Power Converters and AC Electrical Drives

with Linear Neural Networks

Energy, Power Electronics, and Machines

Series Editor
Ali Emadi

Switched Reluctance Motor Drives: Fundamentals of Magnetic Design and Control
Babak Fahimi

Power Converters and AC Electrical Drives with Linear Neural Networks
Maurizio Cirrincione, Marcello Pucci, and Gianpaolo Vitale

Energy Harvesting: Solar, Wind, and Ocean Energy Conversion Systems
Alireza Khaligh and Omer C. Onar

Power Converters and AC Electrical Drives

with Linear Neural Networks

Maurizio Cirrincione
Marcello Pucci
Gianpaolo Vitale

CRC Press
Taylor & Francis Group
Boca Raton London New York

CRC Press is an imprint of the
Taylor & Francis Group, an **informa** business

MATLAB® and Simulink® are trademarks of The MathWorks, Inc. and are used with permission. The MathWorks does not warrant the accuracy of the text or exercises in this book. This book's use or discussion of MATLAB® and Simulink® software or related products does not constitute endorsement or sponsorship by The MathWorks of a particular pedagogical approach or particular use of the MATLAB® and Simulink® software.

CRC Press
Taylor & Francis Group
6000 Broken Sound Parkway NW, Suite 300
Boca Raton, FL 33487-2742

First issued in paperback 2017

© 2012 by Taylor & Francis Group, LLC
CRC Press is an imprint of Taylor & Francis Group, an Informa business

No claim to original U.S. Government works
Version Date: 20120329

ISBN 13: 978-1-138-07746-1 (pbk)
ISBN 13: 978-1-4398-1814-5 (hbk)

Library of Congress Cataloging-in-Publication Data

Cirrincione, Maurizio, 1961-
Power converters and AC electrical drives with linear neutral networks / Maurizio Cirrincione, Marcello Pucci, Gianpaolo Vitale.
p. cm. -- (Energy, power electronics, and machines)
Includes bibliographical references and index.
ISBN 978-1-4398-1814-5 (hardcover : alk. paper)
1. Electric motors, Alternating current--Automatic control. 2. Adaptive control systems. 3. Electric current converters. 4. Neural networks (Computer science) I. Pucci, Marcello. II. Vitale, Gianpaolo. III. Title.

TK2851.C57 2012
621.46--dc23
2012005271

Visit the Taylor & Francis Web site at
http://www.taylorandfrancis.com

and the CRC Press Web site at
http://www.crcpress.com

Contents

Foreword xv
Preface... xvii
Acknowledgments xxi
Authors xxiii

1. Review of Basic Concepts: Space-Vector Analysis 1
   1.1   Introduction 1
   1.2   Space-Vector Definition 2
   1.3   3 → 2 and 2 → 3 Transformations 5
       1.3.1   Non-Power-Invariant Form Ver.1 5
       1.3.2   Power-Invariant Form 6
       1.3.3   Non-Power-Invariant Form Ver.2 7
   1.4   Coordinate Transformation 7
   1.5   Instantaneous Real and Imaginary Powers 9
   References 13

Part I Power Converters

2. Pulsewidth Modulation of Voltage Source Inverters 17
   2.1   Fundamentals of Voltage Source Inverters 17
       2.1.1   Performance Criteria 20
           2.1.1.1   Current Harmonics 20
           2.1.1.2   Harmonic Spectrum 21
           2.1.1.3   Maximum Modulation Index 21
           2.1.1.4   Torque Harmonics 22
           2.1.1.5   Switching Frequency and Switching Losses 22
           2.1.1.6   Common-Mode Voltage (CMV) 24
   2.2   Open-Loop PWM 24
       2.2.1   Carrier-Based PWM 25
           2.2.1.1   Suboscillation Method 25
           2.2.1.2   Modified Suboscillation Method 26
           2.2.1.3   Sampling Techniques 28
           2.2.1.4   Space-Vector Modulation 29
           2.2.1.5   Discontinuous Pulsewidth Modulation 34
           2.2.1.6   General Expression of the Duty Cycle in Carrier-Based PWM Techniques 37
           2.2.1.7   Modified Space-Vector Modulation 38
           2.2.1.8   Synchronized Carrier Modulation 38
       2.2.2   Carrierless PWM 38
       2.2.3   Overmodulation 39
       2.2.4   SV-PWM for the Minimization of the Common-Mode Emissions ... 41
           2.2.4.1   Common-Mode Voltage 41
           2.2.4.2   Switching Strategy 42

        2.2.5   Optimized Open-Loop PWM 43
        2.2.6   Experimental Verification of Open-Loop PWM Techniques.................44
    2.3   Closed-Loop Control of VSIs...51
        2.3.1   Classification of Closed-Loop Control Strategy 51
            2.3.1.1   Hysteresis Current Control...52
            2.3.1.2   Suboscillation Current Control 54
            2.3.1.3   Space-Vector Current Control..55
            2.3.1.4   Closed-Loop PWM with Real-Time Optimization..................57
            2.3.1.5   Predictive Current Control 58
            2.3.1.6   Pulsewidth Control with Field Orientation.....................59
            2.3.1.7   Trajectory Tracking Control...60
        2.3.2   From the Six-Pulse Rectifier to the Active Rectifier 61
        2.3.3   Current Control of VSIs...66
            2.3.3.1   Voltage-Oriented Control (VOC) 66
            2.3.3.2   Virtual Flux-Oriented Control 71
            2.3.3.3   Experimental Results with VOC and VF-OC....................72
        2.3.4   Power Control of VSIs 72
            2.3.4.1   Direct Power Control 72
            2.3.4.2   Switching Table DPC 73
            2.3.4.3   Virtual Flux Direct Power Control 78
            2.3.4.4   Experimental Results with DPC and VF-DPC..................78
            2.3.4.5   DPC EMC (Electromagnetic Compatible).......................79
            2.3.4.6   Simulation Results with DPC-EMC 1 and DPC-EMC 2...........83
            2.3.4.7   Comparative Experimental Analysis of DPC, DPC-EMC 1,
                        and DPC-EMC 2...84
            2.3.4.8   Standard Compliance: Comparative Analysis.........................91
    List of Symbols..92
    References 93
    Further Reading...96

3. Power Quality...97
    3.1   Nonlinear Loads...97
        3.1.1   Current Source Type of Harmonic Sources
                (Harmonic Current Sources) 97
        3.1.2   Voltage Source Type of Harmonic Sources
                (Harmonic Voltage Sources) 97
    3.2   Harmonic Propagation on the Distribution Network 97
    3.3   Passive Filters..102
    3.4   Active Power Filters 105
        3.4.1   Introduction on APFs 105
        3.4.2   Basic Operating Issues of Parallel and Series Filters 107
        3.4.3   Shunt Active Filters...108
            3.4.3.1   Harmonic Current Source Compensation by PAF 108
            3.4.3.2   Harmonic Voltage Source Compensation by PAF.................111
            3.4.3.3   Control of PAF Based on p-q Theory 112
        3.4.4   Series Active Filters 117
            3.4.4.1   Harmonic Current Loads...117
            3.4.4.2   Voltage Harmonic Loads...119
            3.4.4.3   Control of SAF Based on the p-q Theory.........................121

3.4.5 Comparison between PAFs and SAFs..122
3.4.6 Hybrid Active Filters 123
List of Symbols 131
References 131

Part II Electrical Drives

4. Dynamic and Steady-State Models of the Induction Machine 135
4.1 Introduction..135
4.2 Definition of the Machine Space-Vector Quantities....................................135
4.3 Phase Equations of the IM..141
4.4 Space-Vector Equations in the Stator Reference Frame 142
4.5 Space-Vector Equations in the Rotor Reference Frame................................143
4.6 Space-Vector Equations in the Generalized Reference Frame.......................144
4.6.1 Mutually Coupled Magnetic Circuits 146
4.6.2 Space-Vector Equations in the Rotor Flux-Linkage Reference Frame....147
4.6.3 Space-Vector Equations in the Stator Flux-Linkage Reference Frame...149
4.6.4 Space-Vector Equations in the Magnetic Flux-Linkage
Reference Frame..153
4.7 Mathematical Dynamic Model of the IM Taking into Account
Magnetic Saturation..154
4.8 Steady-State Space-Vector Model of the IM..158
4.9 Experimental Validation of the Space-Vector Model of the IM 162
4.10 IM Model Including Slotting Effects...165
4.10.1 Space-Vector Model of the IMs Including Stator
and Rotor Slotting Effects...169
4.10.2 Space-Vector State Model of IM Including Rotor Slotting Effects 171
4.10.3 Space-State Model of IM Including Rotor Slotting Effects 174
4.10.4 Space-State Model of IM Including Stator and Rotor
Slotting Effects 176
4.10.5 Experimental Validation of the Space-Vector Model Considering
the Stator and Rotor Slotting Effects...178
4.10.5.1 Stator Current Harmonics Caused by the Stator
and Rotor Slotting Effect..178
4.10.5.2 Results..179
List of Symbols 186
References 187

5. Control Techniques of Induction Machine Drives..189
5.1 Introduction on Induction Machine (IM) Control..189
5.2 Scalar Control of IMs...190
5.2.1 Scalar Control with Impressed Voltages 190
5.2.1.1 Experimental Results with Open-Loop Scalar Control
with Impressed Voltages...195
5.2.2 Scalar Control with Impressed Currents..196
5.3 FOC of IMs...200
5.3.1 Principle of Field-Oriented Control..200
5.3.2 Rotor Flux-Oriented Control 202

    5.3.3   Rotor Flux-Linkage Acquisition 203
            5.3.3.1   Voltage Flux Models 204
            5.3.3.2   Current Flux Models 206
            5.3.3.3   Rotor Flux-Oriented Control with Impressed Currents 208
            5.3.3.4   Rotor Flux-Oriented Control with Impressed Voltages 213
            5.3.3.5   Experimental Results on a Rotor Flux-Oriented Drive
                      with Impressed Voltages 216
    5.3.4   Stator Flux-Oriented Control 218
            5.3.4.1   Stator Flux-Linkage Acquisition 223
    5.3.5   Magnetizing Flux-Oriented Control 225
5.4   DTC of IM 230
    5.4.1   Electromagnetic Torque Generation in the IM 230
    5.4.2   Relationship between the Stator Flux-Linkage Space-Vector
            and the Inverter Configurations 232
    5.4.3   Criteria for the Selection of the Voltage Space-Vectors
            and Control Strategies 234
    5.4.4   Estimation of the Stator Flux and Electromagnetic Torque 237
    5.4.5   DTC Scheme 240
    5.4.6   DTC EMC 242
            5.4.6.1   Common-Mode Voltage 242
            5.4.6.2   Switching Strategy 242
            5.4.6.3   Common-Mode Voltage Spectrum of DTC EMC 246
    5.4.7   Experimental Results with Classic DTC and DTC EMC 246
    5.4.8   DTC-SVM 249
    5.4.9   Experimental Results on a DTC-SVM Drive 251
    5.4.10 Direct Self-Control 253
    5.4.11 Comparison between FOC and DTC 254
List of Symbols 256
References 258

6. **Sensorless Control of Induction Machine Drives** 261
6.1   Introduction on Sensorless Control 261
6.2   Model-Based Sensorless Control 261
6.3   Anisotropy-Based Sensorless Control 262
6.4   Model-Based Sensorless Techniques 263
    6.4.1   Open-Loop Integration 263
            6.4.1.1   Low-Pass Filter 264
    6.4.2   Inverter Nonlinearity 270
    6.4.3   Machine Parameter Mismatch 273
    6.4.4   Estimators and Observers 275
    6.4.5   Open-Loop Speed Estimators 276
    6.4.6   Model Reference Adaptive Systems 279
            6.4.6.1   Classic MRAS 279
            6.4.6.2   Closed-Loop MRAS 282
    6.4.7   Full-Order Luenberger Adaptive Observer 284
            6.4.7.1   State-Space Model of the IM 284
            6.4.7.2   Adaptive Speed Observer 285

6.4.8   Full-Order Sliding-Mode Observer...289
6.4.9   Reduced-Order Adaptive Observer 291
      6.4.9.1   Reduced-Order Observer Equations 291
      6.4.9.2   Possible Choices of the Observer Gain Matrix.......292
      6.4.9.3   Speed Estimation..295
6.4.10   Extended Kalman Filter 296
6.5   Anisotropy-Based Sensorless Techniques..298
    6.5.1   Signal Injection Techniques..298
      6.5.1.1   Revolving Carrier Techniques....................................298
      6.5.1.2   IM Saliency Analysis under Rotating Carrier by FEA 303
      6.5.1.3   Pulsating Carrier Techniques.....................................308
      6.5.1.4   High-Frequency Excitation Techniques 310
6.6   Conclusions on Sensorless Techniques for IM Drives........................314
References 316

7. **Permanent Magnet Synchronous Motor Drives** 319
7.1   Introduction 319
    7.1.1   DC Brushless Motors...320
    7.1.2   AC Brushless Motors...320
    7.1.3   Permanent Magnets...322
7.2   Space-Vector Model of Permanent Magnet Synchronous Motors 323
7.3   Control Strategies of PMSM Drives...331
    7.3.1   Field-Oriented Control of PMSM Drives..................................331
    7.3.2   Torque-Controlled Drives...333
      7.3.2.1   Surface-Mounted PMSM 333
      7.3.2.2   Interior-Mounted PMSM...336
      7.3.2.3   Feed-Forward Control 339
    7.3.3   Speed-Controlled Drives 339
      7.3.3.1   Experimental Results...340
    7.3.4   Direct Torque Control 341
      7.3.4.1   Electromagnetic Torque Production in the PMSM....343
      7.3.4.2   Criteria for the Selection of the Voltage Space-Vectors
               and Control Strategies 343
      7.3.4.3   Estimation of the Stator Flux and the Electromagnetic
               Torque 345
      7.3.4.4   The Direct Torque Control Scheme 347
7.4   Sensorless Control of PMSM Drives..348
    7.4.1   Anisotropy-Based Sensorless Techniques................................349
      7.4.1.1   Rotating Carrier Signal Injection 349
      7.4.1.2   Alternating Carrier Signal Injection..........................354
      7.4.1.3   INFORM Sensorless Technique 359
    7.4.2   Model-Based Sensorless Techniques..362
      7.4.2.1   Open-Loop Estimators 362
      7.4.2.2   Observer-Based Estimators...363
Appendix: Experimental Test Setup..372
References 372

Part III   Neural-Based Orthogonal Regression

8. Neural-Based Orthogonal Regression 377
  8.1   Introduction: ADALINE and Least Squares Problems...............................377
  8.2   Approaches to the Linear Regression 378
        8.2.1   OLS Problem...378
        8.2.2   DLS Problem...378
        8.2.3   TLS Problem 379
  8.3   Minor Component Analysis and the MCA EXIN Neuron 380
        8.3.1   Some MCA Applications 380
        8.3.2   Neural Approach 380
  8.4   MCA EXIN Neuron 381
        8.4.1   Convergence during the First Transient Phase 381
        8.4.2   Dynamic Behavior of the MCA Neuron.....................................383
        8.4.3   Dynamic Stability and Learning Rate 385
        8.4.4   Numerical Considerations..386
               8.4.4.1   Computational Cost..386
               8.4.4.2   Quantization Errors...386
        8.4.5   Acceleration Techniques 387
        8.4.6   Simulations 387
        8.4.7   Conclusions and Prospects for the MCA Neuron.....................392
  8.5   TLS EXIN Neuron..392
        8.5.1   Stability Analysis (Geometrical Approach) 395
        8.5.2   Convergence Domain..397
        8.5.3   Nongeneric TLS Problem...399
  8.6   Generalization of Linear Least Squares Problems 402
  8.7   GeMCA EXIN Neuron 403
        8.7.1   Qualitative Analysis of the Critical Points of the GeMCA EXIN
               Error Function 405
        8.7.2   Analysis of the Error Function GeTLS (Geometrical Approach)..........405
        8.7.3   Critical Loci: Center Trajectories..406
  8.8   GeTLS EXIN Neuron 408
        8.8.1   GeTLS Domain of Convergence...409
        8.8.2   Scheduling 410
        8.8.3   Accelerated MCA EXIN Neuron (MCA EXIN+)......................412
  References 414

Part IV   Selected Applications

9. Least-Squares and Neural Identification of Electrical Machines...........................419
  9.1   Parameter Estimation of Induction Machines (IMs)..................................419
  9.2   Sensitivity of the Flux Model to Parameter Variations............................420
        9.2.1   Sensitivity of the Current Flux Model.......................................420
        9.2.2   Sensitivity of the Voltage Flux Model 426
  9.3   Experimental Analysis of the Effects of Flux Model Detuning
        on the Control Performance 431
  9.4   Methods for the On-Line Tracking of the Machine Parameter Variations.......432

9.5 On-Line Estimation of the IM Parameters with the Ordinary Least Squares Method 433
  9.5.1 Space-Vector Voltage Equations in the General Reference Frame 434
  9.5.2 Estimation of the Magnetizing Curve 438
  9.5.3 Ordinary Least Squares Identification 439
  9.5.4 RLS Algorithm 440
  9.5.5 Signal Processing System 442
  9.5.6 Description of the Test Setup for the Experimental Application 446
  9.5.7 Simulation and Experimental Results 447
9.6 Constrained Minimization for Parameter Estimation of IMs in Saturated and Unsaturated Conditions 450
  9.6.1 Constrained Minimization: Analytical Solution 451
    9.6.1.1 First Constrained Minimization Method 452
    9.6.1.2 Second Constrained Minimization Method 456
9.7 Parameter Estimation of an IM with the Total Least Squares Method 470
  9.7.1 Simulation and Experimental Results 473
9.8 Application of the RLS-Based Parameter Estimation to Flux Model Adaptation in FOC and DTC IM Drives 479
9.9 Estimation of the IM Parameters at Standstill 483
List of Symbols 488
References 489

10. Neural-Enhanced Single-Phase DG Systems with APF Capability 497
  10.1 Introduction 497
  10.2 General Operating Principle 498
  10.3 ADALINE Design Criteria 499
    10.3.1 Notch Operation 501
    10.3.2 Band Operation 501
    10.3.3 MATLAB®–Simulink® Implementation 503
    10.3.4 Comparison with Traditional Digital Filters 503
    10.3.5 NN Band Filter versus PLL: Theoretical Comparison 505
  10.4 Building the Current Reference 507
  10.5 Multiresonant Current Controller 508
  10.6 Stability Issues 510
  10.7 Test Rig 515
  10.8 Experimental Results 515
    10.8.1 APF Insertion 515
    10.8.2 Power Reference Insertion 518
    10.8.3 Load Variation 519
    10.8.4 NN Filter versus PLL 520
    10.8.5 NN Filter versus p-q Theory 522
    10.8.6 Compliance with International Standards 523
  10.9 APF Connection Procedure 527
  References 528

11. Neural Sensorless Control of AC Drives 531
  11.1 NN-Based Sensorless Control 531
  11.2 BPN-Based MRAS Speed Observer 533
    11.2.1 On-Line Training of the BPN MRAS Observer 534

11.2.2 Implementation Issues of the BPN MRAS Observer 535
11.2.3 Experimental Results with the BPN MRAS Observer 536
11.3 LS-Based MRAS Speed Observer 538
11.3.1 Experimental Results with the OLS MRAS Observer 539
11.3.2 TLS EXIN MRAS Observer 541
11.3.2.1 Neural Adaptive Integrator 544
11.3.2.2 Experimental Results with the TLS EXIN MRAS
Observer 549
11.3.3 Modified Euler Neural Adaptive Model 555
11.3.3.1 Simulation Mode and Prediction Mode in MRAS
Observers: Modified Euler against Simple Euler 557
11.3.4 MCA EXIN + MRAS Observer 561
11.4 TLS EXIN Full-Order Luenberger Adaptive Observer 563
11.4.1 State-Space Model of the IM 563
11.4.2 Adaptive Speed Observer 564
11.4.3 TLS-Based Speed Estimation 564
11.4.4 Stability Issues of the TLS EXIN Full-Order Adaptive Observer 567
11.4.5 Experimental Results with the TLS EXIN Full-Order
Luenberger Adaptive Observer 570
11.4.5.1 Dynamic Performance 571
11.4.5.2 Accuracy in the Low-Speed Ranges 573
11.4.5.3 Accuracy at Very Low Speed (below 2 rad/s) 573
11.4.5.4 Robustness to Load Perturbations 576
11.4.5.5 Regenerative Mode at Very Low Speed 577
11.4.5.6 Field-Weakening Operation 579
11.4.5.7 Zero-Speed Operation 579
11.4.6 Experimental Comparative Tests 581
11.5 MCA EXIN + Reduced-Order Observer 584
11.5.1 Reduced-Order Observer Equations 584
11.5.2 MCA EXIN + Based Speed Estimation 585
11.5.3 Proposed Choice of the Observer Gain Matrix 586
11.5.4 Computational Complexity 587
11.5.5 Experimental Results with the MCA EXIN + Reduced-Order
Adaptive Observer 588
11.5.5.1 Dynamic Performance 588
11.5.5.2 Accuracy at Low Speed 589
11.5.5.3 Zero-Speed Operation 589
Appendix A: Implemented Control Schemes 592
Appendix B: Description of the Test Setup 596
List of Symbols 599
References 600
Index 605

Foreword

It is indeed an honor, privilege, and pleasure to write this foreword at the invitation of the authors of this book. I sincerely hope that this novel and state-of-the-art book on power electronics and motor drives gets wide and enthusiastic acceptance from the professional power electronics community consisting of R&D professionals, practicing engineers, university professors, and even graduate students. I would like to congratulate the authors for writing such an excellent book.

Power electronics and motor drive technology is a very complex and multidisciplinary field, and it has gone through dynamic evolution over the last several decades through many inventions in power semiconductor devices, converters, PWM techniques, electrical machines, motor drives, and advanced control and simulation techniques. In recent years, the frontier of power electronics has advanced further with the advent of artificial intelligence (AI) techniques, such as expert systems, fuzzy logic systems, neural networks, and genetic algorithms (or evolutionary computation). Power electronics has now been established as a major discipline in electrical engineering, and its tremendous impact is obvious not only in global industrialization and general energy systems, but also in energy conservation, renewable energy systems, bulk energy storage, and electric/hybrid vehicles in the twenty-first century. The widespread applications of power electronics in industry have effected an unprecedented revolution in industrial engineering. The role of power electronics in this new era will be as important as, if not more important than, that of computers, communication, and information technology.

I have devoted a number of years of my career to the state-of-the-art development of AI applications in power electronics and motor drives, particularly for expert systems, fuzzy logic, and neural networks. I have also written a number of chapters on these areas for my books. I am truly excited to see the growth and development in these areas and in their applications in industry. Among all the AI techniques, neural networks have emerged as the most important area for complex system identification, control, and estimation in power electronics and motor drives. In the future, they are expected to have widespread applications in industry.

This state-of-the-art book, authored by Maurizio Cirrincione, Marcello Pucci, and Gianpaolo Vitale, is the first book that systematically explores the application of neural networks in the field of power electronics. It emphasizes, particularly, neural network applications in sensorless control of AC drives, including their applications in active power filtering.

Broadly, the content of the book is classified in 4 parts consisting of 11 chapters. Chapter 1, the introductory chapter, presents space-vector theory as well as instantaneous power theory. Suffice it to say that concepts of space-vector theory are very important in modern power electronics and drives. Part I (Chapters 2 and 3) provides a general description of voltage-fed converters and their control (open-loop and closed-loop controls) and deals with PWM algorithms in detail. Both voltage-oriented control (VOC) and power-oriented control (POC) methods are introduced. In addition, power quality control with shunt active and series active filters are discussed. Part II (Chapters 4 through 7) deals with induction and permanent magnet synchronous motor drives. It includes dynamic model descriptions of AC machines, and scalar, vector (or field-oriented), and DTC control of induction and synchronous motors. The sensorless control of induction machine (IM) drives in Chapter 6 is particularly important.

Part III (Chapter 8) and Part IV (Chapters 9 through 11) form the core of this book in that they describe the theoretical aspects of linear neural networks, particularly the EXIN family (jointly developed by one of the authors) and their applications, ranging from neural-based parameter estimation and sensorless control (which includes MRAS observers, full-order Luenberger adaptive observers, and reduced-order observers) to neural-based distributed generation systems from renewable sources and active power filters. Each chapter contains extensive references, including major textbooks in the area. Extensive simulation and experimental results are also provided to validate the theories. As far as I know, this part IV of the book is unique and is not available in any other book.

A reader of this book should have a general background in power electronics and motor drives, including some knowledge of linear algebra. Some background in neural networks is also desirable but not essential.

For whom would this book be useful? In my opinion, the book is suitable primarily for graduate students (as a one-semester course) and researchers. Portions of Parts I and II can also be taught in undergraduate courses. Selected portions of the book could also be useful for practicing engineers.

Finally, I wish the book great success and hope it becomes readily accepted by the professional community.

Dr. Bimal K. Bose
Life Fellow, IEEE
Condra Chair of Excellence/Emeritus in Power Electronics
Department of Electrical Engineering and Computer Science
The University of Tennessee
Knoxville, Tennessee

Preface

Power electronics and electrical drives are heavily characterized by a strong interdisciplinarity. A thorough knowledge of the field involves being proficient in

- Electrical machines
- Circuit theory
- Control system theory
- Signal processing system
- Electronics
- Electromagnetic fields
- Numerical analysis
- Solid-state physics
- Power plants

A recent trend for the development of these disciplines is the application of *artificial intelligence* (AI) tools, such as *expert systems* (ES), *artificial neural networks* (ANN), *fuzzy logic systems* (FLS), *genetic algorithms* (GA), and, more recently, *multi-agent systems* (MAS). These tools have been proven to be able to boost the performance of these systems in real-world and industrial applications thanks to features such as "learning," "self-organization," and "self-adaptation."

With particular regard to ANNs for nonlinear function approximation, in power electronics and electrical drives applications, they are used for control and identification, such as the multilayer perceptron (MLP) or the radial basis function (RBF). Another kind of neuron that has also been applied recently is linear neurons (ADALINE), whose simplicity has given surprisingly good results.

On the other hand, the detailed unified mathematical treatment of space-vectors has made it possible to embed the theory of linear neural networks, resulting in improvements, both theoretical and experimental, of classical approaches in electrical drives and power electronics. This standpoint is the goal of the book: to present in a systematic way the classical theory based on space-vectors in identification, control of electrical drives and of power converters, and the improvements that can be attained when using linear neural networks.

With this outlook, this book is divided into four parts:

- Part I deals specifically with voltage source inverters (VSI) and their control.
- Part II deals with AC electrical drive control, with particular attention to induction and permanent magnet synchronous motor drives.
- Part III deals with theoretical aspects of linear neural networks.
- Part IV deals with specific applications of linear neural networks to electrical drives and power quality.

Outline of the Book

Chapter 1 presents the theory of space-vectors and instantaneous power. This chapter is fundamental for understanding the rest of the book.

Chapter 2 describes the open-loop and closed-loop control of voltage source inverters. With regard to open-loop techniques it also explains the different kinds of pulsewidth modulation (PWM) strategies, and with regard to closed-loop techniques it analyzes both current and power control of VSIs. Voltage-oriented control (VOC) and direct power control (DPC) are also presented. Chapter 3 explains the fundamentals of power quality; parallel active filters (PAFs) and series active filters (SAFs), with reference to their operating principle and control strategies, are investigated. Passive and hybrid filter configurations are also analyzed.

Chapter 4 deals with induction machine (IM) static and dynamic space-vector models. The dynamic model of the IM, including saturation effects, is shown. Finally, the space-vector dynamic model of the IM, including rotor and stator slotting effects, is described. Chapter 5 describes, first, scalar control strategies of IM drives with impressed voltages and currents. It then derives field-oriented control (FOC) strategies, with reference to rotor, stator, and magnetizing flux linkage orientations. Related flux models are also presented. Finally, direct torque control (DTC) strategies are presented, particularly the classic switching table (ST) DTC, the space-vector modulation (SVM) DTC, and the electromagnetically compatible (DTC). The so-called direct self-control (DSC) is also described. Chapter 6 covers sensorless control of IM drives, with particular reference to both model-based and anisotropy-based techniques. With regard to model-based techniques, the following estimators/observers are described: open-loop speed estimators, model reference adaptive systems (MRAS), full-order Luenberger adaptive observer (FOLO), full-order sliding-mode observer, reduced-order adaptive observer (ROO), and, finally, the extended Kalman filter. With reference to anisotropy-based techniques, the following methodologies have been described: revolving Carrier techniques, pulsating carrier techniques, and high-frequency excitation techniques. Chapter 7 derives the permanent magnet synchronous motor space-vector model. Field-oriented control (FOC) with both impressed voltage and currents is described. Various control strategies are presented to maximize the electromagnetic torque production or the drive efficiency. The DTC of the PMSM is also presented. Finally, both anisotropy- and model-based sensorless techniques are explained.

Chapter 8 deals with the theory of linear neural networks, particularly the neural EXIN family. Starting from the adaptive linear neuron (ADALINE) structure, it presents more recent and performing linear neural networks: the TLS EXIN neuron, the Ge-TLS EXIN neuron, the MCA EXIN neuron, and, finally, the MCA EXIN + neuron.

Chapter 9 covers, first, the sensitivity analysis of the classic flux models of IM drives versus parameter variations. It then presents some on-line parameter estimation techniques of IMs by the least-squares (LS) technique, including both unconstrained and constrained estimations. Finally, it shows the neural self-commissioning of IM drives. Chapter 10 deals with the application of neural adaptive filtering to distributed generation (DG) and active power filter (APF) systems. The ADALINE design criteria for the fundamental frequency extraction and the harmonic load current compensation are presented. The stability issues of the entire system are included. Experimental verification of the neural approach is presented in comparison with classic approaches. Chapter 11 presents some applications of LS-based techniques to speed estimation of IMs. In particular, the following neural-based observers are presented and discussed: the MCA EXIN + MRAS observer, the TLS EXIN full-order Luenberger adaptive observer, and, finally, the reduced order adaptive observer. The general approach of this book is presenting initially the theoretical background of each subject immediately followed by a set of simulations of experimental results supporting the analytical part. It is the authors' opinion that the presence of many results should help the reader in understanding better the treated theoretical aspects.

How to Use This Book

This book can be used in different ways, depending on the background of the reader. Instructors for undergraduate students can use Chapters 1, 2, 4, 5, and 7 for a one-semester course in power electronics and electrical drives. Postgraduate and PhD students could also cover Chapters 3 and 6. Researchers involved in applications of ANNs to power electronics and drives would find it more interesting to go through Chapters 9 through 11. Theoreticians will find a comprehensive coverage of linear neural networks in Chapter 8. Figure P.1 suggests different paths for reading this book.

Chapter 1 should be read first, since it provides the basic tools. The power electronics path consists of Chapters 2, 3, and 10. The IM drives path includes Chapters 4 through 6 and 11 (electrical machine approach) or, alternatively, Chapters 2, 5, 6, and 11 (power converter approach). The PMSM drives path includes Chapters 2 and 7 (or directly Chapter 7). The IM identification path includes Chapters 4 and 9. Finally, the ANN path includes Chapters 8 through 11.

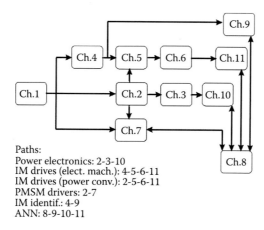

Paths:
Power electronics: 2-3-10
IM drives (elect. mach.): 4-5-6-11
IM drives (power conv.): 2-5-6-11
PMSM drivers: 2-7
IM identif.: 4-9
ANN: 8-9-10-11

FIGURE P.1
Different paths for reading the book.

Prerequisites

To read this book, a basic knowledge of electrical machines and power electronics is required as well as some familiarity with notions of control systems and signal processing. To fully understand the fundamentals of linear neural networks, notions of linear algebra and numerical analysis are necessary; however, no *a priori* knowledge of neural networks is needed.

Remark

Nowadays, writing a book follows a logic contrary to what is usually demanded from researchers. Researchers are increasingly encouraged to submit as many projects as possible and to find money and publish papers at the earliest. They are not expected to dedicate time to writing a systematic treatise of their activities, despite the fact that this is a fundamental task for further improving research on a more solid foundation.

Unfortunately, public institutions find it difficult to understand the importance of devoting time to write. A remark like this amounts to heresy, but it represents a solid base for future generations.

For MATLAB® and Simulink® product information, please contact:

The MathWorks, Inc.
3 Apple Hill Drive
Natick, MA, 01760-2098 USA
Tel: 508-647-7000
Fax: 508-647-7001
E-mail: info@mathworks.com
Web: www.mathworks.com

Acknowledgments

We would like to thank Giuseppe Scordato and Antonio Sauro for their invaluable help with the setup of the experimental rigs at the Institute of Intelligent Systems for the Automation (ISSIA) section of Palermo of the National Research Council of Italy (CNR).

We also would like to thank Nora Konopka and Jennifer Ahringer from Taylor & Francis Group for their friendly help and support throughout the writing of the book.

Authors

Maurizio Cirrincione, PhD, is a full professor of control and signal processing at the University of Technology of Belfort, Montbeliard, France. His current research interests include neural networks, modeling and control, system identification, intelligent control, and electrical machines and drives. E-mail address: maurizio.cirrincione@utbm.fr

Marcello Pucci, PhD, is a senior researcher at the Institute of Intelligent Systems for the Automation (ISSIA) section of Palermo of the National Research Council of Italy (CNR). His current research interests include electrical machines and drives, power converters, wind and photovoltaic generation systems, intelligent control, and neural networks applications. E-mail address: pucci@pa.issia.cnr.it

Gianpaolo Vitale, laurea of electrical engineering, is a senior researcher at the Institute of Intelligent Systems for the Automation (ISSIA) section of Palermo of the National Research Council of Italy (CNR). He is also professor of power electronics and applied electronics at the University of Palermo, Italy. His current research interests include power electronics, generation from renewable energy, and related problems of electromagnetic compatibility. E-mail address: vitale@pa.issia.cnr.it

1

Review of Basic Concepts: Space-Vector Analysis

1.1 Introduction

The development of any control system requires the knowledge of the mathematical model describing it. As for electric machines, before the 1950s, only sinusoidal steady-state models were used [1–3]. In these models, the rotating speed is assumed constant, while the voltage or the current supply is sinusoidally time varying. Afterward, dynamic models have been developed, particularly for AC machines [4–6]. These models permit the analysis of the motor behavior during both steady-state and transients with any time-varying waveform of the supply voltage or current. Thus dynamic models have been playing a key role in properly designing control systems of electric drives so that the stability can be achieved with desired time and frequency performance. At the same time, even the electronic power converters have been initially studied only with reference to their input-output steady-state relationship [7]. Afterwards, their dynamic models have been developed which permitted the analysis and design of suitable high-performance control systems as well as novel pulsewidth modulation (PWM) techniques [8–14]. Such control techniques have been adopted in the area of the *active rectifier* [11–13], the *active power filters* (APF) [14–16], and the *distributed generation systems from renewable sources* [17].

A big step ahead in the dynamic modeling of the systems mentioned earlier has been done made thanks to the space-vector theory. This theory has been initially developed with reference to electric machine dynamic study [4–6], but has been then extended to the study of any three-phase system, among which the power converters, for several applications. It is for this reason that, also in this chapter, the space-vector quantity will be initially defined with reference to an AC electric machine, because of its physical meaning, and then extended to other three-phase variable systems. Space-vector theory has been the core around which the all the vector control techniques for electric machines, like field-oriented control (FOC) and direct torque control (DTC) [18–23], as well as for three-phase active rectifiers, like voltage-oriented control (VOC) and direct power control (DPC) [8], have been developed. Finally, it has been the necessary tool for developing the well-known set of space-vector modulation techniques [24–29].

In the following, the theory of space-vectors is briefly presented, since it is a most useful framework for studying AC machine and power converter dynamic models.

1.2 Space-Vector Definition

Let a generic three-phase system be considered. The system can be represented by a three-phase winding (Figure 1.1), which could be the stator winding of an AC electric machine, whose phases are called sA, sB, and sC. Let $i_{sA}(t)$, $i_{sB}(t)$, and $i_{sC}(t)$ be the instantaneous values of the stator currents for each stator phase, and t is time. No zero-sequence current is assumed to be present, as, for example, if the neutral point of the stator three-phase system is isolated. This means that instantaneously

$$i_{sA}(t) + i_{sB}(t) + i_{sC}(t) = 0 \tag{1.1}$$

The axes of the three phases are mutually displaced in space of 120°, with the sA axis superimposed to the real axis of the complex plane. Each phase winding, if interested by a time-varying sinusoidal current, is assumed to create a space-varying sinusoidal distribution of the magnetomotive force (mmf). It means that the distributed phase windings are assumed to have a sinusoidal winding density. Each phase current component therefore creates a specific sinusoidal mmf distribution, whose amplitude is proportional to the amplitude of the current itself and whose spatial orientation is determined by the spatial position of the winding axis and by the current polarity. If only the phase sA is energized with a positive current i_{sA}, a sinusoidal current density distribution is created which is spatially displaced 90° with respect to the sA winding axis. This representation is sketched in Figure 1.2 [30], where the sinusoidality of the stator winding is represented with a varying cross section of the winding conductors or alternatively with two half-moon-shaped segments.

The global resultant mmf is due to the superposition of the current densities of the three phases. If each phase winding is sinusoidal, even the global current density will present a sinusoidal spatial distribution. The amplitude and spatial orientation of such a distribution will be dependent on the instantaneous amplitude of each phase current i_{sA}, i_{sB}, or i_{sC}. Since a three-phase system is characterized by time-varying current waveforms, the current density field will modify in time both its amplitude and its spatial distribution. Coherently, it is well known from classic electric machine theory that a three-phase set of equilibrated currents circulating on a three-phase symmetric winding generates a sinusoidal mmf of constant amplitude, rotating in time with a constant velocity equal to the supply pulsation (the so-called Galileo Ferraris' field).

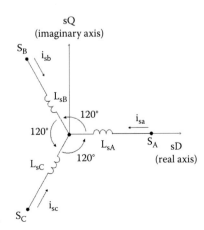

FIGURE 1.1
Sketch of a three-phase winding.

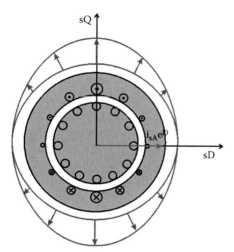

FIGURE 1.2
Current density distribution in a stator winding with only phase sA energized.

The superimposition of the current density profiles can be represented by the spatial addition of each phase current. For this purpose, the space-vector of stator current can be defined in this way:

$$\mathbf{i}_s(t) = k\left[i_{sA}(t) + a i_{sB}(t) + a^2 i_{sC}(t)\right] = |\mathbf{i}_s| e^{j\alpha_s} = i_{sD}(t) + j i_{sQ}(t) \tag{1.2}$$

where
  j is the imaginary unit
  $a = e^{j2\pi/3}$ is a complex operator that makes a vector rotate $2\pi/3$ rad in the counter-clockwise direction
  $|\mathbf{i}_s|$ is the amplitude of the stator current space-vector
  α_s is the angle of the stator current space-vector from the sD axis
  $i_{sD}(t)$ and $i_{sQ}(t)$ are, respectively, the instantaneous values of the sD and sQ components of the stator current space-vector corresponding to the real and imaginary components of the space-vector in the complex plane

In Equation 1.2, since a (a^2) represents a complex vector of unitary amplitude lying in the direction of the sB (sC) axis, then the quantity $a i_{sB}(t)$ $(a^2 i_{sC}(t))$ represents the sinusoidal spatial distribution generated by the phase current i_{sB} (i_{sC}). Figure 1.3 shows the resultant sinusoidal current density distribution in the stator windings of the machine, when all the

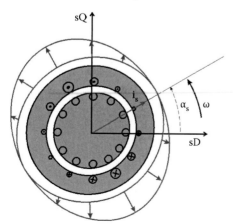

FIGURE 1.3
Resultant current density distribution is a stator winding with all phases energized.

three phases are energized, and consequently, the stator current space-vector presents an angular displacement with respect to the sD axis equal to α_s.

In Equation 1.2, k is a constant factor that can take several values; consequently, the space-vector quantity presents different characteristics. If k = 2/3, then the so-called asymmetrical form (non-power-invariant form) of the stator current space-vector results, which has the property that the amplitude of the space-vector has the same value of the peak sinusoidal value. If $k = \sqrt{2/3}$, then the so-called symmetrical (or power-invariant) form of the space-vector would result, whose property is that no additional factor is used, when instantaneous power is dealt with by means of space-vectors. Finally, if k = 3/2 is adopted, another power invariant form is used.

By assuming a symmetrical three-phase operation in sinusoidal steady-state, whereby currents, voltages, and flux linkages are sinusoids and form a positive sequence, then, for example, the instantaneous stator currents, if I_s is the rms value of the current and ω is its angular frequency and t the time, can be expressed as follows:

$$\begin{cases} i_{sA}(t) = \sqrt{2}I_s \cos(\omega t) \\ i_{sB}(t) = \sqrt{2}I_s \cos\left(\omega t - \frac{2\pi}{3}\right) \\ i_{sC}(t) = \sqrt{2}I_s \cos\left(\omega t - \frac{4\pi}{3}\right) \end{cases} \tag{1.3}$$

By substituting (1.3) into Equation 1.2, the corresponding stator current space-vector is obtained:

$$\mathbf{i}_s(t) = |\mathbf{i}_s|e^{j\omega t} = I_s e^{j\omega t} \tag{1.4}$$

Thus, the space-vector is equal to the time rotating vector (phasor) of phase-A current, and it rotates in space with a constant amplitude with angular speed equal to ω (synchronous speed) in the positive direction. This is coherent with the Galileo Ferraris' field description. Given a set of equilibrated stator currents of unitary amplitude like in (1.3), a graphical description of the space-vector quantity can be given, as in Figure 1.4. If $\omega = 0$, corresponding

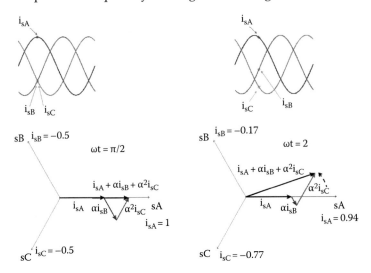

FIGURE 1.4
(See color insert.) Graphical description of the space-vector quantity.

to the positive peak of the sA stator current, then the resultant stator current space-vector lies in the sD axis. Some instants later, when $\omega = \pi/9$, the stator current space-vector has maintained the same amplitude, but its phase angle modified from 0° to 20°, according to the earlier theoretical description.

Because of the earlier emphasized physical meaning, the space-vector quantity has been historically defined with reference to the AC electric machine currents. Its application can, however, be extended to any three-phase time-varying quantity, being it not related to electric machines. Given therefore a set of three-phase quantities $x_A(t)$, $x_B(t)$, and $x_C(t)$, the corresponding space-vector could be defined in this way:

$$\mathbf{x}(t) = k\left[x_A(t) + ax_B(t) + a^2 x_C(t)\right] = |\mathbf{x}|e^{j\alpha_x} = x_D(t) + jx_Q(t) \tag{1.5}$$

1.3 3 → 2 and 2 → 3 Transformations

Let $\mathbf{x}(t) = x_D(t) + jx_Q(t)$ be a generic space-vector. It should be remarked that its direct and quadrature components $x_D(t)$ and $x_Q(t)$ can be directly computed from the three-phase variables $x_A(t)$, $x_B(t)$, and $x_C(t)$ and vice versa. The transformation from the three-phase into the biphase variables is called 3 → 2 transformation, while that from the biphase into the three-phase variables is called 2 → 3. Both these transformations are linear and depend on the constant factor k in the definition of the space-vector (see Equation 1.2). In the following, three different transformations according to the choice of k are described.

1.3.1 Non-Power-Invariant Form Ver.1

In this case, if the D and Q components are to be computed from the A, B, and C ones, the 3 → 2 transformation is the so-called two-axis Park, given by

$$\begin{cases} x_D = \dfrac{2}{3}\left(x_A - \dfrac{1}{2}x_B - \dfrac{1}{2}x_C\right) \\[2mm] x_Q = \dfrac{1}{\sqrt{3}}(x_B - x_C) \end{cases} \tag{1.6}$$

If the A, B, and C components are to be computed from the D and Q ones, the 2 → 3 transformation is the so-called inverse Park, given by

$$\begin{cases} x_A = x_D \\[2mm] x_B = \dfrac{1}{2}\left(-x_D + \sqrt{3}x_Q\right) \\[2mm] x_C = \dfrac{1}{2}\left(-x_D - \sqrt{3}x_Q\right) \end{cases} \tag{1.7}$$

If no zero-sequence is present, the non-power-invariant form permits the instantaneous values of the phase variables to be reconstructed as projection of the corresponding space-vector into the axis of each phase, as

$$\begin{cases} x_A(t) = Re(\mathbf{x}) \\ x_B(t) = Re(a^2\mathbf{x}) \\ x_C(t) = Re(a\mathbf{x}) \end{cases} \tag{1.8}$$

This kind of transformation is more frequently adopted in electric drive control [18–22]. The main characteristic of such a transformation is that the D and Q components of the space-vector present the same amplitude as the A, B, and C phase ones, different from the other forms.

In this book, the asymmetrical form (non-power-invariant form with k = 2/3) will be always used, if not stated otherwise.

1.3.2 Power-Invariant Form

In this case, if the D and Q components are to be computed from the A, B, and C ones, the 3 → 2 transformation is the so-called two-axis Clarke, given by

$$\begin{cases} x_D = \sqrt{\dfrac{2}{3}}\left(x_A - \dfrac{1}{2}x_B - \dfrac{1}{2}x_C \right) \\ x_Q = \dfrac{1}{\sqrt{2}}\left(x_B - x_C \right) \end{cases} \tag{1.9}$$

If the A, B, and C components are to be computed from the D and Q ones, the 2 → 3 transformation is the so-called inverse Clarke, given by

$$\begin{cases} x_A = \sqrt{\dfrac{2}{3}}\,x_D \\ x_B = \sqrt{\dfrac{2}{3}}\left(-\dfrac{1}{2}x_D + \dfrac{\sqrt{3}}{2}x_Q \right) \\ x_C = \sqrt{\dfrac{2}{3}}\left(-\dfrac{1}{2}x_D - \dfrac{\sqrt{3}}{2}x_Q \right) \end{cases} \tag{1.10}$$

If no zero-sequence is present, this form permits the instantaneous values of the phase variables to be reconstructed on the basis of the projection of the corresponding space-vector into the axis of each phase, as

$$\begin{cases} x_A(t) = \sqrt{\dfrac{2}{3}}\,Re(\mathbf{x}) \\ x_B(t) = \sqrt{\dfrac{2}{3}}\,Re(a^2\mathbf{x}) \\ x_C(t) = \sqrt{\dfrac{2}{3}}\,Re(a\mathbf{x}) \end{cases} \tag{1.11}$$

This kind of transformation is more frequently adopted in APF control [14–16].

1.3.3 Non-Power-Invariant Form Ver.2

In this case, if the D and Q components are to be computed from the A, B, and C ones, the $3 \rightarrow 2$ transformation is the following:

$$\begin{cases} x_D = \dfrac{3}{2} x_A \\[2mm] x_Q = \dfrac{\sqrt{3}}{2}(x_B - x_C) \end{cases} \tag{1.12}$$

If the A, B, and C components are to be computed from the D and Q ones, the $2 \rightarrow 3$ transformation is the following:

$$\begin{cases} x_A = \dfrac{2}{3} x_D \\[2mm] x_B = \dfrac{1}{3}\left(-x_D + \sqrt{3} x_Q\right) \\[2mm] x_C = \dfrac{1}{2}\left(-x_D - \sqrt{3} x_Q\right) \end{cases} \tag{1.13}$$

If no zero-sequence is present, this form permits the instantaneous values of the phase variables to be reconstructed on the basis of the projection of the corresponding space-vector into the axis of each phase, as

$$\begin{cases} x_A(t) = \dfrac{2}{3} Re(\mathbf{x}) \\[2mm] x_B(t) = \dfrac{2}{3} Re(a^2 \mathbf{x}) \\[2mm] x_C(t) = \dfrac{2}{3} Re(a \mathbf{x}) \end{cases} \tag{1.14}$$

This kind of transformation is more frequently adopted in electric drive control [23].

1.4 Coordinate Transformation

Another very important feature offered by the space-vector quantity is the possibility to perform a coordinate transformation by means of a vector rotation. This characteristic is particularly important for all the vector-based control techniques. Let us consider a generic space-vector $\mathbf{x}(t) = |\mathbf{x}| e^{j\alpha_x} = x_D(t) + j x_Q(t)$ and let us assume to represent this space-vector quantity in a generic reference frame rotating at the speed $\omega_g = d\theta_g/dt$, where θ_g is the angle between the direct axis x of the generalized reference frame and the direct axis sD (Figure 1.5).

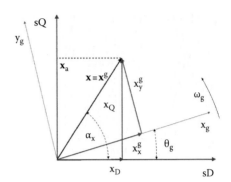

FIGURE 1.5
Generalized reference frame: vector diagram.

A nonlinear transformation is needed to retrieve the space-vector $\mathbf{x}^g = x_x^g + jx_y^g$ expressed in this generic reference frame rotating at the speed ω_g from the corresponding $\mathbf{x}(t) = x_D + jx_Q$ expressed in the stationary reference frame:

$$\mathbf{x}^g = x_x^g + jx_y^g = \mathbf{x}e^{-j\theta_g} \tag{1.15}$$

Decomposing Equation 1.15 in its real and imaginary components, it is possible to compute x_x^g, x_y^g starting from the corresponding components in the stationary reference frame x_D, x_Q as

$$\begin{cases} x_x^g = x_D \cos(\theta_g) + x_Q \, sen(\theta_g) \\ x_y^g = -x_D \, sen(\theta_g) + x_Q \cos(\theta_g) \end{cases} \tag{1.16}$$

If, among the infinite reference frames, that synchronous with the space-vector itself is chosen (the x axis lying in the same direction of $\mathbf{x}$), which implies $\theta_g = \alpha_s$, then

$$\mathbf{x}^g = x_x^g + jx_y^g = \mathbf{x}e^{-j\theta_g} = |\mathbf{x}|e^{j\alpha_s}e^{-j\theta_g} = |\mathbf{x}| \tag{1.17}$$

It means that, in this particular reference frame, the space-vector is a real number coinciding with its amplitude.

If the space-vector in the stationary reference frame $\mathbf{x}(t) = x_D + jx_Q$ has to be computed from that in the generic reference frame $\mathbf{x}^g = x_x^g + jx_y^g$, then the nonlinear transformation inverse of that in Equation 1.15 should be adopted:

$$\mathbf{x} = \mathbf{x}^g e^{-j\theta_g} \tag{1.18}$$

Decomposing Equation 1.18 into its real and imaginary components, it is possible to compute x_D, x_Q starting from the corresponding components in the stationary reference frame as x_x^g, x_y^g:

$$\begin{cases} x_D = x_x^g \cos(\theta_g) - x_y^g \, sen(\theta_g) \\ x_Q = x_x^g \, sen(\theta_g) + x_y^g \cos(\theta_g) \end{cases} \tag{1.19}$$

1.5 Instantaneous Real and Imaginary Powers

In general, the input "instantaneous three power $p_{3\varphi}(t)$" in three-phase systems with a neutral wire is given by Akagi et al., [14]

$$S_{3\varphi}(t) = u_A(t)i_A(t) + u_B(t)i_B(t) + u_C(t)i_C(t) + 3u_0(t)i_0(t) \tag{1.20}$$

where
$i_a(t)$, $i_b(t)$, $i_c(t)$ and $u_a(t)$, $u_b(t)$, $u_c(t)$ are the input phase currents and voltages
$i_0(t)$ and $u_0(t)$ are the zero-sequence current and voltage

Equation 1.20 can be expressed in terms of the corresponding non-power-invariant space-vectors **u** and **i** as follows:

$$p_{3\varphi}(t) = \frac{3}{2}\mathrm{Re}(\mathbf{ui^*}) + 3u_0(t)i_0(t) = p(t) + 3u_0(t)i_0(t) \tag{1.21}$$

where
* symbol stands for the conjugate operator
$p(t) = (3/2)\mathrm{Re}(\mathbf{ui^*})$ is called "instantaneous real power"

If no neutral wire exists, then there is no zero-sequence current component and $p_{3\varphi}(t) = p(t)$, while $s(t) = \mathbf{ui^*}$ is defined as the instantaneous complex power.

In particular, if the space-vector is expressed in the rectangular components D and Q, where D and Q form a monometric orthogonal frame reference (like the one of Figure 1.1 where D and Q coincide with sD and sQ, respectively), then

$$\mathbf{u} = u_D + ju_Q$$
$$\mathbf{i} = i_D + ji_Q \tag{1.22}$$

which, by using (1.21), yields

$$p(t) = \frac{3}{2}(u_D i_D + u_Q i_Q) \tag{1.23}$$

if no zero-sequence is present. The physical result is easy to interpret, since the instantaneous power is given by the sum of the powers of the fictitious D and Q phases.

In a three-phase system in "sinusoidal steady-state," without any zero-sequence component, the complex power can be defined as

$$\mathbf{S} = \frac{3}{2}\mathbf{ui^*} \tag{1.24}$$

By bearing in mind (1.4), since the sinusoidal steady-state is considered, and the analogous for the voltage, the active power P and the reactive power Q can be introduced as

$$\begin{cases} P = \text{Re}(\mathbf{S}) = \dfrac{3}{2}\text{Re}(\mathbf{ui^*}) & (1.25\,\text{a}) \\[4mm] Q = \text{Im}(\mathbf{S}) = \dfrac{3}{2}\text{Im}(\mathbf{ui^*}) & (1.25\,\text{b}) \end{cases}$$

If the current space-vector $\mathbf{i}$ lags the voltage space-vector $\mathbf{u}$ by the angle φ, then (1.25a and 1.25b) become the well-known

$$\begin{cases} P = 3UI\cos\varphi & (1.26\,\text{a}) \\ Q = 3UI\sin\varphi & (1.26\,\text{b}) \end{cases}$$

The earlier definition entails the introduction of a novel variable, called "instantaneous imaginary power" as follows:

$$q(t) = \frac{3}{2}\text{Im}(\mathbf{ui^*}) \qquad (1.27)$$

It should be remarked that the literature also gives the definition of q as

$$q(t) = \frac{3}{2}\text{Re}(-j\mathbf{ui^*}) = \frac{3}{2}\mathbf{u}\wedge\mathbf{i} \qquad (1.28)$$

which is the vector product of the voltage space-vector $\mathbf{u}$ with the space-vector $\mathbf{i}$ in accordance with the definition of the instantaneous real power as their scalar product:

$$p(t) = \frac{3}{2}\text{Re}(\mathbf{ui^*}) = \mathbf{u}\cdot\mathbf{i} = \mathbf{u}^T\mathbf{i} \qquad (1.29)$$

In the following, this definition, which entails a difference in sign with the definition of Q in sinusoidal steady-state, will not be adopted.

To summarize, in a three-phase system, the instantaneous real and imaginary power have been introduced ($p(t)$ and $q(t)$, respectively), along with the instantaneous complex power $s(t)$:

Non-power invariant:
$$\begin{cases} p(t) = \dfrac{3}{2}\text{Re}(\boldsymbol{ui^*}) & (1.30\,\text{a}) \\[4mm] q(t) = \dfrac{3}{2}\text{Im}(\boldsymbol{ui^*}) & (1.30\,\text{b}) \\[4mm] s(t) = \boldsymbol{ui^*} & (1.30\,\text{c}) \end{cases}$$

However, it should be remarked that the earlier equations have been obtained by using the non-power-invariant form of the space-vectors, as defined, for example, by (1.6), where the factor 2/3 is present.

If however the power-invariant form of the space-vector is chosen, characterized by the factor $\sqrt{2/3}$, then the following equations are obtained:

$$\text{Power invariant:} \begin{cases} p(t) = \mathrm{Re}(\mathbf{ui}^*) & (1.31\mathrm{a}) \\ q(t) = \mathrm{Im}(\mathbf{ui}^*) & (1.31\mathrm{b}) \\ s(t) = \mathbf{ui}^* & (1.31\mathrm{c}) \end{cases}$$

where it should be borne in mind that the involved space-vectors are not the same as those of (1.30a through 1.30c), for the reasons explained above. It would be therefore recommended to specify always if the power-invariant or the non-power-invariant form of the space-vector is used.

(1.22), (1.25), and (1.30b) yield for the non-power-invariant space-vectors:

$$q(t) = \frac{3}{2}(u_Q i_D - u_D i_Q) \tag{1.32}$$

while (1.9), (1.28), and (1.32) yield for the power-invariant space-vectors:

$$q(t) = (u_Q i_D - u_D i_Q) \tag{1.33}$$

By using the Clark transformation (1.9), (1.32) can be transformed into

$$q(t) = \frac{1}{\sqrt{3}}\left[(u_b - u_c)i_a + (u_c - u_a)i_b + (u_a - u_b)i_c\right] \tag{1.34}$$

In Chapters 2 and 3, it will be more useful to use the power definitions (1.31) derived from the power-invariant form. For this reason, some further considerations will be developed in the following by using these space-vectors, but simple computations can be easily made to obtain the counterpart in case the non-power-invariant space-vectors were used.

Now (1.31a) and (1.31b) can be written in matrix form as

$$\begin{bmatrix} p \\ q \end{bmatrix} = \begin{bmatrix} u_D & u_Q \\ u_Q & -u_D \end{bmatrix} \begin{bmatrix} i_D \\ i_Q \end{bmatrix} \tag{1.35}$$

Since the matrix is orthogonal, the following relationship can be easily inverted to obtain

$$\begin{bmatrix} i_D \\ i_Q \end{bmatrix} = \frac{1}{u_D^2 + u_Q^2}\begin{bmatrix} u_D & u_Q \\ u_Q & -u_D \end{bmatrix} \begin{bmatrix} p \\ q \end{bmatrix} \tag{1.36}$$

which can be rewritten as

$$\mathbf{i} = \begin{bmatrix} i_D \\ i_Q \end{bmatrix} = \frac{1}{u_D^2 + u_Q^2}\left(\begin{bmatrix} u_D \\ u_Q \end{bmatrix}p + \begin{bmatrix} u_Q \\ -u_D \end{bmatrix}q\right) \overset{\Delta}{=} \begin{bmatrix} i_{Dp} \\ i_{Qp} \end{bmatrix} + \begin{bmatrix} i_{Dq} \\ i_{Qq} \end{bmatrix} = \mathbf{i}_p + \mathbf{i}_q \tag{1.37}$$

where
  $\mathbf{i}_p$ is the instantaneous active current
  $\mathbf{i}_q$ is the instantaneous reactive current

From (1.37), the components of these currents on the axis D and Q can be computed. Specifically,

$$i_{Dp} = \frac{u_D}{u_D^2 + u_Q^2} p \quad \text{instantaneous active current on the D-axis} \tag{1.38}$$

$$i_{Qp} = \frac{u_Q}{u_D^2 + u_Q^2} p \quad \text{instantaneous active current on the Q-axis} \tag{1.39}$$

$$i_{Dq} = \frac{u_Q}{u_D^2 + u_Q^2} q \quad \text{instantaneous reactive current on the D-axis} \tag{1.40a}$$

$$i_{Qq} = \frac{-u_D}{u_D^2 + u_Q^2} q \quad \text{instantaneous reactive current on the Q-axis} \tag{1.40b}$$

Then, since

$$p(t) = u_D i_D + u_Q i_Q = \mathbf{u}^T \mathbf{i} = \mathbf{u}^T (\mathbf{i}_p + \mathbf{i}_q) = \mathbf{u}^T \mathbf{i}_p + \mathbf{u}^T \mathbf{i}_q = u_D i_{Dp} + u_D i_{Dq} + u_Q i_{Qp} + u_Q i_{Qq} \tag{1.41}$$

using (1.38) and (1.39), it results that

$$p(t) = \frac{1}{u_D^2 + u_Q^2} (u_D^2 p + u_Q^2 p + u_D u_Q q - u_D u_Q q) \tag{1.42}$$

From (1.41), it is clear that $p(t)$ is only given by the terms corresponding to $u_D i_{Dp} + u_Q i_{Qp}$, that is, the sum of the product of the D component of the voltage space-vector and the instantaneous active current on the D axis and the product of the Q component of the voltage space-vector and the instantaneous active current on the Q axis. That is why (1.38) and (1.39) are called instantaneous active currents.

The last two terms of the sum of (1.42), $u_D u_Q q - u_D u_Q q$, sum up to zero and correspond to the term $u_Q i_{Qp} + u_Q i_{Qq}$ of (1.42). This means that the terms depending on q (1.40a and 1.40b) do not give any "instantaneous" power contribution between the source and the load. That is why (1.40) are called instantaneous reactive currents. Remark that the instantaneous contribution of these currents in terms of power is null, not the average flow of the power, as in the case of reactive power in sinusoidal steady-state.

On the basis of what was written earlier, the powers can be separated as follows:

$$p_{Dp} = u_D i_{Dp} = \frac{u_D^2}{u_D^2 + u_Q^2} p \quad \text{instantaneous active power on the D axis} \tag{1.43a}$$

$$p_{Dq} = u_Q i_{Qp} = \frac{u_Q^2}{u_D^2 + u_Q^2} p \quad \text{instantaneous active power on the Q axis} \tag{1.43b}$$

$$p_{Dq} = u_D i_{Dq} = \frac{u_D u_Q}{u_D^2 + u_Q^2} q \quad \text{instantaneous reactive power on the D axis} \tag{1.44a}$$

$$p_{Qq} = u_Q i_{Qq} = \frac{-u_D u_Q}{u_D^2 + u_Q^2} q \quad \text{instantaneous reactive power on the Q axis} \tag{1.44b}$$

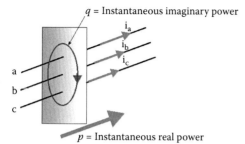

q = Instantaneous imaginary power

i_a
i_b
i_c

a
b
c

p = Instantaneous real power

FIGURE 1.6
Physical meaning of the instantaneous real and imaginary powers.

By inspection of (1.44a and 1.44b) and the last two terms of the sum of (1.42), $u_D u_Q q -u_D u_Q q$, it can be concluded that the instantaneous imaginary power q "is proportional to the quantity of energy that is being exchanged between the phases of the system. It does not contribute to the energy transfer between the source and the load at any time" [14].

Figure 1.6 summarizes the physical meaning of the instantaneous real and imaginary powers.

In a symmetrical three-phase doubly fed electric machine, which is where both the stator and the rotor circuits are supplied, the total instantaneous input power is the sum of the contribution of the stator and the rotor circuits, which gives

$$p(t) = \text{Re}(\mathbf{u}_s \mathbf{i}_s^* + \mathbf{u}_r \mathbf{i}_r^*) \tag{1.45}$$

References

1. G. Kron, *Equivalent Circuits of Electrical Machinery*, New York: John Wiley & Sons Inc., 1951.
2. P. L. Alger, *The Nature of Induction Machine*, New York: Gordon and Breach, 1965.
3. M. Kostenko, L. Piotrovsky, *Electrical Machines*, Moscow, Russia: MIR, 1969 (two volumes).
4. K. P. Kovacs, I. Ràcz, *Transiente Vorgänge in Wechselstrommachinen*, Akadémia Kiadò, Budapest, Hungary, 1954.
5. K. P. Kovacs, *Transient Processes in Electrical Machines (in Hungarian)*, Müszaki Konyvkiadò, Budapest, Hungary, 1970.
6. K. P. Kovacs, Programmierung von Asynchronenmotoren unter Berük-sichtigung der Sättigung, *Archiv für Elektrotechnik*, 47(4), 193–200, 1962.
7. N. Mohan, T. M. Undeland, W. P. Robbins, *Power Electronics, Converters, Applications and Design*, John Wiley & Sons, Inc., New York, 1995.
8. M. P. Kazmierkowski, R. Krishnan, F. Blaabjerg, *Control in Power Electronics*, London, U.K.: Academic Press, 2002.
9. M. P. Kazmierkowski, Current control techniques for three-phase voltage-source PWM converters: A survey, *IEEE Transactions on Industrial Electronics*, 45(5), 691–703, 1998.
10. M. P. Kazmierkowski, H. Tunia, *Automatic Control of Converter-Fed Drives*, Elsevier, Amsterdam, the Netherlands, 1994.
11. M. Malinowski, Sensorless control strategies for three—Phase PWM rectifiers, PhD thesis, University of Warsaw, Warsaw, Poland, 2001.
12. M. Malinowski, M. P. Kazmierkowski, S. Hansen, F. Blaabjerg, Virtual-flux-based direct power control of three-phase PWM rectifiers, *IEEE Transactions on Industrial Applications*, 37(4), 1019–1027, 2001.

13. M. Malinowski, M. P. Kazmierkowski, M. Trzynadlowski, A comparative study of control techniques for PWM rectifiers in AC adjustable speed drives, *IEEE Transactions on Power Electronics*, 18(6), 1390–1396, 2003.

14. H. Akagi, E. H. Watanabe, M. Aredes, *Instantaneous Power Theory and Applications to Power Conditioning*, IEEE Press, Wiley-Interscience, Piscataway, NJ, 2007.

15. H. Akagi, Active harmonic filters, *Proceedings of the IEEE*, 93(12), 2128–2141, 2005.

16. H. Akagi, Trends in active power conditioners, *IEEE Transactions on Power Electronics*, 9(3), 263–268, May 1994.

17. M. Godoy Simões, F. A. Farret, *Alternative Energy Systems: Design and Analysis with Induction Generators*, 2nd edn., CRC Press, Boca Raton, FL.

18. P. Vas, *Electric Machines and Drives, a Space-Vector Theory Approach*, Clarendon Press, Oxford, U.K., 1992.

20. P. Vas, *Sensorless Vector and Direct Torque Control*, Oxford Science Publications, Oxford, U.K., 1998.

22. P. Vas, *Vector Control of AC Machines*, Oxford Science Publications, Oxford, U.K., 1990.

23. W. Leonhard, *Control of Electrical Drives*, Berlin, Germany: Springer-Verlag, 1997.

24. J. Holtz, Optimal pulsewidth modulation for AC servos and low-cost industrial drives, *IEEE Transactions on Industrial Applications*, 30(4), 1039–1047, July/August 1994.

25. B. K. Bose, *Power Electronics and Motor Drives: Advances and Trends*, Academic Press, London, U.K., 2006.

26. B. K. Bose, *Modern Power Electronics and AC Drives*, Prentice Hall, Upper Saddle River, NJ, 2001.

27. B. K. Bose, *Power Electronics and Variable Frequency Drives: Technology and Applications*, Wiley-IEEE Press, New York, 1996.

28. D. G. Holmes, T. A. Lipo, *Pulse Width Modulation for Power Converters: Principles and Practice*, Wiley-IEEE Press, New York, 2003.

29. J. G. Kassakian, M. F. Schlecht, G. C. Verghese, *Principles of Power Electronics*, Prentice Hall, Upper Saddle River, NJ, 1991.

30. J. Holtz, Sensorless control of induction motor drives, *Proceedings of the IEEE*, 90(8), 1359–1394, 2002.

Part I

Power Converters

2

Pulsewidth Modulation of Voltage Source Inverters

2.1 Fundamentals of Voltage Source Inverters

The three-phase voltage source pulsewidth modulation (PWM) converter, also called voltage source inverter (VSI), can perform the connection between the DC side and the AC side of a source/load coupling: if the energy flows from the DC side to AC side, the converter behaves like an inverter; otherwise, it behaves like a rectifier (see Figure 2.1).

The circuit is based on the so-called switch mode operation of semiconductor devices. They are suitably driven to be either in the ON or in the OFF state. In the ON state, each device, in the ideal case, has a null voltage, and its current depends on the external circuit; in the OFF state, the current is null, and the voltage depends on the external circuit. Differently from a classical switch, power electronic devices allow the current to flow only in one direction. An antiparallel diode is therefore added when reverse current flow is necessary; this is current a common situation in supplying inductive loads. This diode placed antiparallel to the power switch is commonly called free-wheeling diode.

No operation in the active region of the power switch occurs; this behavior is similar to that of a conventional switch, and, being nonlinear, the whole circuit is nonlinear.

In real cases, power devices will be characterized by the maximum allowable voltage in the OFF state, the maximum current in the ON state, the turn-on time, that is the time needed to commutate from OFF to ON state, and the turn-off time, that is, the time needed to commutate them from ON to OFF. It is also noteworthy that a circuit (known as driver circuit is further needed), which converts the logic signal into an electric signal is able to drive the power device.

Manufacturers have been trying to improve these parameters by designing devices with high blocking voltage and high conduction current or lower turn-on and turn-off time requiring a cheap driver; these efforts have yielded power devices such as BJT, MOSFET, IGBT, and so on.

The devices in each leg (A, B, and C) in Figure 2.1 must in a complementary way; the ON state of a device must correspond to the OFF state of the complementary device in the same leg and vice versa so as to avoid the short circuit of the DC side. This condition must be valid also during commutation. In order to prevent this undesired condition, known as shoot-through, an additional time interval, known as dead time, before ON commutation of each device, is introduced; during the dead time, both power devices of a leg are in OFF state, and the effects due to the storage charge are definitively canceled. The dead time is small compared to the switching time of the power devices, but it must be a bit greater of the time needed to commutate from ON to OFF and vice versa in the power devices. The contemporary OFF state of two devices on the same leg is admissible; however, such

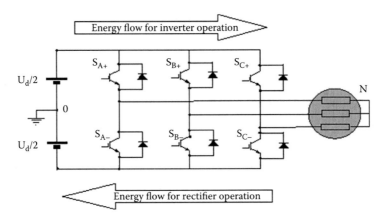

FIGURE 2.1
Schematics of VSI with energy flow direction for inverter and rectifier behavior.

condition does not allow the voltage to be controlled at the correspondent output and is usually avoided for long time intervals. In IGBT-based power converters with power ratings of few kilovolt-amperes, the dead time is usually set between 1 and $4\,\mu s$, depending on the speed of the device itself.

The state of each device is usually indicated by using the binary notation: conventionally the ON state corresponds to "1," while the OFF state to "0." To identify the status of the converter, only three bits are sufficient because, by indicating the status of the upper device, the status of the lower device is implicitly defined (it is its negation as explained above) and so is the inverter output voltage. Table 2.1 gives all possible values of the phase voltage with respect to the middle point 0 of the DC link (see Figure 2.1) and to the neutral point N of the load.

In Table 2.1, the second column represents the status of the upper devices. A gray code (with the less meaningful bit on the left) is recognizable going from u_0 to u_7. The usefulness of this characteristic will be clarified in the section dedicated to the space-vector modulation (SV-PWM).

The voltage u_{N0} is the so-called common-mode voltage common mode voltages (CMVs) and is given by $(u_{sA0} + u_{sB0} + u_{sC0})/3$ if the load is balanced.

TABLE 2.1

Possible Values of the Phase Voltage with Respect to the Middle Point 0 of the DC Link

	$(S_a\,S_b\,S_c)$	u_{sA0}	u_{sB0}	u_{sC0}	u_{sAN}	u_{sBN}	u_{sCN}	u_{N0}
u_0	(000)	$-U_d/2$	$-U_d/2$	$-U_d/2$	0	0	0	$-U_d/2$
u_1	(100)	$U_d/2$	$-U_d/2$	$-U_d/2$	$2U_d/3$	$-U_d/3$	$-U_d/3$	$-U_d/6$
u_2	(110)	$U_d/2$	$U_d/2$	$-U_d/2$	$U_d/3$	$U_d/3$	$-2U_d/3$	$U_d/6$
u_3	(010)	$-U_d/2$	$U_d/2$	$-U_d/2$	$-U_d/3$	$2U_d/3$	$-U_d/3$	$-U_d/6$
u_4	(011)	$-U_d/2$	$U_d/2$	$U_d/2$	$-2U_d/3$	$U_d/3$	$U_d/3$	$U_d/6$
u_5	(001)	$-U_d/2$	$-U_d/2$	$U_d/2$	$-U_d/3$	$-U_d/3$	$2U_d/3$	$-U_d/6$
u_6	(101)	$U_d/2$	$-U_d/2$	$U_d/2$	$U_d/3$	$-2U_d/3$	$U_d/3$	$U_d/6$
u_7	(111)	$U_d/2$	$U_d/2$	$U_d/2$	0	0	0	$U_d/2$

A voltage space-vector* can be associated with each inverter configuration u_k:

$$\mathbf{u_s} = \frac{2}{3}[u_{sA}(t) + au_{sB}(t) + a^2 u_{sC}(t)] = u_{sD}(t) + ju_{sQ}(t) = \mathbf{u_s} e^{j\alpha_s} \tag{2.1}$$

where $u_{sA}(t)$, $u_{sB}(t)$, and $u_{sC}(t)$ are the voltages, with respect to the middle point, of the phase A, B, and C, respectively.

A VSI can generate eight voltage space-vectors as it is clear from Table 2.1. Two of these vectors, corresponding to u_0 (all lower devices in ON state) and u_7 (all upper devices in ON state), give zero voltage with respect to the N-point. For these reasons, they are called zero vectors.

If the N-point is isolated, then the sum of the load currents on the three phases is null, $i_{sA}(t) + i_{sB}(t) + i_{sC}(t) = 0$; the phase voltages are directly linked to the logic state of the upper devices as follows:

$$\begin{cases} u_{sA} = \dfrac{u_d}{3}(2S_a - S_b - S_c) \\[2mm] u_{sB} = \dfrac{u_d}{3}(2S_b - S_a - S_c) \\[2mm] u_{sC} = \dfrac{u_d}{3}(2S_c - S_a - S_b) \end{cases} \tag{2.2}$$

The voltage space-vector is then given by replacing (2.2) into (2.1), which yields the following:

$$\mathbf{u_s} = \frac{2}{3}u_d[S_a + aS_b + a^2 S_c] \tag{2.3}$$

The six active configurations (u_k, k = 1,..., 6) give voltage space-vectors with amplitude equal to $2u_d/3$ and phase equal to $q_k = (k-1)\pi/3$. Then from (2.3), the generable voltage space-vectors therefore are as follows:

$$\mathbf{u_s} = \begin{cases} \dfrac{2}{3}u_d e^{j(k-1)\pi/3} & k = 1,2,...,6 \\[2mm] 0 & k = 0,7 \end{cases} \tag{2.4}$$

The vertices of the nonnull six vectors define the so-called characteristic hexagon of the VSI, as shown in Figure 2.2.

Even if the inverter is a highly nonlinear circuit, it is operated like a linear amplifier: its output must follow an input reference signal with the same shape and magnified amplitude. This task is performed through a suitable control strategy of the power devices, called PWM technique. Several PWM techniques have been proposed in literature whose performance can be assessed using the criteria given in the following section [1,2].

---

* In this chapter, the nonpower invariant definition of the space-vector is used (see Chapter 1).

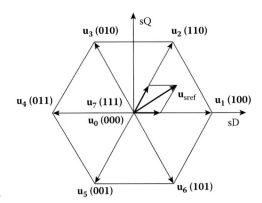

FIGURE 2.2
Characteristic hexagon of a VSI inverter.

2.1.1 Performance Criteria

In real working conditions, the desired output of the inverter is different from the ideal behavior of a linear amplifier: several parameters, like the distortion of the current waveform, the harmonic losses in the power converter and in the load, the oscillation of the electromagnetic torque (if the load is a rotating machine drive), and the common mode voltages (CMVs), are to be considered in the performance criteria, which are therefore important guidelines for the correct choice of the PWM strategy [2,3].

2.1.1.1 Current Harmonics

The rms harmonic current is defined as follows:

$$I_{h\,rms} = \sqrt{\frac{1}{T}\int_T [i(t) - i_1(t)]^2 dt} \tag{2.5}$$

where
   $i_1(t)$ is the time domain fundamental current
   T is the period

It should be noted that this parameter depends on both the modulation algorithm and on the load impedance. This last dependence can be avoided using a normalized harmonic current (distortion factor):

$$d = \frac{I_{h\,rms}}{I_{h\,rms\,six\text{-}step}} \tag{2.6}$$

in which the harmonic current $I_{h\,rms}$ is referred to the current $I_{h\,rms\,six\text{-}step}$ obtained with the same load but operating by repeating the sequence $\mathbf{u}_1 - \mathbf{u}_2 - \mathbf{u}_3 - \mathbf{u}_4 - \mathbf{u}_5 - \mathbf{u}_6$ (see Table 2.1) with the same period of the fundamental output waveform or six-step operation: in this case the obtained line-to-line output voltage waveform exhibits six steps.

   At six-step operation, $d = 1$, however, this parameter can be greater than unity. The harmonic current of a space-vector trajectory can be computed as follows:

$$I_{h\,rms} = \sqrt{\frac{1}{T}\int_T [i(t) - i_1(t)][i(t) - i_1(t)]^* dt} \tag{2.7}$$

where the (*) stands for complex conjugate.

The copper losses in the load are proportional to the square of the harmonic current and, as a consequence d^2 represents the loss factor.

2.1.1.2 Harmonic Spectrum

The synchronized PWM occurs when the switching frequency can be expressed as $f_s = Nf_1$, where f_1 is the fundamental frequency and N is an integer. In this case, the spectrum is composed of discrete current components $h_i(k \cdot f_1)$ with k the order of the harmonic component:

$$h_i(k \cdot f_1) = \frac{I_{h\,rms}(kf_1)}{I_{h\,rms\,six\text{-}step}} \tag{2.8}$$

It should be noted that all components are normalized as in Equation 2.6 so they do not depend on the load.

For nonsynchronized PWM, a continuous amplitude frequency spectrum $h_d(f)$ can be defined by dividing current spectrum amplitude for the harmonic current in six-step mode. The continuous spectrum contains both periodic and nonperiodic components; the $h_d(f)$ function, differently from a discrete spectrum that is dimensional, is measured in $[\mathrm{Hz}^{-1/2}]$.

The normalized harmonic current can be calculated starting from Equation 2.6 for a discrete spectrum and for amplitude density spectrum as follows:

$$d = \frac{I_{h\,rms}}{I_{h\,rms\,six\text{-}step}} = \frac{\sqrt{\dfrac{1}{T}\displaystyle\int_T (i(t)-i_1(t))^2\,dt}}{I_{h\,rms\,six\text{-}step}} = \frac{\displaystyle\sum_{k\neq1} I_{k\,rms}(kf_h)}{I_{k\,rms\,six\text{-}step}} = \sqrt{\sum_{k\neq1} h_k^2(k\cdot f_1)} \tag{2.9}$$

$$d = \frac{\sqrt{\dfrac{1}{T}\displaystyle\int_T (i(t)-i_1(t))^2\,dt}}{I_{h\,rms\,six\text{-}step}} = \sqrt{\frac{1}{T}\int_T \left(\frac{i(t)-i_1(t)}{I_{h\,rms\,six\text{-}step}}\right)^2 dt} = \sqrt{\int_{0,f\neq f_1}^{\infty} h_d^2(f)\,df} \tag{2.10}$$

Moreover, multiplying the distortion factor and the switching frequency of the inverter, another figure of merit is obtained. This parameter allows comparing different PWM algorithms operated at different switching frequency; the pulse number N must be greater than 15, and, for lower values of N, the relation is nonlinear.

2.1.1.3 Maximum Modulation Index

A modulation index can be defined as the normalized fundamental voltage:

$$M = \frac{u_1}{u_{1(six\text{-}step)}} \tag{2.11}$$

where
   u_1 is the amplitude of the fundamental voltage obtained with the selected PWM
   $u_{1(six\text{-}step)}$ is the amplitude of the fundamental voltage in six-step operation

It is verified when the relation $0 \leq M \leq 1$ is valid and $M = 1$ means six-step mode operation. The maximum achievable modulation index M_{max} can vary even of 25%, in dependence of which PWM technique is chosen.

It should be noted that the maximum modulation index is important because the power delivered by a PWM inverter is proportional to the maximum voltage at the AC side.

2.1.1.4 Torque Harmonics

Torque harmonics are responsible for vibrations and noise in electric drives, whose mitigation is important both in civil (e.g., passenger's comfort) and military (e.g., acoustic discretion) applications.

In general, vibrations can have aerodynamic, mechanical, or electromagnetic origin; torque harmonics are mainly tied to electromagnetic causes and particularly to harmonic currents.

The torque ripple can be expressed as

$$\Delta T = \frac{(T_{e\max} - T_{eav})}{T_L} \tag{2.12a}$$

where

T_{emax} is the maximum air-gap torque
T_{eav} is the average air-gap torque
T_L is the rated machine load torque

$$\delta_T \% = \frac{100}{T_{e0}} \sqrt{\frac{1}{N} \sum_{i}^{N} (T_{ei} - T_{e0})^2} \tag{2.12b}$$

where

T_{e0} is the desired torque
T_{ei} is the torque at the ith sample time
N is the number of samples

2.1.1.5 Switching Frequency and Switching Losses

Losses in a power switch occur during the conduction and during commutation.

During conduction, lost energy is given by the product of the conduction current (i_0), the voltage drop of the device (v_{ON}), and the conduction time (T_{ON}). The mean power $\langle P_{ON} \rangle$ during a switching period is given by the following:

$$\langle P_{ON} \rangle = \frac{i_0 v_{ON} T_{ON}}{T_{sw}} = i_0 v_{ON} T_{ON} f_{sw} \tag{2.13}$$

where T_s is the switching time duration whose inverse is the switching frequency $f_{sw} = 1/T_{sw}$.

During the commutation from conduction to interdiction, the voltage across the device rises from v_{ON} to the maximum value, and the current falls to zero. During this transient, both the voltage and the current are nonnull resulting in power losses.

Power losses during commutation can be described considering the triangle defined by the voltage rise, and the current fall, having as the base the sum of the rising time of the

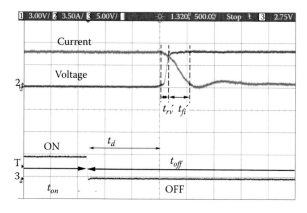

FIGURE 2.3
Time domain waveform in a power device during a commutation from ON to OFF state.

voltage (t_{rv}) and the falling time of the current (t_{fi}). It is shown in Figure 2.3. Commutation is usually delayed respect to the command (t_d) because of the electronic circuitry needed for the command of the power devices.

The energy lost during the transition is given with a good approximation by the area of the triangle:

$$E_{ON \to OFF} = \frac{1}{2} v_0 i_0 (t_{rv} + t_{fi}) \tag{2.14}$$

where
  v_0 is the voltage acting on the power switch during the OFF state
  i_0 represents its current during the ON state

The analysis of this figure shows that the faster the commutation, the lower the area of the aforementioned triangle, the lower the switching losses in the device.

Similar considerations are valid for the transition from interdiction to conduction, considering now the rise time of the current (t_{ri}) and the decay time of the voltage (t_{fv}). In this way the mean power loss $\langle P_{SW} \rangle$ can be computed:

$$\langle P_{SW} \rangle = \frac{v_0 i_0}{2} \frac{[(t_{rv} + t_{fi}) + (t_{ri} + t_{fv})]}{T_{sw}} = \frac{v_0 i_0}{2} \cdot [(t_{rv} + t_{fi}) + (t_{ri} + t_{fv})] f_{sw} \tag{2.15}$$

It should be noted that both (2.13) and (2.15) depend on switching frequency. The higher the f_{sw}, the higher the power losses, and more attention is to be paid to electromagnetic compatibility of the system; on the other hand, the dynamic performance of the converter is improved, and the size of the filtering reactive devices is less bulky.

Other power losses occur during ON state. In particular, the mean power in a switching period is given by $\langle P_{ON} \rangle = i_0 v_{on} t_{on} f_{sw}$, where v_{on} is the drop voltage due to the conduction current i_0 and t_{on} is the time in which the power device is in conduction state. However, this term is usually negligible compared to the switching losses and the power losses during OFF state in which the current is practically null.

Finally, currents depending on switching frequency produce magnetostrictive mechanical deformations in magnetic materials resulting in possible generation of acoustic noise, in the range of audible frequencies, between 50 Hz and 10 kHz, where the peak of human sensitivity is around 1–2 kHz.

2.1.1.6 Common-Mode Voltage (CMV)

The CMV at the output of the inverter, defined as $u_{N0} = (u_{Sa} + u_{sB} + u_{sC})/3$, is one of the effects of the nonideality of the inverter considered as a linear amplifier. Particularly, even if the input references are sinusoidal and represent a symmetrical three-phase system, the CMV will be different from zero due to nonideality of the VSI. For each inverter configuration, a value of u_{N0} is identified in Table 2.1. The spectrum of this waveform depends both on the switching frequency and on the speed of commutations.

High-level CMV variations cause high-frequency common-mode currents flowing to the ground through the parasitic capacitances between the different parts of the VSI load and the ground. In particular, for electric drives with induction machine (IM), these (CMVs) can cause the drive to be less reliable: the ball bearings can deteriorate, unexpected fault relay tripping can occur, or nearby electronic equipments can be disturbed.

Various kinds of solutions have been devised so far to diminish the common-mode emissions. Some of them are hardware solutions based on the application of either passive filters [4] or active cancellers adding additional power devices to the converter structure [5]. Another idea for lessening the common-mode emissions is to employ modified converter structures, for example, four-leg inverters [6], or to properly design the electrical machine [7]. All of the previous solutions imply, however, an increase of the overall cost of the drive and a reduction of its reliability. Another approach consists in limiting the cause of the common-mode disturbance by proper selection of the switching pattern of the inverter [8–10]. This PWM technique is more deeply described in Section 2.2.4.

2.2 Open-Loop PWM

According to the control strategy and particularly the presence or not of a feedback, the modulation techniques can be at first classified on open-loop schemes and closed-loop schemes as shown in Figure 2.5. The open-loop schemes are based on the application of the voltages generated by inverter on the basis of a reference value on a load (see Figure 2.4).

Four main categories can be recognized. The first, named "carrier based", comprises five methods: suboscillation method, modified suboscillation method, sampling techniques, SV-PWM, and modified SV-PVM. The second and the third are known as "carrierless

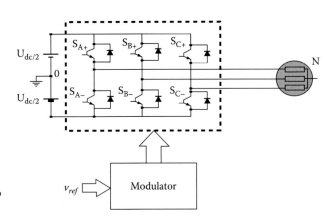

FIGURE 2.4
Representation of a VSI with open-loop control strategy.

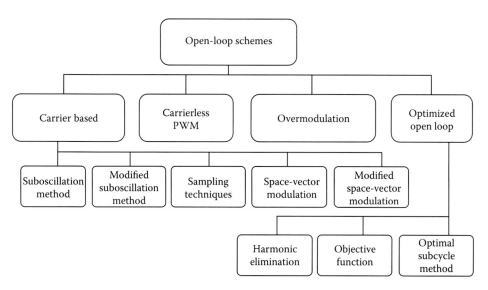

FIGURE 2.5
Classification of open-loop control strategy.

PWM" and "overmodulation" techniques. Finally, the "optimized open loop" is divided into harmonic elimination, objective function, and optimal subcycle methods.

2.2.1 Carrier-Based PWM

2.2.1.1 Suboscillation Method

This method employs an individual modulator for each of the three phases. Three sinusoidal voltage references are needed u_{sAref}, u_{sBref}, and u_{sCref} for phase sA, sB, and sC, respectively, representing a symmetrical three-phase system; they can be obtained from the reference vector u_{sref} using the following equations (see Chapter 1):

$$\begin{cases} u_{sAref} = \Re e[\mathbf{u}_{sref}] \\ u_{sBref} = \Re e[a^2\mathbf{u}_{sref}] \\ u_{sCref} = \Re e[a\mathbf{u}_{sref}] \end{cases} \tag{2.16}$$

These references are compared with a triangular carrier (the same for each phase) whose frequency is much higher than the frequency of the references. The result of the comparison is a logic signal for each leg of the inverter controlling the complementary power switches shown in the block diagram of Figure 2.6. An example of triangular carrier and two reference voltages corresponding to ratios between the carrier and the reference respectively equals to 1 and to 0.5 in Figure 2.7.

This PWM technique does not exploit the DC link voltage at the best. As a matter of fact, the maximum allowed modulation index (see Equation 2.11) is reached when the amplitude of the carriers equals that of the reference that corresponds to $M_{max} = (\pi/4) = 0.785$, obtained with $m = 1$. In this case, $u_{sA0} = U_d/2$; taking into account that for six-step operation, $u_{sA0} = (U_d/2)(4/\pi)$, it follows that $M_{max} = \pi/4 = 0.785$. As a consequence, the DC link is not fully exploited, and the maximum amplitude of the line-to-line voltage is equal to $\sqrt{3}U_d/2 = 0.866\,U_d$. It should be noted that when the switches S_{A+} and S_{B-} are in the ON state, the line-to-line voltage between phases sA and sB is equal to U_d.

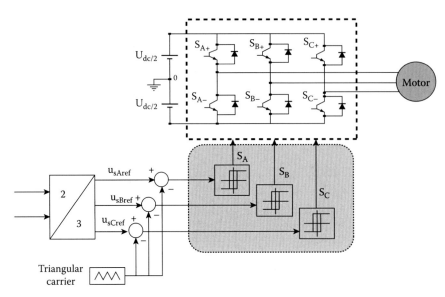

FIGURE 2.6
PWM with suboscillation method.

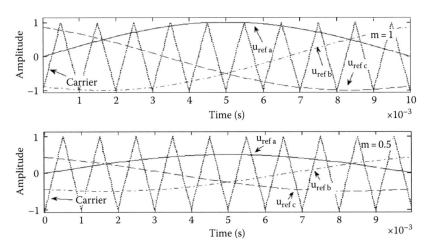

FIGURE 2.7
Triangular carrier and reference voltage signals with m = 1 and m = 0.5.

2.2.1.2 Modified Suboscillation Method

The above method exploits the DC link poorly; a higher maximum value of the fundamental voltage can be achieved by adding to the reference value a waveform with zero sequence. As a matter of fact, in the phase voltage, no triple harmonics are present (this holds for the general case of three-phase symmetrical PWM waveforms), and because all triple harmonics form zero sequences, if the star point of the load is not connected, no corresponding current is produced on the load.

A distorted reference with added zero sequence can be obtained in infinite ways, for example, if a third harmonic with amplitude equal to 25% of the fundamental is added, the maximum modulation index rises to $M'_{max} = 0.882$ [11]. In this case, the technique is called

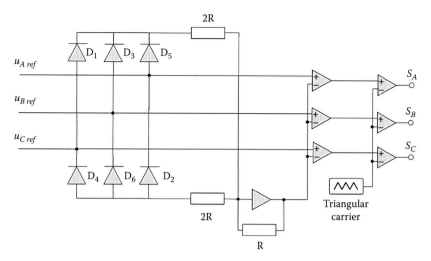

FIGURE 2.8
Circuit for the implementation of analog SV-PWM.

third-harmonic injection PWM (THI-PWM). If a rectangular waveform of triple fundamental frequency is used, the modulation index further increases to $M''_{max} = \sqrt{3}\pi/6 = 0.907$; it represents the maximum value achievable adopting this technique and consequently the best DC link exploitation [12]. If a triangular zero sequence with ¼ amplitude of the fundamental is added, then the obtained PWM corresponds to the conventional (analog) SV-PWM with symmetrical placement of the zero vectors (naturally sampled SV-PWM). Figure 2.8 shows a circuit for the implementation of analog SV-PWM [13]. In this scheme, the three-phase voltages are rectified to produce their envelope magnitude. This magnitude, scaled of a factor 0.5, is added as an offset to each reference voltage.

Figure 2.9 shows the fundamental reference waveform on phase sA, the zero-sequence third-harmonic sinusoidal sequence, and the global reference as obtained when the modulation index m = 0.7 and a 25% sinusoidal third harmonic is added, as obtained with the THI-PWM.

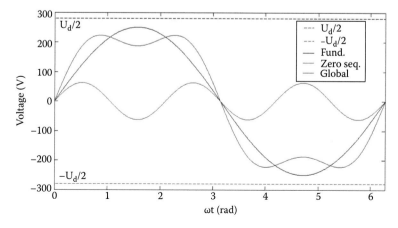

FIGURE 2.9
Reference fundamental, zero-sequence, and global voltages with the THIPWM.

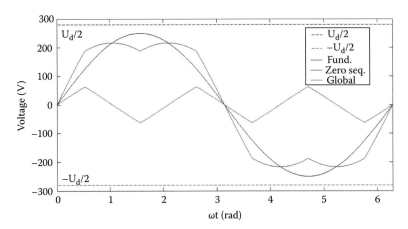

FIGURE 2.10
Reference fundamental, zero-sequence, and global voltages with the SV-PWM with symmetrical placement of
the zero vectors.

Figure 2.10 shows the fundamental reference waveform on phase sA, the zero-sequence
third-harmonic triangular sequence, and the global reference as obtained when the modulation index m = 0.7 and a 25% third harmonic is added, as obtained with the SV-PWM
with symmetrical placement of the zero vectors.

2.2.1.3 Sampling Techniques

In the previous suboscillation method, the reference signal is continuously compared with
the triangular carrier. This is easy to implement by analog circuits but becomes time consuming with microprocessor-based circuits. In the following, it is assumed that the carrier
signal has unitary amplitude while the reference voltage signal is normalized (it takes a
value between 0 and 1).

In Figure 2.11, a representation of the reference and carrier signal is sketched.

Under this assumption, considering that the slope of the triangular carrier is equal to
$2/T_{PWM}$, the equation of the straight line that represents the carrier during $0 < t < T_{PWM}/2$ is
$u_{carrier} = -1 + (4/T_{PWM})t$. If the time t is expressed as $t = (T_{PWM}/4)(u_{carrier}(t) + 1)$, the switching
instants t_{1n} and t_{2n} can be computed by imposing that the carrier is equal to the reference

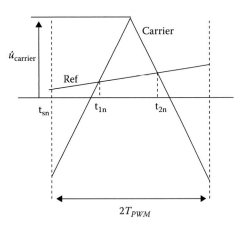

FIGURE 2.11
Representation of the reference and carrier signal.

signal sampled at time t_{sn} (with sampling frequency $f_s = f_{PWM} = 1/T_{PWM}$), that is, $u_{snref}(t_{sn}) = u_{carrier}(t_{1n}) = u_{carrier}(t_{2n})$. To compute t_{2n}, the carrier equation during $T_{PWM}/2 < t < T_{PWM}$ should be considered. As a result, the switching instants are the following:

$$\begin{cases} t_{1n} = \dfrac{1}{4} T_{PWM}(1 + u_{snref}(t_{sn})) \\ t_{2n} = \dfrac{1}{2} T_{PWM} + \dfrac{1}{4} T_{PWM}(1 - u_{snref}(t_{sn})) \end{cases} \tag{2.17}$$

where the reference signal is assumed to be constant during the time interval T_{PWM} and equal to its sampled value; this does not introduce an appreciable error because the frequency of the carrier is typically higher than the frequency of the reference, which means that $u_{sn}(t_{sn}) \approx u_{sn}(t_{1n}) \approx u_{sn}(t_{2n})$.

The switching time interval can be therefore computed as follows:

$$\Delta t = t_{2n} - t_{1n} = \frac{T_{PWM}}{2}(1 - u_{snref}) \tag{2.18}$$

The timing of the command signal to be applied to the power device is the complementary with respect to T_{PWM}. The duty cycle of the phase-n inverter leg $\delta_{sn} = (t_{on}/T_{PWM})$, representing the ratio between the on time t_{on} of its upper device and the T_{PWM}, can be therefore computed as follows:

$$\delta_{sn} = \frac{T_{PWM} - \Delta t}{T_{PWM}} = \frac{1}{2} + \frac{u_{snref}}{2} \tag{2.19}$$

This last case is known as "symmetrical regular sampling." If the reference signal is sampled at double sampling frequency $2f_s$, the method is referred as "asymmetrical regular sampling" whose advantages are an improvement of the dynamic response and of the harmonic distortion content of the load currents.

Equation 2.19 is derived under the assumption of carrier of unitary magnitude and normalized reference voltage. If the reference voltage is not normalized and the carrier presents a peak-to-peak amplitude equal to the DC link voltage U_d, which has precise physical meaning, the following duty cycle expression can be found:

$$\delta_{sn} = \frac{1}{2} + \frac{u_{snref}}{U_d} \tag{2.20}$$

This last expression represents the well-known duty cycle of the so-called sinusoidal PWM (SPWM). This last expression is more suited for digital implementation of SPWM.

2.2.1.4 Space-Vector Modulation

Space-vector modulation (SV-PWM) has been proposed in mid-80s [14–16]. It differs significantly from the previously described SPWM in one main aspect: while the SPWM requires separate modulators for each phase of the VSI, the SV-PWM treats the reference voltage space-vector as a unitary complex quantity. SV-PWM permits, with respect to the

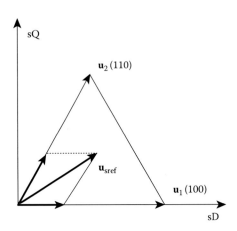

FIGURE 2.12
Decomposition of the reference voltage u_{sref} into two vectors $u_1(100)$ and $u_2(110)$.

classic SPWM, a lower distortion of the stator voltages and currents as well as a higher value of the amplitude of the fundamental component of the stator voltage. This technique can be easily implemented in programmable hardware structures (DSP, ASIC, FPGA). As a result, a higher flexibility of the control system, a reduction of the number of components, and a more compact structure can be achieved. The basic idea of the SV-PWM is to reconstruct in discrete form the space-vector of the reference voltage u_{sref}, which generally does not coincide with any of the six active voltage space-vectors that the VSI can generate, in a time interval equal to T_{PWM}.

As a matter of fact, when the desired output u_s does not coincide with one of the space state vectors of the characteristic hexagon, it can be built during a time interval equal to the modulation time, called T_{PWM}, by using a linear combination of these vectors. In Figure 2.12, the vector u_{sref} is composed as the weighted sum of the space-vectors u_1 and u_2 and a null vector.

The sum of the weighing time intervals is equal to T_{PWM}, and the weights are defined as follows:

$$0 = \int_0^{T_{PWM}} [\mathbf{u_s}(t) - \mathbf{u}_{sref}(t)]dt \tag{2.21}$$

where
  u_{sref} is the reference voltage space-vector
  u_s is the voltage space-vector generated by the inverter: since it comes from the space-vectors generated by the inverter, it is a step constant function

The vector u_{sref} is rebuilt by using two generable space-vectors belonging to the boundaries of the sector in which u_{sref} lies and by null vectors. The crossing from a configuration to the following must take place by commuting only one switch of the inverter, resulting in the least number of commutations. This explains the use of gray code introduced in Section 2.1.

Equation 2.18 can be rewritten as follows:

$$\int_{T_{PWM}} \mathbf{u_{sref}}(t)dt = \int_{T_k} \mathbf{u_k}(t)dt + \int_{T_{k+1}} \mathbf{u_{k+1}}(t)dt + \int_{T_0/2} \mathbf{u_0}(t)dt + \int_{T_0/2} \mathbf{u_7}(t)dt \tag{2.22}$$

where

u_k and u_{k+1} are the nonnull voltage space-vectors to be generated in sequence
u_0 and u_7 are the null space voltage vectors
T_k, T_{k+1}, and T_0 are, respectively, their weights, or the time during while each u_k must be operated with the condition that $T_{PWM} = T_k + T_{k+1} + T_0$

Because u_k and u_{k+1} are constant during their operation time T_k and T_{k+1}, Equation 2.19 can be rewritten as follows:

$$T_{PWM}\mathbf{u_{sref}} = T_k\mathbf{u_k} + T_{k+1}\mathbf{u_{k+1}} \tag{2.23}$$

The Equation 2.18 can be split into the two components on the real and imaginary axes; this, together with Equation 2.4, results in two equations, while T_k and T_{k+1} can be obtained as follows:

$$T_k = \frac{\sqrt{3}T_{PWM}}{U_d}\left[u_{sDref}\,sen\left(\frac{k\pi}{3}\right) - u_{sQref}\cos\left(\frac{k\pi}{3}\right)\right] = \frac{\sqrt{3}T_{PWM}|\mathbf{u_{sref}}|}{U_d}\,sen\left(\frac{k\pi}{3} - \alpha_{sref}\right) \tag{2.24}$$

$$T_{k+1} = \frac{\sqrt{3}T_{PWM}}{U_d}\left[-u_{sDref}\,sen\left(\frac{(k-1)\pi}{3}\right) + u_{sQref}\cos\left(\frac{(k-1)\pi}{3}\right)\right]$$

$$= \frac{3T_{PWM}|\mathbf{u_{sref}}|}{2U_d}\left[\cos\left(\alpha_{sref} - \frac{k\pi}{3}\right) - \frac{1}{\sqrt{3}}\,sen\left(\frac{k\pi}{3} - \alpha_{sref}\right)\right] \tag{2.25}$$

where

u_{sDref} and u_{sQref} are the direct and quadrature components of the reference voltage (in a fixed reference system)
$|\mathbf{u}_{sref}|$ is the module of this vector
α_{ref} is the angle that it forms with the direct-axis sD

Finally, the operation time of the null vector is given by the following:

$$T_0 = T_{PWM} - T_k - T_{k+1} \tag{2.26}$$

The nonnull space voltage vectors are generally used symmetrically in respect to the instant $t = T_{PWM}/2$. In this case, it is called SV-PWM with symmetrical placement of the zero vectors.

Moreover, since the transition from an inverter configuration to the next one must occur with a unique commutation, the null vector u_o is used at the beginning and at the end of the modulation time T_{PWM}, and the null vector u_7 is used in the middle of the modulation time. Finally, between these two null vectors, first, the vector with odd k and the vector with even k are used during the first half time of T_{PWM}; during the second half, they are applied symmetrically. Figure 2.13 shows an example of the driver signal on the upper devices of the inverter, T_a, T_b, and T_c, to obtain a voltage space-vector inside the first sector ($0 < \alpha_{sref} < \pi/3$).

The values of T_a, T_b, and T_c depend on the operation time T_k and T_{k+1}, of nonnull voltage space-vectors u_k and u_{k+1}, and on the time T_0 of null voltage space-vectors u_0 and u_7.

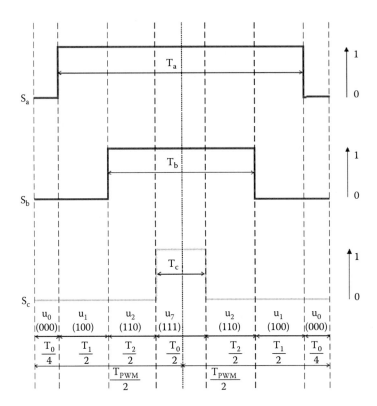

FIGURE 2.13
Time domain waveforms of driver signals of VSI upper devices during T_{PWM}.

Applying the space-vectors symmetrically with respect to the instant $t = T_{PWM/2}$, it can be noted from Figure 2.13 that, if the reference space state vector lies on the first sector, the following equations are valid:

$$\begin{cases} T_a = T_1 + T_2 + \dfrac{T_0}{2} = T_{PWM} - \dfrac{T_0}{2} & (2.27\,a) \\[3mm] T_b = T_2 + \dfrac{T_0}{2} = T_{PWM} - T_1 - \dfrac{T_0}{2} & (2.27\,b) \\[3mm] T_c = \dfrac{T_0}{2} = T_{PWM} - T_1 - T_2 - \dfrac{T_0}{2} & (2.27\,c) \end{cases}$$

In practice, since the Equations 2.24 and 2.25 are computationally cumbersome, simpler equations giving directly the operation time of the upper devices of the inverter, based on the three reference phase voltages, can be utilized under the hypothesis of SV-PWM with symmetrical zero vector placement. During the modulation time T_{PWM}, the expression of the operation time, T_{sn}, of the upper device connected to the phase n (n = A, B, C) is given by the following:

$$T_{sn} = T_{PWM} \left(\frac{1}{2} + \frac{u_{snref} + u^*}{U_d} \right) \tag{2.28}$$

where u_{snref} ($u_{snref} = u_{sAref}, u_{sBref}, u_{sCref}$) is the amplitude of the reference voltage of the phase n and u^* is given by the following:

$$u^* = -\frac{1}{2}\left[\max_n(u_{snref}) + \min_n(u_{snref})\right] \tag{2.29}$$

The zero-sequence voltage u^* corresponds to a triangular third harmonic of amplitude 1/4 of the fundamental, as shown in Figure 2.10.

The previously described algorithm allows any voltage space-vector to be generated with an amplitude lying inside the characteristic hexagon of the inverter.

The corresponding duty cycle of δ_{sn} of the phase-n leg is as follows:

$$\delta_{sn} = \frac{1}{2} + \frac{u_{snref} + u^*}{U_d} \tag{2.30}$$

This last expression is valid in the linear range.

If the reference voltages represent a symmetrical three-phase system, the generated voltage space-vector draws a circle in the complex plane. The maximum amplitude of the phase space voltage vector that can be generated for any value of the angle α_{sref} is given by the radius of the circle inscribed in the hexagon, and it is equal to the following (see Figure 2.2):

$$u_{max} = \frac{2}{\sqrt{3}}\frac{U_d}{2} = 1.15\frac{U_d}{2} = 0.577U_d \tag{2.31}$$

It should be noted that when the reference voltage coincides with one of the space-vectors $u_1 - u_6$, it can be obtained directly, and the obtained voltage amplitude is equal to U_{max}, and when a smaller amplitude is necessary, it is realized by combining it with the null vector.

The maximum amplitude of the achievable fundamental harmonic of the phase voltage is 15% greater than the one obtainable with the traditional suboscillation method (SPWM) described in Section 2.2.1.1.

As for the null vector, two redundant vectors are available. The u_0 (u_7) null vector is used when the previous used vector is u_2, u_4, or u_6 (u_1, u_3, or u_5) so that only half bridge commutation occurs.

The method described earlier strongly depends on the modulation time T_{PWM}, while some further considerations are necessary. A small value of T_{PWM} improves the reconstruction of the reference voltage, as long as it is greater than the sampling time of the control system; this means $1/T_{PWM} = f_{PWM} = f_{sampling}$. In addition, because the modulation frequency $1/T_{PWM} = f_{PWM}$ must be also less than the switching frequency of the power devices, the upper limit for f_{PWM} is given by the minimum value between the sampling frequency of the control system and the switching frequency of the power devices $f_{PWM} < \min(f_{sw1}, f_{fs})$. As a consequence, the vector modulation is suitable for those VSIs with fast-power devices like IGBTs and MOSFETs.

The vector modulation can be further classified into synchronous and asynchronous modulation. With reference to an electric drive, in the asynchronous modulation, f_{PWM} is constant no matter what the value of the supply frequency of the stator is. On the contrary, in the synchronous modulation, f_{PWM} is tied to the supply frequency of the stator f_1 so that the ratio $f_{PWM}/f_1 = p$ is maintained constant and equal to an integer number p.

The synchronous vector modulation is usually employed in high-power electric drives (GTO inverters) where switching losses, which increase with the switching frequency, attain significant values. This is the reason why in these kinds of inverters, the value of p is low (actually for f_1 constant, a low p means a low f_{PWM}, resulting in a possible low f_{sw}).

The asynchronous vector modulation is used in low-power electric drives working at high f_{sw} (IGBT inverters). In this case, p is higher, and f_{PWM} is synchronized with the sampling frequency of the control system. In this case, T_{PWM} is generally chosen to be as much as twice the sampling time ($T_{PWM} = 2T_s$) so that the stator current signals can be sampled in the middle of the interval times in which the null vector is applied. With this strategy, stator current signals need not be filtered analogically before sampling [17,18].

2.2.1.5 Discontinuous Pulsewidth Modulation

In PWM techniques based on zero-sequence signal injection, the voltage linearity, the waveform quality (current ripple), and the switching losses are strongly influenced by the choice of the zero-sequence signal (zero-state partitioning). On the basis of these considerations, many researchers have investigated high-performance PWM techniques, among which the so-called discontinuous PWM (DPWM). DPWM is also called two-phase SV-PWM and is a particular kind of carrier-based technique, where the zero-sequence signals are added in such a way that only two phases are switched at the same time. Each phase is de facto clamped either to the upper or to the lower DC link bus for a certain angle typically (60° or 120°). It adopts only one zero voltage vector for the entire T_{PWM} period. Figure 2.14 shows typical switching patterns of the inverter phase sA leg as obtained with the DPWM when the phase sA is linked to the upper side of the DC link (DPWMMAX).

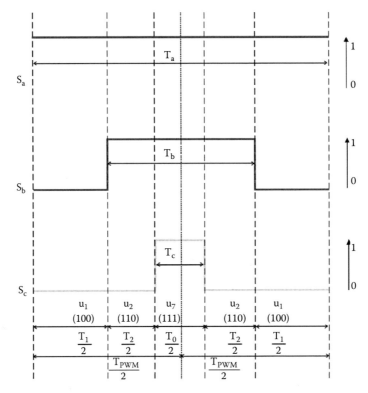

FIGURE 2.14
Switching patterns of the inverter phase sA leg with DSPWM.

As a result, DPWM permits a 33% reduction of the effective switching frequency. However, switching losses depend strongly also on the load power factor, which permits a further reduction of the switching losses.

Depenbrock [17] developed a DPWM technique, later called discontinuous, in the following referred to as DPWM1. Since, during each carrier cycle, one phase is not modulated, and correspondingly it is clamped to the upper or lower DC link terminal, DPWM1 permits a wider linearity range than SPWM, reduced switching losses, and superior high modulation range current waveform quality.

The poor low modulation range performance, like narrow pulse problems and poor current waveform quality as well as a certain complexity in its implementation, limited its practical application. Ogasawara et al. [16] developed a direct digital method with superior high modulation range waveform quality and reduced switching losses. It was later discovered that this space-vector-theory-based method has a triangle-intersection-implementation-based DPWM equivalent [18], in the following referred to as DPWM2. This modulator has been later reinvented and called "minimum switching losses PWM" [19]. These last two DPWMs as well as the main ones in literature [20,21] are described by the voltage waveforms in Figure 2.15.

As far as DPWMMIN and DPWMMAX techniques are concerned, the zero-sequence signal to be added to the sinusoidal reference is described by the following equations:

$$u_0 = -\frac{U_d}{2} - \min_n(u_{snref}) \quad \text{for DPWMMIN} \tag{2.32}$$

$$u_0 = \frac{U_d}{2} - \max_n(u_{snref}) \quad \text{for DPWMMAX} \tag{2.33}$$

It means that with DPWMMIN, the lower sections of the three-phase envelopes are clamped to the lower DC bus; with DPWMAX, the higher sections of the three-phase envelopes are clamped to the upper DC bus. With these DPWMs, all the phases are connected

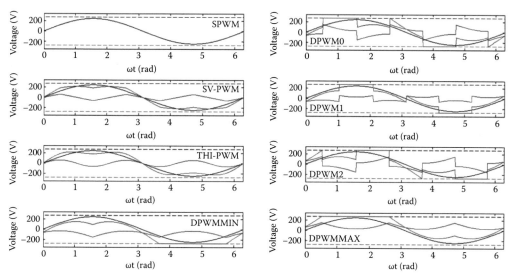

FIGURE 2.15
Modulation waveforms for different PWM techniques.

only to one side of the DC bus, respectively, the upper or the lower. Moreover, each phase is connected to the DC bus for an electrical angle equal to 120°.

As far as DPWM0, DPWM1, and DPWM2 are concerned, the idea is to clamp each phase alternatively on the upper and on the lower DC bus for an electrical angle of 60°. The difference among these techniques lies on the angle in correspondence to which each phase is connected to the DC bus. Specifically, DPWM0 clamps with a leading angle of 30°, DPWM1 clamps with voltage peaks, and DPWM2 clamps with a lagging angle of 30°. Since the main difference between them lies in the angle, an attempt to unify them has led to the development of a generalized DPWM (GDPWM) technique [22,23]. To properly describe GDPWM, a modulator phase angle ψ is defined, whose value zero is defined in correspondence to the intersection point between the two reference sinusoidal modulation waves at $\omega t = \pi/6$. From ψ to $\psi + \pi/3$, the zero sequence is defined as the difference between the saturation voltage ($U_d/2$ or $-U_d/2$) and the reference sinusoidal modulation signal that successfully passes the so-called *maximum magnitude test*. The three-phase modulation signal u_{snref} (n = A, B, C) is phase shifted by $\psi - \pi/6$ giving three new signals u_{snrefx}; the maximum amplitude of these new signals determines the zero-sequence signal in this way:

$$u_0 = sign(u_{snref}^*)\frac{U_d}{2} - u_{snref} \qquad (2.34)$$

where u_{snref}^* is the shifted phase voltage u_{snrefx} currently presenting its maximum value in absolute terms, $|u_{snref}^*| = \max_n (|u_{snrefx}|)$.

Adding this zero-sequence signal to the three original modulation waves u_{snref}, the GDPWM waves are generated. The only control variable is the angle ψ, whose variation range is between 0 and $\pi/3$. Figure 2.16 shows the modulation and zero-sequence waveforms for four values of ψ. It can be noted that DPWM0 corresponds to $\psi = 0$, DPWM1 corresponds to $\psi = \pi/6$, and DPWM2 corresponds to $\psi = \pi/3$.

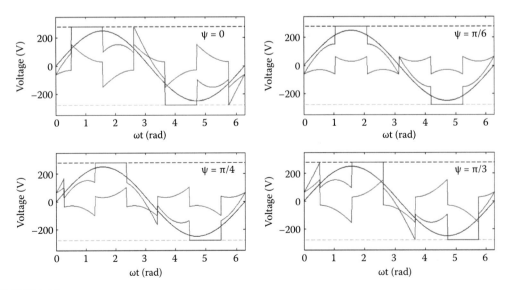

FIGURE 2.16
Modulation and zero-sequence waveforms obtained with GDPWM for four values of ψ.

2.2.1.6 General Expression of the Duty Cycle in Carrier-Based PWM Techniques

Equation 2.20, rewritten here for easy readability, describes the expression of the duty cycle δ_{sn} to be applied to the n-th phase of the inverter leg in the case of SPWM [24]:

$$\delta_{sn} = \frac{1}{2} + \frac{u_{snref}}{U_d} \tag{2.35}$$

THI-PWM and DPWM can be obtained by Equation 2.35 by adding a suitable zero-sequence signal to the reference. The following expression of the duty cycle can be therefore obtained:

$$\delta_{sn} = \frac{1}{2} + \frac{u_{snref} + u_0}{U_d} \tag{2.36}$$

where $u_0 = f(3\omega t)$ is a zero-sequence signal to be determined according to the assigned target conditions. The expression of the duty cycle with the standard SV-PWM is given in Equation 2.30, rewritten here for readability:

$$\delta_{sn} = \frac{1}{2} + \frac{u_{snref} + u^*}{U_d} \tag{2.37}$$

where

$$u^* = -\frac{1}{2}\left[\max_n(u_{snref}) + \min_n(u_{snref}) \right]$$

A general expression for the expression of the duty cycle synthetizing the most carrier-based PWM techniques can therefore be found modifying the value of u* in Equation 2.37 in this way:

$$u^* = g\frac{U_d}{2} + k\left[\min_n(u_{snref}) - \max_n(u_{snref}) \right] - \min_n(u_{snref}) \tag{2.38}$$

where
  k is a parameter taking into consideration the unequal null-state sharing
  g is a parameter permitting to take into consideration even the DPWM techniques

As a matter of fact, the unequal share of null vectors at the beginning and the end of each subcycle allows obtaining different carrier-based PWM techniques [18,25].

If $g = 0$ and k is selected on the basis of this equation

$$k = \frac{\min_n(u_{snref})}{\min_n(u_{snref}) - \max_n(u_{snref})} \tag{2.39}$$

then the classic SPWM is obtained. If $g = 0$ and $k = 0.5$, then the standard SV-PWM is obtained. If $g = 1$ and if $k = 1$, then the DPWMMAX is obtained, while if $g = -1$ and if $k = 0$, then the DPWMMIN is obtained. The GDPWM [26] can be obtained by switching k

TABLE 2.2

Comparison between SV-PWM and Modified SV-PWM Sequences

	SV-PWM							
Vector	u_0	u_1	u_2	u_7	u_7	u_2	u_1	u_0
Time	$T_0/2$	T_1	T_2	$T_0/2$	$T_0/2$	T_2	T_1	$T_0/2$
			$T_{PWM}/2$				$T_{PWM}/2$	

	Modified SV-PWM							
Vector	—	u_0	u_1	u_2	u_2	u_1	u_0	—
Time	—	$T_0/3$	$2T_1/3$	$T_2/3$	$T_2/3$	$2T_1/3$	$T_0/3$	—
			$T_{PWM}/2$				$T_{PWM}/2$	

between 0 and 1 and correspondingly g between –1 and 1 at each interval of $\pi/3$ electrical degrees and changing only the phase with respect to the reference frame.

2.2.1.7 Modified Space-Vector Modulation

As shown in Figure 2.10, the SV-PWM utilizes a vector sequence in which the null vector is present both at the beginning and at the end of each subinterval of time $T_{PWM}/2$.

The modified SV-PWM uses the null vector only at the beginning of the first subinterval of time $T_{PWM}/2$ and at the end of the second subinterval. While the SV-PWM implies three switching states for each subinterval of time $T_{PWM}/2$, the modified SV-PWM has only two switching states; consequently their duration is reduced to maintain the same duration T_{PWM}. A comparison between the two sequences is shown in Table 2.2.

The choice between these two techniques depends on the value of the reference vector and implies different harmonics of the inverter output current. It can be seen that by adopting the modified switching sequence, the harmonic content is lower at a high modulation index; on the contrary, using SV-PWM, a low harmonic current occurs with a low modulation index.

2.2.1.8 Synchronized Carrier Modulation

The synchronized carried modulation is a particular case of the aforementioned modulation techniques and implies an integer value of the ratio between carrier frequency and fundamental inverter output frequency.

In a carrier-based modulation, the carrier frequency is usually fixed, but the frequency of the reference signal can vary; this is identified as asynchronous modulation. As a consequence, the switching sequence can be nonperiodic, and the Fourier spectrum of the output waveform contains also subharmonics whose frequency is lower than the carrier sidebands.

Using this technique, the sampling instants in a period of the fundamental are equally spaced and are given by $t_{sn} = (n/N \cdot f_1)$; $n = 1, 2 \ldots N$, and the reference signal value is in p.u. $u^*(t_{sn}) = m \cdot \sin(2\pi f_1 t_{sm})$, where f_1 is the fundamental output frequency. In a microprocessor-based control system, these values can be calculated, stored in a memory, and then utilized in Equation 2.17 to determine the switching instants.

2.2.2 Carrierless PWM

All the modulation techniques described in the last section are based on the use of a time base. Some of these suboscillation methods, for example, use a triangular carrier; others like sampling techniques adopt directly a set of equations; in any case, the time interval in which the driving pulse is built remains constant (in Figure 2.9, it is indicated as T_{PWM} for SV-PWM). This results in a periodical signal and in the presence of harmonics with frequencies equal to the fundamental of the carrier and its harmonics. This phenomenon has not only electrical consequences, but also effects connected with the use of machines and electric drives in civil (e.g., passenger's comfort) and military (e.g., acoustic discretion) applications.

In order to avoid the harmonic energy to be concentrated around the carrier fundamental frequency, it is possible to vary it in a random manner. With reference to suboscillation method this means keeping constant the triangular shape with a linear slope and modifying the period according to a random number generator. Finally, to avoid anomalous thermal stress to power devices, the average switching frequency must be constant. The resulting spectrum in case of random modulation exhibits the lack of steep peaks on the contrary, the spectrum is almost continuous maintaining the overall energy level. In this way, noise is not concentrated, but it appears as a "white noise" [2].

2.2.3 Overmodulation

In Section 2.2.1.1, which deals with the suboscillation method, it has been shown that when the amplitude of the reference signal is equal to the amplitude of carrier, then $m = 1$ and $u_{SA0} = (U_d/2)$. By increasing the amplitude of reference signal, the six-step operation could be hypothetically reached, to which corresponds a generated voltage equal to $u_{A0} = (U_d/2)(4/\pi)$.

When $m < 1$, the amplitude of the phase voltage increases linearly with m; for $m > 1$, the relationship becomes nonlinear. The operating region in correspondence to which $m > 1$ is called overmodulation region.

As far as SV-PWM is concerned, the overmodulation behavior can be considered in this way [27,28]. It is almost apparent from averaging the voltage space-vector to be synthesized, that the higher the modulation index M, the lower the application times T_0 and T_7 of the zero voltage vectors $\mathbf{u}_0$ and $\mathbf{u}_7$. The overmodulation region starts when the circular path of the reference voltage vector $\mathbf{u}_{sref}$ coincides with the circle inscribed in the inverter hexagon. The condition $T_0 = 0$ is reached when $M = M_{max2}$, which happens when the circular path of $\mathbf{u}_{sref}$ touches the inverter hexagon itself (see Figure 2.17). At this point, the controllable region of linear modulation finishes. On the other hand, the six-step mode corresponds to the maximum generable voltage, to which corresponds $M = 1$. The control of the voltage in

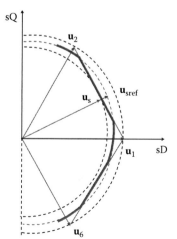

FIGURE 2.17
Overmodulation mode 1: trajectory of the output average voltage vector.

the region $M_{max} < M < 1$ is achieved by overmodulation. Overmodulation techniques are subdivided into two modes, called modes 1 and 2. In mode 1, the trajectory of the average voltage space-vector follows a circle of radius $M > M_{max2}$ as long as the circle arc is located inside the hexagon, tracking the hexagon itself in the remaining part of the path.

In this case, if u_k and u_{k+1} are the voltage vectors that should synthesize the reference voltage vector u_{sref}, whose angular position is α_{ref}, then the respective application times T_k and T_{k+1}, when the reference trajectory lies on the inverter hexagon, are computed as follows:

$$
\begin{cases}
T_k = T_{PWM} \dfrac{\sqrt{3}\cos(\alpha_{ref}) - \sin(\alpha_{ref})}{\sqrt{3}\cos(\alpha_{ref}) + \sin(\alpha_{ref})} \\[3mm]
T_{k+1} = T_{PWM} - T_k \\[2mm]
T_0 = 0
\end{cases}
\tag{2.40}
$$

Overmodulation mode 2 is reached when $M > M_{max3} = 0.952$, in correspondence to which the length of the arcs reduces to zero and the trajectory of the average stator voltage vector becomes purely hexagonal. In this mode, the speed of the average voltage vector is controlled along the linear trajectory by varying the duty cycle. The higher the duty cycle, the higher the speed of the voltage vector in the middle of the central part of the hexagon path and the lower the speed in correspondence to the corners. In synthesis, the higher the duty cycle, the closer the behavior of the SV-PWM to the six-step operation. Ideally, when six-step operation is reached, the speed at the center is infinite while the speed at the corners is null.

A way to perform overmodulation mode 2 has been proposed in Ref. [29]. In this case, the trajectory of the average voltage vector changes smoothly between the inverter hexagon and the six-step operation. To achieve that, both the magnitude and the phase of the reference voltage vector are to be modified. The current angular position α_{ref} becomes α_{refp}, which can be computed as follows:

$$
\alpha_{refp} =
\begin{cases}
0 & \text{for } 0 \le \alpha \le \alpha_h \\[3mm]
\dfrac{\alpha - \alpha_h}{\pi/6 - \alpha_h} \dfrac{\pi}{6} & \text{for } \alpha_h \le \alpha \le \dfrac{\pi}{3} - \alpha_h \\[3mm]
\pi/3 & \text{for } \pi/3 - \alpha_h \le \alpha \le \dfrac{\pi}{3}
\end{cases}
\tag{2.41}
$$

where α_h is defined as in Figure 2.18. In a nutshell, the modified vector is held at the vertexes of the hexagon for a certain angle α_h, while it tracks the hexagon itself for the remaining part of the switching period T_{PWM}. α_h, defines the time inside which the voltage vector remains at the vertexes, and is a nonlinear function of the modulation index M, which can be piecewise linearized as follows [29]:

$$
\alpha_h =
\begin{cases}
6.4M - 6.09 & \text{for } 0.952 \le M \le 0.98 \\
11.75M - 11.34 & \text{for } 0.98 \le M \le 0.9975 \\
48.96M - 48.43 & \text{for } 0.9875 \le M \le 1
\end{cases}
\tag{2.42}
$$

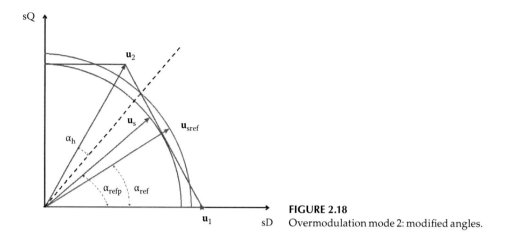

FIGURE 2.18
Overmodulation mode 2: modified angles.

2.2.4 SV-PWM for the Minimization of the Common-Mode Emissions

The fast switching frequency of PWM techniques when used in variable speed drives has, among other detrimental effects, the drawback of generating high-level CMV variations. As a result, high-frequency common-mode currents flowing to the ground through the parasitic capacitances between the different parts of the drive and the ground occur, resulting in the drive itself to be less reliable. The ball bearings can deteriorate, unexpected fault relay tripping can occur, or nearby electronic equipments can be disturbed. In general, the faster the variation of the CMV, the smaller the machine-to-ground common-mode capacitive impedance: in this respect, the trend to increase the switching frequency of the power devices causes an unavoidable increase of these effects.

A possible way to reduce the variation of the CMV and consequently of the common-mode currents is to act directly on the switching pattern of the VSI and then on the PWM strategy. In this regard, two PWM strategies have been devised for the reduction of the common-mode emissions [8]. The first PWM technique, although effectively reducing the common-mode currents, is affected by a heavy limitation on the maximum generable stator voltage. The second PWM technique permits an increase of the maximum allowable stator voltage at the cost of a slight increase of the common-mode currents.

2.2.4.1 Common-Mode Voltage

In a star-connected three-phase electric machine, the CMV u_{N0} is given by the following:

$$u_{N0} = \frac{u_{sA0} + u_{sB0} + u_{sC0}}{3} \tag{2.43}$$

where u_{sA0}, u_{sB0}, and u_{sC0} are the inverter output phase voltages referred to the medium point of the DC link bus, which in this case is assumed to be connected to ground. The lack of this connection to ground causes an additional voltage term. If the machine is supplied with a symmetric sinusoidal three-phase voltage, u_{N0} is instantaneously equal to zero. However, when the machine is supplied by an inverter, the CMV is always different from zero, and its instantaneous value can be computed on the basis of the DC link voltage (U_d) and the switching pattern of the inverter, as shown in Table 2.3 ($\mathbf{u}_i$ stands for the ith stator voltage space-vector).

TABLE 2.3

Inverter States and CMVs

State	u_{sA0}	u_{sB0}	u_{sC0}	u_{N0}
$\mathbf{u}_0\,(0, 0, 0)$	$-\mathbf{u}_d/2$	$-\mathbf{u}_d/2$	$-\mathbf{u}_d/2$	$-\mathbf{u}_d/2$
$\mathbf{u}_1\,(1, 0, 0)$	$\mathbf{u}_d/2$	$-\mathbf{u}_d/2$	$-\mathbf{u}_d/2$	$-\mathbf{u}_d/6$
$\mathbf{u}_2\,(1, 1, 0)$	$\mathbf{u}_d/2$	$\mathbf{u}_d/2$	$-\mathbf{u}_d/2$	$\mathbf{u}_d/6$
$\mathbf{u}_3\,(0, 1, 0)$	$-\mathbf{u}_d/2$	$\mathbf{u}_d/2$	$-\mathbf{u}_d/2$	$-\mathbf{u}_d/6$
$\mathbf{u}_4\,(0, 1, 1)$	$-\mathbf{u}_d/2$	$\mathbf{u}_d/2$	$\mathbf{u}_d/2$	$\mathbf{u}_d/6$
$\mathbf{u}_5\,(0, 0, 1)$	$-\mathbf{u}_d/2$	$-\mathbf{u}_d/2$	$\mathbf{u}_d/2$	$-\mathbf{u}_d/6$
$\mathbf{u}_6\,(1, 0, 1)$	$\mathbf{u}_d/2$	$-\mathbf{u}_d/2$	$\mathbf{u}_d/2$	$\mathbf{u}_d/6$
$\mathbf{u}_7\,(1, 1, 1)$	$\mathbf{u}_d/2$	$\mathbf{u}_d/2$	$\mathbf{u}_d/2$	$\mathbf{u}_d/2$

Source: Cirrincione, M. et al., New direct power control strategies of three-phase VSIs for the minimization of common-mode emissions in distributed generation systems, in *The Forty-Second Annual Meeting of the IEEE Industry Applications Society (IAS 2007)*, New Orleans, LA, September 23–27, 2007.

2.2.4.2 Switching Strategy

Table 2.3 shows that if only even or only odd active voltage vectors are used (u_k, with k, respectively, even or odd), no CMV variation is generated. If a transition from an even voltage vector to an odd one (or vice versa) occurs, a common-mode variation of amplitude $U_d/3$ is generated. If a transition from an odd (even) voltage vector to the zero (seventh) voltage vector occurs, a common-mode variation of amplitude $U_d/3$ is generated. Finally, if a transition from an odd (even) voltage vector to the seventh (zero) voltage vector occurs, a common-mode variation of amplitude $2U_d/3$ is generated. Thus, from the point of view of common-mode emissions, the worst case is a transition from an odd (even) voltage vector to the seventh (zero) voltage vector. For this reason, whatever inverter control technique is devised to minimize the common-mode emissions of the drive, the exploitation of both null voltage vectors (zero and seventh) should be avoided.

The best way to minimize the variations of the CMV is to devise a PWM technique in which either only an odd or only an even voltage space-vector is used. In this way, basically no variation of the CMV occurs, as clearly apparent from Table 2.3. In this case, if a space-vector averaging approach is adopted and the odd vectors are chosen, the following expression holds:

$$T_{PWM}\mathbf{u}_{ref} = T_1\mathbf{u}_1 + T_3\mathbf{u}_3 + T_5\mathbf{u}_5 \tag{2.44}$$

with

$$T_{PWM} = T_1 + T_3 + T_5 \tag{2.45}$$

With such a PWM technique, the CMV is maintained constant at the value $-U_d/6$. Alternatively, the same kind of equations could be written with the even voltage space-vectors, and the CMV would be maintained at the value $U_d/6$. With such an approach, the linear region achievable with this PWM technique is defined by the circle inscribed in the triangle, as shown in Figure 2.19. The maximum sinusoidal phase voltage that can be synthesized is $U_d/3$, which is quite limited.

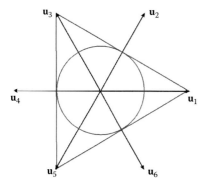

FIGURE 2.19
Linear region obtainable with the first modified PWM strategy.

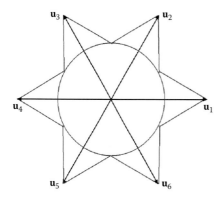

FIGURE 2.20
Linear region obtainable with the second modified
PWM strategy.

In order to increase the maximum value of generable voltage, a second PWM strategy has been developed. The complex plane has been divided into six sectors of $\pi/3$ amplitude, which, however, has been rotated of an angle $\pi/6$ in the positive angular direction. Whenever the voltage reference vector lies in the kth sector, the kind of vectors to be used to synthesize it is defined by the sector itself. In details, when the voltage reference vector lies in the first sector, then only odd voltage vectors are used, while if it lies in the second sector, then only even voltage vectors are used. In this way, only six pulses of the CMV are generated inside a period of the fundamental waveform. Equations 2.44 and 2.45, and the corresponding ones written for the even space-vectors are thus used when the reference vector lies, respectively, in an odd or in an even sector. As a result, the average voltage space-vector is limited by the six-point star as in Figure 2.20. Correspondingly, the maximum sinusoidal phase voltage that can be synthesized is increased to $(2\sqrt{3}U_d)/9$.

Figure 2.21 shows the switching pattern of the inverter when the voltage reference lies in the first sector. It could be noted that, differently from the other PWM techniques, there is never a time interval in which two legs present the same switching configuration.

2.2.5 Optimized Open-Loop PWM

For high-power inverters, some precautions are necessary for avoiding high-power losses. Firstly, low switching frequencies are used, (see Equation 2.5); then only synchronized pulse schemes are adopted to minimize the generation of subharmonic components, and there are some switching instants that are generally computed off-line to create a priori optimal switching patterns. These patterns differ depending on the optimization objective.

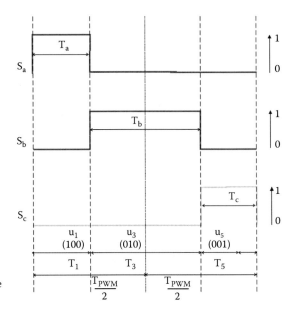

FIGURE 2.21
Switching pattern of the VSI when the voltage reference vector lies in the first sector.

The performance is usually optimized in steady-state conditions because transients seldom occur. Three main techniques can be recognized: harmonic elimination, objective functions, and optimal subcycle method.

The harmonic elimination technique focusses on a specific harmonic value to be eliminated. This is mandatory, for example, in case of mechanical resonance in electric drives excited by torque harmonic induced by the inverter supply [30].

The objective function techniques try to minimize a target parameter: it can be the loss factor d^2 defined in Equation 2.6 and (2.9) [31] or the highest pulse of the phase current [32] or the maximization of the efficiency of inverter/machine system [33].

The optimal subcycle method is based on the variation of T_{PWM} to build the output voltage in SV-PWM. The target is to nullify the instantaneous distortion current at the beginning and at the end of each $T_{PWM}/2$ in which T_k, T_{k+1}, and T_0 are calculated. Some sets of optimal sequences for T_k, T_{k+1}, and T_0 are calculated in advance and utilized depending on the reference signal [3].

2.2.6 Experimental Verification of Open-Loop PWM Techniques

This section shows some simulation and experimental results of two PWM techniques, the SPWM, explained in Section 2.2.1.1, and the SV-PWM explained in Section 2.2.1.4.

Simulations have been performed in MATLAB®–Simulink® environment; experimental results have been carried out on a 7 kW three-phase IGBT inverter prototype. The PWM frequency has been set to 5 kHz. The inverter supplies an IM, and is driven by a simple scalar control based on impressed voltage.

Figure 2.22 shows the numerical simulation of the logic command signals, in one PWM period, of the upper IGBTs of the three inverter legs when the SV-PWM with symmetrical zero vectors is used. Figure 2.23 shows the waveforms of the reference voltage and of the phase voltage synthesized by the inverter when a reference voltage of amplitude equal to 103 V corresponding to a rotating speed of the machine of 50 rad/s has been given, obtained, respectively, with the SPWM and the SV-PWM. Figure 2.24 shows the experimental waveforms of the logic command signals, in one PWM period, of the upper IGBTs

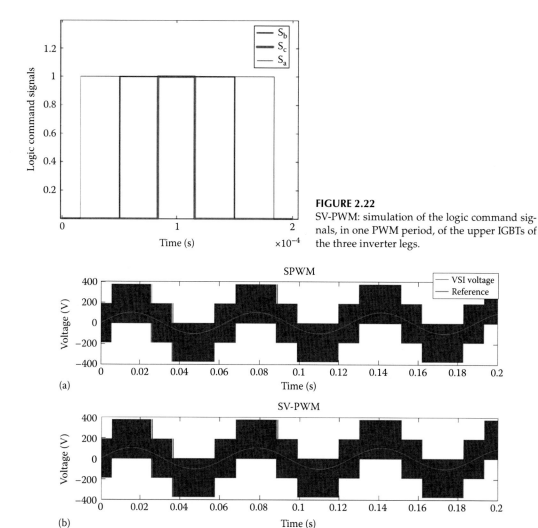

FIGURE 2.22
SV-PWM: simulation of the logic command signals, in one PWM period, of the upper IGBTs of the three inverter legs.

FIGURE 2.23
Simulation waveform of the phase voltage obtained with SPWM (a) and SV-PWM (b).

of the three inverter legs when the SV-PWM is used. It should be noted a good correspondence between the simulated and the experimental waveforms of the command signals.

A comparison between Figures 2.22 and 2.24 shows that the SV-PWM is characterized by a time interval, centered in the middle of T_{PWM}, in which the zero vector u_7 is applied.

Figure 2.25 shows the experimental waveforms of both the reference and measured phase voltages obtained, respectively, with SPWM and SV-PWM. The tests have been done, as in simulation, by imposing a reference voltage of amplitude equal to 103 V corresponding to a rotating speed of the machine of 50 rad/s. By observing these waveforms in time domain, any difference is hardly noticeable; on the contrary, in frequency domain, the spectral composition (obtained with FFT) exhibits the presence of harmonics around the switching frequency (5 kHz) whose amplitudes are lower in the case of SV-PWM both in the low and in the high frequency range, see Figures 2.26 and 2.27. Moreover, the SV-PWM offers lower frequency fundamental harmonic and correspondingly a lower THD% (8.56% for SV-PWM against 11.94% for SPWM).

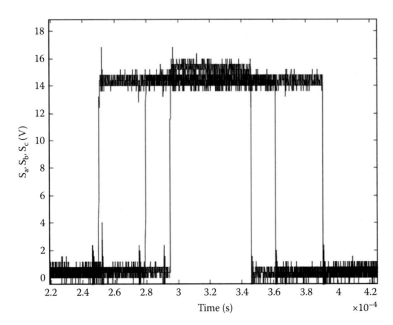

FIGURE 2.24
SV-PWM: experimental waveforms of the logic command signals, in one PWM period, of the upper IGBTs of the three inverter legs.

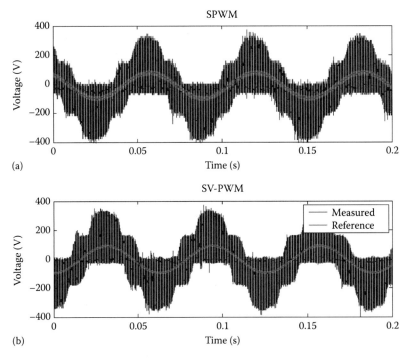

FIGURE 2.25
Experimental waveform of the phase voltage obtained with sinusoidal PWM (a) and SV-PWM (b).

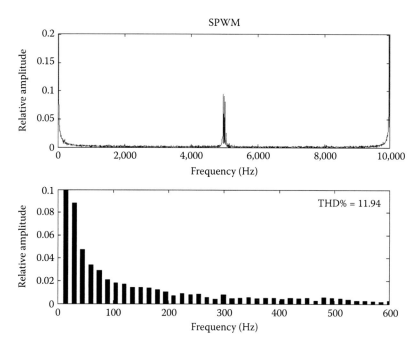

FIGURE 2.26
Spectrum of phase voltage obtained with sinusoidal PWM.

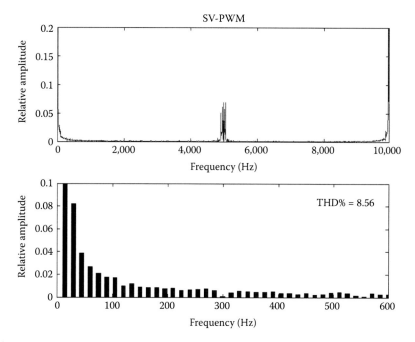

FIGURE 2.27
Spectrum of phase voltage obtained with SV-PWM.

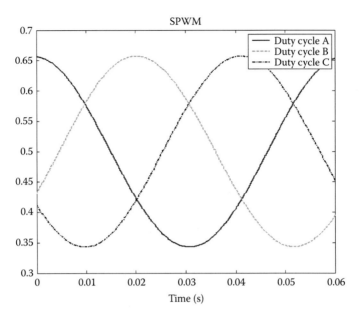

FIGURE 2.28
Duty cycles for SPWM.

Figures 2.28 and 2.29 show the duty cycles, obtained with both SPWM and SV-PWM, under the same working condition. It should be noted that as far as SPWM is concerned, the duty cycles are sinusoidal waveforms, while as far as SV-PWM is concerned, the duty cycles present the typical triangular triple harmonic content added to the fundamental.

Finally, Figures 2.30 and 2.31 show the reference and measured phase voltage obtained with the SV-PWM in the overmodulation mode 1 working region, obtained, respectively,

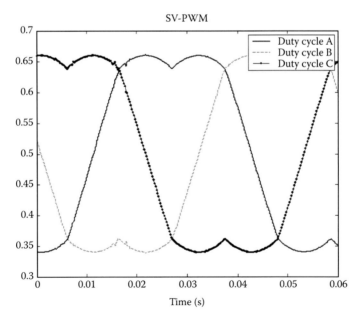

FIGURE 2.29
Duty cycles for SV-PWM.

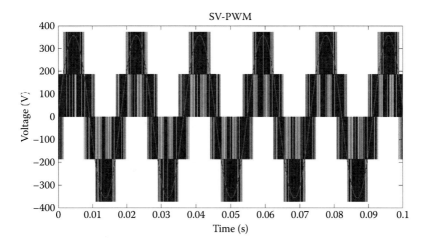

FIGURE 2.30
Reference and measured phase voltage obtained with the SV-PWM in the overmodulation working region (simulation).

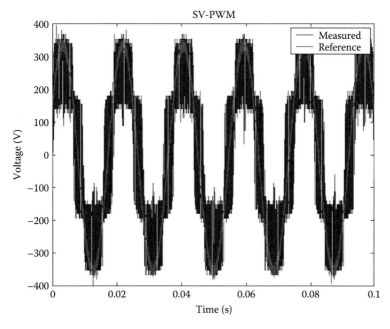

FIGURE 2.31
Reference and measured phase voltage obtained with the SV-PWM in the overmodulation working region (experiment).

in numerical simulation and experimentally. In particular, they refer to a reference voltage of amplitude equal to 357 V corresponding to a working speed of 172 rad/s; the test corresponds to a voltage space-vector whose amplitude is comprised between the circle inscribed in the inverter hexagon and the inverter hexagon itself. This is confirmed by Figure 2.32, showing the reference and real average voltage space-vectors loci during this test.

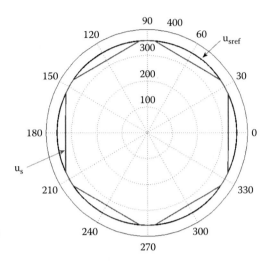

FIGURE 2.32
Loci of the reference and average voltage space-vector
in overmodulation.

It can be observed that the average voltage vector coincide in some parts with the reference while in others lies on the inverter hexagon, as expected.

Figure 2.33 shows the spectrum obtained with the FFT of the inverter phase voltage. The THD% in this case is equal to 4.56. Finally, Figure 2.34 shows the duty cycles obtained with SV-PWM in the same working condition. It is clearly observable the saturation duty cycle to one in a long time interval as expected.

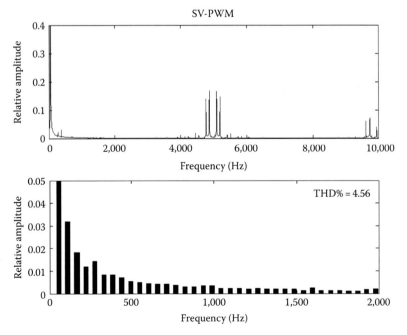

FIGURE 2.33
Spectrum of phase voltage obtained with SV-PWM in overmodulation condition.

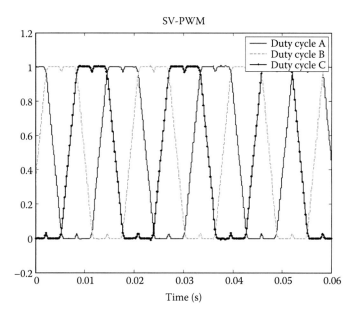

FIGURE 2.34
Duty cycle obtained with SV-PWM in overmodulation mode 1 region.

2.3 Closed-Loop Control of VSIs

Closed-loop schemes, differently from open-loop ones, inject currents; this approach requires a feedback in which load currents are sensed and compared with the desired ones.

In the case of grid inverter connection, when a renewable supply is available, the grid represents a load that imposes the voltage. The inverter is required to inject current that depends on the power given by the source.

A schematic representation of inverter with current feedback is given in Figure 2.35. The currents are measured and compared with reference ones; the error signal is given to the controller/modulator block whose outputs are the driving power switch signal. The structure of this block is connected with the loop control strategy.

It should be noted that in any case the resulting current is obtained on the basis of a voltage applied to the load impedance; as a consequence, a dependence on this impedance is expected. This situation is different from current source inverters (CSIs); however, the most applications are based on VSIs rather than CSIs mainly because of the characteristics of power switches.

2.3.1 Classification of Closed-Loop Control Strategy

Figure 2.36 shows a classification of closed-loop control strategies. Closed-loop PWM control can be divided into nonoptimal methods and closed-loop PWM with real-time optimization. Nonoptimal methods include hysteresis current control, suboscillation current control, and space-vector current control. Closed-loop PWM with real-time optimization methods contains predictive current control, pulsewidth control with field orientation, and trajectory tracking control.

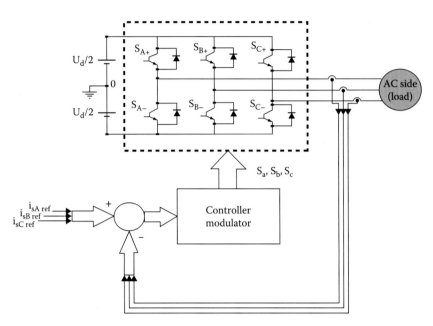

FIGURE 2.35
Representation of a VSI with closed-loop control strategy.

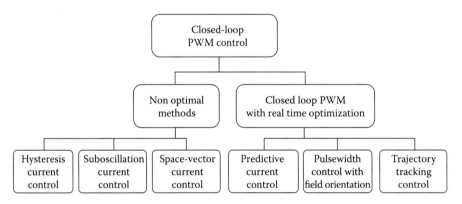

FIGURE 2.36
Classification of closed-loop control strategy.

2.3.1.1 Hysteresis Current Control

Hysteresis current control belongs to the so-called on-off controllers; each measured current is compared to the desired one, and the error signal is fed to a two-level hysteresis comparator whose output is the driving signal of the corresponding inverter leg. The diagram of this control strategy is shown in Figure 2.37. By comparing this last figure with Figure 2.35, it can be noted that the controller/modulator block is represented by the hysteresis comparators.

Three hysteresis comparators are present, one for each phase. The switching status of each leg does not depend on the other legs. The error of each current is maintained within the hysteresis bandwidth. Among advantages of these strategies, there are the inherent simplicity and the good dynamic performance. On the other hand, some disadvantages are

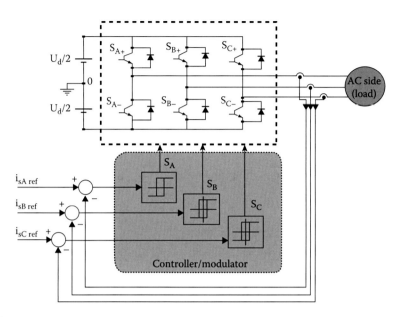

FIGURE 2.37
Schematic representation of hysteresis current controller.

that the switching frequency varies with load parameters and AC voltage, and the operation is influenced by the randomness of hysteresis controllers that are independent to each other. Due to these reasons, the current error in isolated neutral system can be twice of the hysteresis band. Finally, subharmonics can be generated. Figure 2.38 shows the waveforms of the reference and measured stator currents, as well as the bandwidth limits of the hysteresis controller, obtained in numerical simulation on an FOC drive based on injected currents where current controllers are based on hysteresis comparators. The parameters of the machine under tests are given in Table 4.1. The upper and lower limits of the hysteresis controllers have been set equal to almost 10% of the reference current amplitude. It can be

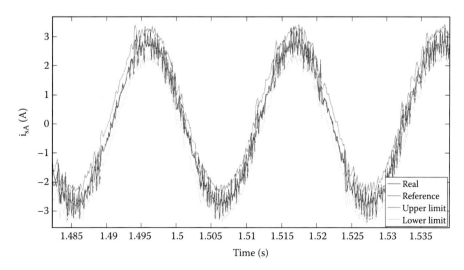

FIGURE 2.38
Hysteresis current control in an FOC IM drive based on impressed currents.

observed that the real current is inside the limits of the bandwidth controller, as expected. It should be also noted that the reference current is not perfectly sinusoidal but more irregular, as expected since it is generated by the output of the flux and speed controllers.

2.3.1.2 Suboscillation Current Control

In this strategy, the error signal is processed by a proportional integral (PI) controller, and then the driving signals are obtained by a comparison with a triangular carrier. The PWM is the one described in the suboscillation method; however, in this case, the current ripple is fed back, and it influences the switching times (Figure 2.39). This interaction can be minimized by a high gain in the PI controller while the integral part reduces the error at low frequency. A steady-state tracking error is to be expected; the higher the proportional gain of the PI, the lower the tracking error.

For correct operation, the slope of the error signal must be lower than the slope of the carrier. Considering the representation for a phase as in Figure 2.40, where the connection to an IM is accounted for, the motor is modelled by a series of the transient stator inductance L_s with rotor resistance R_s, the PWM inverter is represented by a constant gain K, the counter electromotive force by E_s, and the PI controller has a gain K_I and a time constant T.

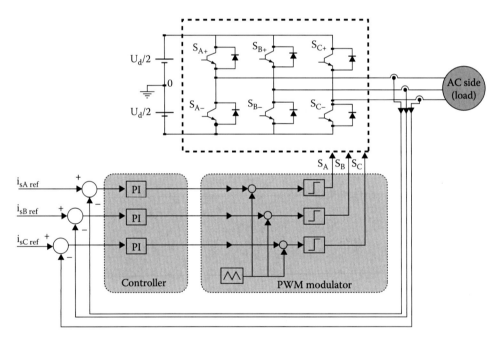

FIGURE 2.39
Schematic representation of the suboscillation current controller.

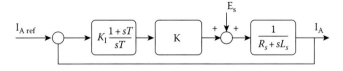

FIGURE 2.40
One phase representation for the control of the inverter with suboscillation current controller.

As it will be explained in Chapter 10, the same consideration holds for the connection of the inverter to a grid. In this case, instead of the stator inductance, the interconnecting inductance with its parasitic resistance while, instead of the counter electromotive force, the grid voltage have to be considered.

The condition that ties on the slope current error and the carrier slope is as follows:

$$\frac{d}{dt}(i_{sA\,ref} - i_A)\bigg|_{max} < 4\hat{V}_t f_t \tag{2.46}$$

where $\hat{V}_t$ and f_t are the carrier amplitude and frequency, respectively.

The maximum slope of the reference current is given by the following:

$$\frac{d}{dt}(i_{sA\,ref})\bigg|_{max} = \hat{I}_{sA\,ref}\,\omega\cos(\omega t)\big|_{\omega=0} = 2\pi f\hat{I}_{sA\,ref} \tag{2.47}$$

The maximum slope of inverter current is obtained when the voltage across the inductance is maximum:

$$\frac{d}{dt}(i_{sA})\bigg|_{max} = \frac{E_s + U_d/2}{L_s} \tag{2.48}$$

The condition (2.28) becomes

$$\frac{E_s + U_d/2}{L_s} + 2\pi f\hat{I}_{sA\,ref} < 4\hat{V}_t f_t \tag{2.49}$$

Once the grid inductance is known, (2.31) allows the triangular carrier parameters, $\hat{V}_t$ and f_t, to be chosen.

2.3.1.3 Space-Vector Current Control

As explained in Section 2.2.1.4, the SV-PWM utilizes the null vector at the beginning, at the end, and at the center of T_{PWM}. In simple words, if the current signal is always sampled at $t = T_{PWM}/2$ (see Figure 2.13), one is sure that the null vector is applied, and it implies that harmonic current is not present in the signal to be compared with the reference one. This overcomes one of the main limits of the other two techniques in which the harmonic current can influence the control strategy performance.

Among PI controllers that utilize space-vector theory, three techniques are recognizable: stationary vector controller, synchronous vector controller, and stationary resonant controller.

2.3.1.3.1 Stationary Vector Controller

In a three-phase system with isolated neutral, the knowledge of only two currents is necessary; in this case, the controller can be based on two PIs and a 3 → 2 coordinate transformation as described in (1.24). A scheme of the stationary PI controller is drawn in Figure 2.41. Comparing this scheme with that of Figure 2.39, it should be noted that only two PIs are used;

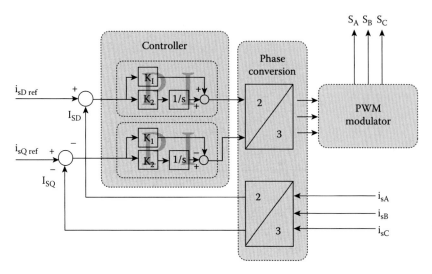

FIGURE 2.41
Stationary PI controller.

on the other hand, two coordinate transformation blocks are necessary. Moreover, this strategy acts on AC current components; therefore a steady-state tracking error is to be expected.

2.3.1.3.2 Synchronous Vector Controller

The steady-state tracking errors can be eliminated by using a synchronous vector controller. Differently from the stationary one, currents are converted in a rotating synchronous reference frame; as a consequence, they are DC quantities, and the PI controllers are able to control the steady-state error to zero. On the other hand, this scheme is more complicated than the stationary controller and requires the knowledge of the angle needed for the coordinate transformation. A scheme of the synchronous vector controller is drawn in Figure 2.42.

An equivalent synchronous controller in stationary coordinates (sD and sQ) has been proposed in Ref. [34]; this solution generates the reference voltages even if the corresponding

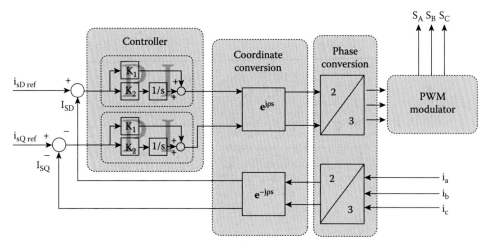

FIGURE 2.42
Synchronous PI controller.

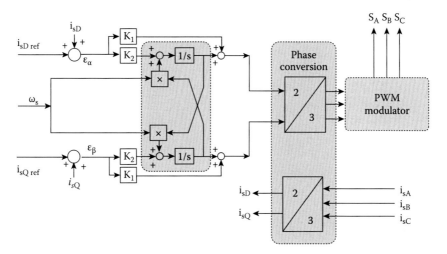

FIGURE 2.43
Synchronous PI controller working in stationary coordinates.

error is null, thus avoiding steady-state error; however, it has a dynamic performance worse than that of the synchronous controller due to the cross coupling of sD and sQ components. The corresponding scheme is shown in Figure 2.43.

2.3.1.3.3 Stationary Resonant Controller

In the stationary reference frame, the PI compensators of Figure 2.41 are substituted by a controller that has the same DC control response but centered on the control frequency.

The transfer function of PI controller in a synchronous reference frame is given by the following:

$$G(s) = K_1 + \frac{K_2}{s} \tag{2.50}$$

if it is converted from the synchronous to the stationary frame the transfer function of the resonant controller is obtained.

$$G(s) = K_1 + \frac{sK_2}{s^2 + \omega_s^2} \tag{2.51}$$

Equation 2.33 exhibits an infinite gain at the resonant frequency defined by ω_s; this is the condition that gives a zero steady-state tracking error [35]. The resonant controller scheme is shown in Figure 2.44.

2.3.1.4 Closed-Loop PWM with Real-Time Optimization

Closed-loop PWM techniques with real-time optimization tend to be used mainly in high-power industrial and traction AC drives [36], where power devices are predominantly thyristors or, at increasing powers, even GTOs. For such applications, power converter switching losses play a relevant part among the overall losses. For such a reason, switching frequency is typically maintained low, around hundreds of Hertz. The reduction of the switching frequency is, on the other hand, the cause of increase of the stator current harmonics and

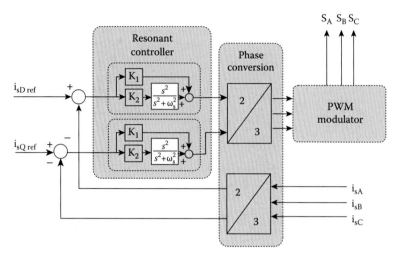

FIGURE 2.44
Resonant controller.

correspondingly of the torque ripple as well as of further machine losses. A suitable trade-off between the reduction of the switching losses and the current harmonics is the adoption of optimal switching strategies for the inverter control. A simple way to do that is to determine the optimal switching patterns off-line [37]. The algorithm, in this case, optimizes the switching angles (see Figure 2.45) under the hypothesis of steady-state behavior of the machine. The switching angles are stored on a look-up tables (in a memory of the drive system) and then recalled on-line by the inverter control system. This method obviously provides very good results at steady-state and quasi steady-state, but presents several limitations during dynamic behavior. Better results can be achieved with the predictive control, the pulsewidth control with field orientation, and the trajectory tracking control, as shown in the following [37].

2.3.1.5 Predictive Current Control

Among the closed-loop PWM techniques with real-time optimization, an interesting solution is the so-called predictive current control [38–40]. Predictive current control presents some common features with methodologies based on look-up tables. These methodologies have been applied mainly in electric drive control. In both techniques, the underlying idea

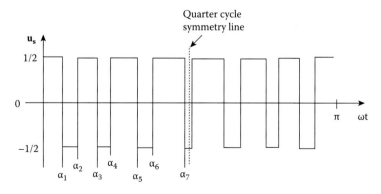

FIGURE 2.45
Phase voltage under optimal switching angles of the inverter.

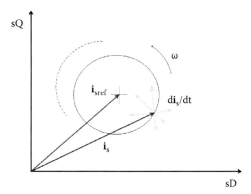

FIGURE 2.46
Reference and actual stator current space-vector and relative boundary.

is to control the stator current by suitably limiting it within some boundaries. One way is to select a boundary of circular shape centered on the apex of the reference stator current space-vector i_{sref} (see Figure 2.46). Whenever, during the operation of the drive, the stator current space-vector i_s touches or overcomes instantaneously its circular boundary, then the subsequent switching configuration of the inverter is determined in two phases: a prediction and an optimization phase. During the prediction phase, the loci of the stator current space-vector for each switching pattern of the inverter are computed, and, correspondingly, the time needed for reaching again the circular boundary is predicted. The prediction is made on the basis of a simplified mathematical model of the machine. The control system takes into consideration the movement of the circular boundary in the sD–sQ plane due to the modification of the reference current. The prediction of the times in which the stator current space-vector lies inside its boundary is made for all the inverter switching patterns. Finally, the switching pattern ensuring the maximum crosswalk time inside the current error boundary is selected as the optimal choice. This kind of selection, providing the longest possible duration between two subsequent switching instants, permits the switching frequency to be minimized. This optimization can be performed, taking into consideration the next two inverter switching patterns. The entire algorithm, run on a DSP, typically requires a sampling time of about 20 μs. This delay time can be dealt with by the system only at low switching frequencies. At increasing frequencies, a so-called double prediction method should be used. The prediction of trajectories is made by adopting a lower sampling time on the basis of the prediction of the switching pattern at the following boundary intersection. Figure 2.47 shows the block diagram of the predictive current control system of an AC drive.

2.3.1.6 Pulsewidth Control with Field Orientation

To further reduce the inverter switching frequency, which could be needed for very high-power drives, a rectangular current error boundary could be devised instead of a circular one [36,41]. The rectangular boundary should be aligned with the rotor flux space-vector of the machine (see Figure 2.48). The idea is to transfer the biggest part of the stator current harmonic distortion on the rotor flux axis, where it does not provide a significant influence on the machine electromagnetic torque. In this way, the harmonic content of the stator quadrature component $i_{sy}^{\psi_r}$ is negligible. At the same time, since in the rotor flux-oriented reference frame the system is represented on the direct x axis by a first-order system with time constant equal to the machine rotor time constant T_r, its typical big value is able to filter with a low pass behavior the residual effect caused by the harmonics in the stator direct current component $i_{sx}^{\psi_r}$ on the rotor flux amplitude. On the basis of the earlier considerations, a current trajectory-oriented control scheme can be devised which is able to force

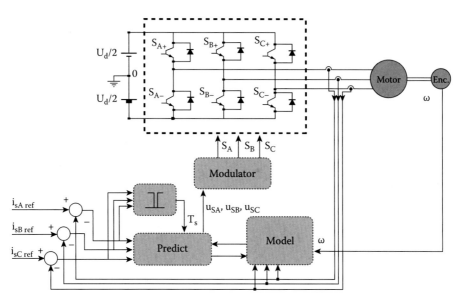

FIGURE 2.47
Block diagram of the predictive current control scheme.

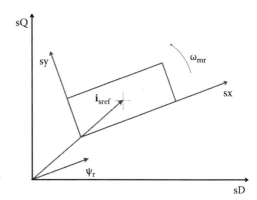

FIGURE 2.48
Current rectangular boundary and its alignment with the
rotor flux space-vector.

a small tracking error on $i_{sy}^{\psi_r}$, while permitting a high tracking error on $i_{sx}^{\psi_r}$. As a matter of fact, to do that, the error boundary on the x axis (aligned with the rotor flux space-vector) is much longer than that on the y axis. In this way, the system is devised so that the switching on the direct x axis will be avoided as long as possible.

Also in this case, the choice of the switching patterns is made on the basis of a prediction, satisfying the criterion of minimization of the switching frequency. It has been demonstrated that the use of a rectangular boundary instead of a circular one leads to a further reduction of the switching frequency of the inverter [36].

2.3.1.7 Trajectory Tracking Control

The idea of trajectory tracking control arises from the following considerations. The off-line optimization method [36] imposes a priori the entire set of switching patterns of the inverter within a period of the fundamental. This feature makes the dynamic performance of the system poor, so poor to be sometimes unfeasible. This is due to the fact that, during transient operation, the VSI reference voltage becomes nonsinusoidal. The switching pattern is, even in these conditions, assembled on the basis of precalculated switching patterns, which are, however, correctly associated only with steady-state working conditions. As a result, the optimization process fails, and the so-called dynamic modulation error arises, which can take values as big as the stator current fundamental component, resulting in big torque pulsations. Since the assumption that a PWM system behaves as a linear amplifier is valid only if the switching frequency is much higher than the bandwidth of the voltage reference signal, high-power drives supplied by VSI with low switching frequency typically do not match this condition. In such cases, the so-called dynamic modulation error occurs [42], deriving from the difference between the reference and actual integrals of the inverter voltage space-vectors: $\int \mathbf{u}_{sref}(t)dt \neq \int \mathbf{u}_s(t)dt$. As a result, a stator current error term, called dynamic modulation error $\delta(t)$, arises, which is a component of the stator current space-vector that is generated by the PWM in transient operation.

All the on-line optimization methods [38–40] take as basis of the optimization process the next or the next two switching patterns. This limiting assumption does not permit the optimization process to be properly fulfilled and, in addition, does not allow a synchronous switching.

The above considerations call for a proper integration of off-line methods, to cope for the steady-state behavior, with on-line methods, to cope for transient operation. A potential solution can exploit the advantages of both methods, as proposed in [42], which adopts the trajectory tracking method as a solution for eliminating the dynamic modulation error. The trajectory tracking approach in [42] utilizes the steady-state trajectories of the stator current space-vector, computed from the optimal switching patterns, as template. A tracking controller then reacts if the actual current vector deviates from the given path during transient operation. Whenever a dynamic modulation error occurs, the VSI switching pattern is modified so that the modulation law is respected. Several methods for pattern modification could be used, among which the step-by-step minimization method, the dead-beat control, and the pulse insertion method, each of them presenting advantages and disadvantages on the basis of the modulator working operating condition. In [42], a generalized algorithm, of which the above algorithms can be considered subsets, has been adopted. In addition, the decomposition of the stator current space-vector into its components enables the identification of the fundamental component of the stator current, permitting the implementation of a fast current control system.

2.3.2 From the Six-Pulse Rectifier to the Active Rectifier

The operation of the active rectifier can be conceptually considered as an extension of the classical six-pulse rectifier.

A great part of the early efforts in power electronics have focused on the conversion from a sinusoidal to a DC voltage. One of the reasons was the necessity to utilize the energy produced by synchronous generators to supply load that required constant voltage, as DC motors, particularly for railway applications.

First applications were based on motor-generator (an M-G set or a dynamotor for dynamo-motor) in which a sinusoidal voltage supplied an IM coaxial with a DC motor utilized as generator. Its brushes give a rectification effect allowing the output voltage to become unipolar.

The same concept is utilized in static rectifiers where no moving parts are present. First high-power rectifiers were mercury arc rectifiers in which the mercury was heated to become a gas in a glass tube: a unidirectional current could therefore flow from a steel (or coal) electrode to the remaining part of the liquid mercury. These devices were utilized for about 40 years starting from 1920. To obtain a low ripple at the output voltage,

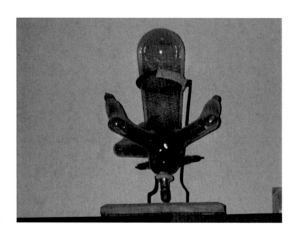

FIGURE 2.49
Three-phase mercury arc rectifier.

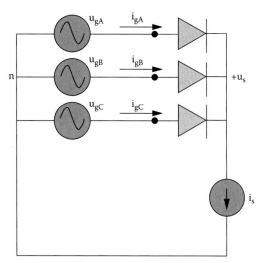

FIGURE 2.50
Equivalent scheme of three-phase mercury arc rectifier.

several electrodes have been used that utilize three-, six-, or polyphase supply obtained with suitable transformers. Figure 2.49 shows an example of a mercury arc rectifier and Figure 2.50 its equivalent circuit. It should be noted that it is a common cathode circuit, because there can be several steel anodes but only one liquid mercury cathode.

Due to the evolution of semiconductor power diodes, the common cathode rectifier has been replaced by the bridge topology, which ensures a lower voltage ripple. Figure 2.51 shows the electric scheme of a traditional three-phase uncontrolled six-pulse rectifier supplied by the grid and with a current generator as a load. This generator represents the behavior of a high inductive load, which is a common situation in power electronics applications.

It should be borne in mind that all rectifier solutions described earlier are based on elements that allow the current to flow only in one direction and are nonlinear elements. As a consequence, the circuit should be studied as a nonlinear circuit, and because of the presence of a sinusoidal supply, harmonics are to be expected. Section 3.1 explains the consequence of nonlinear loads.

To understand the operation of the circuit drawn in Figure 2.51, it should be noted that, among the top devices (D1, D3, D5), the diode with anode higher potential will conduct, and, among the bottom devices (D2, D4, D6), the diode with cathode lower potential will conduct.

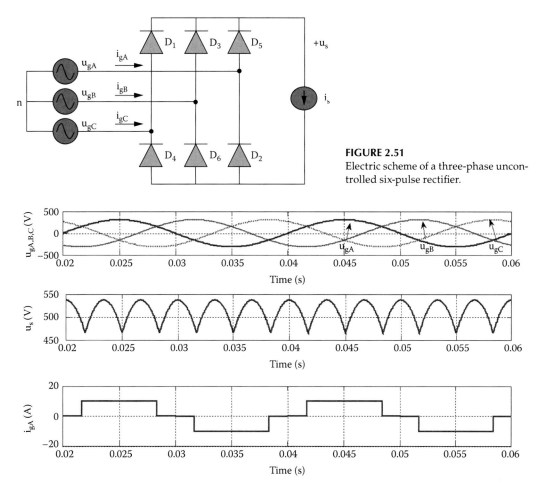

FIGURE 2.51
Electric scheme of a three-phase uncontrolled six-pulse rectifier.

FIGURE 2.52
Phase to neutral supply voltage, load voltage and current supplied by the grid obtained with the six-pulse rectifier shown in Figure 2.51.

In Figure 2.52, the characteristic waveforms obtained by a six pulse rectifier are shown, in particular, the phase supply voltage (referred to neutral), the load voltage, and the current supplied by the grid. The fundamental frequency has a period of 20 ms. It should be noted that the load voltage is composed of six-pulse-shaped elements for each period of the fundamental. This is the reason why the rectifier is called six-pulse rectifier. The load voltage u_s has a frequency that is six times the fundamental frequency, and its instantaneous value is given by ($u_{gx = A,B,C} - u_{gy = A,B,C}$), thus representing a line-to-line voltage. If this maximum value is $U_{LL(max)}$, then its mean value is given by the following:

$$U_{mean} = \frac{1}{\pi/3} \int_{\theta=-\pi/6}^{\theta=\pi/6} U_{LL(max)} \cos(\theta) d\theta = \frac{3}{\pi} U_{LL(max)} = \frac{3\sqrt{2}}{\pi} U_{LL(rms)} = U_{do} \qquad (2.52)$$

where
  $U_{LL(rms)}$ is its rms value
  $\theta = \omega t$ where t is the time and ω the pulsation of the supply

As for the current supplied by the grid, since the load current is constant, each diode allows the current to flow for $2\pi/3$, and the line current is given by the following:

$$i_{gA} = \begin{cases} i_s \to D_5 \ on \\ 0 \to D_5, D_2 \ off \\ -i_s \to D_2 \ on \end{cases} \tag{2.53}$$

If the parasitic elements are neglected, the commutation can be considered instantaneous. The rms value of the current supplied by each phase is as follows:

$$I_{g(rms)} = \sqrt{\frac{2}{3}} I_s = 0.816 I_d \tag{2.54}$$

By Fourier analysis, the fundamental value of the current can be evaluated; its fundamental value and its rms value are, respectively, as follows:

$$I_{g1(max)} = \frac{2\sqrt{3}}{\pi} I_s \tag{2.55}$$

$$I_{g1(rms)} = \frac{\sqrt{6}}{\pi} I_s \tag{2.56}$$

The spectrum has no even harmonics and multiple of the triple because of the symmetry of the time waveform; the other harmonics have amplitudes $I_{gh} = I_{g1}/N$, being h the order of the harmonic.

The time domain current with its fundamental harmonic and phase voltage are shown in Figure 2.53. It should be noted that, when the load is a pure current generator, the fundamental harmonic of the current delivered by the grid is in phase with the grid voltage. However, many current harmonics are present; the corresponding spectrum is shown in Figure 2.54.

If the diodes of the circuit of Figure 2.51 are replaced with thyristors, a controlled rectifier is obtained (the term controlled rectifier is also used for a single thyristor). By controlling the firing angle α, it is possible to vary the average value of the load voltage, which is then given by the following:

$$u_{s\alpha} = \frac{3\sqrt{2}}{\pi} U_{LL(rms)} \cos(\alpha) = 1.35 U_{LL(rms)} \cos\alpha = U_{do} \cos(\alpha) \tag{2.57}$$

where
$U_{LL(rms)}$ is the rms value of the line-to-line voltage
U_{do} is the mean voltage obtained with six-pulse diode rectifier (see (2.34))

Consequently the average power supplied to the load is as follows:

$$P = u_{s\alpha} I_s = 1.35 U_{LL} I_s \cos(\alpha) \tag{2.58}$$

The (2.57) gives (2.52) for $\alpha = 0$; moreover, it can be noted that the power that flows from the grid to the load for $0 < \alpha < \pi/2$ is null for $\alpha = \pi/2$ and flows from the load to the grid for $\pi/2 < \alpha < \pi$.

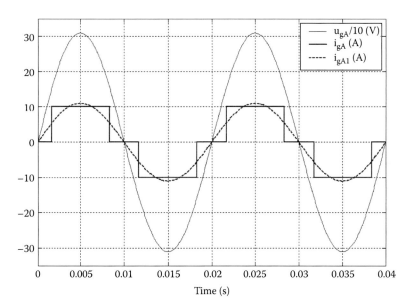

FIGURE 2.53
Time domain current waveform with its fundamental harmonic and phase voltage.

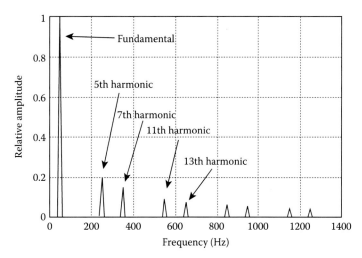

FIGURE 2.54
Spectrum of the current supplied by the grid to the six-pulse rectifier loaded by a current generator.

With a controlled rectifier, it is possible the inversion of power flow. Since only the voltage can change its sign, it corresponds to a two-quadrant operation.

It should be borne in mind that this analysis is valid for a load that can supply power as in this case. With a generic passive load, the inversion of the voltage sign is possible only instantaneously, but the mean value in a period of the fundamental remains positive (from the grid to the load).

A further improvement has been devised with the controlled rectifier that has the same topology of the circuit of Figure 2.51 adopting controlled power switches (BJT, MOSFET, and IGBT). More precisely in this case, the circuit becomes an active rectifier.

Active rectifiers present this additional advantage of their bidirectional power flow. They can be properly used also when a renewable source is utilized to supply the grid; this is the case of distributed generation (DG).

Most common control strategies of active rectifiers are based on current control of VSIs [43]. Among these, one of the most adopted is the so-called voltage-oriented control (VOC), where the current control is performed in the grid voltage space-vector oriented reference frame [44,45]. Another approach is based on the idea of directly controlling the active and reactive powers by choosing the proper switching patterns on the basis of the instantaneous position of the grid voltage space-vector [46]. This technique has been called direct power control (DPC). On the basis of a parallelism between the electrical grid and an electrical machine, both VOC and DPC have been further improved in their virtual flux (VF)-based versions, called, respectively, virtual flux-oriented control (VF-OC) and virtual flux direct power control (VF-DPC) [47,48].

2.3.3 Current Control of VSIs

2.3.3.1 Voltage-Oriented Control (VOC)

The VOC has been directly derived from the field-oriented control (VOC), formerly devised for the control of electric drives. Like in the drive control counterpart, the VOC is based on the idea to find a rotating reference frame in which the current control corresponds to the active and reactive power control. On this basis, first, the VOC has been developed, where the direct axis lies in the direction of the grid voltage space-vector, and second, the VF-OC has been developed, where the direct axis lies in the direction of a virtual flux, obtained on the basis of the time integration of the grid voltage components. Obviously, since the virtual flux lies in quadrature with respect to the grid voltage, the direct and quadrature components of the injected currents are displaced of 90° with respect to the VOC. In the following, it is assumed that the power is positive when it is absorbed by the DC source from the grid: therefore, powers generated by renewable sources are assumed negative (the load is studied with the user's sign convention).

A voltage-controlled bridge converter is shown in Figure 2.55a. It can be seen that the inverter is connected on the DC side with a load and via an inductance with its parasitic resistance (sketched in the dotted box) to the grid. A different solution could be to connect the VSI to the power grid via an LC filter (Figure 2.55b) or by an LCL filter (Figure 2.55c). The first solution is the simplest solution as for the number of components; on the other hand, to reduce the current harmonics around the switching frequency, a high value of input inductance should be selected. This, for applications above several kilowatts, implies a high cost and poor system dynamical performance.

The solution with an LC filter allows lower values to be adopted and improves the low-pass behavior of the filter that becomes a second-order filter. With the use of LCL filter, the main inductance can be lowered, and a better decoupling of the inverter behavior from the grid impedance is achieved. With both the LC and LCL solutions, however, particular care has to be taken to avoid resonances [49].

The operation can be explained with the aid of the single-phase representation of Figure 2.56. The inverter acts as a current source because it is obtained as the difference between the generated voltage, depending of the power switch states, and the grid voltage applied to the inductance L. The resulting current can be injected both to the grid and to the load.

Some characteristic situations are sketched in Figure 2.57 using a vector diagram valid in sinusoidal steady-state. In particular, Figure 2.57a shows a generic operating condition. The grid voltage u_g is obtained as the sum of inverter voltage phasor u_s, the drop on inductance

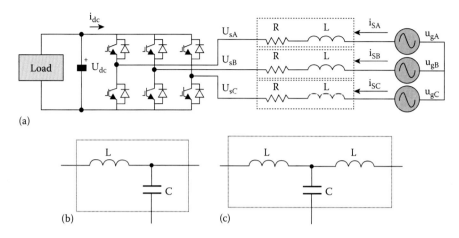

(a)

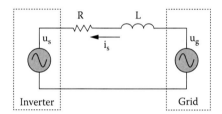

(b)   (c)

FIGURE 2.55
(a) Three legs bridge converter connecting the grid with a load by an interconnecting inductance, (b) by an LC filter, and (c) by an LCL filter.

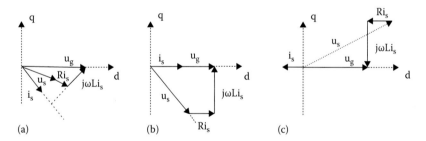

FIGURE 2.56
One-phase representation of the scheme of Figure 2.25.

(a)   (b)   (c)

FIGURE 2.57
Operating condition of an active rectifier: (a) generic condition, (b) rectification at unity power factor, and (c) inversion at unity power factor.

$j\omega L i_s$, and the drop on the parasitic resistance $R i_s$. Figure 2.57b represents a rectification at unity power factor. With the imposed inverter voltage u_s, the sum of the drop on inductance and of the drop on resistance makes the current i_s in phase with the grid voltage u_g. The power flow goes from the grid to inverter representing the rectifier behavior. Finally, Figure 2.57c shows the inversion at unity power factor. In this last case, the grid voltage space-vector and current space-vector are in opposition. It should be borne in mind that in Figure 2.54, the resistive drop has been exaggerated for the sake of clarity.

2.3.3.1.1 Circuital Analysis of the Active Rectifier

With reference to Figure 2.55, three main parts are recognizable: the inverter, the grid, and the RL circuit. The analysis can be performed by writing the grid voltage, the inverter voltage, and the Kirchhoff's voltage law equations (KVL) from which the current supplied by the active rectifier can be calculated. The equations can be written for each phase, in the stationary and synchronous reference frame.

The phase grid voltages and currents are as follows:

$$\begin{cases} u_{gA} = E_m \cos(\omega t) \\ u_{gB} = E_m \cos\left(\omega t + \frac{2}{3}\pi\right) \\ u_{gC} = E_m \cos\left(\omega t - \frac{2}{3}\pi\right) \end{cases} \tag{2.59}$$

$$\begin{cases} i_{sA} = I_m \cos(\omega t + \varphi) \\ i_{sB} = I_m \cos\left(\omega t + \frac{2}{3}\pi + \varphi\right) \\ i_{sC} = I_m \cos\left(\omega t - \frac{2}{3}\pi + \varphi\right) \end{cases} \tag{2.60}$$

The phase voltage generated by the active rectifier depends on the state of its power devices as expressed by (2.2) here rewritten for the sake of clarity:

$$\begin{cases} u_{sA} = \dfrac{2S_a - (S_b + S_c)}{3} \cdot U_d \\ u_{sB} = \dfrac{2S_b - (S_a + S_c)}{3} \cdot U_d \\ u_{sC} = \dfrac{2S_c - (S_a + S_b)}{3} \cdot U_d \end{cases} \tag{2.61}$$

The KVL are as follows:

$$\begin{bmatrix} u_{gA} \\ u_{gB} \\ u_{gC} \end{bmatrix} = R \begin{bmatrix} i_{sA} \\ i_{sB} \\ i_{sC} \end{bmatrix} + L\frac{d}{dt}\begin{bmatrix} i_{sA} \\ i_{sB} \\ i_{sC} \end{bmatrix} + \begin{bmatrix} u_{sA} \\ u_{sB} \\ u_{sC} \end{bmatrix} \tag{2.62}$$

Finally, a further equation for the DC link voltage and current can be added:

$$C\frac{du_d}{dt} = S_A \cdot i_{sA} + S_B \cdot i_{sB} + S_C \cdot i_{sC} - i_{dc} \tag{2.63}$$

2.3.3.1.2 Stationary Reference Frame Analysis

In the stationary reference frame, the grid voltages are obtained by (2.59) with the Clarke transformation (1.9):

$$\begin{cases} u_{gD} = \sqrt{\dfrac{3}{2}} E_m \cos(\omega t) \\ u_{gQ} = \sqrt{\dfrac{3}{2}} E_m \sin(\omega t) \end{cases} \tag{2.64}$$

The inverter voltages can be deduced by applying (1.9) to the vector of inverter voltages written for each phase:

$$
\begin{cases}
u_{sD} = \sqrt{\dfrac{2}{3}}\left(u_{sA} - \dfrac{1}{2}u_{sB} - \dfrac{1}{2}u_{sC} \right) \\
u_{sQ} = \dfrac{1}{\sqrt{2}}(u_{sB} - u_{sC})
\end{cases}
\tag{2.65}
$$

The KVL are as follows:

$$
\begin{bmatrix} u_{gD} \\ u_{gQ} \end{bmatrix} = R\begin{bmatrix} i_{sD} \\ i_{sQ} \end{bmatrix} + L\frac{d}{dt}\begin{bmatrix} i_{sD} \\ i_{sQ} \end{bmatrix} + \begin{bmatrix} u_{sD} \\ u_{sQ} \end{bmatrix}
\tag{2.66}
$$

The DC link Equation 2.63 becomes the following:

$$
C\frac{du_d}{dt} = (i_{sD}S_D + i_{sQ}S_Q) - i_{dc}
\tag{2.67}
$$

It should be noted that in the stationary reference, the quantity $(S_A \cdot i_{sA} + S_B \cdot i_{sB} + S_C \cdot i_{sC})$ corresponds to $(i_{sD}S_D + i_{sQ}S_Q)$, where (S_D, S_Q) are the Clarke transformation of the switch status considered as space-vectors.

The active and reactive power are given by the following:

$$
\begin{cases}
p = u_{gD} \cdot i_{SD} + u_{gQ} \cdot i_{SQ} \\
q = u_{gQ} \cdot i_{SD} - u_{gD} \cdot i_{SQ}
\end{cases}
\tag{2.68}
$$

2.3.3.1.3 Synchronous Reference Frame Analysis

In the synchronous reference frame, the grid voltage can be deduced by the equation written in stationary reference frame (2.64) after a vector rotation:

$$
\begin{bmatrix} u_{gd} \\ u_{gq} \end{bmatrix} = \begin{bmatrix} \sqrt{(3/2)}E_m \\ 0 \end{bmatrix} = \begin{bmatrix} \sqrt{u_{gD}^2 + u_{dQ}^2} \\ 0 \end{bmatrix}
\tag{2.69}
$$

The inverter voltage can be obtained by the transformation from stationary to synchronous frame:

$$
\begin{bmatrix} u_{sd} \\ u_{sq} \end{bmatrix} = \begin{bmatrix} \cos\gamma & \sin\gamma \\ -\sin\gamma & \cos\gamma \end{bmatrix}\begin{bmatrix} u_{sD} \\ u_{sQ} \end{bmatrix}
\tag{2.70}
$$

where

$$
\begin{cases}
\sin\gamma = \dfrac{u_{sQ}}{\sqrt{(u_{sD})^2 + (u_{sQ})^2}} \\[4mm]
\cos\gamma = \dfrac{u_{sD}}{\sqrt{(u_{sD})^2 + (u_{sQ})^2}}
\end{cases}
\tag{2.71}
$$

In synchronous coordinates, the DC vector component i_{sd} is responsible of the active power flow; the quadrature current vector component i_{sq} determinates the reactive power. These two terms can be controlled independently. If the vector i_{sd} is aligned with the grid voltage u_g, then the unitary power factor condition is reached. By imposing the direct-axis of the rotating coordinates on the grid voltage vector, a simplified dynamic model can be obtained.

The KVL voltage space-vector equations of the system on the direct (d) and quadrature (q) axis give the following:

$$\begin{cases} u_{gd} = u_{sd} + Ri_{sd} + L\dfrac{di_{sd}}{dt} - \omega L i_{sq} \\ u_{gq} = u_{sq} + Ri_{sq} + L\dfrac{di_{sq}}{dt} + \omega L i_{sd} \end{cases} \tag{2.72}$$

Equation 2.72 shows that the direct (quadrature) component of the injected currents depends on the direct (quadrature) component of the inverter voltages. However, like in the electric drive counterpart, there are some coupling terms on both axis equations, which should be compensated with feed-forward control terms. The same consideration is true for the direct component of the grid voltage, which should be compensated with a suitable feed-forward term.

Finally, the active and reactive powers can be written as follows:

$$\begin{cases} P = \dfrac{3}{2} u_{gd} i_{sd} \\ Q = -\dfrac{3}{2} u_{gd} i_{sq} \end{cases} \tag{2.73}$$

If the target is to control directly the active and reactive power, the current references can be obtained on the basis of the reference active and reactive power.

A control scheme is shown in Figure 2.58. The grid voltages (u_{gA}, u_{gB}, and u_{gB}) and inverter currents (i_{sA}, i_{sB}, and i_{sC}) are first sampled by analog/digital converters; they are converted

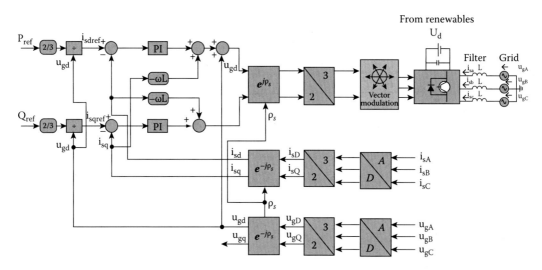

FIGURE 2.58
Control scheme of a controlled rectifier in the synchronous reference frame. (From Pucci, M. et al., *Electric Power Syst. Res.*, 81(4), 830, April 2011, doi: 10.1016/j.epsr.2010.11.007.)

in stationary D–Q coordinates and then in rotating d–q coordinates. The reference values of direct, i_d, and quadrature current, i_q, are obtained on the basis of active and reactive reference power P_{ref} and Q_{ref} by using (2.73). The error signal is processed by PI regulators; in particular, the error on DC processed by PI regulator is added to the feed-forward term, $-\omega L i_{sq}$, and to the direct component of grid voltage, the error on quadrature current processed by PI regulator is added to the feed-forward term $\omega L i_{sd}$.

The direct and quadrature components of the inverter voltages are then computed as follows:

$$u_{sd} = \omega L i_{sq} + u_{gd} + \Delta u_{sd}$$
$$u_{sq} = -\omega L i_{sd} + \Delta u_{sq} \tag{2.74}$$

where the parasitic resistance of the inductance grid is neglected, and PI controllers are used for controlling the current error to zero in this way, forcing the current error on both axes to zero:

$$\Delta u_{sd} = k_p(i_{dref} - i_d) + k_1 \int (i_{dref} - i_d)dt$$
$$\Delta u_{sq} = k_p(i_{qref} - i_q) + k_1 \int (i_{qref} - i_q)dt \tag{2.75}$$

The entire control scheme is shown in Figure 2.58, which exhibits the well-known decoupling terms. However it is slightly different from [46–50] where the direct component of the reference current is computed as the output of a DC link voltage controller. Current control is performed in the grid voltage reference frame by means of PI controllers.

2.3.3.2 Virtual Flux-Oriented Control

To use the control techniques typical the electric motors, deeply studied in literature, also in the sectors of the DG and the active rectifiers has lead to the concept of "virtual motor." It is possible to consider the electric grid as an equivalent virtual motor characterized by a transient leakage inductance (motor) equivalent to the connecting filter inductance of the inverter (active rectifier) and the back-electromotive force (motor) equivalent to the grid voltages (active rectifier). On the basis of this abstraction, the concept of virtual flux, equivalent to the magnetic flux of an electric motor, can be defined as the time integral of the grid voltages. The virtual flux is phase shifted of 90° with respect to the grid voltage. The direct and quadrature components of the virtual flux in the stationary reference frame are as follows:

$$\begin{cases} \psi_{gD} = \int \left(u_{sD} + L \dfrac{di_{sD}}{dt} \right) dt \\ \psi_{gQ} = \int \left(u_{sQ} + L \dfrac{di_{sQ}}{dt} \right) dt \end{cases} \tag{2.76}$$

The use of the virtual flux in the VOC technique is based on the orientation of the rotating reference frame on the vector of the virtual flux rather than on the vector of the grid voltage. The estimation of the virtual flux for experimental applications requires an open-loop integration. A simple way to solve this problem is to approximate the

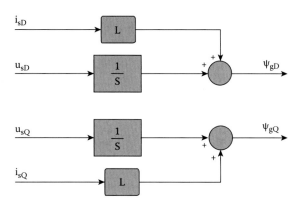

FIGURE 2.59
Block diagram of the VF estimator. (From Giglia, G. et al., Comparison of control techniques for three-phase distributed generation based on VOC and DPC, in *International Conference on Renewable Energy and Power Quality (ICREPQ'08)*, Santander, Spain, March 12–14, 2008.)

pure integrator with a first-order low-pass filter (see Chapter 6). The practical implementation of (2.76) is shown in Figure 2.59.

2.3.3.3 Experimental Results with VOC and VF-OC

The VOC control technique with its virtual flux corresponding version has been implemented in MATLAB®–Simulink® environment. In particular, for the circuital scheme analysis, the PLECS® has been adopted. The PLECS® software has been used for the IGBT three-phase inverter, the interconnecting inductance, and the voltage and current sensors. The control algorithm has been implemented in classic Simulink®, in the discrete domain. A sampling frequency of 10 kHz with a SV-PWM frequency of 5 kHz has been adopted both for the VOC and VF-OC. Figures 2.60 and 2.61 show the steady-state inverter current time waveforms and their corresponding spectra calculated with the FFT (Fast Fourier Transform) up to the 40th harmonics, obtained, respectively, with the VOC and VF-OC when Pref = −2 kW, Qref = 0 VAR have been given as references. As expected, VF-OC exhibits a slightly better harmonic content of the injected current, both considering a harmonic-by-harmonic analysis and considering the %THD equal, respectively, to 0.77% for the VOC and 0.72% for the VF-OC.

Figure 2.62, referring to an experimental test, shows a set of transients of the inverter-generated direct and quadrature current, obtained with a VOC-based wind generator under a set of step wind variations corresponding to active power variations. The quadrature current component has been controlled to zero, as usual, to avoid reactive power exchange with the grid. Figure 2.63 shows the corresponding waveforms of the generated active and reactive powers.

2.3.4 Power Control of VSIs

2.3.4.1 Direct Power Control

DPC has been directly derived from its counterparts devised for the control of electric drives called direct torque control (DTC). DPC is based on the idea to find instantaneously a switching pattern of the inverter permitting to increase or decrease directly, without current control, and in a decoupled way the active and reactive power exchanges between the DC stage and the grid. Even in this case, the VF DPC has been developed as a further improvement, where active and reactive powers are estimated on the basis of the virtual

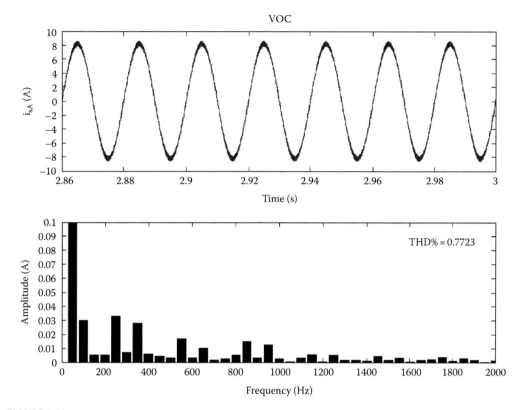

FIGURE 2.60
Inverter current and its FFT at Pref = –2 kW, Qref = 0 VAR with VOC. (From Giglia, G. et al., Comparison of control techniques for three-phase distributed generation based on VOC and DPC, in *International Conference on Renewable Energy and Power Quality (ICREPQ'08)*, Santander, Spain, March 12–14, 2008.)

flux components instead of the voltage ones. Like Malinowski et al., [48] and [50] only the behavior as generator is addressed, since the behavior of the control systems as a controlled load or a generation system is not perfectly symmetric, particularly in the DPC case.

2.3.4.2 Switching Table DPC

DPC is based on the idea to control directly and in a decoupled way the active and reactive power exchanged by the inverter with the electrical grid, avoiding any current control. The optimal switching pattern is to be selected on the basis of the active and reactive power demand and depends on the instantaneous position of the grid voltage space-vector. It can be shown that if the grid voltage vector u_g lies instantaneously in the sector k, the effect on the active and reactive power, P and Q, exchanged with the grid caused by the application of any VSI voltage vector can be summarized in Table 2.4, which takes also into consideration each subsector inside a sector (A is the first and B the second subsector in the rotating sense of the grid voltage vector [24]). In the table, a single arrow means a small variation while a double arrow a big variation.

Figure 2.64 shows the effects of the application of any active voltage vector on the active and reactive powers exchanged with the grid after a sampling time of the control system T_s. This figure is drawn when the grid voltage vector $u_g(t)$ lies in the first sector and each inverter active voltage vector is applied. Under the assumption that the control system is

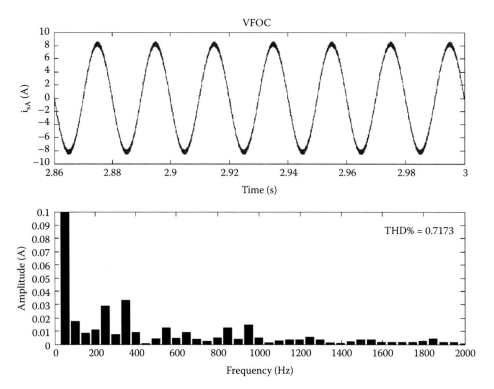

FIGURE 2.61
Inverter current and its FFT at Pref = –2 kW, Qref = 0 VAR with VF-OC. (From Giglia, G. et al., Comparison of control techniques for three-phase distributed generation based on VOC and DPC, in *International Conference on Renewable Energy and Power Quality (ICREPQ'08)*, Santander, Spain, March 12–14, 2008.)

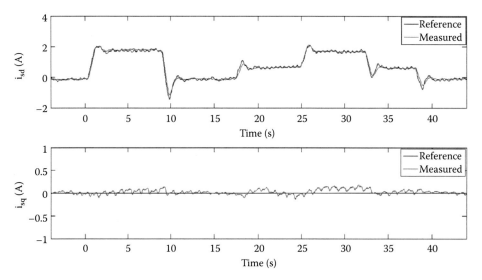

FIGURE 2.62
Grid side i_{sd}, i_{sq} reference and measured currents.

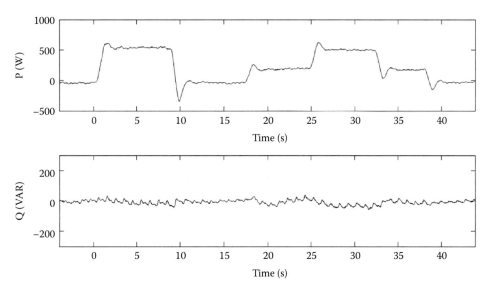

FIGURE 2.63
Active (*P*) and reactive (*Q*) power flowing to the power grid.

TABLE 2.4

Effect of the VSI Voltage Vectors on P and Q

	$\mathbf{u}_0$	$\mathbf{u}_k$	$\mathbf{u}_{k+1}$	$\mathbf{u}_{k+2}$	$\mathbf{u}_{k+3}$	$\mathbf{u}_{k-1}$	$\mathbf{u}_{k-2}$
P	⇑⇑	⇓⇓	⇑⇑	⇑⇑	⇑⇑	⇑	⇑⇑
Q	⇑A ⇓B	⇑A ⇓B	⇑⇑	⇑	⇓A⇑ B	⇓⇓	⇓

Source: Cirrincione, M. et al., New direct power control strategies of three-phase VSIs for the minimization of common-mode emissions in distributed generation systems, in *The Forty-Second Annual Meeting of the IEEE Industry Applications Society (IAS 2007)*, New Orleans, LA, September 23–27, 2007.

correctly working and therefore the injected current $i_s(t)$ in the time instant t is in phase with the grid voltage, the voltage drop on the series inductance L, neglecting its parasitic resistance, is the vector difference $\mathbf{u}_g(t) - \mathbf{u}_s(t)$. The current vector $i_s(t + T_s)$ after a sampling time of the control system is thus obtained as the vector sum of $i_s(t)$ and an additional term in phase with the inductance voltage drop.

On the basis of Table 2.4, the optimal switching table (ST) proposed by Noguchi et al., [46] can be directly deduced. This ST, shown in Table 2.5, has been used here for the experimental application of the methodology.

The active and reactive power for control feedback is estimated instantaneously on the basis of the following equations:

$$P = u_{sA}i_{sA} + u_{sB}i_{sB} + u_{sC}i_{sC} \tag{2.77}$$

$$Q = \frac{1}{\sqrt{3}}((u_{sB} - u_{sC})i_{sA} + (u_{sC} - u_{sA})i_{sB} + (u_{sA} - u_{sB})i_{sC}) \tag{2.78}$$

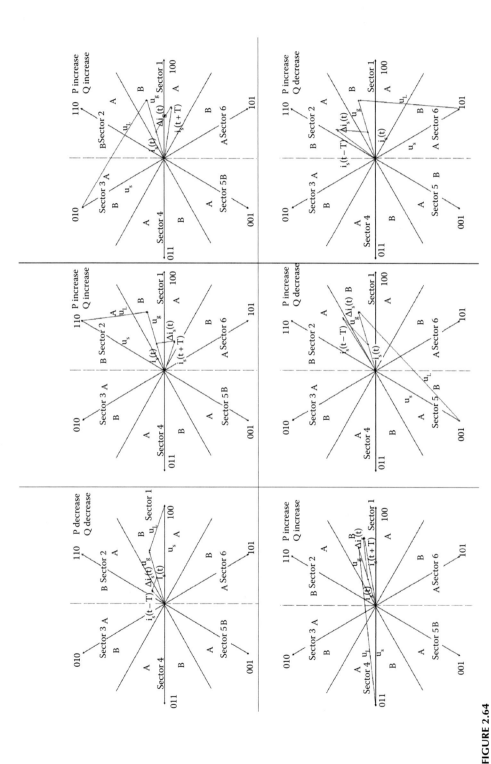

FIGURE 2.64

(See color insert.) Effect of the application of any voltage vector on the active and reactive power exchanged with the grid. (From Pucci, M. et al., *Electric Power Syst. Res.*, 81(4), 830, April 2011, doi: 10.1016/j.epsr.2010.11.007.)

TABLE 2.5

Optimal ST

		Sub A	Sub B
P ⇑	Q ⇑	u_0	u_0
	Q ⇓	u_{k-1}	u_0
P ⇓	Q ⇑	u_k	u_{k+1}
	Q ⇓	u_{k-1}	u_k

Source: Cirrincione, M. et al., New direct power control strategies of three-phase VSIs for the minimization of common-mode emissions in distributed generation systems, in *The Forty-Second Annual Meeting of the IEEE Industry Applications Society (IAS 2007)*, New Orleans, LA, September 23–27, 2007.

The three-phase quantities of the inverter voltages and currents have been used for the computation of the active and reactive powers, and so no coordinate variation is needed differently from the VOC. The entire control scheme is shown in Figure 2.65. Active and reactive power control is performed by two-level hysteresis controllers. A sector and subsector finding algorithm permits computing where the grid voltage vector instantaneously lies, with the approximation of $\pi/6$ rad.

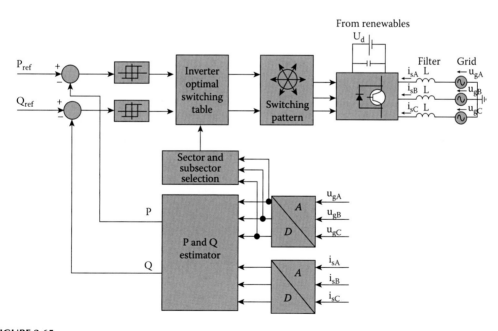

FIGURE 2.65
Block diagram of the DPC control schemes. (From Pucci, M. et al., *Electric Power Syst. Res.*, 81(4), 830, April 2011, doi: 10.1016/j.epsr.2010.11.007.)

2.3.4.3 Virtual Flux Direct Power Control

The use of the virtual flux in the DPC technique allows the active and reactive powers to be calculated using the virtual flux rather than the grid voltage that is more noisy. The estimation of the power based on the virtual flux is based on the following equations:

$$\begin{cases} P = \dfrac{3}{2}\omega(\psi_{gD}i_{sQ} - \psi_{dQ}i_{sD}) \\[2mm] Q = \dfrac{3}{2}\omega(\psi_{gD}i_{sD} + \psi_{gQ}i_{sQ}) \end{cases} \tag{2.79}$$

The advantages introduced with the virtual flux are a reduction of the harmonics in the grid current and in the ripples of the active and reactive powers.

2.3.4.4 Experimental Results with DPC and VF-DPC

The two control techniques described above—DPC with its virtual flux corresponding version—have been implemented in MATLAB–Simulink and PLECS environment for the IGBT three-phase inverter, the interconnecting inductance, and the voltage and current sensors. All the control algorithms have been implemented in classic Simulink, in the discrete domain. In this case, a sampling frequency of 15 kHz has been adopted.

Figures 2.66 and 2.67 show the steady-state inverter current time waveforms and their corresponding spectra under the same working conditions of Section 2.3.3.2

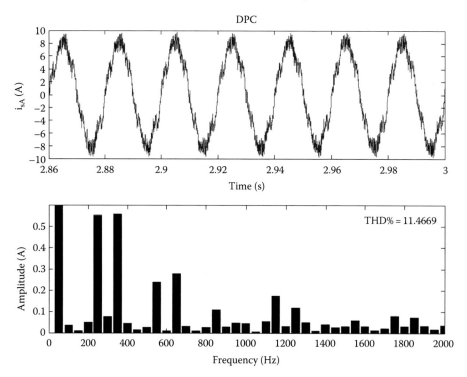

FIGURE 2.66
Inverter current and its FFT at Pref = −2 kW, Qref = 0 VAR with DPC. (From Giglia, G. et al., Comparison of control techniques for three-phase distributed generation based on VOC and DPC, in *International Conference on Renewable Energy and Power Quality (ICREPQ'08)*, Santander, Spain, March 12–14, 2008.)

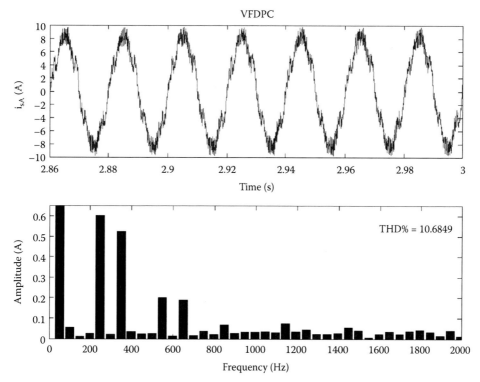

FIGURE 2.67
Inverter current and its FFT at Pref = −2 kW, Qref = 0 VAR with VF-DPC. (From Giglia, G. et al., Comparison of control techniques for three-phase distributed generation based on VOC and DPC, *International Conference on Renewable Energy and Power Quality (ICREPQ'08)*, Santander, Spain, March 12–14, 2008.)

(P_{ref} = −2 kW, Q_{ref} = 0 VAR), respectively, with the DPC and the VF-DPC. As expected, VF-DPC exhibits a slightly better harmonic content of the injected current, both considering a harmonic-by-harmonic analysis and considering the %THD equal, respectively, to 11.47% for the DPC and 10.68% for the VF-DPC.

2.3.4.5 DPC EMC (Electromagnetic Compatible)

In Section 2.1, Table 2.1 shows the CMV for each inverter state. It can be seen that, if only even or only odd active voltage vectors are used (u_k, with k, respectively, even or odd), no CMV variation is generated. If a transition from an even voltage vector to an odd one (or vice versa) occurs, the variation of the CMV is equal to $U_d/3$ is generated. If a transition from an odd (even) voltage vector to the zero (seventh) voltage vector occurs, a common-mode variation of amplitude $U_d/3$ is generated. Finally, if a transition from an odd (even) voltage vector to the seventh (zero) voltage vector occurs, a common-mode variation of amplitude $2U_d/3$ is generated.

The step shape exhibited by CMV variation causes the presence of harmonics that propagate toward the grid. The frequency of these harmonics depends both on the switching time of power devices and on the repetition of step variation of CMV.

Thus, from the point of view of common-mode emissions, the worst case is a transition from an odd (even) voltage vector to the seventh (zero) voltage vector.

For this reason, whatever inverter control technique is devised, to minimize the generated common-mode emissions of the drive, the exploitation of both null voltage

TABLE 2.6

Optimal ST of DPC-EMC 1

		Sect. k
P ⇑	Q ⇑	u_{k+2}
	Q ⇓	u_{k-2}
P ⇓	Q ⇑	u_k
	Q ⇓	u_k

Source: Cirrincione, M. et al., New direct power control strategies of three-phase VSIs for the minimization of common-mode emissions in distributed generation systems, in *The Forty-Second Annual Meeting of the IEEE Industry Applications Society (IAS 2007)*, New Orleans, LA, September 23–27, 2007.

vectors (zero and seventh) should be avoided. If a DPC technique is used, this consideration is helpful also from the control point of view. As a matter of fact, the original DPC [46] has been devised so that the zero voltage vector is adopted when a *P* increase is required when both a *Q* increase and decrease are needed. In Table 2.6, a first version of this technique, called DPC-EMC 1, is shown [51–54]; it can be noted that, when the grid voltage vector lies in the *k-th* sector, the application of the *k-th* voltage vector produces a high decrease of the absorbed active power and a low increase (subsector A) or decrease (subsector B) of the reactive power. On the contrary, the application of the u_{k+2} voltage vector produces a slight increase both of the active and reactive power while the u_{k-2} voltage vector produces a slight increase of the active power and a decrease of the reactive power.

In case of VOC, the harmonic content of the CMV is basically in the $f_{PWM} = 1/T_{PWM}$ frequency range, in particular, a strong component at f_{PWM} is present and its odd multiples, and its amplitude decreases with a $sin(x)/x$ law with frequency. Figure 2.68 shows the amplitude of the CMV spectrum versus frequency for different values of the duty cycle (δ_1) and $f_{PWM} = 4\,kHz$ obtained with VOC. It should be noted that the harmonics starting from the f_{PWM} frequency can be found up to 100 kHz with significant amplitude.

The spectrum of the CMV that is generated by DPC-EMC 1 is easily predictable; as a matter of fact, the theoretical steady-state CMV, u_{N0}, neglecting all parasitic effects and the rise/fall time, is a periodic waveform, with a period equal to 1/3 of the period of the fundamental of the grid voltage. Figure 2.69 shows the theoretical time domain waveform of the CMV obtainable with the DPC-EMC 1. Its harmonic spectrum can be analytically inferred by computing the coefficients of its Fourier series expansion as functions of the DC link voltage U_d and the fundamental pulsation of the grid voltage ω:

$$u_{N0} = \sum_{n=1}^{\infty} \frac{U_d}{3\pi n} \sin(3n\omega t) \tag{2.80}$$

On this basis, the harmonic content of the CMV is basically in the low frequency region, in particular, the amplitudes of the third harmonic of the fundamental (150 Hz in Europe)

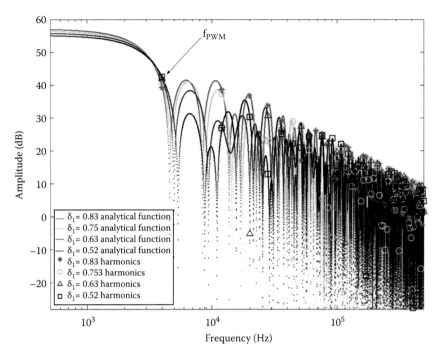

FIGURE 2.68
Amplitude of the CMV spectrum versus frequency for different values of the duty cycle (δ1) in VOC. (From Pucci, M. et al., *Electric Power Syst. Res.*, 81(4), 830, April 2011, doi: 10.1016/j.epsr.2010.11.007.)

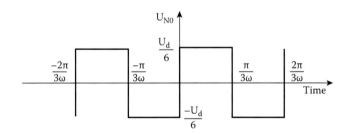

FIGURE 2.69
Theoretical waveform of the CMV for DPC_EMC1. (From Pucci, M. et al., *Electric Power Syst. Res.*, 81(4), 830, April 2011, doi: 10.1016/j.epsr.2010.11.007.)

and its odd multiples (3rd, 9th, 15th, 25th, etc., of the fundamental) decrease with inverse proportionality with the frequency itself. Such a controlled generator, therefore, shows a common-mode harmonic content that is quite low; its spectrum presents a low harmonic content compared to traditional modulation technique.

On the other hand, this technique has the drawback to consider only the common voltage spectrum optimization, which implies a high ripple of P and Q waveforms and of injected current, and the presence of a bias in the controlled reactive power in generating mode. As a matter of fact, the poor control of the reactive power in generating mode is due to the fact that both the u_{k+2} and u_{k-2} voltage vectors cause small variations of the reactive power. On the contrary, u_{k+1} and u_{k-1} voltage vectors cause high variations of the reactive power (see Table 2.4).

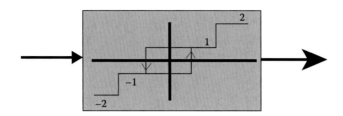

FIGURE 2.70
Four-level hysteresis controller. (From Pucci, M. et al., *Electric Power Syst. Res.*, 81(4), 830, April 2011, doi: 10.1016/j.epsr.2010.11.007.)

A trade-off between these two constraints is achieved with a modified version of DPC-EMC 1 called DPC-EMC 2.

The DPC-EMC 2 technique, has been devised to improve the drawbacks of DPC-EMC 1 by using a four-level hysteresis controller (Figure 2.70) for Q control, instead of a two-level one. In this way, when the Q error is low, the output of the controller is ±1 and thus vectors $u_{k\pm2}$ are used (the controller coincides with that of the DPC-EMC 1). On the contrary, when the Q error is high, the output of the controller is ±2 and thus vectors $u_{k\pm1}$ are used. The control strategy is summarized in Table 2.7.

The result is a better capability to control the Q error to zero (no reactive power exchanged with the grid), lower ripple in the P and Q waveforms, and lower harmonic content of the injected currents. With regard to the CMV, the result is a waveform, which is the six-step one of DPC-EMC 1 with few additional spikes due to the rare application of vectors $u_{k\pm1}$. However, the harmonic content of the CMV at frequencies about some kilohertz slightly increases with respect to that of DPC-EMC 1, but this deterioration is negligible, and in any case, much lower than that of classic DPC.

Differently from DPC-EMC 1, DPC-EMC 2 presents a CMV waveform, which is a square wave at 150 Hz with some internal spikes; as a result, its spectrum presents lower values of the low frequency harmonics and slightly higher values of the harmonics around higher frequencies than DPC-EMC 1.

TABLE 2.7

Optimal ST of DPC-EMC 2

		Sect. k
$P\Downarrow$	$Q\Uparrow$	u_k
	$Q\Downarrow$	u_k
$P\Uparrow$	$Q\Uparrow\Uparrow$	u_{k+1}
	$Q\Downarrow\Downarrow$	u_{k-1}
$P\Uparrow$	$Q\Uparrow$	u_{k+2}
	$Q\Downarrow$	u_{k-2}

Source: Cirrincione, M. et al., New direct power control strategies of three-phase VSIs for the minimization of common-mode emissions in distributed generation systems, in *The Forty-Second Annual Meeting of the IEEE Industry Applications Society (IAS 2007)*, New Orleans, LA, September 23–27, 2007.

2.3.4.6 Simulation Results with DPC-EMC 1 and DPC-EMC 2

The DPC-EMC 1, the DPC-EMC 2, and their virtual flux versions VF-DPC-EMC 1 and VF-DPC-EMC 2 control techniques have been implemented in MATLAB–Simulink and PLECS environment for the IGBT three-phase inverter, the interconnecting inductance, and the voltage and current sensors. All the control algorithm has been implemented in classic Simulink, in the discrete domain.

Figures 2.71 and 2.72 show the steady-state inverter current time waveforms and their corresponding spectra under the same working conditions Section 2.3.3.2 (P_{ref} = −2 kW, Q_{ref} = 0 VAR), respectively, with the DPC-EMC 1 and the VF-DPC-EMC 1. Also in this case, VF-DPC-EMC 1 exhibits a slightly better harmonic content of the injected current, both considering a harmonic-by-harmonic analysis and considering the %THD equal, respectively, to 8.93% for the DPC-EMC 1 and 8.36% for the VF-DPC-EMC 1. Finally, Figures 2.73 and 2.74 show the steady-state inverter current time waveforms and their corresponding spectra under the same working conditions, respectively, with the DPC-EMC 2 and the VF-DPC-EMC 2. Also in this case, VF-DPC-EMC 2 exhibits a slightly better harmonic content of the injected current, both considering a harmonic-by-harmonic analysis and considering the %THD equal, respectively, to 7.77% for the DPC-EMC 2 and 6.70% for the VF-DPC-EMC 2. As a global comparative analysis, the %THD of the injected current versus the generated power has been drawn for all control techniques.

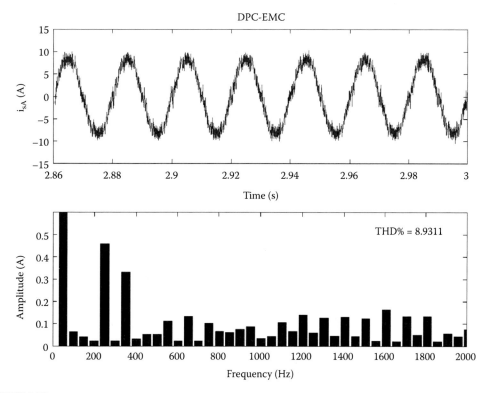

FIGURE 2.71
Inverter current and FFT at Pref = −2 kW, Qref = 0 VAR with DPC-EMC1. (From Giglia, G. et al., Comparison of control techniques for three-phase distributed generation based on VOC and DPC, in *International Conference on Renewable Energy and Power Quality (ICREPQ'08)*, Santander, Spain, March 12–14, 2008.)

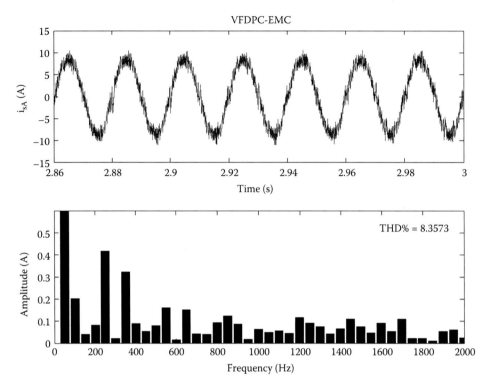

FIGURE 2.72

Inverter current and FFT at Pref = −2 kW, Qref = 0 VAR with VF-DPC-EMC1. (From Giglia, G. et al., Comparison of control techniques for three-phase distributed generation based on VOC and DPC, in *International Conference on Renewable Energy and Power Quality (ICREPQ'08)*, Santander, Spain, March 12–14, 2008.)

2.3.4.7 Comparative Experimental Analysis of DPC, DPC-EMC 1, and DPC-EMC 2

All DPC control techniques have been tested experimentally on a properly devised test setup. It is composed of the following items:

- A 7.5 kVA, three-phase VSI
- An electronic card with voltage and current sensors (model LEM CV3-1000 and LEM LA-55P)
- A dSPACE card (DS1103) with a PowerPC 604e at 400 MHz and a floating-point DSP TMS320F240
- A DC voltage source of 560 V, emulating a voltage renewable source
- A 100 V sinusoidal grid
- An interconnection series inductance of 12 mH with a parasitic resistance of 0.9 Ω

The sampling frequency of all control strategies has been set to f_s = 20 kHz. Figures 2.75 and 2.76 show, respectively, the electrical scheme and the photograph of the test setup. The CMV has been measured with a capacitor divider (C_m = 1 nF in Figure 2.75).

The DC supply configuration corresponds to U_d = 560 V. Figures 2.77 through 2.79 show, respectively, with DPC-EMC 1, DPC-EMC 2, and the classic DPC, the reference and estimated P and Q when steps P_{ref} = −1000 W and Q_{ref} = 0 VAR are given and the

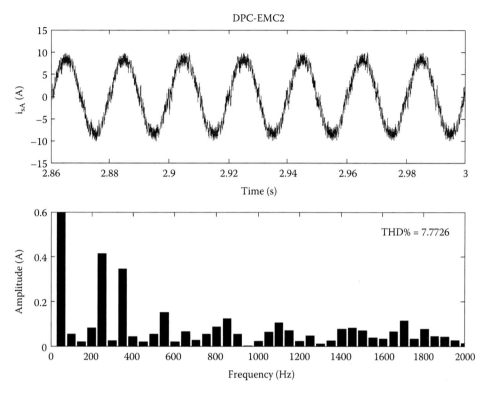

FIGURE 2.73
Inverter current and FFT at Pref = –2 kW, Qref = 0 VAR with DPC-EMC2. (From Giglia, G. et al., Comparison of control techniques for three-phase distributed generation based on VOC and DPC, in *International Conference on Renewable Energy and Power Quality (ICREPQ'08),* Santander, Spain, March 12–14, 2008.)

corresponding time waveforms of the three injected currents are also given. Experimental results are consistent with those obtained with the simulations. Particularly, all DPC strategies present a good dynamic performance as for the P step command, with no appreciable differences. Between the DPC-EMC 1 and the DPC-EMC 2, there is no appreciable difference in controlling the reactive power due to the high DC link voltage value of U_d = 560 V. It should be borne in mind that, if the DC link voltage is below a certain threshold, the system starts to have a bias in the controlled Q when DPC-EMC 1 is used, which is not present with DPC-EMC 2. In the following, only tests with a higher DC link voltage value are shown for stability reasons of the control system. On the contrary, the classic DPC exhibits a significant ripple with negative peaks on the estimated Q, which are present also in simulation. With regard to the harmonic content of the injected currents, results are summarized in Figure 2.80, which shows the %THD versus the generated power with all DPC techniques including, for comparison, also the VOC results and the limit of 5% required by the international IEEE Standard [55]. As expected, VOC outperforms all the other techniques, always remaining below the prescribed limit, except at very low generated power. All of the DPC techniques present THDs decreasing for increasing generated power, with the DPC-EMC 1 and the DPC-EMC 2 with very similar results.

In particular, the DPC-EMC 1 is significantly worse for low generated power levels (THD% almost equal to 70%) while it is almost the same at increasing power. The classic DPC has a better behavior among DPC techniques for low generated power almost

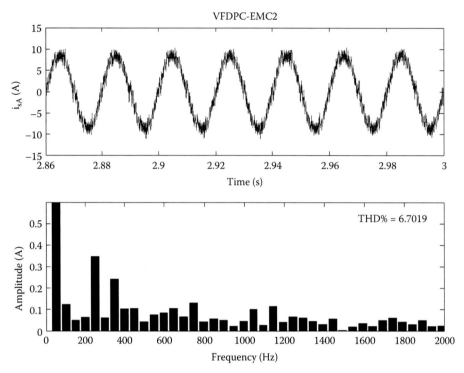

FIGURE 2.74
Inverter current and FFT at Pref = −2 kW, Qref = 0 VAR with VF-DPC-EMC2. (From Giglia, G. et al., Comparison of control techniques for three-phase distributed generation based on VOC and DPC, in *International Conference on Renewable Energy and Power Quality (ICREPQ'08)*, Santander, Spain, March 12–14, 2008.)

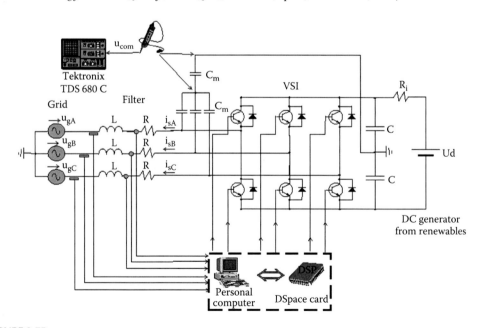

FIGURE 2.75
Electrical scheme of the DG generation unit. (From Pucci, M. et al., *Electric Power Syst. Res.*, 81(4), 830, April 2011, doi: 10.1016/j.epsr.2010.11.007.)

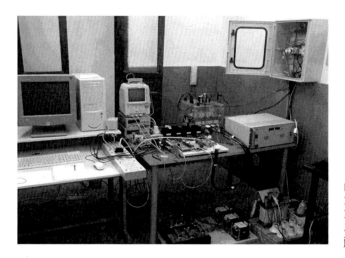

FIGURE 2.76
Photograph of the test setup. (From Pucci, M. et al., *Electric Power Syst. Res.*, 81(4), 830, April 2011, doi: 10.1016/j.epsr.2010.11.007.)

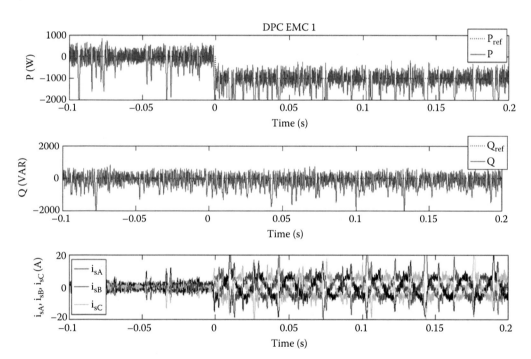

FIGURE 2.77
P_{ref}, P, Q_{ref}, Q, i_{sA}, i_{sB}, and i_{sC} during a step Pref = −1000 W and Q_{ref} = 0 at U_d = 560 V with DPC-EMC 1. (From Pucci, M. et al., *Electric Power Syst. Res.*, 81(4), 830, April 2011, doi: 10.1016/j.epsr.2010.11.007.)

complying with the standard for increasing value of generator power. It is evident that none of the DPC techniques fully complies with the standard; however, it should be borne in mind that the high value of THD is mainly caused by the presence in the current spectrum by high-frequency harmonics. Since the THD is computed considering up to 40th harmonics and all ST-DPC techniques present harmonics in the range from 20th to 40th, this effect is expected. In a real world application, this limitation could be coped with, filtering the line currents by a low-pass power line filter at the output of the inverter.

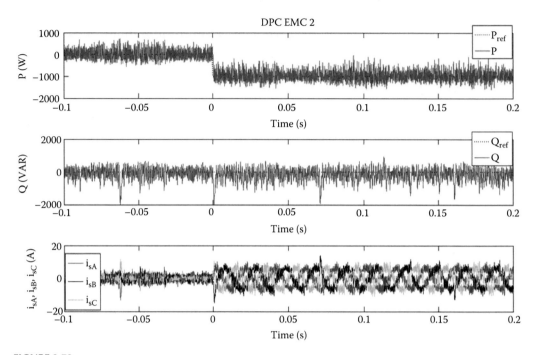

FIGURE 2.78

P_{ref}, P, Q_{ref}, Q, i_{sA}, i_{sB}, and i_{sC} during a step Pref = –1000 W and Q_{ref} = 0 at U_d = 560 V with DPC-EMC 2. (From Pucci, M. et al., *Electric Power Syst. Res.*, 81(4), 830, April 2011, doi: 10.1016/j.epsr.2010.11.007.)

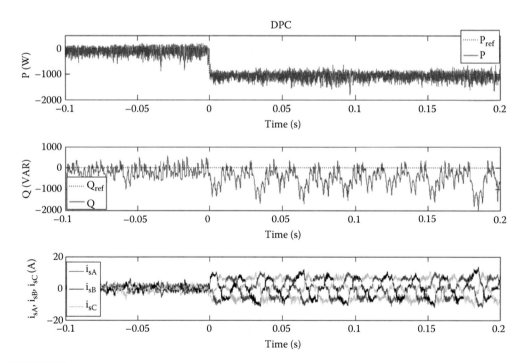

FIGURE 2.79

P_{ref}, P, Q_{ref}, Q, i_{sA}, i_{sB}, and i_{sC} during a step P_{ref} = –1000 W and Q_{ref} = 0 at U_d = 560 V with classic DPC. (From Pucci, M. et al., *Electric Power Syst. Res.*, 81(4), 830, April 2011, doi: 10.1016/j.epsr.2010.11.007.)

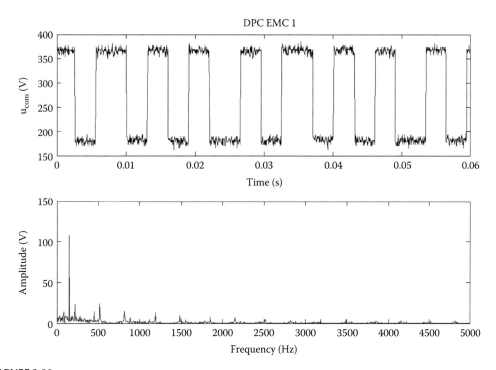

FIGURE 2.80
CMV and its spectrum obtained with FFT with DPC-EMC 1. (From Pucci, M. et al., *Electric Power Syst. Res.*, 81(4), 830, April 2011, doi: 10.1016/j.epsr.2010.11.007.)

It should be noted that this filter makes all DPC techniques equivalent from the point of view of standard compliance; it means that the proposed DPC techniques can abate the common-mode spectrum content without worsening the THD in compliance with the standard.

Moreover, it is to be expected that increasing the sampling frequency of the control system by optimizing its software implementation or the use of a more performing programmable hardware can further contribute to reduce the harmonic content of the line current. As a matter of fact, the current variation **Δi** at each sampling time of the system is proportional to the sampling time itself. The lower the sampling time, the lower the **Δi** with ripple reduction and reduced harmonic content. In this case, a sampling frequency of 20 kHz has been used but a higher sampling frequency is expected to be implementable with more dedicated hardware structures like FPGAs.

It should be noted that even if the VOC has a better performance in terms of %THD of the inverter current, it presents the worst behavior in terms of common-mode emissions. On the other hand, DPC techniques would be able to minimize the common-mode emissions and also have the well-known advantages of lower complexity and computational demand. This is particularly interesting for the exploitation of renewable sources.

Finally, Figures 2.80 through 2.82 show the steady-state CMV time waveforms and their FFT obtained, respectively, with the DPC-EMC 1, the DPC-EMC 2, and the classic DPC. Experimental results are in agreement with the simulations and with the theory. In particular, it is to be expected that the Fourier series expansion of the CMV obtained with DPC-EMC 1 shows harmonics basically in the low frequency region, in particular, third harmonic of the fundamental (150 Hz in Europe) and its odd multiples (3rd, 9th, 15th, 25th, etc., of the fundamental), with amplitudes decreasing with inverse proportionality with the frequency. DPC-EMC 2 is expected to present a harmonic content that is slightly higher in

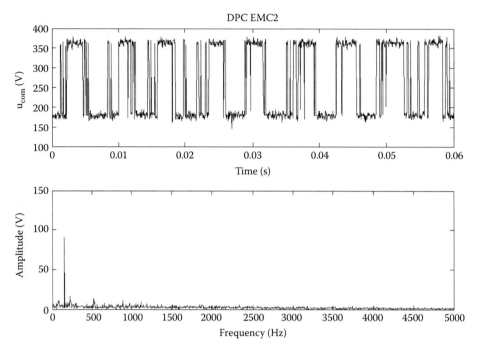

FIGURE 2.81
CMV and its spectrum obtained with FFT with DPC-EMC 2. (From Pucci, M. et al., *Electric Power Syst. Res.*, 81(4), 830, April 2011, doi: 10.1016/j.epsr.2010.11.007.)

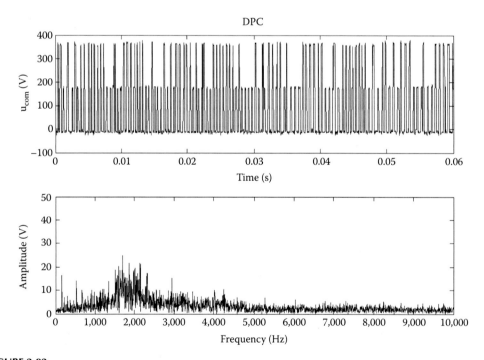

FIGURE 2.82
CMV and its spectrum obtained with FFT with classic DPC. (From Pucci, M. et al., *Electric Power Syst. Res.*, 81(4), 830, April 2011, doi: 10.1016/j.epsr.2010.11.007.)

the frequency range of kilohertz, remaining, however, very close to that of DPC-EMC 1. On the contrary, in the classic DPC, the switching pattern of the power devices of the inverter is commanded by the control law itself, and therefore the CMV variation is unpredictable.

The analysis of results shows that the classic DPC presents the worst behavior with spectral lines of high magnitude in a frequency range from 2 to 10 kHz. On the contrary, the best behavior is shown by the DPC-EMC 1, whose CMV waveform is a square wave with the fundamental frequency at 150 Hz and harmonics only at low frequency, followed by DPC-EMC 2, which is slightly worse [56].

2.3.4.8 Standard Compliance: Comparative Analysis

A comparative analysis of all previously presented techniques has been made. The operating condition is $P_{ref} = -2\,kW$ and $Q_{ref} = 0$ VAR. In particular, in Figure 2.83, the %THD of the injected current versus generated power is shown.

It should be noted that, in general, VOC and VF-OC present better performances, especially for low values of the generated power. Among the different DPC techniques, the worst is the DPC while the best is the DPC-EMC 2. Each of them presents an improvement in its VF version. In general, whatever technique is used, the lower the generated power, the higher the harmonic content. It should be noted that only VOC and VF-OC satisfy the requirements of the American [55] and European standards [57], equal in both cases to 5%.

Figure 2.84 shows %THD of the injected current versus DC link voltage. The knowledge of the power quality issues related to the DC link value is particularly important, especially when this value cannot be considered constant. This figure shows that only VOC and VF-OC always respect the standards' limit. Other techniques are not complying with it for all values of voltage. With regard to DPC-EMC 1 and DPC-EMC 2, the trend is a significant increase of the THD for decreasing values of U_d. Same considerations are true for VOC, which, however, present a slight increase of the THD at lower values of U_d. Finally, DPC does not present significant variations of the THD for the different values of U_d.

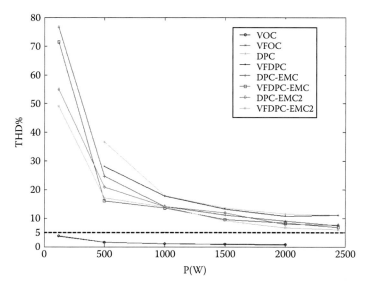

FIGURE 2.83
%THD of the injected current versus generated power. (From Giglia, G. et al., Comparison of control techniques for three-phase distributed generation based on VOC and DPC, in *International Conference on Renewable Energy and Power Quality (ICREPQ'08)*, Santander, Spain, March 12–14, 2008.)

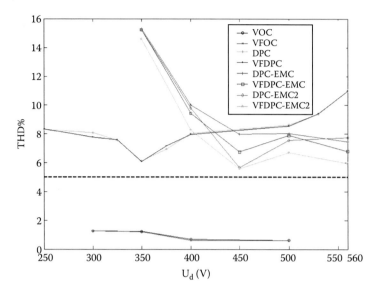

FIGURE 2.84
%THD of the injected current versus DC link voltage. (From Giglia, G. et al., Comparison of control techniques for three-phase distributed generation based on VOC and DPC, in *International Conference on Renewable Energy and Power Quality (ICREPQ'08)*, Santander, Spain, March 12–14, 2008.)

List of Symbols

f_s	switching frequency
f_1	fundamental frequency
f_{switch}	switching frequency
$I_{h\ rms}$	rms harmonic current
$I_{h\ rms\ six\ step}$	rms harmonic current in six-step operation
i_{sA}, i_{sB}, i_{sC}	stator phase currents
$\mathbf{i}_s = i_{sD} + j i_{sQ}$	space-vector of the stator currents in the stator reference frame
$\mathbf{i}_s = i_{sx} + j i_{sy}$	space-vector of the stator currents in a generic rotating reference frame
$\mathbf{i}'_r = i_{rd} + j i_{rq}$	space-vector of the rotor currents in the stator reference frame
$\langle P_{ON} \rangle$	mean power loss during a switching period
$\langle P_{sw} \rangle$	mean power loss during a commutation
S_a, S_b, S_c	command signals of the VSI legs
T_s	sampling time of the control system
T_{ON}	conduction time of the power switch
t_{ri}	rise time of the power switch current
t_{rv}	rise time of the power switch voltage
t_{fi}	fall time of the power switch current
t_{fv}	fall time of the power switch current
$T_{PWM} =$	pulsewidth modulation period
u_{N0}	common mode voltage (CMV)
u_{sA}, u_{sB}, u_{sC}	stator phase voltages

$\mathbf{u}_s = u_{sD} + ju_{sQ}$ space-vector of the stator voltages in the stator reference frame
$\mathbf{u}_s = u_{sx} + ju_{sy}$ space-vector of the stator voltages in a generic rotating reference frame
U_d DC link voltage
v_{ON} voltage drop of the power switch

All quantities with *ref* in pedex are reference quantities.

References

1. N. Mohan, T. Underland, W. Robbins, *Power Electronics*, Wiley, New York, 1995.
2. M. P. Kazmierkowski, R. Khrishnan, F. Blaabjerg, *Control in Power Electronics*, Academic Press, Waltham, MA, 2002.
3. J. Holtz, Pulsewidth modulation for electronic power conversion, *Proceedings of the IEEE*, 82(8), 1194–1214, August 1994.
4. S. Ogasawara, H. Ayano, H. Akagi, Measurement and reduction of EMI radiated by a PWM inverter-fed AC motor drive system, *IEEE Transactions on Power Electronics*, 33(4), 1019–1026, July–August 1997.
5. S. Ogasawara, H. Ayano, H. Akagi, An active circuit for cancellation of common-mode voltage generated by a PWM inverter, *IEEE Transactions on Power Electronics*, 13(5), 835–841, September 1998.
6. A. L. Julian, G. Oriti, T. A. Lipo, Elimination of common-mode voltage in three-phase sinusoidal power converters, *IEEE Transactions on Power Electronics*, 14(5), 982–989, September 1999.
7. J. Adabi, F. Zare, Analysis, calculation and reduction of shaft voltage in induction generators, *International Conference on Renewable Energies and Power Quality (ICREPQ'09)*, Valencia, Spain, April 2009.
8. M. Cacciato, A. Consoli, G. Scarcella, A. Testa, Reduction of common-mode currents in PWM inverter motor drives, *IEEE Transactions on Industry Applications*, 35(2), 469–476, March–April 1999.
9. M. Cirrincione, M. Pucci, G. Vitale, G. Cirrincione, A new direct torque control strategy for the minimization of common-mode emissions, *IEEE Transactions on Industry Applications*, 42(2), 504–517, March–April 2006.
10. M. Cirrincione, M. Pucci, G. Vitale, Direct power control of three-phase VSIs for the minimization of common-mode emissions in distributed generation systems, *Electric Power Systems Research*, 81, 830–839, 2011.
11. D. A. Grant, J. A. Houldsworth, PWM ac motor drive employing ultrasonic carrier, *IEE Conference on Power Electronics and Variable, Speed Drives*, London, U.K., 1984.
12. J. W. Kolar, H. Ertl, F. C. Zach, Influence of the modulation method on the conduction and switching losses of a PWM converter system, *IEEE Transactions on Industry Applications*, 27(6), 1063–1075, November–December 1991.
13. M. Depenbrock, Pulsewidth control of a 3-phase inverter with non-sinusoidal phase voltages, *IEEE Industry Applications International Semiconductor Power Converter Conference*, 1997.
14. D. G. Holmes, T. A. Lipo, *Pulse Width Modulation for Power Converters*, Wiley, New York, 2003.
15. H. W. Van der Broek, H. Skudelny, G. Stanke, Analysis and realization of a pulse width modulator based on voltage space-vectors, *IEEE Transactions on Industry Applications*, 24(1), 142–150, January–February 1988.
16. O. Ogasawara, H. Akagi, A. Nabae, A novel PWM scheme of voltage source inverters based on space-vector theory, *European Power Electronics Conference (EPE)*, Aachen, Germany, 1989.

17. V. Blasko, Analysis of a hybrid PWM based on modified space-vector and triangle-comparison methods, *IEEE Transactions on Industry Applications*, 33(3), 756–764, May–June 1997.
18. D. G. Holmes, The significance of zero space-vector placement for carrier based PWM schemes, *IEEE Transactions on Industry Applications*, 32(5), 1122–1129, September–October 1996.
19. A. M. Trzynadlowsky, S. Legowsky, Minimum-loss vector PWM strategy for three-phase inverters, *IEEE Transactions on Power Electronics*, 9, 26–34, January 1994.
20. H. W. Van der Broek, Analysis of the harmonics in voltage fed inverter drives caused by PWM schemes with discontinuous switching operation, *European Power Electronics Conference (EPE)*, Florence (Italy) 1991.
21. K. Taniguchi, Y. Ogino, H. Irie, PWM technique for power MOSFET inverter, *IEEE Transactions on Power Electronics*, 3, 328–334, July 1988.
22. A. M. Hava, R. J. Kerkman, T. A. Lipo, A high performance generalized discontinuous PWM algorithm, *Presented at IEE APEC'97 Conference*, Atlanta, GA, February 23–27, 1997.
23. A. M. Hava, R. J. Kerkman, T. A. Lipo, A high performance generalized discontinuous PWM algorithm, *IEEE Transactions on Industry Applications*, 34(5), 1059–1071, September–October 1998.
24. W. Chung, J. S. Sheok, S. K. Sul, Unified voltage modulation technique for real time three phase power conversion, *IEEE Transactions on Industry Applications*, 34(2), 374–380, March–April 1998.
25. D. G. Holmes, A unified modulation algorithm for voltage and current source inverters based on AC–AC matrix converter theory, *IEEE Transactions on Industry Applications*, 28(1), 31–40, January/February 1992.
26. O. Ojo, The generalized discontinuous PWM scheme for three-phase voltage source inverters, *IEEE Transactions on Industry Electronics*, 51(6), 1280–1289, December 2004.
27. W. Lotzkat, Aufwandarme und netzausfallsichere Frequenzum-richter zur parameterunempfindlichen Regelung von Asynchron-maschinen fur industrielle Standardantriebe, PhD dissertation, Wuppertal University, Wuppertal, Germany, 1991.
28. J. Holtz, W. Lotzkat, A. Khambadkone, On continuous control of PWM inverters in the over-modulation range with transition to the six-step mode, *IECON*, San Diego, CA, 1992.
29. D. C. Lee, G. M. Lee, A novel overmodulation technique for space-vector PWM inverters, *IEEE Transactions on Power Electronics*, 1144–1151, 13, 1998.
30. H. S. Patel, R. G. Hoft, Generalized techniques of harmonic elimination and voltage control in thyristor inverters, *IEEE Transactions on Industry Applications*, IA-9(3), 310–317, May–June 1973.
31. G. S. Buja, G. B. Indri, Optimal pulsewidth modulation for feeding AC motors, *IEEE Transactions on Industry Applications*, IA-13(1), 3844, January–February 1977.
32. J. Holtz, S. Stadtfeld, H.-P. Wurm, A novel PWM technique minimizing the peak inverter current at steady-state and transient operation, *Elektrische Bahnen*, 81, 5541, 1983.
33. F. C. Zach, H. Ertl, Efficiency optimal control for AC drives with PWM inverters, *IEEE Transactions on Industry Applications*, IA-21(4), 987–1000, July–August 1985.
34. T. M. Rowan, R. J. Kerkman, A new synchronous current regulator and analysis of current regulated inverter, *IEEE Transactions on Industry Applications*, IA-22, 163–171, 1982.
35. D. N. Zmood, D. G. Holmes, Frequency-domain analysis of three-phase linear current regulators, *IEEE Transactions on Industry Applications*, 37, 601–610, 2001.
36. A. Khambadkone, J. Holtz, Low switching frequency and high dynamic pulsewidth modulation based on field orientation for high-power inverter drive, *IEEE Transactions on Power Electronics*, 7(4), October 1992.
37. J. Holtz, Pulsewidth modulation—A survey, *IEEE Transactions on Industry Electronics*, 39(5), 410–420, December 1992.
38. J. Holtz, S. Stadtfeld, A predictive controller for the stator current vector of AC machines fed from a switched voltage source, *International Power Electronics Conference*, Tokyo, Japan, 1983.

39. U. Boelkens, Comparative study on trajectory-based pulsewidth modulation methods for three-phase converters feeding induction machines, PhD thesis (Dissertation in German), Wuppertal University, Wuppertal, Germany, 1989.

40. J. Holtz, S. Stadtfeld, A PWM inverter drive system with on-line optimized pulse patterns, *EPE Conference*, Brussels, Belgium, 1985.

41. J. Holtz, E. Bube, Field-oriented asynchronous pulse-width modulation for high-performance AC machine drives operating at low switching frequency, *IEEE Transactions on Industry Applications*, 27(3), May–June 1991.

42. J. Holtz, B. Beyer, The trajectory tracking approach—A new method for minimum distortion PWM in dynamic high-power drives, *IEEE Transactions on Industry Applications*, 30(4), July–August 1994.

43. M. P. Kazmierkowski, Current control techniques for three-phase voltage-source PWM converters: A survey, *IEEE Transactions on Industrial Electronics*, 45(5), 691–703, October 1998.

44. M. P. Kazmierkowski, H. Tunia, *Automatic Control of Converter-Fed Drives*, Elsevier, Amsterdam, the Netherlands, 1994.

45. M. Liserre, A. Dell'Aquila, F. Blaabjerg, An overview of the three-phase voltage source active rectifiers interfacing the utility, *Powertech Conference*, Bologna, Italy, June 23–26, 2003.

46. T. Noguchi, H. Tomini, S. Kondo, Direct power control of PWM converter without power-source voltage sensor, *IEEE Transactions on Industry Applications*, 34(3), 473–479, May–June 1998.

47. M. Malinowski, Sensorless control strategies for three—Phase PWM rectifiers, PhD thesis, University of Warsaw, Warsaw, Poland, 2001.

48. M. Malinowski, M. P. Kazmierkowski, S. Hansen, F. Blaabjerg, Virtual-flux-based direct power control of three-phase PWM rectifiers, *IEEE Transactions on Industry Applications*, 37(4), 1019–1027, July–August 2001.

49. M. Liserre, F. Blaabjerg, S. Hansen, Design and control of an LCL-filter-based three-phase active rectifier, *IEEE Transactions on Industry Applications*, 41(5), 1281–1291, September–October 2005.

50. M. Malinowski, M. P. Kazmierkowski, M. Trzynadlowski, A comparative study of control techniques for PWM rectifiers in AC adjustable speed drives, *IEEE Transactions on Power Electronics*, 18(6), 1390–1396, November 2003.

51. M. Pucci, G. Vitale, M. Cirrincione, Direct power control of three-phase VSIs for the minimization of common-mode emissions in distributed generation systems, *Electric Power Systems Research*, 81(4), 830–839, April 2011, doi: 10.1016/j.epsr.2010.11.007.

52. M. Cirrincione, M. Pucci, G. Vitale, Direct power control of three-phase VSIs for the minimization of common-mode emissions in distributed generation systems, *ISIE 2006 IEEE International Symposium on Industrial Electronics*, Vigo, Spain, June 4–7, 2007.

53. G. Giglia, C. Serporta, M. Pucci, G. Vitale, Comparison of control techniques for three-phase distributed generation based on VOC and DPC, *International Conference on Renewable Energy and Power Quality (ICREPQ'08)*, Santander, Spain, March 12–14, 2008.

54. M. Cirrincione, M. Pucci, G. Vitale, New direct power control strategies of three-phase VSIs for the minimization of common-mode emissions in distributed generation systems, *The Forty-Second Annual Meeting of the IEEE Industry Applications Society (IAS 2007)*, New Orleans, LA, September 23–27, 2007.

55. IEEE 1547-2003, Standard for interconnecting distributed resources with electric power systems, 2003.

56. G. Giglia, C. Serporta, M. Pucci, G. Vitale, Experimental comparison of three-phase distributed generation systems based on VOC and DPC control techniques, *EPE 2007 European Conference on Power Electronics and Applications*, Aalborg, Denmark, September 2–5, 2007.

57. Standard IEC/EN61727, Photovoltaic (PV) Systems—Characteristics of the utility interface, IEC Std, 1997.

Further Reading

Bose, B. K., *Power Electronics and Variable Frequency Drives: Technology and Applications*, 1 edn., Wiley, New York, September 21, 1996.

Bose, B. K., *Power Electronics and Variable Speed Drives*, IEEE Press, New York, 1997.

Bose, B. K., *Modern Power Electronics and AC Drives*, 1 edn., *Prentice Hall*, Upper Saddle River, NJ, October 22, 2001.

Bose, B. K., *Power Electronics and Motor Drives: Advances and Trends*, Academic Press, Waltham, MA, August 11, 2006.

Holmes, D. G., T. A. Lipo, *Pulse Width Modulation for Power Converters: Principle and Practice*, IEEE Press, Piscataway, NJ, 2003.

Kassakian, J. G., *Principles of Power Electronics*, Facsimile edn., Prentice Hall, Upper Saddle River, NJ, July 11, 1991.

Rashid, M., *Power Electronics Handbook*, 3 edn., Butterworth-Heinemann, Burlington, MA, December 23, 2010.

Trzynadlowski, A., *Introduction to Modern Power Electronics*, 2 edn., Wiley, New York, March 15, 2010.

3

Power Quality

3.1 Nonlinear Loads

3.1.1 Current Source Type of Harmonic Sources (Harmonic Current Sources)

The classic harmonic current source (CS) is the thyristor rectifier with a series inductance, which maintains the load current constant, as shown in Figure 3.1. The switching operation of the diodes makes the source current periodic (three-step square wave) but nonsinusoidal; as a consequence, a distortion on voltage source (VS) occurs due to the voltage drop on its impedance. Since this kind of load imposes the distorted current to the source, it is called harmonic CS. In this case, the load current is not significantly dependent on the AC side, so it can be considered a CS. Figure 3.2 shows the typical time waveform of the load voltage and current obtained with a three-phase diode rectifier connected to a highly inductive load. The voltage waveform appears quasi-sinusoidal because of the reduced voltage drop on the line impedance represented by the series R_g, L_g. Figure 3.3 shows the corresponding spectrum of the current, obtained with the fast Fourier transform (FFT). The spectrum exhibits the presence of only odd harmonics with amplitudes decreasing with the frequency, as theoretically expected.

3.1.2 Voltage Source Type of Harmonic Sources (Harmonic Voltage Sources)

When a diode rectifier is connected to a smoothing capacitance, as shown in Figure 3.4, the required current is highly distorted, as in the case of a harmonic current load. Differently from it, dually, the magnitude of the load current is significantly dependent on the source impedance and the distortion of the voltage source if present. On the other hand, the voltage on the AC side is less dependent on the source: that is why it can be considered a harmonic voltage load. Figure 3.5 shows the typical time waveforms of the load voltage and current obtained with a three-phase rectifier connected to a highly capacitive load. As expected, both the load voltage and current present significant distortion. Figure 3.6 shows the voltage spectrum obtained with the FFT. As in the harmonic current load counterpart, this spectrum exhibits only odd harmonics, with amplitudes decreasing with frequency.

3.2 Harmonic Propagation on the Distribution Network

Harmonics started being present in distribution networks in the 1920s due to the necessity to convert electric energy delivered by three-phase AC systems to supply DC motors. At that time, rectifiers were built with mercury arc rectifier. With the use of thyristors

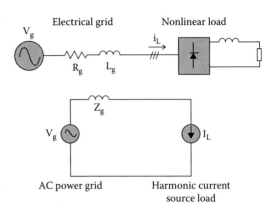

FIGURE 3.1
Typical harmonic current load and corresponding circuit scheme.

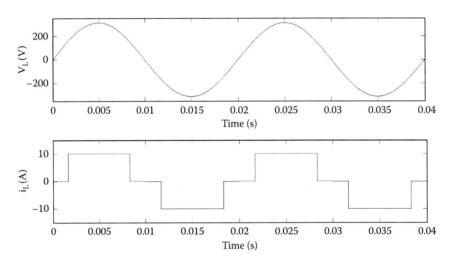

FIGURE 3.2
Voltage and current on a typical harmonic current load.

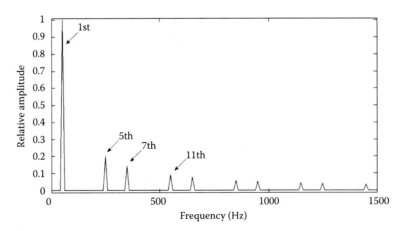

FIGURE 3.3
Spectrum of the current in a current harmonic load.

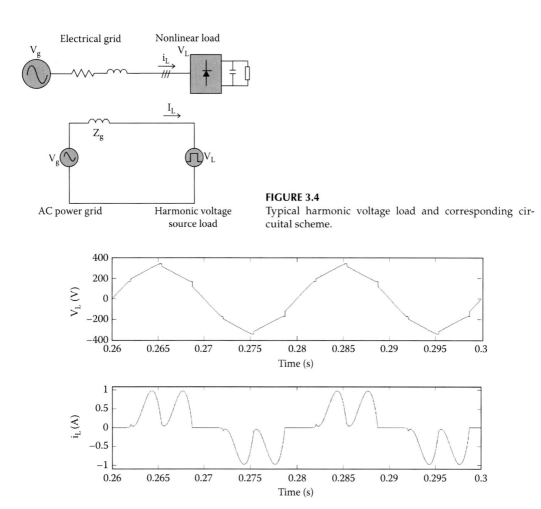

FIGURE 3.4
Typical harmonic voltage load and corresponding circuital scheme.

FIGURE 3.5
Voltage and current on a typical harmonic voltage load.

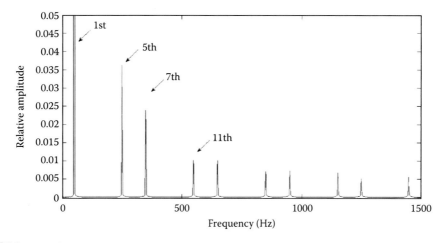

FIGURE 3.6
Spectrum of the voltage in a voltage harmonic load.

since the 1950s and then with modern switching power devices, harmonic propagation has become more and more important. Nowadays, about 80% of the generator output supplies nonlinear converters.

In general, the operation of switching power devices implies the presence of harmonics due to their intrinsic nonlinearity, so in a distribution network, the presence and mutual interaction of several harmonic generators have to be accounted for. Among these, besides rectifiers, considered previously, and inverters, there are also static VAR compensators and power transformers.

The most common source of harmonics and harmonic problems in power systems is the three-phase Graetz bridge. It produces harmonics of order $n = 6k \pm 1$ where n is the harmonic order and k is an integer (see Figures 3.3 and 3.6). Such a rectifier can supply the DC side of an inverter, for example, in an induction drive, or can generate DC voltage for a high voltage direct current (HVDC) conversion system. In this last case, harmonics can be reduced by increasing the pulse number of the rectifier as in the 12-pulse rectifier where two 6-pulse rectifiers are used connected in series on the DC side and in parallel on the AC side. In this way, some of the produced harmonics can be canceled.

Pulsewidth modulated (PWM) three-phase inverters commonly used in variable speed drives are supplied by a DC link, which is the output of a rectifier described earlier.

Static VAR compensators are used in power systems to control voltage at the end of long transmission lines: if well designed, they should have little impact on the grid.

Power transformers generate low harmonic levels at steady-state; however, when they are initially energized, the presence of the so-called inrush current implies the presence of a harmonic current, including a DC term, which can be as high as 60% of the transformer rated current of the transformer.

Finally, several kinds of low-power sources are commonly present, whose cumulative effect can be relevant: for example, ballast inductors for fluorescent light and converters for consuming electronics like PCs and TVs.

Besides harmonics multiple of the fundamental harmonic, some additional terms are commonly present, known as interharmonics. The term "interharmonic" covers a wide range of frequencies generated by nonlinear static conversion systems, that are not integer multiples of the fundamental frequency [1].

Interharmonics often appear where two AC systems that are joined by the same DC link operate at different frequencies. In case of electric variable speed drives, both the power supplying AC system and the motor will experience interharmonics, including frequencies that are below the fundamental value. This can be explained considering that the DC link does not decouple the two systems perfectly: its voltage is not constant and contains a superimposed ripple.

Harmonic pollution results both from large nonlinear loads such as variable speed drives and power converter and from a great number of small nonlinear loads due to their cumulative effect.

Three main groups of small nonlinear loads can be identified. The first group contains loads that utilize the single-phase capacitor-filtered diode bridge rectifier (DBR): these circuits require a pulsed current that is rich in harmonics. In this group, there are computers, TVs, battery chargers, and small adjustable speed drives for heat pumps. The second group contains loads that use phase angle voltage controllers in which the input voltage is controlled by using a thyristor. Light dimmers, heaters, and single-phase induction machines (IMs) belong to this group. Finally, the third group contains the compact fluorescent lamps employing magnetic ballast. The contribution, in this case, results from the non linearity of gas discharges.

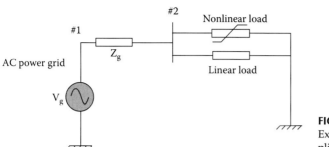

FIGURE 3.7
Example of linear and nonlinear load supplied by the same power grid.

The cumulative effect of the loads mentioned earlier is made more complicated by their interconnection to the power grid.

With reference to Figure 3.7, in which a linear load and a nonlinear one are parallely connected and supplied by the grid generator V_g by its impedance Z_g, it can be noticed that if the grid impedance is negligible, the voltage at bus #2 is sinusoidal, and the nonlinear load represents the unique source of current harmonics. On the contrary, if the inductive impedance Z_g is not neglected, the distorted current due to the nonlinear load creates a harmonic voltage drop on Z_g. As a consequence, the voltage on bus #2 is distorted as well; the linear load is then supplied by a distorted voltage and a resulting distorted current and the total current supplied by the grid generator is highly polluted as for its harmonic contents [2].

A real distribution system should be considered as made up of a large number of converters coupled through DC links, transformers, and bus-bars. Figure 3.8 shows a typical real system in an industry. There are several PCCs (points of common coupling). PCCI represents the connection of the main transformer to the distribution network. At PCC2, nonlinear loads are connected to a step-down distribution transformer. At PCC3, a DC motor converter and an IM drive are connected in parallel to the other step-down transformer. The separate effects of each of these two loads can be compensated (or calculated) at the corresponding PCCs. It is apparent that such a situation requires appropriated methods to foresee the harmonic pollution and providing remedies to avoid harmonic propagation [3].

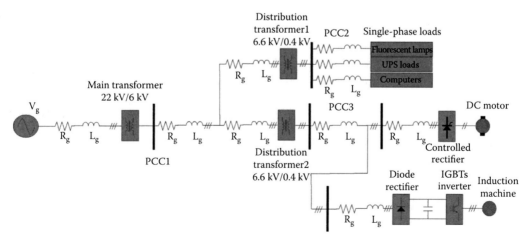

FIGURE 3.8
Example of industrial network with linear and nonlinear loads.

Many techniques for harmonic analysis have been devised. The simplest is the "frequency scan": it is based on the response of a network in a particular bus or node: a 1 per unit sinusoidal current (or voltage) is injected into the bus, and the corresponding response is calculated. The injection is repeated using discrete frequency step in the interval of interest. Mathematically, it is equivalent to solve the network equation

$$[Y_n] \cdot [V_n] = [I_n] \tag{3.1}$$

where
  I_n is the vector of injected currents
  V_n is the bus voltage vector to be computed
  Y_n is the admittance matrix, while n indicates the number of considered harmonic

This technique allows resonances to be detected, and it is useful for filter design.

The "harmonic iteration" method considers a harmonic-producing device that acts as a supply voltage–dependent CS following a function in the form

$$I_n = f(V_1, V_2, \ldots, V_n, c) \quad n = 1, 2, \ldots, h \tag{3.2}$$

where
  $V_1, V_2, \ldots, V_n$ are the harmonic phasors of the supply voltage
  c is a set of control variables as converter firing angle or output powers

At the first iteration, (3.2) is solved with an estimated supply voltage; then the result is used as a CS in (3.1), which gives voltage harmonics. They are introduced in (3.2), and the process is repeated until the convergence is reached.

Another method consists in solving simultaneously (3.1) and (3.2) using a Newton-based algorithm. In this case, generally, the devices have to be modeled in a closed form or in a form that allows derivatives to be calculated. If the starting point is chosen close to the final point, the convergence is better than that obtained with the previous method.

Finally, the analysis can be performed in time domain by electromagnetic transient programs such as EMTP [4,5].

Limits to the distortion at the PCC are specified by the IEEE standard [6]. To comply with this standard, compensating circuits have to be connected to the PCC of the distorting system. These circuits can be passive or active or both; they are illustrated in the following sections.

3.3 Passive Filters

Passive filters are built by passive components such as capacitors, inductors, and/or resistors. Their aim is to block a single harmonic (tuned filter) or a set of harmonics whose frequency is higher than a threshold. In general, they are connected in parallel with the nonlinear load [7]: in this case, they are called parallel passive filters (PPFs). Figure 3.9 shows the electric schemes of some passive filters. Figure 3.9a, in particular, shows a single-tuned filter: the frequency to be eliminated is given by the values of C and L, while R is the parasitic inductor resistance. Figure 3.9b shows a double-tuned filter. In Figure 3.9c through e, HPFs of the first, second, and third orders are sketched, respectively.

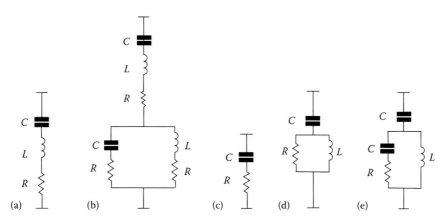

FIGURE 3.9
Electric schemes of some passive filters: (a) single-tuned filter, (b) double-tuned filter, (c) first order high-pass filter, (d) second order high-pass filter, (e) third order high-pass filter.

Figure 3.10 shows the frequency responses of each component of a bank of PPF, composed of filters tuned at the 5th harmonic, a filter tuned at the 7th harmonic, and an HPF tuned to let harmonics above the 11th pass. Figure 3.11 shows the overall frequency response of the bank of filters, presenting the expected resonances at the 5th and 7th harmonics, while the 11th HPF acts above the 11th harmonic. Table 3.1 shows the parameters of the filter components [7].

In practice, for several reasons, the characteristic parameters may differ from the theoretical ones. With reference to the circuit of Figure 3.9a, in which the resistance R represents the parasitic resistance of the inductor, the impedance is given by

$$Z = \sqrt{R^2 + \left(\omega L - \frac{1}{\omega C}\right)^2} \tag{3.3}$$

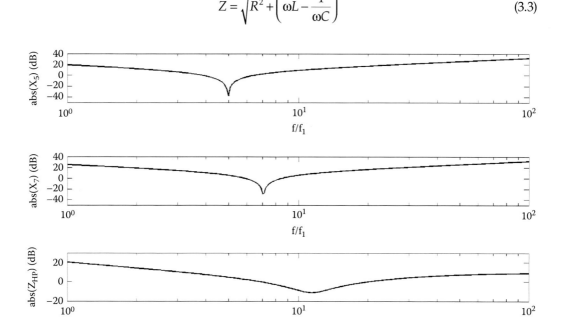

FIGURE 3.10
Frequency response of the single component of a PPF.

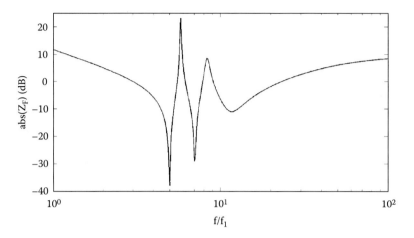

FIGURE 3.11
Frequency response of the overall PPF.

TABLE 3.1

Parameters of the PPF Components

	Inductance (mH)	Capacitance (μF)	
Fifth-order filter	1.2	340	Q = 14
Seventh-order filter	1.2	170	Q = 14
HPF	0.26	300	R = 3 Ω

It presents a minimum value at the resonance pulsation

$$\omega_r = \frac{1}{\sqrt{LC}} \tag{3.4}$$

by using the definition of "quality factor of the inductor" as follows:

$$Q_r = \frac{\omega_r L}{R} \tag{3.5}$$

the 3 dB bandwidth can be expressed as

$$B_{3dB} = \frac{\omega_r}{2\pi Q_r} \tag{3.6}$$

where reasonable values for Q_r range from 20 to 100.

In practice, both ω_r and Q_r may vary for several reasons. Considering, for example, a situation in which the series-tuned filter is connected with a voltage generator, which has an equivalent series internal impedance with a resistor R_g and an inductance L_g, the new values of ω_r and Q_r are modified as follows:

$$\omega_r^* = \frac{1}{\sqrt{(L+L_g)\cdot C}} \tag{3.7}$$

$$Q_r^* = \frac{\omega_r^*(L+L_g)}{(R+R_g)} = \frac{\sqrt{(L+L_g)}}{(R+R_g)\sqrt{C}} \tag{3.8}$$

It should be noted that the presence of a real voltage generator implies that the resonant pulsation is decreased and the band of the filter is broadened. Moreover, the filter has to be connected to a load, which further contributes to variation in its performance.

Finally, the resonant pulsation is affected by the variation of the parameters. Indeed, the variation of ω_r due to the tolerance of the inductor and the capacitor yields

$$\Delta\omega_r = \frac{\partial\omega_r}{\partial L}\cdot\Delta L + \frac{\partial\omega_r}{\partial C}\cdot\Delta C = \left(-\frac{1}{2}\frac{C}{L}\omega_r\right)\Delta L + \left(-\frac{1}{2}\frac{L}{C}\omega_r\right)\Delta C \tag{3.9}$$

Therefore the relative variation of ω_r is proportional to the relative variation of the capacitor and to the inductance value, as follows:

$$\frac{\Delta\omega_r}{\omega_r} = -\frac{1}{2}\left(C\frac{\Delta L}{L} + L\frac{\Delta C}{C}\right) \approx -\frac{L}{2}\frac{\Delta C}{C} \tag{3.10}$$

Similar consideration can be extended to the other circuits of Figure 3.8b through e.

3.4 Active Power Filters

3.4.1 Introduction on APFs

As known, harmonics have a negative impact on distribution networks and influence the behavior of system components and loads, for example, conductors suffer from losses and skin effects; eddy current losses can have detrimental effects on transformers, with consequent equipment overheating; capacitors may be affected by resonance phenomena with potential breakdown; and machines can suffer from vibration phenomena.

Over the last few years, active power filters (APFs) have drawn great attention and are expected to be a suitable remedy for the problem of harmonic pollution [7–9], thanks to the recent advances in semiconductor technology.

The basic theory on active filters was proposed about 40 years ago [10–15]; researchers have been applying them ever increasingly, thanks to the progress of technology both of power electronic devices and microprocessors for signal processing and control algorithm implementation.

In particular, power devices like insulated gate bipolar transistors (IGBTs) and power MOSFETs with fast switching capability and insulated gate structure have allowed the use of fast modulation frequencies and simple and low-power-consuming driving circuits. Moreover, by using new less expensive microprocessors like digital signal processors (DSPs) and field-programmable gate arrays (FPGAs), the algorithm's performance has been increased as well [15–20].

Nowadays, the trend is to substitute traditional passive harmonic filters with active harmonic filters.

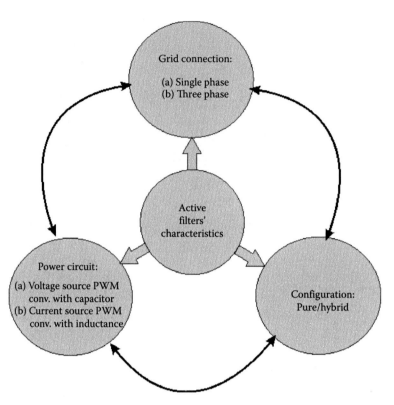

FIGURE 3.12
Classification of filtering strategies.

"Active filtering" is something included inside the more general term of "power conditioning" used to take into consideration several features in addition to harmonic filtering like harmonic damping, harmonic isolation, harmonic termination, reactive power control for power factor correction and voltage regulation, load balancing, and voltage-flicker reduction.

The main features of active filters are sketched in Figure 3.12. First, they can be divided into single-phase and three-phase active filters. The first typology is used for low-power applications and for electric traction and rolling stock. The second has been recently drawing more attention from industries and researchers due to their wide range of applications. The active filter configurations can be further divided into pure or hybrid. This last configuration is obtained by a pure active filter adopted in combination with traditional passive filters. Finally, the power circuit can be a PWM converter in a VS configuration adopting a capacitor on the DC side or a CS with a series inductance.

The widest spread active filters employ a PWM in VS configuration due to its advantages like lower cost, physical size, and efficiency; hybrid configuration is preferred for both viability and economical reasons as explained in the following.

Two main categories of pure APFs exist: shunt filters, also called parallel active filters (PAFs), and series active filters (SAFs). PAFs are effective for those nonlinear loads that can be considered as current harmonic sources (see Section 3.1.1); they are therefore used to generate harmonic currents to compensate load harmonic currents. SAFs are effective in generating harmonic voltages to compensate load harmonics and grid voltage harmonics and are suitable for compensating voltage harmonic sources [7].

Three-phase APFs have historically drawn more attention. In this case, three "optimal" compensation targets can be addressed by the APF controller, from which three control strategies derive [7]:

- Constant instantaneous source power control
- Sinusoidal current control
- Generalized Fryze current control

As demonstrated by the instantaneous active-reactive (p-q) power theory (see Chapter 1) [7,8], only under three-phase undistorted balanced voltages is it possible to simultaneously perform these three control actions, otherwise, one out of the three is to be chosen.

Active filters can be installed by individual consumers or in substations and/or distribution feeders [19]. In the first case, the consumer aims to compensate harmonics produced by its own loads so as to present the grid an equivalent load complying with the standard. Due to harmonic propagation through the grid, the presence of an active filter in a substation and/or power feeders can perform harmonic damping and compensate voltage imbalance.

3.4.2 Basic Operating Issues of Parallel and Series Filters

Figure 3.13 shows a typical connection of a PPF to a CS nonlinear load. In this case, tuned filters is devised to present a low impedance to the harmonic to be suppressed; the filter would then represent a sink for the correspondent harmonic current. In practice, the suppression acts on a frequency range which depends on the quality factor of the inductor as explained above [22]. Second-order HPFs exhibit good filtering performance but cause higher fundamental frequency losses compared to single-tuned ones.

It should be noted that the PPF is composed of two single-tuned filters and an HPF [23]. The first two filters aim to block a single frequency by having a low impedance at those harmonics (e.g., the 5th and the 7th). The HPF exhibits a low impedance for harmonics higher than the 11th.

It should be borne in mind that the presence of a tuned passive filter can imply some problems. With reference to Figure 3.13, the first problem is that the filter impedance depends

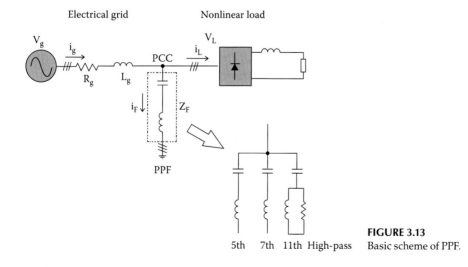

FIGURE 3.13
Basic scheme of PPF.

5th   7th   11th   High-pass

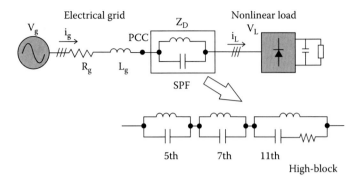

FIGURE 3.14
Basic scheme of SPF.

on the load impedance which is not constant. Then, the filter can become a sink also for frequencies different from the tuned ones; correspondingly, a current, whose frequency is ω_i, will flow through the loop composed of the filter, the grid impedance $\dot{Z}_g$, and the grid voltage generator. This will cause a distorted voltage drop on the source impedance with a consequent load voltage distortion. Finally, the contemporary presence of both the filter inductance and the grid impedance can cause a parallel resonance, with a consequent high voltage at frequency ω_i appearing on the load at frequency ω_i.

On the basis of what is mentioned earlier it is clear that the grid impedance plays an important role in harmonic compensation. Its value should be high at frequency values corresponding to the harmonics to be compensated; in addition, it should have a low value at the fundamental frequency to avoid the voltage drop at the fundamental frequency. Both requirements can be fulfilled with a hybrid filter, as shown in the following.

Figure 3.14 shows the basic configuration of a series passive filter (SPF) [18,24]. The filter is connected in series with the power grid. Differently from the PPF, the SPF acts as a block versus the voltage grid harmonics. SPFs are typically composed of parallel-resonant circuits and high-block (HB) circuits. The tuned filters are designed to block specific harmonics, while the HB filter blocks harmonics above a certain order.

3.4.3 Shunt Active Filters

A shunt active filter or a PAF is placed in parallel with a load considered as the harmonic generator [7,8]. The PAF injects a harmonic current whose amplitude is the same as that of the load with opposite phase: for this purpose, the load harmonic current has to be detected through a dedicated circuit. As a consequence, an inner loop is present in this kind of filter.

The basic diagram is shown in Figure 3.15. In this figure, the PAF is represented as a CS $\dot{I}_C$; the injected current depends on the load current $\dot{I}_L$ and a gain factor G. In case of a harmonic current load, it can be represented by the Norton equivalent circuit as in Figure 3.15 [25,26]. If the load is considered as a voltage harmonic generator, it can be represented by the Thévenin equivalent circuit as in Figure 3.16. The two circuital representations are equivalent; however, their adoption depends on the load characteristics, as explained later in the text. The following analysis has been done under the hypothesis of sinusoidal steady-state for the generic harmonic of order h; in this way, phasor quantities are associated with each electric variable.

3.4.3.1 Harmonic Current Source Compensation by PAF

With reference to Figure 3.15, the current supplied by the filter can be expressed as

$$\dot{I}_C = G\dot{I}_L \tag{3.11}$$

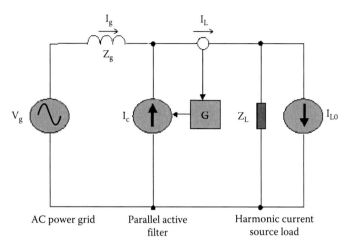

FIGURE 3.15
Basic principle of PAF for harmonic current load.

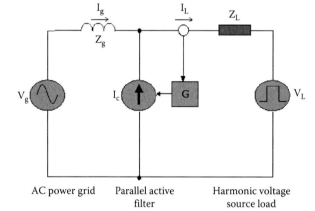

FIGURE 3.16
Basic principle of PAF for harmonic voltage load.

while the grid current can be written as

$$\dot{I}_g = \frac{\dot{Z}_L}{\dot{Z}_g + \dot{Z}_L/(1-G)} \dot{I}_{L0} + \frac{\dot{V}_g}{\dot{Z}_g + \dot{Z}_L/(1-G)} \tag{3.12}$$

and the load current as

$$\dot{I}_L = \frac{\dot{Z}_L/(1-G)}{\dot{Z}_g + \dot{Z}_L/(1-G)} \dot{I}_{L0} + \frac{1}{(1-G)} \cdot \frac{\dot{V}_g}{\dot{Z}_g + \dot{Z}_L/(1-G)} \tag{3.13}$$

As for the generic harmonic of order h, the $\dot{I}_{gh}$ (the hth order harmonic of the grid current) can be made null if the following equation holds:

$$\left| \frac{\dot{Z}_L}{1-G} \right|_h \gg \left| \dot{Z}_g \right|_h \tag{3.14}$$

In this case, Equations 3.11 through 3.13 can be rewritten as

$$\dot{I}_C = \dot{I}_{Lh} \tag{3.15}$$

and

$$\dot{I}_{gh} \approx (1-G)\big|_h \, \dot{I}_{L0h} + (1-G)\big|_h \, \frac{\dot{V}_{gh}}{\dot{Z}_L} \approx 0 \tag{3.16}$$

$$\dot{I}_{Lh} = \dot{I}_{L0h} + \frac{\dot{V}_{gh}}{\dot{Z}_L} \tag{3.17}$$

Note that if $(1-G)\big|_h = 0$, the condition (3.14) is always verified and the source current $\dot{I}_g$ is sinusoidal. $(1-G)\big|_h = 0$ or (3.14) can thus be considered the basic conditions of operation of a PAF. However remark that (3.14) contains the grid impedance $\dot{Z}_g$ and the load impedance: the ratio between these two quantities contributes to the filter performance (as in the case of passive filter). It should be noticed also that for a pure current source where $|\dot{Z}_L| \gg |\dot{Z}_g|$, Equations 3.12 and 3.13 are reduced to

$$\frac{\left|\dot{I}_g\right|}{\left|\dot{I}_{L0}\right|} = (1-G) \tag{3.18}$$

$$\frac{\left|\dot{I}_L\right|}{\left|\dot{I}_{L0}\right|} \approx 1 \tag{3.19}$$

$$(1-G)\big|_h \ll 1 \tag{3.20}$$

From what was shown earlier, the PAF performance improves when the load impedance is bigger than the grid impedance as, for example, for highly inductive thyristor rectifiers. This last condition depends on the grid and load configuration. Moreover, the contribution of the PAF consists of a gain G able to discriminate between the fundamental and harmonic currents. For practical cases, $(1-G)\big|_h$ ranges from 0.1 to 0.3.

In summary, the capability of the APF to cancel the load current harmonics depends therefore not only on the filter itself (G) but also on the load ($\dot{Z}_L$) and source ($\dot{Z}_g$) impedances. However, if $(1-G)\big|_h \approx 0$, the harmonic compensation capability is independent from the system impedances. On the contrary, if $(1-G)\big|_h \neq 0$, the ratio $|\dot{Z}_L|/|\dot{Z}_g|$ must be taken into account.

Figure 3.17 shows a set of curves representing the low frequency harmonic amplitudes of the line current, in percent of the fundamental of the load current (assumed to be a square waveform), versus the active filter gain for a fixed value of the ratio $|\dot{Z}_L|/|\dot{Z}_g| = 30$. It shows that with $G = 0$, the harmonic content of the line current is obviously the greatest, and no current harmonic compensation occurs. According to Equation 3.14, the higher the gain G, the lower the harmonic of the line current. When G approaches 1, all the harmonics of the line current become close to 0, independent of the harmonic order and the ratio $|\dot{Z}_L|/|\dot{Z}_g|$. Figure 3.18 shows a set of curves representing the third harmonic amplitude of the grid current versus the active filter gain for different values of the ratio $|\dot{Z}_L|/|\dot{Z}_g|$. It shows that, for lower values of G, there is a dependency of the third harmonics on the grid impedance. The higher the ratio $|\dot{Z}_L|/|\dot{Z}_g|$, the lower this dependency as explained earlier. For values of G close to 1, the third harmonic amplitude is almost independent from the impedance ratio.

On the other hand, some common situations occur in which the load impedance is low for some harmonics. It is the case of loads that present a highly capacitive bank or with a passive filter. As a consequence, the PAF performance will depend on the grid

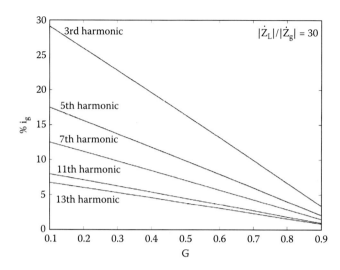

FIGURE 3.17
Percent harmonic current compensation versus active filter gain for $|\dot{Z}_L|/|\dot{Z}_g| = 30$.

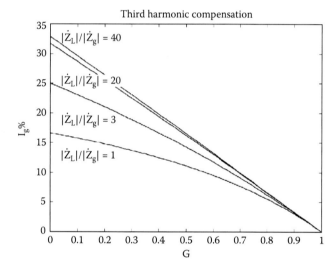

FIGURE 3.18
Percent third harmonic current compensation versus active filter gain for different values of $|\dot{Z}_L|/|\dot{Z}_g|$.

characteristics; in addition, from (3.17), it is evident that the difference between the hth harmonic of the load current $\dot{I}_{Lh}$ and the hth harmonic of the Norton equivalent generator load current $\dot{I}_{L0h}$ depends strongly on the load impedance. In particular, if the load impedance tends to become infinite, then $\dot{I}_{L0h} \cong \dot{I}_{Lh}$. On the other hand, for low values of the load impedance (stiff grid), $\dot{I}_{Lh}$ becomes high, and the filter must be correspondingly designed.

3.4.3.2 Harmonic Voltage Source Compensation by PAF

Given the Thevenin equivalent circuit in Figure 3.16 [27,28], the equations describing the circuit are

$$\dot{I}_C = G\dot{I}_L \tag{3.21}$$

$$\dot{I}_g = \frac{\dot{V}_g - \dot{V}_L}{\dot{Z}_g + \dot{Z}_L/(1-G)} \tag{3.22}$$

$$\dot{I}_L = \frac{1}{(1-G)} \cdot \frac{\dot{V}_g - \dot{V}_L}{\dot{Z}_g + \dot{Z}_L/(1-G)} = \frac{\dot{V}_g - \dot{V}_L}{\dot{Z}_g(1-G) + \dot{Z}_L} \tag{3.23}$$

If the following condition is satisfied,

$$\left| \dot{Z}_g + \frac{\dot{Z}_L}{(1-G)} \right|_h \gg 1pu \tag{3.24}$$

the grid current is sinusoidal. In this case, it happens that

$$\dot{I}_C = \dot{I}_{Lh} \tag{3.25}$$

$$\dot{I}_{gh} \approx 0 \tag{3.26}$$

$$\dot{I}_{Lh} = \frac{\dot{V}_{gh} - \dot{V}_{Lh}}{\dot{Z}_L} \tag{3.27}$$

Equation 3.24 is the dual equation of (3.14); it should be noted, however, that a harmonic VS has usually a low internal impedance, and the operation of PAF makes null the grid impedance seen by the load side (see Equation 3.23 when $(1 - G) \approx 0$). From Equation 3.25, it is clear that the harmonic current injected by PAF flows into the load and, finally from (3.27), that the distortion of the grid voltage $\dot{V}_{gh}$ will cause harmonic current into the load, independent of the filter itself. As a consequence, especially for low load impedance, the required volt-ampere rating of the filter has to be increased.

3.4.3.3 Control of PAF Based on p-q Theory

Among the different control methods for APFs, the technique based on the p-q power theory is one of the most performing [7,8]. The complete scheme of a PAF based on the p-q controller is sketched in Figure 3.19.

Two main blocks can be identified: the controller that calculates the compensating current reference and the power converter that amplifies it and performs the injection into the power grid.

Harmonic currents can be generated both by load nonlinearity and by harmonic voltages (see Sections 3.1 and 3.2) in the power system. In such a case, the grid voltage is composed of the term at fundamental frequency and additional terms with frequencies multiple of the fundamental.

Since the shunt active filter usually is used to compensate the harmonic current introduced by the load, it gives a compensating current with the same amplitude and opposite phase of the harmonic current of the load. However, the voltage at the PCC can be distorted by the presence of the harmonic voltage generated by the grid. When the filter compensates both the grid and load harmonic components, the voltage at the PCC becomes sinusoidal.

It should be remarked that if the load impedance is low (or the power system has a high short-circuit capacity), the filter should deliver a high current, which could be unfeasible. In such a case, other solutions such as series filters could be exploited.

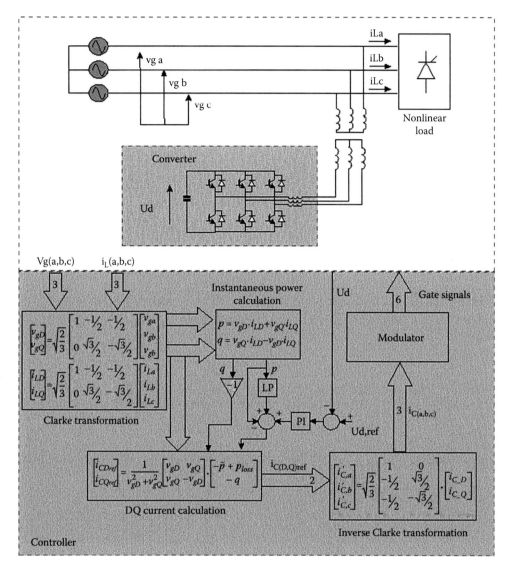

FIGURE 3.19
PAF scheme based on a p-q controller.

3.4.3.3.1 PWM Converters for Shunt Active Filters

The power converter shown in Figure 3.19 is a voltage source converter (VSC); however, other topologies can be adopted such as the current source converter (CSC) shown in Figure 3.20; both of the converters can behave as controlled CSs. The storage element (a capacitor for the VSC, an inductance for the CSC) is sufficient to supply the inverter. Ideally, the exchanged energy is null; in reality, inverter losses require that the storage element voltage be controlled to keep the DC voltage (current) constant.

It should be noted that since all inverters are sources of harmonics, in order to avoid harmonic propagation through the grid, a passive RC wye-connected low-pass filter (LPF) is usually added at the inverter's output. Usually, a small passive filter is required because the switching frequency of the inverter is high. If, instead of a three leg inverter, three

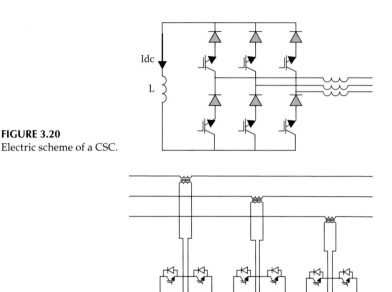

FIGURE 3.20
Electric scheme of a CSC.

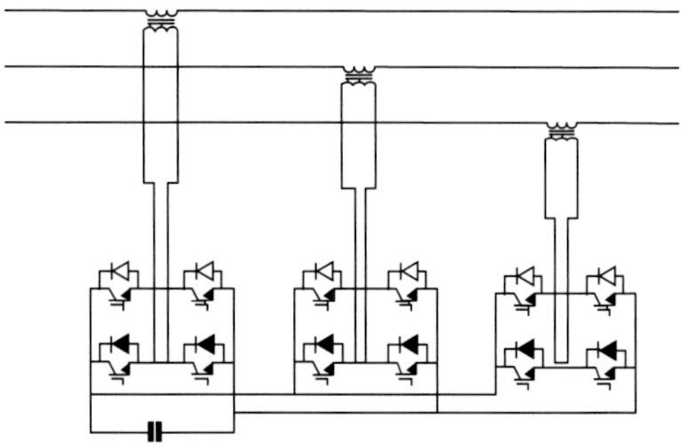

FIGURE 3.21
Electric scheme of the H bridge configuration.

H bridges are adopted, they should have in common the storage element as shown in Figure 3.21. This last configuration is employed in the SAFs.

The CSC exhibits a high robustness [27–28], but the VCS has a low initial cost, high efficiency, and smaller size [8,9]. At present, manufacturers provide six IGBT modules with antiparallel free-wheeling diode, and as a consequence, the VCS is more practical and cheaper to realize.

3.4.3.3.2 Active Filter Controllers Using p-q Theory

The use of p-q theory is a very satisfactory strategy to design an active filter controller; however, it is demonstrated [7] that when the power system contains voltage harmonics and/or imbalances at the fundamental frequency, the three main targets of compensation—(a) to draw a constant instantaneous active power from the grid, (b) to draw a sinusoidal current from the grid, and (c) to draw the minimum rms value of the current that transports the same energy to the load with minimum losses along the transmission line (unit power factor)—cannot be simultaneously satisfied.

For this reason, three fundamental compensation strategies can be identified [7]:

1. Constant instantaneous power control strategy

2. Sinusoidal current control strategy

3. Generalized Fryze current control strategy

Under sinusoidal, balanced system voltages, the three control strategies can produce the same results.

Figure 3.19 shows the most important parts of a three-phase, three-wire shunt active filter for current compensation. The input variables of the controller are the phase voltages $v_g(a,b,c)$, measured at the PCC, and the load current $i_L(a,b,c)$ flowing into the load to be compensated. The DC link voltage U_d is measured in order to compensate the inverter losses. The controller outputs six gate signals to drive the inverter power switches.

The sampled input signals are processed as follows: voltages and currents are preliminarily converted from the three-phase to the biphase system by means of the Clarke transformation to obtain the corresponding variables in the D-Q reference frame. Afterward, both the instantaneous active power p and reactive power q are calculated. The compensating current is finally obtained on the basis of the grid voltage components v_{gD}, v_{gQ} in the D-Q reference, the reactive power q, the oscillating active power $\tilde{p}$, and the power term due to the losses p_{loss}. The obtained compensating current is finally converted by the inverse Clarke transformation into the three reference values. A current controller is always present, which commands a PWM providing the six signals to drive the inverter power switches. Several current control methods can be devised, working in both the stationary and rotating reference frames [29]. Finally, an LPF (not shown in Figure 3.19) could be placed before the voltage grid measurement in order to avoid problems of instability or resonance; it is adopted with the constant instantaneous power control strategy method explained in the following text.

3.4.3.3.3 Constant Instantaneous Power Control Strategy

The goal of this strategy consists in drawing a constant instantaneous power from the source. The filter has to be connected as close as possible to the load. In terms of p-q theory, this means that the oscillating real power $\tilde{p}$ will be compensated by the filter.

The current supplied by the filter is the oscillating part of the active current on the D and Q axis computed on the basis of $\tilde{p}$ as follows:

$$i_{CD\tilde{p}} = \frac{v_{gD}}{v_{gD}^2 + v_{gQ}^2}(-\tilde{p}) \tag{3.28}$$

$$i_{CQ\tilde{p}} = \frac{v_{gQ}}{v_{gD}^2 + v_{gQ}^2}(-\tilde{p}) \tag{3.29}$$

where the sign of the power is positive when it flows into the load such as in the "load current convention."

When the filter operates, the supply source (the grid) gives the constant value of the real power $\bar{p}$. The compensation of $\tilde{p}$ by the filter implies exploiting the capacitor supplying the inverter DC link. Its value should be designed so as to avoid large DC voltage, which however should be always higher than the peak value of the grid voltage. Moreover, because of the presence of a small amount of average real power (P_{loss}), due to switching and ohmic inverter losses, the DC voltage must be controlled with an additional control loop to maintain the DC link voltage constant.

The current reference calculated by Equation 3.28 and (3.29) can be added with a further term to compensate the instantaneous reactive current on D and Q axis:

$$i_{CDq} = \frac{v_{gQ}}{v_{gD}^2 + v_{gQ}^2}(-q) \tag{3.30}$$

$$i_{CQq} = \frac{-v_{gD}}{v_{gD}^2 + v_{gQ}^2}(-q) \tag{3.31}$$

The separation between the average $\bar{p}$ and the oscillating part $\tilde{p}$ of the power is performed by an LPF (Figure 3.19). It influences the active filter dynamical performance because it introduces a delay [7], so its cutoff frequency has to be chosen carefully.

3.4.3.3.4 Sinusoidal Current Control Strategy

The goal of this strategy is to make the grid current sinusoidal and balanced. This require-ment cannot be accomplished with a compensated constant power supplied to the load in the presence of a voltage imbalance or voltage distortion. This is the reason why, in this last case, a choice between the two described methods must be made.

In this method, the filter has to compensate the currents that differ from the fundamen-tal positive-sequence current i_{+1}. The strategy consists in determining the fundamental positive sequence of the load current. In this sense, the scheme of Figure 3.19 needs to be modified by inserting a positive-sequence detector block [7] to process the voltage before the Clarke transformation. On the other hand, the presence of the low-pass filter on the phase voltage measurement is not necessary.

3.4.3.3.5 Generalized Fryze Current Control Strategy

The Fryze method aims to minimize the rms value of the compensated currents; in this case, the ohmic losses are minimized too, and the average active power supplied by the grid remains the same as the power given by the original current. Moreover, this method does not utilize the p-q theory, but directly exploits the three grid voltages $v_g(a,b,c)$ and load currents $i_L(a,b,c)$, and, consequently, the calculation effort is lower than the other methods.

The scheme of the Fryze controller is shown in Figure 3.22. The ratio between voltage and current is maintained equal to a conductance defined as the mean instantaneous power on the three phases divided by the square of the sum of the rms values of the phase voltages:

$$G_e = \frac{(1/T)\int_0^T \left(v_{gA}i_{LA} + v_{gB}i_{LB} + v_{gC}i_{LC}\right)}{\sqrt{(1/T)\left(\int_0^T v_{gA}^2(t)dt + \int_0^T v_{gB}^2(t)dt + \int_0^T v_{gC}^2(t)dt\right)}} \tag{3.32}$$

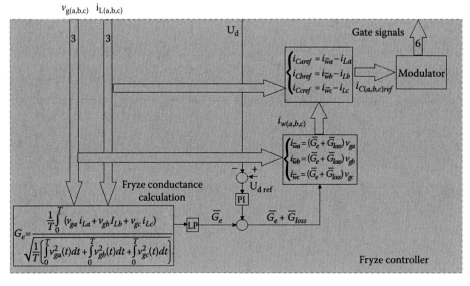

FIGURE 3.22
PAF scheme based on the Fryze controller.

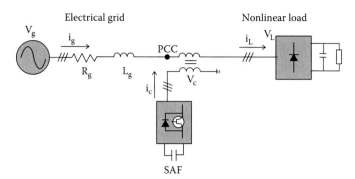

FIGURE 3.23
Basic schematics of the SAF.

If Equation 3.32 is used, the speed of the method is affected by the need of calculation in one period T; however, an LPF can be utilized as shown in Figure 3.22. Moreover, a further term to take into account losses on the DC link has to be added. This last term is obtained by the error values between the desired DC link voltage and the real one processed with a PI.

The obtained quantity $\bar{G}_e + \bar{G}_{loss}$ multiplied to the phase voltages gives the generalized Fryze currents $i_{\bar{w}}(A,B,C)$. Then the reference currents $i_{Cref}(A,B,C)$ are obtained by subtracting the load current.

3.4.4 Series Active Filters

The series active power filter (SAF) is an active system which should be placed between the AC source and the nonlinear load (either current or voltage type), in series with both of them [7,24]. Figure 3.23 shows the basic diagram of an SAF, with a typical voltage harmonic nonlinear load, which can be valid for both three-phase (three wires) systems and single-phase ones. The three-phase SAF is composed of a suitably controlled three-phase PWM VS inverter (or three single-phase inverters), connected in series with the load and the power line by a three-phase transformer (or three single-phase transformers). The SAF is basically a controlled voltage generator, and therefore PWM inverter typically does not present any inner current loop, different from the PAF.

The basic idea of the SAF is to block harmonics by generating proper compensating voltages. To avoid the current harmonics from flowing between the AC source and the nonlinear load, the SAF must present a high impedance at the harmonic frequencies, while presenting a zero impedance at the fundamental frequency.

As in the case of the PAF, the effectiveness of the SAF in cases of current harmonic loads and voltage harmonic loads should be evaluated. The SAF can be more properly used for compensating harmonic voltage nonlinear loads, as in the case of a three-phase rectifier with a capacitive load.

3.4.4.1 Harmonic Current Loads

As in the case of the PAF, the complete system composed by the AC source, the nonlinear current harmonic load, and the SAF can be represented by the corresponding Norton equivalent circuit, as in Figure 3.24 [25,26]. The power grid is represented by a voltage generator $\dot{V}_g$ in series with an impedance $\dot{Z}_g$, the current type load with a current generator providing $\dot{I}_{L0}$ in parallel with its equivalent impedance $\dot{Z}_L$, and finally, the SAF is represented by a controlled voltage generator $\dot{V}_C$.

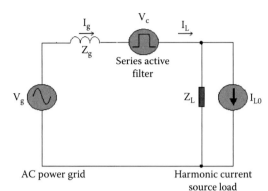

FIGURE 3.24
Schematics of an SAF for current harmonic compensation.

The voltage equation of the system at any frequency (fundamental or harmonic of order h) can be written as

$$\dot{V}_g = \dot{I}_g \dot{Z}_g + \dot{V}_c + \dot{V}_L \tag{3.33}$$

where

$$\dot{V}_c = KG\dot{I}_g \tag{3.34}$$

with G representing the equivalent transfer function of the current harmonic detection system, including its dynamics and delay times, and K is a gain quantity (dimensionally in $[\Omega]$). The value of G should be almost 0 at the fundamental frequency, $G|_1 = 0$, while its value should be almost equal to 1 at any harmonic h to be compensated, $G|_h = 1$.

The load voltage can be written as

$$\dot{V}_L = (\dot{I}_g - \dot{I}_{L0})\dot{Z}_L \tag{3.35}$$

After substituting Equations (3.33) and (3.34) in (3.32), the following expression of the grid current $\dot{I}_g$ can be obtained:

$$\dot{I}_g = \frac{\dot{V}_g + \dot{I}_{L0}\dot{Z}_L}{\dot{Z}_g + \dot{Z}_L + KG} \tag{3.36}$$

If the value of K satisfies the following hypothesis for each harmonic of order h to be compensated,

$$K|_h \gg \left| \dot{Z}_g + \dot{Z}_L \right|_h \tag{3.37 a}$$

$$K|_h \gg \left| \dot{Z}_L \right|_h \tag{3.37 b}$$

then the component of the source current at harmonic h, $\dot{I}_{gh}$, from Equation 3.36 becomes

$$\dot{I}_{gh} \cong 0 \qquad (3.38)$$

while the component of the voltage generated by the SAF for the harmonic h, $\dot{V}_{Ch}$, becomes

$$\dot{V}_{Ch} \cong \dot{Z}_{Lh}\dot{I}_{L0h} + \dot{V}_{gh} \qquad (3.39)$$

where
   $\dot{V}_{gh}$ is the hth harmonic component of the source voltage
   $\dot{I}_{L0h}$ is the hth harmonic component of the load current

Equations 3.37a and 3.37b are therefore the required operating conditions of the SAF, which demands that the value of K should be high enough, while the load impedance for the hth harmonic $|\dot{Z}_L|_h$ should be sufficiently low. If these requirements are fulfilled, Equation 3.38 confirms that the source (grid) current becomes almost equal to 0, which means that the SAF prevents the load harmonic current from flowing into the grid, and Equation 3.39 confirms that the voltage generated by the SAF properly compensates both the voltage drop of the load harmonic current into the load impedance and the VS harmonics, that will not be transmitted on the load. Unfortunately, real cases of typical current loads, like conventional phase-controlled thyristor rectifiers, have a very high value of $|\dot{Z}_L|_h$, which makes the requirements in Equation 3.26 not properly satisfied with resulting bad operation of the SAF. This is confirmed by Equation 3.31, which shows that, in this case, the voltage required to the SAF tends to become infinite. The pure SAF is therefore not suitable for real current type nonlinear loads. A solution for using the SAF in case of these kinds of loads is to supplement the SAF with a passive filter placed in parallel with the load, trying to reduce its equivalent impedance, shown in the following.

3.4.4.2 Voltage Harmonic Loads

To analyze the behavior of the SAF in case of harmonic voltage nonlinear loads, the complete system composed by the AC source, the nonlinear voltage harmonic load, and the SAF can be represented by the corresponding Thévenin equivalent circuit, as shown in Figure 3.25 [25,26].

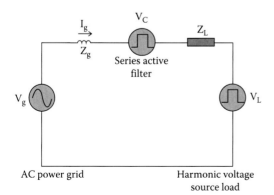

AC power grid   Harmonic voltage
                                 source load

FIGURE 3.25
Schematics of an SAF for voltage harmonic compensation.

The voltage equation of the system at any frequency, fundamental or harmonic of order h, can be written as

$$\dot{V}_g = \dot{I}_g(\dot{Z}_g + \dot{Z}_L) + \dot{V}_c + \dot{V}_L \tag{3.40}$$

where

$$\dot{V}_c = KG\dot{I}_g \tag{3.41}$$

with G representing the equivalent transfer function of the current harmonic detection system, including its dynamics and delay times, and K is a gain quantity (dimensionally in [Ω]). Substituting Equation 3.41 into 3.40, the grid current can be obtained as

$$\dot{I}_g = \frac{\dot{V}_g - \dot{V}_L}{\dot{Z}_g + \dot{Z}_L + KG} \tag{3.42}$$

Equation 3.42 confirms that if K is sufficiently high at the hth harmonics, the resulting source current becomes null:

$$\dot{I}_{gh} \cong 0 \tag{3.43}$$

and the voltage $\dot{V}_{Ch}$ generated by the SAF at the hth harmonics is

$$\dot{V}_{Ch} \cong \dot{V}_{gh} - \dot{V}_{Lh} \tag{3.44}$$

A sufficiently high value of K, ideally infinite, is the required operating condition of the SAF in the presence of nonlinear voltage loads. There are several ways to try and obtain such a condition. The easiest is to control the current with a hysteresis controller. In this case, sketched schematically in Figure 3.26, the value of the gain $K \to \infty$. Another way is to directly control the SAF hth harmonic voltage $\dot{V}_{Ch}$, trying to compensate the load voltage $\dot{V}_{Lh}$ on the basis of the source current $\dot{I}_{gh}$, as shown in Figure 3.27.

From Figure 3.27, the SAF reference voltage at the $\dot{V}_{Cref}$ can be written as

$$\dot{V}_{Cref} = G(K\dot{I}_g - \dot{V}_L) \tag{3.45}$$

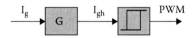

FIGURE 3.26
Hysteresis control of the SAF.

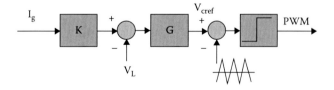

FIGURE 3.27
Carrier-based control of the SAF.

Assuming the voltage is properly controlled, that is, $\dot{V}_C = \dot{V}_{Cref}$, and recalling Equation 3.41, the grid current can be computed as

$$\dot{I}_g = \frac{\dot{V}_g - \dot{V}_L(1-G)}{\dot{Z}_g + \dot{Z}_L + KG} \tag{3.46}$$

Under the hypothesis of sufficiently low source (grid) voltage harmonics, that is, $\dot{V}_{gh} \cong 0$, if the value of G at the hth harmonic satisfies the following condition,

$$(1-G)\big|_h \cong 0 \tag{3.47}$$

then the source current hth harmonic is almost null, independent of the ratio between K and $|\dot{Z}_g + \dot{Z}_L|_h$, that is, independent of the value of the load, that is:

$$\dot{I}_{gh} = -\frac{(1-G)\dot{V}_{Lh}}{\left|\dot{Z}_g + Z_L\right| + KG} \cong 0 \tag{3.48}$$

If the VS harmonics are nonnull, then an uncompensated term appears which, on the contrary, depends on the load impedance. Equation 3.47 is then the operating condition of the SAF for nonlinear voltage loads. It can be observed that the behavior and the effectiveness of the SAF in case of nonlinear voltage loads are dually equivalent to those of the PAF in case of nonlinear current loads.

3.4.4.3 Control of SAF Based on the p-q Theory

The classic control strategy for SAFs is based on the p-q theory [7]. The complete control scheme with the electric scheme typical of a three-phase three-wire system is drawn in Figure 3.28. It is based on three single-phase converters with common DC link capacitor. The underlying assumption is that no zero-sequence current component exists. The goal of the control action is to generate a set of voltages to compensate the harmonic load voltage components which are the cause of oscillating active and reactive power components on the load side. Such a control scheme permits the grid voltages and currents to have pure sinusoidal waveforms.

In this system, the instantaneous values of the load currents $i_L(a,b,c)$ and voltages $v_L(a,b,c)$ are measured and are then given as inputs to the SAF controller. Starting from the three-phase variables, exploiting the Clarke transformation, the corresponding D-Q variables are obtained, respectively (i_{LD}, i_{LQ}) and (v_{LD}, v_{LQ}). On this basis, the instantaneous active (p) and reactive (q) power components are computed. The oscillating active $\tilde{p}$ and reactive $\tilde{q}$ power components are then computed after subtracting their average values obtained by a low-pass (LP) filter. The zero-sequence power components are assumed equal to 0, since the zero-sequence current is equal to 0 as seen before. Given the oscillating power components, the instantaneous voltages to be generated by the SAF to compensate the load voltage harmonics can be computed and given as references to the power converter. Another important underlying assumption is that the power converter is ideal, that is, no power losses exist, which implies no theoretical need to control the DC link voltage. In real-world application, an additional term $\Delta\bar{p}$ should be added to $\tilde{p}$ for suitably compensating the losses. Furthermore, any displacement between load voltages and currents, resulting in a nonnull average reactive power $\bar{q}$, cannot be taken into consideration, or else an additional term should be added.

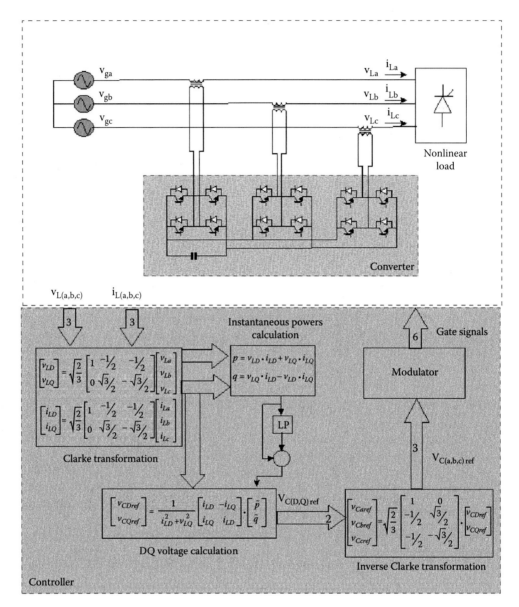

FIGURE 3.28
Control scheme of SAF based on the p-q theory.

3.4.5 Comparison between PAFs and SAFs

The analysis of PAFs and SAFs has shown that each of them works for a certain kind of load, with a precise required operating condition, and that between them, a dual relation exists. The main characteristics of both of them can be summarized in the following:

- *Basic operating principle*: PAF works as a CS while SAF works as a VS.
- *Nonlinear load type*: PAF works for inductive or harmonic CSs, like phase controller rectifiers, while SAF works for capacitive or harmonic VSs, like diode rectifiers with smoothing capacitors.

- *Operating condition*: PAF works properly with high values of the load impedance and when $(1 - G)|_h \ll 1$, while SAF works properly with low values of the load impedance and when $(1 - G)|_h \ll 1$.
- *Compensation properties*: PAF presents an excellent compensation for current harmonic loads, independent from the grid impedance only when the load impedance is high, while SAF presents an excellent compensation independent from both the grid and the load impedances only for harmonic voltage loads, depending on the load impedance for harmonic current loads.
- *Applicative issues*: PAF injects current on the load, which can cause overcurrents when applied to a voltage harmonic load, while SAF needs a low impedance parallel branch when used with a harmonic current load.

3.4.6 Hybrid Active Filters

There are some conditions which make the use of an active filter impractical or not proper [7]. The reasons are the following:

- Sometimes the design and construction of an active filter of high power rating is not reasonable from the technical point of view.
- The cost of an active filter is high compared to that of a passive filter.

With specific regard to the PPFs, some considerations should be further done:

- Their filtering performance depends on the power grid impedance, which could not be properly known in advance and could be time-varying, due to the modifications of the power grid structure.
- The higher the power grid impedance at the given current harmonic, the better the PPF filtering performance; it is, however, in contrast with the need to have a low value of the grid impedance at the fundamental frequency to limit the voltage drops on the power line.

The earlier considerations suggest the idea of combining properly active and passive filters. The resulting filter is typically called hybrid active. Hybrid active filters are the combinations of active and passive filters, with the goal of increasing the performance of the compensation and reducing the power rating of the active part [7, 21 and 26].

Figure 3.29 shows the parallel combination of a PAF with a PPF, which is suitable for harmonic current loads. In this case, the PAF is used to compensate low-order current harmonics, typically fifth and seventh, because of the limitations coming from the switching frequency of the inverter and its rating, while the PPF is used to compensate current harmonics above the 11th, permitting a more compact design of it. In general, in such a configuration, a good rule is to adopt the PAF to compensate low-order harmonics and the SAF to compensate high-order ones. PAF is then adopted to avoid resonance phenomena between the power grid and the PPF.

The dual configuration is shown in Figure 3.30 and is obtained by combining in series an SAF with an SPF, suitable for harmonic voltage loads. Even in this case, SAF and SPF share the compensation tasks, working the SAF for low-order voltage harmonics and the SPF for high-order voltage harmonics.

Figure 3.31 shows the scheme of a combination of an SAF with a PPF [7,25,26,30]. In this case, the main filter is the PPF, which works as a current harmonic sink, while the SAF is

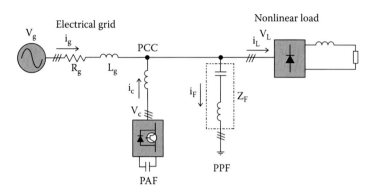

FIGURE 3.29
Parallel combination of a PAF with a PPF.

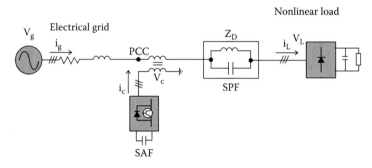

FIGURE 3.30
Series combination of an SAF with an SPF.

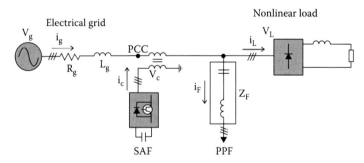

FIGURE 3.31
Combination of an SAF with a PPF.

designed with a smaller rating with respect to it. The SAF is basically adopted to avoid resonance phenomena between the power grid and the PPF and also to enhance the performance of the PPF.

The basic idea in this case is not to directly compensate the harmonic voltages but to help the PPF by properly blocking harmonics. The SAF should thus present a high equivalent impedance at high frequencies and a low (ideally 0) impedance at fundamental frequency. Practically, the SAF behaves as an "active impedance," offering zero impedance at fundamental frequency and an impedance K at load or grid harmonic frequencies. To evaluate the effectiveness of this kind of hybrid filters, the ratio between the grid current harmonic amplitude and the load current harmonic amplitude is usually chosen [7,30]:

$$\frac{\dot{I}_{gh}}{\dot{I}_{Lh}} = \frac{\dot{Z}_F}{\dot{Z}_g + \dot{Z}_F + K} \tag{3.49}$$

This ratio, depending on the active resistance K, is computed under the assumption of null grid voltage harmonics ($V_{gh} = 0$) and is called *distribution factor*. This parameter is called distribution factor. Figures 3.32 and 3.33 show the distribution factor versus the normalized frequency for different values of K (0, 1, 2 Ω) and for two different values of grid inductances, respectively $L_g = 0.04\,H$ and $L_g = 0.112\,H$. These graphs have been drawn with the PPF made up of two filters, tuned, respectively, for the fifth and seventh fundamental harmonics, and an HPF as shown in Figure 3.32 (the parameters of the PPF are shown in Table 3.1). The condition $K = 0$ corresponds to the absence of the SAF, which

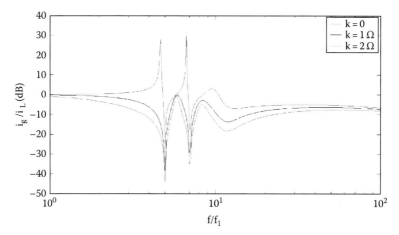

FIGURE 3.32
Distribution factor for the grid inductance of $L_g = 0.04\,H$.

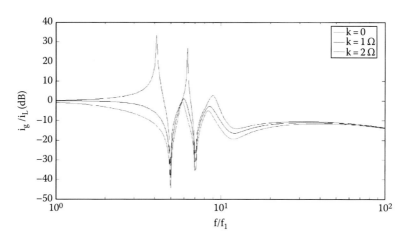

FIGURE 3.33
Distribution factor for the grid inductance of $L_g = 0.112\,H$.

means that only the PPF is working. It can be observed that, in this condition, resonances occur at the fourth and sixth fundamental harmonics, with amplification factors of more than 22 dB. When the SAF is working, with K equal respectively to 1 or 2, the distribution factor can be significantly reduced over the entire frequency range. The best results are obtained with $K = 2$.

The dual configuration is shown in Figure 3.34, referring to the combination of a PAF with an SPF. In this case, the main filter is the SPF, blocking the current harmonics, while the low rating PAF is used to enhance the SPF performance, as well as to avoid resonance phenomena between the power grid and the SPF.

Figure 3.35 shows the series combination of a PAF with a PPF. In this case, the PAF injects current harmonics to compensate the load current ones. It is also controlled to generate a fundamental current component permitting the fundamental grid voltage to be applied only on the PPF and not at all on the PAF. This feature permits reducing the power rating of the PAF significantly, still maintaining its excellent compensation performance.

Figure 3.36 presents the dual circuit, composed of the parallel combination of an SAF with an SPF. In this case, the power rating of the SAF can be reduced by permitting the fundamental current to flow in the SPF, since the passive filter is designed to be a short circuit at the fundamental frequency.

The ideal configuration for the compensation of a harmonic current load is sketched in Figure 3.37. The PAF is used to inject harmonic currents to compensate the load ones, while the SAF is used to block the harmonic currents from flowing through the power grid.

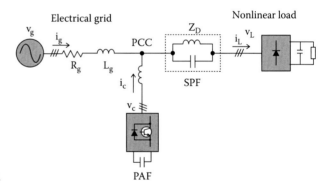

FIGURE 3.34
Combination of a PAF with an SPF.

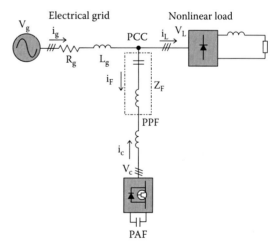

FIGURE 3.35
Series combination of a PAF with a PPF.

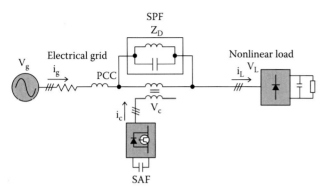

FIGURE 3.36
Parallel combination of an SAF with an SPF.

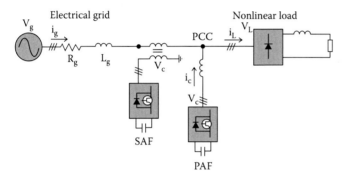

FIGURE 3.37
Combination of a PAF with an SAF for current harmonic loads.

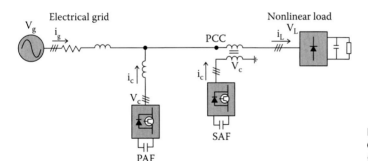

FIGURE 3.38
Combination of an SAF with a PAF for voltage harmonic loads.

In practice, the SAF creates a distorted voltage waveform to make the power line see a linear sinusoidal load. This system is potentially able not only to cancel the effect of the load harmonic currents but also to provide a pure sinusoidal voltage on the load, independent of the harmonics present in the power grid voltage or its fluctuations.

The dual configuration is shown in Figure 3.38, which is based on the combination of an SAF with a PAF for a typical harmonic voltage load. In this case, the system can block any current harmonic, ensuring at the same time a pure sinusoidal voltage on the load.

Configurations for the ideal compensation of current or voltage harmonic loads can also be obtained also with passive filter configurations. Figures 3.39 and 3.40 show, respectively, the two systems.

Since active filter applications to high-power systems present some problems, one of the targets is to reduce their power rating. To this aim, several solutions have been devised, whose schemes are drawn in Figures 3.41 through 3.48. In particular,

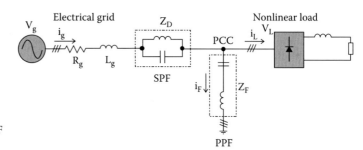

FIGURE 3.39
Combination of a PPF with an SPF for current harmonic loads.

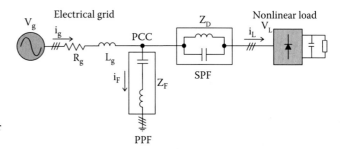

FIGURE 3.40
Combination of an SPF with a PPF for voltage harmonic loads.

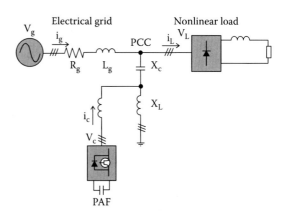

FIGURE 3.41
PAF with fundamental voltage reduction: scheme 1.

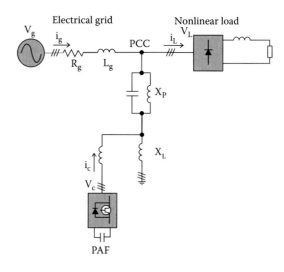

FIGURE 3.42
PAF with fundamental voltage reduction: scheme 2.

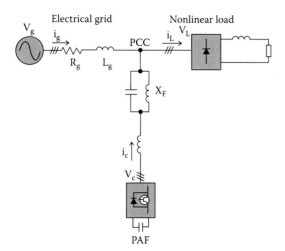

FIGURE 3.43
PAF with fundamental voltage reduction: scheme 3.

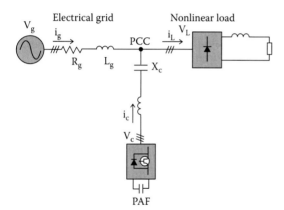

FIGURE 3.44
PAF with fundamental voltage reduction: scheme 4.

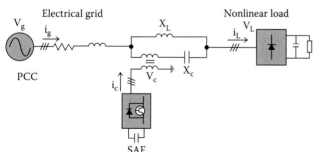

FIGURE 3.45
SAF with fundamental current reduction:
scheme 1.

Figures 3.41 through 3.44 refer to PAFs, while Figures 3.45 through 3.48 refer to SAFs. In the scheme in Figure 3.41, the reduction of the voltage acting on the PAF is obtained with an L/C voltage divider; the voltage across the PAF, depending on the ratio between the capacitive and inductive reactances X_L and X_C, can be reduced by the ratio $X_L/(X_L + X_C)$. The dual scheme for the SAF is sketched in Figure 3.45, where L and C determine a current divider; also in this case, the current flowing in the SAF can be reduced by the ratio $X_L/(X_L + X_C)$. In the scheme in Figure 3.45, an LC circuit

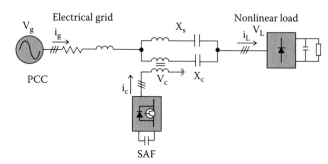

FIGURE 3.46
SAF with fundamental current reduction:
scheme 2.

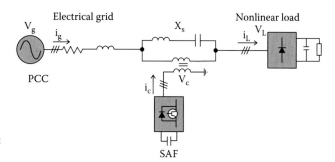

FIGURE 3.47
SAF with fundamental current reduction:
scheme 3.

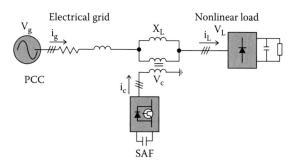

FIGURE 3.48
SAF with fundamental current reduction: scheme 4.

with resonance tuned at the grid fundamental frequency is used to increase the impedance ratio, so as to further decrease the fundamental voltage across the PAF by the factor $X_L/(X_P + X_C)$. The dual scheme for the SAF is sketched in Figure 3.46, where the current divider depends on a series LC circuit resonating at the grid fundamental frequency; also in this case, the current flowing in the SAF can be reduced by the factor $X_S/(X_S + X_C)$. Figures 3.43 and 3.44 show schemes permitting to reduce the fundamental voltage across the PAF to be reduced to 0, with a suitable control of the fundamental current generated by the PAF. The drawback of such a scheme is the voltage drop caused by the harmonic current generated by the PAF on the reactance X_P or X_C. The dual schemes for the SAF are shown in Figures 3.47 and 3.48, and they both permit a zero fundamental current to flow through the SAF, with a proper control of the SAF fundamental voltage. The drawback, in these cases, is that the harmonic voltages produced by the SAF cause harmonic currents to flow in X_P or X_C.

List of Symbols

$\dot{I}_g$	phasor of the power grid current
$\dot{I}_{L0}$	phasor of the current of the ideal current generator
$\mathbf{i}_C = i_{CD} + ji_{CQ}$	space-vector of the converter current
$\mathbf{i}_g = i_{gD} + ji_{gQ}$	space-vector of the power grid current
$\mathbf{i}_L = i_{LD} + ji_{LQ}$	space-vector of the load current
p	instantaneous active power
$\tilde{p}$	instantaneous oscillating active power
$\bar{p}$	instantaneous average active power
q	instantaneous reactive power
$\tilde{q}$	instantaneous oscillating reactive power
$\bar{q}$	instantaneous average reactive power
$\dot{V}_C$	phasor of the voltage generated from the APF
$\dot{V}_g$	phasor of the AC power grid voltage
$\mathbf{v}_g = v_{gD} + jv_{gQ}$	space-vector of the power grid voltage
$\mathbf{v}_L = v_{LD} + jv_{LQ}$	space-vector of the load voltage
$\dot{Z}_F$	filter impedance
$\dot{Z}_g$	impedance of the AC power grid
$\dot{Z}_L$	equivalent impedance of the load

Pedex h is given to variables, referring to any harmonic of order h.
Pedex *ref* is given to reference control variables.

References

1. R. Yacamini, Power system harmonics—Part 1 harmonic sources, *Power Engineering Journal*, 8(4), 193–198, August 1994.
2. E. F. El-Saadany, Parameters affecting harmonic propagation and distortion levels in non-linear distribution systems, *Power Engineering Society Summer Meeting, 2002 IEEE*, Chicago, IL, Vol. 2, pp. 1010–1016, 2002.
3. A. I. Maswood, J. Zhu, Attenuation and diversity effect in harmonic current propagation study, *Power Engineering Society General Meeting, 2003*, Vol. 3, pp. 1480–1485, 2003.
4. Task force on harmonics modelling and simulation, Modeling and simulation of the propagation of harmonics in electric power networks, *IEEE Transactions on Power Delivery*, 11(1), 452–465, January 1996.
5. J. Arrillaga, N. E. Watson, *Power System Harmonics*, Wiley, New York, 2007.
6. *IEEE Recommended Practices and Requirements for Harmonic Control in Electric Power Systems*, IEEE Std. 519-1993, IEEE, New York, 1993.
7. H. Akagi, E. H. Watanabe, M. Aredes, *Instantaneous Power Theory and Applications to Power Conditioning*, IEEE Press, Wiley Interscience, Piscataway, NJ, 2007.
8. H. Akagi, Active harmonic filters, *Proceedings of the IEEE*, 93(12), 2128–2141, December 2005.
9. H. Akagi, Trends in active power conditioners, *IEEE Transactions on Power Electronics*, 9(3), 263–268, May 1994.
10. B. M. Bird, J. F. Marsh, P. R. McLellan, Harmonic reduction in multiple converters by triple-frequency current injection, *IEE Proceedings*, 116(10), 1730–1734, 1969.

11. H. Sasaki, T. Machida, A new method to eliminate ac harmonic currents by magnetic compensation—Consideration on basic design, *IEEE Transactions on Power Apparatus Systems*, PAS-90, 2009–2019, 1971.

12. A. Ametani, Harmonic reduction in thyristor converters by harmonic current injection, *IEEE Transactions on Power Apparatus Systems*, PAS-95, 441–449, 1976.

13. L. Gyugyi, E. C. Strycula, Active ac power filters, *Proceedings of 1976 IEEE IAS Annual Meeting*, Orlando, FL, pp. 529–535, 1976.

14. N. Mohan, H. A. Peterson, W. F. Long, G. R. Dreifuerst, J. J. Vithaythil, Active filters for ac harmonic suppression, *Presented at the 1977 IEEE PES Winter Meeting*, A77026-8, 1977.

15. J. Uceda, F. Aldana, P. Martinez, Active filters for static power converters, *IEE Proceedings*, 130(Pt B, 5), 347–354, 1983.

16. H. Akagi, New trends in active filters for power conditioning, *IEEE Transactions on Industry Applications*, 32(6), 1312–1322, November/December 1996.

17. W. M. Grady, M. J. Samotyj, A. H. Noyola, Survey of active power line conditioning methodologies, *IEEE Transactions on Power Delivery*, 5(3), 1536–1542, July 1990.

18. H. Akagi, N. Watanabe, Present state and future trend of active power filters in Japan, *Conference Record International Symposium on Power Electronics*, pp. 445–452, 1992.

19. H. Akagi, New trends in active filters, *Conference Record EPE 1995*, vol. 0, pp. 17–26, keynote papers, 1995.

20. M. Routimo, M. Salo, H. Yuusa, Comparison of voltage source and current source shunt active power filters, *Conference Record IEEE PESC 2005*, pp. 2571–2577, 2005.

21. B. Singh, K. Al-Haddad, A. Chandra, A review of active filters for power quality improvements, *IEEE Transactions on Industrial Electronics*, 46(6), 960–971, December 1999.

22. E. B. Makram, E. V. Subramaniam, A. A. Girgis, R. Catoe, Harmonic filter design using actual recorded data, *IEEE Transactions on Industry Applications*, 29(6), 1176–1193, November/December 1993.

23. F. Z. Peng, G. J. Su, A series LC filter for harmonic compensation of ac drives, *Proceedings of IEEE Power Electronics Specialist Conference*, Charleston, SC, pp. 213–218, August 1999.

24. F. Z. Peng, M. Kohata, Series-active filter, National Convention Record IEE of Japan, 1991, Paper #568.

25. F. Z. Peng, Application issues of active power filters, *IEEE Industry Applications Magazine*, 4(5), 21–30, September/October 1998.

26. F. Z. Peng, Harmonic sources and filtering approaches, *IEEE Industry Applications Magazine*, 7(4), 18–25, July/August 2001.

27. L. Gyugyi, E. C. Strycula, Active AC power filters, *Proceedings of IEEE Industry Applications Annual Meeting*, Chicago, Vol. 19-C, pp. 529–535, 1976.

28. S. Fukuda, T. Endoh, Control method for a combined active filter system employing a current source converter and a high pass filter, *IEEE Transaction on Industrial Applications*, 31(3), 590–595, May/June 1995.

29. M. P. Kazmierkowski, R. Krishnan, F. Blaabjerg, *Control in Power Electronics*, Academic Press, London, U.K., 2002.

30. F. Z. Peng, H. Akagi, A. Nabae, Compensation characteristics of the combined system of shunt passive and series active filters, *IEEE Transactions on Industry Applications*, 29(1), 144–152, January/February 1993.

Part II

Electrical Drives

4

Dynamic and Steady-State Models
of the Induction Machine

4.1 Introduction

Scientific literature about induction machine (IM) modeling is huge [1–15]. This chapter describes the dynamic and steady-state models of the IM, which consists of the differential equations describing the electromagnetic relationships of the stator and of the rotor as well as the equation of motion.

These equations will be written by using the space-vector theory [1–3,11–14] described in Chapter 1 and under the same assumptions. It should be remarked that these equations are of course valid for any three-phase AC machine, but in this chapter, they will be oriented to IMs. These equations will be presented in the stator reference frame, in the rotor reference frame, and in a general reference frame. By using this last general reference frame, the equations will be also written in the stator flux-linkage reference frame, the rotor flux-linkage reference frame, and the magnetizing flux-linkage reference frame.

4.2 Definition of the Machine Space-Vector Quantities

Let a three-phase electric machine be considered with cylindrical structure, smooth air-gap, and one pole pair, with three-phase windings both on the stator and the rotor, whose cross surface is shown in Figure 4.1. The following assumptions are made:

- Infinite permeability of iron.
- Radial direction of the flux density in the air-gap.
- Stator and rotor slotting effects are neglected.
- Stator and rotor losses are neglected.
- Each stator and rotor three-phase winding is considered as a full-pitch multi-turn winding.
- The axes of each stator three-phase winding (s_A, s_B, and s_C) are displaced by the angle $2\pi/3$ from each other in space around the stator periphery.

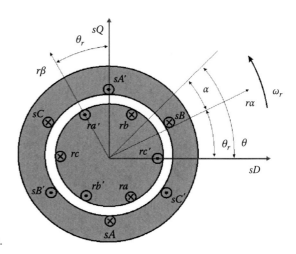

FIGURE 4.1
Cross section of a three-phase machine.

- The axes of each rotor three-phase winding (r_a, r_b, and r_c) are displaced by the angle $2\pi/3$ from each other in space around the rotor periphery. The electric angle between the axis of the rotor a phase (r_a) and the stator A phase (s_A) is θ_r, and the rotor speed in electric rad/s is given by $\omega_r = d\theta_r/dt$.

No zero-sequence stator current is present, as, for example, if the neutral point of the stator three-phase system is isolated. This means that

$$i_{sA}(t) + i_{sB}(t) + i_{sC}(t) = 0 \tag{4.1}$$

where
- t is the time variable
- $i_{sA}(t)$, $i_{sB}(t)$, and $i_{sC}(t)$ are the instantaneous values of the stator currents for each stator phase

Under these assumptions, the time and space waveform $f_s(\theta,t)$ of the resultant total mmf (magnetomotive force) caused by the three stator phases can be expressed as follows [11]:

$$f_s(\theta,t) = N_{se}\left[i_{sA}(t)\cos(\theta) + i_{sB}(t)\cos\left(\theta - \frac{2\pi}{3}\right) + i_{sC}\cos\left(\theta - \frac{4\pi}{3}\right) \right] \tag{4.2}$$

where
- θ is the electric angle coordinate along the stator periphery from the "sD" axis, coincident with the axis of phase s_A (Figure 4.1)
- N_{se} is the number of effective turns of each stator winding (given by the number of turns N_s multiplied by the winding factor k_{ws} of the stator winding)

If complex numbers are used, (4.2) can be rewritten as follows:

$$f_s(\theta,t) = \frac{3}{2}N_{se}\,\mathrm{Re}\left\{ \frac{2}{3}\left[i_{sA}(t) + a i_{sB}(t) + a^2 i_{sC}(t) \right]e^{-j\theta} \right\} \tag{4.3}$$

where
- Re means "real part"
- j is the imaginary unit
- $a = e^{j2\pi/3}$ is a complex operator that makes a vector rotate of $2\pi/3$ rad

From (4.3), it is possible to define the stator current space-vector in the stationary reference frame, coincident with the stator reference frame "*sD-sQ*," as follows:

$$\mathbf{i_s}(t) = \frac{2}{3}\left[i_{sA}(t) + ai_{sB}(t) + a^2 i_{sC}(t)\right] = |\mathbf{i_s}|e^{j\alpha_s} = i_{sD}(t) + ji_{sQ}(t) \tag{4.4a}$$

where
  $|\mathbf{i_s}|$ is the amplitude of the stator current space-vector
  α_s is the angle of the stator current space-vector from the *sD* axis
  $i_{sD}(t)$ and $i_{sQ}(t)$ are, respectively, the instantaneous values of the *sD* and *sQ* components of the stator current space-vector

It should be, however, remarked that $i_{sD}(t)$ and $i_{sQ}(t)$ can be directly computed from the phase variables by the so-called two-axis Park transformation (also called 3 → 2 transformation), given by the following:

$$\begin{cases} i_{sD} = \dfrac{2}{3}\left(i_{sA} - \dfrac{1}{2}i_{sB} - \dfrac{1}{2}i_{sC}\right) \\[3mm] i_{sQ} = \dfrac{1}{\sqrt{3}}(i_{sB} - i_{sC}) \end{cases} \tag{4.4b}$$

It should be emphasized that since the current is time varying, also the amplitude and the angle of the stator current space-vector are time varying.

The stator current space-vector then encompasses, thanks to its complex representation, the instantaneous amplitude and angle of the sinusoidal mmf distribution produced by the three stator currents.

The previous considerations permit the definition of the stator mmf space-vector as follows:

$$\mathbf{f_s}(t) = N_{se}\mathbf{i_s}(t) \tag{4.5}$$

By analogy with the stator currents and mmf, the rotor current space-vector and the rotor mmf space-vector can be defined in a similar way.

Actually if $i_{ra}(t)$, $i_{rb}(t)$, and $i_{rc}(t)$ are the instantaneous values of the rotor currents in the three rotor phases and if N_{re} is the number of effective turns of each rotor winding (given by the number of turns N_r multiplied by the winding factor k_{wr} of the rotor winding), under the assumption that there are no zero-sequence rotor currents, the space and time mmf distribution produced by the rotor currents is given by the following:

$$f_r(\theta,t) = N_{re}\left[i_{ra}(t)\cos(\alpha) + i_{rb}(t)\cos\left(\alpha - \frac{2\pi}{3}\right) + i_{rc}\cos\left(\alpha - \frac{4\pi}{3}\right)\right] \tag{4.6}$$

where α, which is a function of the angle θ, is the electric angle coordinate along the rotor periphery from the "*ra*" axis, coincident with the axis of phase "*ra*" (Figure 4.1). If complex numbers are used, (4.6) can be rewritten as follows:

$$f_r(\theta,t) = \frac{3}{2}N_{re}\,\mathrm{Re}\left\{\frac{2}{3}\left[i_{ra}(t) + ai_{rb}(t) + a^2 i_{rc}\right]e^{-j\alpha}\right\} \tag{4.7}$$

It is possible to define the rotor current space-vector in the rotor reference frame, coincident with the reference frame "$r\alpha$-$r\beta$," as follows:

$$\mathbf{i}_r(t) = \frac{2}{3}\left[i_{ra}(t) + ai_{rb}(t) + a^2 i_{rc}(t)\right] = |\mathbf{i}_r|e^{j\alpha_r} = i_{r\alpha}(t) + ji_{r\beta}(t) \qquad (4.8)$$

where
$|\mathbf{i}_r|$ is the amplitude of the rotor current space-vector
α_r is the angle of the rotor current space-vector from the $r\alpha$ axis
$i_{r\alpha}(t)$ and $i_{r\beta}(t)$ are, respectively, the instantaneous values of the $r\alpha$ and $r\beta$ components of the rotor current space-vector

With a rotational transformation of θ_r, the rotor current space-vector can be expressed in the stationary reference frame as follows:

$$\mathbf{i}'_r(t) = \mathbf{i}_r(t)e^{j\theta_r} = |\mathbf{i}_r|e^{j(\theta_r + \alpha_r)} = |\mathbf{i}_r|e^{j\alpha'_r} = i_{rd}(t) + ji_{rq}(t) \qquad (4.9)$$

where
$\alpha'_r = \alpha_r + \theta_r$ is the angle of the rotor current space-vector from the direct sD axis of the stationary reference frame
$i_{rd}(t)$ and $i_{rq}(t)$ are, respectively, the instantaneous sD and sQ components of the rotor current space in the stationary reference frame

The rotor mmf space-vector can be likewise defined as follows:

$$\mathbf{f}_r(t) = N_{re}\mathbf{i}_r(t) \qquad (4.10)$$

In the same way, the stator flux-linkage space-vector and the rotor flux-linkage space-vector can be defined. Particularly the stator flux-linkage space-vector is defined from the instantaneous value of the total flux linkage in the stator phases s_A, s_B, and s_C, that is, $\psi_{sA}(t)$, $\psi_{sB}(t)$, and $\psi_{sC}(t)$ in the stationary reference frame, as follows:

$$\mathbf{\psi_s} = \frac{2}{3}(\psi_{sA}(t) + a\psi_{sB}(t) + a^2\psi_{sC}(t)) \qquad (4.11)$$

The total flux linkage in the stator phases due to the combined effect of stator and rotor currents is given by the following relationships:

$$\begin{cases} \psi_{sA} = \bar{L}_s i_{sA} + \bar{M}_s i_{sB} + \bar{M}_s i_{sC} + \bar{M}_{sr}\cos\theta_r i_{ra} + \bar{M}_{sr}\cos\left(\theta_r + \frac{2\pi}{3}\right)i_{rb} + \bar{M}_{sr}\cos\left(\theta_r + \frac{4\pi}{3}\right)i_{rc} & (4.12\text{a}) \\[2mm] \psi_{sB} = \bar{L}_s i_{sB} + \bar{M}_s i_{sA} + \bar{M}_s i_{sC} + \bar{M}_{sr}\cos\left(\theta_r + \frac{4\pi}{3}\right)i_{ra} + \bar{M}_{sr}\cos\theta_r i_{rb} + \bar{M}_{sr}\cos\left(\theta_r + \frac{2\pi}{3}\right)i_{rc} & (4.12\text{b}) \\[2mm] \psi_{sC} = \bar{L}_s i_{sC} + \bar{M}_s i_{sB} + \bar{M}_s i_{sA} + \bar{M}_{sr}\cos\left(\theta_r + \frac{2\pi}{3}\right)i_{ra} + \bar{M}_{sr}\cos\left(\theta_r + \frac{4\pi}{3}\right)i_{rb} + \bar{M}_{sr}\cos\theta_r i_{rc} & (4.12\text{c}) \end{cases}$$

where

$\bar{L}_s$ is the self-inductance of each stator phase (assumed equal for each phase for symmetry reasons)

$\bar{M}_s$ is the mutual inductance between one stator winding and one of the other two stator windings (also in this case assumed equal for each phase for symmetry reasons)

$\bar{M}_{sr}$ is the maximum value of the mutual inductance between the stator and the rotor windings

Due to symmetry reasons, this value is constant whatever stator and rotor winding is considered. However, the mutual inductance between the stator and the rotor windings varies with the rotor angle.

By using (4.12a through 4.12c) and (4.11) and some algebra, the stator flux-linkage space-vector can be expressed as a function of the stator current space-vector and the rotor current space-vector in the stator reference frame:

$$\mathbf{\psi_s} = L_s \mathbf{i_s} + L_m \mathbf{i'_r} = L_s \mathbf{i_s} + L_m \mathbf{i_r} e^{j\theta_r} = \psi_{sD}(t) + j\psi_{sQ}(t) \tag{4.13}$$

where

$L_s = \bar{L}_s - \bar{M}_s$ is the total three-phase stator self-inductance

$L_m = 3/2\bar{M}_{sr}$ is the three-phase magnetizing inductance

$\psi_{sD}(t)$ and $\psi_{sQ}(t)$ are, respectively, the instantaneous values of the sD and sQ components of the stator flux-linkage space-vector

Equation 4.13 shows that the stator flux linkage is made up of two terms: one is the stator self-flux linkage $L_s \mathbf{i_s}$ due only to the stator currents and the other is the mutual flux component $L_m \mathbf{i'_r}$ due only to the rotor currents which link the stator winding.

Likewise, the rotor flux-linkage space-vector is defined from the instantaneous value of the total flux linkage in the rotor phases r_a, r_b, and r_c, that is, $\psi_{ra}(t)$, $\psi_{rb}(t)$, and $\psi_{rc}(t)$, in the rotor reference frame as follows:

$$\mathbf{\psi}_r = \frac{2}{3}(\psi_{ra}(t) + a\psi_{rb}(t) + a^2\psi_{rc}(t)) \tag{4.14}$$

Also in this case, the total flux linkage in the rotor phases due to the combined effect of stator and rotor currents is given by the following relationships:

$$
\begin{cases}
\psi_{ra} = \bar{L}_r i_{ra} + \bar{M}_r i_{rb} + \bar{M}_r i_{rc} + \bar{M}_{sr}\cos\theta_r i_{sA} + \bar{M}_{sr}\cos\left(\theta_r + \frac{4\pi}{3}\right)i_{sB} + \bar{M}_{sr}\cos\left(\theta_r + \frac{2\pi}{3}\right)i_{sC} & (4.15\,\text{a}) \\[2mm]
\psi_{rb} = \bar{L}_r i_{rb} + \bar{M}_r i_{ra} + \bar{M}_r i_{rc} + \bar{M}_{sr}\cos\left(\theta_r + \frac{2\pi}{3}\right)i_{sA} + \bar{M}_{sr}\cos\theta_r i_{sB} + \bar{M}_{sr}\cos\left(\theta_r + \frac{4\pi}{3}\right)i_{sC} & (4.15\,\text{b}) \\[2mm]
\psi_{rc} = \bar{L}_r i_{rc} + \bar{M}_r i_{ra} + \bar{M}_r i_{rb} + \bar{M}_{sr}\cos\left(\theta_r + \frac{4\pi}{3}\right)i_{sA} + \bar{M}_{sr}\cos\left(\theta_r + \frac{2\pi}{3}\right)i_{sB} + \bar{M}_{sr}\cos\theta_r i_{sC} & (4.15\,\text{c})
\end{cases}
$$

where

$\bar{L}_r$ is the self-inductance of each rotor phase (assumed equal for each phase for symmetry reasons)

$\bar{M}_r$ is the mutual inductance between one rotor winding and one of the other two rotor windings (also in this case assumed equal for each phase for symmetry reasons)

By using (4.15a through 4.15c) and Equation 4.14 and some algebra, the rotor flux-linkage space-vector can be expressed as a function of the stator current space-vector and the rotor current space-vector in the rotor reference frame:

$$\boldsymbol{\psi}_r = L_r \mathbf{i}_r + L_m \mathbf{i}_s' = L_r \mathbf{i}_r + L_m \mathbf{i}_s e^{-j\theta_r} = \psi_{r\alpha}(t) + j\psi_{r\beta}(t) \tag{4.16}$$

where
   $L_r = \overline{L}_r - \overline{M}_r$ is the total three-phase rotor self-inductance
   $\mathbf{i}_s' = \mathbf{i}_s e^{-j\theta_r}$ is the stator current space-vector in the rotor reference frame
   $\psi_{r\alpha}(t)$ and $\psi_{r\beta}(t)$ are, respectively, the instantaneous values of the $r\alpha$ and $r\beta$ components of the rotor flux-linkage space-vector

The rotor flux-linkage space-vector in the stator reference frame can be expressed by the following:

$$\boldsymbol{\psi}_r' = \boldsymbol{\psi}_r e^{j\theta_r} = L_r \mathbf{i}_r' + L_m \mathbf{i}_s = \psi_{rd}(t) + j\psi_{rq}(t) \tag{4.17}$$

where $\psi_{rd}(t)$ and $\psi_{rq}(t)$ are, respectively, the instantaneous values of the sD and sQ components of the rotor flux-linkage space-vector.

Similar to the previous definition, the stator voltage space-vector and the rotor voltage space-vector can be introduced. Thus, in the absence of zero-sequence stator voltages, the stator voltage space-vector in the stator reference frame is defined as follows:

$$\mathbf{u}_s = \frac{2}{3}\left[u_{sA}(t) + au_{sB}(t) + a^2 u_{sC}(t)\right] = u_{sD}(t) + ju_{sQ}(t) \tag{4.18}$$

where
   $u_{sA}(t)$, $u_{sB}(t)$, and $u_{sC}(t)$ are the instantaneous values of the voltages in the stator phases sA, sB, and sC
   $u_{sD}(t)$ and $u_{sQ}(t)$ are, respectively, the instantaneous values of the sD and sQ components of the stator voltage space-vector

Likewise, in the absence of zero-sequence rotor voltages, the rotor voltage space-vector in the rotor reference frame is defined as follows:

$$\mathbf{u}_r = \frac{2}{3}\left[u_{ra}(t) + au_{rb}(t) + a^2 u_{rc}(t)\right] = u_{r\alpha}(t) + ju_{r\beta}(t) \tag{4.19}$$

where
   $u_{ra}(t)$, $u_{rb}(t)$, and $u_{rc}(t)$ are the instantaneous values of the voltages in the rotor phases r_a, r_b, and r_c
   $u_{r\alpha}(t)$ and $u_{r\beta}(t)$ are, respectively, the instantaneous values of the $r\alpha$ and $r\beta$ components of the rotor voltage space-vector

The rotor voltage space-vector in the stator reference frame is given by the following:

$$\mathbf{u}_r' = \mathbf{u}_r e^{j\theta_r} = u_{rd}(t) + ju_{rq}(t) \tag{4.20}$$

where $u_{rd}(t)$ and $u_{rq}(t)$ are, respectively, the instantaneous values of the sD and sQ components of the rotor voltage space-vector.

4.3 Phase Equations of the IM

In the stator reference frame, the equations describing the AC machine for each stator phase are the following:

$$\left\{\begin{array}{l} u_{sA}(t) = R_s i_{sA}(t) + \dfrac{d\psi_{sA}(t)}{dt} \\[2mm] u_{sB}(t) = R_s i_{sB}(t) + \dfrac{d\psi_{sB}(t)}{dt} \\[2mm] u_{sC}(t) = R_s i_{sC}(t) + \dfrac{d\psi_{sC}(t)}{dt} \end{array}\right.$$

(4.21a)

(4.21b)

(4.21c)

where

R_s is the resistance of each stator winding and the stator flux linkages

$\psi_{sA}, \psi_{sB},$ and ψ_{sC} are functions of stator and rotor currents (see Equations 4.12a through 4.12c)

Similarly, in the rotor reference frame, the equations describing the AC machine for each stator phase are the following:

$$\left\{\begin{array}{l} u_{ra}(t) = R_r i_{ra}(t) + \dfrac{d\psi_{ra}(t)}{dt} \\[2mm] u_{rb}(t) = R_r i_{rb}(t) + \dfrac{d\psi_{rb}(t)}{dt} \\[2mm] u_{rc}(t) = R_r i_{rc}(t) + \dfrac{d\psi_{rc}(t)}{dt} \end{array}\right.$$

(4.22a)

(4.22b)

(4.22c)

where

R_r is the resistance of each rotor winding and the rotor flux linkages

$\psi_{ra}, \psi_{rb},$ and ψ_{rc} are functions of stator and rotor currents (see Equations 4.15a through 4.15c)

The equation of motion is given by the following:

$$t_e(t) - t_l(t) = J_m \frac{d\omega_m(t)}{dt} + D\omega_m(t)$$

(4.23)

where

t_e is the electromagnetic torque

t_l the load torque

J_m is the inertia of the rotating masses

D is the damping constant which accounts for losses due to friction and windage

ω_m is the mechanical rotor speed expressed in mechanical angles*

---

* The rotor speed expressed in electrical angles ω_r is related to the rotor speed expressed in mechanical angles ω_m by the relationship $\omega_r = p\omega_m$, where p is the number of pair poles of the machine.

Equations 4.21a through 4.21c and 4.22a through 4.22c together with Equations 4.12a through 4.12c and 4.15a through 4.15c as well as the equation of the motion (4.23) constitute the so-called "phase-variable" dynamic mathematical model of the IM. In this model, the input-output relationship between currents and voltages requires a matrix impedance of 36 terms, half of which are time-dependent because of the presence of cosine functions of the rotor position. However, this model can be significantly simplified by using the theory of space-vectors described in Chapter 1.

4.4 Space-Vector Equations in the Stator Reference Frame

The differential Equations 4.21a through 4.21c and 4.22a through 4.22c can be easily rewritten by using the space-vector theory in the stator or stationary reference frame, which yields the following two equations, written for the stator and the rotor, respectively:

$$\mathbf{u_s} = R_s \mathbf{i_s} + \frac{d\mathbf{\psi_s}}{dt} \tag{4.24}$$

$$\mathbf{u'_r} = R_r \mathbf{i'_r} + \frac{d\mathbf{\psi'_r}}{dt} - j\omega_r \mathbf{\psi'_r} \tag{4.25}$$

The stator and rotor flux linkages can be expressed as a function of the rotor and stator currents by using Equations 4.13 and 4.17 rewritten as follows for convenience:

$$\mathbf{\psi_s} = L_s \mathbf{i_s} + L_m \mathbf{i'_r} = L_s \mathbf{i_s} + L_m \mathbf{i_r} e^{j\theta_r} = \psi_{sD}(t) + j\psi_{sQ}(t) \tag{4.26}$$

$$\mathbf{\psi'_r} = \mathbf{\psi_r} e^{j\theta_r} = L_r \mathbf{i'_r} + L_m \mathbf{i_s} = \psi_{rd}(t) + j\psi_{rq}(t) \tag{4.27}$$

By replacing the flux linkages of (4.26) and (4.27) into Equations 4.24 and 4.25 results in the differential equations relating rotor and stator voltage space-vectors to rotor and stator current space-vectors:

$$\mathbf{u_s} = R_s \mathbf{i_s} + \frac{d(L_s \mathbf{i_s})}{dt} + \frac{d(L_m \mathbf{i'_r})}{dt} \tag{4.28}$$

$$\mathbf{u'_r} = R_r \mathbf{i'_r} + \frac{d(L_r \mathbf{i'_r})}{dt} + \frac{d(L_m \mathbf{i_s})}{dt} - j\omega_r (L_r \mathbf{i'_r} + L_m \mathbf{i_s}) \tag{4.29}$$

The previous two equations (4.28) and (4.29) together with the equation of the motion (4.23) constitute the dynamic model of the IM expressed with space-vectors in the stator reference frame.

The electromagnetic torque t_e created by the IM can be written, by using space-vectors in the stator reference frame, as a function of the stator (or rotor) flux linkage and the rotor or stator current, or directly as a function of only the stator and rotor currents, as shown in the following:

$$t_e = -\frac{3pL_m}{2L_s}\,\mathbf{\psi_s} \wedge \mathbf{i_r'} = \frac{3pl_m}{2L_r}\,\mathbf{\psi_r'} \wedge \mathbf{i_s} = -\frac{3}{2}pL_m\mathbf{i_s} \wedge \mathbf{i_r'} \tag{4.30}$$

where

p is the number of pair poles of the machine
"$\wedge$" is the symbol of vector product for complex numbers*

4.5 Space-Vector Equations in the Rotor Reference Frame

The differential equations (4.21a through 4.21c) and (4.22a through 4.22c) or (4.24) and (4.25) can be easily rewritten by using the space-vector theory in the rotor reference frame, which yields the following two equations, written for the stator and the rotor, respectively:

$$\mathbf{u_s'} = R_s\mathbf{i_s'} + \frac{d\mathbf{\psi_s'}}{dt} + j\omega_r\mathbf{\psi_s'} \tag{4.31}$$

$$\mathbf{u_r} = R_r\mathbf{i_r} + \frac{d\mathbf{\psi_r}}{dt} \tag{4.32}$$

The stator and rotor flux linkages can be expressed in the rotor reference frame as a function of the rotor and stator currents as follows:

$$\mathbf{\psi_s'} = \mathbf{\psi_s}e^{-j\theta_r} = L_s\mathbf{i_s'} + L_m\mathbf{i_r} = \psi_{sd}(t) + j\psi_{sq}(t) \tag{4.33}$$

$$\mathbf{\psi_r} = L_r\mathbf{i_r} + L_m\mathbf{i_s'} = \psi_{r\alpha}(t) + j\psi_{r\beta}(t) \tag{4.34}$$

By replacing the flux linkages of (4.33) and (4.34) into Equations 4.31 and 4.32 results in the differential equations relating rotor and stator voltage space-vectors to rotor and stator current space:

$$\mathbf{u_s} = R_s\mathbf{i_s} + \frac{d(L_s\mathbf{i_s})}{dt} + \frac{d(L_m\mathbf{i_r'})}{dt} \tag{4.35}$$

$$\mathbf{u_r'} = R_r\mathbf{i_r'} + \frac{d(L_r\mathbf{i_r'})}{dt} + \frac{d(L_m\mathbf{i_s})}{dt} - j\omega_r(L_r\mathbf{i_r'} + L_m\mathbf{i_s}) \tag{4.36}$$

---

* The vector product between two complex numbers **x** and **y** is given by the scalar **x** $\wedge$ **y** = Re(j **x y***) = −Im (**x y***) and is not obviously commutative.

The previous two equations (4.35) and (4.36) together with the equation of the motion (4.23) constitute the dynamic model of the IM expressed with space-vectors in the rotor reference frame.

The electromagnetic torque t_e created by the IM can be written, by using space-vectors in the rotor reference frame, as a function of the stator (or rotor) flux-linkage and the rotor or stator current, or directly as a function of only the stator and rotor currents, as shown below:

$$t_e = -\frac{3pL_m}{2L_s}\,\mathbf{\psi}'_s \wedge \mathbf{i_r} = \frac{3pL_m}{2L_r}\,\mathbf{\psi_r} \wedge \mathbf{i'_s} = -\frac{3}{2}\,pL_m\mathbf{i'_s} \wedge \mathbf{i_r} \qquad (4.37)$$

4.6 Space-Vector Equations in the Generalized Reference Frame

If a general reference frame is adopted, rotating at a speed $\omega_g = d\theta_g/dt$, where θ_g is the angle between the direct axis x of the generalized reference frame and the direct axis sD of the stator reference frame (Figure 4.2), the corresponding differential equations for the stator and the rotor can be found once the expressions of voltages, currents, and flux linkages are found in this reference frame.

Particularly it is possible to express the stator current space-vector in the generalized reference frame as a function of the stator current space-vector in the stator reference frame as follows:

$$\mathbf{i}_s^g = \mathbf{i}_s e^{-j\theta_g} = |\mathbf{i}_s|e^{j\alpha_s}e^{-j\theta_g} = i_{sx}^g(t) + ji_{sy}^g(t) \qquad (4.38)$$

where
   $\mathbf{i}_s^g$ is the stator current space-vector in the generalized reference frame
   $\mathbf{i_s}$ is the stator current space-vector in the stator reference frame
   α_s is the angle of the stator current space-vector with respect to the real axis of the stator reference frame, while i_{sx}^g and i_{sy}^g are, respectively, the components of the vector $\mathbf{i}_s^g$ on the direct axis x_g and the quadrature axis y_g in the generalized reference frame (Figure 4.2)

In the same fashion, the stator voltage and stator flux-linkage space-vectors in the generalized reference frame can be obtained:

$$\mathbf{u}_s^g = \mathbf{u}_s e^{-j\theta_g} = u_{sx}^g(t) + ju_{sy}^g(t) \qquad (4.39)$$

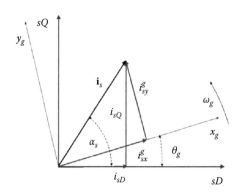

FIGURE 4.2
Generalized reference frame: vector diagram.

$$\boldsymbol{\psi}_s^g = \boldsymbol{\psi}_s e^{-j\theta_g} = \psi_{sx}^g(t) + j\psi_{sy}^g(t) \tag{4.40}$$

where
  $\mathbf{u}_s^g$ is the stator voltage space-vector in the generalized reference frame
  $\mathbf{u_s}$ is the stator voltage space-vector in the stator reference frame
  u_{sx}^g and u_{sy}^g are, respectively, the components of the vector $\mathbf{u}_s^g$ on direct axis x_g and the quadrature axis y_g in the generalized reference frame
  $\boldsymbol{\psi}_s^g$ is the stator flux-linkage space-vector in the generalized reference frame
  $\boldsymbol{\psi_s}$ is the stator flux-linkage space-vector in the stator reference frame
  ψ_{sx}^g and ψ_{sy}^g are, respectively, the components of the vector $\boldsymbol{\psi}_s^g$ on the direct axis x_g and the quadrature axis y_g in the generalized reference frame

It is possible to express the rotor variable space-vectors in the generalized reference frame, considering that the angle between the direct axis x_g of the generalized reference frame and the direct axis rα of the rotor reference frame is equal to $\theta_g - \theta_r$. This yields the following:

$$\mathbf{i}_r^g = \mathbf{i}_r e^{-j(\theta_g - \theta_r)} = |\mathbf{i}_r| e^{j\alpha_r} e^{-j(\theta_g - \theta_r)} = i_{rx}^g(t) + ji_{ry}^g(t) \tag{4.41}$$

$$\mathbf{u}_r^g = \mathbf{u}_r e^{-j(\theta_g - \theta_r)} = u_{rx}^g(t) + ju_{ry}^g(t) \tag{4.42}$$

$$\boldsymbol{\psi}_r^g = \boldsymbol{\psi}_r e^{-j(\theta_g - \theta_r)} = \psi_{rx}^g(t) + j\psi_{ry}^g(t) \tag{4.43}$$

where
  $\mathbf{i}_r^g$ is the rotor current space-vector in the generalized reference frame
  $\mathbf{i_r}$ is the rotor current space-vector in the rotor reference frame
  α_r is the angle of the rotor current space-vector with respect to the real axis of the rotor reference frame
  i_{rx}^g and i_{ry}^g are, respectively, the components of the vector $\mathbf{i}_r^g$ on the direct axis x_g and the quadrature axis y_g in the generalized reference frame
  $\mathbf{u}_r^g$ is the rotor voltage space-vector in the generalized reference frame
  $\mathbf{u_r}$ is the rotor voltage space-vector in the rotor reference frame
  u_{rx}^g and u_{ry}^g are, respectively, the components of the vector $\mathbf{u}_r^g$ on the direct axis x_g and the quadrature axis y_g in the generalized reference frame
  $\boldsymbol{\psi}_r^g$ is the rotor flux-linkage space-vector in the generalized reference frame
  $\boldsymbol{\psi_r}$ is the rotor flux-linkage space-vector in the rotor reference frame
  ψ_{rx}^g and ψ_{ry}^g are, respectively, the components of the vector $\boldsymbol{\psi}_r^g$ on the direct axis x_g and the quadrature axis y_g in the generalized reference frame

The vector Equations 4.24 and 4.25 can be therefore written in the generalized reference frame as follows:

$$\mathbf{u}_s^g = R_s \mathbf{i}_s^g + \frac{d\boldsymbol{\psi}_s^g}{dt} + j\omega_g \boldsymbol{\psi}_s^g \tag{4.44}$$

$$\mathbf{u}_r^g = R_r \mathbf{i}_r^g + \frac{d\boldsymbol{\psi}_r^g}{dt} + j(\omega_g - \omega_r)\boldsymbol{\psi}_r^g \tag{4.45}$$

where the stator and rotor flux linkages can be expressed as a function of the stator and rotor current space-vectors in the generalized reference frame by using the following equations:

$$\boldsymbol{\psi}_s^g = L_s \mathbf{i}_s^g + L_m \mathbf{i}_r^g \qquad (4.46)$$

$$\boldsymbol{\psi}_r^g = L_r \mathbf{i}_r^g + L_m \mathbf{i}_s^g \qquad (4.47)$$

By replacing Equations 4.46 and 4.47 into Equations 4.44 and 4.45, the following differential equations can be obtained between the stator and rotor currents and voltages:

$$\mathbf{u}_s^g = R_s \mathbf{i}_s^g + \frac{d(L_s \mathbf{i}_s^g)}{dt} + \frac{d(L_m \mathbf{i}_r^g)}{dt} + j\omega_g (L_s \mathbf{i}_s^g + L_m \mathbf{i}_r^g) \qquad (4.48)$$

$$\mathbf{u}_r^g = R_r \mathbf{i}_r^g + \frac{d(L_r \mathbf{i}_r^g)}{dt} + \frac{d(L_m \mathbf{i}_s^g)}{dt} + j(\omega_g - \omega_r)(L_r \mathbf{i}_r^g + L_m \mathbf{i}_s^g) \qquad (4.49)$$

Equations 4.48 and 4.49 together with the equation of the motion (4.23) constitute the dynamic model of the IM in the general reference frame.

The electromagnetic torque t_e of the machine can be expressed also either by using the stator or rotor flux-linkage space-vectors or by using the stator or rotor current space-vectors or both, as follows in the generalized reference frame:

$$t_e = -\frac{3pL_m}{2L_s} \boldsymbol{\psi}_s^g \wedge \mathbf{i}_r^g = \frac{3pL_m}{2L_r} \boldsymbol{\psi}_r^g \wedge \mathbf{i}_s^g = -\frac{3}{2} pL_m \mathbf{i}_s^g \wedge \mathbf{i}_r^g \qquad (4.50)$$

In the following, these equations in the generalized reference frame will be rewritten when using either the stator flux-linkage, or the rotor flux-linkage, or the magnetizing flux-linkage reference frame.

4.6.1 Mutually Coupled Magnetic Circuits

Whatever reference frame is chosen, in general Equations 4.46 and 4.47 represent in general two magnetic circuits mutually coupled by the total mutual inductance L_m. Note that three stator circuits along with three rotor circuits are represented, thanks to the use of space-vectors. These equations are written again in the following for convenience:

$$\boldsymbol{\psi}_s^g = L_s \mathbf{i}_s^g + L_m \mathbf{i}_r^g \qquad (4.51)$$

$$\boldsymbol{\psi}_r^g = L_r \mathbf{i}_r^g + L_m \mathbf{i}_s^g \qquad (4.52)$$

If the parameter n is introduced, these equations can be rewritten as follows:

$$\boldsymbol{\psi}_s^g = L_s \mathbf{i}_s^g + L_m \mathbf{i}_r^g = (L_s - nL_m)\mathbf{i}_s^g + nL_m \left(\mathbf{i}_s^g + \frac{\mathbf{i}_r^g}{n} \right) \qquad (4.53)$$

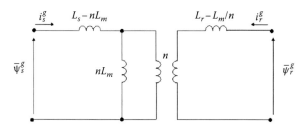

FIGURE 4.3
Equivalent circuit of mutually coupled magnetic circuits.

$$n\boldsymbol{\psi}_r^g = nL_r\mathbf{i}_r^g + nL_m\mathbf{i}_s^g = n^2\left(L_r - \frac{L_m}{n}\right)\frac{\mathbf{i}_r^g}{n} + nL_m\left(\mathbf{i}_s^g + \frac{\mathbf{i}_r^g}{n}\right) \tag{4.54}$$

This means that the equivalent circuit of these equations can be represented as in Figure 4.3.

The parameter n can have various values according to the assumptions made for describing the AC machine. If $n = N_{se}/N_{re}$, where N_{se} is the number of stator effective turns and N_{re} is the number of rotor effective turns, then $L_s - nL_m$ and $L_r - L_m/n$ represent, respectively, the stator and rotor leakage inductances $L_{s\sigma}$ and $L_{r\sigma}$, nL_m the magnetic inductance, and $\mathbf{i}_m^g = \left(\mathbf{i}_s^g + \mathbf{i}_r^g/n\right)$ the magnetizing current.

If $n = L_m/L_r$, then the "rotor magnetizing current" can be defined, referred to the stator, as follows:

$$\mathbf{i}_{mr}^g = \frac{\boldsymbol{\psi}_r^g}{L_m} = \mathbf{i}_s^g + \frac{L_r}{L_m}\mathbf{i}_r^g \tag{4.55}$$

In this case, $L_s - nL_m = L_s\left(1 - \dfrac{L_m^2}{L_sL_r}\right) = \sigma L_s = L_s'$, where $\sigma = 1 - L_m^2/L_sL_r$ is the "total leakage factor" or Blondel coefficient and $L_s' = \left(L_s - L_m^2/L_r\right) = \sigma L_s$ is the stator transient inductance.

In particular, if $N_{se} = N_{re}$, then Equation 4.55 can be written as $\mathbf{i}_{mr}^g = \mathbf{i}_s^g + \mathbf{i}_r^g L_r/L_m$ $= \mathbf{i}_s^g + \mathbf{i}_r^g(1 + \sigma_r)$, where $\sigma_r = L_{r\sigma}/L_m$ is the "rotor leakage factor" since in this case $L_r = L_{r\sigma} + L_m$.

If $n = L_s/L_m$, then the "stator magnetizing current" can be defined, referred to the rotor, as follows:

$$\mathbf{i}_{ms}^g = \frac{\boldsymbol{\psi}_s^g}{L_m} = \frac{L_s}{L_m}\mathbf{i}_s^g + \mathbf{i}_r^g \tag{4.56}$$

In particular, if $N_{se} = N_{re}$, then Equation 4.55 can be written as $\mathbf{i}_{ms}^g = \mathbf{i}_s^g L_s/L_m + \mathbf{i}_r^g$ $= \mathbf{i}_s^g(1 + \sigma_s) + \mathbf{i}_r^g$, where $\sigma_s = L_{s\sigma}/L_m$ is the "stator leakage factor" since in this case $L_s = L_{s\sigma} + L_m$.

4.6.2 Space-Vector Equations in the Rotor Flux-Linkage Reference Frame

This paragraph describes the equations of the IM in the rotor flux-linkage reference frame. This reference frame is also called "field reference frame" or simply "rotor-flux reference frame." This reference frame is a key issue for explaining the "rotor-flux-oriented vector control" for AC machines, which is one of the versions of the so-called field-oriented control (FOC). This kind of control strategy will be deeply described in Chapter 5. This reference frame is defined as having the quadrature-axis component of the rotor flux-linkage space-vector equal to 0 $\left(\psi_{ry}^{\psi_r} = 0\right)$, that is, the x-axis is parallel to the direct-axis component the rotor flux-linkage space-vector.

Let ρ_r be the angle between the direct axis x of the rotor flux-linkage reference frame and the direct axis sD of the stator reference frame and let $\omega_{mr} = d\rho_r/dt$ be the angular speed of the rotor flux-linkage space-vector under the assumption that the machine has a linear magnetic behavior. Then Equations 4.44 and 4.45 can be rewritten in the rotor flux-linkage reference frame as follows:

$$\mathbf{u}_s^{\psi_r} = R_s \mathbf{i}_s^{\psi_r} + L_s \frac{d\mathbf{i}_s^{\psi_r}}{dt} + L_m \frac{d\mathbf{i}_r^{\psi_r}}{dt} + j\omega_{mr}L_s\mathbf{i}_s^{\psi_r} + j\omega_{mr}L_m\mathbf{i}_r^{\psi_r} \tag{4.57}$$

$$\mathbf{u}_r^{\psi_r} = R_r \mathbf{i}_r^{\psi_r} + L_r \frac{d\mathbf{i}_r^{\psi_r}}{dt} + L_m \frac{d\mathbf{i}_s^{\psi_r}}{dt} + j(\omega_{mr} - \omega_r)L_r\mathbf{i}_r^{\psi_r} + j(\omega_{mr} - \omega_r)L_m\mathbf{i}_s^{\psi_r} \tag{4.58}$$

where
  $\mathbf{u}_s^{\psi_r}$ and $\mathbf{u}_r^{\psi_r}$ are, respectively, the stator and rotor voltage space-vectors in the rotor flux-linkage reference frame
  $\mathbf{i}_s^{\psi_r}$ and $\mathbf{i}_r^{\psi_r}$ are the stator and rotor current space-vectors in the rotor flux-linkage reference frame

Let $\mathbf{i}_{mr}$ be the rotor magnetizing current space-vector in the rotor flux-linkage reference frame defined as in Equation 4.55 and rewritten here for convenience in the case $N_{se} = N_{re}$, where the apex g is replaced with ψ_r to indicate the rotor flux-linkage reference frame:

$$\mathbf{i}_{mr} = \frac{\mathbf{\Psi}_r^{\psi_r}}{L_m} = \mathbf{i}_s^{\psi_r} + (1 + \sigma_r)\mathbf{i}_r^{\psi_r} \tag{4.59}$$

As explained earlier, this selection of reference frame means that the rotor magnetizing current space-vector is coaxial with the direct axis of the rotor flux-linkage reference frame. This means that $\mathbf{i}_{mr}$ coincides with a real number whose value is equal to its modulus, that is, $\mathbf{i}_{mr} = |\mathbf{i}_{mr}|$.

Equations 4.57 and 4.58 can be therefore written in the following form, considering $\mathbf{u}_r^{\psi_r} = \mathbf{0}$, and after some manipulation:

$$T_s' \frac{d\mathbf{i}_s^{\psi_r}}{dt} + \mathbf{i}_s^{\psi_r} = \frac{\mathbf{u}_s^{\psi_r}}{R_s} - j\omega_{mr}T_s'\mathbf{i}_s^{\psi_r} - (T_s - T_s')\left(j\omega_{mr}\mathbf{i}_{mr} + \frac{d\mathbf{i}_{mr}}{dt}\right) \tag{4.60a}$$

$$T_r \frac{d\mathbf{i}_{mr}}{dt} + \mathbf{i}_{mr} = \mathbf{i}_s^{\psi_r} - j(\omega_{mr} - \omega_r)T_r\mathbf{i}_{mr} \tag{4.60b}$$

where
  $T_s = L_s/R_s$ is the stator time constant
  $T_s' = L_s'/R_s$ is the stator transient time constant
  $T_r = L_r/R_r$ is the rotor time constant

It should be remarked that this choice of reference frame results in the appearance of the stator transient time constant and the stator transient inductance, that is, the assumption $n = L_m/L_r$ as explained in Section 4.6.1.

Equations 4.60a and b, if projected onto the direct x-axis and quadrature y-axis of the rotor flux-linkage reference frame, yield the following:

$$\begin{cases} T_s' \dfrac{di_{sx}^{\psi_r}}{dt} + i_{sx}^{\psi_r} = \dfrac{u_{sx}^{\psi_r}}{R_s} + \omega_{mr} T_s' i_{sy}^{\psi_r} - \left(T_s - T_s'\right) \dfrac{d\left|\mathbf{i}_{mr}\right|}{dt} & (4.61\,a) \\[3ex] T_s' \dfrac{di_{sy}^{\psi_r}}{dt} + i_{sy}^{\psi_r} = \dfrac{u_{sy}^{\psi_r}}{R_s} - \omega_{mr} T_s' i_{sx}^{\psi_r} - \left(T_s - T_s'\right) \omega_{mr}\left|\mathbf{i}_{mr}\right| & (4.61\,b) \end{cases}$$

$$\begin{cases} T_r \dfrac{d\left|\mathbf{i}_{mr}\right|}{dt} + \left|\mathbf{i}_{mr}\right| = i_{sx}^{\psi_r} & (4.62\,a) \\[3ex] \omega_{mr} = \omega_r + \dfrac{i_{sy}^{\psi_r}}{T_r \left|\mathbf{i}_{mr}\right|} & (4.62\,b) \end{cases}$$

where

$u_{sx}^{\psi_r}$ and $u_{sy}^{\psi_r}$ are the x-axis and y-axis components of the stator voltage space-vector
$i_{sx}^{\psi_r}$ and $i_{sy}^{\psi_r}$ are the x-axis and y-axis components of the stator current space-vector in the rotor flux-linkage reference frame

The electromagnetic torque t_e can be then easily written from (4.50) in the following form by using the rotor flux-linkage space-vector and the stator current space-vector in the field:

$$t_e = \frac{3pL_m}{2L_r}\, \psi_{rx}^{\psi_r} i_{sy}^{\psi_r} = \frac{3pL_m^2}{2L_r}\left|\mathbf{i}_{mr}\right| i_{sy}^{\psi_r} = \frac{3pL_m}{2(1+\sigma_r)}\left|\mathbf{i}_{mr}\right| i_{sy}^{\psi_r} \tag{4.63}$$

In this reference frame, the torque is the given only by the product of the direct-axis component of the rotor flux-linkage space-vector and the quadrature-axis component of the stator current space-vector: it resembles the torque equation of a separately excited DC current machine, and it should be noted that these components are constant in steady-state. This is the key factor of the field oriented control of IMs (see Chapter 5).

Figure 4.4 shows the block diagram of the dynamic model of the IM in the rotor flux-linkage reference by using (4.61a and 4.61b), (4.62a and 4.62b), and the equation of their motion (4.23). In Figure 4.4, s is the complex Laplace variable.

4.6.3 Space-Vector Equations in the Stator Flux-Linkage Reference Frame

This paragraph describes the equations of the IM in the stator flux-linkage reference frame and is the basis for the "stator-flux-oriented vector control." This reference frame is defined as having the quadrature-axis component of the stator flux-linkage space-vector equal to 0 $\left(\psi_{ry}^{\psi_s} = 0\right)$, that is, the x-axis is parallel to the direct-axis component of the stator flux-linkage space-vector.

Let ρ_s be the angle between the direct axis x of the stator flux-linkage reference frame and the direct axis sD of the stator reference frame and let $\omega_{ms} = d\rho_s/dt$ be the angular speed of the stator flux-linkage space-vector under the assumption that the machine has a linear magnetic behavior. Then Equations 4.44 and 4.45 can be rewritten in the stator flux-linkage reference frame as follows:

$$\mathbf{u}_s^{\psi_s} = R_s \mathbf{i}_s^{\psi_s} + L_s \frac{d\mathbf{i}_s^{\psi_s}}{dt} + L_m \frac{d\mathbf{i}_r^{\psi_s}}{dt} + j\omega_{ms} L_s \mathbf{i}_s^{\psi_s} + j\omega_{ms} L_m \mathbf{i}_r^{\psi_s} \tag{4.64}$$

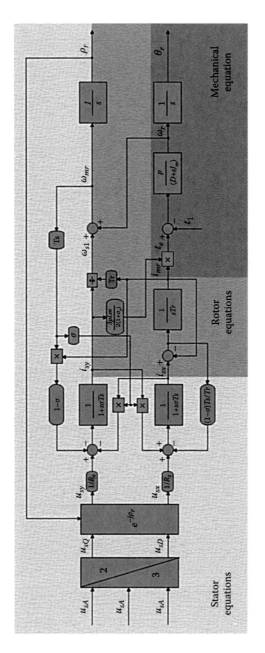

FIGURE 4.4

(See color insert.) Block diagram of the dynamical model of the IM in the rotor flux-linkage reference frame.

$$\mathbf{u}_r^{\psi_s} = R_r\mathbf{i}_r^{\psi_s} + L_r\frac{d\mathbf{i}_r^{\psi_s}}{dt} + L_m\frac{d\mathbf{i}_s^{\psi_s}}{dt} + j(\omega_{ms} - \omega_r)L_r\mathbf{i}_r^{\psi_s} + j(\omega_{ms} - \omega_r)L_m\mathbf{i}_s^{\psi_s} \qquad (4.65)$$

where

$\mathbf{u}_s^{\psi_s}$ and $\mathbf{u}_r^{\psi_s}$ are, respectively, the stator and rotor voltage space-vectors in the stator flux-linkage reference frame

$\mathbf{i}_s^{\psi_s}$ and $\mathbf{i}_r^{\psi_s}$ are the stator and rotor current space-vectors in the stator flux-linkage reference frame

Let $\mathbf{i}_{ms}$ be the stator magnetizing current space-vector in the stator flux-linkage reference frame defined as in (4.56) and rewritten here for convenience in the case $N_{se} = N_{re}$, where the subscript g is replaced with ψ_s to indicate the stator flux-linkage reference frame:

$$\mathbf{i}_{ms} = \frac{\boldsymbol{\psi}_s^{\psi_s}}{L_m} = \frac{L_s}{L_m}\mathbf{i}_s^{\psi_s} + \mathbf{i}_r^{\psi_s} = \mathbf{i}_s^{\psi_s}(1+\sigma_s) + \mathbf{i}_r^{\psi_s} \qquad (4.66)$$

As explained earlier, this selection of reference frame means that the stator magnetizing current space-vector is coaxial with the direct axis of the stator flux-linkage reference frame. This means that $\mathbf{i}_{ms}$ coincides with a real number whose value is equal to its modulus, that is, $\mathbf{i}_{ms} = |\mathbf{i}_{ms}|$.

Now if a **squirrel-cage IM** is considered, for which the rotor current is not accessible, the rotor current space-vector should be eliminated by using Equation 4.32, which yields the following:

$$\mathbf{i}_r^{\psi_s} = |\mathbf{i}_{ms}| - \frac{L_s\mathbf{i}_s^{\psi_s}}{L_m} \qquad (4.67)$$

which substituted into (4.30) gives

$$\mathbf{u}_s^{\psi_s} = R_s\mathbf{i}_s^{\psi_s} + L_m\frac{d|\mathbf{i}_{ms}|}{dt} + j\omega_{ms}L_m|\mathbf{i}_{ms}| \qquad (4.68)$$

Notice that this equation could be also derived from Equation 4.44 expressed in the stator flux-linkage reference frame. By resolving (4.68) into the direct and quadrature components, the following equations are formed:

$$\begin{cases} i_{sx}^{\psi_s} = \dfrac{u_{sx}^{\psi_s}}{R_s} - \dfrac{L_m}{R_s}\dfrac{d|\mathbf{i}_{ms}|}{dt} & (4.69\,a) \\[3mm] i_{sy}^{\psi_s} = \dfrac{u_{sy}^{\psi_s}}{R_s} - \omega_{ms}\dfrac{|\mathbf{i}_{ms}|}{R_s} & (4.69\,b) \end{cases}$$

Equation 4.32 can be written in the following form considering $\mathbf{u}_r^{\psi_s} = \mathbf{0}$ in the squirrel-cage IM, and after some manipulation:

$$0 = R_r\left[|\mathbf{i}_{ms}| - \frac{L_s}{L_m}\mathbf{i}_s^{\psi_s}\right] + L_r\frac{d|\mathbf{i}_{ms}|}{dt} - \frac{L_s'L_r}{L_m}\frac{d\mathbf{i}_s^{\psi_s}}{dt} + j\omega_{sl}\left[L_r|\mathbf{i}_{ms}| - \frac{L_s'L_r}{L_m}\mathbf{i}_s^{\psi_s}\right] \qquad (4.70)$$

It should be noted that on the one hand, (4.68) is much simpler than Equation 4.60a, but on the other hand, (4.70) is more involved than the counterpart Equation 4.60b. Thus, the choice of the stator flux-linkage reference frame for the control of the IM results in a simplification of the stator voltage equation (no derivative of the stator current, no stator leakage inductance, further simplification if the stator magnetizing current is kept constant); however, a decoupling circuit is necessary to eliminate the coupling between $|\mathbf{i}_{ms}|$ and the torque-producing current component $i_{sy}^{\psi_s}$.

This is apparent if Equation 4.70 is resolved into its direct and quadrature components:

$$
\begin{cases}
\dfrac{L_m}{L_s'}\dfrac{d|\mathbf{i}_{ms}|}{dt} + \dfrac{L_m}{L_s T_r'}|\mathbf{i}_{ms}| = \dfrac{di_{sx}^{\psi_s}}{dt} + \dfrac{1}{T_r'}i_{sx}^{\psi_s} - \omega_{sl}i_{sy}^{\psi_s} & (4.71\,\mathrm{a})\\[3mm]
\omega_{sl}\left(\dfrac{L_m}{L_s'}|\mathbf{i}_{ms}| - i_{sx}^{\psi_s}\right) = \dfrac{di_{sy}^{\psi_s}}{dt} + \dfrac{1}{T_r'}i_{sy}^{\psi_s} & (4.71\,\mathrm{b})
\end{cases}
$$

where
$T_r' = L_r'/R_s$ is the transient rotor time constant
$L_r' = \sigma L_r$

The electromagnetic torque t_e can be then easily written in the following form (4.50) by using the stator flux-linkage space-vector and the stator current space-vector in the field:

$$
t_e = \frac{3p}{2}\psi_{sx}^{\psi_s}i_{sy}^{\psi_s} = \frac{3p}{2}L_m|\mathbf{i}_{ms}|i_{sy}^{\psi_s} \tag{4.72}
$$

In the reference frame, the torque is given only by the product of the direct-axis component of the stator flux-linkage space-vector and the quadrature-axis component of the stator current space-vector: in a way, it resembles the torque equation of an independent excitation DC current machine. This is the key factor of the stator flux oriented control of IMs (see Chapter 5).

If a **doubly fed IM** is considered, then it is more convenient to express $\mathbf{i}_s^{\psi_s}$ in terms of $|\mathbf{i}_{ms}|$ and $\mathbf{i}_r^{\psi_s}$, which yields the following:

$$
\mathbf{i}_s^{\psi_s} = \frac{L_m}{L_s}\left(|\mathbf{i}_{ms}| - \mathbf{i}_r^{\psi_s}\right) \tag{4.73}
$$

Then the stator voltage Equation 4.30 becomes as follows:

$$
\mathbf{u}_s^{\psi_s} = R_s\frac{L_m}{L_s}\left(|\mathbf{i}_{ms}| - \mathbf{i}_r^{\psi_s}\right) + L_m\frac{d|\mathbf{i}_{ms}|}{dt} + j\omega_{ms}L_m|\mathbf{i}_{ms}| \tag{4.74}
$$

This gives the following equation by resolving it into its direct and quadrature components and after some algebra:

$$
T_s\frac{d|\mathbf{i}_{ms}|}{dt} + |\mathbf{i}_{ms}| = \frac{L_s}{R_s L_m}u_{sx}^{\psi_s} + i_{rx}^{\psi_s} \tag{4.75\,a}
$$

$$
\omega_{ms}T_s|\mathbf{i}_{ms}| = \frac{L_s}{R_s L_m}u_{sy}^{\psi_s} + i_{sy}^{\psi_s} \tag{4.75\,b}
$$

These equations, in respect with Equation 4.69a and 4.69b, contain the rotor currents, which can be controlled in stator-flux-oriented control.

By substituting (4.65) into the stator voltage Equation 4.74 and by defining the angular slip frequency as $\omega_{sl} = \omega_{ms} - \omega_r$ yields the following:

$$\mathbf{u}_r^{\psi_s} = R_r \mathbf{i}_r^{\psi_s} + L_r' \frac{d\mathbf{i}_r^{\psi_s}}{dt} + \frac{L_m^2}{L_s} \frac{d|\mathbf{i}_{ms}|}{dt} + j\omega_{sl} \left[\frac{L_m^2}{L_s} \mathbf{i}_r^{\psi_s} + L_r' \mathbf{i}_r^{\psi_s} \right] \tag{4.76}$$

and by resolving (4.76) into its direct and quadrature components, the following equations are obtained:

$$T_r' \frac{di_{rx}^{\psi_s}}{dt} + i_{rx}^{\psi_s} = \frac{u_{rx}^{\psi_s}}{R_r} + \omega_{sl} T_r' i_{ry}^{\psi_s} - (T_r - T_r') \frac{d|\mathbf{i}_{ms}|}{dt} \tag{4.77 a}$$

$$T_r' \frac{di_{ry}^{\psi_s}}{dt} + i_{ry}^{\psi_s} = \frac{u_{ry}^{\psi_s}}{R_r} + \omega_{sl} T_r' i_{rx}^{\psi_s} - (T_r - T_r') \frac{d|\mathbf{i}_{ms}|}{dt} \tag{4.77 b}$$

which are very similar to Equations 4.69a and 4.69b and can be used for the stator-flux-oriented control if the rotor currents are not impressed.

4.6.4 Space-Vector Equations in the Magnetic Flux-Linkage Reference Frame

This paragraph describes the equations of the IM in the magnetic flux-linkage reference frame and is the basis for the "magnetizing-flux-oriented" vector control. This reference frame is defined as having the quadrature-axis component of the magnetizing flux-linkage space-vector equal to 0 $\left(\psi_{my}^{\psi_m} = 0\right)$, that is, the x-axis is parallel to the direct-axis component the magnetizing flux-linkage space-vector. In this case, in this reference frame, the magnetizing current space-vector is given by $\mathbf{i}_{mm} = |\mathbf{i}_{mm}|$.

From (4.50) and the assumption that $N_{se} = N_{re}$, the expression of the torque in the magnetizing reference frame can be obtained, given by

$$t_e = \frac{3p}{2} \psi_{mx}^{\psi_m} i_{sy}^{\psi_m} = \frac{3p}{2} L_m |\mathbf{i}_{mm}| i_{sy}^{\psi_m} \tag{4.78}$$

And this formula is the basis for the so-called magnetizing-flux-oriented vector control.

Similar to the earlier two paragraphs, let ρ_m be the angle between the direct axis x of the magnetizing flux-linkage reference frame and the direct axis sD of the stator reference frame and let $\omega_{mm} = d\rho_m/dt$ be the angular speed of the magnetizing flux-linkage space-vector under the assumption that the machine has a linear magnetic behavior. Then Equations 4.48 and 4.49 can be rewritten in the stator flux-linkage reference frame as follows:

$$\mathbf{u}_s^{\psi_m} = R_s \mathbf{i}_s^{\psi_m} + L_s \frac{d\mathbf{i}_s^{\psi_m}}{dt} + L_m \frac{d\mathbf{i}_r^{\psi_m}}{dt} + j\omega_{mm} L_s \mathbf{i}_s^{\psi_m} + j\omega_{mm} L_m \mathbf{i}_r^{\psi_m} \tag{4.79}$$

$$\mathbf{u}_r^{\psi_m} = R_r \mathbf{i}_r^{\psi_m} + L_r \frac{d\mathbf{i}_s^{\psi_m}}{dt} + L_m \frac{d\mathbf{i}_s^{\psi_m}}{dt} + j(\omega_{mm} - \omega_r) L_r \mathbf{i}_r^{\psi_m} + j(\omega_{ms} - \omega_r) L_m \mathbf{i}_s^{\psi_m} \tag{4.80}$$

where

$\mathbf{u}_s^{\psi_m}$ and $\mathbf{u}_r^{\psi_m}$ are, respectively, the stator and rotor voltage space-vectors in the magnetizing flux-linkage reference frame

$\mathbf{i}_s^{\psi_m}$ and $\mathbf{i}_r^{\psi_m}$ are the stator and rotor current space-vectors in the magnetizing flux-linkage reference frame

Let $\mathbf{i}_{mm}$ be the magnetizing current space-vector in the magnetizing flux-linkage reference frame in the case $N_{se} = N_{re}$, in accordance with Section 4.6.1 where the apex g is replaced with ψ_m to indicate the magnetizing flux linkage reference frame:

$$\left|\mathbf{i}_{mm}\right| = \frac{\left|\mathbf{\Psi}_m\right|}{L_m} = \mathbf{i}_r^{\psi_m} + \mathbf{i}_s^{\psi_m} \tag{4.81}$$

As explained earlier, this selection of reference frame means that the stator magnetizing current space-vector is coaxial with the direct axis of the stator flux-linkage reference frame. This means that $\mathbf{i}_{mm}$ coincides with a real number whose value is equal to its modulus, that is, $\mathbf{i}_{mm} = \left|\mathbf{i}_{mm}\right|$.

From (4.80), the rotor current can be eliminated, and considering also Section 4.6.1, (4.79) becomes as follows:

$$\mathbf{u}_s^{\psi_m} = R_s \mathbf{i}_s^{\psi_m} + L_{s\sigma} \frac{d\mathbf{i}_s^{\psi_m}}{dt} + L_m \frac{d\left|\mathbf{i}_{mm}\right|}{dt} + j\omega_{mm}\left(L_{s\sigma}\mathbf{i}_s^{\psi_m} + L_m\left|\mathbf{i}_{mm}\right|\right) \tag{4.82}$$

while the rotor equations, in case a squirrel-cage motor is considered, are as follows:

$$R_r\left[\left|\mathbf{i}_{mm}\right| - \mathbf{i}_s^{\psi_m}\right] + L_r \frac{d\left|\mathbf{i}_{mm}\right|}{dt} = L_{r\sigma}\frac{d\mathbf{i}_s^{\psi_m}}{dt} - j(\omega_{mm} - \omega_r)\left[L_r\left|\mathbf{i}_{mm}\right| - L_{r\sigma}\mathbf{i}_s^{\psi_m}\right] \tag{4.83}$$

And by resolving Equations 4.48 and 4.49 into their direct and quadrature components, the following equations are obtained. For the stator equation,

$$i_{sx}^{\psi_m} = \frac{u_{sx}^{\psi_m}}{R_s} - T_{s\sigma}\frac{di_{sx}^{\psi_m}}{dt} - \frac{L_m}{R_s}\frac{d\left|\mathbf{i}_{mm}\right|}{dt} + \omega_{mm}T_{s\sigma}i_{sy}^{\psi_m} \tag{4.84a}$$

$$i_{sy}^{\psi_m} = \frac{u_{sx}^{\psi_m}}{R_s} - T_{s\sigma}\frac{di_{sy}^{\psi_m}}{dt} - \frac{\omega_{mm}\left(L_{s\sigma}i_{sx}^{\psi_m} + L_m\left|\mathbf{i}_{mm}\right|\right)}{R_s} \tag{4.84b}$$

and for the rotor equation,

$$\frac{\left|\mathbf{i}_{mm}\right| + T_r\, d\left|\mathbf{i}_{mm}\right|/dt}{T_{rl}} = \frac{di_{sx}^{\psi_m}}{dt} + \frac{i_{sx}^{\psi_m}}{T_{r\sigma}} - \omega_{sl}i_{sy}^{\psi_m} \tag{4.85a}$$

$$\omega_{sl}\left(\left|\mathbf{i}_{mm}\right|\frac{T_r}{T_{r\sigma}} - i_{sx}^{\psi_m}\right) = \frac{di_{sy}^{\psi_m}}{dt} + \frac{i_{sy}^{\psi_m}}{T_{r\sigma}} \tag{4.85b}$$

Both (4.84) and (4.85) have a coupling term, which requires a decoupling circuit in the magnetizing flux-linkage vector control.

4.7 Mathematical Dynamic Model of the IM Taking into Account Magnetic Saturation

This section shows the mathematical dynamic model of the IM which takes into account the magnetic saturation of the iron paths in the rotor flux-linkage reference frame [16–18]. As a matter of fact, some electric parameters of the IM, as the magnetizing inductance, as well as the stator and rotor inductances, and the leakage factors σ_s, σ_r, and σ vary with the magnetic saturation. In the following, the effect of the magnetizing inductance is considered by using the relationship linking the three-phase magnetizing inductance and the rotor magnetizing current, as defined in Section 4.6.1. The analysis is therefore made in the rotor flux-linkage reference frame, but it could not be made in any reference frame.

For this purpose, Equations 4.48 and 4.49 are used, that is, the machine stator and rotor equations after expressing them in the rotor flux-linkage reference frame. In Equations 4.48 and 4.49, the three-phase magnetizing inductance L_m varies with the magnetic saturation as well as $L_s = L_m + L_{s\sigma}$ and $L_r = L_m + L_{r\sigma}$. The stator and rotor leakage inductances $L_{s\sigma}$ and $L_{r\sigma}$ are considered constant and not affected therefore by the magnetic saturation.

The magnetizing inductance is assumed to be a nonlinear function of the rotor magnetizing current $\mathbf{i}_{mr}$ through the relationship $L_m = L_m(|\mathbf{i}_{mr}|)$. From Equation 4.55, L_m can be defined as the ratio between the amplitude of the rotor flux-linkage vector space and the amplitude of the rotor magnetizing current space-vector $\mathbf{i}_{mr}$. This results in the following:

$$L_m = \frac{|\boldsymbol{\psi}_r|}{|\mathbf{i}_{mr}|} \tag{4.86}$$

Remark that L_m is also called *static inductance*.

The *dynamic or incremental inductance L* is the derivative of the amplitude of the rotor flux-linkage vector space with respect to the amplitude of the rotor magnetizing current space-vector $\mathbf{i}_{mr}$, that is,

$$L = \frac{d|\boldsymbol{\psi}_r|}{d|\mathbf{i}_{mr}|} = \frac{d(L_m|\mathbf{i}_{mr}|)}{d|\mathbf{i}_{mr}|} \tag{4.87}$$

From (4.87), the following relationship results:

$$L = L_m + |\mathbf{i}_{mr}| \frac{dL_m}{d|\mathbf{i}_{mr}|} \tag{4.88}$$

From (4.88), it is apparent that under linear conditions, L_m and L coincide and are both independent of time.

Let also T_r^*, the "modified rotor time constant," be defined as follows:

$$T_r^* = T_r \frac{L}{L_m} = \frac{L}{R_r}\left(1 + \frac{L_{r\sigma}}{L_m}\right) \tag{4.89}$$

Equation 4.89 shows that under the linearity assumption, the modified rotor time constant coincides with the rotor time constant.

After algebraic manipulation of Equations 4.48 and 4.49, it is possible to obtain equations describing the behavior of the machine in magnetic saturation. Thanks to the definition of dynamic inductance and modified rotor time constant, these equations are formally analogous to Equations 4.60a and 4.60b, which are valid under the linear magnetic assumption. The following vector equations are therefore obtained [11–14]:

$$T_s'\frac{d\mathbf{i}_s^{\psi_r}}{dt} + \mathbf{i}_s^{\psi_r} = \frac{\mathbf{u}_s^{\psi_r}}{R_s} - \left\{ \frac{L - L_m}{R_s}\left[1 - \frac{L_{r\sigma}^2}{L_r^2\left(1 - \mathbf{i}_s^{\psi_r}/|\mathbf{i}_{mr}|\right)}\right] + \frac{L_m^2}{L_r^2 R_s}\right\}\frac{d|\mathbf{i}_{mr}|}{dt} +$$

$$- j\omega_{mr}\left[T_s'\mathbf{i}_s^{\psi_r} + \frac{L_m^2}{L_r R_s}|\mathbf{i}_{mr}|\right] \tag{4.90}$$

$$T_r^* \frac{d\left|\mathbf{i}_{mr}\right|}{dt} + \left|\mathbf{i}_{mr}\right| = \mathbf{i}_s^{\psi_r} - j(\omega_{mr} - \omega_r)T_r\left|\mathbf{i}_{mr}\right| \tag{4.91}$$

Stator Equation 4.90 shows that the magnetic saturation results not only in the variation of the transient time constant T_s' with respect to the rotor magnetizing current but also in an additive term given by

$$\Delta\mathbf{i}_s^{\psi_r} = \left\{ \frac{(L - L_m)}{R_s}\left[1 - \frac{L_{r\sigma}^2}{L_r^2\left(1 - \mathbf{i}_s^{\psi_r}/\left|\mathbf{i}_{mr}\right|\right)} \right] \right\} \frac{d\left|\mathbf{i}_{mr}\right|}{dt} \tag{4.92}$$

Under linearity conditions, this term $\Delta\mathbf{i}_s^{\psi_r}$ is null, and Equation 4.90 coincides with Equations 4.60a and b.

Rotor Equation 4.91 shows that magnetic saturation effect results not only in the variation of the rotor time constant T_r with respect to the rotor magnetizing current but also in the appearance of the modified rotor time constant T_r^*, which also varies with the rotor magnetizing current but differently from how T_r does. Actually T_r^* is linked both to the dynamic inductance L and the static inductance L_m, while T_r is only linked to the static inductance L_m. Under magnetic linearity conditions, $T_r^* = T_r$, and (4.91) coincides with Equation 4.60b.

Equations 4.91 and 4.92, resolved into the axes x and y of the rotor flux-linkage reference frame, yield the following:

$$\begin{cases} T_s'\dfrac{di_{sx}^{\psi_r}}{dt} + i_{sx}^{\psi_r} = \dfrac{u_{sx}^{\psi_r}}{R_s} + \omega_{mr}T_s'i_{sy}^{\psi_r} - (T_s - T_s')\dfrac{d\left|\mathbf{i}_{mr}\right|}{dt} + \Delta i_{sx}^{\psi_r} \tag{4.93a} \\[4mm] T_s'\dfrac{di_{sy}^{\psi_r}}{dt} + i_{sy}^{\psi_r} = \dfrac{u_{sy}^{\psi_r}}{R_s} - \omega_{mr}T_s'i_{sx}^{\psi_r} - (T_s - T_s')\omega_{mr}\left|\mathbf{i}_{mr}\right| + \Delta i_{sy}^{\psi_r} \tag{4.93b} \end{cases}$$

$$\begin{cases} T_r^*\dfrac{d\left|\mathbf{i}_{mr}\right|}{dt} + \left|\mathbf{i}_{mr}\right| = i_{sx}^{\psi_r} \tag{4.94a} \\[4mm] \omega_{mr} = \omega_r + \dfrac{i_{sy}^{\psi_r}}{T_r\left|\mathbf{i}_{mr}\right|} \tag{4.94b} \end{cases}$$

where

$$\begin{cases} \Delta i_{sx}^{\psi_r} = \left\{ (L_m - L)\dfrac{\left[1 - \left(L_{r\sigma}^2/L_r^2\right)\left(1 - i_{sx}^{\psi_r}/\left|\mathbf{i}_{mr}\right|\right)\right]}{R_s} \right\}\dfrac{d\left|\mathbf{i}_{mr}\right|}{dt} \tag{4.95a} \\[4mm] \Delta i_{sy}^{\psi_r} = \dfrac{(L_m - L)L_{r\sigma}^2 i_{sy}^{\psi_r}}{L_r^2 R_s\left|\mathbf{i}_{mr}\right|}\dfrac{d\left|\mathbf{i}_{mr}\right|}{dt} \tag{4.95b} \end{cases}$$

The electromagnetic torque in magnetic saturation conditions is the same as the one given by Equation 4.63, taking into account of the variations of L_m and L_r with $\mathbf{i}_{mr}$.

Figure 4.5 shows the block diagram of the mathematical model of the IM including iron path saturation effects. It includes Equations 4.93a and 4.93b, 4.94a and 4.94b), and the

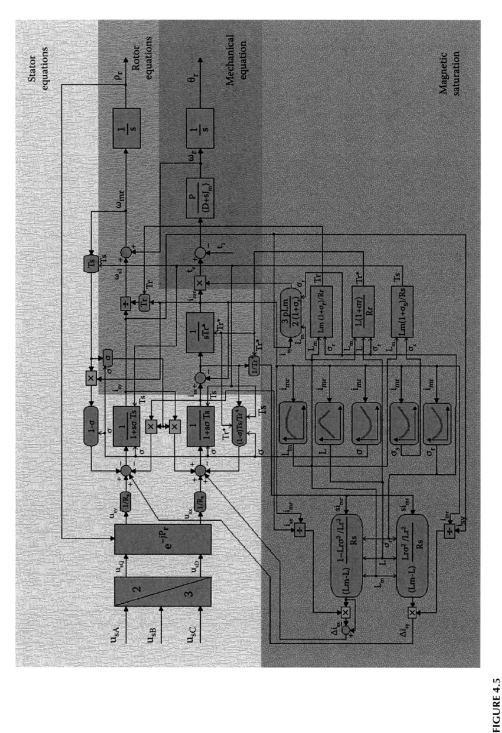

FIGURE 4.5
(See color insert.) Block diagram of the mathematical model of the IM including iron path saturation effects. (From Cirrincione, M. et al., *IEEE Trans. Ind. Electron.*, 52(5), October 2005.)

equation of motion (4.23). For numerical simulation purposes the effect of the magnetic saturation can be taken into account by lookup tables linearly interpolated, which give the nonlinear relationship between the electric parameters L_m, L, σ, σ_s, and σ_r and the rotor magnetizing current space-vector.

4.8 Steady-State Space-Vector Model of the IM

The steady-state stator and rotor equations of the IM and the corresponding space-vector equivalent circuit, describing the steady-state operation of the machine under symmetrical three-phase sinusoidal supply at constant pulsation ω_1, can be deduced from the previously illustrated space-vector dynamic model.

Starting from the space-vector dynamic equations written in the stationary reference frame in Equations 4.24 and 4.25, the following steady-state equations can be deduced. Passing from the dynamic to the steady-state analysis, the derivation operator becomes $d(\ )/dt = j\omega_1(\ )$. On this basis, the stator and rotor equations become as follows [11]:

$$\begin{cases} \mathbf{U}_s = R_s \mathbf{I}_s + j\omega_1 \Psi_s \\ \mathbf{0} = R_r \mathbf{I}'_r + j(\omega_1 - \omega_r)\Psi'_r = R_r \mathbf{I}'_r + js\omega_1 \Psi'_r \end{cases} \tag{4.96}$$

where $s = (\omega_1 - \omega_r)/\omega_1$ is the slip and all the symbols in upper case mean the steady-state values of the corresponding variables written in lower case.

Correspondingly, the relationships between the flux linkages and the stator and rotor currents are as follows:

$$\begin{cases} \Psi_s = L_s \mathbf{I}_s + L_m \mathbf{I}'_r \\ \Psi'_r = L_m \mathbf{I}_s + L_r \mathbf{I}'_r \end{cases} \tag{4.97}$$

Equations 4.96 and 4.97 can be implemented by the space-vector equivalent circuit in Figure 4.6, once the stator and rotor leakage inductances are defined as $L_{s\sigma} = L_s - L_m$ and $L_{r\sigma} = L_r - L_m$.

The equivalent circuit drawn in Figure 4.6 can be suitably used to obtain the expression of the steady-state electromagnetic torque as well as some power components. To do that,

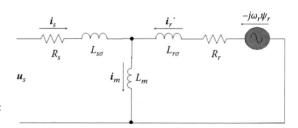

FIGURE 4.6
Steady-state space-vector equivalent circuit of the IM.

the air-gap power P_{gap} is expressed as the difference between the input power P_1 and the stator joule losses P_s:

$$P_{gap} = P_1 - P_s = 3\Re(\mathbf{U}_s^*\mathbf{I}_s) - 3R_s|\mathbf{I}_s|^2 \tag{4.98}$$

where * denotes the complex conjugate operator. The air-gap power is balanced by the sum of the mechanical power P_{mech} and the rotor joule losses P_r:

$$P_{gap} = P_{mech} + P_r = P_{mech} + 3R_r|\mathbf{I}_r'|^2 \tag{4.99}$$

The steady-state electromagnetic torque T_e can be expressed on the basis of the air-gap power as follows:

$$T_e = \frac{P_{gap}}{\omega_1} \tag{4.100}$$

The air-gap power and the rotor joules losses are in relationship as follows:

$$P_r = 3R_r|\mathbf{I}_r'|^2 = T_e s\omega_1 = sP_{gap} \tag{4.101}$$

The electromagnetic torque T_e can be therefore written as follows:

$$T_e = \frac{3R_r|\mathbf{I}_r'|^2}{s\omega_1} \tag{4.102}$$

To express the electromagnetic torque as a function of the supply voltage, the following relationship between the rotor current and the stator voltage can be deduced from the equivalent circuit in Figure 4.6:

$$\mathbf{I}_r' = \frac{\mathbf{U}_s}{(1+\sigma_s)\left\{(R_s + R_r/s) + j\left(\dfrac{X_{s\sigma}}{(1-\sigma_s)} + X_{r\sigma}'\right)\right\}} \tag{4.103}$$

where $X_{s\sigma} = j\omega_1 L_{s\sigma}$ and $X_{r\sigma}' = j\omega_1 L_{r\sigma}$ are, respectively, the stator and rotor leakage reactances and $\sigma_s = X_{s\sigma}/X_m$ is the stator leakage factor.

The following expression can finally be found [11]:

$$T_e = \frac{3}{\omega_1} \frac{|\mathbf{U}_1|^2}{(1+\sigma_s)^2} \frac{R_r/s}{(R_s + R_r/s)^2 + X'^2} \tag{4.104}$$

where $X' = X_{r\sigma}' + (X_{s\sigma}/1+\sigma_s)$ is the transient rotor reactance.

The maximum air-gap electromagnetic torque can be developed by the machine at a specific pulsation slip s_{max}, equal to the following:

$$s_{max} = \frac{\pm R_r}{\left(R_s^2 + X'^2\right)^{1/2}} \tag{4.105}$$

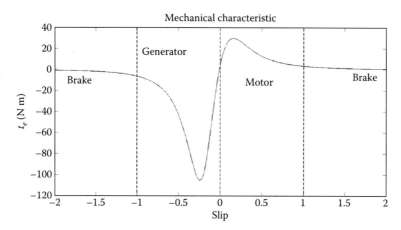

FIGURE 4.7
Steady-state characteristic torque versus slip of a 2.2 kW IM.

In correspondence to such a pulsation slip, the maximum electromagnetic torque is equal to the following:

$$T_{e\,max} = \frac{(3/\omega_1)\left(|U_s|/1+\sigma_s\right)^2}{2\left[R_s \pm \left(R_s^2 + X'^2\right)^{1/2}\right]} \tag{4.106}$$

where the positive (negative) sign is valid in case of motoring (generating) operation.

From the earlier discussion, it can be deduced that the maximum value of the electromagnetic torque is proportional to the square of the stator voltage amplitude. Moreover, in absolute terms, the maximum torque as generator is bigger than that as motor. The physical reason is that in generating operation, because of the voltage drop of the stator current on the stator impedance, the resulting back emf is higher. Correspondingly, even the stator flux amplitude in generating mode will be higher. This difference, which is due to the non-null value of the stator resistance, can be usually neglected in large machines.

Figure 4.7 shows the steady-state torque versus slip characteristic of a 2.2 kW, 4 poles IM (parameters in Table 4.1) obtained under a 220 V RMS, 50 Hz supply. The three slip

TABLE 4.1

Parameters of the 2.2 kW IM

Rated power, P_{rated} [kW]	2.2
Rated voltage, U_{rated} [V]	220
Rated frequency, f_{rated} [Hz]	50
Rated speed [rad/s]	149.75
Pole pairs	2
Stator resistance, R_s [Ω]	3.88
Stator inductance, L_s [mH]	252
Rotor resistance, R_r [Ω]	1.87
Rotor inductance, L_r [mH]	252
Three-phase magnetizing inductance, L_m [mH]	236
Moment of inertia, J [kg · m²]	0.0266

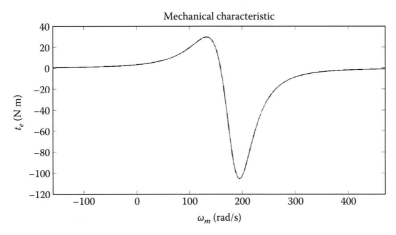

FIGURE 4.8
Steady-state characteristic torque versus speed of a 2.2 kW IM.

intervals corresponding to the behavior of the machine as motor, generator, and brake can be clearly observed. Moreover, the difference between the maximum torques as generator and motor could be noted as well. Figure 4.8 shows the corresponding characteristic torque versus rotational speed.

Figure 4.9 shows the steady-state current versus speed characteristic of the same machine. It can be observed that the maximum current is absorbed when the machine is at standstill, decreasing with the speed down to the no-load current at slip equal to 0.

Figure 4.10 shows the steady-state percent efficiency versus slip of the same machine. As expected, the maximum efficiency (about 70%) is at slip null, decreasing with the slip down to about 45% at the pullout torque slip.

Finally, Figure 4.11 shows the *Heyland* diagram of the same machine, describing the steady-state trajectory of the stator current space-vector in the complex plane, according to the variation of the slip.

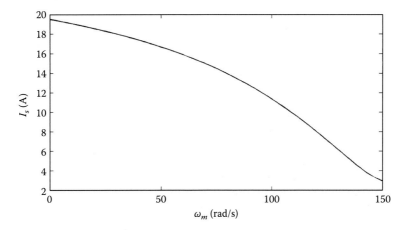

FIGURE 4.9
Steady-state characteristic stator current versus speed of a 2.2 kW IM.

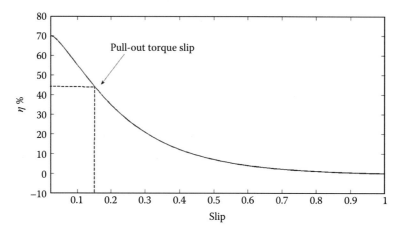

FIGURE 4.10
Steady-state characteristic efficiency versus speed of a 2.2 kW IM.

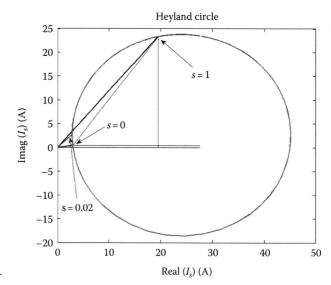

FIGURE 4.11
Heyland diagram of a 2.2 kW IM.

4.9 Experimental Validation of the Space-Vector Model of the IM

In this section, the results of the numerical simulation of the previously described space-vector model of the IM are shown, in comparison with the corresponding experimental results in order to validate the numerical model [19]. The software adopted for the numerical simulation is MATLAB®–Simulink® of Mathworks. Both the classic dynamic model considering the machine linear from the point of view of its magnetic characteristic (Figure 4.4) and the model taking into consideration the magnetic saturation in the iron core (Figure 4.5) have been implemented and tested.

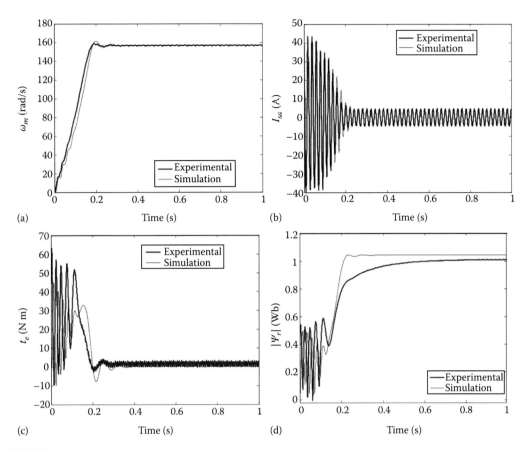

FIGURE 4.12
(a) Speed of the IM during the start-up at no load. (b) Stator current of the IM during the start-up at no load. (c) Electromagnetic torque of the IM during the start-up at no load. (d) Rotor flux-linkage amplitude of the IM during the start-up at no load.

With regard to the classic dynamic model considering the machine linear from the point of view of its magnetic characteristic, the parameters of the machine under test are shown in Table 4.1.

A start-up of the IM under a sinusoidal voltage supply 220 V RMS, 50 Hz at no load has been simulated. Figure 4.12a shows the time simulated and experimental waveform of the machine speed during the start-up. Figure 4.12b through d shows the corresponding time waveforms of the absorbed stator phase current of the electromagnetic torque and of the rotor flux-linkage amplitude. All these figures show a good matching between the simulated and experimental waveforms. Finally, Figure 4.13 shows the dynamic torque versus speed characteristics of the machine during the start-up. Even in this case, there is a good matching between the two curves.

With regard to the dynamic model taking into consideration the magnetic saturation in the iron core, the parameters of the machine under test are shown in Table 4.2. Figure 4.14 shows the magnetic characteristic of the machine, rotor magnetizing current i_{mr} versus rotor flux-linkage amplitude ψ_r. Figure 4.15 shows the corresponding waveforms of the static L_m and dynamic L inductances of the machine, as well as the stator σ_s, rotor σ_r, and global σ leakage factors as a function of the rotor magnetizing current. In the following, the numerical simulations of the model taking into consideration the magnetic saturation

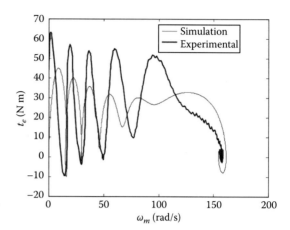

FIGURE 4.13
Torque versus speed characteristic of the IM during the start-up at no load.

TABLE 4.2

Parameters of the 22 kW IM

Rated power, P_{rated} [kW]	22
Rated voltage, U_{rated} [V]	220
Rated frequency, f_{rated} [Hz]	50
Rated speed [rad/s]	150
Pole pairs	2
Stator resistance, R_s [Ω]	0.18
Stator inductance, L_s [mH]	44.8×10^{-3}
Rotor resistance, R_r [Ω]	0.26
Rotor inductance, L_r [mH]	45.6×10^{-3}
Three-phase magnetizing inductance, L_m [mH]	43.7×10^{-3}
Moment of inertia, J [kg·m²]	0.19

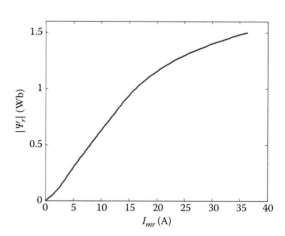

FIGURE 4.14
Magnetization curve of the IM.

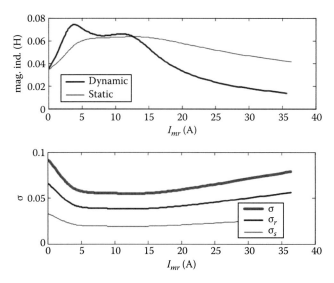

FIGURE 4.15

L_m, L, σ_s, σ_r, σ versus i_{mr}.

and the model not taking the saturation into consideration are compared. With regard to the linear model, the parameters of the machine are assumed to be constant and equal to those in Table 4.2.

A start-up of the IM under a sinusoidal voltage supply 220 V RMS, 50 Hz at no load has been simulated, with a step torque insertion of 90 N m at 0.7 s. Figure 4.16a shows the time simulated waveforms of the machine speed during the start-up as obtained with the two models. Figure 4.16b through e shows the corresponding time waveforms of the absorbed stator phase current, of the electromagnetic torque, of the rotor flux-linkage amplitude, and of the rotor magnetizing current amplitude. Figure 4.17 shows the dynamic torque versus speed characteristics of the machine during the start-up. A good matching between the two models could be noted; it could also be observed how the magnetic saturation of the iron core is cause of a delay in the time response of the machine. Moreover, while the rotor flux amplitude is equal with both models, the rotor magnetizing current obtained with the model taking into consideration the saturation is lower. This is due to fact that, for the adopted supply voltage, the steady-state working point on the magnetic characteristic corresponds lower than the intersection point between the nonlinear and linear magnetic characteristic (Figure 4.18).

4.10 IM Model Including Slotting Effects

The previously described dynamic model of the IM is based on the basic assumption that only the fundamental of the mmf is taken into consideration. This model cannot thus consider several aspects of a real IM, among which the distributed windings, stator and rotor slotting effects, slot shapes, etc. [22–24]. There is, on the contrary, a need of models dealing with some magnetic saliencies of the machine. It arises from the increasing trend to try to estimate the machine speed/position by tracking one or some of its magnetic

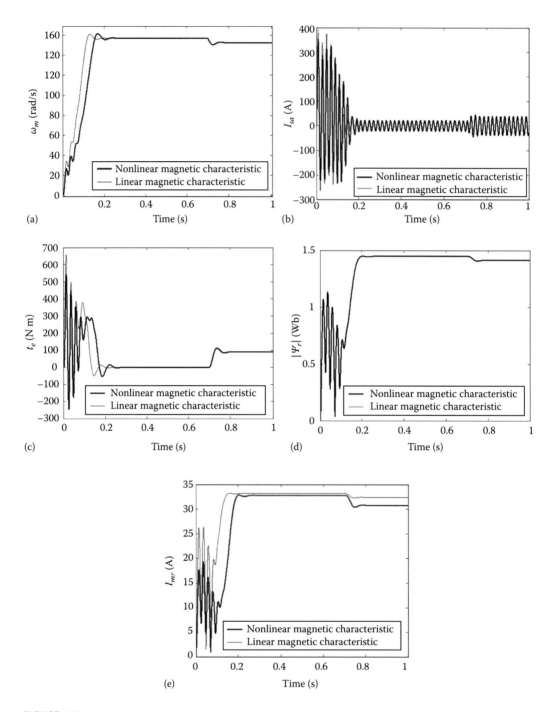

FIGURE 4.16
(a) Speed of the IM during the start-up at no load. (b) Stator current of the IM during the start-up at no load. (c) Electromagnetic torque of the IM during the start-up at no load. (d) Rotor flux-linkage amplitude of the IM during the start-up at no load. (e) Rotor magnetizing current amplitude of the IM during the start-up at no load.

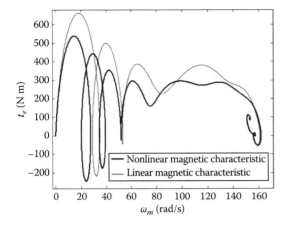

FIGURE 4.17
Torque versus speed characteristic of the IM during the start-up at no load.

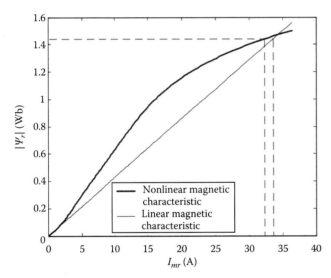

FIGURE 4.18
Linear and nonlinear magnetization curve of the IM.

saliencies [25–27]. Among the others, the main ones are certainly the saturation of the main flux and the so-called rotor slotting effects. For this purpose, the principal slot harmonics (PSH) of some electric variables of the machine (stator current, common-mode voltages or currents) should be detected and suitably processed. However, to take into consideration the slotting effects, only simplified models have been used so far for control applications; in any case, these models are valid only under a high frequency supply (signal injection techniques or test vectors); they consider only the variation of the leakage stator inductance with the rotor slotting and cannot be used at the fundamental frequency supply. They therefore present some important limitations if used to test sensorless algorithms. An attempt to develop a space-vector dynamic model of the IM considering the slotting effect has been done in Refs [28,29]: this model considers a variation of the mutual inductance with the stator/rotor slotting but does not develop a space-vector state formulation, useful for control applications.

This paragraph presents, in fact, a space-vector dynamic model of the IM including stator and rotor effects [20,30]. In accordance with Refs. [26,27], the slotting effect has been considered only in the stator and rotor leakage variations.

Model assumptions

The IM space-vector model including slotting effects is based on the following assumptions, which extend those used in Section 4.2:

- Infinite permeability of iron
- Radial direction of flux density in the air-gap
- Concentrated stator and rotor windings, shifted 120° from one another
- Stator windings with a star connection
- Stator and rotor slotting effects taken into consideration in the stator and rotor leakage inductance variations

The model has been developed for a machine with one pole pair, but obviously the results can be easily extended to multi-pole-pair machines.

Comments on the model assumptions

This model is first derived considering both the stator and rotor slotting effects. Then, for the space-state representation, only the rotor slotting effects have been considered (scheme in Figure 4.19a). This first simplification arises from the practical consideration that IMs are usually designed and constructed with a number of stator slots per pole pair which is a multiple of three, and therefore, no stator slot harmonics (SSHs) are observable on the stator current.

The second simplification is that the stator and rotor slotting effects have been considered only in the stator and rotor leakage inductance variations. More rigorously, also the mutual inductance should vary according to the rotor slotting effect. However, this simplification has been suggested by the fact that it permits a significant reduction of the complexity of the resulting space formulation without implying an important reduction of the model accuracy. As a matter of fact, it can be observed that, if the rotor slotting effect is considered also in the mutual inductance, a set of terms appear which are proportional to the second and third powers of the slotting inductance L_h: all these terms are practically negligible, and in any case, the result of this simplification can be easily evaluated. The comparison of simulation and experimental results in Ref. [20] shows that the rotor slot harmonics (RSHs) obtained by adopting the model are exactly those obtainable theoretically

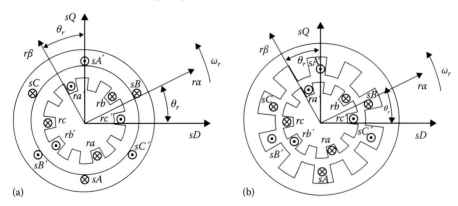

FIGURE 4.19
Schematics of the IM including rotor slotting effects (a) and stator and rotor slotting effects (b) with concentrated windings. (From Cirrincione, M. et al., *IEEE Trans. Ind. Appl.*, 44, 1683, 2008; Cirrincione, M. et al., Space-vector state model of induction machines including rotor and stator slotting effects, in *IEEE IEMDC '07*, Antalya, Turkey, pp. 673–682, May 3–5, 2007.)

from literature and in agreement with those measured experimentally. Moreover, these harmonics shift with frequency as expected. This means that, as a worst case, a slight phase angle error in the stator slotting current could be caused by the adopted simplification, which is an acceptable compromise to avoid a huge increase in the complexity of the model, especially in view of devising an observer based on this model.

4.10.1 Space-Vector Model of the IMs Including Stator and Rotor Slotting Effects

If the slotting effects influence only the stator and rotor leakage inductances, the AC motor inductances can be written as follows for each phase:

$$
\begin{cases}
\bar{\mathbf{L}}_s = \bar{L}_{\sigma s}\mathbf{U} + \bar{L}_{sm}\mathbf{S} + \bar{L}_{mh}\mathbf{C}(q_r\theta_r) + \bar{L}_{mh}\mathbf{C}(q_s\theta_r) \\
\bar{\mathbf{L}}_r = \bar{L}_{\sigma r}\mathbf{U} + \bar{L}_{rm}\mathbf{S} + \bar{L}_{mh}\mathbf{C}^T(q_r\theta_r) + \bar{L}_{mh}\mathbf{C}^T(q_s\theta_r) \\
\bar{\mathbf{L}}_{sr} = \bar{L}_{sr}\mathbf{C}^T(\theta_r) \\
\bar{\mathbf{L}}_{rs} = \bar{L}_{sr}\mathbf{C}(\theta_r)
\end{cases}
\tag{4.107}
$$

where
$\bar{\mathbf{L}}_s$ and $\bar{\mathbf{L}}_r$ are, respectively, the vector of the stator and of the rotor self-inductances of the three phases of the motor
$\bar{\mathbf{L}}_{rs}$ and $\bar{\mathbf{L}}_{sr}$ are the mutual inductance matrices
$\bar{L}_{\sigma s}$ is the mean stator leakage inductance
$\bar{L}_{\sigma r}$ is the mean rotor leakage inductance
$\bar{L}_{sm}$ is the maximum mutual inductance between two stator phases
$\bar{L}_{rm}$ is the maximum mutual inductance between two rotor phases
$\bar{L}_{sr}$ is the maximum mutual inductance between one phase of the rotor and one phase of the stator
$\bar{L}_{mh}$ is the slot inductance
q_s is the number of stator slots per pole pair
q_r is the number of rotor slots per pole pair
θ_r is the angular position of rotor a-phase with respect to stator a-phase (positive if counterclockwise)

$$
\mathbf{S} = \begin{bmatrix} 1 & -1/2 & -1/2 \\ -1/2 & 1 & -1/2 \\ -1/2 & -1/2 & 1 \end{bmatrix}
\quad
\mathbf{U} = \begin{bmatrix} 1 & 0 & 0 \\ 0 & 1 & 0 \\ 0 & 0 & 1 \end{bmatrix}
$$

$$
\mathbf{C}(k\vartheta_r) = \begin{bmatrix} \cos(k\vartheta_r) & \cos(k(\vartheta_r - 4\pi/3)) & \cos(k(\vartheta_r - 2\pi/3)) \\ \cos(k(\vartheta_r - 2\pi/3)) & \cos(k\vartheta_r) & \cos(k(\vartheta_r - 4\pi/3)) \\ \cos(k(\vartheta_r - 4\pi/3)) & \cos(k(\vartheta_r - 2\pi/3)) & \cos(k\vartheta_r) \end{bmatrix}
\quad \text{where } k = 1, q_s, q_r
$$

The values that can be taken on by q_s and q_r are $3n$, $3n + 1$, or $3n - 1$ with $n \in N$. Thus, nine possible different models in function of the stator and rotor slots per pole are possible.

It should be noted that in Equation 4.107, the mutual inductance matrices $\bar{\mathbf{L}}_{rs}$ and $\bar{\mathbf{L}}_{sr}$ vary with the rotor position for the normal coupling due to the rotation (no q_r term appears in the mutual inductance).

If the following phase vectors are introduced for the stator flux linkages, rotor flux linkages, stator currents, and rotor currents

$$\boldsymbol{\psi}_{fs} = \begin{pmatrix} \psi_{sa} \\ \psi_{sb} \\ \psi_{sc} \end{pmatrix}, \quad \boldsymbol{\psi}_{fr} = \begin{pmatrix} \psi_{ra} \\ \psi_{rb} \\ \psi_{rc} \end{pmatrix}, \quad \boldsymbol{i}_{fs} = \begin{pmatrix} i_{sa} \\ i_{sb} \\ i_{sc} \end{pmatrix}, \quad \boldsymbol{i}_{fr} = \begin{pmatrix} i_{ra} \\ i_{rb} \\ i_{rc} \end{pmatrix}$$

then the following flux-linkage equations can be written in phase coordinates:

$$\begin{cases} \boldsymbol{\psi}_{fs} = \overline{\mathbf{L}}_s \boldsymbol{i}_{fs} + \overline{\mathbf{L}}_{rs} \boldsymbol{i}_{fr} \\ \boldsymbol{\psi}_{fr} = \overline{\mathbf{L}}_r \boldsymbol{i}_{fr} + \overline{\mathbf{L}}_{sr} \boldsymbol{i}_{fs} \end{cases} \tag{4.108}$$

Now it should be recalled that, to obtain the space-vector form of a variable from its phase coordinates, it is enough to make the scalar product between the phase vector and the following complex vector:

$$\mathbf{a} = \frac{2}{3} \begin{pmatrix} 1 \\ a \\ a^2 \end{pmatrix} \quad \text{where } a = e^{j(2\pi/3)} \text{ and } a^2 = e^{j(4\pi/3)}$$

Thus, by pre-multiplying (4.108) by $\mathbf{a}^T$ and remembering that

$$\text{If } k = 3n: \mathbf{a}^T \mathbf{C}(k\vartheta_r) = \mathbf{a}^T \mathbf{C}(k\vartheta_r)^T = 0$$

$$\text{If } k = 3n+1: \mathbf{a}^T \mathbf{C}(k\vartheta_r) = \frac{3}{2} e^{jk\vartheta_r} \mathbf{a}^T$$

$$\mathbf{a}^T \mathbf{C}(k\vartheta_r)^T = \frac{3}{2} e^{-jk\vartheta_r} \mathbf{a}^T$$

$$\text{If } k = 3n-1: \mathbf{a}^T \mathbf{C}(k\vartheta_r) = \frac{3}{2} e^{-jk\vartheta_r} \mathbf{a}^T$$

$$\mathbf{a}^T \mathbf{C}(k\vartheta_r)^T = \frac{3}{2} e^{jk\vartheta_r} \mathbf{a}^T$$

Then the following equations are obtained:

$$\boldsymbol{\psi}_s = (L_s + L_h e^{\mp jq_r \theta_r} + L_h e^{\mp jq_s \theta_r}) \mathbf{i}_s + L_m \mathbf{i}'_r$$

$$\boldsymbol{\psi}_r = L_m \mathbf{i}'_s + (L_r + L_h e^{\pm jq_r \theta_r} + L_h e^{\pm jq_s \theta_r}) \mathbf{i}_r \tag{4.109}$$

where

the upper sign refers to the case $q_r = 3n - 1$, $q_s = 3m - 1$ and the lower one to the case $q_r = 3n + 1$. $q_s = 3m + 1$, with $n,m \in N$

$\mathbf{i}'_r$ and $\mathbf{i}_r$ are the rotor current space-vectors, respectively, in the stator and rotor reference frame

$\mathbf{i}_s$ and $\mathbf{i}'_s$ are the stator current space-vectors, respectively, in the stator and rotor reference frame

$\boldsymbol{\Psi}_s$ is the stator flux-linkage space-vector in the stator reference frame

$\boldsymbol{\Psi}'_r$ and $\boldsymbol{\Psi}_r$ are the rotor flux-linkage space-vectors, respectively, in the stator and rotor reference frame

Note that $\mathbf{i}'_r = e^{j\vartheta_r}\mathbf{i}_r$, $\mathbf{\Psi}'_r = e^{j\vartheta_r}\mathbf{\Psi}_r$, and $\mathbf{i}'_s = e^{-j\vartheta_r}\mathbf{i}_s$

while

$L_s = \bar{L}_{\sigma s} + \dfrac{3}{2}\bar{L}_{sm}$ is the total three-phase stator inductance without slotting

$L_r = \bar{L}_{\sigma r} + \dfrac{3}{2}\bar{L}_{rm}$ is the total three-phase rotor inductance without slotting

$L_m = \dfrac{3}{2}\bar{L}_{sr}$ is the three-phase magnetizing inductance

$L_h = \dfrac{3}{2}\bar{L}_{mh}$ is the three-phase slot inductance

4.10.2 Space-Vector State Model of IM Including Rotor Slotting Effects

Starting from Equations 4.109, if only the rotor slotting effects are considered ($q_s = 3m$), which is the most usual case, the following equations can be deduced where all rotor quantities are transformed into the stator reference frame:

$$\mathbf{\Psi}_s = (L_s + L_h e^{\mp jq_r\theta_r})\mathbf{i}_s + L_m\mathbf{i}'_r$$
$$\mathbf{\Psi}'_r = L_m\mathbf{i}_s + (L_r + L_h e^{\pm jq_r\theta_r})\mathbf{i}'_r \tag{4.110}$$

After substituting the expression of $\mathbf{i}'_r$ from the equation of the rotor flux linkage into that of the stator flux linkage, the following equation is deduced:

$$\mathbf{\Psi}_s = \left(L_s + L_h e^{\mp jq_r\theta_r} - \frac{L_m^2}{L_r + L_h e^{\pm jq_r\theta_r}}\right)\mathbf{i}_s + \left(\frac{L_m}{L_r + L_h e^{\pm jq_r\theta_r}}\right)\mathbf{\Psi}'_r \tag{4.111}$$

A modified global leakage factor including rotor slotting effects can be defined in this way:

$$\sigma^{slot} = \Re_e\left(1 - \frac{L_m^2}{\left(L_r + L_h e^{\pm jq_r\theta_r}\right)\left(L_s + L_h e^{\mp jq_r\theta_r}\right)}\right) \tag{4.112}$$

After computing the time derivative of Equation 4.111, considering that $d\vartheta_r/dt = \omega_r$ and substituting this derivative into the stator equation of the IM

$$\mathbf{u_s} = R_s\mathbf{i_s} + \frac{d\mathbf{\Psi}_s}{dt} \tag{4.113}$$

where R_s is the resistance of a stator winding, the following expression can be deduced:

$$\frac{d\mathbf{\Psi}'_r}{dt} = \frac{L_r + L_h e^{\pm jq_r\theta_r}}{L_m}\left\{\mathbf{u_s} - \left[R_s \pm j\omega_r\left(\frac{q_r L_m L_h}{\left(L_r + L_h e^{\pm jq_r\theta_r}\right)^2}e^{\pm jq_r\theta_r} \mp q_r L_h e^{\mp jq_r\theta_r}\right)\right]\mathbf{i_s}\right.$$

$$\left. - \left[L_s + L_h e^{\mp jq_r\theta_r} - \frac{L_m^2}{L_r + L_h e^{\pm jq_r\theta_r}}\right]\frac{d\mathbf{i_s}}{dt} \pm j\omega_r\frac{q_r L_m L_h}{\left(L_r + L_h e^{\pm jq_r\theta_r}\right)^2}e^{\pm jq_r\theta_r}\mathbf{\Psi}'_r\right\} \tag{4.114}$$

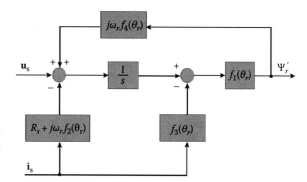

FIGURE 4.20
Block diagram of the "voltage" flux model including rotor slotting effects (plotted for the case $q_r = 3n - 1$). (From Cirrincione, M. et al., *IEEE Trans. Ind. Appl.*, 44, 1683, 2008.)

which can be also written in a simplified way:

$$\frac{d\mathbf{\Psi}'_r}{dt} = f_1(\vartheta_r)\left\{\mathbf{u_s} - [R_s \pm j\omega_r f_2(\vartheta_r)]\mathbf{i_s} - f_3(\vartheta_r)\frac{d\mathbf{i_s}}{dt} \pm j\omega_r f_4(\vartheta_r)\mathbf{\Psi}'_r\right\} \tag{4.115}$$

where the complex functions $f_i(\vartheta_r)$, with $i = 1, 2, 3, 4$, are obviously defined. Equation 4.115 gives a new definition of the so-called *"voltage"* flux model of the IM in the stator reference frame, including the rotor slotting effects. Figure 4.20 shows the block diagram of this flux model. It should be observed that in the classical "voltage" flux model, no feedback exists between the output and the input of the model (open-loop integration) and the rotor flux does not depend on the rotor speed; in contrast to it, the model including rotor slotting effects presents both a feedback term depending on the rotor speed and on the vector function $f_4(\vartheta_r)$ and a forward term depending also on the rotor speed and the function $f_2(\vartheta_r)$. If $L_h = 0$, this model coincides with the classical one.

The *"current"* flux model, based on the rotor equations of the machine in the stator reference frame including the rotor slotting effects, can be similarly derived as follows:

$$\frac{d\mathbf{\Psi}'_r}{dt} = \frac{R_r L_m}{L_r + L_h e^{\pm jq_r\theta_r}}\mathbf{i_s} + \left(j\omega_r - \frac{R_r}{L_r + L_h e^{\pm jq_r\theta_r}}\right)\mathbf{\Psi}'_r \tag{4.116}$$

Figure 4.21 shows the block diagram of the corresponding "current" flux model.

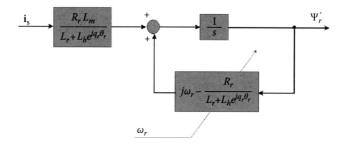

FIGURE 4.21
Block diagram of the "current" flux model in the stator reference frame including rotor slotting effects (plotted for the case $q_r = 3n - 1$). (From Cirrincione, M. et al., *IEEE Trans. Ind. Appl.*, 44, 1683, 2008.)

A modified rotor time constant including rotor slotting effects can be defined as follows:

$$T_r^{slot} = \Re_e \left(\frac{L_r + L_h e^{\pm jq_r\theta_r}}{R_r} \right) = \frac{L_r + L_h}{R_r}$$

The *"current"* flux model including the rotor slotting effects can be derived also in the rotor-flux-oriented reference frame. This last model is particularly employed for flux estimation in FOC since it does not require open-loop integration. Since $\Psi_r^{\psi_r} = \Psi_r' e^{-\rho_r}$, where $\Psi_r^{\psi_r}$ is the rotor flux vector in the rotor-flux-oriented reference frame and ρ_r is its angle with respect to the sD axis, from (4.116), the following space-vector rotor equation is obtained:

$$\frac{d\Psi_r^{\psi_r}}{dt} = \frac{R_r L_m}{L_r + L_h e^{\pm jq_r\theta_r}} i_s^{\psi_r} - \frac{R_r}{L_r + L_h e^{\pm jq_r\theta_r}} \Psi_r^{\psi_r} - j\omega_{sl}\Psi_r^{\psi_r} \tag{4.117}$$

where $i_s^{\psi_r}$ is the stator current space-vector in the rotor-flux-oriented reference frame and $\omega_{sl} = \omega_{mr} - \omega_r$ is the slip pulsation, the difference between the speed of the rotor flux vector and the rotor speed (in electric angles).

Equation 4.117 can be split into the direct (x) and quadrature (y) axis so that the following equation can be obtained:

$$\begin{cases} \dfrac{d\left|\Psi_r^{\psi_r}\right|}{dt} = \dfrac{R_r L_m}{L_r + L_h e^{\pm jq_r\theta_r}} i_{sx}^{\psi_r} - \dfrac{R_r}{L_r + L_h e^{\pm jq_r\theta_r}} \left|\Psi_r^{\psi_r}\right| & \text{(4.118a)} \\[4mm] \omega_{sl} = \dfrac{R_r L_m}{L_r + L_h e^{\pm jq_r\theta_r}} \dfrac{i_{sy}^{\psi_r}}{\left|\Psi_r^{\psi_r}\right|} & \text{(4.118b)} \end{cases}$$

Figure 4.22 shows the block diagram of the corresponding "current" flux model in the rotor-flux-oriented reference frame.

Equations 4.118a and 4.118b and Figure 4.22 show that, even if the rotor flux vector depends directly on the stator and rotor currents, it usually does not exhibit significant PSH components caused by the slotting effect because the amplitude of the rotor flux vector is low pass filtered by a first-order system with a time constant equal to T_r^{slot}, whose inverse is much lower than the lower PSH.

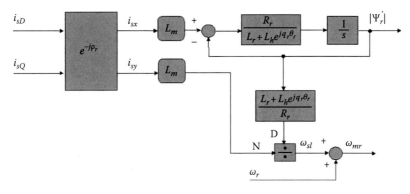

FIGURE 4.22
Block diagram of the "current" flux model in the rotor-flux-oriented reference frame including rotor slotting effects (plotted for the case $q_r = 3n - 1$). (From Cirrincione, M. et al., *IEEE Trans. Ind. Appl.*, 44, 1683, 2008.)

Equations 4.118a and 4.118b show also that, in rotor flux coordinates, field orientation always causes decoupling of the flux and torque current components. Remark, however, that, differently from the classical model, in steady-state, the amplitude of rotor fluxes is not constant any longer because of the occurrence of slotting higher order harmonics varying in time with the position of the rotor.

4.10.3 Space-State Model of IM Including Rotor Slotting Effects

To retrieve the full-state space-vector model of the IM including rotor slotting effects, the first complex equation is derived from Equation 4.114, rewritten as a function of di_s/dt, where $d\Psi'_r/dt$ is obtained from the *"current"* model (4.116). The second complex equation is the *"current"* model. The obtained full-state space-vector equation is therefore as follows:

$$\frac{d}{dt}\begin{bmatrix} \mathbf{i}_s \\ \mathbf{\psi}'_r \end{bmatrix} = \begin{bmatrix} \mathbf{A}_{11} & \mathbf{A}_{12} \\ \mathbf{A}_{21} & \mathbf{A}_{22} \end{bmatrix}\begin{bmatrix} \mathbf{i}_s \\ \mathbf{\psi}'_r \end{bmatrix} + \begin{bmatrix} \mathbf{B}_1 \\ 0 \end{bmatrix}\mathbf{u}_s = \mathbf{Ax} + \mathbf{Bu}_s \tag{4.119a}$$

$$\mathbf{i}_s = \mathbf{Cx} \tag{4.119b}$$

where $\mathbf{C} = [\mathbf{I} \quad 0]$ and

$$\mathbf{A}_{11} = \frac{1}{L_s + L_h e^{\mp jq_r\theta_r} - (L_m^2/L_r + L_h e^{\pm jq_r\theta_r})}$$
$$\times \left\{ -R_s - \frac{R_r L_m^2}{(L_r + L_h e^{\pm jq_r\theta_r})^2} \pm j\omega_r\left[q_r L_h e^{\mp jq_r\theta_r} - \frac{q_r R_r L_m L_h}{(L_r + L_h e^{\pm jq_r\theta_r})^2} e^{\pm jq_r\theta_r} \right] \right\} \tag{4.120a}$$

$$\mathbf{A}_{12} = \frac{1}{L_s + L_h e^{\mp jq_r\theta_r} - (L_m^2/L_r + L_h e^{\pm jq_r\theta_r})}$$
$$\times \left\{ \frac{R_r L_m}{(L_r + L_h e^{\pm jq_r\theta_r})^2} + j\omega_r\left[\pm \frac{q_r L_m L_h}{(L_r + L_h e^{\pm jq_r\theta_r})^2} e^{\pm jq_r\theta_r} - \frac{L_m}{L_r + L_h e^{\pm jq_r\theta_r}} \right] \right\} \tag{4.120b}$$

$$\mathbf{A}_{21} = \frac{R_r L_m}{L_r + L_h e^{\pm jq_r\theta_r}} \tag{4.120c}$$

$$\mathbf{A}_{12} = j\omega_r - \frac{R_r}{L_r + L_h e^{\pm jq_r\theta_r}} \tag{4.120d}$$

$$\mathbf{B}_1 = \frac{1}{L_s + L_h e^{\mp jq_r\theta_r} - (L_m^2/L_r + L_h e^{\pm jq_r\theta_r})} \tag{4.120e}$$

Remark that the coefficients of matrix $\mathbf{A}$ and vector $\mathbf{B}$ are complex numbers and depend not only on the parameters of the machine but also on the rotor speed and angle. In particular, $\mathbf{A}_{11}$ and $\mathbf{A}_{12}$ present additional terms depending on the rotor speed which are not present in the classical fundamental harmonic model. If $L_h = 0$, all the coefficients become real numbers and coincide with the parameters of the classic full-state model of the IM. The previous full-order space-state model is particularly interesting since it can be employed as a full-order state observer in high-performance IM drives, by adding the correction Luenberger term driven by the difference between the measured and estimated stator currents. In this case, the estimated current would take into consideration also the RSH terms present in the stator current. Moreover, this model can also be used for implementing position sensorless techniques of IM drives. Indeed, this space-vector state model encompasses, in addition to the information about the rotor speed like the classical model, also the information about the rotor angle, which can be retrieved by adopting proper numerical techniques.

Most terms of the space-vector state model are combinations of exponential terms depending on the rotor angle θ_r and the number of rotor slots per pole pair q_r, and this could result in a high computational burden. However, since they are all periodic complex functions with the same period 2π, they can be straightforwardly implemented on a DSP for real-time applications by employing linearly interpolated look-up tables, having in input the rotor angle, which would require just a slightly higher increase of the complexity of the observer.

Because of the rotor slotting effect, the eigenvalues of the **A** matrix of the state representation modify when changing from the classical model. Figure 4.23 shows the eigenvalue locus of matrix **A**, according to the variation of the rotor speed ω_r, obtained with the classic space-state model neglecting the slotting effects and the locus of the eigenvalues of matrix **A** obtained with the proposed model according to the variation of the rotor angle θ_r for a single value of ω_r (85 rad/s). It can be observed that the eigenvalue locus for each rotor speed value modifies from a single point into a helix turning around the value obtained with the classic model. The number of turns of this helix is equal to the number of the rotor slots for pole pair q_r. This means that in this case, each eigenvalue of the classical model becomes a branchpoint of q_r Riemann surfaces in the complex plane when the rotor effects are accounted for in the model. In addition, the presence of this helix also implies that, even if the locus of the eigenvalues of the classical model lies on the left s-plane, the actual path of the pole in a Riemann surface can go into the right s-plane, with resulting instability.

This remark is of course a theoretical possibility: this instability phenomenon can happen if the eigenvalues of the classic model are very close to the imaginary axis, which is not the case in real-world IMs.

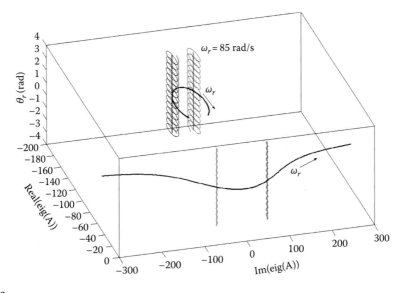

FIGURE 4.23
(See color insert.) Eigenvalue locus of matrix **A** obtained with the fundamental mmf model and the model taking into consideration the rotor slotting effects. (From Cirrincione, M. et al., *IEEE Trans. Ind. Appl.*, 44, 1683, 2008.)

4.10.4 Space-State Model of IM Including Stator and Rotor Slotting Effects

Starting from Equation 4.109, if both the stator and the rotor slotting effects are considered (Figure 4.19b), a procedure similar to that followed in Section 4.10.2 can be followed to retrieve the state space-vector model of the IM [30]. In this case, if L_{sh} and L_{rh} are, respectively, the three-phase stator and rotor slot inductances, the equations describing the "voltage model" in the stationary reference frame are the following:

$$
\frac{d\mathbf{\Psi}_r'}{dt} = \frac{L_r + L_{rh}e^{\pm jq_r\theta_r} + L_{sh}e^{\pm jq_s\theta_r}}{L_m}\left\{ \mathbf{u_s} - \left[R_s + j\omega_r \left(\pm \frac{L_m^2\left(q_rL_{rh}e^{\pm jq_r\theta_r} + q_sL_{sh}e^{\pm jq_s\theta_r}\right)}{\left(L_r + L_{rh}e^{\pm jq_r\theta_r} + L_{sh}e^{\pm jq_s\theta_r}\right)^2} \mp q_rL_{rh}e^{\mp jq_r\theta_r} \right.\right.
$$

$$
\left.\left. \mp q_sL_{sh}e^{\mp jq_s\theta_r} \right)\right]\mathbf{i_s} - \left[L_s + L_{rh}e^{\mp jq_r\theta_r} + L_{sh}e^{\mp jq_s\theta_r} - \frac{L_m^2}{L_r + L_{rh}e^{\pm jq_r\theta_r} + L_{sh}e^{\pm jq_s\theta_r}} \right]\frac{d\mathbf{i_s}}{dt}
$$

$$
\left. + j\omega_r \frac{L_m\left(\pm q_rL_{rh}e^{\pm jq_r\theta_r} \pm q_sL_{sh}e^{\pm jq_s\theta_r}\right)}{\left(L_r + L_{rh}e^{\pm jq_r\theta_r} + L_{sh}e^{\pm jq_s\theta_r}\right)^2}\mathbf{\Psi}_r' \right\} \tag{4.121}
$$

where the + and − signs follow the same logic explained in Section 4.10.2.

In the same manner, the equation of the "current model" in the stationary reference frame is the following:

$$
\frac{d\mathbf{\Psi}_r'}{dt} = \frac{R_rL_m}{L_r + L_{rh}e^{\pm jq_r\theta_r} + L_{sh}e^{\pm jq_s\theta_r}}\mathbf{i_s} + \left(j\omega_r - \frac{R_r}{L_r + L_{rh}e^{\pm jq_r\theta_r} + L_{sh}e^{\pm jq_s\theta_r}} \right)\mathbf{\Psi}_r' \tag{4.122}
$$

With regard to the state formulation, following the same procedure of Section 4.10.3, these are the parameters of the space-state model:

$$
\mathbf{A_{11}} = \frac{1}{L_s + L_{rh}e^{\mp jq_r\theta_r} + L_{sh}e^{\mp jq_s\theta_r} - \left(L_m^2/L_r + L_{rh}e^{\pm jq_r\theta_r} + L_{sh}e^{\pm jq_s\theta_r}\right)}
$$

$$
\times \left\{ -R_s - \frac{R_rL_m^2}{\left(L_r + L_{rh}e^{\pm jq_r\theta_r} + L_{sh}e^{\pm jq_s\theta_r}\right)^2} \right.
$$

$$
\left. + j\omega_r\left[\pm q_rL_{rh}e^{\mp jq_r\theta_r} \pm q_sL_{sh}e^{\mp jq_s\theta_r} - \frac{R_rL_m^2\left(\mp q_rL_{rh}e^{\pm jq_r\theta_r} \mp q_sL_{sh}e^{\pm jq_s\theta_r}\right)}{\left(L_r + L_{rh}e^{\pm jq_r\theta_r} + L_{sh}e^{\pm jq_s\theta_r}\right)^2} \right] \right\}
$$

$$\mathbf{A}_{12} = \frac{1}{L_s + L_{rh}e^{\mp jq_r\theta_r} + L_{sh}e^{\mp jq_s\theta_r} - \left(L_m^2/L_r + L_{rh}e^{\pm jq_r\theta_r} + L_{sh}e^{\pm jq_s\theta_r}\right)}$$

$$\times \left\{ \frac{R_r L_m}{\left(L_r + L_{rh}e^{\pm jq_r\theta_r} + L_{sh}e^{\pm jq_s\theta_r}\right)^2} \right.$$

$$\left. + j\omega_r \left[\frac{L_m\left(\pm q_r L_{rh}e^{\pm jq_r\theta_r} \pm q_s L_{sh}e^{\pm jq_s\theta_r}\right)}{\left(L_r + L_{rh}e^{\pm jq_r\theta_r} + L_{sh}e^{\pm jq_s\theta_r}\right)^2} - \frac{L_m}{L_r + L_{rh}e^{\pm jq_r\theta_r} + L_{sh}e^{\pm jq_s\theta_r}} \right] \right\}$$

$$\mathbf{A}_{21} = \frac{R_r L_m}{L_r + L_{rh}e^{\pm jq_r\theta_r} + L_{sh}e^{\pm jq_s\theta_r}}$$

$$\mathbf{A}_{12} = j\omega_r - \frac{R_r}{L_r + L_{rh}e^{\pm jq_r\theta_r} + L_{sh}e^{\pm jq_s\theta_r}}$$

$$\mathbf{B}_1 = \frac{1}{L_s + L_{rh}e^{\mp jq_r\theta_r} + L_{sh}e^{\mp jq_s\theta_r} - \left(L_m^2/L_r + L_{rh}e^{\pm jq_r\theta_r} + L_{sh}e^{\pm jq_s\theta_r}\right)} \tag{4.123}$$

It should be remarked that, even in this case, the coefficients of matrix $\mathbf{A}$ and vector $\mathbf{B}$ are complex numbers and depend not only on the parameters of the machine but also on the rotor speed and angle. In particular, $\mathbf{A}_{11}$ and $\mathbf{A}_{12}$ present additional terms depending on the rotor speed which are not present in the classical fundamental harmonic model. If $L_{sh} = 0$, the parameters of the state model coincide with those in Equation 4.120. If $L_{rh} = 0$, all the coefficients become real numbers and coincide with the parameters of the classic full-state model of the IM. Figure 4.24 shows the eigenvalue locus of matrix $\mathbf{A}$, according to

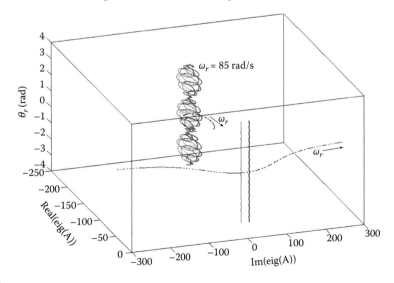

FIGURE 4.24
(See color insert.) Eigenvalue locus of matrix $\mathbf{A}$ obtained with the fundamental mmf model and the model taking into consideration the stator and rotor slotting effects. (From Cirrincione, M. et al., Space-vector state model of induction machines including rotor and stator slotting effects, *IEEE IEMDC '07*, Antalya, Turkey, pp. 673–682, May 3–5, 2007.)

the variation of the rotor speed ω_r, obtained with the classic space-state model neglecting the slotting effects and the locus of the eigenvalues of matrix **A** obtained with the model considering both the stator and rotor slotting effects, according to the variation of the rotor angle θ_r for a single value of ω_r (85 rad/s). It can be observed that the eigenvalue locus for each rotor speed value modifies from a single point into a helix turning around the value obtained with the classic model. The same considerations written in Section 4.10.3 could be made in this case.

4.10.5 Experimental Validation of the Space-Vector Model Considering the Stator and Rotor Slotting Effects

4.10.5.1 Stator Current Harmonics Caused by the Stator and Rotor Slotting Effect

In an IM, the RSHs and the static and dynamic eccentricity-related harmonics are essentially a function of the number of pole pairs, the number of rotor slots per pole pair, and the speed. In a compact form, RSH is given by the following [21–24]:

$$f_{RSH} = \left(\frac{rq_r \mp n_d}{p}\right)(1-s)f_1 \mp \nu f_1 \tag{4.124}$$

where
f_1 is the fundamental harmonic of the supply voltage
s is the slip
p is the number of pole pairs
q_r is the number of rotor slots for pole pair
n_d is the eccentricity order ($n_d = 0$ in case of static eccentricity and $n_d = 1, 2, 3 \ldots$ in case if dynamic eccentricity)
r is any positive integer which gives the order of the space harmonic
$\nu = 1, 3, 5 \ldots$ is the order of the stator time harmonics present in the power supply driving the motor

If the time harmonics of the stator and rotor currents as well as the static and dynamic eccentricities are neglected, the RSHs are obtained by Equation 4.124 with $n_d = 0$, $\nu = 1$, and $k = 1$. In this case, the rotor slotting effects are at frequencies:

$$f_{RSH} = q_r f_1(1-s) \mp f_1 \tag{4.125}$$

Therefore, in a healthy machine supplied at frequency f_1, two slotting frequencies appear in the stator current spectrum. Alger [5] gives a proof of the creation of the previous frequency components caused by rotor slotting.

Identically, the stator slotting effects are at frequencies:

$$f_{SSH} = q_s f_1(1-s) \mp f_1 \tag{4.126}$$

It should be noted, however, that the harmonics as described by the former equations are not present in a real machine for any combination of the *number of rotor slots* and *pole pairs* [31–33]. These time harmonics result from the corresponding space harmonics of the resulting mmf which are of order $q_r \pm 1$ and $q_s \pm 1$. Since $q_r = 3n \pm 1$ and $q_s = 3m \pm 1$ (with n and m integer), this also implies that, both for RSH and SSH, one of the two space harmonics is always a multiple of three, and therefore, it never induces a time harmonic in a healthy machine (e.g., balanced three-phase winding).

As far as the machine under test in the following section is considered, since the adopted motors have 2 pole pairs, 36 stator slots ($q_s = 18 = 3m$), and 28 rotor slots ($q_r = 14 = 3n − 1$), at no load (s = 0) with a pure sinusoidal supply and under the assumption of the model, only the 13th and 15th time harmonics should be expected, corresponding to spectral lines at 650 and 750 Hz under a 50 Hz fundamental supply. These two time harmonics in the current spectrum are originated from the two 13th and 15th space harmonics of the mmf, generated by the slots in the rotor. The 15th space harmonic cannot, in any case, induce any time harmonic because of the balanced three-phase winding [5]. On the contrary, since $q_s = 18$ no SSH can be observed.

Furthermore, it should be noted that, while at no load, the time harmonic components caused by the slotting effect lie at frequencies which are integral multiple of the fundamental ($q_r \pm 1$ and $q_s \pm 1$) and that in loaded conditions, the same harmonic components lie at frequencies which are nonintegral multiple of the fundamental, which affects in a significant way also the locus of the stator currents, as will be shown in the following.

To validate the model and to make the stator slotting effect observable in the simulation test, a motor with $q_s = 17$ has been considered. Therefore, in simulation at no load (s = 0), the 16th and 18th harmonics should be expected, corresponding to spectral lines at 800 and 900 Hz. As explained before, the 900 Hz harmonic is not present.

4.10.5.2 Results

With regard to the numerical simulation, the model has been implemented in the MATLAB®–Simulink® environment. With regard to the experimental verification of the model, three motor prototypes have been built with the same stator and different rotors. All the three motors have 2 pole pairs, 36 stator slots, and 28 rotor slots. Particularly, *motor 1* has a skewed rotor, *motor 2* has an unskewed rotor, and *motor 3* has an unskewed rotor with open rotor slots. Figure 4.25 shows the photograph of the three rotors.

FIGURE 4.25
Photograph of the three rotors, skewed (left), unskewed (center), and unskewed with open slots (right). (From Cirrincione, M. et al., *IEEE Trans. Ind. Appl.*, 44, 1683, 2008.)

TABLE 4.3

Parameters of the 2.2 kW IM with the Three Different Rotor Configurations

	Motor 1	Motor 2	Motor 3
Rated power, P_{rated} [kW]	2.2	2.2	2.2
Rated voltage, U_{rated} [V]	220	220	220
Rated frequency, f_{rated} [Hz]	50	50	50
Pole pairs	2	2	2
Stator resistance, R_s [Ω]	2.9	2.9	2.9
Stator inductance, L_s [mH]	275.1	270.0	203.0
Stator leakage inductance, $L_{\sigma s}$ [mH]	10.7	8.2	6.2
Stator leakage factor, σ_s	0.0406	0.0314	0.0317
Rotor resistance, R_r [Ω]	2.7	1.54	1.54
Rotor inductance, L_r [mH]	285.8	278.2	209.3
Rotor leakage inductance, $L_{\sigma r}$ [mH]	21.4	16.4	12.5
Rotor leakage factor, σ_r	0.0811	0.0628	0.0634
Three-phase magnetizing inductance, L_m [mH]	264.4	261.8	196.8
Global leakage factor, σ	0.1111	0.0877	0.0886
Slotting inductance, L_h [mH]	1.32	1.30	0.98
Moment of inertia, J [kg·m²]	0.0048	0.0048	0.0048

Source: Cirrincione, M. et al., Space-vector state model of induction machines including rotor and stator slotting effects, in *IEEE IEMDC '07*, Antalya, Turkey, pp. 673–682, May 3–5, 2007.

The parameters of the motor are shown in Table 4.3, as measured with the usual no-load and locked-rotor tests.

To assess the proposed model, two kinds of representation have been made. At first, after supplying the motor both in numerical simulation and experimentally with a sinusoidal waveform at the rated frequency of 50 Hz and voltage of 220 V RMS, in the first representation, the steady-state stator current spectrum has been computed. In the second, the stator current space-vector locus at steady-state has been drawn (i_{sQ} versus i_{sD}). If supplied with a pure sinusoidal voltage, neglecting all the spatial harmonics in the machine, the stator current locus would be a circle. In the presence of the RSH and the SSH, the stator current locus deviates from a circle. However, while on the basis of the model assumptions, only the RSHs and SSHs are to be expected as well as the fundamental in the stator current spectrum of the simulated machine; in the real machine, also other harmonics are to be expected, in particular, those generated by the distributed stator winding, which have usually a higher amplitude than those of RSH and SSH. For this reason, the experimentally obtained stator current locus is expected to be quite different from the simulated one. To make the experimentally obtained stator current locus comparable with the simulated one, the stator current signals have been properly filtered and reconstructed to contain only the fundamental and the RSH components (no SSH is present in the experimental current signature). This has been obtained with the scheme shown in Figure 4.26. The three current signals have been acquired by a Dspace DS1103 board, then converted from three to two phases in the stator reference frame. The obtained i_{sD} and i_{sQ} components have been processed by two select-band digital filters, tuned, respectively, on the fundamental (50 Hz) and on the RSH (only 650 Hz at no load), and then summed. In this way, the filtered current sD and sQ components contain only the fundamental and the RSH components. It should be noted that the employed filters do not cause any phase distortion of the signal, since the phase characteristics of the filter is 0 at the selected band.

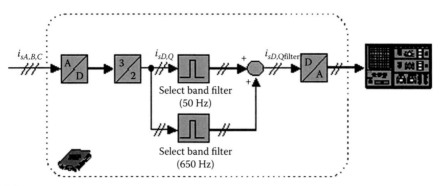

FIGURE 4.26
Schematics for the filtering of the stator current components for drawing the stator current locus. (From Cirrincione, M. et al., Space-vector state model of induction machines including rotor and stator slotting effects, in *IEEE IEMDC '07*, Antalya, Turkey, pp. 673–682, May 3–5, 2007.)

In the numerical test, $L_{sh} = L_{rh}$ have been assigned the value of 0.5% L_m; with such a value, the simulation results match the experimental ones well. Figure 4.27 shows the steady-state stator current on phase sA obtained in numerical simulation and its spectrum, obtained with the fast Fourier transform (FFT), at steady-state at no load. The time waveform of the stator current shows a distortion which is caused by the RSH and SSH. As written earlier, since $q_r = 14$ and $q_s = 17$ (only in simulation), at no load (s = 0), the 13th (RSH) and 16th (SSH) harmonics are present, corresponding to spectral lines at 650 and 800 Hz. This is confirmed by the results in Figure 4.27, and it is in agreement with literature [22–24,34]. The same test

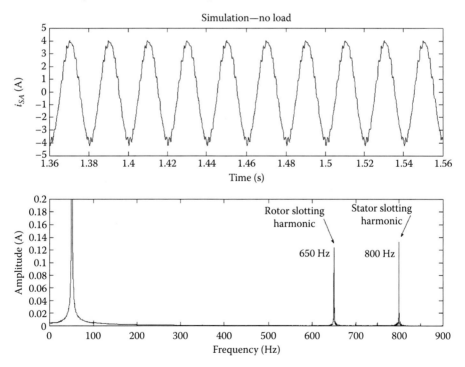

FIGURE 4.27
Stator current waveform and its spectrum at no load (numerical simulation). (From Cirrincione, M. et al., Space-vector state model of induction machines including rotor and stator slotting effects, in *IEEE IEMDC '07*, Antalya, Turkey, pp. 673–682, May 3–5, 2007.)

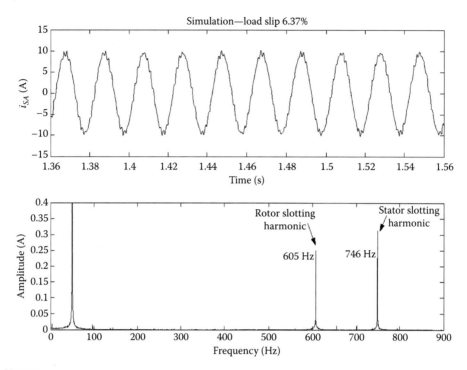

FIGURE 4.28

Stator current waveform and its spectrum at 6.37% load slip (numerical simulation) and stator current waveform and its spectrum obtained with motor 1 (experiment). (From Cirrincione, M. et al., Space-vector state model of induction machines including rotor and stator slotting effects, in *IEEE IEMDC '07*, Antalya, Turkey, pp. 673–682, May 3–5, 2007.)

has been made by applying a constant load torque of 10 N m, corresponding to a slip of 6.37%. Figure 4.28 shows the steady-state stator current on phase sA obtained in numerical simulation and its spectrum at steady-state at the earlier load. As expected, both the RSH and the SSH have a frequency shift caused by the slip equal, respectively, to $q_r f_1 s$ and $q_s f_1 s$: as a result, the line at 650 Hz moves to 605 Hz, and the line at 800 Hz moves to 746 Hz.

To check the accordance of the proposed model with the different rotor constructions, the same test with no load has been done with the three rotors. Figures 4.29 through 4.31 show the steady-state stator current on phase sA and its spectrum obtained experimentally with the FFT with the three motors. All spectra show that in the experimental case, only the 13th at 650 Hz is present.

It should be remarked that the lowest slotting harmonic is obtained with *motor 1* (skewed rotor), then with *motor 2* (unskewed rotor), and finally with *motor 3* (unskewed rotor with open slots), as expected: the proposed model matches the real motor construction better with unskewed rotors than with skewed ones and, even better, with open rotor slots. Obviously, while in the simulated test, only the rotor slotting harmonics are present as well as the fundamental; in the experimental one, other harmonics are also present. As a matter of fact, in a real machine the stator and rotor windings are distributed while in the proposed model they are concentrated, a real machine can have some small static or dynamic eccentricities, and finally, the supply voltage from the grid is never purely sinusoidal with consequent time harmonics in the stator currents. All these reasons account for the presence of some other harmonics in the stator current signature of a real machine, in particular, the 5th, 7th, 11th, and 13th, as explained in Cameron et al. [22].

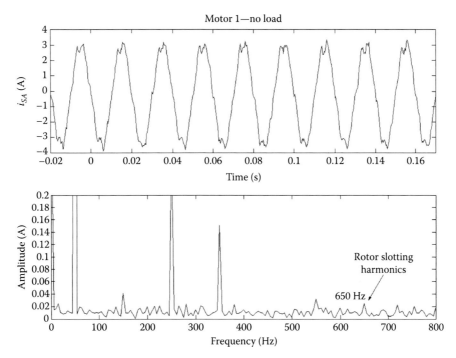

FIGURE 4.29
Stator current waveform and its spectrum obtained with motor 1 (experiment). (From Cirrincione, M. et al., *IEEE Trans. Ind. Appl.*, 44, 1683, 2008.)

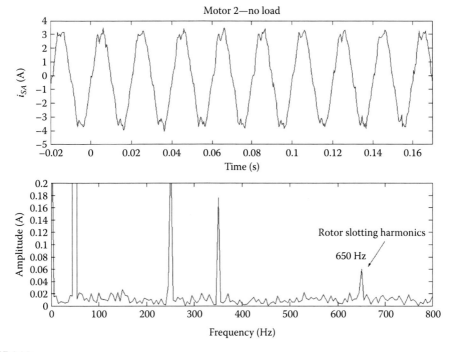

FIGURE 4.30
Stator current waveform and its spectrum obtained with motor 2 (experiment). (From Cirrincione, M. et al., *IEEE Trans. Ind. Appl.*, 44, 1683, 2008.)

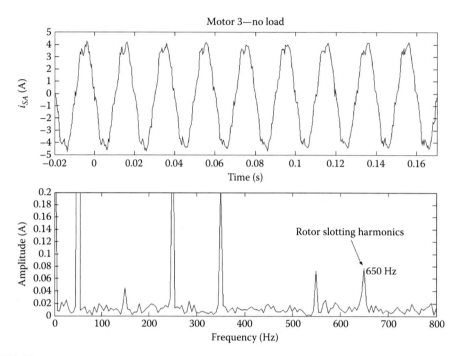

FIGURE 4.31
Stator current waveform and its spectrum obtained with motor 3 (experiment). (From Cirrincione, M. et al., *IEEE Trans. Ind. Appl.*, 44, 1683, 2008.)

Figure 4.32 shows the steady-state current components of *motors 2* and *3* supplied at the rated frequency of 50 Hz and voltage of 220 V, respectively, and the measured and the filtered one so to show only the RSH. Measurements have been done only on *motors 2* and *3* since *motor 1* does not exhibit any significant PSH (see Figure 4.29). Figure 4.32 shows that filters do not modify the phase position of the notch frequency, thus avoiding undesirable phase distortion. By this filtering process, any effect on the stator current locus of the significant 5th, 7th, 11th, and 13th has been canceled. The stator current locus obtained in numerical simulation is not a circle but presents some lobes caused by the 650 Hz (605 Hz) RSH and 800 Hz (756 Hz) SSH at no load (10 N m load), as shown in Figure 4.33a and b. The locus has been drawn three times: considering the effect only of the stator slotting effects, only of the rotor slotting effects, and both the stator and rotor slotting effect. The presence of slotting in all three cases can be observed by the lobes in the current locus. However, when considering either the stator or the rotor slotting, the current locus is regular in steady-state presenting a polar symmetry; when both stator and rotor slotting are considered, the locus presents lobes but is not regular anymore and does not present a polar symmetry. Figure 4.34a and b shows the steady-state stator current space-vector locus obtained experimentally on *motors 2* and *3* under the same supply at no load. Two loci are shown that obtained with the measured currents and that obtained with the filtered currents. The loci obtained with the measured currents are quite irregular and different from the theoretical ones (obtained in simulation). The lobes of the dominant harmonics hide those due to the slotting. The loci of the filtered currents, however, which contain only the information on the fundamental and the RSH show a shape quite similar to that obtained in numerical simulation and confirm that the two motors with an unskewed rotor exhibit a significant slotting effect.

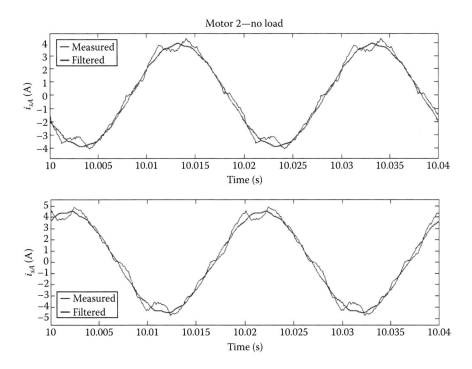

FIGURE 4.32
Measured and filtered i_{sD} components (experiment). (From Cirrincione, M. et al., *IEEE Trans. Ind. Appl.*, 44, 1683, 2008.)

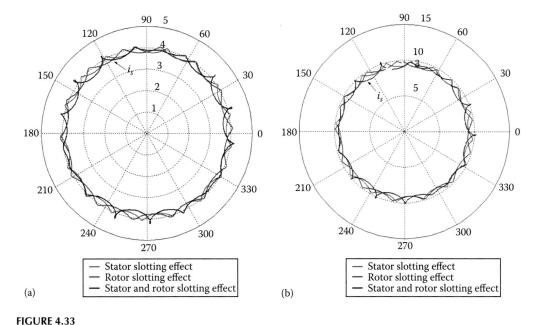

FIGURE 4.33
(See color insert.) (a) Locus of the stator current at no load (numerical simulation). (b) Locus of the stator current at 6.37% load slip (numerical simulation). (From Cirrincione, M. et al., Space-vector state model of induction machines including rotor and stator slotting effects, in *IEEE IEMDC '07*, Antalya, Turkey, pp. 673–682, May 3–5, 2007.)

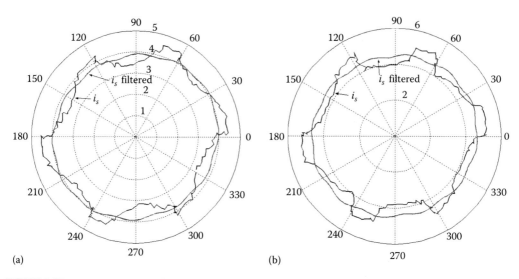

(a) (b)

FIGURE 4.34
(a) Locus of the measured and filtered stator current at no load with motor 2 (experiment). (b) Locus of the measured and filtered stator current at no load with motor 3 (experiment). (From Cirrincione, M. et al., *IEEE Trans. Ind. Appl.*, 44, 1683, 2008.)

List of Symbols

i_{sA}, i_{sB}, i_{sC}	stator phase currents
$\mathbf{i}_s = i_{sD} + ji_{sQ}$	space-vector of the stator currents in the stator reference frame
$\mathbf{i}_s^g = i_{sx}^g + ji_{sy}^g$	space-vector of the stator currents in a generic rotating reference frame
$\mathbf{i}_r' = i_{rd} + ji_{rq}$	space-vector of the rotor currents in the stator reference frame
$\mathbf{i}_r^g = i_{rx}^g + ji_{ry}^g$	space-vector of the rotor currents in a generic rotating reference frame
$\mathbf{i}_{mr} = i_{mrD} + ji_{mrQ}$	space-vector of the rotor magnetizing current in the stator reference frame
$\mathbf{i}_{ms} = i_{msD} + ji_{msQ}$	space-vector of the stator magnetizing current in the stator reference frame
$\mathbf{i}_{mm} = i_{mmD} + ji_{mmQ}$	space-vector of the magnetizing current in the stator reference frame
L_s	stator inductance
L_s'	stator transient inductance
L_r	rotor inductance
L_m	total static magnetizing inductance
$L_{s\sigma}$	stator leakage inductance
$L_{r\sigma}$	rotor leakage inductance
p	number of pole pairs
R_s	resistance of a stator phase winding
R_r	resistance of a rotor phase winding
t_e	electromagnetic torque
T_s	stator time constant

T_r	rotor time constant
T_s'	stator transient time constant
T_r'	rotor transient time constant
$T_{r\sigma}$	rotor leakage time constant
u_{sA}, u_{sB}, u_{sC}	stator phase voltages
$\mathbf{u}_s = u_{sD} + ju_{sQ}$	space-vector of the stator voltages in the stator reference frame
$\mathbf{u}_s^g = u_{sx}^g + ju_{sy}^g$	space-vector of the stator voltages in a generic rotating reference frame
ρ_r	phase angle of the rotor flux-linkage space-vector with respect to the sD axis
ρ_s	phase angle of the stator flux-linkage space-vector with respect to the sD axis
ρ_m	phase angle of the magnetizing flux-linkage space-vector with respect to the sD axis
σ	$1 - L_m^2/(L_s L_r)$ total leakage factor
σ_r	rotor leakage factor
σ_s	stator leakage factor
ϑ_r	angular position of the rotor with respect to the sD axis
$\boldsymbol{\psi}_r' = \psi_{rd} + j\psi_{rq}$	space-vector of the rotor flux linkages in the stator reference frame
$\boldsymbol{\psi}_s = \psi_{sD} + j\psi_{sQ}$	space-vector of the stator flux linkages in the stator reference frame
ω_{mr}	angular speed of the rotor flux space-vector
ω_{ms}	angular speed of the stator flux space-vector
ω_{mm}	angular speed of the magnetizing flux space-vector
ω_{sl}	angular slip speed
ω_r	angular rotor speed (in electric angles per second)
ω_m	angular rotor speed (in mechanical angles per second)

The symbols ψ_r', ψ_s, ψ_m in apex mean the reference frame in which the variables are expressed.

References

1. K. P. Kovacs, I. Ràcz, *Transiente Vorgänge in Wechselstrommachinen*, Akadémia Kiadò, Budapest, Hungary, 1954.
2. K. P. Kovacs, *Transient Processes in Electrical Machines (in Hungarian)*, Müszaki Konyvkiadò, Budapest, Hungary, 1970.
3. K. P. Kovacs, Programmierung von Asynchronnmotoren unter Berük-sichtigung der Sättigung, *Archiv für Elektrotechnik*, 47, 193–206, 1962.
4. G. Kron, *Equivalent Circuits of Electrical Machinery*, John Wiley & Sons Inc., New York, 1951.
5. P. L. Alger, *The Nature of Induction Machine*, Gordon and Breach, New York, 1965.
6. M. Kostenko, L. Piotrovsky, *Electrical Machines*, MIR, Moscow, Russia, 1969, (two volumes).
7. P. C. Krause, *Analysis of Electrical Machinery*, McGraw Hill, New York, 1986.
8. C. V. Jones, *The Unified Theory of Electrical Machines*, Butterworth, London, U.K., 1967.
9. A. E. Fitzgerald, C. Kingsley, A. Kusko, *Electric Machinery*, McGraw Hill, New York, 1971.
10. S. A. Nasar, I. Boldea, *The Induction Machine Handbook*, CRC Press, Boca Raton, FL, 2001.
11. P. Vas, *Electric Machines and Drives, a Space-Vector Theory Approach*, Clarendon Press, Oxford, U.K., 1992.
12. P. Vas, *Sensorless Vector and Direct Torque Control*, Oxford Science Publications, Oxford, U.K., 1998.

13. P. Vas, *Vector Control of AC Machines*, Oxford Science Publications, Oxford, U.K., 1990.

14. W. Leonhard, *Control of Electrical Drives*, Springer-Verlag, Berlin, Germany, 1997.

15. J. Holtz, The representation of AC machine dynamics by complex signal flow graphs, *IEEE Transactions on Industrial Electronics*, 42(3), 263–271, June 1995.

16. P. Vas, Generalized analysis of saturated AC machines, *Archiv für Elektrotechnik*, 64, 57–62, 1981.

17. J. A. Melkebeek, Magnetizing field saturation and dynamic behaviour of induction machines. Part I: An improved calculation method for induction machine dynamics, *IEE Proceedings B*, 30(1), pp. 218–224, 1–9, January 1983.

18. P. Vas, M. Alaküla, Field-oriented control of saturated induction machines, *IEEE Transactions on Energy Conversion*, pp. 218–224, March 1990.

19. M. Pucci, Novel numerical techniques for the identification of induction motors for the control of AC drives: Simulations and experimental implementations (in Italian), PhD thesis, University of Palermo, Palermo, Italy, January 2001.

20. M. Cirrincione, M. Pucci, G. Cirrincione, A. Miraoui, Space-vector state model of induction machines including rotor slotting effects: Towards a new category of observers, *IEEE Transactions on Industry Applications*, 44, Vol. 44, n. 6, pp. 1683–1692, 2008.

21. P. Vas, *Parameter Estimation, Condition Monitoring, and Diagnosis of Electrical Machines*, Oxford Science Publications, Oxford, U.K., 1993.

22. J. R. Cameron, W. T. Thomson, A. B. Dow, Vibration and current monitoring for detecting airgap eccentricity in large induction motors, *IEE Proceedings B Electric Power Applications*, Vol. 133(Pt. B, 3), 155–163, May 1986.

23. S. Nandi, Modeling of induction machines including stator and rotor slot effects, *IEEE Transactions on Industry Applications*, 40(4), 1058–1065. July/August 2004.

24. S. Nandi, Slot permeance effects on rotor slot harmonics in induction machines, *IEEE IEMDC '03*, Vol. 3, pp. 1633–1639, June 1–4, 2003.

25. K. Rajashekara, A. Kawamura, K. Matsuse, *Sensorless Control of AC Motor Drives*, IEEE Press, Piscataway, NJ, 1996.

26. J. Holtz, Sensorless control of induction motor drives, *Proceedings of the IEEE*, 90(8), 1359–1394, August 2002.

27. P. L. Jansen, R. D. Lorenz, Transducerless position and velocity estimation in induction and salient AC machines, *IEEE Transactions on Industry Applications*, 31(2), 240–247, March/April 1995.

28. J. M. Aller, J. A. Restrepo, A. Bueno, M. I. Giménez, G. Pesse, Squirrel cage induction machine model for the analysis of sensorless speed measurement methods, *Second IEEE International Caracas Conference on Devices, Circuits and Systems*, Isla de Margarita, Venezuela, pp. 243–248, March 2–4, 1998.

29. J. M. Aller, J. A. Restrepo, A. Bueno, M. I. Giménez, V. M. Guzmàn, Induction machine model for sensorless speed measurement systems, *IEEE Power Engineering Review*, 18(7), pp. 53–54, July 1998.

30. M. Cirrincione, M. Pucci, G. Cirrincione, A. Miraoui, Space-vector state model of induction machines including rotor and stator slotting effects, *IEEE IEMDC '07*, Antalya, Turkey, pp. 673–682, May 3–5, 2007.

31. S. Nandi, S. Amhed, H. Toliyat, Detection of rotor slot and other eccentricity related harmonics in a three phase induction motor with different cages, *IEEE Transactions on Energy Conversion*, 16(3), pp. 253–260, September 2001.

32. S. Nandi, S. Amhed, H. Toliyat, M. Bharadwaj, Selection criteria of induction machines for speed-sensorless drive applications, *IEEE Transactions on Industry Applications*, 39(3), May/June 2003.

33. S. Nandi, H. Toliyat, Novel frequency-domain-based technique to detect stator interturn faults in induction machines using stator-induced voltages after switch-off, *IEEE Transactions on Industry Applications*, 38(1), pp. 101–109, January/February 2002.

34. M. M. Liwschitz, Field harmonics in induction motors, *Transactions of the American Institute of Electrical Engineers*, 61, 797–803, November 1942.

5

Control Techniques of Induction Machine Drives

5.1 Introduction on Induction Machine (IM) Control

This chapter mainly deals with high-performance control techniques of IM drives. Control techniques of IMs can be divided into two main categories: *scalar* and *vector controls*. *Scalar control* is based on the steady-state model of the IM and therefore permits regulating at steady-state only the magnitudes and frequency of the stator voltages, currents, flux linkages, and electromagnetic torque. Since it does not act on the angular position of the space-vectors of the control variables, it does not permit the best dynamic performance to be achieved. On the contrary, *vector controls* are based on the dynamic model of the machine; they permit the drive to achieve its best dynamic performance in terms of electromagnetic torque control, thanks to their feature to take into consideration the instantaneous angular position of the stator voltages, currents as well as of the flux linkages.

According to these characteristics, *vector controls* can be obtained in several ways. The most popular method is well-known as field-oriented control (FOC), alternatively called vector control, and has been proposed by Hasse [1] and Blaschke [2]. Other methods based on modern nonlinear control [3–5] have been more recently proposed. Marino et al. [6] have proposed a nonlinear transformation of the motor state variables so that, in the new coordinates, the speed and rotor flux amplitude can be decoupled by feedback; the method is called feedback linearization control (FLC) or input–output decoupling [6]. A similar approach has been proposed by Krzeminski [7], which is based on a multiscalar model of the IM. A different possible approach, called passivity-based control (PBC), is based on the variation theory and energy shaping and has been only recently focused [8]. To do that, the machine is modeled in terms of the Euler–Lagrange equations in generalized coordinates.

During the mid-1980s, while FOC philosophy was gradually becoming the industrial standard for industrial high-performance drive control, completely new ideas came up from Takahashi and Noguchi [9] and Depenbrock [10–12]. The underlying idea was always in the direction to develop a suitable coordinate transformation permitting the drive to be controlled in a decoupled way similarly with DC motor control. The decoupling control typical of FOC has been replaced by a bang-bang control, which suitably matches the typical on–off command of the inverter power devices. This control strategy has been called direct torque control (DTC), and since 1985, it has been continuously developed and improved. Figure 5.1 shows the classification of the most common control techniques for IMs [13].

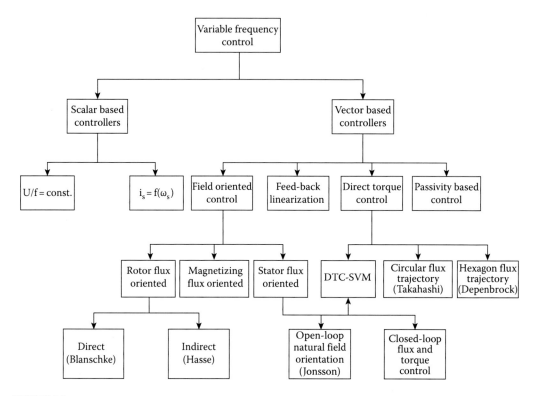

FIGURE 5.1
Classification of IM control techniques (NFO = natural field orientation).

5.2 Scalar Control of IMs

Some kinds of mechanical loads exist which do not require a high dynamic performance. Typical examples are fans and pumps where it is sufficient to regulate the speed of the IM with adequate efficiency over a wide speed range. This implies that it is sufficient to use the steady-state model of the IM instead of the dynamic one, as far as the control system design is concerned. The machine is supposed to be supplied by a pulsewidth modulation (PWM) voltage source inverter (VSI), able to generate a set of three-phase voltages whose fundamental component is characterized by the desired amplitude and frequency. Scalar control of IMs was born with the idea to have at disposal as a simple methodology for regulating the speed of an AC machine as that adopted for controlling the DC machine.

5.2.1 Scalar Control with Impressed Voltages

Starting from the steady-state space-vector equations of the IM, the air-gap electromagnetic torque can be written as follows [14]:

$$T_e = \frac{3}{\omega_1} \frac{|\mathbf{U}_1|^2}{(1+\sigma_s)^2} \frac{R_r/s}{(R_s + R_r/s)^2 + X'^2} \tag{5.1}$$

where

$\mathbf{U}_1$ is the steady-state space-vector of the stator voltages

ω_1 is the fundamental supply pulsation

R_s and R_r are, respectively, the stator and rotor resistances

$\sigma_s = X_{s\sigma}/X_m$ is the stator leakage factor, $X_{s\sigma}$ and X_m are the stator leakage and magnetizing reactances

$X' = X'_{r\sigma} + (X_{s\sigma}/1 + \sigma_s)$ is the transient rotor reactance, $X_{r\sigma}$ is the rotor leakage reactance

s = the expression of the pulsation is the pulsation slip

The maximum air-gap electromagnetic torque can be developed by the machine at a specific pulsation slip $s_{\max}$ equal to the following:

$$s_{\max} = \frac{\pm R_r}{\left(R_s^2 + X'^2\right)^{1/2}} \tag{5.2}$$

In correspondence to such a pulsation slip, the maximum electromagnetic torque is equal to the following:

$$T_{e\max} = \frac{(3/\omega_1)\left(|\mathbf{U}_s|/(1+\sigma_s)\right)^2}{2\left[R_s \pm \left(R_s^2 + X'^2\right)^{1/2}\right]} \tag{5.3}$$

where the positive (negative) sign is valid in case of motoring (generating) operation.

If the VSI is commanded so as to keep a constant ratio between the stator voltage amplitude and the stator fundamental pulsation

$$\frac{|\mathbf{U}_s|}{\omega_1} = \frac{|\mathbf{U}_{srat}|}{\omega_{1rat}} \tag{5.4}$$

where the "rat" pedex is referred to the rated values of the corresponding variables; then so long as the stator resistance R_s is neglected, the stator flux amplitude at steady-state is kept constant as well, since

$$\mathbf{\Psi}_s \cong \frac{\mathbf{U}_s}{j\omega_1} \tag{5.5}$$

where $\mathbf{\Psi}_s$ is the steady-state space-vector of the stator flux linkage.

At the same time, the maximum electromagnetic air-gap torque $T_{e\max}$ is kept constant as apparent from Equation 5.3, while the electromagnetic air-gap torque can be rewritten as follows:

$$\frac{T_e}{T_{e\max}} \cong \frac{2}{s/s_{\max} + s_{\max}/s} \tag{5.6}$$

Equation 5.6 implies that the steady-state torque, for different values of the supply frequency, depends only on the slip, being the relationship almost linear for small values of the speed itself (close to the synchronous speed). In this case [14],

$$\frac{T_e}{T_{e\max}} \cong 2\frac{s}{s_{\max}} \tag{5.7}$$

On the basis of above, the easiest way to control the speed of the IM is to open-loop regulate its fundamental supply pulsation ω_1 simultaneously keeping constant the

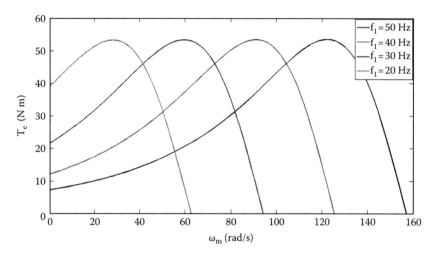

FIGURE 5.2
Set of ideal steady-state speed versus torque characteristics for different supply frequencies.

ratio $|\mathbf{U}_s|/\omega_1$. This corresponds to varying only the synchronous speed of the drive, without the need of measuring the machine speed on the one hand but without the possibility to compensate any variation of the speed caused by the load torque on the other hand. Figure 5.2 shows a set of ideal steady-state torque versus speed character-istics of a 2.2 kW machine, obtained for different values of the supply frequency under the assumption of null stator resistance. It clearly shows that the higher the supply fre-quency, the higher the synchronous speed of the machine, as expected. Moreover, the lower the supply frequency, the higher the starting torque. Finally, the maximum torque is maintained constant, while varying the supply frequency.

Since scalar control does not take into consideration the dynamic relationships of the machine, it suffers besides the above mentioned limited dynamic performance, from other drawbacks. In particular, sudden frequency variations could be the cause of the following phenomena:

- High overcurrents in the motor
- Oscillations of the stator currents, fluxes, and torque amplitudes
- Potential instability phenomena since the slip can overcome that corresponding to the maximum torque

To properly attenuate these drawbacks, frequency variations can be applied with a ramp instead of with a step. A ramp with a time constant close to the electromechanical one of the motor permits its speed to track the frequency changes more closely, with consequent reduction of the oscillations. Moreover, a better behavior changes more closely from the point of view of the stability of the system can be further achieved.

Figure 5.3a shows the block diagram of the open-loop scalar control scheme [15]. In this scheme, the gradient limiter reduces the bandwidth of the stator frequency command. The band-limited stator frequency signal then generates the stator voltage reference mag-nitude while its integral determines the phase angle. The amplitude and phase of the reference stator voltage space-vector constitute the input of the PWM system that, in turn, establishes the switching pattern of the inverter synthesizing the reference voltages.

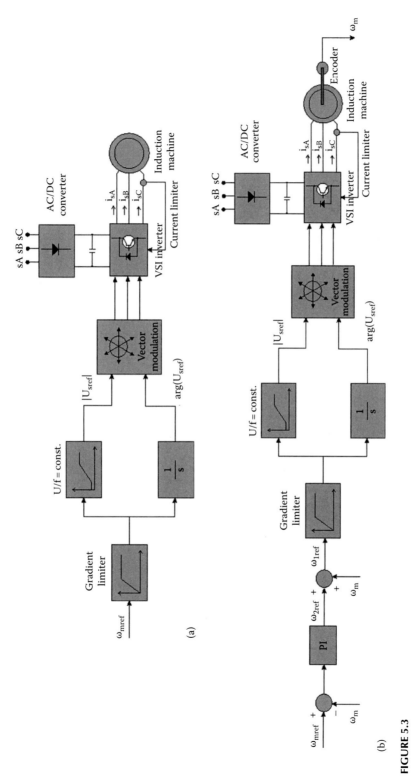

FIGURE 5.3
(a) Block diagram of the open-loop scalar control. (b) Block diagram of the closed-loop scalar control with impressed voltages.

Even if theoretically no stator current sensor is needed, since no direct current control is performed, in practical terms, it is frequently mounted to inhibit the switching of the inverter power devices for overload protection in the presence of overcurrents. Since this control system behaves as a feed-forward action, the mechanical speed of the motor will differ from the reference one in the presence of a load torque. The maximum speed tracking error depends on the nominal slip of the machine, almost equal to $3 \div 5\%$ for low-power machines and even lower for higher ratings. A way to overcome this disadvantage is to implement a load-current-dependent slip compensation scheme [16].

Alternatively, a closed-loop control of the rotor speed can be achieved with the scheme in Figure 5.3b [17]. In this case, the reference speed ω_{mref} is compared with the measured one ω_m, being the error processed by a PI controller. The output of such a controller is the reference slip speed ω_{2ref} which, added to the measured speed, provides the stator pulsation reference ω_{1ref}. The reference slip speed must be properly limited to the range where the speed/torque relationship is almost linear, to avoid pull-out phenomena.

With reference to the voltage-frequency relationship, some further remarks should be made. The $|\mathbf{U}_s|/\omega_1$ ratio defines the rate of change of the linear function in Figures 5.3 and 5.4 and is usually set equal to the rated stator flux amplitude of the machine, $(|\mathbf{U}_s|/\omega_1) = const = |\Psi_{srat}|$, when the motor speed remains below the rated one. Above the rated speed, field weakening can be simply achieved by limiting the voltage amplitude to the rated voltage of the machine, $|\mathbf{U}_s| = |\mathbf{U}_{srat}|$. At very low stator frequency, there is a preset minimum value of the stator voltage programmed to account for the resistive stator voltage drop, $|\mathbf{U}_s| = |\mathbf{U}_{s\,min}|$. This is due to the fact that, because of a nonnull value of the stator resistance, as long as the voltage-frequency reduces, the stator flux amplitude reduces too. This can be clearly observed from Figure 5.4, showing the stator flux amplitude versus of the rotor speed of a 2.2 kW machine for a fixed value of the ratio $(|\mathbf{U}_s|/\omega_1) = 0.7$ at slip equal to 0.

This implies a reduced torque capability of the drive at decreasing supply frequencies, as apparent from Figure 5.5, showing the real (with nonnull stator resistance) torque versus speed characteristics of a 2.2 kW machine for different values of the supply frequency. The lower the supply frequency, the lower the maximum electromagnetic torque of the machine. Instead of setting a minimum value of the stator voltage at low speed, a more

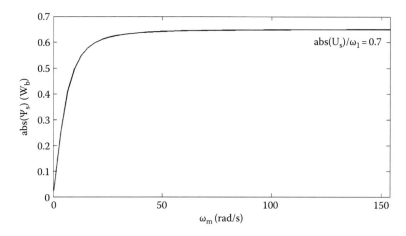

FIGURE 5.4
Stator flux amplitude versus rotor speed.

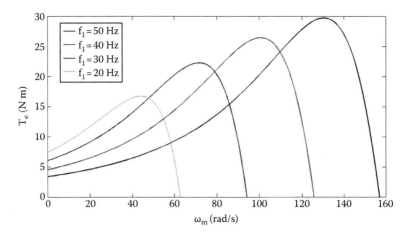

FIGURE 5.5
Set of ideal steady-state speed versus torque characteristics for different supply frequencies.

sophisticated method [17] is to correct the stator voltage reference on the basis of such a relationship, compensating the stator resistance ohmic drop:

$$\left|\mathbf{U}_s\right| = \left|j\frac{\omega_1}{\omega_{1rat}}\mathbf{U}_{srat} + R_s\mathbf{I}_s\right| \tag{5.8}$$

or, if the operating range has to be extended at low speeds,

$$\left|\mathbf{U}_s\right| = \left|\frac{\omega_1}{\omega_{1rat}} + \frac{R_s}{j\omega_{1rat}L_s}\frac{1+j\omega_2 T_r}{1+j\omega_2\sigma T_r}\right|\mathbf{U}_{srat} \tag{5.9}$$

Figure 5.6a shows two voltage versus pulsation frequency curves: the ideal one and that obtained with the R_s compensation method in Equation 5.9, respectively.

Finally, the stator flux amplitude decreases also, for a given supply frequency, for increasing values of the slip as observable from Figure 5.6b, showing the stator flux amplitude versus the slip for a fixed value of the voltage-frequency ratio. The slip variation during typical operation is limited, but its increase with the load is always a cause of decreased flux amplitude with consequent torque capability reduction. This further complicates the behavior of such a control system at low speeds.

The open-loop scalar control ensures high robustness at the expense of reduced dynamic performance. The absence of closed-loop control and the restriction to low dynamic performance make controlled drives very robust. For very high-speed applications like centrifuges and grinders, open-loop control can also be an advantage: as a matter of fact, the current control system of closed-loop schemes tends to destabilize when operated at field weakening up to 5–10 times the nominal speed [15].

The particular attraction of scalar-controlled drives is their extremely simple control structure which favours an implementation with few highly integrated electronic components. These cost-saving aspects are most important for applications at low power below 5 kW.

5.2.1.1 Experimental Results with Open-Loop Scalar Control with Impressed Voltages

In the following, some experimental results of open-loop scalar-controlled IM drive with impressed voltages are presented. The scalar control scheme in Figure 5.3a has been

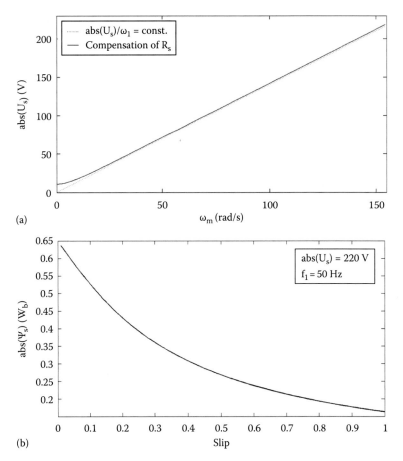

FIGURE 5.6
(a) Voltage versus pulsation curve: standard and with R_s compensation. (b) Stator flux amplitude versus slip.

adopted where, however, no limitation either of the rate of change of the reference speed or of the stator current amplitude has been implemented to show the corresponding detrimental effects. A test setup with a 2.2 kW IM has been used for this test (see Appendix 11.A). Figure 5.7a shows the reference and measured speed during a speed reversal of the type $100 \to -100$ rad/s. Figure 5.7b shows the corresponding waveform of the electromagnetic torque and rotor flux-linkage amplitude. Figure 5.7c shows the corresponding waveform of the three-phase stator currents. These figures confirm the previous considerations. Firstly, the speed reversal is followed by significant uncontrolled oscillations of the speed, which die out in a time interval longer than 0.5 s. The oscillations in the speed result in oscillations of the electromagnetic torque and rotor flux amplitude, both of which are uncontrolled variables. Finally, the three-phase uncontrolled currents attain very high values (over 50 A), significantly higher than the rated ones (around 10 A). Some oscillations in the stator current amplitude occur in correspondence of the speed oscillations.

5.2.2　Scalar Control with Impressed Currents

Open-loop scalar control with impressed voltages is sensitive to stator resistance variations, occurring with temperature changes. The dependence of the control performance on the stator resistance and leakage reactance can be eliminated using a scheme with

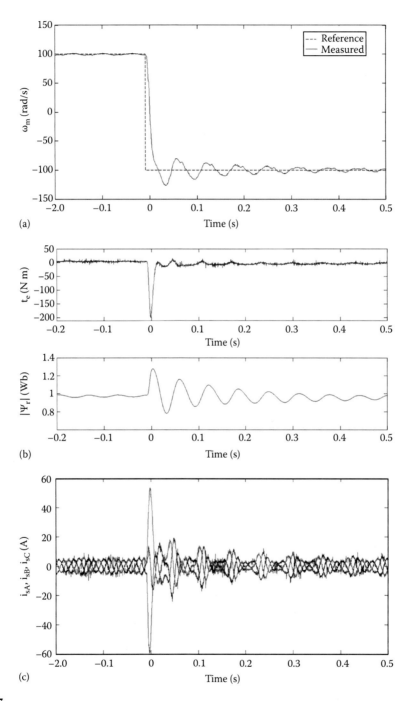

FIGURE 5.7
(a) Reference and measured rotor speed with open-loop scalar control. (b) Electromagnetic torque and rotor flux-linkage amplitude with open-loop scalar control. (c) Three-phase stator currents with open-loop scalar control.

impressed stator currents instead of voltages. Moreover, the assumption of sinusoidal supply is certainly more realistic with currents than with voltages for a switched converter.

If the motor is assumed to be supplied with sinusoidal currents, it is preferable to control the rotor flux-linkage amplitude, instead of the stator or magnetizing one. The reason is that, in this way, any variation of both the stator resistance and leakage inductance does not affect the control action at all. From the equivalent circuit of the IM, it is possible to compute the stator current steady-state space-vector needed to maintain constant the magnitude of the rotor flux linkage, resulting in [14] the following:

$$\mathbf{I}_s = \left[1 + \frac{j\omega_2(1-\sigma)L_s(1-\sigma_r)}{L_m R_r}\right]\mathbf{\Psi}_r' = \frac{(1+j\omega_2 T_r)}{L_m}\mathbf{\Psi}_r' \tag{5.10}$$

To maintain the rotor flux amplitude constant, the stator current amplitudes must follow a function depending on the slip speed ω_2 and on rotor parameters only

$$\left|\mathbf{I}_s\right| = \frac{\sqrt{(1+(\omega_2 T_r)^2)}}{L_m}\left|\mathbf{\Psi}_r'\right| \tag{5.11}$$

This can be obtained with a function generator having as input the slip speed ω_2. Figure 5.8 shows the current versus slip speed curves for two values of T_r.

The steady-state torque can be expressed as a function of the stator currents in this form [14]:

$$T_e = \frac{3\left|\mathbf{I}_s\right|^2 R_r L_m^2 \omega_2}{R_r^2 + (\omega_2 L_r)^2} = \frac{3\omega_2}{R_r}\left|\mathbf{\Psi}_r'\right|^2 \tag{5.12}$$

Equation 5.12 confirms that, as long as the rotor magnitude is maintained constant with the technique described above the steady-state torque depends linearly on the slip speed, and therefore, torque control can be directly performed by acting on the slip speed itself. Figure 5.9 shows the block diagram of a closed-loop speed scalar control with impressed currents [17]. The reference speed ω_{mref} is compared with the measured one ω_m, and the error is processed by a PI controller. The output of such a controller is the reference slip

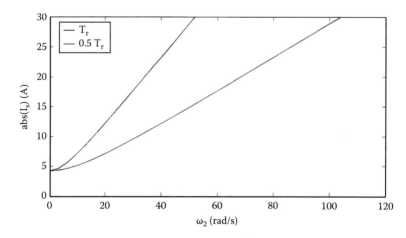

FIGURE 5.8
Stator current amplitude versus slip speed curves.

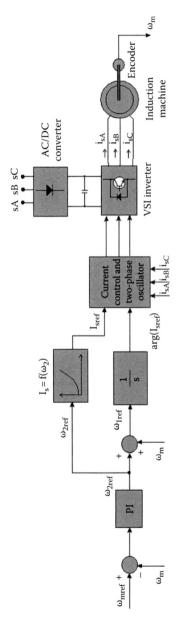

FIGURE 5.9
Block diagram of the closed-loop scalar control with impressed currents.

speed ω_{2ref} which, added to the measured speed, provides the stator pulsation reference ω_{1ref}. The integral of the stator pulsation reference gives the stator angle. The ω_{2ref} signal is fed to the function generator implementing the stator current versus slip speed relationship, whose output is the stator current reference. Both the stator current amplitude and the angle are then fed to a two-phase oscillator, at the output of which the two instantaneous reference currents are obtained. A current control system permits the instantaneous currents to be controlled to their reference values.

The scalar control scheme with impressed currents is certainly more complex than the corresponding with impressed voltages. It requires the knowledge of the parameters contained in the current versus slip speed relationship (L_m and T_r). Moreover, it requires the motor to be controlled in current, which offers the advantages of a quicker response in the current and torque commands, since the dynamic relationship of the stator is not accounted for.

5.3 FOC of IMs

Scalar control of IMs, although yet successfully employed in industry, is not adoptable for those application requiring high dynamic performance (e.g., servo drives, flying shears, rolling mills, robotic manipulators). The open-loop control of the magnetic flux linkage, typical of scalar control, makes the generation of the rated electromagnetic torque of the machine basically impossible at very low and zero speed.

FOC of IM drives was introduced almost 30 years ago [1,2], but it has been intensively studied and over the last few years [17–33], becoming nowadays the industrial standard as far as high-performance IM drives are concerned. Its development has been a significant breakthrough in the field of control of electrical drives with IM since it has permitted the use of this kind of motor for applications where only DC motors offered adequate dynamic performance. FOC permits the IMs to be controlled with dynamic performance comparable with that achievable with DC motor drives, but without the drawbacks caused by the brushes.

The application of FOC has been possible, even at an industrial level, thanks to the development of power electronics, resulting in reliable, cheap and fast-switching off-the-shelf power devices along with powerful low cost digital programmable architectures. As a matter of fact, today, IM drives are a valid alternative to DC motor drives, also from an economical point of view.

5.3.1 Principle of Field-Oriented Control

The DC motor with separate excitation is easily controllable with high dynamic performance, thanks to its structure that is naturally decoupled. As a matter of fact, in the DC machine, acting on the excitation current, it is possible to control the magnetic flux, while acting on the armature current, it is possible to control the electromagnetic torque and so the angular speed and the angular position.

FOC permits the IM to be controlled as a DC motor with separate excitation, with equivalent dynamic performance. It can be demonstrated, by adopting the space-vector theory explained in Chapter 1, that the choice of a suitable rotating reference frame with the direct axis lying on the direction of the rotor flux-linkage space-vector (or equivalently the stator or the magnetization flux linkage) makes it possible to write the instantaneous value of the electromagnetic torque as the product of the amplitude of the rotor (or stator or magnetization) flux linkage and the quadrature component of the stator current expressed in the same reference frame. The electromagnetic torque can be therefore expressed, in these reference frames, in the following ways [18]:

$$t_e = c_{1r} |\mathbf{\psi}_r| i_{sy}^{\psi_r} \quad \text{rotor flux-linkage-oriented reference frame} \tag{5.13}$$

$$t_e = c_{1s} |\mathbf{\psi}_s| i_{sy}^{\psi_s} \quad \text{stator flux-linkage-oriented reference frame} \tag{5.14}$$

$$t_e = c_{1m} |\mathbf{\psi}_m| i_{sy}^{\psi_m} \quad \text{magnetization flux-linkage-oriented reference frame} \tag{5.15}$$

where

c_{1r}, c_{1s}, and c_{1m} are constants depending on the machine parameters

$|\mathbf{\psi}_r|$, $|\mathbf{\psi}_s|$, and $|\mathbf{\psi}_m|$ are the amplitudes of, respectively, the rotor, stator, and magnetization flux linkages

$i_{sy}^{\psi_r}$, $i_{sy}^{\psi_s}$, and $i_{sy}^{\psi_m}$ are the quadrature components of the stator current space-vector in the rotor flux-oriented, stator flux-oriented, and magnetization flux-oriented reference frames (in the following just called field reference frame)

Equations 5.13 through 5.15 are similar to the electromagnetic torque equation of the DC motor with independent excitation. This suggests that the control of the electromagnetic torque should be performed, once the amplitude of the rotor (stator or magnetization) flux-linkage amplitude is fixed, by acting on the quadrature components of the stator current space-vector in the chosen field reference frame.

The control of the rotor (stator or magnetization) flux-linkage amplitude can be performed by acting, on the contrary, on the direct component of the stator current in the field reference frame. Decoupled control of the magnetic flux and electromagnetic torque can then be obtained in FOC, by acting, respectively, on the direct and quadrature components of the stator currents in one of the field reference frames. To implement FOC the stator current space-vector can be thus transformed into the corresponding reference frame. In any of the field reference frames, the direct and quadrature components of the stator current play the same role of the excitation and armature currents in the DC motor. FOC is usually implemented in the rotor flux-linkage reference frame. As a matter of fact, in this reference frame, the rotor flux linkage is dependent only on the direct component of the stator current, with a consequent natural decoupling in current. If any of the other two field reference frames is adopted, the flux linkage does not depend only on the direct component of the stator current, with consequent coupling terms. This calls for feed-forward control terms, called decoupling circuits, which however make the overall control system more complex. The implementation of the FOC, with orientation in any field reference frame, requires information on the electric angle between the direct-axis sD in the stationary reference frame and the space-vector of the adopted flux linkage. As far as rotor flux-oriented control is concerned, there are two ways to obtain the amplitude and phase of the rotor flux space-vector. When the so-called direct field-oriented control (DFOC) is used, also called flux feedback, these quantities are either directly measured (by Hall effect sensors, additional windings, etc.) or more commonly they are computed on the basis of the flux models. When the so-called indirect field-oriented control (IFOC) is used, the amplitude and phase of the rotor flux space-vector are obtained, respectively, by the reference values of the direct and quadrature components of the stator current and by the machine angular speed measured on the shaft. In particular, in the IFOC, the phase of the rotor flux space-vector is obtained as the sum of the measured rotor position θ_r and the computed slip angle θ_{sl}. The slip angle θ_{sl} gives the angular position of the rotor flux space-vector with respect to the direct-axis ra in the rotor reference frame. This angle can be computed on the basis of the direct and quadrature components of the reference stator current space-vector in the field reference frame, whose accuracy in the estimation strongly depends on the correct knowledge of the rotor time constants. The incorrect knowledge of this parameters leads to a wrong field orientation.

5.3.2 Rotor Flux-Oriented Control

The rotor flux-oriented control is suited for motors directly supplied in current. If the machine is supplied by a current generator, the dynamic model of the system is significantly simplified. Supplying the machine with impressed currents, it is possible to take into consideration, in the control system, simply the rotor space-vector equations of the machine, without considering the stator ones.

Adopting the rotor flux-oriented reference frame, the rotor space-vector equation becomes:

$$T_r \frac{d|\mathbf{i}_{mr}|}{dt} + |\mathbf{i}_{mr}| = \mathbf{i}_s^{\psi_r} - j(\omega_{mr} - \omega_r)T_r|\mathbf{i}_{mr}| \tag{5.16}$$

where
$\mathbf{i}_s^{\psi_r}$ is the space-vector of the stator current in the field reference frame
$\mathbf{i}_{mr}$ is the space-vector of the rotor magnetization current defined as $\mathbf{i}_{mr} = \boldsymbol{\psi}_r/L_m$, where L_m
   is the three-phase magnetization inductance, $\boldsymbol{\psi}_r$ is the rotor flux-linkage space-vector
T_r is the rotor time constant
ω_r is the speed of the machine in electrical angles
ω_{mr} is the speed of the rotor flux linkage in electrical angles

The space-vector Equation 5.16 can be decomposed into the direct x and quadrature y axes of the field reference frame, giving the following scalar equations:

$$\begin{cases} T_r \dfrac{d|\mathbf{i}_{mr}|}{dt} + |\mathbf{i}_{mr}| = i_{sx}^{\psi_r} \\[3mm] \omega_{mr} = \omega_r + \dfrac{i_{sy}^{\psi_r}}{T_r|\mathbf{i}_{mr}|} \end{cases} \tag{5.17a,b}$$

where $i_{sx}^{\psi_r}$ and $i_{sy}^{\psi_r}$ are the direct and quadrature components of the stator current space-vector in the field reference frame.

The expression of the electromagnetic torque in the rotor flux-linkage reference frame, as a function of $\mathbf{i}_{mr}$ and $i_{sy}^{\psi_r}$, is as follows:

$$t_e = \frac{3pL_m}{2(1+\sigma_r)}|\mathbf{i}_{mr}|i_{sy}^{\psi_r} \tag{5.18}$$

where σ_r is the rotor leakage factor.

Figure 5.10 shows the vector diagram with the different variables.

Equation 5.17a suggests that since the rotor magnetizing current depends only on the direct component of the stator current $i_{sx}^{\psi_r}$, it is possible to control the rotor flux linkage acting directly on this last component $i_{sx}^{\psi_r}$. It should be noted that $\mathbf{i}_{mr}$ and $i_{sx}^{\psi_r}$ are related by a dynamic relationship of the first order, governed by the rotor time constant T_r.

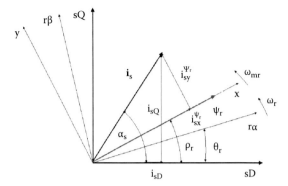

FIGURE 5.10
Representation of the rotor flux-linkage space-vector and the stator current.

Equation 5.18 highlights that the electromagnetic torque can be controlled by acting either on the rotor magnetization current $\mathbf{i}_{mr}$ or on the quadrature component of the stator current $i_{sy}^{\psi_r}$. To obtain the best dynamic performance, it is more convenient, when the machine works below the rated speed, to maintain at a constant value the amplitude of the rotor flux linkage (or equally of the rotor magnetizing current), that is, to maintain at a constant value the direct component of the stator current $i_{sx}^{\psi_r}$. The electromagnetic torque, in the same range of speed, can be controlled by acting on the quadrature component of the stator current $i_{sy}^{\psi_r}$. To obtain the correct decomposition of the stator current space-vector into its components $i_{sx}^{\psi_r}$ and $i_{sy}^{\psi_r}$, it is necessary to know the instantaneous value of the angular position of the rotor flux-linkage space-vector ρ_r with respect to the direct-axis sD of stationary reference frame. In the DFOC, the value of ρ_r is provided by the flux model. It should be noted that, in the rotor flux-linkage reference frame, at steady-state, the direct and quadrature components of the stator current, $i_{sx}^{\psi_r}$ and $i_{sy}^{\psi_r}$, are constant variables.

5.3.3 Rotor Flux-Linkage Acquisition

As far as a DFOC with rotor flux orientation is concerned, a distinction should be made between the case where the rotor flux is measured and the case where the rotor flux is estimated by a flux model.

A direct measurement of the magnetic flux can be made by inserting on the stator inner periphery properly displaced on it, some magnetogalvanic or magnetoresistive sensors like those based on the Hall effect. By interpolating the local measurements of the flux density, it is possible to have the amplitude and angular position of the magnetic flux density at the air-gap. On the basis of this measurement, measuring also the stator currents, it is possible to compute the rotor flux-linkage space-vector. The direct measurement of the rotor flux presents, however, some drawbacks [17]:

1. It is not possible to use an off-the-shelf motor, since the sensors must be inserted during the construction of the machine.

2. The Hall effect sensors are fragile from both a mechanical and thermal point of view and present an additive cost besides the assembling problems.

3. There exist problems connected to the thermal offset of the Hall effect sensors and consequent processing of the signals at their output because of the harmonics caused by the rotor slots. This implies the need of adopting a proper filtering.

4. The electromagnetic torque estimation by the direct measurement of the flux with few measurement points on the stator periphery is not very reliable.

Another possibility is to use some additional windings on the stator, to determine the magnetic flux on the basis of the integration of the induced electromotive forces. Such a solution implies a structural modification of the motor and causes problems at low speed because of the offset of the integrators and the variations of the resistance with the temperature. Moreover, the measurement of the induced electromotive forces requires reliable and accurate sensors, besides a suitable signal filtering.

The most common solution is the adoption of the so-called flux models. Flux models are mathematical models based on the equations describing the behavior of the machine, and they permit the estimation of the magnetic flux on the basis of the measurement of some electrical (e.g., stator voltages and currents) or mechanical (e.g., rotor speed) variables.

5.3.3.1 Voltage Flux Models

Flux models are typically divided into voltage and current flux models. The first are based on the stator equations, while the second are based on the rotor equations of the machine. Flux models are open-loop estimators.

The voltage flux model, as recalled earlier, is based on the stator equations of the machine. After processing Equation (4.35), expressing the rotor current space-vector as a function of the stator current and the rotor magnetizing current space-vectors, the following space-vector equation can be obtained:

$$(1-\sigma)T_s \frac{d\mathbf{i}_{mr}}{dt} = \frac{\mathbf{u_s}}{R_s} - \mathbf{i_s} - T_s' \frac{d\mathbf{i_s}}{dt} \tag{5.19}$$

where
T_s is the stator time constant
T_s' is the transient time constant of the machine
σ is the global leakage factor
$\mathbf{u_s}$ is the stator voltage space-vector in the stationary reference frame
$\mathbf{i_s}$ is the stator current space-vector in the same reference frame

The space-vector equation (5.19) can be decomposed into the direct and quadrature components in the stationary reference frame in this way:

$$\begin{cases} (1-\sigma)T_s \dfrac{di_{mrD}}{dt} = \dfrac{u_{sD}}{R_s} - i_{sD} - T_s' \dfrac{di_{sD}}{dt} \\[3mm] (1-\sigma)T_s \dfrac{di_{mrQ}}{dt} = \dfrac{u_{sQ}}{R_s} - i_{sQ} - T_s' \dfrac{di_{sQ}}{dt} \end{cases} \tag{5.20a,b}$$

where
i_{mrD} and i_{mrQ} are the direct and quadrature components of the rotor magnetization space-vector in the stationary reference frame
i_{sD} and i_{sQ} are the direct and quadrature components of the stator current
u_{sD} and u_{sQ} are the direct and quadrature components of the stator voltage in the same reference frame

Figure 5.11a shows the block diagram of the voltage flux model described by Equations 5.20a and 5.20b.

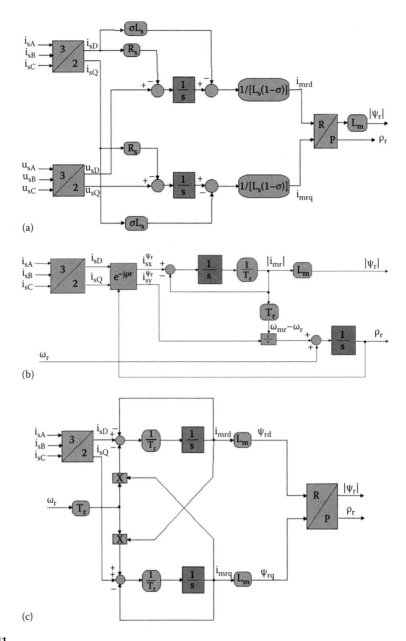

FIGURE 5.11
(a) Block diagram of the voltage flux model. (b) Block diagram of the current flux model in the rotor flux-oriented reference frame. (c) Block diagram of the current flux model in the stationary reference frame.

In Figure 5.11a, s denotes the Laplace variable. This flux model requires the measurement of the stator voltages and currents, while it does not need the measurement of the angular speed of the machine. The electrical parameters of the machine to be known for the implementation of such flux model are the stator resistance, the stator inductance, and the global leakage factor. The block identified as 3/2 in Figure 5.11a permits the three-phase variables to be transformed into the 2–phase stationary reference frame. The equations describing such transformations are as follows:

$$\begin{cases} u_{sD} = \left(\dfrac{2}{3}\right)u_{sA} - \left(\dfrac{1}{3}\right)u_{sB} - \left(\dfrac{1}{3}\right)u_{sC} \\[4mm] u_{sQ} = \left(\dfrac{1}{\sqrt{3}}\right)u_{sB} - \left(\dfrac{1}{\sqrt{3}}\right)u_{sC} \end{cases} \tag{5.21a,b}$$

$$\begin{cases} i_{sD} = \left(\dfrac{2}{3}\right)i_{sA} - \left(\dfrac{1}{3}\right)i_{sB} - \left(\dfrac{1}{3}\right)i_{sC} \\[4mm] i_{sQ} = \left(\dfrac{1}{\sqrt{3}}\right)i_{sB} - \left(\dfrac{1}{\sqrt{3}}\right)i_{sC} \end{cases} \tag{5.22a,b}$$

where u_{sA}, u_{sB}, u_{sC} and i_{sA}, i_{sB}, i_{sC} are the phase voltages and currents of the machine.

The block identified as R/P (rectangular-polar) in Figure 5.11a performs the coordinate transformation from Cartesian to polar, by means of the following equations:

$$\begin{cases} |\mathbf{i_{mr}}| = \sqrt{i_{mrD}^2 + i_{mrQ}^2} \\[4mm] \rho_r = \arctan\left(\dfrac{i_{mrQ}}{i_{mrD}}\right) \end{cases} \tag{5.23a,b}$$

This model presents the following drawbacks [18,19]:

1. Offset of the integrators caused by the open-loop integration of the flux
2. Sensitivity of the model at low speed to the variation of R_s with the temperature, which requires a compensation of the voltage drop on R_s before the integration
3. Need of the measurement of the stator voltages with proper sensors, which are typically expensive, and need of filtering voltage signals before sampling (anti-aliasing filters)

On the contrary, it presents the following advantages:

1. No need of speed measurement
2. Robustness to parametric variations at high speeds
3. Dependence of the estimation only on the stator parameters of the machine

5.3.3.2 Current Flux Models

Current flux models are, on the contrary, described by the rotor equations of the machine. Several current flux models exist, depending on how the equations are rewritten and on which reference frame is adopted. The two most common current flux models are those adopting, respectively, the rotor flux-linkage reference frame and the stator reference frame.

The flux model based on the rotor equations in the rotor flux-oriented reference frame is described by Equation 5.16, given by:

$$T_r \frac{d|\mathbf{i}_{mr}|}{dt} + |\mathbf{i}_{mr}| = \mathbf{i}_s^{\psi_r} - j(\omega_{mr} - \omega_r)T_r |\mathbf{i}_{mr}| \tag{5.24}$$

This equation gives rise to the following scalar equations on the direct x and quadrature y axes:

$$\begin{cases} T_r \dfrac{d|\mathbf{i}_{mr}|}{dt} + |\mathbf{i}_{mr}| = i_{sx}^{\psi_r} \\[2mm] \omega_{mr} = \omega_r + \dfrac{i_{sy}^{\psi_r}}{T_r |\mathbf{i}_{mr}|} \end{cases} \tag{5.25a,b}$$

Equations 5.25a and 5.25b are valid under the hypothesis of linearity from the point of view of the magnetic behavior of the machine. They highlight that, in the rotor flux reference frame, the quadrature component of the stator current acts only on the slip speed $\omega_{mr} - \omega_r$, while the rotor magnetizing current (rotor flux) is linked only to the direct component of the stator current. Figure 5.11b shows the block diagram of the flux model described by Equations 5.25a and 5.25b.

The block identified as $e^{-j\rho_r}$ permits the transformation of the stator current components i_{sD} and i_{sQ} in the stationary reference frame into the corresponding $i_{sx}^{\psi_r}$ and $i_{sy}^{\psi_r}$ in the rotor flux-oriented reference frame on the basis of the knowledge of the angle ρ_r.

The equations describing this transformation are as follows:

$$\begin{cases} i_{sx}^{\psi_r} = i_{sD} \cos(\rho_r) + i_{sQ} \, sen(\rho_r) \\[2mm] i_{sy}^{\psi_r} = -i_{sD} \, sen(\rho_r) + i_{sQ} \cos(\rho_r) \end{cases} \tag{5.26a,b}$$

The block diagram in Figure 5.11b highlights that the direct component of the stator current $i_{sx}^{\psi_r}$ is linked to the rotor magnetizing current $\mathbf{i}_{mr}$ by a dynamic relationship of the first order. At steady-state, therefore, when $\mathbf{i}_{mr}$ is constant, the rotor magnetizing current coincides with the direct component of the stator current in the field reference frame.

The main advantages of such a flux model are as follows:

- Simple equations
- Closed-loop integration for the flux computation
- Constant variables at steady-state

It should, however, be noted that in the model described by Equations 5.25a and 5.25b, both the amplitude and angle of the rotor magnetizing current depend significantly on the correct knowledge of the rotor time constant of the machine. A wrong knowledge of it causes undesired couplings between variables on the x and y axes. This coupling can result in the reduction of the dynamic performance of the drive, the reduction of its load capability and can finally lead the drive to an unstable behavior.

The flux model based on the rotor equations in the stationary reference frame is described by the following vector equation:

$$T_r \frac{d\mathbf{i}'_{mr}}{dt} = \mathbf{i}_s - \mathbf{i}'_{mr} + j\omega_r T_r \mathbf{i}'_{mr} \tag{5.27}$$

where $\mathbf{i}'_{mr}$ is the rotor magnetizing current space-vector in the stationary reference frame. From this equation, the two scalar equations on the direct sD and quadrature sQ axes can be obtained:

$$\begin{cases} T_r \dfrac{di'_{mrD}}{dt} = i_{sD} - i''_{mrD} - \omega_r T_r i''_{mrQ} \\[2mm] T_r \dfrac{di'_{mrQ}}{dt} = i_{sQ} - i''_{mrQ} + \omega_r T_r i''_{mrD} \end{cases} \tag{5.28a,b}$$

The block diagram of the flux model described by Equations 5.28a and 5.28b is drawn in Figure 5.11c. Even in this case, the block 3/2 performs the transformation of the stator currents i_{sA}, i_{sB}, and i_{sC} from the three-phase system to the biphase one in the stationary reference frame, providing i_{sD} and i_{sQ}, while the block R/P performs the transformation from Cartesian to polar coordinates. The correct estimation of the amplitude and phase of the rotor magnetizing current depends, even in this case, on the correct knowledge of the rotor time constant of the machine.

Different from the model described by Equations 5.25a and 5.25b, this flux model presents the following disadvantages:

- Open-loop integration for the flux computation
- AC electrical variables
- Higher estimation errors with respect to those obtainable with the current model in the field reference frame

As a matter of fact, manipulating Equation 5.27 and considering the steady-state values of all variables, the rotor magnetizing current $\mathbf{I}'_{mr}$ can be computed as a function of the stator current $\mathbf{I}_s$ and the pulsation of the stator current ω_1:

$$\mathbf{I}'_{mr} = \frac{\mathbf{I}_s}{1 + j(\omega_1 - \omega_r)T_r} \tag{5.29}$$

Equation 5.29 confirms that, for high values of the rotor speed and low values of the slip, a slight error in the rotor speed measurement can lead to a big error in the estimation of $|\mathbf{i}'_{mr}|$ and mainly of ρ_r. Some flux models based on a proper combination of the current and voltage models of the IM have been then devised, which are not very sensitive to errors on the measurement of the rotor speed [18,19].

5.3.3.3 Rotor Flux-Oriented Control with Impressed Currents

Rotor flux-oriented control with impressed currents can be obtained by employing a current-controlled cycloconverter, a current source converter (CSI) supplied by a controllable AC/DC converter, or adopting a current-controlled VSI, CRPWM (current-regulated PWM) supplied by an AC/DC converter.

The current-controlled cycloconverter is typically used only for high-power drives. This converter requires, in fact, the use of at least 36 thyristors. Moreover, its big limit is the maximum value of the fundamental frequency f_1 that is able to generate. The maximum achievable value is usually variable in the range $0 < f_1 < p_u \cdot f_{1max}/15$, where p_u is the number of pulses of the converter [17]. In case of a power grid at fundamental frequency of 50 Hz with a three-phase bridge converter ($p_u = 6$), then $f_{1max} \approx 20$ Hz. If a higher rated frequency power grid is available, as in the case of ship propulsion supplied by turbines or diesel engines, the frequency range at the converter output consequently increases.

Also the CSI supplied by an AC/DC converter with a smoothing inductor is usually adopted in high-power drives. In this case, the control of the magnitude of the stator current is performed by acting on the current at the DC level with a controlled AC/DC converter. The control of the fundamental frequency of the stator current is performed by acting on the CSI. The CSI requires the adoption of 12 thyristors, not necessarily of high switching frequency. As a matter of fact, the switching frequency of such a converter cannot be in any case very high. The dynamic response of the converter to any current reference variation strongly depends on the smoothing inductance on the DC side, on the machine inductance, on the gain of the current control loop on the DC side, on the rated value of the output voltage of the AC/DC converter, and, finally, on the load torque [18]. The commutation time can be suitably reduced by choosing a motor with a low leakage inductance. It should be noted that, with such a converter, the current absorbed by the motor has a square waveform, while the voltage on the motor windings exhibits a quasi-sinusoidal waveform, except for some spikes generated in the commutation instants of the current. The harmonic content of the stator current is therefore very high and this is further cause for a significant ripple present in the electromagnetic torque, which can be problematic at low rotating speed. This is the reason why this converter is not suitable for electrical drives to be used in machine tools where a continuous precise control of the angular position is required. When high dynamic performance are required to the drive, a CRPWM-controlled VSI is the best solution. If the DC link voltage is sufficiently high, it is possible to design the current control loop with a very high gain, permitting the measured currents to properly track their references. As recalled earlier, the current control permits avoiding the use of the dynamic relationship described by the stator equations of the machine. The control system is therefore highly simplified with respect to the case of voltage controlled VSI. A simple way to control the three stator currents is based on the hysteresis comparators. Figure 5.12 shows the block diagram of a stator current control system based on hysteresis comparators.

The phase stator currents i_{sA}, i_{sB}, and i_{sC} are instantaneously compared with the corresponding reference values i_{sAref}, i_{sBref}, and i_{sCref} provided by the control system, and the error is processed by hysteresis comparators. If the positive error is higher than half of the hysteresis band, then the upper device of the inverter leg is commanded; otherwise, if the negative error is lower than half of the hysteresis band, then the lower device of the inverter leg is commanded. The adoption of hysteresis comparators with reduced band permits the stator currents to be shaped with a very low harmonic content. On the contrary, the disadvantage is that the switching frequency of the converter is variable, which can be problematic from the point of view of a potential interference with other systems. The switching frequency depends on how quickly the stator current goes from the lower to the higher limit of the hysteresis band and vice versa. It depends on the magnitude of the DC link voltage, on the value of the back electromotive force offered by the motor and by the inductance of the motor itself. The switching frequency is, however, limited to few kilohertz if bipolar devices are used (e.g., BJT), while it can overcome even 16 kHz, above the acoustic frequencies, in case IGBTs or MOSFETs are used [17]. The use of MOSFETs permits the significant

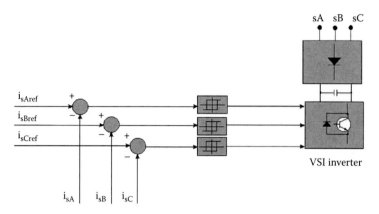

FIGURE 5.12
Block diagram of stator current control by hysteresis comparators.

increase of the switching frequency of the converter, from which a stator current waveform with lower harmonic content results, but it imposes an upper limit to the maximum power of the drive. For this reason, FOC with impressed current by a CRPWM converter can be a solution only in low-/medium-power drives. A suitable way to maintain constant the switching frequency is to adopt, instead of the hysteresis comparators, simple on–off comparators and to consequently command the device commutation at constant frequency. The constant frequency command of the converter is obviously the cause of an increase of the harmonic content of the stator current. Finally, another way to control the VSI with constant switching frequency is sketched in Figure 5.13 [19]. The actual value of each phase current is compared with its corresponding reference. The current error is processed by a PI controller, whose output is compared with a triangular carrier waveform (u_{carr} in Figure 5.13) with fixed frequency. The error is processed by a comparator which, on the basis of its sign, commands the upper or lower devices of the inverter leg. This technique, compared to the simple use of on–off comparators, permits the reduction of the harmonic content of the stator current, maintaining a constant switching frequency of the converter.

Figure 5.14 shows the block diagram of a rotor flux-oriented drive with impressed currents IM. The IM is supplied by a frequency converter, composed by a diode three-phase

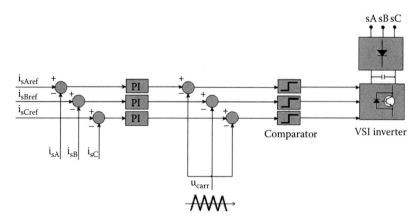

FIGURE 5.13
Block diagram of a current control system with constant switching frequency.

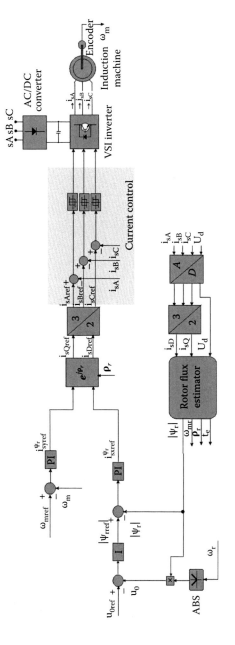

FIGURE 5.14
Block diagram of an IM drive with FOC based on current control with rotor flux orientation.

rectifier plus an IGBT-based VSI. The VSI is current controlled by high gain control loops. The block "current control" describes a current control system based on hysteresis comparators. This block could be substituted with the block diagram of the current control with constant switching frequency shown in Figure 5.13.

It can be observed that a closed-loop control of both the machine speed ω_m and the amplitude of the rotor flux ψ_r is made. The motor speed, assumed to be measured with an incremental encoder, is compared with its reference ω_{mref}. The speed error is then processed by a PI controller including an anti-wind-up system.* The output of the speed controller is the reference value of the quadrature component of the stator current space-vector $i_{syref}^{\psi_r}$ in the rotor flux-oriented reference frame. The control of the reference flux, below and above the rated speed of the motor, is made by the closed-loop control of the voltage u_0. Given the voltage reference u_{0ref}, this is compared with the actual voltage u_0, computed as the product of the estimated value of the amplitude of the rotor flux for the absolute value of the measured rotor speed. The tracking error is processed by an integrator, whose output is the rotor reference flux ψ_{rref}. The integrator output is limited to the value of the rotor flux in which the machine is assumed to work for rotating speeds below the rated one. This value is typically chosen in correspondence of the knee of the magnetic characteristic of the machine.† Below the rated speed of the machine, therefore, the rotor flux linkage is kept constant and equal to the limit value provided to the integrator. Above the rated speed, on the contrary, the control system reduces the reference rotor flux linkage so as to maintain the value of the product $|\omega_r| \cdot \psi_r$ constant and equal to the reference voltage u_{0ref}.

An alternative method, even if less efficient for the exploitation of the inverter is to open-loop control the reference rotor flux linkage. The reference flux linkage is, in this case, selected by a function generator having as input the current value of the rotor speed.

The reference rotor flux linkage ψ_{rref} is then compared with the estimated rotor flux linkage ψ_r, computed by the so-called flux model. The tracking error is processed by a PI controller. The output of the flux controller is the reference value of the direct component of the stator current space-vector $i_{sxref}^{\psi_r}$ expressed in the rotor flux-linkage reference frame.

The block "flux model" in Figure 5.14 can represent any of the flux models previously described. Typically, the current flux model based on the rotor equations in the rotor flux-linkage reference frame is used, as described by the block diagram in Figure 5.11b. The flux model has as output the angular position ρ_r, representing the angular displacement between the rotor flux-linkage space-vector and the direct axis in the stationary reference frame sD. The instantaneous value of ρ_r is necessary to correctly perform the field orientation and is the input of the block $e^{j\rho_r}$ in Figure 5.14. This block performs the transformation of the reference direct and quadrature components of the stator current space-vector $i_{sxref}^{\psi_r}$ and $i_{syref}^{\psi_r}$ in the rotor flux-linkage reference frame into the corresponding components i_{sDref} and i_{sQref} in the stationary reference frame. This transformation is described by the following equations:

$$\begin{cases} i_{sDref} = i_{sxref}^{\psi_r} \cos(\rho_r) - i_{syref}^{\psi_r} sen(\rho_r) \\ i_{sQref} = i_{sxref}^{\psi_r} sen(\rho_r) + i_{syref}^{\psi_r} \cos(\rho_r) \end{cases} \qquad (5.30a,b)$$

---

* The anti-wind-up system permits the action of the PI integrator to be mullified in the time interval when the output of the controller is limited by a saturator. This system permits the overshoot of the output of the control system to be significantly reduced.
† The magnetic characteristic of the machine is intended for the curve which gives the rotor flux linkage as a function of the rotor magnetizing current.

Finally, the block 2/3 in Figure 5.14 permits the reference current components i_{sDref} and i_{sQref} in the stationary reference frame to be transformed into the reference currents i_{sAref}, i_{sBref}, and i_{sCref} in the three-phase system.

5.3.3.4 Rotor Flux-Oriented Control with Impressed Voltages

FOC with impressed voltages is typically adopted in servo drives of rated power higher than 100 kW, requiring very high dynamic performance. At these power levels, VSIs with BJTs or, at increasing powers, GTOs or even thyristors could be adopted. The maximum switching frequency of these devices is usually between 100 Hz and some kilohertz. If the VSI were controlled in current at so low switching frequency, the harmonic content of the stator current would be too high. For such application, then, the VSI should be controlled in voltage by a proper PWM technique. This implies the necessity to take into consideration, in the control system, even the stator equations of the machine. Current control is therefore performed in the rotor flux-linkage space-vector reference frame, in correspondence to which the steady-state values of the currents are constant.

The relationship between the stator voltages and currents in the field-oriented reference frame is described by the vector equation (4.49) of Chapter 4, rewritten in the following for the sake of simplicity:

$$T_s' \frac{d\mathbf{i}_s^{\psi_r}}{dt} + \mathbf{i}_s^{\psi_r} = \frac{\mathbf{u}_s^{\psi_r}}{R_s} - j\omega_{mr} T_s' \mathbf{i}_s^{\psi_r} - (T_s - T_s') \left(j\omega_{mr} |\mathbf{i}_{mr}| + \frac{d|\mathbf{i}_{mr}|}{dt} \right) \tag{5.31}$$

See the list of symbols for their interpretation.

If Equation 5.31 is decomposed into the direct x and quadrature y axes in the field-oriented reference frame, the following equations can be obtained:

$$\begin{cases} T_s' \dfrac{di_{sx}^{\psi_r}}{dt} + i_{sx}^{\psi_r} = \dfrac{u_{sx}^{\psi_r}}{R_s} + \omega_{mr} T_s' i_{sy}^{\psi_r} - (T_s - T_s') \dfrac{d|\mathbf{i}_{mr}|}{dt} \\[4mm] T_s' \dfrac{di_{sy}^{\psi_r}}{dt} + i_{sy}^{\psi_r} = \dfrac{u_{sy}^{\psi_r}}{R_s} - \omega_{mr} T_s' i_{sx}^{\psi_r} - (T_s - T_s') \omega_{mr} |\mathbf{i}_{mr}| \end{cases} \tag{5.32a,b}$$

Equations 5.32a and 5.32b show that, with respect to the current components $i_{sx}^{\psi_r}$ and $i_{sy}^{\psi_r}$, the machine behaves as a first-order system with a time constant equal to the transient time constant of the machine and a gain equal to the inverse of the stator resistance. It should be noted, however, that an undesired coupling between the dynamic systems on the x and y axes exists [17,18]. This means that a variation of the voltage $u_{sx}^{\psi_r}$ determines the desired variation of the current $i_{sx}^{\psi_r}$ but also an undesired variation of the current $i_{sy}^{\psi_r}$. The two control variables $u_{sx}^{\psi_r}$ and $u_{sy}^{\psi_r}$ cannot thus be considered decoupled for the control of the rotor flux linkage and the electromagnetic torque. The current components $i_{sx}^{\psi_r}$ and $i_{sy}^{\psi_r}$ can be controlled independently only if the coupling terms of Equation 5.32 are eliminated. The coupling term in Equation 5.32a is $\omega_{mr} T_s' i_{sy}^{\psi_r}$, while that one in Equation 5.32b is $-\omega_{mr} T_s' i_{sx}^{\psi_r} - (T_s - T_s') \omega_{mr} |\mathbf{i}_{mr}|$.

If the delay times of the signal processing are neglected and if the system is assumed to work at constant rotor flux amplitude, the decoupling of the direct and quadrature axis

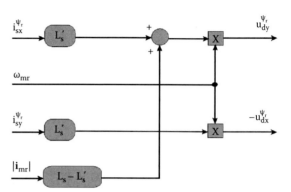

FIGURE 5.15
Block diagram of the decoupling circuit.

systems can be achieved by adding at each sampling time of the control system the following voltage terms $u_{dx}^{\psi_r}$ and $u_{dy}^{\psi_r}$ to the outputs $\hat{u}_{sx}^{\psi_r}$ and $\hat{u}_{sy}^{\psi_r}$ of the current controllers:

$$\begin{cases} u_{dx}^{\psi_r} = -\omega_{mr} L_s' i_{sy}^{\psi_r} \\ u_{dy}^{\psi_r} = \omega_{mr} L_s' i_{sx}^{\psi_r} + (L_s - L_s')\omega_{mr} |\mathbf{i}_{mr}| \end{cases} \tag{5.33a,b}$$

The voltage components $u_{sxref}^{\psi_r}$ and $u_{syref}^{\psi_r}$ which the control system gives to the PWM are then as follows:

$$\begin{cases} u_{sxref}^{\psi_r} = \hat{u}_{sx}^{\psi_r} + u_{dx}^{\psi_r} = R_s i_{sx}^{\psi_r} + L_s' \dfrac{di_{sx}^{\psi_r}}{dt} \\ u_{syref}^{\psi_r} = \hat{u}_{sy}^{\psi_r} + u_{dy}^{\psi_r} = R_s i_{sy}^{\psi_r} + L_s' \dfrac{di_{sy}^{\psi_r}}{dt} \end{cases} \tag{5.34a,b}$$

Figure 5.15 shows the block diagram of the decoupling circuit described by Equations 5.33.

The values of $i_{sx}^{\psi_r}$ and $i_{sy}^{\psi_r}$ to be adopted by the decoupling circuit are obtained by the measurements of the real currents of the machine, while the values of $\mathbf{i}_{mr}$ and ω_{mr} are computed by the flux model. Other decoupling circuits, different from that in Figure 5.18, can be derived having as input the stator voltages $\hat{u}_{sx}^{\psi_r}$ and $\hat{u}_{sy}^{\psi_r}$ instead of the stator currents $i_{sx}^{\psi_r}$ and $i_{sy}^{\psi_r}$ [18]. Furthermore, some decoupling circuits taking into consideration also the delay times introduced by the power converter and the signal processing system have been introduced [18]. It is noteworthy that an undesired coupling between the x and y axis systems could be a cause of instability of the drive and, in any case, of a significant worsening of the dynamic performance of the FOC. The effectiveness of the decoupling circuit is high in dependance of the accuracy in the knowledge of the parameters L_s and σ of the motor. These parameters can vary during the normal operation of the drive, because of the variations of the rotor flux-linkage amplitude (e.g., in a field-weakening region). The output of the control system is the value of the voltage that the inverter should generate. For this purpose several PWM techniques could be used, among which the carrier-based and the space-vector techniques (see Chapter 2 for more details).

Figure 5.16 shows the block diagram of a rotor flux-oriented IM drive with impressed voltages. The control scheme is, in this case, much more complex than that with impressed currents. In particular, it should be noted that current control is performed here in the field-oriented reference frame, and a decoupling circuit is present (see Figure 5.15). A control of both the angular speed of the drive ω_m and the amplitude of its rotor flux linkage ψ_r is adopted in this scheme.

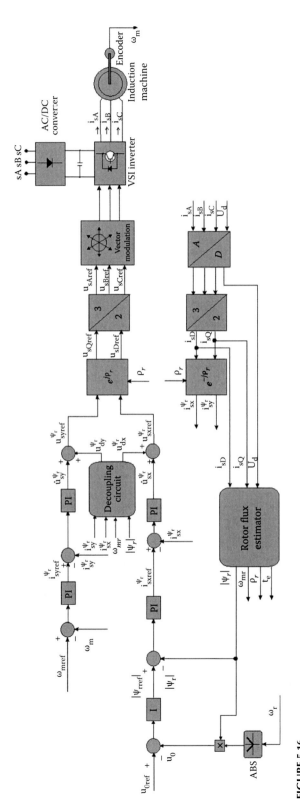

FIGURE 5.16
Block diagram of an IM drive with FOC based on voltage control with rotor flux orientation.

On the direct axis, three controllers are present. The voltage u_0, equal to the product between the absolute value of the machine speed ω_r and the amplitude of the rotor flux linkage ψ_r, is kept constant by an integral (I) controller. The output of this controller is the reference value of the rotor flux-linkage amplitude ψ_{rref}. The rotor flux-linkage amplitude ψ_r is closed-loop controlled by a PI controller, whose output is the reference value of the direct component of the stator current in the field-oriented reference frame $i_{sxref}^{\psi_r}$. The rotor flux amplitude ψ_r is estimated by the block "flux model," which implements any of the flux models previously described. The flux model computes also the angle ρ_r, expressing the angular position of the rotor flux-linkage space-vector, necessary for performing the field orientation by the coordinate rotations. The direct component of the stator current $i_{sx}^{\psi_r}$ is closed-loop controlled by a PI controller, whose output is the reference value of the direct component of the stator voltage in the field-oriented reference frame $\hat{u}_{sx}^{\psi_r}$. For the reasons explained above to this last term, a voltage component $u_{dx}^{\psi_r}$ coming from the decoupling circuit is added to obtain the voltage term $u_{sxref}^{\psi_r}$, that is, the reference direct component of the stator voltage in the field-oriented reference frame.

On the quadrature axis, two controllers are present as well. The motor speed ω_m is closed-loop controlled to its reference ω_{mref} by a PI controller. The output of this controller is the reference value of the quadrature component of the stator current in the field-oriented reference frame $i_{syref}^{\psi_r}$. The quadrature component of the stator current $i_{sy}^{\psi_r}$ is closed-loop controlled by a PI controller, whose output is the reference value of the quadrature component of stator voltage in the field-oriented reference frame $\hat{u}_{sy}^{\psi_r}$. For the reasons explained above to this last term, a voltage component $u_{dy}^{\psi_r}$ coming from the decoupling circuit is added to obtain the voltage term $u_{syref}^{\psi_r}$, that is, the reference quadrature component of the stator voltage in field-oriented reference frame.

The block $e^{j\rho_r}$ performs a vector rotation from the field-oriented to the stationary reference frame on the basis of the instantaneous knowledge of the rotor flux-linkage angle ρ_r, provided by the flux model. This transformation, described by the following equations, permits the voltage components $u_{sxref}^{\psi_r}$ and $u_{syref}^{\psi_r}$ to be transformed into u_{sDref} and u_{sQref}:

$$\begin{cases} u_{sDref}^{\psi_r} = u_{sxref}^{\psi_r}\cos(\rho_r) - u_{syref}^{\psi_r}\,sen(\rho_r) \\ u_{sQref}^{\psi_r} = u_{sxref}^{\psi_r}\,sen(\rho_r) + u_{syref}^{\psi_r}\cos(\rho_r) \end{cases} \qquad (5.35\text{a,b})$$

Finally, the block 2/3 permits the voltage components u_{sDref} and u_{sQref} to be transformed into the corresponding three-phase components u_{sAref}, u_{sBref}, and u_{sCref}.

5.3.3.5 Experimental Results on a Rotor Flux-Oriented Drive with Impressed Voltages

In the following, some experimental results of a rotor flux-oriented IM drive with impressed voltages are presented. The rotor flux-oriented scheme in Figure 5.16 has been adopted.

Firstly, a test setup with a 2.2 kW IM has been adopted (see Table 4.1 for the parameters of the machine). The test is composed of the following speed step references $0 \rightarrow 100 \rightarrow -100 \rightarrow 0$ rad/s. Correspondingly, the following step load torques have been applied: at t = 0.5 s, a step load torque of 6 N m has been applied; at t = 1 s, the load torque has been set to 0; at t = 2 s, a step load torque of −6 N m has been applied; and at t = 2.5 s, the load torque has been set to 0.

Figure 5.17 shows the reference and measured speed during this test. It can be observed that the measured speed property tracks its reference, with a quick response to any sudden application of the load torque. Figure 5.18 shows the reference and measured i_{sx} and i_{sy} current components. It can be observed that the direct component of the stator current

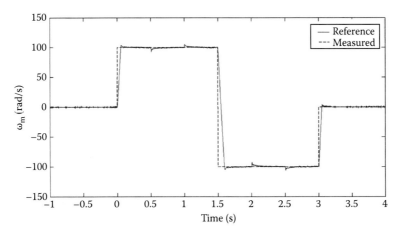

FIGURE 5.17
Reference and measured speed of the FOC drive.

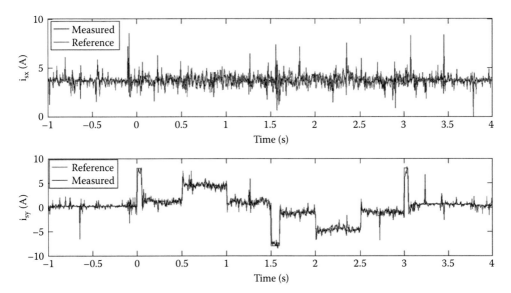

FIGURE 5.18
Reference and measured i_{sx}, i_{sy} of the FOC drive.

is controlled to a constant value, according to a constant magnetic rotor flux amplitude command. The quadrature component of the stator current presents a step variation at each speed transient or at any sudden application of the load torques, as expected. This is confirmed by Figure 5.19, showing the load and electromagnetic torque of the drive.

Secondly, a test setup with a 22 kW IM has been adopted (see Table 4.2 for the parameters of the machine). The test is composed of the following speed step references, including the field-weakening region: $0 \rightarrow -188 \rightarrow -63 \rightarrow 63 \rightarrow 188 \rightarrow 0$ rad/s. Correspondingly, the following step load torques have been applied: at t = 5 s, a step load torque of −60 N m has been applied; at t = 6 s, the load torque has been set to 0; at t = 8 s, a step load torque o 60 N m has been applied; and at t = 9 s, the load torque has been set to 0.

Figure 5.20 shows the reference and measured speed during this test. It can be observed that the measured speed property tracks its reference, with a quick response to any sudden

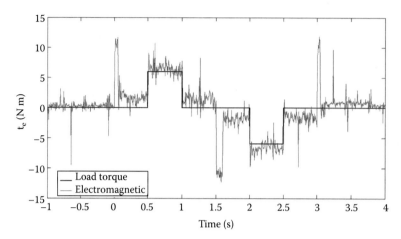

FIGURE 5.19
Load and electromagnetic torque of the FOC drive.

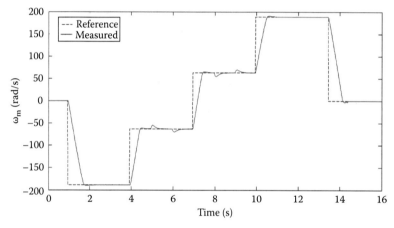

FIGURE 5.20
Reference and measured speed of the FOC drive.

application of the load torque. Figure 5.21 shows the measured i_{sx} and i_{sy} current components. It can be observed that the direct component of the stator current is controlled to a constant value, according to a constant magnetic rotor flux amplitude command, except when the machine works in field-weakening region where the i_{sx} is reduced. The quadrature component of the stator current presents a step variation at each speed transient or at any sudden application of the load torque, as expected. This is confirmed by Figure 5.22, showing the load and electromagnetic torque of the drive.

5.3.4 Stator Flux-Oriented Control

This section describes the stator flux-linkage-oriented control of the IM. Since there are several similarities between rotor flux and stator flux orientation, only the significant differences are emphasized.

Equation 5.14, rewritten in the following for the sake of simplicity, shows that the electromagnetic torque is proportional to the product between the stator flux-linkage

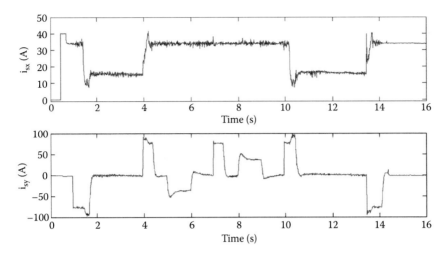

FIGURE 5.21
Measured i_{sx}, i_{sy} of the FOC drive.

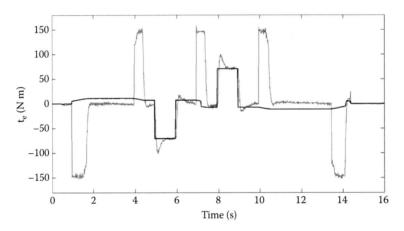

FIGURE 5.22
Load and electromagnetic torque of the FOC drive.

amplitude and the quadrature component of the stator current, expressed in the stator flux-oriented reference frame:

$$t_e = \frac{3}{2} p |\boldsymbol{\psi}_s| i_{sy}^{\psi_s} = \frac{3}{2} p L_m |\mathbf{i}_{ms}| i_{sy}^{\psi_s} \tag{5.36}$$

where

$\boldsymbol{\psi}_s$ is the stator flux-linkage space-vector

$i_{sy}^{\psi_s}$ is the quadrature component of the stator current in the stator flux-oriented reference frame

$\mathbf{i}_{ms}$ is the stator magnetizing current space-vector, defined as follows:

$$|\mathbf{i}_{ms}| = \frac{|\boldsymbol{\psi}_s|}{L_m} = \mathbf{i}_r^{\psi_s} + \frac{L_s}{L_m} \mathbf{i}_s^{\psi_s} = \left(i_{rx}^{\psi_s} + j i_{ry}^{\psi_s} \right) + \frac{L_s}{L_m} \left(i_{sx}^{\psi_s} + j i_{sy}^{\psi_s} \right) \tag{5.37}$$

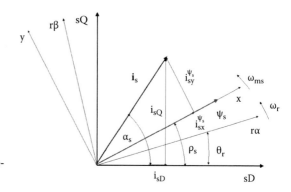

FIGURE 5.23
Representation of the stator flux-linkage space-vector and the stator current.

The apex term ψ_s denotes the stator flux reference frame in which all the variables are expressed. Figure 5.23 shows the vector diagram with the different variables.

Even in this case, if the machine is supplied by a current generator, the dynamic model of the system is significantly simplified. Supplying the machine with impressed currents, it is possible to take into consideration, in the control system, simply the rotor space-vector equations of the machine, without considering the stator ones.

Adopting the stator flux-oriented reference frame, the rotor space-vector equation is the following:

$$R_r\left[|\mathbf{i}_{ms}| - \frac{L_s}{L_m}\mathbf{i}_s^{\psi_s}\right] + L_r\frac{d|\mathbf{i}_{ms}|}{dt} = \frac{L_s'L_r}{L_m}\frac{d\mathbf{i}_s^{\psi_s}}{dt} - j\omega_{sl}\left[L_r|\mathbf{i}_{ms}| - \frac{L_s'L_r}{L_m}\mathbf{i}_s^{\psi_s}\right] \tag{5.38}$$

where $\omega_{sl} = \omega_{ms} - \omega_r$ is the slip pulsation, equal to the difference between the stator flux-linkage speed ω_{ms} and the rotor speed ω_r.

The space-vector Equation 5.38 can be split into the direct x and quadrature y axes of the field reference frame, giving the following scalar equations:

$$\begin{cases} \dfrac{L_m}{L_s'}\dfrac{d|\mathbf{i}_{ms}|}{dt} + \dfrac{L_m}{L_sT_r'}|\mathbf{i}_{ms}| = \dfrac{di_{sx}^{\psi_s}}{dt} + \dfrac{1}{T_r'}i_{sx}^{\psi_s} - \omega_{sl}i_{sy}^{\psi_s} \\[4mm] \omega_{sl}\left(\dfrac{L_m}{L_s'}|\mathbf{i}_{ms}| - i_{sx}^{\psi_s}\right) = \dfrac{di_{sy}^{\psi_s}}{dt} + \dfrac{1}{T_r'}i_{sy}^{\psi_s} \end{cases} \tag{5.39a,b}$$

where T_r' is the rotor transient time constant. A comparison between Equations 5.38 and 5.16 shows that the rotor equation in the stator flux reference frame is quite more complicated than the corresponding one in the rotor flux reference frame. In particular, there exists a coupling between the two axes: a variation occurring on the torque current component $i_{sy}^{\psi_s}$ will cause a transient variation on the flux component $i_{sx}^{\psi_s}$. This undesired coupling is to be compensated with a decoupling circuit that, differently from the rotor flux orientation, acts on the current components control instead of on the voltage references. Under the assumption that the stator flux magnitude is fed back in the control system, a current component $i_{dx}^{\psi_s}$ should then be added to the output of the flux controller $\hat{i}_{sx}^{\psi_s}$ to provide the reference direct component of the stator current $i_{sxref}^{\psi_s}$. It can be therefore written as follows:

$$i_{sx}^{\psi_s} = \hat{i}_{sx}^{\psi_s} + i_{dx}^{\psi_s} \tag{5.40}$$

The $i_{dx}^{\psi_s}$ term can be easily computed after substituting Equation 5.40 in Equation 5.39a, obtaining the following equation:

$$i_{dx}^{\psi_s} = \omega_{sl} \frac{T_r'}{1+sT_r'} i_{sy}^{\psi_s} \tag{5.41}$$

where s is the Laplace variable. Equation 5.41 requires the knowledge of the slip pulsation ω_{sl}, that can be obtained from Equation 5.39b as follows:

$$\omega_{sl} = \frac{(1+sT_r')L_s}{T_r \left(L_m |\mathbf{i}_{ms}| - L_s' i_{sx}^{\psi_s} \right)} i_{sy}^{\psi_s} \tag{5.42}$$

Equation 5.41, together with Equation 5.42, defines the decoupling circuit to be adopted in a stator flux-oriented control with impressed currents. Figure 5.24 shows the block diagram of a stator flux-oriented IM drive with impressed currents. It is basically the same scheme of Figure 5.17, with few differences. The main difference is that, being the orientation on the stator flux vector, all vector rotations need the knowledge of the stator flux space-vector angle ρ_s instead of the rotor one. This angle is provided by the flux model, estimating, in this case, both the amplitude and the angular position of the stator flux space-vector. Moreover, the output of the PI flux controller $\hat{i}_{sx}^{\psi_s}$ is added to the decoupling term $i_{dx}^{\psi_s}$, defined as in Equation 5.41.

If the machine is supplied by a voltage generator, also the stator equations of the IM should be considered. If the stator equation (4.48) in Chapter 4 (written in the general reference frame) is expressed in the stator flux-linkage reference frame, the following space-vector equations can be obtained:

$$\mathbf{u}_s^{\psi_s} = R_s \mathbf{i}_s^{\psi_s} + L_m \frac{d|\mathbf{i}_{ms}|}{dt} + j\omega_{ms} L_m |\mathbf{i}_{ms}| \tag{5.43}$$

obtained by rewriting the rotor vector as a function of the stator current and the stator magnetizing space-vectors. It can easily be observed that Equation 5.43 is much simpler than the corresponding Equation 5.31, written in the rotor flux reference frame. Equation 5.43 does not contain the derivative of the stator current, differently from Equation 5.31. If the stator magnetizing current amplitude is kept constant, Equation 5.43 is significantly simplified; decomposing it into the direct and quadrature axes, the following two scalar equations are obtained:

$$\begin{cases} i_{sx}^{\psi_s} = \dfrac{u_{sx}^{\psi_s}}{R_s} - \dfrac{L_m}{R_s} \dfrac{d|\mathbf{i}_{ms}|}{dt} \\[2mm] i_{sy}^{\psi_s} = \dfrac{u_{sy}^{\psi_s}}{R_s} - \omega_{ms} \dfrac{|\mathbf{i}_{ms}|}{R_s} \end{cases} \tag{5.44a,b}$$

Equations 5.44a and 5.44b are much simpler than Equations 5.32a and 5.32b written in the rotor flux reference frame. Like in the rotor flux orientation counterpart, $u_{sx}^{\psi_s}$ and $u_{sy}^{\psi_s}$ cannot be considered as decoupled variables for the control of the stator flux and electromagnetic torque. As a matter of fact, some coupling terms appear, which are, however, much

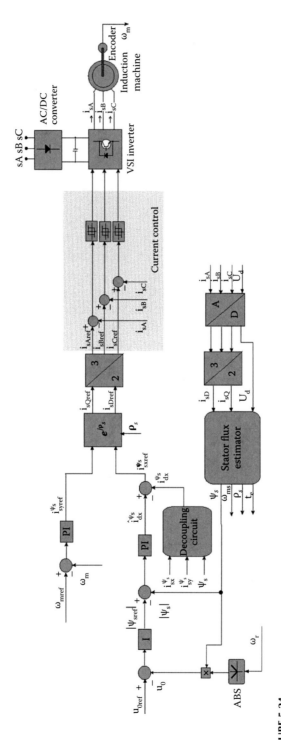

FIGURE 5.24
Block diagram of an IM drive with FOC based on current control with stator flux orientation.

simpler than the corresponding counterparts in the rotor flux-linkage orientation scheme. The decoupling voltage terms that must be added at the output of the current controllers are the following:

$$
\begin{cases}
u_{dx}^{\psi_s} = 0 \\
u_{dy}^{\psi_s} = \omega_{ms} L_m \left| \mathbf{i}_{ms} \right|
\end{cases}
\tag{5.45}
$$

A simple decoupling circuit can therefore be used, having at input just the amplitude of the stator magnetizing current space-vector $\left| \mathbf{i}_{ms} \right|$, or in the same manner the stator flux amplitude, and its rotational speed ω_{ms}. Both of these terms are provided by the flux model. Figure 5.25 shows the block diagram of a stator flux-oriented IM drive with impressed voltages. It can be observed that some decoupling terms at the output of the flux controller (Equation 5.45) and at the output of the quadrature axis current controller (Equation 5.41) are present, as expected on the basis of the earlier considerations. Besides the different decoupling terms, the control scheme is basically the same as the one obtained with the rotor flux orientation, with the main difference that all vector rotations are based on the knowledge of the stator flux-linkage angle ρ_s.

5.3.4.1 Stator Flux-Linkage Acquisition

Like rotor flux orientation, the stator flux linkage is hardly measured by proper sensors. It is usually estimated by the so-called flux models. Even in this case, there are several possible solutions. The main solutions are the so-called voltage and current models. The first is based on the stator equations, while the second on the rotor equations of the IM.

With regard to the voltage model, after processing Equation 4.48, expressing the rotor current space-vector as a function of the stator current and the stator magnetizing current space-vectors, the following space-vector equation can be obtained based on the stator equations of the machine, written in the stationary reference frame:

$$
\frac{L_m}{R_s} \frac{d\mathbf{i}_{ms}}{dt} = \frac{\mathbf{u}_s}{R_s} - \mathbf{i}_s
\tag{5.46}
$$

The space-vector equation (5.46) can be decomposed into the direct and quadrature components in the stationary reference frame:

$$
\begin{cases}
\dfrac{L_m}{R_s} \dfrac{di_{msD}}{dt} = \dfrac{u_{sD}}{R_s} - i_{sD} \\
\dfrac{L_m}{R_s} \dfrac{di_{msQ}}{dt} = \dfrac{u_{sQ}}{R_s} - i_{sQ}
\end{cases}
\tag{5.47a,b}
$$

Figure 5.26 shows the block diagram of the voltage model, representing Equations 5.47a and 5.47b. Advantages and disadvantages of such model are the same of those summarized in Section 5.3.3.1.

Another model can be deduced, which is based on the stator equations of the machine, but written in the stator flux-linkage reference frame. Figure 5.27 shows the block diagram of this flux model. In this case, the stator voltage components in stator reference frame are subtracted of the ohmic drops, obtaining the terms $d\psi_{sD}/dt$ and $d\psi_{sQ}/dt$. These two

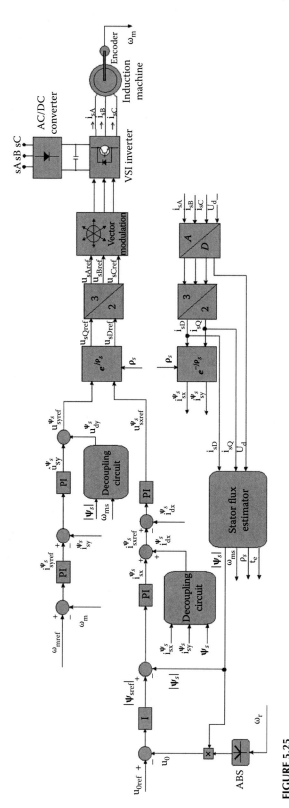

FIGURE 5.25

Block diagram of an IM drive with FOC based on voltage control with stator flux orientation.

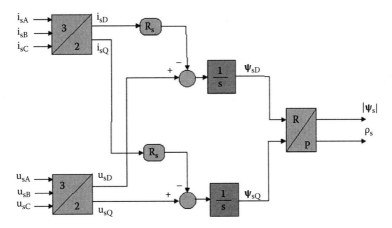

FIGURE 5.26
Block diagram of the voltage model.

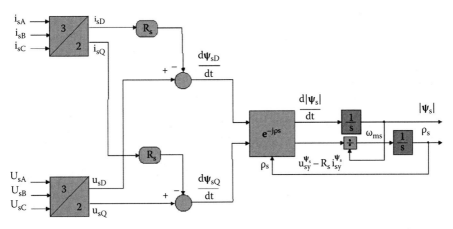

FIGURE 5.27
Block diagram of the voltage model in the stator flux-oriented reference frame.

terms are then transformed in the stator flux reference frame by means of the vector rotation $e^{-j\rho_s}$. The resulting direct component is further integrated to obtain the amplitude of the stator flux space-vector. The resulting quadrature component, equal to $u_{sy}^{\psi_s} - R_s i_{sy}^{\psi_s}$, is further divided by $|\psi_s|$ to obtain the angular speed of the stator flux vector ω_{ms}. Its integration provides the angular position of the stator flux vector ρ_s. The main advantage of this scheme is the closed-loop integration of the stator flux with consequent absence of the problems of DC drift and open-loop integration. This is obviously paid back with an increase of the complexity and the computational demand of such a model.

5.3.5 Magnetizing Flux-Oriented Control

This section describes the magnetizing flux-linkage-oriented control of the IM [18]. Since there are several similarities between rotor flux and magnetizing flux orientation, only the significant differences are emphasized.

Equation 5.15, rewritten in the following for the sake of simplicity, shows that the electromagnetic torque is proportional to the product between the magnetizing flux-linkage amplitude and the quadrature component of the stator current, expressed in the magnetizing flux-oriented reference frame:

$$t_e = \frac{3}{2} p \left| \mathbf{\Psi}_m \right| i_{sy}^{\psi_m} = \frac{3}{2} p L_m \left| \mathbf{i}_{mm} \right| i_{sy}^{\psi_m} \tag{5.48}$$

where

$\mathbf{\Psi}_m$ is the magnetizing flux-linkage space-vector
$i_{sy}^{\psi_m}$ is the quadrature component of the stator current in the magnetizing flux-oriented reference frame
$\mathbf{i}_{mm}$ is the magnetizing current space-vector, defined as follows:

$$\left| \mathbf{i}_{mm} \right| = \frac{\left| \mathbf{\Psi}_m \right|}{L_m} = \mathbf{i}_r^{\psi_m} + \mathbf{i}_s^{\psi_m} = \left(i_{rx}^{\psi_m} + j i_{ry}^{\psi_m} \right) + \left(i_{sx}^{\psi_m} + j i_{sy}^{\psi_m} \right) \tag{5.49}$$

The apex term $\mathbf{\psi}_m$ denotes the magnetizing flux reference frame in which all the variables are expressed. Figure 5.28 shows the vector diagram with the different variables.

If the machine is supplied by a current generator, the dynamic model of the system is significantly simplified. Supplying the machine with impressed currents, it is possible to take into consideration, in the control system, simply the rotor space-vector equations of the machine, without considering the stator ones.

Adopting the magnetizing flux-oriented reference frame, the rotor space-vector equation is the following:

$$R_r \left[\left| \mathbf{i}_{mm} \right| - \mathbf{i}_s^{\psi_m} \right] + L_r \frac{d \left| \mathbf{i}_{mm} \right|}{dt} = L_{r\sigma} \frac{d \mathbf{i}_s^{\psi_m}}{dt} - j(\omega_{mm} - \omega_r) \left[L_r \left| \mathbf{i}_{mm} \right| - L_{r\sigma} \mathbf{i}_s^{\psi_m} \right] \tag{5.50}$$

where

ω_{mm} is rotational speed of the magnetizing space-vector
$L_{r\sigma}$ is the rotor leakage inductance

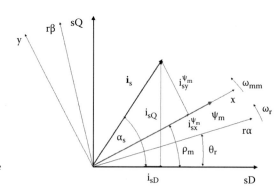

FIGURE 5.28
Representation of the magnetizing flux-linkage space-vector and the stator current.

The space-vector equation (5.50) can be split on the direct x and quadrature y axes of the field reference frame, giving the following scalar equations:

$$\begin{cases} \dfrac{|\mathbf{i}_{mm}| + T_r d|\mathbf{i}_{mm}|/dt}{T_{r\sigma}} = \dfrac{di_{sx}^{\psi m}}{dt} + \dfrac{1}{T_{r\sigma}} i_{sx}^{\psi m} - \omega_{sl} i_{sy}^{\psi m} \\ \omega_{sl}\left(\dfrac{T_r}{T_{r\sigma}}|\mathbf{i}_{mm}| - i_{sx}^{\psi m}\right) = \dfrac{di_{sy}^{\psi m}}{dt} + \dfrac{1}{T_{r\sigma}} i_{sy}^{\psi m} \end{cases}$$ (5.51a,b)

where

$T_{r\sigma} = L_{r\sigma}/R_r$ is the rotor leakage time constant

$\omega_{sl} = \omega_{mm} - \omega_r$ is the slip speed, equal to the difference between the magnetizing flux space-vector and the rotor speed

Equations 5.51a and 5.51b shows that there is an undesired coupling between the two equations: as a consequence, $|\mathbf{i}_{mm}|$ will be modified by any change of the torque current component $i_{sy}^{\psi m}$. This undesired coupling is to be compensated with a decoupling circuit that acts on the current components. Under the assumption that the magnetizing flux magnitude is fed back in the control system, a current component $i_{dx}^{\psi m}$ should thus be added to the output of the flux controller $\hat{i}_{sx}^{\psi m}$ to provide the reference direct component of the stator current $i_{sxref}^{\psi m}$. It can be thus written as follows:

$$i_{sx}^{\psi m} = \hat{i}_{sx}^{\psi m} + i_{dx}^{\psi m}$$ (5.52)

The $i_{dx}^{\psi m}$ term can be easily computed after substituting Equation 5.52 in Equation 5.51a, obtaining the following:

$$\frac{|\mathbf{i}_{mm}| + T_r d|\mathbf{i}_{mm}|/dt}{T_{r\sigma}} = \frac{d\hat{i}_{sx}^{\psi m}}{dt} + \frac{1}{T_{r\sigma}} \hat{i}_{sx}^{\psi m} + \frac{di_{dx}^{\psi m}}{dt} + \frac{1}{T_{r\sigma}} i_{dx}^{\psi m} - \omega_{sl} i_{sy}^{\psi m}$$ (5.53)

The correct decoupling term to be added is thus as follows:

$$i_{dx}^{\psi m} = \frac{T_{r\sigma}\omega_{sl}}{1 + sT_{r\sigma}} i_{sy}^{\psi m}$$ (5.54)

where s is the Laplace variable. Equation 5.54 requires the knowledge of the slip pulsation ω_{sl}, that can be obtained from Equation 5.51b as follows:

$$\omega_{sl} = \frac{(1 + sT_{r\sigma})}{T_r|\mathbf{i}_{mm}| - T_{r\sigma}i_{sx}^{\psi m}} i_{sy}^{\psi m}$$ (5.55)

Equation 5.54, together with Equation 5.55, defines the decoupling circuit to be adopted in a magnetizing flux-oriented control with impressed currents. Figure 5.29 shows the block diagram of a magnetizing flux-oriented IM drive with impressed currents. It is basically the same scheme of Figure 5.17, with few differences. The main difference is that, being the orientation on the magnetizing flux vector, all vector rotation needs the knowledge of the magnetizing flux space-vector angle ρ_m instead of the rotor one. This angle is provided by the flux model, estimating, in this case, both the amplitude and the angular position

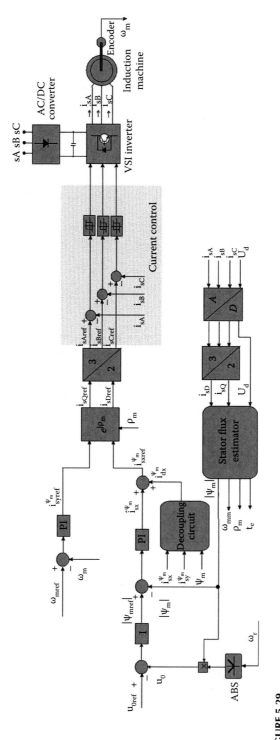

FIGURE 5.29

Block diagram of an IM drive with FOC based on current control with magnetizing flux orientation.

of the magnetizing flux space-vector. Moreover, the output of the PI flux controller $\hat{i}_{sx}^{\psi m}$ is added the decoupling term $i_{dx}^{\psi m}$, defined as in Equation 5.54.

If the machine is supplied by a voltage generator, also the stator equations of the IM should be considered. If the stator equation (4.48) in Chapter 4 (written in the general reference frame) is particularized in the magnetizing flux-linkage reference frame, the following space-vector equations can be obtained:

$$\mathbf{u}_s^{\psi m} = R_s \mathbf{i}_s^{\psi m} + L_{s\sigma} \frac{d\mathbf{i}_s^{\psi m}}{dt} + L_m \frac{d|\mathbf{i}_{mm}|}{dt} + j\omega_{mm}\left(L_{s\sigma}\mathbf{i}_s^{\psi m} + L_m|\mathbf{i}_{mm}|\right) \qquad (5.56)$$

where $L_{s\sigma} = L_s - L_m$ is the stator leakage inductance. This last equation is obtained by rewriting the rotor vector as a function of the stator and magnetizing current space-vectors. Comparing Equation 5.56 with Equation 5.43, it is noteworthy that the stator equation written in the magnetizing flux reference frame is much more complicated than the one written in the stator flux one. As a matter of fact, Equation 5.56 contains the derivative of the stator current space-vector and a separate stator leakage voltage drop, whereas the corresponding equation in the stator flux reference frame does not.

After decomposing Equation 5.56 into the direct and quadrature axes, the following scalar equations are obtained:

$$\begin{cases} i_{sx}^{\psi m} = \dfrac{u_{sx}^{\psi m}}{R_s} - T_{s\sigma}\dfrac{di_{sx}^{\psi m}}{dt} - \dfrac{L_m}{R_s}\dfrac{d|\mathbf{i}_{mm}|}{dt} + \omega_{mm}T_{s\sigma}i_{sy}^{\psi m} \\[3mm] i_{sy}^{\psi m} = \dfrac{u_{sy}^{\psi m}}{R_s} - T_{s\sigma}\dfrac{di_{sy}^{\psi m}}{dt} - \omega_{mm}\dfrac{\left(L_{s\sigma}i_{sx}^{\psi m} + L_m|\mathbf{i}_{mm}|\right)}{R_s} \end{cases} \qquad (5.57a,b)$$

where $T_{s\sigma} = L_{s\sigma}/R_s$ is the stator leakage time constant. Like in the rotor flux orientation counterpart, $u_{sx}^{\psi m}$ and $u_{sy}^{\psi m}$ cannot be considered as decoupled variables for the control of the magnetizing flux and electromagnetic torque. As a matter of fact, some undesired coupling terms appear. The decoupling voltage terms that must be added at the output of the current controllers are the following, under the assumption that the magnetizing flux amplitude is kept constant by the controller:

$$\begin{cases} u_{dx}^{\psi m} = -\omega_{mm}L_{s\sigma}i_{sy}^{\psi m} \\[3mm] u_{dy}^{\psi m} = \omega_{mm}\left(L_{s\sigma}i_{sx}^{\psi m} + L_m|\mathbf{i}_{mm}|\right) \end{cases} \qquad (5.58)$$

These equations are very similar to Equations 5.33a and 5.33b, written for the rotor flux-linkage orientation, and the corresponding block diagram describing the decoupling circuit is sketched in Figure 5.30. This decoupling circuit has as input the amplitude of the magnetizing current space-vector $|\mathbf{i}_{mm}|$ and its rotational speed ω_{mm} as well as the direct and quadrature components of the stator current $i_{sx}^{\psi m}$ and $i_{sy}^{\psi m}$. Both of these terms are provided by the flux model. Figure 5.31 shows the block diagram of a stator flux-oriented IM drive with impressed voltages. It can be observed that some decoupling terms at the output of the flux controller (Equation 5.54) and at the output of the quadrature axis current controller (Equation 5.58) are present, as expected on the basis of the earlier considerations.

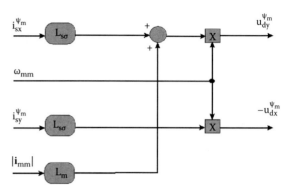

FIGURE 5.30
Block diagram of the decoupling circuit.

Besides the different decoupling terms, the control scheme is basically the same as the one obtained with the rotor flux orientation, with the main difference that all vector rotations are based on the knowledge of the magnetizing flux-linkage angle ρ_m.

5.4 DTC of IM

DTC has been developed around mid-80's [9–13,18]. The ABB company has been, however, the first and unique company which, in 1995, introduced in the marked an electrical drive based on a DTC.

By means of the DTC, it is possible, like the FOC, to control instantaneously and in a decoupled way the magnetic flux of the machine and its electromagnetic torque. The control actions, in this case, aim to track the references of magnetic flux and electromagnetic torque, by means of the application of suitable values of stator voltages of the machine. On this basis, DTC can be implemented by adopting a simple VSI and does not require the adoption either of hysteresis comparators for current control (typical of FOC with impressed currents) or of PWMs (typical of FOC with impressed voltages). This is the reason for the significant simplification of the control system, with the corresponding advantages in terms of cost reduction and increase of the system reliability.

In DTC electrical drives with IM supplied by VSIs, it is possible to directly control the stator (or rotor or even magnetizing) flux linkage and the electromagnetic torque by the selection of optimal switching configurations of the inverter. The selection of such switching configurations is made so as to maintain instantaneously the loci of the stator flux linkage and electromagnetic torque inside hysteresis bands centered on the relative reference values. This kind of control results in high dynamic performance in the torque response, low switching frequency of the inverter, and low losses due to the harmonics.

5.4.1 Electromagnetic Torque Generation in the IM

The electromagnetic torque developed by an IM can be expressed as a function of the stator and rotor flux-linkage space-vectors as follows [18]:

$$t_e = \frac{3}{2}p\frac{L_m}{L_s'L_r}\boldsymbol{\psi}_r' \wedge \boldsymbol{\psi}_s = \frac{3}{2}p\frac{L_m}{L_s'L_r}|\boldsymbol{\psi}_r'||\boldsymbol{\psi}_s|sin(\rho_s-\rho_r) = \frac{3}{2}p\frac{L_m}{L_s'L_r}|\boldsymbol{\psi}_r'||\boldsymbol{\psi}_s|sin\gamma \tag{5.59}$$

See the list of symbols for their interpretation. Equation 5.59 shows that the electromagnetic torque depends on the product between the amplitudes of the stator and

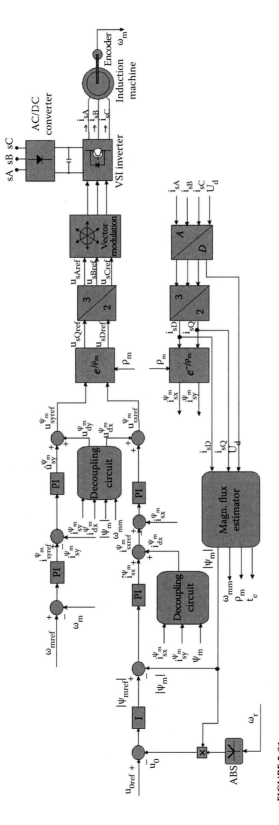

FIGURE 5.31
Block diagram of an IM drive with FOC based on voltage control with magnetizing flux orientation.

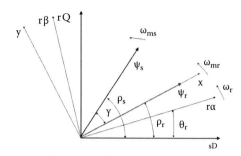

FIGURE 5.32
Reciprocal position of the stator and rotor flux linkages.

rotor flux linkages as well as on the *sin* of their angular displacement γ (cross-product between the space-vector flux linkages).

Figure 5.32 shows the reciprocal position of both the stator and rotor flux linkages as well as their position with respect to the stationary reference frame.

After expressing the stator and rotor flux-linkage space-vectors in the rotor reference frame, the following relationship between the same space-vectors can be found:

$$\boldsymbol{\psi}_r = \frac{L_m/L_s}{1 + s\sigma T_r} \boldsymbol{\psi}'_s \tag{5.60}$$

where s is the Laplace variable. Equation 5.60 clearly shows that the rotor flux linkage adapts itself to the variation of the stator flux linkage by means of a dynamic relationship typical of first-order systems. The rotor flux linkage, therefore, changes more slowly than the stator one.

The analysis of Equations 5.59 and 5.60 clearly shows that it is possible to make the machine generate an electromagnetic torque by modifying the instantaneous angular position of the stator flux-linkage space-vector. As a matter of fact, the modification of the angular position of the stator flux linkage creates an instantaneous variation of the electromagnetic torque. This variation is limited by the fact that the rotor flux-linkage space-vector tends to adapt itself to the variations of the stator flux-linkage space-vector, with a delay governed by the dynamic relationship (type low-pass) described by Equation 5.60. The angle γ between the two space-vectors tends to become zero in time. If the stator flux-linkage space-vector is therefore accelerated along the positive direction (direction of motion of the rotor) to amplify the value of $d\rho_s/dt$ and if, at the same time, the amplitude of the stator flux-linkage space-vector is kept constant, it is possible to achieve the fastest torque response of the machine. On the contrary, if the stator flux-linkage space-vector is accelerated along the negative direction, an electromagnetic torque of opposite sign is then generated, which is able to brake the machine.

5.4.2 Relationship between the Stator Flux-Linkage Space-Vector and the Inverter Configurations

If the stator equation of the IM in the stationary reference frame is considered and the ohmic drop on the stator resistance is neglected, then the following equation is obtained:

$$\mathbf{u}_s = \frac{d\boldsymbol{\psi}_s}{dt} \quad \Rightarrow \quad d\boldsymbol{\psi}_s = \mathbf{u}_s \, dt \tag{5.61}$$

Since the voltage space-vector $\mathbf{u}_s$ is applied for a finite time interval Δt, then Equation 5.61 can be rewritten in the following way:

$$\Delta\boldsymbol{\psi}_s = \mathbf{u}_s\Delta t \qquad (5.62)$$

Equation 5.62 shows that, neglecting the ohmic drops on the stator resistance and considering sufficiently small time intervals, the space-vector of the variation of the stator flux linkage $\Delta\psi_s$ has the same direction and the same orientation as the stator voltage space-vector.

If the stator voltage space-vector is expressed as a function of the logic command signals to the upper devices of the three legs of the inverter, it is possible to obtain the relationship between the space-vector of the variation of the stator flux linkage and the inverter state configurations:

$$\Delta\boldsymbol{\psi}_s = \frac{2}{3}U_d[S_a + aS_b + a^2S_c]\Delta t \qquad (5.63)$$

where
S_a, S_b, and S_c are the logic command signals to the upper devices of the three legs of the inverter
U_d is the DC link voltage

On this basis, the expression linking the space-vector at the time instant $t + \Delta t$ to that at the time instant t and to the inverter state configurations is as follows:

$$\boldsymbol{\psi}_s(t+\Delta t) = \boldsymbol{\psi}_s(t) + \frac{2}{3}U_d[S_a + aS_b + a^2S_c]\Delta t \qquad (5.64)$$

The stator flux-linkage space-vector can be therefore maintained in its position (neglecting the ohmic drop on the stator resistance that tends to reduce the flux) by applying a null voltage vector; alternatively, it can be driven in one of the six directions corresponding to the six nonnull stator voltage space-vectors that can be generated by the inverter.

The application of any nonnull stator voltage space-vector modifies, in general, both the amplitude and the initial angular position of the stator flux-linkage space-vector. In details, to modify the amplitude of the stator flux-linkage vector, it is necessary to apply stator voltage vectors with adequate radial components (with respect to the direction of the stator flux-linkage space-vector), while to modify its angular position, it is necessary to apply voltage vectors with adequate tangential components. A proper selection of the most suitable voltage space-vector permits modifying both the amplitude and the angular position of the stator flux vector in the desired way. The decoupled control of the magnetic flux and the electromagnetic torque is achieved by separately acting on the radial and tangential components of the stator flux-linkage space-vector.

Since both the amplitude and the angular position of the stator flux vector are directly linked to the supply voltage and the stator current (once the stator resistance value is known), in classic DTC, the stator flux-linkage vector is chosen as the controlled variable. In this way, a low sensitivity of the control action versus the parameter variations of the machine is achieved. It is, however, possible to perform the control of either the rotor or the magnetization flux vector. It can be demonstrated that, by controlling the rotor flux-linkage space-vector, a higher torque overloading of the drive can result even if at the expenses of a higher sensitivity to the machine parameter variation [34]. The control of the magnetizing flux-linkage vector, which offers intermediate characteristics in terms of torque overloading and parameter mismatch sensitivity with respect to stator or rotor flux vector controls, is usually adopted when a direct measurement of the air-gap magnetic flux is made by means of Hall effect sensors.

5.4.3 Criteria for the Selection of the Voltage Space-Vectors and Control Strategies

The control of the stator flux amplitude and the electromagnetic torque, performed on the basis of the criteria described in Section 5.4.2, is not intrinsically decoupled since the application of any nonnull voltage vector modifies at the same time both the amplitude and angular position of the stator flux vector itself. The suitable selection, at each sampling time of the control system, of a voltage vector able to minimize the tracking errors on both the flux amplitude and the electromagnetic torque, permits implementing de facto a decoupled control of the two variables. The voltage vector to be applied depends on the instantaneous angular position of the stator flux-linkage space-vector.

For determining the control strategy, it is necessary to ideally divide the complex plane on which the voltage space-vectors lie into sectors of amplitude $\pi/3$, each centered on one of the nonnull voltage space-vectors. If, in a generic instant, the stator flux vector lies inside a generic sector k (k = 1, 2,..., 6), the applicable voltage vectors produce the following effects [34]:

- The voltage space-vector $\mathbf{u}_k$ increases the amplitude of the stator flux linkage and maintains almost unchanged the angular position (radial positive configuration).
- The voltage space-vector $\mathbf{u}_{k+1}$ increases the amplitude of the stator flux linkage and rotates it in the counterclockwise direction (direct positive configuration).
- The voltage space-vector $\mathbf{u}_{k+2}$ reduces the amplitude of the stator flux linkage and rotates it in the counterclockwise direction (direct negative configuration).
- The voltage space-vector $\mathbf{u}_{k+3}$ reduces the amplitude of the stator flux linkage and maintains almost unchanged the angular position (radial negative configuration).
- The voltage space-vector $\mathbf{u}_{k-1}$ increases the amplitude of the stator flux linkage and rotates it in the clockwise direction (inverse positive configuration).
- The voltage space-vector $\mathbf{u}_{k-2}$ reduces the amplitude of the stator flux linkage and rotates it in the clockwise direction (inverse negative configuration).
- The voltage space-vectors $\mathbf{u}_0$ and $\mathbf{u}_7$ maintain almost unchanged the amplitude and the angular position of the stator flux linkage.

On this basis, given the reference values of the stator flux amplitude and electromagnetic torque, four cases can be encountered:

1. *The electromagnetic torque should be increased, and the flux should be increased*: a voltage vector with positive tangential component and positive radial component should be chosen.
2. *The electromagnetic torque should be increased, and the flux should be reduced*: a voltage vector with positive tangential component and negative radial component should be chosen.

3. *The electromagnetic torque should be reduced, and the flux should be increased*: a voltage vector with negative or null tangential component and positive radial component should be chosen.

4. *The electromagnetic torque should be reduced, and the flux should be reduced*: a voltage vector with negative or null tangential component and negative radial component should be chosen.

In case the torque should be increased, the configuration to be applied is always the one corresponding to a voltage vector with positive tangential and nonnull radial components. On the contrary, in case the torque should be reduced, the choice is not unique since, if the rotational speed of the machine is nonnull, a reduction of the angular displacement between the stator and rotor flux-linkage space-vectors can be obtained by applying either the configurations with null voltage space-vectors or those implementing inverse configurations. Several strategies for the control of the electromagnetic torque and flux can be devised. With reference to a stator flux vector lying in the sector k at a certain time instant, Table 5.1 synthesizes the most usual strategies for the control of the torque and stator flux [34].

Figure 5.33 shows the voltage space-vectors to be applied, in the four possible cases, when the stator flux space-vector lies in each of the six quadrants, under the hypothesis to adopt the control strategy called D in Table 5.1.

Different characteristics of the drive can be achieved in terms of torque and flux ripple, switching frequency, harmonic content of the stator currents, DC link current, behavior in regenerative phase, behavior at low speed and dynamic performance according to the choice of the strategy. In particular, the following could be observed:

- *Strategy A* utilizes the direct configurations for the increase of the torque and null configurations for the reduction of the torque. When a reduction of the torque is to be obtained, the application of the null voltage vector blocks the position of the stator flux linkage. The braking action varies with the rotational speed of the motor and is less efficient at lower rotating speed. This strategy presents a good behavior at high speed and is characterized by switching frequencies lower than other possible strategies.

- *Strategy B* utilizes the radial positive configurations, instead of the null ones, for the reduction of the electromagnetic torque. In this way, the flux reduction of the machine at low speed can be avoided. The switching frequency is higher than that achievable with *strategy A*, and the locus of the stator flux space-vector is in general more irregular, because of the adoption of the radial positive components. Torque control does not present significant differences with respect to *strategy A*.

TABLE 5.1

Possible Control Strategies for DTC

	$t_e \Uparrow \; |\psi_s| \Uparrow$	$t_e \Uparrow \; |\psi_s| \Downarrow$	$t_e \Downarrow \; |\psi_s| \Uparrow$	$t_e \Downarrow \; |\psi_s| \Downarrow$
Strategy A	u_{k+1}	u_{k+2}	u_0, u_7	u_0, u_7
Strategy B	u_{k+1}	u_{k+2}	u_k	u_0, u_7
Strategy C	u_{k+1}	u_{k+2}	u_k	u_{k+3}
Strategy D	u_{k+1}	u_{k+2}	u_{k-1}	u_{k-2}

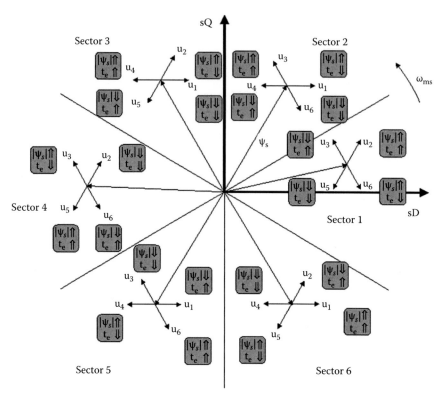

FIGURE 5.33
Selection of the voltage vectors to be applied with strategy D.

- *Strategy C* does not adopt null configurations. Torque reduction is obtained applying only radial configurations (positive or negative), which block the stator flux space-vector. The adoption of only radial configurations influences significantly the shape of the stator flux vector locus that must be strongly corrected to lie inside the hysteresis band. This is a cause of an increase of the switching frequency.

- *Strategy D* adopts the direct configurations to increase the electromagnetic torque and only the inverse configurations to reduce them. Adopting such strategy, the reduction of the electromagnetic torque is performed by a rotation of the stator flux vector in an opposite direction with respect to the rotation of the machine. In this way, it is possible to achieve fast torque reduction even at low speed. On the contrary, this strategy causes a significant increase of the switching frequency.

By a proper selection of the voltage vectors to be generated, DTC permits the stator flux and torque references to be tracked with the bang-bang type technique which is very similar to a sliding-mode control [38]. In order to properly limit the switching frequency, hysteresis bands centered on the reference values of the stator flux and torque are usually adopted.

Figure 5.34 shows how the stator flux vector modifies, in each quadrant, because of the application of the suitable voltage space-vectors (strategy D), while always keeping inside the hysteresis band $2\Delta B\psi_s$ centered on the reference flux ψ_{sref}. The underlying assumption is that the control system always requires an increase of the electromagnetic torque.

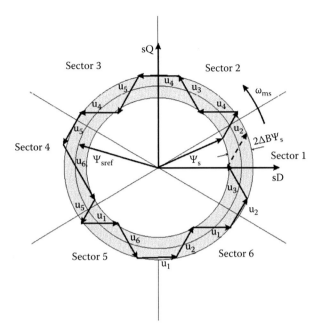

FIGURE 5.34
Locus of the stator flux space-vector.

If the control system is made analogically, in the absence of hysteresis, the switching frequency can take very high values (up to tens of kilohertz). If the control system is made digitally, on the contrary, the maximum switching frequency coincides with the sampling frequency of the control system. If a hysteresis control is adopted, however, there is a direct correspondence between the hysteresis bands and the switching frequency. The frequency of the commutation driven by the flux and torque control system is inversely proportional to the amplitudes of the corresponding hysteresis bands ($2\Delta B_{te}$, $2\Delta B_{\psi s}$).

5.4.4 Estimation of the Stator Flux and Electromagnetic Torque

In DTC, to properly track the stator flux and electromagnetic torque references, it is necessary to know the instantaneous values of these variables. The direct measurement of such variables by means of suitable sensors (e.g., Hall effect sensors for the magnetic flux and torquemeters for the electromagnetic torque) is very rarely accepted at an industrial level, not matching the requirements in terms of economizing, simplicity of construction, and reliability.

It is therefore more common to estimate such variables by the so-called flux and torque models. They are mathematical models based on the differential equations of the machine, permitting the estimation of the magnetic flux and the electromagnetic torque on the basis of the measurements of some electrical (stator voltages and/or currents) or mechanical (angular speed) variables.

In DTC, the amplitude of the stator flux linkage is usually obtained by the so-called voltage model based on the integration of the stator voltages, taking into account the ohmic drop on the stator resistance.

From Equation 4.24, stator flux-linkage space-vector can be obtained from the measurements of the stator voltages and currents as follows:

$$\boldsymbol{\psi}_s = \int (\mathbf{u}_s - R_s \mathbf{i}_s)dt \tag{5.65}$$

After decomposing the vector equation (5.65) into its real sD and quadrature sQ components in the stationary reference frame, the following scalar equations can be obtained:

$$\begin{cases} \psi_{sD} = \int (u_{sD} - R_s i_{sD}) \, dt \\ \psi_{sQ} = \int (u_{sQ} - R_s i_{sQ}) \, dt \end{cases} \tag{5.66a,b}$$

For a correct interpretation of the symbols, see the list. The direct and quadrature components of the stator voltages and currents can be obtained from the three-phase values by means of the following transformations:

$$\begin{cases} u_{sD} = \left(\dfrac{2}{3}\right) u_{sA} - \left(\dfrac{1}{3}\right) u_{sB} - \left(\dfrac{1}{3}\right) u_{sC} \\ u_{sQ} = \left(\dfrac{1}{\sqrt{3}}\right) u_{sB} - \left(\dfrac{1}{\sqrt{3}}\right) u_{sC} \end{cases} \tag{5.67a,b}$$

$$\begin{cases} i_{sD} = \left(\dfrac{2}{3}\right) i_{sA} - \left(\dfrac{1}{3}\right) i_{sB} - \left(\dfrac{1}{3}\right) i_{sC} \\ i_{sQ} = \left(\dfrac{1}{\sqrt{3}}\right) i_{sB} - \left(\dfrac{1}{\sqrt{3}}\right) i_{sC} \end{cases} \tag{5.68a,b}$$

The electromagnetic torque can be computed from the estimated values of the stator flux-linkage components and from the stator currents, by means of the following relationship:

$$t_e = \frac{3}{2} p \boldsymbol{\psi}_s \wedge \mathbf{i}_s = \frac{3}{2} p (\psi_{sD} i_{sQ} - \psi_{sQ} i_{sD}) \tag{5.69}$$

Figure 5.35 shows the block diagram of the flux and torque model described by Equations 5.66 and 5.69. In this figure, s represents the Laplace variable.

The flux model sketched in Figure 5.35 presents, besides a significant simplicity, a weak robustness to the parameter variations of the machine, especially at low speed. As a matter of fact, the value of the stator resistance is not properly known at each time instant since it is

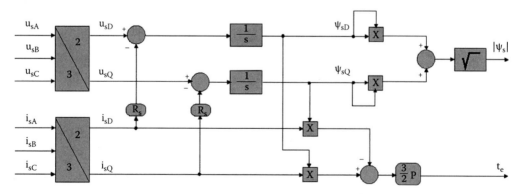

FIGURE 5.35
Block diagram of the stator flux and torque model.

variable with the temperature of the stator winding. When the machine works at high speed, the ohmic drop is negligible with respect to the supply voltage and an error in the knowledge of R_s does not significantly affect the stator flux estimation. At low speed, on the contrary, the ohmic drop on the stator resistance is comparable with the supply voltage, and so an error in the knowledge of R_s influences strongly the estimation of both the stator flux and the torque. The sensitivity analysis of such a flux model versus the variations of the R_s is shown in Chapter 9. The scheme in Figure 5.35 can be modified by adopting a low-pass filter in place of pure integrators. The blocks with transfer function equal to $1/s$ in Figure 5.35 would then be substituted with blocks having transfer function equal to $T/(1+sT)$, where T is a properly chosen time constant. This solution permits partially solving the problems of DC drift and initial conditions of the integrator (detailed analysis and solutions in Chapter 6).

The flux model in Figure 5.35 requires the adoption of voltage sensors for the measurement of the stator voltage, which is not typically accepted at an industrial level. It is, however, possible to modify such a scheme by, computing the stator voltage on the basis of the measurement of the DC link voltage U_d and of the knowledge of the working configuration of the inverter, represented by the logic command signals S_a, S_b, and S_c.

Equations 5.70a through 5.70c permit the stator voltages to be computed on the basis of the switching configuration of the inverter:

$$
\begin{cases}
u_{sA} = \dfrac{U_d}{3}\left(2S_a - S_b - S_c\right) \\[2mm]
u_{sB} = \dfrac{U_d}{3}\left(2S_b - S_a - S_c\right) \\[2mm]
u_{sC} = \dfrac{U_d}{3}\left(2S_c - S_a - S_b\right)
\end{cases}
\qquad (5.70a,b,c)
$$

It is noteworthy that a scheme where the phase voltages are computed on the basis of the measurement of the DC link voltage and the knowledge of the logic command signals to the upper devices of the inverter legs is sensitive to the following:

- Errors on the estimation of the voltage, due to the protection time to be considered (the effect is more significant at low speed)
- The voltage drops on the power devices of the inverter
- Fluctuations of the DC link voltage U_d
- Stator resistance variations (this variation can identically occur in case of direct measurement of the phase voltages)

On the contrary, the adoption of such a solutin requires only one voltage sensor, instead of two as in the case of direct measurement of the stator voltage, with lower bandwidth since the DC link voltage presents slight variations because of the presence of the capacitors. The phase voltages are characterized by a very high dynamics and so require more sophisticated and expensive sensors. This solution is globally cheaper than that with the direct measurement of the stator voltages.

While, in FOC, the knowledge of the instantaneous angular position of the rotor flux-linkage space-vector ρ_r is required, in DTC, the knowledge of the instantaneous angular position of the stator flux-linkage space-vector ρ_s is needed. The knowledge of ρ_s is necessary to recognize in which quadrant the stator flux-linkage vector instantaneously lies, in order to choose the optimal command pattern to drive the inverter devices.

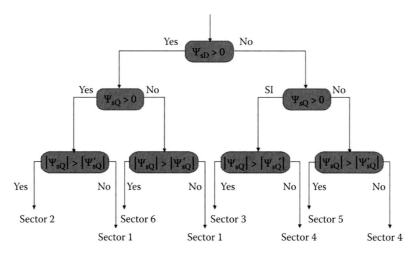

FIGURE 5.36
Flow chart of the algorithm for the sector determination.

The value of ρ_s can be computed on the basis of the knowledge of the direct and quadrature components of the stator flux space-vectors ψ_{sD} and ψ_{sQ} as follows:

$$\rho_s = \tan^{-1}\left(\frac{\psi_{sQ}}{\psi_{sD}}\right) \tag{5.71}$$

As a matter of fact, classic DTC does not require an accurate knowledge of the angular position of the stator flux space-vector, because it is sufficient to find of the quadrant in which this vector lies. It is then possible to characterize the sector in which the stator flux vector lies on the basis of the sign of the components ψ_{sD} and ψ_{sQ}. This permits avoiding the use of transcendent functions as the $\tan^{-1}$. As far as only the signs of ψ_{sD} and ψ_{sQ} are considered, it is not possible to discriminate between sectors n° 1 and n° 6, between n° 1 and n° 2, between n° 3 and n° 4, and finally between n° 4 and n° 5. To distinguish, for example, if the stator flux vector lies in the sector n° 1 or n° 2, it must be checked if ρ_s is higher or lower than $\pi/6$ rad; if $\rho_s > \pi/6$ rad, then the stator flux vector lies in sector n° 2, otherwise in sector n° 1. Practically, after defining $\psi'_{sQ} = \psi_{sD} \tan(\pi/6)$, $|\psi_{sQ}|$ is compared with $|\psi'_{sQ}|$; if $|\psi_{sQ}| > |\psi'_{sQ}|$, then the stator flux vector lies in sector n° 2, if $|\psi_{sQ}| < |\psi'_{sQ}|$ in sector n° 1.

In the same way, the indetermination problem in the other sectors is solved. Figure 5.36 shows the flow chart of the algorithm for the sector determination.

5.4.5 DTC Scheme

Figure 5.37 shows the block diagram of DTC IM drive. In this case, the closed-loop control of the machine speed ω_m and the stator flux ψ_s is performed. The machine measured speed ω_m is compared with its reference ω_{mref}, and the tracking error is processed by the PI controller. The output of the speed controller is the torque reference t_{eref} which is compared with the estimated torque t_e, computed by the flux and torque model, being the tracking error processed by a hysteresis comparator, whose output is one of the inputs of the block "inverter optimal switching table."

The stator flux reference ψ_{sref} is compared with the estimated one ψ_s, computed also in this case with the flux and torque model, and the tracking error is processed by a hysteresis

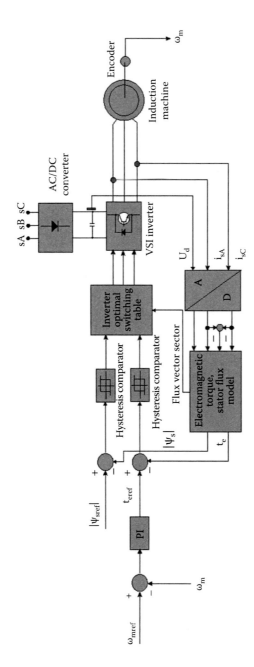

FIGURE 5.37
Block diagram of the DTC IM drive.

comparator, whose output is the other input of the block "inverter optimal switching table." The flux and torque model is represented in Figure 5.35. In this scheme, the phase voltages have been reconstructed by the knowledge of the DC link voltage U_d and the switching patterns of the inverter S_a, S_b, and S_c. The block "inverter optimal switching table" chooses, at each sampling time of the control system, the optimal configuration of the inverter on the basis of the angular position of the stator flux-linkage space-vector. The algorithm adopted for the quadrant identification is shown in Figure 5.36. The control strategy is usually the D, as defined in Table 5.1, since it permits the best dynamic response.

5.4.6 DTC EMC

Recently, [35] a new switching-table DTC strategy has been devised, called DTC EMC (electro-magnetically compatible), with the aim to reduce the common-mode emissions of the drive. In general, the inherent fast switching frequency of both PWM fed and DTC drives has, among other detrimental effects, the drawback of generating high-level common-mode voltage variations with resulting high-frequency common-mode currents flowing to the ground through the parasitic capacitances between the different parts of the drive and the ground, thus causing the drive itself to be less reliable; for example, the ball bearings can deteriorate, unexpected fault relay tripping can occur, or nearby electronic equipments can be disturbed.

DTC EMC reduces the common-mode emissions of the drive by alternatively using only even or only odd voltage vectors in each of the six sectors of the inverter hexagon, without using any null vector. This approach permits the common-mode emissions to be reduced in comparison with the classic DTC algorithm, at the expense of a slight increase of the torque and flux ripples as well as in the harmonic content of the voltage and current waveforms.

5.4.6.1 Common-Mode Voltage

In a star-connected three-phase electric machine, the common-mode voltage u_{com} is given by the following:

$$u_{com} = \frac{u_{sA0} + u_{sB0} + u_{sC0}}{3} \tag{5.72}$$

where u_{sA0}, u_{sB0}, and u_{sC0} are the inverter output phase voltages referred to the medium point of the DC link bus, which in this case is assumed to be connected to ground. The lack of this connection to ground results in an additional voltage term. If the machine is supplied with a symmetric sinusoidal three-phase voltage, u_{com} is instantaneously equal to zero. However, when the machine is supplied with an inverter, the common-mode voltage is always different from zero, and its instantaneous value can be computed on the basis of the DC link voltage (U_d) and the switching pattern of the inverter, as shown in Table 5.2 ($\mathbf{u}_i$ stands for the ith stator voltage space-vector).

5.4.6.2 Switching Strategy

Table 5.2 shows that, if only even or only odd active voltage vectors are used ($\mathbf{u}_k$, with k, respectively, even or odd), no common-mode voltage variation is generated. If a transition from an even voltage vector to an odd one (or vice versa) occurs, a common-mode variation of amplitude $U_d/3$ is generated. If a transition from an odd (even) voltage vector to the zero (seventh) voltage vector occurs, a common-mode variation of amplitude

TABLE 5.2

Inverter States and Common Voltages

State	u_{sA0}	u_{sB0}	u_{sC0}	u_{com}
$\mathbf{u_0}$ $(0, 0, 0)$	$-U_d/2$	$-U_d/2$	$-U_d/2$	$-U_d/2$
$\mathbf{u_1}$ $(1, 0, 0)$	$U_d/2$	$-U_d/2$	$-U_d/2$	$-U_d/6$
$\mathbf{u_2}$ $(1, 1, 0)$	$U_d/2$	$U_d/2$	$-U_d/2$	$U_d/6$
$\mathbf{u_3}$ $(0, 1, 0)$	$-U_d/2$	$U_d/2$	$-U_d/2$	$-U_d/6$
$\mathbf{u_4}$ $(0, 1, 1)$	$-U_d/2$	$U_d/2$	$U_d/2$	$U_d/6$
$\mathbf{u_5}$ $(0, 0, 1)$	$-U_d/2$	$-U_d/2$	$U_d/2$	$-U_d/6$
$\mathbf{u_6}$ $(1, 0, 1)$	$U_d/2$	$-U_d/2$	$U_d/2$	$U_d/6$
$\mathbf{u_7}$ $(1, 1, 1)$	$U_d/2$	$U_d/2$	$U_d/2$	$U_d/2$

Source: Cirrincione, M. et al., *IEEE Trans. Ind. Appl.*, 42(2), 504, 2006.

$U_d/3$ is generated. Finally, if a transition from an odd (even) voltage vector to the seventh (zero) voltage vector occurs, a common-mode variation of amplitude $2U_d/3$ is generated. Thus, from the point of view of common-mode emissions, the worst case is the transition from an odd (even) voltage vector to the seventh (zero) voltage vector. For this reason, whatever inverter control technique is devised to minimize the common-mode emissions of the drive, the exploitation of both null voltage vectors (zero and seventh) should be avoided. If a DTC technique is used, this consideration is helpful also from the control point of view.

DTC EMC has been devised on the basis of the earlier considerations, and its basic idea is shown in the flow chart of Figure 5.38. When the stator flux space-vector lies in the *k*th

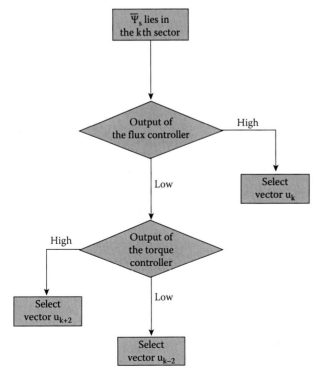

FIGURE 5.38
Flow chart of the DTC EMC algorithm. (From Cirrincione, M. et al., *IEEE Trans. Ind. Appl.*, 42(2), 504, 2006.)

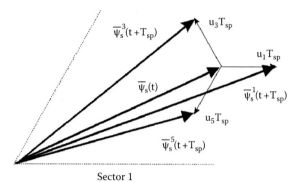

FIGURE 5.39
Effect of the application of the vectors $\mathbf{u}_1$, $\mathbf{u}_3$, and $\mathbf{u}_5$ in the first sector. (From Cirrincione, M. et al., *IEEE Trans. Ind. Appl.*, 42(2), 504, 2006.)

sector, if an increase in the flux is required, then the kth voltage vector is applied; otherwise, if an increase (decrease) in the torque is required, then the (k + 2)th (respectively, the (k − 2)th) voltage vector is applied.

Actually, when the stator flux linkage lies in the kth sector, the application of the kth voltage vector produces an increase in the flux amplitude and a slight decrease of the electromagnetic torque, the application of the (k + 2)th voltage vector produces a slight decrease in the flux amplitude and an increase in the electromagnetic torque, while the application of the (k − 2)th voltage vector produces a slight decrease in the flux amplitude and a decrease in the electromagnetic torque. As an example, Figure 5.39 shows that when the stator flux vector $\boldsymbol{\psi}_s(t)$ lies in the first sector and one of the voltage vectors $\mathbf{u}_1$, $\mathbf{u}_3$, and $\mathbf{u}_5$ is applied, the flux vector resulting after a sampling time of the control system T_{sp} becomes, respectively, $\boldsymbol{\psi}_s^1(t + T_{sp})$, $\boldsymbol{\psi}_s^3(t + T_{sp})$, or $\boldsymbol{\psi}_s^5(t + T_{sp})$. In particular, the application of $\mathbf{u}_1$ causes a strong increase of the flux amplitude and a slight decrease of the torque, the application of $\mathbf{u}_3$ causes a slight decrease of the flux amplitude and a strong increase of the torque, and, finally, the application of $\mathbf{u}_5$ causes a strong decrease of the flux amplitude and a strong decrease of the torque. With such a control law, flux control should be performed before torque control (see Figure 5.38); otherwise, the machine cannot be magnetized at zero speed without load torque (zero speed operation), or at steady-state gets demagnetized rapidly after a series of torque commands. Moreover, it is apparent that a reduction of the flux always results in torque variation.

This means that when the flux lies in the odd (even) sector, only odd (even) voltage vectors are employed, as summarized in Figure 5.40 where all the possible configurations as obtained both with the classic DTC and with the DTC EMC are shown.

In the end, it is clear that, as long as the stator flux linkage lies in one sector, no common-mode voltage variation occurs. Each commutation of the common-mode voltage appears only when the stator flux linkage goes from one sector to the adjacent one. Moreover, at each sector crossing, the common-mode voltage variation is the minimum achievable, equal in magnitude to $U_d/3$. Thus, in steady-state, theoretically only six variations of the common-mode voltage of amplitude $U_d/3$ appear in each period of the fundamental of the voltage waveform. It should be noted, however, that in a real-world application of this control strategy, some undesirable further commutations of the common-mode voltage, with consequent spikes of the common-mode current, can occur during each sector crossing. It can happen that if a torque decrease or even simply a flux increase is commanded by the control system when the stator flux vector lies at the beginning of a sector, the stator flux itself goes back in the antecedent sector. In this case, instead of a simple commutation of the common-mode voltage, at least three commutations occur with consequent common-mode current spikes, as it will be shown both in the simulation and experimental results.

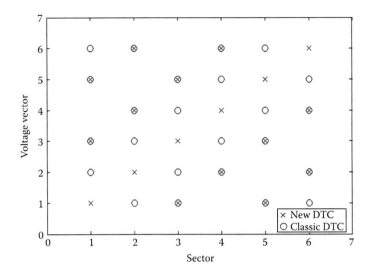

FIGURE 5.40
Map of the voltage vectors with DTC EMC and classic DTC. (From Cirrincione, M. et al., *IEEE Trans. Ind. Appl.*, 42(2), 504, 2006.)

However, the benefits of a significant reduction of the common-mode emissions of the drive are paid back with a poorer exploitation of the DC link capability of the inverter (no zero voltages are employed), with higher ripples both in the flux and torque waveforms and finally with higher harmonic contents of the stator voltages and currents. The resulting increase of the stator flux and torque ripples can be easily deduced also from Figure 5.41 which shows the locus of the stator flux vector during a positive torque

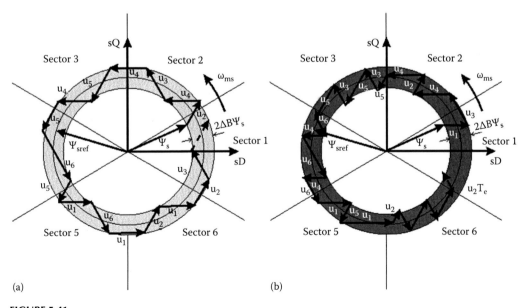

(a)   (b)

FIGURE 5.41
Locus of the stator flux vector during a positive torque command with classic DTC (a) and DTC EMC (b). (From Cirrincione, M. et al., *IEEE Trans. Ind. Appl.*, 42(2), 504, 2006.)

command with the classic DTC and the DTC EMC. It highlights that the flux locus with the DTC EMC is sharper than that obtained with the classic DTC, given the sampling time of the control system and the amplitude of the hysteresis bands of the flux controller.

5.4.6.3 Common-Mode Voltage Spectrum of DTC EMC

In DTC EMC, the theoretical steady-state common-mode voltage u_{com}, neglecting the unde-sired commutation of the common-mode voltage at sector crossings, all parasitic effects, and the rise/fall time, is a periodic waveform (Figure 5.42), with a period equal to 1/3 of the period of the fundamental harmonic of the supply voltage of the drive.

Thus it exhibits a harmonic spectrum which can be analytically inferred by computing the coefficients of its Fourier series expansion as functions of the DC link voltage U_d and the fundamental pulsation of the supply voltage ω:

$$u_{com} = \sum_{n}^{\infty} \frac{2U_d}{3\pi n} \sin(3n\omega t) \tag{5.73}$$

where
   n is the harmonic order (only odd in this case)
   t is time

On this basis, the harmonic content of the common-mode voltage is basically in the low frequency region, in particular, the 3rd harmonic of the fundamental and its odd multiples (3rd, 9th, 15th, 25th, etc., of the fundamental), and its amplitude decrease with inverse pro-portionality with the frequency itself. Such a drive, therefore, exhibits a common-mode harmonic content which is quite low and depends on the rotor speed of the drive.

5.4.7 Experimental Results with Classic DTC and DTC EMC

In the following, some comparative experimental results between the classic DTC [9] and the DTC EMC are presented [35]. These tests have been done on an experimental rig with a 2.2 kW IM (details regarding the test setup in Appendix 11.A of Chapter 11).

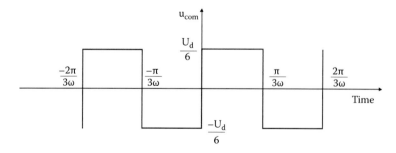

FIGURE 5.42
Theoretical waveform of the common-mode voltage. (From Cirrincione, M. et al., *IEEE Trans. Ind. Appl.*, 42(2), 504, 2006.)

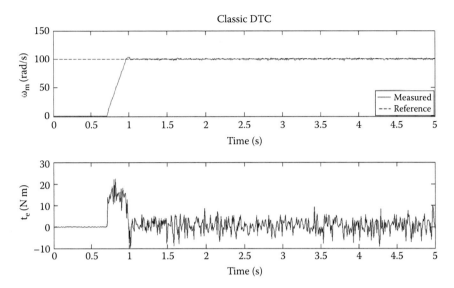

FIGURE 5.43
Rotor speed and electromagnetic torque with classic DTC. (From Cirrincione, M. et al., *IEEE Trans. Ind. Appl.*, 42(2), 504, 2006.)

Figures 5.43 and 5.44 show the rotor speed and the electromagnetic torque with the classic DTC and the DTC EMC, when a step speed reference of 100 rad/s at no load is given. It is apparent the increase of the torque ripple with DTC EMC, as expected.

Figures 5.45a and b and 5.46a and b show the common-mode voltage waveforms and the corresponding spectra, computed with the FFT, obtained at constant speeds of 110 and

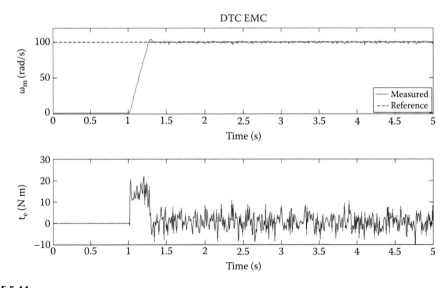

FIGURE 5.44
Rotor speed and electromagnetic torque with DTC EMC. (From Cirrincione, M. et al., *IEEE Trans. Ind. Appl.*, 42(2), 504, 2006.)

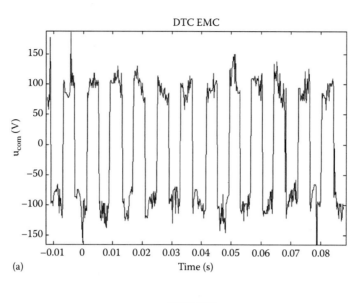

(a)

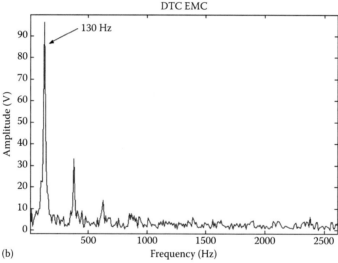

(b)

FIGURE 5.45

Common-mode voltage waveform (a) and its FFT (b) at the reference speed of 110 rad/s with the DTC EMC (0.12 slip). (From Cirrincione, M. et al., *IEEE Trans. Ind. Appl.*, 42(2), 504, 2006.)

2 rad/s with the DTC EMC technique. These figures clearly confirm that the main harmonic component of the common-mode voltage is the third harmonic of the supply frequency of the machine, around 130 and 3 Hz, respectively, at the reference speeds of 110 and 2 rad/s with a 12% slip. Figure 5.47a and b show the common-mode voltage waveforms and the corresponding spectrum obtained with the FFT, in the case of the classic DTC. The spectra present a series of harmonic clusters. A comparison between Figures 5.45b and 5.46b with 5.47b shows a strong reduction of the common-mode voltage harmonic content, achievable with the DTC EMC with respect to the classic DTC.

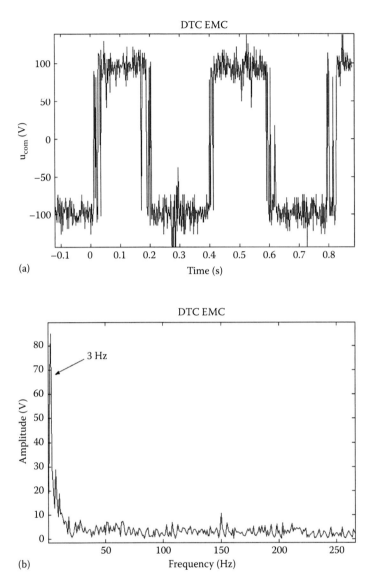

FIGURE 5.46
Common-mode voltage waveform (a) and its FFT (b) at the reference speed of 2 rad/s with the DTC EMC (0.12 slip). (From Cirrincione, M. et al., *IEEE Trans. Ind. Appl.*, 42(2), 504, 2006.)

5.4.8 DTC-SVM

The classic DTC consists essentially in the bang-bang control of the electromagnetic torque and flux, and it is therefore characterized by a fast response to the control commands. However, in steady-state operation especially at low speeds, the DTC control results in chaotic switching patterns of the inverter with resulting torque ripple and undesirable acoustic and vibration effects associated with it. To optimize the steady-state switching process, various methods have been devised [36,37], among which one of the most

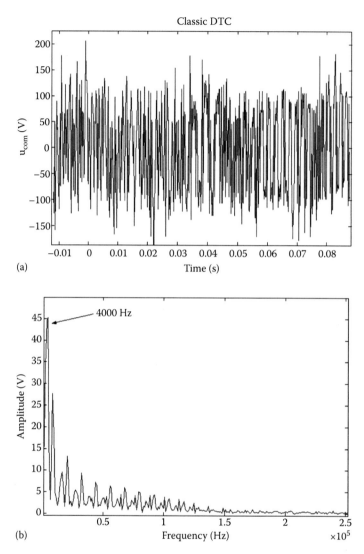

FIGURE 5.47
Common-mode voltage waveform (a) and its FFT (b) at the reference speed of 2 rad/s with the classic DTC (0.12 slip). (From Cirrincione, M. et al., *IEEE Trans. Ind. Appl.*, 42(2), 504, 2006.)

interesting employs the well-known space-vector modulation (SVM) associated with the DTC strategy [37–39]. Such switching patterns result in low-ripple stator currents, as well as smooth flux and torque waveforms. On this basis, the classical DTC has been improved so that the control system can generate a stator voltage reference, instead of directly indicating the next switching pattern of the inverter. The stator voltage is then synthesized with the SVM.

The control law comes directly from the direct and quadrature stator equations of the IM in the stator flux-oriented reference frame:

$$u_{sx}^{\psi_s} = R_s i_{sx}^{\psi_s} + \left(\frac{d|\psi_s|}{dt}\right) \tag{5.74}$$

$$u_{sy}^{\psi_s} = R_s i_{sy}^{\psi_s} + \omega_{ms} \left| \mathbf{\psi}_s \right| \tag{5.75}$$

For the symbols, see the list. The last equation can be rewritten as a function of the electromagnetic torque as follows:

$$u_{sy}^{\psi_s} = \frac{3}{2p} R_s \frac{t_e}{\left| \mathbf{\psi}_s \right|} + \omega_{ms} \left| \mathbf{\psi}_s \right| \tag{5.76}$$

where the angular speed of the stator flux ω_{ms} has been estimated in a discrete form from the last two temporal samples k and $k-1$ of the stator flux components ψ_{sD} and ψ_{sQ} in the stator reference frame, as follows:

$$\omega_{ms} = \frac{\psi_{sQ}(k)\psi_{sD}(k-1) - \psi_{sD}(k)\psi_{sQ}(k-1)}{\left| \mathbf{\psi_s} \right|^2 T_s} \tag{5.77}$$

with T_s sampling time of the control system.

From Equation 5.75, it can be seen that the $u_{sx}^{\psi_s}$ component of the stator voltage can control the stator flux amplitude, while from Equation 5.76, that $u_{sy}^{\psi_s}$ can control the electromagnetic torque.

Figure 5.48 shows the block diagram of the DTC-SVM IM drive.

In this scheme, a closed-loop control of both the stator flux-linkage amplitude and the rotor speed is performed. Speed control is achieved by employing a PI controller for processing the speed error resulting from the comparison between the reference and the measured speed. The output of the speed controller is the reference torque, which is compared with the estimated one, being the tracking error processed by a PI controller. The output of the torque controller is further added to the decoupling term in Equation 5.76, yielding the quadrature axis reference voltage $u_{sy}^{\psi_s}$. On the direct axis, the stator flux reference is compared with the estimated flux, being the tracking error processed by a PI controller. The stator flux amplitude and the electromagnetic torque are estimated by a flux and torque estimator, as described earlier. The output of the flux controller is the direct-axis reference voltage $u_{sx}^{\psi_s}$. The reference voltages are then transformed from the stator flux linkage to the stationary reference frame by means of a vector rotation on the basis of the knowledge of the stator flux angle ρ_s. The stator voltages are finally synthesized by a PWM VSI.

5.4.9 Experimental Results with a DTC-SVM Drive

In the following, some experimental results of a DTC-SVM motor drive are presented. The rotor flux-oriented scheme in Figure 5.48 has been adopted. A test setup with a 2.2 kW IM has been adopted (details regarding the test setup in Appendix 11.A of Chapter 11). The drive has been given a speed step reference of 100 rad/s at no load. Figure 5.49 shows the reference and measured speed obtained during this test, while Figure 5.50 shows the corresponding waveform of the reference and estimated electromagnetic torque. Comparing Figure 5.50 with Figure 5.43, it can be clearly observed how DTC-SVM permits a significant reduction of the torque ripple with respect to classic DTC.

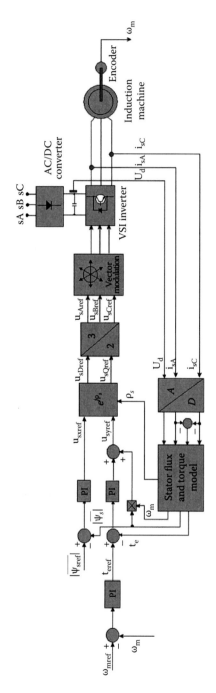

FIGURE 5.48
Block diagram of the DTC-SVM scheme.

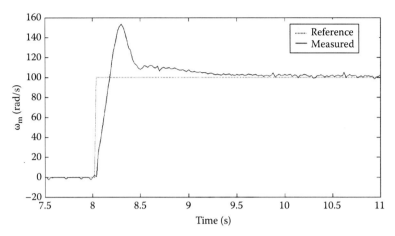

FIGURE 5.49
Reference and measured speed with DTC-SVM.

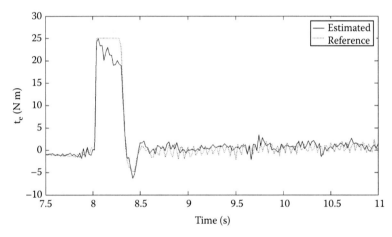

FIGURE 5.50
Reference and estimated speed with DTC-SVM.

5.4.10 Direct Self-Control

The direct self-control (DSC) [10–12] has been proposed by Depenbrock in 1985, almost contemporary with DTC. DSC is more suited for high-power drives supplied by VSIs, where slow-power devices are adopted (e.g., GTOs) and consequently low switching frequencies. In DSC, the VSI is commanded in a quasi-square waveform, with occasional zero states. Zero states are not adopted at all when the drive operates above the rated speed, in field weakening. Even if DSC is frequently presented as a subcase of DTC, its working principle is different. The underlying idea is that, even if the time waveform of the voltage generated by a VSI is discontinuous, its time integral is continuous [37]. These time integrals are called *virtual fluxes* and can be controlled in a feedback way by hysteresis comparators.

In this way, the square-way operation of the inverter can be obtained without any external signal (from which the word "self" in its name). Based on stator virtual flux components $\psi_{s\beta A}$, $\psi_{s\beta B}$, and $\psi_{s\beta C}$, the flux comparators generate the digitized variables d_A, d_B, and d_C, which correspond to active voltage vectors for six-step operation. The hysteresis torque controller, on the other hand, generates the digitized signal d_0 that determines the zero states duration. Figure 5.51 shows the block diagram of a DSC IM drive.

The main characteristics of DSC are as follows [13]:

1. PWM operation in the constant flux region and six-step operation in the field-weakening region.
2. Nonsinusoidal stator flux and current waveforms that, with the exception of the harmonics, are identical for both PWM and six-step operation.
3. Stator flux vector moves along a hexagon path also under PWM operation.
4. No voltage supply reserve is necessary, and the inverter capability is fully utilized.
5. The inverter switching frequency is lower than in the ST-DTC scheme.
6. Excellent torque dynamics in constant and field-weakening regions.

The behavior of a DSC scheme can be reproduced by a ST-DTC scheme when the hysteresis band of the stator flux comparator is properly set [40].

5.4.11 Comparison between FOC and DTC

FOC and DTC are today the most important control techniques, capable of ensuring high dynamic performance of IM drives. For this reasons, most drive manufacturers adopt one of the two solutions. Both industrial and academic environment have been discussing for many years which of the two techniques is the best. Plenty of scientific papers have been written about the comparison between such techniques. It is certainly true that each of the two presents some features making itself preferable to the other. In the following, the main advantages and disadvantages of each control technique are summarized.

The main characteristics of DTC are as follows [18]:

1. Direct control of magnetic flux and electromagnetic torque by the selection of the optimal configuration of the inverter
2. Indirect control of the stator voltages and currents
3. Stator fluxes and currents almost sinusoidal
4. Possibility to reduce the torque ripple depending on the time interval of the null voltage vectors
5. High dynamic performance
6. Variability of the switching frequency of the inverter, depending on the amplitude of the hysteresis bandwidth and on the flux and torque comparators

The main advantages of DTC are as follows [18]:

1. Absence in the control algorithm of coordinate transformation, required on the contrary in FOC
2. Absence of PWM blocks, required on the contrary in FOC

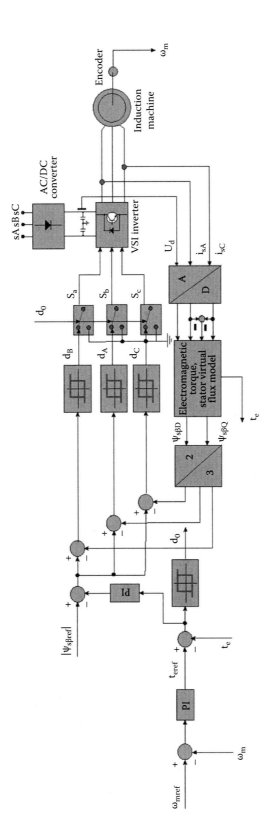

FIGURE 5.51
Block diagram of the DSC IM drive.

3. Absence of voltage decoupling circuits, required, on the contrary, in FOC with impressed voltages

4. Presence of few controllers in the control scheme, while in FOC with impressed voltages, at least four controllers are required

5. Necessity of determining only the sector where the stator flux linkage lies, and not necessarily the actual instantaneous position of this space-vector

6. Low rise time in the response of the torque loop

On the contrary, the main disadvantages of DTC are as follows:

1. Potential problems at the start-up, at low rotating speeds, and during fast variations in torque command

2. Necessity of flux and torque models (same problems occur in FOC)

3. Variability of the inverter switching frequency

4. High torque ripple

It is, however, possible to overcome many of the previously described drawbacks of DTC, like the problems at start-up, at low speed, and the torque ripple. Torque ripple in DTC is caused, from one side, by the fact that normally none of the voltage vectors that the inverter can generate exactly coincide with the voltage vector required by the control system to perform the suitable variation of magnetic flux and torque and, from the other side, that the voltage vector is applied for a time interval equal to the sampling time of the control system, and consequently the control action has a duration equal to this entire period. As a result, for low values of the torque tracking errors, the developed torque overcomes its reference value in a very brief time interval, much lower than the sampling time of the control system, tending to move away from the reference for the remaining duration of this period.

A solution could be obtained applying the stator voltage space-vector not for the entire sampling time period, but for a fraction of it (called in relative terms duty ratio), applying for the remaining part of the period the null voltage vector. The value of the duty ratio should be selected, at each time sample, on the basis of the value of the torque tracking error and of the actual angular position of the stator flux-linkage space-vector. To this aim, some algorithms based on Fuzzy logic have been proposed [36].

Torque ripple can be limited by adopting high switching frequencies of the inverter or different inverter structures. The switching frequency increase results in, however, an increase of the switching losses of the converter, and so this technique cannot be used in high-power drives. Moreover, a high value of the switching frequency requires the adoption of powerful but more expensive microprocessors, reducing the sampling time of the control system at increasing switching frequency.

List of Symbols

$\mathbf{i}_s = i_{sD} + ji_{sQ}$	space-vector of the stator currents in the stator reference frame
$\mathbf{i}_s = i_{sx} + ji_{sy}$	space-vector of the stator currents in a generic rotating reference frame
i_{sA}, i_{sB}, i_{sC}	stator phase currents
$\mathbf{i}'_r = i_{rd} + ji_{rq}$	space-vector of the rotor currents in the stator reference frame

$\mathbf{i_r} = i_{rx} + ji_{ry}$ space-vector of the rotor currents in a generic rotating reference frame

$\mathbf{i}_{mm} = i_{mmD} + ji_{mmQ}$ space-vector of the magnetizing current in the stator reference frame

$\mathbf{i}_{mr} = i_{mrD} + ji_{mrQ}$ space-vector of the rotor magnetizing current in the stator reference frame

$\mathbf{i}_{ms} = i_{msD} + ji_{msQ}$ space-vector of the stator magnetizing current in the stator reference frame

L_m total static magnetizing inductance

L_r rotor inductance

$L_{r\sigma}$ rotor leakage inductance

L_s stator inductance

L'_s stator transient inductance

$L_{s\sigma}$ stator leakage inductance

p number of pole pairs

R_r resistance of a rotor phase winding

R_s resistance of a stator phase winding

S_a, S_b, and S_c command signals of the VSI legs

t_e electromagnetic torque

T_r rotor time constant

T'_r rotor transient time constant

$T_{r\sigma}$ rotor leakage time constant

T_s stator time constant

T_{sp} sampling time of the control system

T'_s stator transient time constant

u_{com} common-mode voltage

$\mathbf{u}_s = u_{sD} + ju_{sQ}$ space-vector of the stator voltages in the stator reference frame

$\mathbf{u}_s = u_{sx} + ju_{sy}$ space-vector of the stator voltages in a generic rotating reference frame

u_{sA}, u_{sB}, u_{sC} stator phase voltages

U_d DC link voltage

ρ_r phase angle of the rotor flux-linkage space-vector with respect to the sD axis

ρ_s phase angle of the stator flux-linkage space-vector with respect to the sD axis

ρ_m phase angle of the magnetizing flux-linkage space-vector with respect to the sD axis

σ $1 - L_m^2/(L_sL_r)$ = total leakage factor

σ_r rotor leakage factor

σ_s stator leakage factor

ϑ_r angular position of the rotor with respect to the sD axis

$\psi_{s\beta A}, \psi_{s\beta B}, \psi_{s\beta C}$ stator phase virtual fluxes

$\boldsymbol{\psi}'_r = \psi_{rd} + j\psi_{rq}$ space-vector of the rotor flux linkages in the stator reference frame

$\boldsymbol{\psi}_s = \psi_{sD} + j\psi_{sQ}$ space-vector of the stator flux linkages in the stator reference frame

ω_{mr} angular speed of the rotor flux space-vector

ω_{ms} angular speed of the stator flux space-vector

ω_{mm} angular speed of the magnetizing flux space-vector

ω_{sl} angular slip speed

ω_r angular rotor speed (in electrical angles per second)

All quantities with *ref* in pedex are reference quantities.

The symbols ψ'_r, ψ_s, and ψ_m in apex mean the reference frame in which the variables are expressed.

References

1. K. Hasse, Drehzahlgelverfahren für schnelle umkehrantriebe mit stromrichtergespeisten asynchron-kurzschlusslaufer-motoren, *Reglungstechnik*, 20, 60–66, 1972.
2. F. Blaschke, The principle of field-orientation as applied to the transvector closed-loop control system for rotating-field machines, *Siemens Review*, 34, 217–220, 1972.
3. M. Bodson, J. Chiasson, R. Novotnak, High performance induction motor control via input-output linearization, *IEEE Control Systems Magazine*, 14, 25–33, August 1994.
4. M. Pietrzak-David, B. de Fornel, Non-Linear control with adaptive observer for sensorless induction motor speed drives, *European Power Electronics and Drives Journal*, 11(4), 7–13, 2001.
5. D. G. Taylor, Nonlinear control of electric machines: An overview, *IEEE Control Systems Magazine*, 14, 41–51, December 1994.
6. R. Marino, Output feedback control of current-fed induction motors with unknown rotor resistance, *IEEE Transactions on Control Systems Technology*, 4, 336–347, July 1996.
7. Z. Krzeminski, Nonlinear control of induction motors, *Proceedings of 10th IFAC World Congress*, Munich, Germany, 1987, Vol. IA-22, pp. 349–354, 820–827.
8. R. Ortega, A. Loria, P. J. Nicklasson, H. Sira-Ramirez, *Passivity-Based Control of Euler-Lagrange Systems*, Springer-Verlag, London, U.K., 1998.
9. I. Takahashi, T. Noguchi, A new quick-response and high efficiency control strategy of an induction machine, *IEEE Transactions on Industry Applications*, IA-22, 820–827, September/October 1986.
10. M. Depenbrock, Direct self-control of the flux and rotary moment of a rotary-field machine, U.S. Patent 4 678 248, July 7, 1987.
11. M. Depenbrock, Direct self control of inverter-fed induction machines, *IEEE Transactions on Power Electronics*, 3, 420–429, October 1988.
12. U. Baader, M. Depenbrock, G. Gierse, Direct self control (DSC) of inverter-fed-induction machine—A basis for speed control without speed measurement, *IEEE Transactions on Industry Applications*, 28, 581–588, May/June 1992.
13. G. S. Buja, M. Kazmierkowski, Direct torque control of PWM inverter-fed AC motors—A survey, *IEEE Transactions on Industrial Electronics*, 51(4), 744–756, August 2004.
14. P. Vas, *Electrical Machines and Drives*, Oxford Science Publications, New York, 1992.
15. J. Holtz, Sensorless control of induction motor drives, *Proceedings of the IEEE*, 90(8), 1359–1394, August 2002.
16. A. Abbondanti, M. B. Brennen, Variable speed induction motor drives use electronic slip calculator based on motor voltages and currents, *IEEE Transactions on Industry Applications*, 29, 344–348, March/April 1993.
17. W. Leonhard, *Control of Electrical Drives*, Springer-Verlag, Berlin, Germany, 1997.
18. P. Vas, *Sensorless Vector and Direct Torque Control*, Oxford Science Publications, New York, 1998.
19. P. Vas, *Vector Control of AC Machines*, Oxford Science Publications, Oxford, U.K., 1990.
20. P. Vas, *Artificial-Intelligence-Based Electrical Machines and Drives*, Oxford Science Publications, Oxford, U.K., 1999.
21. I. Boldea, S. A. Nasar, *Electric Drives*, 2nd edn., Taylor & Francis, New York, 2005.
22. B. K. Bose, *Power Electronics and Variable Speed Drives*, IEEE Press, New York, 1997.
23. D. W. Novotny, T. A. Lipo, *Vector Control and Dynamics of AC Drives*, Oxford Science Publications, Oxford, U.K., 1996.
24. W. Leonhard, 30 Years space vectors, 20 years field orientation, 10 years digital signals processing with controlled AC-drives, a review (part 1), *European Power Electronics and Drives Journal*, 1(1), 13–19, July 1991.
25. W. Leonhard, 30 Years space vectors, 20 years field orientation, 10 years digital signals processing with controlled AC-drives, a review (part 2), *European Power Electronics and Drives Journal*, 1(2), 89–101, October 1991.

26. W. Leonhard, Microcomputer control of high dynamic performance AC-drives, a survey, *Automatica*, 22(1), 1–19, 1986.
27. W. Leonhard, Controlled AC drives, a successful transition from ideas to industrial practice, *Control Engineering Practice*, 4(7), 897–908, 1996.
28. R. Lessmeier, W. Schumacher, W. Leonhard, Microprocessor-controlled AC-servo drives with synchronous or induction motors: Which is preferable?, *IEEE Transactions on Industry Applications*, IA-22(5), 812–819, September/October 1986.
29. R. D. Lorenz, T. A. Lipo, D. W. Novotny, Motion control with induction motors, *Proceedings of the IEEE*, 82(8), 1215–1240, August 1994.
30. H. Theuerkauft, Zur digitalen Nachbildung von Antriebsschaltungen mit umrichtergespeisen Asynchronmachinen, PhD dissertation, Technischen Universitaet Carolo Wihelmina zu Braunschweig, Brunswick, Germany, 1975.
31. R. Gabriel, Feldorientierte Regelung einer Asynchronmachine mit einem Mikrorechner, PhD dissertation, Technischen Universitaet Carolo Wihelmina zu Braunschweig, Brunswick, Germany, 1982.
32. W. Schumacher, Microrechner-Geregelter Asynchron-Steallantrieb, PhD dissertation, Technischen Universitaet Carolo Wihelmina zu Braunschweig, Brunswick, Germany, 1985.
33. N. Mohan, T. M. Undeland, W. P. Robbins, *Power Electronics, Converters, Applications and Design*, John Wiley & Sons, Inc., Hoboken, NJ, 2003.
34. G. Buja, D. Casadei, J. Lalu, G. Serra, Il controllo diretto di coppia negli azionamenti elettrici con motore asincrono, Specialistic Course ANAE-UCIREV, Milano, Italy, November 7–8, 1996.
35. M. Cirrincione, M. Pucci, G. Vitale, G. Cirrincione, A new direct torque control strategy for the minimization of common-mode emissions, *IEEE Transactions on Industry Applications*, 42(2), 504–517, March/April 2006.
36. A. Mir, M. E. Elbuluk, D. S. Zinger, Fuzzy implementation of direct self control of induction motors, *IEEE Transactions on Industry Applications*, 30, 729–735, 1994.
37. A. M. Trzynadlowski, *Control of Induction Motors*, Academic Press, New York, 2001.
38. C. Lascu, I. Boldea, F. Blaabjerg, A modified direct torque control for induction motor sensorless drive, *IEEE Transactions on Industry Applications*, 36(1), 122–130, 2000.
39. T. G. Habetler, F. Profumo, M. Pastorelli, L. M. Tolbert, Direct torque control of induction machines using space vector modulation, *IEEE Transactions on Industry Applications*, 28(5), September/October 1992.
40. G. Buja, D. Casadei, G. Serra, DTC-based strategies for induction motor drives, *Proceedings of IEEE IECON'97*, New Orleans, LA, 1997, Vol. 4, pp. 1506–1516.

6

Sensorless Control of Induction Machine Drives

6.1 Introduction on Sensorless Control

In the last decade, there has been a strong interest in eliminating the speed or position sensor in the shaft of induction machine (IM) drives, while preserving, however, the high dynamic performance of vector-controlled drives. The techniques for the speed or position estimation of the rotor are called in literature sensorless or encoderless [1–5]. The main advantages of sensorless-controlled drives are the reduced hardware complexity, the lower cost, the reduced size of the drive machine, the elimination of the sensor cables, the better noise immunity, the increased reliability, and the lower maintenance requirements [6]. Moreover, a motor without a speed sensor is suitable for dangerous or even hostile environments. Actually, high-performance applications require high accuracy in speed estimation, wide speed ranges, high bandwidth of the speed control loop, robustness to load torque perturbations, and correct zero-speed operation both at no load and at load. A great deal of schemes have been proposed for sensorless control of IMs so far [1–6]; among these, the two main categories are those based on magnetic saturation or in general on machine anisotropies and those based on the fundamental magnetomotive (mmf) force dynamic model of the IM. Table 6.1 summarizes a schematic overview of the different techniques applied to speed sensorless control [6]. Fundamental model techniques are based either on open-loop estimators or on closed-loop observers and are limited by zero stator frequency. In these conditions, the back electromotive force (emf) is zero and the machine speed is unobservable. However, in the low-speed range anisotropies of the machine can provide useful information on the main field angle or on the position of the rotor. An accurate speed estimation can be given, exploiting the voltages induced in the stator windings by the spatial rotor slot harmonics. Proper excitation signals can be either injected signals at frequencies higher than the fundamental or transients is caused by the switching of the inverter. The response of the motor is used to identify either the field angle or the position angle.

6.2 Model-Based Sensorless Control

A great deal of model-based sensorless techniques have devised over the last few years. Scientific literature regarding these methods is huge; thus, in the following, only the milestone ideas are given. The first attempt to estimate the rotor speed is probably proposed by Joetten and Maeder [7], where the slip speed is estimated on the basis of the

TABLE 6.1

Issues of the Different Sensorless Techniques

	Fundamental Model			Exploited Anisotropies		
Additional signal injection	No	No	No	Yes	Yes	Yes/no
Principle	Open-loop models	Observers	Rotor slot harmonics	Main inductance saturation	Artificial saliency	Rotor slot leakage
Minimum frequency	Close to or temporarily zero	Close to or temporarily zero	Below 1 Hz	Theoretically zero	Theoretically zero	Zero
Maximum speed error	Half-rated speed	Half-rated speed	Theoretically zero	Half-rated speed	Small	Theoretically zero
Position error	—	—	—	—	—	Theoretically zero

state equations of the IM. Afterward, [8] adopted for the first time an *MRAS (model reference adaptive system)* scheme. Then, [9] adopted a combination of a *reduced-order observer* (ROO), used as reference model, and the simple current model, used as adaptive model, to estimate the rotor speed. The proposed adaptation mechanism was, however, exactly the same as what is proposed in [8]. A suitable gain matrix choice was proposed for the reduced-order poles allocation. Other possible choices for the poles allocation of ROOs have been analyzed in [10,11]. Almost at the same time [12] proposed a *full-order Luenberger adaptive observer* for the simultaneous estimation of both the rotor flux and the rotor speed of the machine. In [12], an on-line estimation of the stator and rotor resistances was also presented. This kind of observer presents, however, some drawbacks about the stability of its adaptation law, especially in regenerative mode low-speed operation, and about the accuracy in low-speed regions and at zero speed. To overcome the first kind of difficulties, some methods have been come up with [13–16]. Another approach by [17] proposes a speed sensorless control using parameter identification for the speed estimation. A design method of robust adaptive sliding observers is further proposed.

6.3 Anisotropy-Based Sensorless Control

Anisotropy-based sensorless techniques exploit those machine properties which are not reproduced in the fundamental mmf model. As far as signal injection methods are considered, the injected signals excite the machine at frequencies higher than the fundamental. The resulting high-frequency currents generate flux linkages which close through the leakage paths both in the stator and in the rotor, leaving the fundamental flux of the fundamental wave almost unaffected. The high-frequency effects can be therefore considered superimposed to the fundamental behavior of the machine. A magnetic anisotropy can be caused by the saturation of the leakage path through the fundamental field. Other anisotropic structures are the discrete rotor bars in the cage motor. Otherwise, a motor can be also custom designed so as to emphasize some saliencies or exhibit periodic variations within a pole pitch of local magnetic or electrical properties. Typical cases are the variations of the widths of the rotor slot openings [18], of the depths at which the rotor bars are buried below the level of the rotor, or of the resistance of the outer bars of a double cage or deep bar rotor [19]. There are, in general, more anisotropies in the motor. The various anisotropies present different spatial orientations such as the angular position of the fundamental field, the position of the rotor bars within a rotor bar pitch, and if possible, the position of a custom-designed rotor within a fundamental pole pair. The response to an injected high-frequency signal necessarily reflects all the anisotropies, dependant on or position. The injected signals can be periodic, creating either a high-frequency revolving field or an alternative field in a specific predetermined spatial direction. Such signals are called carriers. The carrier signals, mostly created by additional components of the stator voltages, get modulated by the actual orientation in space of the machine anisotropies.

Other kinds of excitation signals are the PWM-switching signals of the inverter. Each commutation of the power devices of the inverter excites repetitive transients in the machine. All flux components, apart from the fundamental one, present high-frequency content and therefore do not penetrate sufficiently into the rotor to create a mutual flux linkage. These fluxes, on the contrary, contribute to the total leakage flux. Among the techniques for retrieving the rotor position from anisotropies, the most important are the so-called *INFORM* (*INdirect Flux detection by On-line Reactance Measurements*) and the *instantaneous rotor position measurement* [6]. The *INFORM* method, developed by Schroedl [20], is based on the measurement of the rate of change of the stator current during a short interval. Two opposite voltage vectors are applied in two consecutive short intervals to compensate the ohmic voltage drop on the stator resistance and the back emf. The estimated field angle can be retrieved on the basis of the rate of change of the stator current measured on the two consecutive short intervals. The *instantaneous rotor position measurement* method exploits the rotor slot anisotropy to identify the rotor position angle, and the magnetic saturation is considered as a disturbance. The methods developed by Jiang [21], Holtz [22,23] [21–23] are based on the instantaneous measurements of anisotropy signals from the terminal voltage between the feeding inverter and the star point of the machine [23].

6.4 Model-Based Sensorless Techniques

Before starting with the description of the most important model-based sensorless techniques, the main problematic issues related to the implementation of model-based techniques will be treated. In particular, the open-loop integration, the inverter nonlinearity, and the machine parameter mismatch will be addressed, with some proposals for solving the related problems.

6.4.1 Open-Loop Integration

It is well known that the main problem of the numerical integration in some flux estimators, when used in high-performance electrical drives, is the presence of DC biases at the input of the integrators themselves. The speed observers suffering from this problem are those employing open-loop flux estimators, for example, open-loop speed estimators and *MRAS* systems, while speed estimators employing closed-loop flux integration, like the ROO and the *full-order Luenberger adaptive observer*, do not have this problem. In particular,

DC drifts are always present in the signal before they are integrated, which causes the integrator to saturate with a resulting inadmissible estimation error, and also after the integration because of the initial conditions [24,25]. The open-loop integration problem at low speed can become even more crucial than the possible detuning of the estimator at the same speeds. Some alternative solutions have been conceived to solve this problem, for example, the integrator with saturation feedback [24], the integrator based on cascaded low-pass filters (LPFs) [26,27], the integrator based on the offset vector estimation, and compensation of residual estimation error [28–30]. In the following, these solutions proposed by literature are analyzed.

6.4.1.1 Low-Pass Filter

In general, first-order LPFs are used instead of a pure integrator, with very low cutoff frequency. However, the behavior of any estimator in low-frequency ranges, close to the cutoff frequency of the LPF, is greatly affected by errors in the amplitude and the phase of the estimated flux [3]. Figure 6.1 shows the Bode diagram of a first-order LPF with cutoff frequency equal to 15 rad/s. It can be observed that the phase angle error becomes significant just below 10 Hz. This makes the LPF solution unsuitable for low-speed operation of sensorless drives.

Moreover, the cutoff frequency of the LPF cannot be reduced too much, since at lower cutoff frequencies, the DC drift filtering effect reduces. It can be easily shown by computing, in the Laplace domain s, the steady-state value of the output of the LPF $G_{LP}(s)$ to a step input of amplitude E_{dr}:

$$\lim_{s \to 0} sG_{LP}(s)\frac{E_{dr}}{s} = \lim_{s \to 0} s\frac{1}{s + \omega_c}\frac{E_{dr}}{s} = \frac{E_{dr}}{\omega_c} \tag{6.1}$$

Equation 6.1 clearly shows that, the lower the cutoff frequency ω_c of the LPF, the higher the undesired effect of the DC drift at the output.

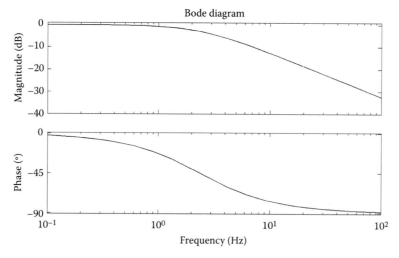

FIGURE 6.1
Bode diagram of a first-order LPF with cutoff frequency of 15 rad/s.

As explained in [25], to solve the problems connected with the use of pure integrators and LPFs, new integration methods have been developed which can be generally expressed as [24]

$$\Psi = \frac{1}{s + \omega_c} e + \frac{\omega_c}{s + \omega_c} \Psi \tag{6.2}$$

where
e and ψ are, respectively, the input and the output of the integrator
ω_c is the cutoff frequency of the LPF

In the case under study, e represents the emf and ψ the corresponding flux linkage. To obtain the approximate z-domain transfer function from the s-domain, the approximation

$$s = \frac{z-1}{T_{sp}} \tag{6.3}$$

has been used, where T_{sp} is the sampling time of the control system. Then (6.2) can be written in the z-domain in the following way:

$$\Psi = \underbrace{\frac{T_{sp}}{z-\beta}}_{\text{feed-forward}} e + \underbrace{\frac{1-\beta}{z-\beta}}_{\text{feed-back}} \Psi \tag{6.4}$$

where $\beta = e^{-T_{sp}\omega_c}$. The idea is to express the integrator as the sum of two terms: a feed-forward and a feed-back term. The block diagram of the integrator in the z-domain, sketched in Figure 6.2, is thus obtained. From this block diagram, two algorithms are derived [24], as explained in the following.

Algorithm 1
If the ψ signal in the second term of the right-hand side of (6.4) is followed by a saturation function of amplitude L, the first algorithm of [24] is obtained. In the following, it will be simply called *algorithm 1*, and its block diagram is shown in Figure 6.3 in the z-domain.

Figure 6.4 shows the locus of the rotor flux linkage experimentally measured on a 2.2 kW IM drive (see Chapter 11 for details on the adopted test setup) when the *algorithm 1* has been adopted in open-loop flux integration (no flux feedback). Results show the correct behavior of the algorithm.

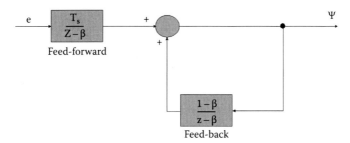

FIGURE 6.2
Block diagram of the integrator in the z-domain. (From Cirrincione, M. et al., *IEEE Trans. Power Electron.*, 19(1), 25, 2004.)

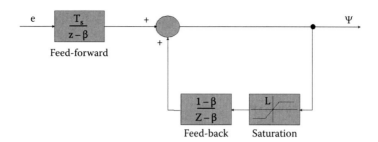

FIGURE 6.3
Block diagram of the integration *algorithm 1*. (From Cirrincione, M. et al., *IEEE Trans. Power Electron.*, 19(1), 25, 2004.)

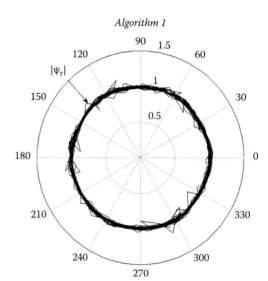

FIGURE 6.4
Locus of the rotor flux linkage with *algorithm 1*.

Algorithm 2
An improvement to *algorithm 1* can be obtained by limiting only the magnitude of the flux as shown in Figure 6.4. Further details can be found in [24]. In the following, it is simply called *algorithm 2*. Figure 6.4 shows the block diagrams of the *algorithm 2* in the z-domain. This algorithm is specifically targeted to the integration of the emf space-vector to obtain the flux-linkage space-vector, so in this case, the direct and quadrature components (subscripts D and Q) of both space-vectors are the output of the system.

In *algorithms 1* and *2*, as explained in [24], the saturation limit L must be chosen close to the reference value (Figure 6.5).

Figure 6.6 shows the locus of the rotor flux linkage experimentally measured on a 2.2 kW IM drive (see Chapter 11 for details on the adopted test setup) when the *algorithm 2* has been adopted in open-loop flux integration (no flux feedback). Results show the correct behavior of the algorithm, with performance very close to those obtainable with *algorithm 1*.

Algorithm 3
In this algorithm, the flux linkage ψ is obtained from emf e by an integration method accomplished by a *programmable cascaded low-pass filter* (PCLPF) implemented by a hybrid neural network consisting of a *recurrent neural network* (RNN) and a *feed-forward artificial neural network* (FFANN). Here, only the fundamentals are described, in accordance with [26,27].

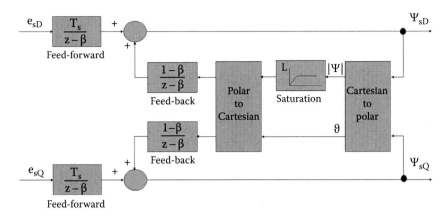

FIGURE 6.5
Block diagram of the integration *algorithm 2*. (From Cirrincione, M. et al., *IEEE Trans. Power Electron.*, 19(1), 25, 2004.)

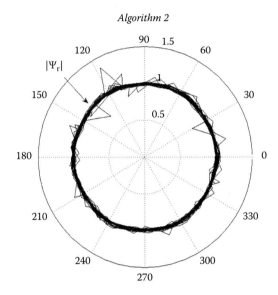

FIGURE 6.6
Locus of the rotor flux linkage with *algorithm 2*.

If two identical LPFs are cascaded (*PCLPF*) with transfer functions in the z-domain given by

$$\frac{\Psi_1(z)}{E(z)} = \frac{Kz^{-1}}{1 - \alpha z^{-1}} \quad \text{and} \quad \frac{\Psi(z)}{\Psi_1(z)} = \frac{Kz^{-1}}{1 - \alpha z^{-1}} \tag{6.5}$$

where

$E(z)$, $\Psi_1(z)$, and $\Psi(z)$ are, respectively, the z-transforms of the input signal $e(k)$, the output of the first filter $\psi_1(k)$, and the output of the second filter $\psi(k)$

$\alpha = (1 - T_{sp}/\tau)$, T_s is the sampling time, τ is the time constant of each component filter

$K = (T_{sp}/\sqrt{G})/\tau$, G is the amplitude of the compensation gain of the *PCLPF*

Then the following discrete-time equation results, expressed in matrix notation:

$$\begin{bmatrix} \Psi_1(k+1) \\ \Psi_2(k+1) \end{bmatrix} = \begin{bmatrix} \alpha & 0 \\ K & \alpha \end{bmatrix} \begin{bmatrix} \Psi_1(k) \\ \Psi_2(k) \end{bmatrix} + \begin{bmatrix} \alpha \\ 0 \end{bmatrix} e(k) \tag{6.6}$$

An equivalent *RNN* is then suggested which results in the following matrix equation:

$$\begin{bmatrix} \Psi_1(k+1) \\ \Psi_2(k+1) \end{bmatrix} = \begin{bmatrix} W_{11} & 0 \\ W_{21} & W_{22} \end{bmatrix} \begin{bmatrix} \Psi_1(k) \\ \Psi_2(k) \end{bmatrix} + \begin{bmatrix} W_{13} \\ 0 \end{bmatrix} e(k) \tag{6.7}$$

where W_{11}, W_{21}, W_{22}, and W_{13} are the weights of the *RNN*, which is shown in Figure 6.7. Although $W_{11} = W_{22} = \alpha$ and $W_{21} = W_{13} = K$, all the weights are considered as independent variables and a function of frequency. These weights can be supplied, for each frequency, by a look-up table with interpolation properties, which can be implemented by an *FFANN*, for example, a multilayer perceptron. To obtain the training set for the FFANN, the weights W_{11}, W_{21}, W_{22}, and W_{13} of the *RNN* at every supply frequency are tuned so that the input voltage wave and the corresponding flux output wave match precisely with a very low error. In this way, the input (frequency)—output (weights of the *RNN*) pairs for training *FFANN* are obtained. During the integration process, the *FFANN* is fed by the estimated frequency of the input signal *e* and produces the corresponding set of weights for the *RNN*, so that this will integrate the input signal *e* correctly. However, the *RNN* weights are dependent on the sampling frequency of the control system: this means that a retraining of the RNN is required if the sampling frequency varies.

Different ways can be followed to train the RNN: either a first- or second-order method or an extended Kalman filter (EKF) [26,27,31]. Figure 6.8 shows the locus of the rotor flux linkage experimentally measured on a 2.2 kW IM drive (see Chapter 11 for details on the adopted test setup) when the *algorithm 3* has been adopted in open-loop flux integration (no flux feedback). Results show the correct behavior of the algorithm, with performance slightly worse than both *algorithms 1* and *2*. As a matter of fact, the locus in this case tends to degenerate on a hexagon instead of a circle.

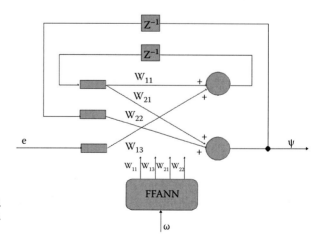

FIGURE 6.7
Block diagram of the integration *algorithm 3*. (From Cirrincione, M. et al., *IEEE Trans. Power Electron.*, 19(1), 25, 2004.)

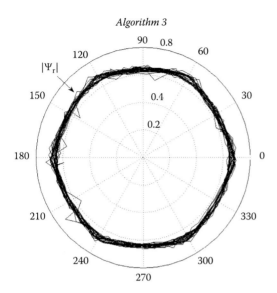

FIGURE 6.8
Locus of the rotor flux linkage with *algorithm 3*.

Algorithm 4

A different approach has been proposed by [28–30], which is based on the estimation of the offset vector and compensation of residual estimation error. In this case, the integrand emf e expressed in the stationary reference frame is integrated to obtain the flux (in [28,29], it is applied to stator flux, but the considerations are general). The resulting flux $\boldsymbol{\psi}$ is limited inside a circular boundary of radius $\boldsymbol{\psi}^*$, providing $\hat{\boldsymbol{\psi}}$. The contribution to the offset voltage term to be subtracted to the emf e is computed as

$$e_{off} = \frac{\hat{\boldsymbol{\psi}}_{max} + \hat{\boldsymbol{\psi}}_{min}}{\Delta t} \tag{6.8}$$

where
Δt is the time interval between zero crossings of the flux components $\hat{\boldsymbol{\psi}}$, equal to the fundamental period
$\hat{\boldsymbol{\psi}}_{max}$, $\hat{\boldsymbol{\psi}}_{min}$ are, respectively, the maximum and minimum values of the flux $\hat{\boldsymbol{\psi}}$ direct and quadrature components in the time interval Δt

In practice, at each period of the fundamental, the average offset voltage causing a drift in the flux is estimated and then subtracted. To improve the behavior of the system, the offset voltage term is low-pass filtered before being subtracted from the emf e. A further noise compensator term e_{hf} is subtracted from the emf e to cope the nonlinearities resulting from switching harmonics, inverter dead-time effects, and, in general, all system nonlinearities characterized by a frequency much higher than the fundamental. An efficient way to compensate these effects is to force the radial component of $\hat{\boldsymbol{\psi}}$ to go close to its reference $\boldsymbol{\psi}_{ref}$ by a fast proportional closed-loop control. Since the tangential component of $\hat{\boldsymbol{\psi}}$ is not altered at all, there is no interaction of this compensation with the flux estimator. Figure 6.9 shows the block diagram of such a compensation scheme.

Figure 6.10 shows the locus of the stator flux linkage experimentally measured on a 2.2 kW IM drive (see Chapter 11 for details on the adopted test setup) when the *algorithm 4* has been adopted in open-loop flux integration (no flux feedback). Results show the correct behavior of the algorithm, with performance similar to both *algorithms 1* and *2*.

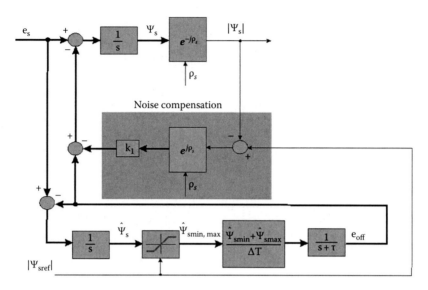

FIGURE 6.9
Block diagram of the *algorithm 4*.

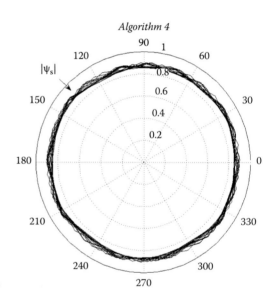

FIGURE 6.10
Locus of the stator flux linkage with *algorithm 4*.

6.4.2 Inverter Nonlinearity

The power devices of the inverter present a finite voltage drop in "on state," due to their forward nonlinear characteristics. This voltage drop has to be taken into consideration at low frequency (low fundamental voltage amplitude) where it becomes comparable with the stator voltage itself, giving rise to distortion and discontinuities in the voltage waveform. An interesting method for the compensation of the inverter device voltage drops has been proposed in [28–30]. This technique is based on modeling the forward characteristics of each power device with an average threshold voltage u_{th} and an average differential resistance R_d. Figure 6.11 shows, for example, the characteristic i_c versus v_{ce} of an IGBT

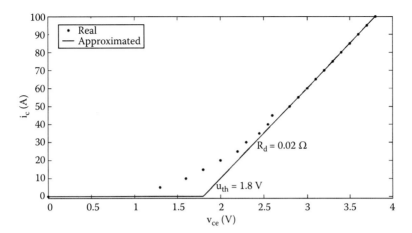

FIGURE 6.11
Forward characteristics of an IGBT Semikron SMK 50 GB 123 at the working temperature of 25°C.

module (model *Semikron* SMK 50 GB 123), the measured (dotted) and the modeled 11 (continuous). This characteristic has been used in the experimental tests shown in Chapter 11. In this case, the characteristic has been approximated with a threshold voltage of $u_{th} = 1.8\,\text{V}$ and a differential resistance of $R_d = 0.02\,\Omega$.

It could be easily demonstrated that in each leg of the inverter, the effect of the threshold voltage depends on the sign of the phase current, since there is always a power device that is forward biased. The threshold voltage on phase sA can be therefore written as $u_{th}sign(i_{sA})$. The threshold voltage space-vector can be then defined as

$$\mathbf{u}_{th} = \frac{2}{3}\left(u_{th}sign(i_{sA}) + \alpha u_{th}sign(i_{sB}) + \alpha^2 u_{th}sign(i_{sC})\right) = \frac{4}{3}u_{th}\mathbf{sec}(\mathbf{i}_s) \qquad (6.9)$$

where

$$\mathbf{sec}(\mathbf{i}_s) = \frac{1}{2}\left(sign(i_{sA}) + \alpha sign(i_{sB}) + \alpha^2 sign(i_{sC})\right) \qquad (6.10)$$

$\mathbf{sec}(\mathbf{i}_s)$ is the function defining the unity vector marking the 60° sector in which the stator current lies.

In general, the stator voltage space-vector $\mathbf{u}_s^*$ can be obtained as the difference between the reference stator voltage $\mathbf{u}_{sref}$, given to the PWM modulator, the forward voltage space-vector $\mathbf{u}_{th}$, and the ohmic drop $R_d\mathbf{i}_s$:

$$\mathbf{u}_s^* = \mathbf{u}_{sref} - \mathbf{u}_{th} - R_d\mathbf{i}_s \qquad (6.11)$$

Figure 6.12a and b shows the locus of the stator voltage space-vector $\mathbf{u}_s^*$, obtained as the difference between $\mathbf{u}_{sref}$ and $\mathbf{u}_{th}$ during both a motoring (a) and a generating (b) phases. Due to the nonlinearity of the inverter, the trajectories of the stator voltage space-vector $\mathbf{u}_s^*$ are distorted and discontinuous. Figure 6.13 shows the block diagram of the compensation methodology proposed in Refs [28–30]. Finally, Figure 6.14 shows the experimental time

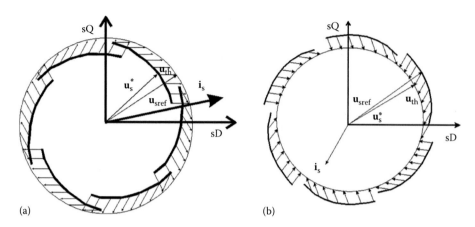

FIGURE 6.12
Locus of the real stator voltage space-vector in motoring (a) and generating (b) phase. (From Cirrincione, M. et al., *IEEE Trans. Ind. Appl.*, 42(1), 89, 2006.)

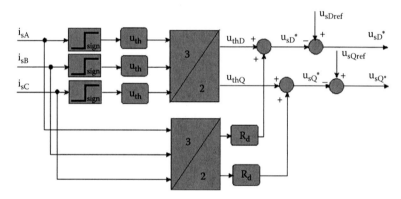

FIGURE 6.13
Block diagram of the forward voltage compensation scheme. (From Cirrincione, M. et al., *IEEE Trans. Ind. Appl.*, 42(1), 89, 2006.)

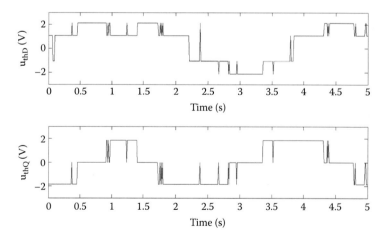

FIGURE 6.14
Direct and quadrature components of the $\mathbf{u}_{th}$ voltage space-vector during a constant speed reference of $1\,\text{rad/s}$. (From Cirrincione, M. et al., *IEEE Trans. Ind. Appl.*, 42(1), 89, 2006.)

waveforms, obtained on a 2.2 kW IM drive, of the direct and quadrature components of the $\mathbf{u}_{th}$ voltage space-vector during a constant speed reference of 1 rad/s.

6.4.3 Machine Parameter Mismatch

A further source of error in flux estimation is the mismatch of the stator and rotor resistances of the observer with their real values because of heating/cooling of the machine. The load-dependent variations of the winding temperature may lead up to 50% error in the modeled resistance. Stator and rotor resistances should be therefore estimated on-line and tracked during the operation of the drive. A great deal of on-line parameter estimation algorithms have been devised, requiring low complexity and computational burden when used in control systems. In any case, it should be emphasized that steady-state estimation of the rotor resistance cannot be performed in sensorless drives; thus, rotor resistance variations must be deduced from stator resistance estimation.

Literature offers several methodologies for the estimation of the electrical parameters of IM. Chapter 9 is devoted to the description of the simultaneous estimation of all the electrical parameters of the IM by means of linear neural networks, in particular the so-called TLS EXIN neuron. In the following, three recently proposed efficient algorithms for the on-line stator resistance estimation are briefly described [12,28,29].

Algorithm 1
Algorithm 1 is based on the concept that at steady-state, the stator flux vector and the induced voltage are orthogonal [28]:

$$\boldsymbol{\psi}_s \cdot \mathbf{e}_s = \boldsymbol{\psi}_s \cdot (\mathbf{u}_s - R_s \mathbf{i}_s) = 0 \tag{6.12}$$

Equation 6.12 intrinsically depends on the stator resistance and can therefore be exploited to estimate on-line the stator resistance itself. To make the estimator less cumbersome from the computational point of view, Equation 6.12 is transformed from the stationary reference frame into a reference frame aligned with the stator current space-vector, on the basis of the vector rotation $e^{-j\alpha_s}$, where α_s is the stator current space-vector angle. From this standpoint, the stator resistance can be estimated on the basis of this equation:

$$R_s = \frac{u_{sx}^{i_s} - (\psi_{sy}^{i_s}/\psi_{sx}^{i_s})u_{sy}^{i_s}}{|\mathbf{i}_s|} = \frac{u_{sx}^{i_s} - \omega_s |\boldsymbol{\psi}_s| \sin(\alpha_s - \rho_s)}{|\mathbf{i}_s|} \tag{6.13}$$

where

$$\frac{\psi_{sy}^{i_s}}{\psi_{sx}^{i_s}} = \tan(\alpha_s - \rho_s) \tag{6.14}$$

and $u_{sy}^{i_s} = e_i \cos(\alpha_s - \rho_s)$ with $e_i = \omega_s |\boldsymbol{\psi}_s|$ at steady-state.
Figure 6.15 shows the block diagram of the stator resistance estimator.

Algorithm 2
Algorithm 2 permits the on-line stator resistance R_s estimation on the basis of the minimization of the error between the measured $\mathbf{i}_{sx}^{\psi_s}$ and estimated $\hat{\mathbf{i}}_{sx}^{\psi_s}$ direct components of the stator current space-vector expressed in the stator flux reference frame [29].

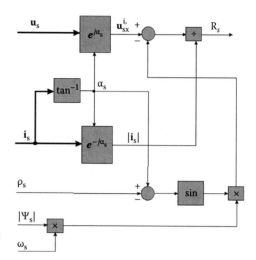

FIGURE 6.15
Block diagram of the *algorithm 1* for the stator resistance
on-line estimation algorithm.

If the stator equations of the IM in the stator flux linkage reference frame are considered, after considering them at steady-state ($d/dt \cong 0$), the following set of equations can be obtained:

$$\begin{cases} \left|\boldsymbol{\psi}_s\right| = -\omega_r T_r \sigma L_s i_{sy}^{\psi_s} + L_s i_{sx}^{\psi_s} & (6.15\,\text{a}) \\ L_s i_{sy}^{\psi_s} = \omega_r T_r \left(\left|\boldsymbol{\psi}_s\right| - \sigma L_s i_{sx}^{\psi_s}\right) & (6.15\,\text{b}) \end{cases}$$

In particular, after eliminating ω_r from the above equations, $\hat{i}_{sx}^{\psi_s}$ is estimated on the basis of the following relationship:

$$\hat{\mathbf{i}}_{sx}^{\psi_s} = \frac{\left|\boldsymbol{\psi}_s\right|}{L_s(1+\sigma)} + \frac{L_s^2\sigma}{L_s(1+\sigma)\left|\boldsymbol{\psi}_s\right|}\left|\mathbf{i}_s^{\psi_s}\right|^2 \tag{6.16}$$

Equation 6.16 does not depend on the stator resistance, and it can be therefore used as reference model of a proper *MRAS*. $\mathbf{i}_{sx}^{\psi_s}$ is obtained by the current measurements on the basis of the transformation from the stationary reference frame to the stator flux reference frame (block $e^{-j\rho_s}$ in Figure 6.16), according to the knowledge of the stator flux angle ρ_s, given by the flux model. The tracking error between the two currents is then processed by an adaptive controller or more simply by a proportional integral (PI) controller. Figure 6.16 shows the block diagram of the previously described algorithm.

Algorithm 3
Algorithm 3 permits the stator resistance to be computed, by exploiting the estimation of the stator current space-vector in the stator reference frame. This method can suitably be integrated with the full-order Luenberger observer (FOLO) in this way, as proposed in Ref. [12].

In particular, R_s is estimated on the basis of the measured i_{sD} and i_{sQ} and estimated $\hat{i}_{sD}$ and $\hat{i}_{sQ}$ stator current components in the stationary reference frame by means of the following update law:

$$\frac{d\hat{R}_s}{dt} = -\lambda\left(\left(i_{sD} - \hat{i}_{sD}\right)\hat{i}_{sD} + \left(i_{sQ} - \hat{i}_{sQ}\right)\hat{i}_{sQ}\right) \tag{6.17}$$

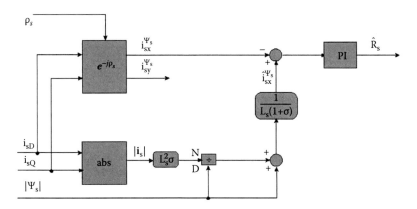

FIGURE 6.16
Block diagram of the *algorithm 2* for the stator resistance on-line estimation algorithm. (From Cirrincione, M. et al., *IEEE Trans. Ind. Appl.*, 40(4), 1116, 2004.)

where λ is a properly chosen positive constant. In practice, the rate of change of the estimated stator resistance is proportional to the scalar product between the stator current estimation error and the estimated current space-vectors. On the contrary, since the same estimation scheme cannot be applied to rotor resistance estimation in sensorless drives, R_r can be estimated considering its variation proportional to that of the R_s on the basis of the following law:

$$\hat{R}_r = R_{srn}\hat{R}_s \qquad (6.18)$$

where R_{srn} is the ratio between the rated values of the stator and rotor resistances. It is therefore assumed that both the resistances vary with the operating temperature of the machine.

6.4.4 Estimators and Observers

In general, an estimator is defined as a dynamic system whose state variables are estimates of some other system (in the case under study, an electrical machine). There are basically two forms of implementation of an estimator: open loop and closed loop, the distinction between the two being whether or not a correction term, involving an estimation error term, is exploited to adjust on-line the response of the estimator [1]. A closed-loop estimator is called an observer. In general, observers are preferable to open-loop estimators, since they permit to robustness to parameter variations and noise.

Observers can be classified in accordance with the kind of representation used for the plant to be observed. If the plant is considered deterministic, correspondingly, the observer is a deterministic observer; otherwise, it must be considered a stochastic observer. As a matter of fact, the FOLO and the ROO are deterministic, while the Kalman Filter (KF) is stochastic. The classic KF can be applied only to linear stochastic systems, while the EKF should be used if a nonlinear stochastic system has to be dealt with. On the other hand, the basic Luenberger observer can be applied only to linear time-invariant deterministic systems, while the extended Luenberger observer (ELO) should be used if a nonlinear time-variant deterministic system has to be dealt with. In summary, both the ELO and the EKF are nonlinear estimators, to be applied, respectively, to deterministic and stochastic systems. The ELO represents a good alternative for the real-time implementation in industrial drive systems because of the simple algorithm and its ease of tuning, making it preferable to EKF [1].

6.4.5 Open-Loop Speed Estimators

Various rotor speed and slip speed open-loop estimators can be obtained by rewriting the stator and rotor equations of the IM. A pioneering work about open-loop speed estimators has been made in [7], where a speed estimator simply based on the voltages and currents measurements has been presented. In [1], five open-loop sensorless schemes are described, which are all based on the stator and rotor equations of the IM, differing from one another by the reference frame in which the equations are expressed. All of these schemes estimate the speed on the basis of the rotor flux-linkage estimation. Some of them are also currently employed in commercial sensorless drives. Analogous open-loop speed estimators can be devised with the speed estimation on the stator flux linkage, rather than on the rotor one [3]. Consequently, equations can be written in the rotor-flux- or stator-flux-oriented reference frames.

If the stator flux linkage is estimated, the rotor speed can be computed as the difference of the stator flux-linkage speed ω_{ms} and the slip speed ω_{sls}, as follows:

$$\omega_r = \omega_{ms} - \omega_{sls} = \underbrace{\left(\psi_{sD}\,d\psi_{sD}/dt - \psi_{sQ}d\psi_{sQ}/dt\right)/\left|\mathbf{\Psi}_s\right|^2}_{\omega_{ms}} - \underbrace{\frac{L_s}{T_r}\,\frac{\sigma T_r\,di_{sy}^{\psi_s}/dt + i_{sx}^{\psi_s}}{\left|\mathbf{\Psi}_s\right| - \sigma L_s i_{sx}^{\psi_s}}}_{\omega_{sls}} \tag{6.19}$$

The stator flux-linkage speed is obtained with the division between the vector product of the stator flux-linkage vector and its derivative and the square of the stator flux amplitude itself. The slip speed is obtained on the basis of the direct and quadrature components of the stator current in the stator flux–oriented reference frame, $i_{sx}^{\psi_s}, i_{sy}^{\psi_s}$; for this reason, a coordinate transformation is needed for this estimator. Figure 6.17 shows the block diagram of the open-loop speed estimator based on Equation 6.19. The correct field orientation is influenced by the accurate estimation of the angle ρ_s that, depending on the open-loop flux estimation, suffers from both the integration problem and the sensitivity to the stator resistance variation. Moreover, the computation of the slip speed needs the time derivative $di_{sy}^{\psi_s}/dt$; it should be noted, however, that the derivative operation amplifies the high-frequency noise in the signal.

The open-loop speed estimator in Figure 6.17 has been experimentally implemented in a 2.2 kW flux-oriented controlled (FOC) IM drive (see Appendix 11.A) on a test setup

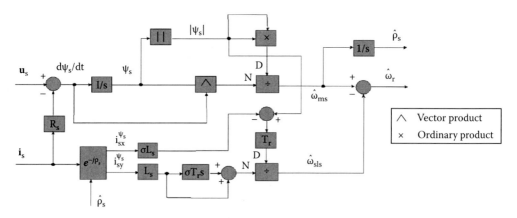

FIGURE 6.17
Open-loop speed estimator based on the stator flux linkage.

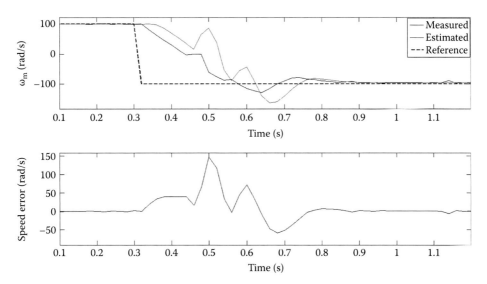

FIGURE 6.18
Reference, measured, and estimated speed with the open-loop speed estimator in Figure 6.17.

(see Appendix 11.B). Figure 6.18 shows the experimental time waveform of the reference, measured, and estimated speed as well as the speed estimation error. In this case, open-loop flux integration has been performed by adopting an LPF (see Figure 6.1). The test refers to a speed reversal from 100 to −100 rad/s at no load. Results show a good tracking of the estimated speed at steady-state, but big estimation errors occur during the transient phase, mainly caused by the presence of the time derivative $di_{sy}^{\psi_s}/dt$ term.

If the rotor flux linkage is estimated, the rotor speed can be computed as the difference of the rotor flux-linkage speed ω_{mr} and the slip speed ω_{slr} as follows:

$$\omega_r = \omega_{mr} - \omega_{slr} = \underbrace{\left(\psi_{rd}\, d\psi_{rq}/dt - \psi_{rq}\, d\psi_{rd}/dt\right)/|\mathbf{\Psi}_r|^2}_{\omega_{mr}} - \underbrace{\left(-\psi_{rq}i_{sD} + \psi_{rd}i_{sQ}\right)L_m/\left(T_r\,|\mathbf{\Psi}_r|^2\right)}_{\omega_{slr}} \quad (6.20)$$

The rotor flux-linkage speed is obtained with the division between the vector product of the rotor flux-linkage vector and its derivative and the square of the rotor flux amplitude. The slip speed is obtained on the basis of the vector product of the rotor flux and the stator current vectors. Figure 6.19 shows the block diagram of the open-loop speed estimator based on Equation 6.20. Unlike the scheme based on the stator flux estimation, here no coordinate transformation is needed, with resulting reduction of the computational demand of the estimator. Also in this case, the open-loop flux estimation suffers from both the integration problem and the sensitivity to the stator resistance variation. On the contrary, different from the scheme based on the stator flux estimation, here, no current derivative is necessary, while the rotor flux derivative is needed, which is less problematic than the current one since the flux is an integral signal, and thus this derivative term is an available signal.

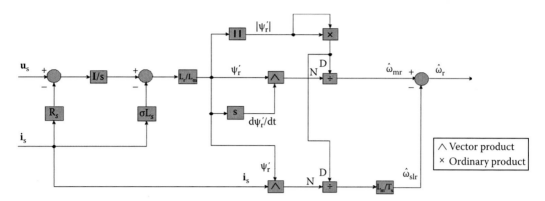

FIGURE 6.19
Open-loop speed estimator based on the rotor flux linkage.

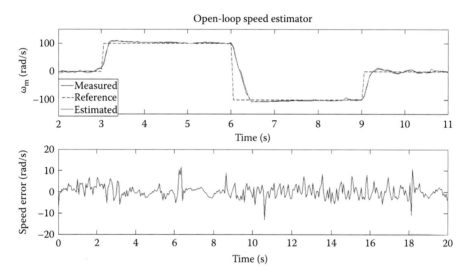

FIGURE 6.20
Reference, measured, and estimated speed with the open-loop speed estimator in Figure 6.19.

Figure 6.20 shows the experimental time waveform of the reference, measured, and estimated speed as well as the speed estimation error, obtained on the same 2.2 kW FOC IM drive where the open-loop speed estimator in Figure 6.19 has been implemented. In this case, open-loop flux integration has been done by adopting an LPF (see Figure 6.1). The test refers to a set of speed reference step variations with this sequence: $0 \rightarrow 100 \rightarrow -100 \rightarrow 0$ rad/s at no load. Results show a good tracking of the estimated speed both during the steady-state and the transient phase, with a peak transient estimation error of 10 rad/s.

In general, however, the effectiveness of any open-loop estimator highly depends on the accuracy in the knowledge of the machine parameters. At low speed, in particular, a reduction of the speed estimation accuracy is to be expected in all these schemes due to a mismatch between the real and estimated flux linkage caused by a wrong model of the stator resistance. The misknowledge of the rotor time constant, on the contrary, mainly influences the estimation of the slip speed and therefore is critical at high loads. This is confirmed by the sensitivity analysis of the flux models to parameter mismatch [32] (see the sensitivity analysis in Chapter 9). Parameter mismatch has, in these schemes, a big influence on the steady-state and transient behavior of the sensorless drive.

6.4.6 Model Reference Adaptive Systems

Both the steady-state and transient accuracy of the speed estimation can be significantly increased by adopting closed-loop speed estimators (observers) instead of the open-loop ones.

An important category of observers is that of *MRAS*s. In this framework, several speed observers have been developed.

6.4.6.1 Classic MRAS

All of the MRAS observers are based on the idea of a *reference model* and an *adaptive model*. Some state variables of the IM are estimated at the same time with both the *reference* and the *adaptive models*; the difference between these two estimations is then processed by a adaptation mechanism, which yields the estimated speed by adjusting the *adaptive model* so that its output coincides with that of the *reference model*. The various MRAS schemes differ from one another by the chosen state variables. From this standpoint, some schemes have been developed, either based on the rotor flux ψ'_r [8] or on the back emf **e** [33]. The appropriate adaptation law can be derived by the Popov's hyperstability criterion [1]. The differences between the state variables estimated, respectively, with the *reference* and *adaptive models* are fed to a speed tuning signal ε, and then processed by a PI (proportional integral) controller, whose output is the rotor speed. Three main schemes have been developed, where the speed tuning signals are, respectively, $\varepsilon_\omega = \mathrm{Im}(\psi'_r \hat{\psi}'^{*}_r)$ [8], $\varepsilon_e = \mathrm{Im}(\mathbf{e}\hat{\mathbf{e}}^{*})$ [33], and $\varepsilon_{\Delta e} = \mathrm{Im}((\mathbf{e}-\hat{\mathbf{e}})\mathbf{i}^{*}_s)$ [33]: all the quantities with $\wedge$ are related to the adaptive model and the * operator denotes the complex conjugate. The speed tuning signals are, in these cases, respectively, the vector product between the two estimated fluxes (ε_ω), the vector product between the two estimated back emfs (ε_e), and finally the vector product between the back emf error and the stator current $(\varepsilon_{\Delta e})$. The angular difference of the outputs from the two estimators is then fed to a PI controller, which gives a nonlinear stable feedback system. Basically, the PI controller drives the phase errors between the two state estimates to zero by aligning the *adaptive model* estimate with the *reference model* one. The pioneering work has been [8], while [33] improves [8] because, thanks to the employment of the back emf instead of the flux, the speed estimation does not suffer from the problem of the open-loop flux integration, with resulting wider bandwidth of the speed loop. It should be noted that, in any case, all the control schemes where the flux loop is closed must deal with the open-loop integration problem. Figure 6.21 shows the block diagram of the classic MRAS scheme [8] (in brackets the variables adopted in the other schemes [33]). The tuning of the PI influences the dynamics of the estimation and, therefore, the bandwidth of the observer; optimal PI tuning can be done analytically so as to achieve the desired observer bandwidth. In other words, the estimated speed is modified in the adjustable model in such a way that the difference between the outputs of the reference and adaptive models becomes zero; only in this condition does the estimated speed coincide with the real one.

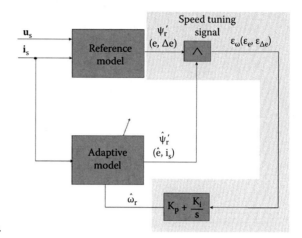

FIGURE 6.21
Block diagram of the classic MRAS observer.

If the classic MRAS scheme based on the rotor flux error is considered, the *reference model* is described by the stator voltage equations in stator reference frame, rewritten here for the sake of simplicity:

$$\begin{cases} \dfrac{d\psi_{rd}}{dt} = \dfrac{L_r}{L_m}\left(u_{sD} - R_s i_{sD} - \sigma L_s \dfrac{di_{sD}}{dt}\right) \\[3mm] \dfrac{d\psi_{rq}}{dt} = \dfrac{L_r}{L_m}\left(u_{sQ} - R_s i_{sQ} - \sigma L_s \dfrac{di_{sQ}}{dt}\right) \end{cases} \tag{6.21}$$

while the *adaptive model* is based on the rotor equations in the stator reference frame, which is the so-called current model:

$$\begin{cases} \dfrac{d\hat{\psi}_{rd}}{dt} = \dfrac{1}{T_r}\left(L_m i_{sD} - \hat{\psi}_{rd} - \omega_r T_r \hat{\psi}_{rq}\right) \\[3mm] \dfrac{d\hat{\psi}_{rq}}{dt} = \dfrac{1}{T_r}\left(L_m i_{sQ} - \hat{\psi}_{rq} + \omega_r T_r \hat{\psi}_{rd}\right) \end{cases} \tag{6.22}$$

In this case, the speed is estimated as

$$\hat{\omega}_r = K_p\left(\psi_{rq}\hat{\psi}_{rd} - \psi_{rd}\hat{\psi}_{rq}\right) + K_i \int\left(\psi_{rq}\hat{\psi}_{rd} - \psi_{rd}\hat{\psi}_{rq}\right)dt \tag{6.23}$$

Figure 6.22, which shows the block diagrams of the *reference* and *adaptive* models, clearly highlights that the reference model suffers from the open-loop integration problem. This problem

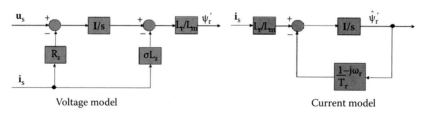

Voltage model   Current model

FIGURE 6.22
Block diagrams of the voltage and current models.

has been solved in Ref. [8] by adopting an LPF instead of a pure integrator, which causes, however, a poor flux amplitude and angle estimation as well as a poor speed estimation at low frequency, around the cutoff frequency of the LPF (usually a few Hertz). This consideration highly limits the minimum working speed of the drive and the correct field orientation, with consequent reduction of the torque performances at low speed. Alternative solutions to be adopted for the open-loop flux integration have been shown in Section 6.4.1. Furthermore, at low speeds, the stator voltage amplitude is small, and thus, an accurate value of the stator resistance is required by the model to have a satisfactory response.

If the MRAS scheme based on the back emf error is considered, the *reference model* can be written in this way:

$$
\begin{cases}
e_d = \dfrac{L_m}{L_r}\dfrac{d\psi_{rd}}{dt} = u_{sD} - R_s i_{sD} - L_s'\dfrac{di_{sD}}{dt} \\[4mm]
e_q = \dfrac{L_m}{L_r}\dfrac{d\psi_{rq}}{dt} = u_{sQ} - R_s i_{sQ} - L_s'\dfrac{di_{sQ}}{dt}
\end{cases}
\tag{6.24}
$$

The corresponding back emf equations for the adaptive model can be written in this way:

$$
\begin{cases}
\hat{e}_d = \dfrac{L_m}{L_r}\dfrac{d\hat{\psi}_{rd}}{dt} = \dfrac{L_m}{L_r}\dfrac{\left(L_m i_{sD} - \hat{\psi}_{rd} - \omega_r T_r \hat{\psi}_{rq}\right)}{T_r} \\[4mm]
\hat{e}_q = \dfrac{L_m}{L_r}\dfrac{d\hat{\psi}_{rq}}{dt} = \dfrac{L_m}{L_r}\dfrac{\left(L_m i_{sQ} - \hat{\psi}_{rq} + \omega_r T_r \hat{\psi}_{rd}\right)}{T_r}
\end{cases}
\tag{6.25}
$$

In this case, the speed is estimated as

$$
\hat{\omega}_r = K_p\left(e_q\hat{e}_d - e_d\hat{e}_q\right) + K_i\int\left(e_q\hat{e}_d - e_d\hat{e}_q\right)dt
\tag{6.26}
$$

As expected, the back emf components can be obtained without any integration, different from the previous scheme. When this last MRAS observer is used in sensorless drives, satisfactory performance can be achieved even at low speeds [1]. What is critical in this scheme is the correct knowledge of the stator resistance, which can influence significantly both the stability and the performance of the entire system at low speed.

An MRAS scheme which is insensitive to the stator resistance variations can be obtained by using a speed tuning signal which depends on a quantity that does not contain the stator resistance term. To achieve that, the speed tuning signal is deliberately chosen as $\mathrm{Im}((\mathbf{e} - \hat{\mathbf{e}})\mathbf{i}_s^*) = \mathrm{Im}(\mathbf{ei}_s^* - \hat{\mathbf{ei}}_s^*)\cdot$

From this standpoint, the following variable is estimated by the *reference model*:

$$
y = \mathbf{i}_s \wedge \mathbf{e} = u_{sQ}i_{sD} - u_{sD}i_{sQ} - L_s'\left(i_{sD}\dfrac{di_{sQ}}{dt} - i_{sQ}\dfrac{di_{sD}}{dt}\right)
\tag{6.27}
$$

while the equivalent one is estimated by the *adaptive model*:

$$
\hat{y} = \mathbf{i}_s \wedge \hat{\mathbf{e}} = \hat{e}_q i_{sD} - \hat{e}_d i_{sQ} = \dfrac{L_m}{L_r}\left[\dfrac{1}{T_r}(\psi_{rd}i_{sQ} - \psi_{rq}i_{sD}) + \omega_r(\psi_{rd}i_{sD} - \psi_{rq}i_{sQ})\right]
\tag{6.28}
$$

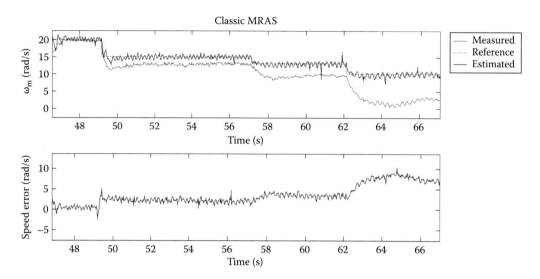

FIGURE 6.23
Reference, measured, and estimated speed with the MRAS observer in Figure 6.21.

In this case, the speed is estimated as:

$$\hat{\omega}_r = K_p \left(y - \hat{y}\right) + K_i \int \left(y - \hat{y}\right) dt \tag{6.29}$$

It could be observed that the tuning signal is independent from the stator resistance, and thus, the speed estimation is insensitive to its variations. It permits the drive to obtain a satisfactory performance even at low speeds [33].

The classic MRAS scheme based on the flux error (Figure 6.21) has been experimentally implemented in a 2.2 kW FOC IM drive (see Appendix 11.A) on a test setup (see Appendix 11.B). Figure 6.23 shows the experimental time waveform of the reference, measured, and estimated speed as well as the speed estimation error. In this case, open-loop flux integration has been done by adopting an LPF (see Figure 6.1). The test refers to a set of speed reference step variations with this sequence: 20 → 15 → 13 → 10 rad/s at no load. Results show a good tracking of the estimated speed at 10 rad/s. Below 10 rad/s, the speed estimation error increases dramatically, mainly because of the open-loop integration performed by LP filtering.

6.4.6.2 Closed-Loop MRAS

The CL-MRAS (Closed Loop) integrates the characteristics of a closed-loop flux observer (CLFO) with those of an MRAS, including also a mechanical system model [34–36]. The CLFO, proposed in Refs [32,37], combines the voltage and the current models of the IM, but rather than proposing an abrupt switching between the two models at a certain speed, it provides a smooth and continuous transition from one model to the other. Figure 6.24 shows the block diagram of the CL-MRAS observer. The rotor flux is estimated as the sum of high-pass-filtered and low-pass-filtered flux, estimated, respectively, by the voltage and the current models. This leads to a correction term which depends on the difference between the two estimated fluxes, subsequently processed by a PI controller. The resulting observer presents a smooth transition between current and voltage model flux

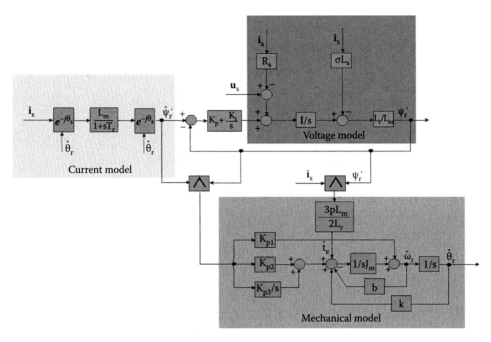

FIGURE 6.24
Block diagram of the CL-MRAS observer.

estimation which is ruled by the closed-loop eigenvalues of the observer, determined by the parameters of the PI controller. At rotor speeds below the bandwidth of the observer, its sensitivity to the parameters corresponds to that of the current model, while at high speeds, its sensitivity corresponds to that of the voltage model. [35,36] applies an MRAS CLFO based on a PI controller–based minimization of the cross product between the rotor fluxes, estimated, respectively, by the voltage model and the current model. In this observer, which is an evolution of [34], the closed-loop topology provides the necessary feedback for the voltage model integration, so that in the voltage model an LPF is not required to cancel the DC drift, different from [8]. However, this results in a bad behavior of the observer at low speeds: in fact, as clearly written in [35,36], on the one hand, the current model has no DC constant output, and therefore, the flux-coupling PI controller ensures zero DC level at the output. On the other hand, as frequency approaches to zero, the cross product also approaches to zero, and speed estimate forcing is lost. A mechanical model of the machine for compensating this aspect is then to be used, which permits the flux and the speed to be estimated also when the vector product between the fluxes approaches to zero. The structure of the mechanical model is such that the electromagnetic torque is estimated, and it is fed to a first-order mechanical model. This is also driven by the fluxes vector product through a PI controller, helping compensate model errors. A feed-forward term (K_{p1} in Figure 6.24) weights the effect of the CLFO and the mechanical model on the speed estimation. If the mechanical model is not properly tuned, simply a reduction of the performance of the observer is obtained, but in any case, there is still an improvement over the case where no mechanical model is used at all [35,36]. Obviously, this is paid with an increase of the complexity of the observer. Moreover, this observer requires two coordinate transformations, respectively, from the stator- to the rotor-oriented reference frame and vice versa, and the flux estimation is dependent on the speed estimation, different from [8].

6.4.7 Full-Order Luenberger Adaptive Observer

To develop the structure of the full-order Luenberger adaptive observer, firstly, a state estimator has been developed, which can be used to estimate the state variables of the IMs. Among the various choices, the rotor flux linkage and the stator current components have been chosen in Ref. [12] since they are the most suitable for integrating the observer in rotor flux–oriented control. The state estimator can be further modified so as to include the speed estimation, giving rise to an adaptive speed estimator (in general a speed-adaptive flux observer). The adaptation mechanism has been derived so as to guarantee the stability of the system, exploiting the state-error dynamic equations of the machine together with the Lyapunov stability theorem.

6.4.7.1 State-Space Model of the IM

If the stator current and the rotor flux-linkage space-vectors are chosen as state variables, the state equations of the IM in the stationary reference frame can be written as [12]

$$\frac{d}{dt}\begin{bmatrix} \mathbf{i}_s \\ \mathbf{\psi}'_r \end{bmatrix} = \frac{d\mathbf{x}}{dt} = \begin{bmatrix} \mathbf{A}_{11} & \mathbf{A}_{12} \\ \mathbf{A}_{21} & \mathbf{A}_{22} \end{bmatrix}\begin{bmatrix} \mathbf{i}_s \\ \mathbf{\psi}'_r \end{bmatrix} + \begin{bmatrix} \mathbf{B}_1 \\ \mathbf{0} \end{bmatrix}\mathbf{u}_s = \mathbf{A}\mathbf{x} + \mathbf{B}\mathbf{u}_s \tag{6.30a}$$

$$\mathbf{i}_s = \mathbf{C}\mathbf{x} \tag{6.30b}$$

where

$$\begin{cases} \mathbf{A}_{11} = -\left\{ R_s/(\sigma L_s) + (1-\sigma)/(\sigma T_r) \right\}\mathbf{I} \times a_{11} & \text{(6.31a)} \\[2mm] \mathbf{A}_{12} = L_m/(\sigma L_s L_r)\left\{ (1/T_r)\mathbf{I} - \omega_r\mathbf{J} \right\} = a_{12}\left\{ (1/T_r)\mathbf{I} - \omega_r\mathbf{J} \right\} & \text{(6.31b)} \\[2mm] \mathbf{A}_{21} = \left\{ L_m/T_r \right\}\mathbf{I} = a_{21}\mathbf{I} & \text{(6.31c)} \\[2mm] \mathbf{A}_{22} = -(1/T_r)\mathbf{I} + \omega_r\mathbf{J} = a_{22}\left\{ (1/T_r)\mathbf{I} - \omega_r\mathbf{J} \right\} & \text{(6.31d)} \\[2mm] \mathbf{B}_1 = 1/(\sigma L_s)\mathbf{I} = b\mathbf{I} & \text{(6.31e)} \end{cases}$$

with

$$\mathbf{i}_s = \begin{bmatrix} i_{sD} & i_{sQ} \end{bmatrix}^T, \quad \mathbf{u}_s = \begin{bmatrix} u_{sD} & u_{sQ} \end{bmatrix}^T, \quad \mathbf{\Psi}'_r = \begin{bmatrix} \Psi_{rd} & \Psi_{rq} \end{bmatrix}^T,$$

$$\mathbf{C}' = \begin{bmatrix} \mathbf{0} & \mathbf{I} \end{bmatrix}, \quad \mathbf{C} = \begin{bmatrix} \mathbf{I} & \mathbf{0} \end{bmatrix}, \quad \mathbf{I} = \begin{bmatrix} 1 & 0 \\ 0 & 1 \end{bmatrix}, \quad \mathbf{J} = \begin{bmatrix} 0 & -1 \\ 1 & 0 \end{bmatrix}$$

In the above state representation, $\mathbf{x} = \begin{bmatrix} \mathbf{i}_s, \mathbf{\psi}'_r \end{bmatrix}$ is the state vector, composed of the stator current and rotor flux-linkage direct and quadrature components in the stationary reference frame, $\mathbf{u}_s$ is the input vector composed of the stator voltage direct and quadrature components in the stationary reference frame, $\mathbf{A}$ is the state matrix (4×4 matrix) depending on the rotor speed ω_r, $\mathbf{B}$ is the input matrix, and finally $\mathbf{C}$ is the output matrix.

6.4.7.2 Adaptive Speed Observer

The full-order Luenberger state observer can be obtained from Equations 6.30 and 6.31 if a correction term is added, containing the difference between the actual and estimated states. In particular, since the only measurable state variables are the stator currents, the correction term involves only the error vector on the stator current $e_{rr} = (i_s - \hat{i}_s)$, as in the following:

$$\frac{d\hat{x}}{dt} = \hat{A}x + Bu_s + G(\hat{i}_s - i_s)$$ (6.32)

where
  ^ means the estimated values
  G is the observer gain matrix which is designed so that the observer is stable

To ensure the stability in the whole speed range, it is convenient to select the observer poles proportional to the motor poles [12] by a suitable constant $k \geq 1$. In this way, the observer dynamics can be made faster than that of the motor, even if small values of k are usually chosen to make the observer more robust against noise. Adopting the classic pole-placement technique, the gain matrix **G** (2×4) can be selected as

$$G = -\begin{bmatrix} g_1 I + g_2 J \\ g_3 I + g_4 J \end{bmatrix}$$ (6.33)

where the four gain terms can be obtained from the eigenvalues of the IM as

$$\begin{cases} g_1 = -(k-1)\left(\frac{1}{T_s'} + \frac{1}{T_r'}\right) \\ g_2 = (k-1)\hat{\omega}_r \\ g_3 = (k^2-1)\left\{-\left[\frac{1}{T_s'} + \frac{(1-\sigma)}{T_r'}\right]\frac{L_s'L_m}{L_r} + \frac{L_m}{T_r}\right\} + L_s'\frac{L_m}{L_r}(k-1)\left(\frac{1}{T_s'} + \frac{1}{T_r'}\right) \\ g_4 = -(k-1)\hat{\omega}_r\frac{L_s'L_m}{L_r} \end{cases}$$ (6.34)

From Equation 6.34, it can be observed that the gain matrix **G** depends on the estimated speed $\hat{\omega}_r$. Such a choice guarantees that the estimated states converge toward the actual ones in the whole speed range.

Figure 6.25a shows the loci of the machine and observer poles, according to the variation of the machine speed, when the observer poles have been selected twice as much as the machine ones.

A slightly different approach for the gain matrix selection has been proposed by Maes and Melkebeek [38], where the observer poles s_{obs} have been selected by shifting the machine

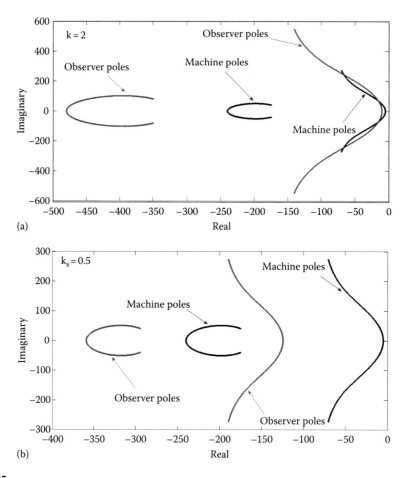

FIGURE 6.25
(a) Machine and observer poles with a proportionality factor equal to 2. (b) Machine and observer poles with $k_s = 0.5$.

poles $\mathbf{s}_{mot}$ to the left in the s plane of an amount equal to $k_s(a_{11} + a_{22})$. Since a_{11} and a_{22} are negative quantities, the observer poles are

$$\mathbf{s}_{obs} = k_s(a_{11} + a_{22}) + \mathbf{s}_{mot} \tag{6.35}$$

with k_s, a positive quantity. The resulting gain matrix components are

$$\begin{cases} g_1 = -k_s R_s \left(1 + \dfrac{a_{22}}{a_{11}}\right)(1 - \sigma a_{22}f) \\[2ex] g_2 = -k_s R_s \hat{\omega}_r \left(1 + \dfrac{a_{22}}{a_{11}}\right)f \\[2ex] g_3 = k_s R_s \dfrac{L_r}{L_m}\left(1 + \dfrac{a_{22}}{a_{11}}\right)(1 + \sigma a_{22}f) \\[2ex] g_4 = -k_s R_s \hat{\omega}_r \dfrac{L_r}{L_m}\left(1 + \dfrac{a_{22}}{a_{11}}\right) \end{cases} \tag{6.36}$$

with

$$f = \frac{(\sigma a_{22} - (k_s + 1)(a_{11} + a_{22}))}{(\hat{\omega}_r^2 + (\sigma a_{22})^2)}$$

The gains, depending also in this case from the estimated speed, can be either computed on-line or off-line and stored in the memory of the microprocessor, to be used during operation. Figure 6.25b shows the loci of the machine and observer poles, according to the variation of the machine speed, when the observer poles have been shifted with $k_s = 0.5$ with respect to the machine ones.

As highlighted in Ref. [38], the following main advantages can be achieved by adopting such gain matrix choice:

- At low speed, the slowest pole of the classical method is closer to the imaginary axis, thereby giving slower error decrease.
- At high speed, the classical pole-placement method results in poles with a very large imaginary part, and they can cause instability when representing the observer in discrete time.

It can be further observed that the state matrix $\hat{\mathbf{A}}$ presents the $^\wedge$ symbol, since it is a function of the estimated speed $\hat{\omega}_r$. In this case, the estimated speed is considered as a parameter of $\hat{\mathbf{A}}$, different from the EKF where it is considered as a state variable [1].

In order to find the speed-adaptive law, the Lyapunov stability theorem has been used. After defining the following Lyapunov function V,

$$V = (\mathbf{x} - \hat{\mathbf{x}})^T (\mathbf{x} - \hat{\mathbf{x}}) + \frac{1}{\lambda}(\hat{\omega}_r - \omega_r)^2 \tag{6.37}$$

where λ is a positive constant. From the time derivative of (6.37), it is possible to find the following speed tuning signal:

$$e_\omega = \frac{d\hat{\omega}_r}{dt} = \lambda \frac{L_m}{\sigma L_s L_r} \Delta \mathbf{i}_s {}^\wedge \hat{\boldsymbol{\psi}}_r' = \lambda \frac{L_m}{\sigma L_s L_r} (\mathbf{i}_s - \hat{\mathbf{i}}_s) {}^\wedge \hat{\boldsymbol{\psi}}_r' = \lambda \frac{L_m}{\sigma L_s L_r} \left[\psi_{rq}(i_{sD} - \hat{i}_{sD}) - \psi_{rd}(i_{sQ} - \hat{i}_{sQ}) \right]$$

$$\tag{6.38}$$

In order to improve the dynamics of the speed estimation, the error term e_ω can be corrected by a PI controller as

$$\hat{\omega}_r = K_p \left(\psi_{rq}(i_{sD} - \hat{i}_{sD}) - \psi_{rd}(i_{sQ} - \hat{i}_{sQ}) \right) + K_i \int \left(\psi_{rq}(i_{sD} - \hat{i}_{sD}) - \psi_{rd}(i_{sQ} - \hat{i}_{sQ}) \right) dt \tag{6.39}$$

The block diagram of the full-order Luenberger adaptive observer is shown in Figure 6.26. It has been experimentally implemented in a 2.2 kW FOC IM drive (see Appendix 11.A) on a test setup (see Appendix 11.B). The observer has been integrated with the R_s and R_r estimation algorithm called *algorithm 3* in Section 6.4.3.

The first test refers to a set of speed reference step variations with this sequence: $0 \rightarrow 100 \rightarrow -100 \rightarrow 0$ rad/s at no load. Figure 6.27a shows the experimental time waveform of the reference, measured, and estimated speed as well as the speed estimation error. Figure 6.27b and c shows

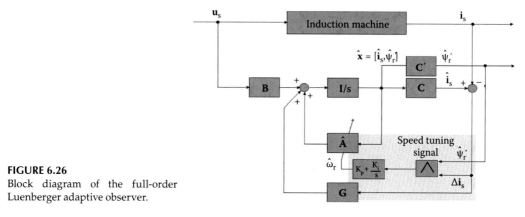

FIGURE 6.26
Block diagram of the full-order
Luenberger adaptive observer.

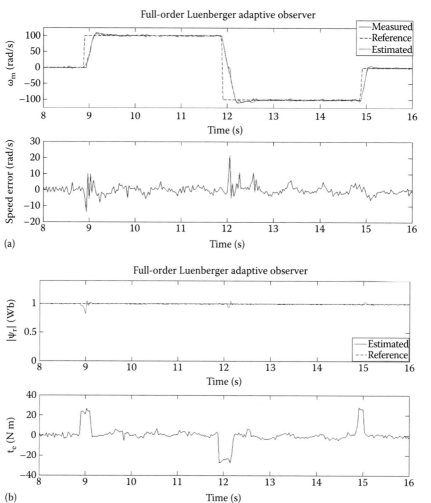

FIGURE 6.27
(a) Reference, measured, and estimated speed with the full-order Luenberger observer. (From Cirrincione, M. et al., *IEEE Trans. Ind. Appl.*, 42(1), 89, 2006.) (b) Rotor flux amplitude and electromagnetic torque with the full-order Luenberger observer. (From Cirrincione, M. et al., *IEEE Trans. Ind. Appl.*, 42(1), 89, 2006.)

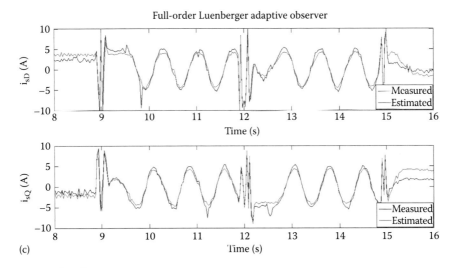

FIGURE 6.27 (continued)
(c) Measured and estimated i_{sD}, i_{sQ} current components with the full-order Luenberger.

the corresponding waveforms of the rotor flux amplitude, electromagnetic torque, as well as the estimated and measured, i_{sD} and i_{sQ}, current components. Results show a very good tracking of the estimated speed both during the steady-state and the transient phase, with a peak transient estimation error lower than 20 rad/s and a good matching between the estimated and measured current components.

The second test refers to a constant speed reference at low speed at no load. The drive has been run at 5 rad/s at no load. Figure 6.28a shows the experimental time waveform of the reference, measured, and estimated speed as well as the speed estimation error. Figure 6.27b shows the corresponding waveforms of the i_{sD} and i_{sQ} current components as well as the estimated R_s and R_r. These figures highlight that the speed estimation error is more than 26% at 5 rad/s, even if the stator current components are properly estimated and even if the stator and rotor resistance are well estimated. Moreover, the estimated speed presents a significant ripple.

6.4.8 Full-Order Sliding-Mode Observer

A sliding-mode controller can be suitably adopted to tune the observer for speed and rotor flux estimation, permitting the increase of the effective gains of the error compensator. A full-order sliding-mode observer has been proposed in Ref. [17]. Figure 6.29 shows the block diagram of this observer. The sliding hyperplane is defined in this case on the basis of the current estimation error $\mathbf{e}_1 = \Delta \mathbf{i}_s = (\hat{\mathbf{i}}_s - \mathbf{i}_s)$, as

$$\xi = \left[\xi_1 \xi_2 \right]^T = \frac{d\hat{\mathbf{i}}_s}{dt} - \frac{d\mathbf{i}_s}{dt} = 0 \tag{6.40}$$

The switching signal z of the of the sliding-mode observer is defined as

$$\mathbf{z} = -K_1 sign(\mathbf{e}_1) \tag{6.41}$$

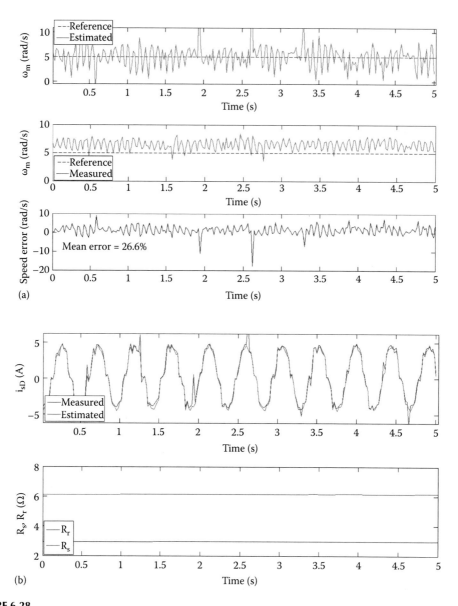

FIGURE 6.28
(a) Reference, measured, and estimated speed with the full-order Luenberger observer. (b) Measured and estimated i_{sD}, i_{sQ} current components and estimated R_s, R_r with the full-order Luenberger observer. (From Cirrincione, M. et al., *IEEE Trans. Ind. Appl.*, 42(1), 89, 2006.)

The estimation error is forced to zero by the high-frequency nonlinear switching controller of Equation 6.41. The output of the switching controller is directly used to tune on-line the observer matrix gain, while it is low pass filtered before being processed by the speed estimation algorithm. The robustness offered by the sliding-mode approach guarantees a null error in the estimation of the stator current. Furthermore, an H_∞ approach has been adopted in Ref. [17] for the design of the observer pole placement in order to maximize the robustness of the observer itself to parameter variations.

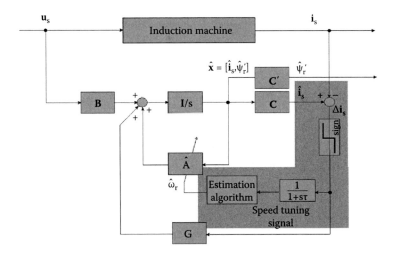

FIGURE 6.29
Block diagram of the full-order sliding-mode observer.

This has been done in evaluating the H_∞ norm of the transfer function starting from the disturbances to the flux estimation error.

The speed estimation has been done in [17], adopting the following adaptation law:

$$\frac{d\hat{\omega}_r}{dt} = -g\frac{\sigma L_s L_r}{pL_m}\mathbf{z}^T \mathbf{J}\hat{\boldsymbol{\psi}}_r' \tag{6.42}$$

with g being positive adaptive gain.

A further improvement of [17] has been proposed by [39], adopting a sliding-mode observer which, however, completely avoids the use of any speed adaptation scheme. In [39] the speed is computed in an open-loop way as the difference between the rotor flux-linkage speed and the slip speed, being the last one computed on the basis of the estimated torque.

6.4.9 Reduced-Order Adaptive Observer

The ROO permits the rotor flux-linkage components of the induction to be estimated by exploiting an observer of reduced dynamic order (2 instead of 4). The consequent main advantage is obviously the reduction of the complexity and the computational demand required for its implementation.

6.4.9.1 Reduced-Order Observer Equations

Starting from the state-space representation of the IM model, the matrix equations of the reduced-order flux observer, with a *voltage error* used for corrective feedback, can be deduced [9–11,40–43]:

$$\frac{d}{dt}\hat{\boldsymbol{\psi}}_r' = \mathbf{A}_{22}\hat{\boldsymbol{\psi}}_r' + \mathbf{A}_{21}\mathbf{i}_s + \mathbf{G}\left(\frac{d}{dt}\mathbf{i}_s - \mathbf{A}_{12}\hat{\boldsymbol{\psi}}_r' - \mathbf{A}_{11}\mathbf{i}_s - \mathbf{B}_1\mathbf{u}_s\right)$$

$$= \left(\mathbf{A}_{22} - \mathbf{G}\mathbf{A}_{12}\right)\hat{\boldsymbol{\psi}}_r' + \left(\mathbf{A}_{21} - \mathbf{G}\mathbf{A}_{11}\right)\mathbf{i}_s - \mathbf{G}\mathbf{B}_1\mathbf{u}_s + \mathbf{G}\frac{d}{dt}\mathbf{i}_s \tag{6.43}$$

where

$$\begin{cases} \mathbf{A_{11}} = -\left\{ R_s / (\sigma L_s) + (1-\sigma)/(\sigma T_r) \right\} \mathbf{I} = a_{11}\mathbf{I} & (6.44\,\mathrm{a}) \\[2mm] \mathbf{A_{12}} = L_m / (\sigma L_s L_r) \left\{ (1/T_r)\mathbf{I} - \omega_r \mathbf{J} \right\} = a_{12} \left\{ (1/T_r)\mathbf{I} - \omega_r \mathbf{J} \right\} & (6.44\,\mathrm{b}) \\[2mm] \mathbf{A_{21}} = \left\{ L_m / T_r \right\} \mathbf{I} = a_{21}\mathbf{I} & (6.44\,\mathrm{c}) \\[2mm] \mathbf{A_{22}} = -(1/T_r)\mathbf{I} + \omega_r \mathbf{J} = a_{22} \left\{ (1/T_r)\mathbf{I} - \omega_r \mathbf{J} \right\} & (6.44\,\mathrm{d}) \\[2mm] \mathbf{B_1} = 1/(\sigma L_s)\mathbf{I} = b\mathbf{I} & (6.44\,\mathrm{e}) \end{cases}$$

where all space-vectors are in the stator reference frame:

$\mathbf{i}_s = [i_{sD} \quad i_{sQ}]^T$ is the stator current vector
$\mathbf{u}_s = [u_{sD} \quad u_{sQ}]^T$ is the stator voltage vector
$\hat{\boldsymbol{\psi}}_r' = \left[\hat{\psi}_{rd} \quad \hat{\psi}_{rq} \right]^T$ is the rotor flux vector
$\mathbf{I} = \begin{bmatrix} 1 & 0 \\ 0 & 1 \end{bmatrix}, \ \mathbf{J} = \begin{bmatrix} 0 & -1 \\ 1 & 0 \end{bmatrix}$
ω_r is the rotor speed
$\mathbf{G}$ is the observer gain matrix

6.4.9.2 Possible Choices of the Observer Gain Matrix

The choice of a suitable gain matrix $\mathbf{G}$ of the observer has been a problem largely discussed in literature [9,10,41–43]. It is well known [9] that the poles of the ROOs are the pair of eigenvalues $\alpha \pm j\beta$ of the matrix $(\mathbf{A_{22}} - \mathbf{GA_{12}})$, where

$$\alpha = -p_r - \frac{L_m}{\sigma L_s L_r} p_r g_{re} - \frac{L_m}{\sigma L_s L_r} \omega_r g_{im}$$

$$\beta = \omega_r + \frac{L_m}{\sigma L_s L_r} \omega_r g_{re} - \frac{L_m}{\sigma L_s L_r} p_r g_{im}$$

and

$$\mathbf{G} = \mathbf{G}_{re} + \mathbf{G}_{im} = g_{re}\mathbf{I} + g_{im}\mathbf{J}$$

Figure 6.30 shows the observer pole locus, the amplitude of poles versus the rotor speed, and the damping factor ζ versus the rotor speed, obtained with five different gain choices of the matrix gain.

Choice 1: In Ref. [44] a criterion for choosing the locus of the observer poles is to make their amplitude constant, in respect to the rotor speed. This criterion leads, as one possible solution if $\alpha^2 + \beta^2 = constant$, to a semicircle pole locus with center in the origin, with radius R, and lying in the complex semiplane with negative real part. In this last case, then the poles placement varies in this locus with the rotor speed, to avoid instability phenomena, a maximum rotor speed ω_{rx} must be properly chosen, in correspondence to which the poles

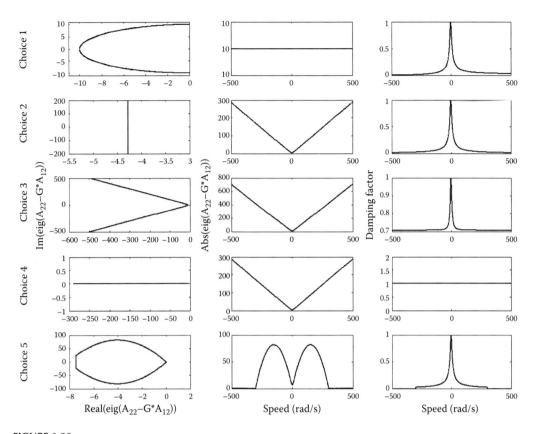

FIGURE 6.30
Pole locus, amplitude versus speed, and ζ versus speed with five matrix gain choices. (From Cirrincione, M. et al., *IEEE Trans. Ind. Electron.*, 54(1), February 2007.)

of the observer lie on the imaginary axis. The matrix gain choice which guarantees this condition is the following:

$$
\mathbf{G} = \mathbf{G}_{re} + \mathbf{G}_{im} = \frac{\sigma L_s L_r}{L_m}\left(-1 + \frac{\omega_{rx}}{\sqrt{\omega_{rx}^2 + p_r^2}}\frac{R}{\sqrt{p_r^2 + \omega_r^2}}\right)\mathbf{I} + \frac{\sigma L_s L_r}{L_m}\frac{p_r}{\sqrt{\omega_{rx}^2 + p_r^2}}\frac{R}{\sqrt{p_r^2 + \omega_r^2}}\mathbf{J} \quad (6.45)
$$

This matrix gain is dependent on the rotor speed, and therefore, the observer requires the correction term d**G**/dt. With such a matrix gain choice, the poles are complex with a constant amplitude *R*, but with a damping factor ζ which drastically decreases, from 1 at zero speed to about 0 at rated speed and above.

Choice 2: Ref. [43] proposes the following matrix gain choice:

$$
\mathbf{G} = \mathbf{G}_{re} = -\frac{\sigma L_s L_m p_r}{R_s}\mathbf{I} \quad (6.46)
$$

With such a matrix gain choice, the poles of the observer are imaginary with magnitude increasing with the rotor speed and the damping factor ζ drastically reducing at increasing speed, from 1 at zero speed to about 0 at rated speed and above. This choice cancels from the observer in Equation 6.43 the contribution of the stator current. It has the advantage that the gain matrix is not dependent on the rotor speed and therefore it is simpler than other solutions; for the same reason, it does not even require the correction term dG/dt.

Choice 3: Ref. [9] proposes the following matrix gain choice:

$$\mathbf{G} = \mathbf{G}_{re} + \mathbf{G}_{im} = \frac{\sigma L_s L_r p_r}{L_m} \left(\frac{\sqrt{2p_r^2 + \omega_r^2} - p_r}{p_r^2 + \omega_r^2} \right) \mathbf{I} + \frac{\sigma L_s L_r \omega_r}{L_m} \left(\frac{\sqrt{2p_r^2 + \omega_r^2} - p_r}{p_r^2 + \omega_r^2} \right) \mathbf{J} \qquad (6.47)$$

This matrix gain is dependent on the rotor speed, and therefore, the observer requires the correction term dG/dt. With such a matrix gain choice, the poles of the observer are complex with magnitude increasing with the speed and the damping factor ζ reducing at increasing speed, from 1 at zero speed to about 0.7 at rated speed and above. However, in [9], it is claimed that this matrix gain choice reduces the sensitivity of the observer to rotor resistance variations.

Choice 4: Ref. [43] proposes also the following matrix gain choice:

$$\mathbf{G} = \mathbf{G}_{re} + \mathbf{G}_{im} = \frac{\sigma L_s L_r}{L_m} \left(-1 + \frac{R p_r}{\sqrt{p_r^2 + \omega_r^2}} \right) \mathbf{I} + \frac{\sigma L_s L_r}{L_m} \frac{R \omega_r}{\sqrt{p_r^2 + \omega_r^2}} \mathbf{J} \qquad (6.48)$$

This matrix gain is dependent on the rotor speed, and therefore, the observer requires the correction term dG/dt. With such a matrix gain choice, the poles of the observer are real and lie on the negative real semiaxis with magnitude increasing with the speed and a damping factor ζ constant with rotor speed and always equal to 1.

Choice 5: Ref. [10] proposes the following matrix gain choice:

$$\begin{cases} \mathbf{G} = 0 \quad \text{for } |\omega_r| \leq \omega_1 \\[2ex] \mathbf{G} = \mathbf{G}_{re} = -\frac{\sigma L_s L_r}{L_m} \frac{\left(|\omega_r| - \omega_1 \right)}{\left(\omega_2 - \omega_1 \right)} \mathbf{I} \quad \text{for } \omega_1 < |\omega_r| < \omega_2 \\[2ex] \mathbf{G} = \mathbf{G}_{re} = -\frac{\sigma L_s L_r}{L_m} \mathbf{I} \quad \text{for } |\omega_r| \geq \omega_2 \end{cases} \qquad (6.49)$$

Assigned two threshold values ω_1 and ω_2 to the rotor speed, the gain matrix $\mathbf{G}$ has three discrete values. Below ω_1, no correction feedback is given to the observer, and it behaves as the simple "current" model of the IM, based on its rotor equations. Above ω_2, the correction feedback given to the observer is a constant multiplied with identity matrix, and it behaves as the simple "voltage" model of the IM, based on its stator equations. Between ω_1 and ω_2, the gain matrix $\mathbf{G}$ linearly varies from the two limit conditions. For this reason, it has been called current voltage model (CVM), since it gives rise to a smooth transition from the "current" to the "voltage" model according to the increase of the rotor speed. With such a choice, the poles of the observer are complex, with magnitude firstly increasing and then decreasing with the rotor speed, and a damping factor ζ drastically reducing at increasing speed, from 1 at zero speed to about 0 at rated speed and above. As mentioned earlier, however, this solution makes the observer work as a simple open-loop estimator both at low and high speeds, with the consequent DC drift integration problems. This is not the case in the other four ones. See [10] for the choice of ω_1 and ω_2.

TABLE 6.2

Issues of 5 Selected Matrix Gain Choices

	Pole Amplitude Varying with ω_r	ζ Decreasing with ω_r	G Depending on ω_r	DC Drift Integration Problems
Choice 1	No	Yes (high)	Yes	No
Choice 2	Yes (low)	Yes (high)	No	No
Choice 3	Yes (high)	Yes (low)	Yes	No
Choice 4	Yes (low)	No	Yes	No
Choice 5	Yes (low)	Yes (high)	Yes	Yes

Source: From Cirrincione, M. et al., *IEEE Trans. Ind. Electron.*, 54(1), 2007.

A slightly different approach is the one developed in the framework of the CL-MRAS [37] (see Section 6.4.6), which proposes an observer where the rotor flux is estimated as the sum of high-pass-filtered and low-pass-filtered flux, estimated, respectively, by the "voltage" and the "current" models. This leads to a correction term that depends, different from the other choices discussed earlier, on the difference between the two estimated fluxes, which are subsequently processed by a PI controller. The resulting observer presents a smooth transition between "current" and "voltage" model flux estimation which is ruled by the closed-loop eigenvalues of the observer, determined by the parameters of the PI controller. At rotor speeds below the bandwidth of the observer, its sensitivity to the parameters corresponds to that of the "current" model, while at high speeds, its sensitivity corresponds to that of the "voltage model." In this sense, it behaves like *choice 5*.

Table 6.2 summarizes the features of all five choices, mainly focusing on the variations of the observer pole amplitude with the rotor speed, the variation of the damping factor ζ with the rotor speed, the dependence on the matrix gain **G** by the rotor speed, and the DC drift integration problems. From the standpoint of the pole amplitude variation, *choice 1* is the best since it permits the amplitude to be constant, *choices 2, 4,* and *5* permit a low variation of the pole amplitudes, and *choice 3* causes a high variation. As for the damping factor variation, *choice 4* is the best since it keeps ζ always equal to 1, *choice 3* permits a low decrease of ζ at increasing rotor speeds, and *choices 1, 2,* and *5* cause a strong reduction of ζ. As for the dependence of **G** on the rotor speed, all the choices except *choice 2* suffer from this variation. As for the DC drift integration problems, only *choice 5* presents this negative issue, especially at low and high rotor speeds.

6.4.9.3 Speed Estimation

The reduced-order adaptive observer proposed in [9] adopts an MRAS system based on flux error processing to estimate the rotor speed. As a matter of fact, it adopts the ROO as reference model and a current model, like in [8], as adaptive model. In this way, the reference model is a combination of the voltage and current models, depending on the rotor speed estimation, while the adaptive model is the current model itself.

In this case, the speed is therefore estimated as

$$\hat{\omega}_r = K_p \left(\hat{\psi}_{rq} \hat{\hat{\psi}}_{rd} - \hat{\psi}_{rd} \hat{\hat{\psi}}_{rq} \right) + K_i \int \left(\hat{\psi}_{rq} \hat{\hat{\psi}}_{rd} - \hat{\psi}_{rd} \hat{\hat{\psi}}_{rq} \right) dt \qquad (6.50)$$

Figure 6.31 shows the block diagram of the reduced-order adaptive observer.

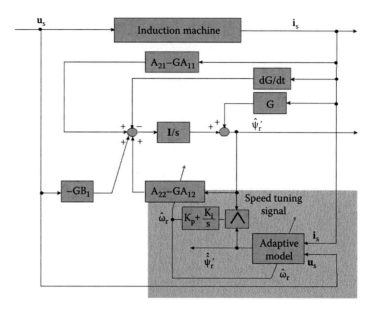

FIGURE 6.31
Block diagram of the reduced-order adaptive observer.

6.4.10 Extended Kalman Filter

The use of the EKF is a further approach for the speed estimation of IM drives [45]. EKF is suitable for high-performance drive, permitting a good accuracy in the speed estimation in a wide speed range, including low speeds. Moreover, it can be further adopted for joint state and parameter estimation. As a counterpart, it is computationally more demanding than the full-order adaptive observer and obviously also than the reduced-order adaptive observer. The EKF is a recursive optimum stochastic state estimator, which can be used for the joint state and parameter estimation of nonlinear dynamic systems in real time by using noisy monitored signals that are affected by random noise. The underlying assumption is that the measurement noise and the disturbance noise are uncorrelated, taking the noise sources into account including measurement and modeling inaccuracies. Different from the full-order adaptive observer, it permits the noise to be taken into consideration in the estimation process, since it is a stochastic estimator. Furthermore, the EKF assumes the speed as a state variable, whereas the full-order adaptive observer assumes it as a parameter.

The main steps for the design of a speed sensorless IM drive adopting an EKF are [1]

1. Selection of the time-domain IM model
2. Discretization of the IM model
3. Determination of the noise and state covariance matrices $\mathbf{Q}$, $\mathbf{R}$, and $\mathbf{P}$
4. Implementation of the discretized EKF and tuning

With regard to the time-domain model, the most adopted is that assuming the stator currents and the rotor flux linkages as state variables, expressed in the stationary reference frame (see Equations 6.30a and 6.30b). If this model is augmented with the estimated quantity, the rotor speed in this case, the obtained state vector equation is the following:

$$\begin{cases} \dfrac{d\mathbf{x}}{dt} = \mathbf{A}\mathbf{x} + \mathbf{B}\mathbf{u} \\ \mathbf{y} = \mathbf{C}\mathbf{x} \end{cases} \tag{6.51}$$

where

$$\mathbf{A} = \begin{bmatrix} -1/T_s'^* & 0 & L_m/(L_s'L_rT_r) & \omega_r L_m/(L_s'L_r) & 0 \\ 0 & -1/T_s'^* & -\omega_r L_m/(L_s'L_r) & L_m/(L_s'L_rT_r) & 0 \\ L_m/T_r & 0 & -1/T_r & -\omega_r & 0 \\ 0 & L_m/T_r & \omega_r & -1/T_r & 0 \\ 0 & 0 & 0 & 0 & 0 \end{bmatrix} \tag{6.52a}$$

$$\mathbf{B} = \begin{bmatrix} 1/L_s' & 0 \\ 0 & 1/L_s' \\ 0 & 0 \\ 0 & 0 \\ 0 & 0 \end{bmatrix} \tag{6.52b}$$

$$\mathbf{C} = \begin{bmatrix} 1 & 0 & 0 & 0 & 0 \\ 0 & 1 & 0 & 0 & 0 \end{bmatrix} \tag{6.52c}$$

where

$$\frac{1}{T_s'^*} = \frac{1}{T_s'} + \frac{1-\sigma}{T_r'}, \quad \mathbf{x} = \begin{bmatrix} i_{sD} & i_{sQ} & \psi_{rd} & \psi_{rq} & \omega_r \end{bmatrix}^T, \quad \mathbf{u} = \begin{bmatrix} u_{sD} & u_{sQ} \end{bmatrix}^T$$

A is the system matrix, and **C** is the output matrix. It should be further noted that the system matrix **A** is nonlinear, depending it on the state variable ω_r, that is, $\mathbf{A} = \mathbf{A}(\mathbf{x})$.

The EKF estimation $\hat{\mathbf{x}}$ is obtained by the predicted values of the states **x**, and this is corrected recursively by using a correction term, which is the product of the Kalman gain **K** and the deviation of the estimated measurement output vector from the actual one $(\mathbf{y} - \hat{\mathbf{y}})$. The Kalman gain has to be chosen to provide the optimal estimated states. The filter algorithm contains two main stages, the prediction stage and the filtering stage [1]. During the prediction stage, the next predicted values of the states $\mathbf{x}(k+1)$ are obtained by using the machine mathematical model (state equations) and the previous values of the estimated states. Furthermore, the predicted state covariance matrix **P** is obtained before the new measurements are made; for this purpose, the mathematical model and the covariance matrix of the system **Q** are adopted. During the filtering stage, the next estimated states $\hat{\mathbf{x}}(k+1)$ are obtained from the predicted estimates $\mathbf{x}(k+1)$ by adding a correction term $\mathbf{K}(\mathbf{y} - \hat{\mathbf{y}})$ to the predicted value. This correction term is a weighted difference between the actual output vector **y** and the predicted output vector $\hat{\mathbf{y}}$ where **K** is the Kalman gain. The predicted state estimation (and the covariance matrix) is corrected by a correction scheme, making use of the real measurement quantities. The Kalman gain is chosen to minimize the estimation error variances of the states to be estimated.

The EKF vector equation is the following:

$$\frac{d\hat{\mathbf{x}}}{dt} = \mathbf{A}(\hat{\mathbf{x}})\hat{\mathbf{x}} + \mathbf{B}\mathbf{u} + \mathbf{K}(\mathbf{i}_s - \hat{\mathbf{i}}_s) \tag{6.53}$$

6.5 Anisotropy-Based Sensorless Techniques

6.5.1 Signal Injection Techniques

6.5.1.1 Revolving Carrier Techniques

In this kind of sensorless technique, a polyphase rotating carrier at pulsation ω_c is usually added to the fundamental voltage generated by the pulsewidth modulation (PWM) system [3,6,18,46], as shown in Figure 6.32. This term is of the type

$$\mathbf{u}_c = u_c e^{j\omega_c t} \tag{6.54}$$

where u_c is the amplitude of the revolving carrier.

The interaction of such a voltage component with the machine anisotropies causes the presence of a current space-vector $\mathbf{i}_c$ at carrier frequency ω_c appearing as a component of

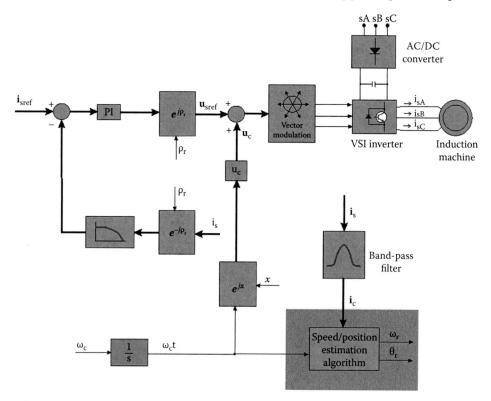

FIGURE 6.32
Current control and signal injection with a revolving carrier.

the stator current space-vector $\mathbf{i}_s$. To be further processed by the speed estimation algorithm, the $\mathbf{i}_c$ component is extracted by a band-pass filter centered at the carrier frequency, which separates it from both the fundamental current component and the high-frequency components due to the switching. Moreover, the stator current $\mathbf{i}_s$ is low pass filtered before being fed to the current controllers, to cutoff the $\mathbf{i}_c$ component.

As far as the anisotropies are concerned, a single anisotropy with a spatial cycle per pole pitch is typical of saturation effects, both of the main flux and slotting ones as well as of custom-designed machines. If a system of coordinates x–y rotating at the speed of the anisotropy ω_x to be detected is considered, to compute the response of the motor to the carrier voltage, the voltage term must be transformed into the reference frame rotating at speed ω_x by multiplying the voltage in (6.54) with the term $e^{-j\omega_x}$. The resulting voltage equation, valid only for studying the high-frequency behavior of the machine, is the following:

$$\mathbf{u}_c^x = u_c e^{j(\omega_c - \omega_x)t} = \mathbf{L}_\sigma^x \frac{d\mathbf{i}_c^x}{dt} \tag{6.55}$$

where
the apex x means that the corresponding terms are expressed in the reference frame rotating at speed ω_x
$\mathbf{L}_\sigma^x$ is the total leakage inductance tensor
$\mathbf{i}_c^x$ is the stator current space-vector due to the corresponding voltage

The total leakage inductance tensor can be written as

$$L_\sigma^x = \begin{bmatrix} L_{\sigma d} & 0 \\ 0 & L_{\sigma q} \end{bmatrix} \tag{6.56}$$

If Equation 6.55 is solved for $\mathbf{i}_c^x$, considering that $\omega_c \gg \omega_x$, the following solution can be obtained [3]:

$$\mathbf{i}_c^x = \frac{-j u_c}{2\omega_c L_{\sigma d} L_{\sigma q}} \left[(L_{\sigma d} + L_{\sigma q}) e^{j(\omega_c - \omega_x)t} + (L_{\sigma d} - L_{\sigma q}) e^{-j(\omega_c - \omega_x)t} \right] \tag{6.57}$$

where $L_{\sigma d}$ and $L_{\sigma q}$ are the direct and quadrature leakage inductance components in the ω_x reference frame. Equation 6.57 shows that, as effect of the application of a rotating voltage carrier, two sidebands appear centered around ω_c at distance ω_x from it. If Equation 6.57 is transformed back in the stationary reference frame:

$$\mathbf{i}_c = \frac{-j u_c}{2\omega_c L_{\sigma d} L_{\sigma q}} \left[(L_{\sigma d} + L_{\sigma q}) e^{j\omega_c t} + (L_{\sigma d} - L_{\sigma q}) e^{-j(\omega_c - 2\omega_x)t} \right] = \mathbf{i}_p + \mathbf{i}_n \tag{6.58}$$

As a result, a first current space-vector rotating at carrier frequency ω_c (positive sequence) appears, as well as a second space-vector that rotates at the angular velocity $-\omega_c + 2\omega_x$ (negative sequence). This last component has the information on the speed ω_x of the anisotropy to be detected. The space-vector given by Equation 6.58 describes an elliptical path trajectory, being the axis ratio of such ellipse $L_{\sigma q}/L_{\sigma d}$ (ranging typically between 0.9 and 0.96). It is very difficult to determine the angular inclination of the ellipse and, therefore, the angular orientation of the anisotropy. As a matter of fact, the component $\mathbf{i}_n$ to be retrieved is very small, significantly smaller than the positive one $\mathbf{i}_p$ and further contaminated by the presence of other anisotropies and noisy signals. Finally, it is literally hidden inside the fundamental component of the stator current $\mathbf{i}_{s1}$ [3].

If the anisotropy to be detected is the saturation of the main flux, then $\omega_x = \omega_r$, where ω_r is the speed of the machine in electrical angles; if it is the rotor slot saliency, then $2\omega_x = q_r\omega_r$, where q_r is the number of rotor slots for pole pair.

The original idea of this technique has been proposed in [47], where a spatial modulation of the rotor leakage inductance had been obtained by creating a periodic variation of the rotor slot opening width. The wide slot openings create high-reluctance flux paths (low inductance), while narrow openings create low-reluctance paths (high inductance). Closed-slot rotor bridges are considered undesirable in this case because of the saturation effects. In [48], the possibility to track saturation-induced saliencies has been explored. Saturation effects are usually associated with the main flux created by the magnetizing current or with localized leakage flux created by slot currents. Both forms of saturations are capable of spatially modulating the stator transient inductance. A further improvement has been proposed in [18] in which a custom-designed anisotropy of one pole pair periodicity is tracked, taking into consideration the rotor slotting–modulated harmonics as disturbance. The adopted carrier frequency is 250 Hz. Figure 6.33 shows the block diagram of such a system, which implements a so-called phase-locked loop (PLL).

The input of the system is the carrier-based component of the stator current $\mathbf{i}_c$, extracted by the band-pass filter in Figure 6.32. It is firstly transformed into the $+\omega_c$ reference frame, in which the $\mathbf{i}_p$ component appears as a constant. $\mathbf{i}_p$ is canceled by an integrator in feedback configuration. The remaining part of the signal, theoretically only $\mathbf{i}_n$ containing all negative-sequence components, is transformed into a $-\omega_c$ reference frame (the frequency origin is so shifted to $-\omega_c$). The unbalance disturbance is compensated by an estimated current vector $\hat{\mathbf{i}}_u = \hat{i}_u e^{j\varphi_u}$. After such a coordinate transformation, $\mathbf{i}_n$ contains signals at frequencies $2\omega_r$ and $q_r\omega_r$ (slotting modulated harmonic). Two anisotropy models are then adopted to generate a synchronization signal:

$$\mathbf{s}_n(\hat{\vartheta}_r) = \hat{\mathbf{i}}_2 e^{j(2\hat{\vartheta}_r + \hat{\varphi}_2)} + \hat{\mathbf{i}}_{slot} e^{j(q_r\hat{\vartheta}_r + \hat{\varphi}_{slot})} \tag{6.59}$$

This synchronization signal depends on the estimated rotor position $\hat{\vartheta}_r$. $\mathbf{s}_n(\hat{\vartheta}_r)$ is phase locked with the transformed negative component $\mathbf{i}_n$. The error signal is the vector product $\varepsilon = \mathbf{i}_n \wedge \mathbf{s}_n(\hat{\vartheta}_r)$, which is forced to zero by a proportional integral derivative (PID) controller, which provides the estimated acceleration of the motor. A mechanical model of the system

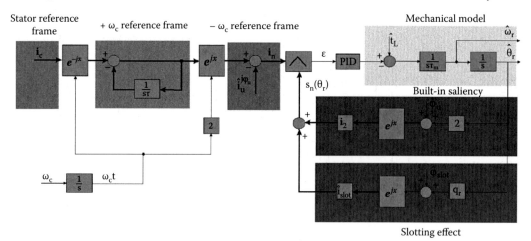

FIGURE 6.33
(See color insert.) Block diagram of the rotor position estimator based on PLL for anisotropy tracking.

is then used to compute both the angular speed and position. An estimated torque signal $\hat{t}_e$ can be used to further improve the dynamics of the observer. This approach does not take into consideration the saturation caused by the main flux, and therefore, it can be suitably used in unsaturated machines. As a matter of fact, it would be difficult to separate custom-made and saturation-induced harmonics, considering that they present very close harmonics ($\omega_s \cong \omega_r$). To overcome this problem, [49] proposes a methodology based on the tracking of the rotor slotting anisotropy, trying to compensate the saturation effects considered as disturbances. A saturation model of the machine is suitably developed, generating a signal, taking into consideration the effects of both the excitation and load current components. Figure 6.22 shows the block diagram of such a scheme.

The input of the system is the carrier-based component of the stator current $\mathbf{i}_c$, extracted by the band-pass filter in Figure 6.32. It is firstly transformed into the $-\omega_c$ reference frame, where an LPF permits the retrieval of the negative current component $\mathbf{i}_n$, containing all negative-sequence components as $\hat{\mathbf{i}}_{sat}$, $\hat{\mathbf{i}}_{slot}$, and $\hat{\mathbf{i}}_u$. The saturation current component $\hat{\mathbf{i}}_{sat}$ is treated as a disturbance, and thus, it is attenuated. It is estimated by two complex functions $f_2(\mathbf{i}_{s1})$ and $f_4(\mathbf{i}_{s1})$, generating, respectively, the second and fourth spatial harmonics, both referred to the fundamental field. The input signal of such functions is the fundamental stator current in field coordinates, characterizing in its direct component the main flux and in its quadrature component the load. As a matter of fact, the saturation of the machine is influenced by both current components. These two functions are machine dependent and must determined off-line by an identification process [49]. Since the current harmonics are difficult to be compensated by working in the frequency domain, a suitable off-line identification method has been developed [50], permitting the generation of time waveforms of this functions in a time interval equal to one electrical revolution: one waveform for each value of excitation and load current. If saturation effects, inverter nonlinearities, and signal unbalances, represented by the corresponding space-vectors $\hat{\mathbf{i}}_{sat}$, $\hat{\mathbf{i}}_{inv}$, and $\hat{\mathbf{i}}_u$, are properly compensated, the remaining signal should depend only on the rotor slotting:

$$\hat{\mathbf{i}}_{slot} = \hat{i}_{slot}e^{j(q_r\hat{\vartheta}_r + \hat{\varphi}_{slot})} \tag{6.60}$$

If this signal does not contain many other disturbing harmonics, instead of using a PLL, a direct computation of the phase angle of $\hat{\mathbf{i}}_{slot}$ can be obtained, as shown in Figure 6.34. Since, to obtain the rotor speed, the corresponding position should in any case be differentiated, a PLL is, in general, preferable.

As an example, Figure 6.35a and b shows, respectively, the steady-state stator current space-vector locus and its relative spectrum (computed with a fast Fourier transform [FFT] and referred to the fundamental component amplitude) of a 2.2 kW motor. These results have been obtained by a finite-element analysis (FEA) (see Section 6.5.1.2 for details) under a working fundamental frequency of 1 Hz and a carrier frequency of 1500 Hz at no load with an open-slots machine. The inverter switching effects have not been considered in this analysis. Figure 6.35a clearly shows that the stator current locus, ideally a circle, presents several additional lobes due to the carrier excitation. Figure 6.35b shows the two harmonic components of the stator current caused by the rotor slotting effect ($q_r = 14$), precisely $\mathbf{i}_p$ at 1500 Hz (ω_c) and $\mathbf{i}_n$ at 1486 Hz ($\omega_c - q_r\omega_r$), as expected. It further shows that while the $\mathbf{i}_p$ amplitude is almost 16% of the fundamental, the $\mathbf{i}_n$ amplitude is less than 0.68%, making its detection quite difficult.

A completely different approach has been proposed by [51], where the rotating carrier is used to modulate the saturation level of the air-gap flux. The underlying idea is that,

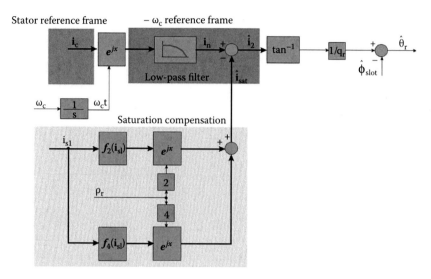

FIGURE 6.34
(See color insert.) Block diagram of the rotor position estimator based on rotor slotting anisotropy tracking.

since the leakage inductances are affected by the main flux by saturation, an unbalance of the phase leakage voltages contains the information of the air-gap flux angle. Since the unbalance of the leakage voltages reflects in the existence of zero sequences in the stator voltage, this quantity is measured either between the midpoint of the DC link of the inverter and the neutral of the stator winding or with an artificially constructed neutral (e.g., resistors). For this reason, this technique has been called ZST (zero-sequence technique). In this case, the carrier frequency is not constant but variable as $\omega_c = \omega_0 + \omega_s$, where $\omega_0 = 2\pi500\,\text{rad/s}$ and ω_s is the stator fundamental frequency. The zero-sequence voltage presents, thus, a constant frequency equal to $\omega_c - \omega_s = \omega_0$. The flux amplitude is computed on the basis of the precommissioned look-up table. It is even possible to eliminate the use of current sensor, replacing them with current estimators as demonstrated in Ref. [51].

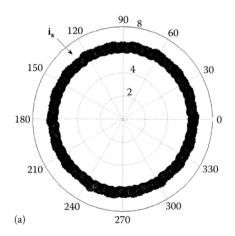

FIGURE 6.35
(a) Locus of the stator current with FEA under a rotating carrier excitation.

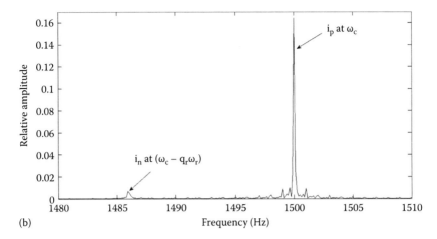

(b)

FIGURE 6.35 (continued)
(b) FFT of the stator current with carrier frequency at 1500 Hz and fundamental frequency at 1 Hz at no load.

A further paper of the same authors [52] focuses on the issues related to the presence of the additional high-frequency field and predicts the resulting harmonic content of the zero-sequence voltage by using three different approaches such as analytical calculations, computer simulation, and FEA. Moreover, an expression is introduced for generalizing the results in the presence of different kinds of injections, both pulsating and rotating carriers, demonstrating the different nature of the ZST.

6.5.1.2 IM Saliency Analysis under Rotating Carrier by FEA

How significantly the machine saliency is excited by the carrier supply is a crucial issue for the estimation of the machine speed/position. The correct choice of the parameter of the machine giving proper information on the saliency is a key point. As far as sensorless control based on high-frequency voltage carrier injection is concerned, the amplitude of the carrier voltage is the cause of the excitation, while the corresponding current harmonic content is the effect. None of the two is, thus, the best parameter to be chosen.

One of the parameters to consider is the space harmonic of the machine stator inductance, in particular its space harmonic responsible for the saliency excited by the carrier. In the following, it is the rotor slotting saliency of the machine which is to be tracked.

The machine stator inductance under fundamental sinusoidal supply has a spatial harmonic due to the rotor slotting effect at frequency $\omega_r(q_r - 1)$. The existence of the high-frequency voltage carrier in addition to the fundamental frequency, however, modifies the magnetic structure of the machine, generating new space harmonics of the stator inductance. In particular, if the machine rotates at steady-state at the speed ω_r, a spatial harmonic of the stator inductance at $\omega_c \pm \omega_r q_r$ appears (the sign depending on the q_r equal to $3n + 1$ or $3n - 1$).

The ratio between the amplitude of this spatial harmonic and the amplitude of the fundamental spatial harmonic of the stator inductance, defined L_{hexc}, can be an interesting index for evaluating the response of the machine to the carrier excitation [53,54]. The best way to obtain this parameter is by the FEA of the machine, which permits finding the exact stator spatial flux distribution at speed steady-state ($\vartheta_r = \omega_r t$) in the presence of only the fundamental and the carrier frequency supply. Figure 6.36a and b shows the magnetic flux density lines obtained with the FEA (software Flux-2D®) of two IMs of rated power 2.2 kW, suitably constructed with identical stators and two rotor configurations, respectively, in case

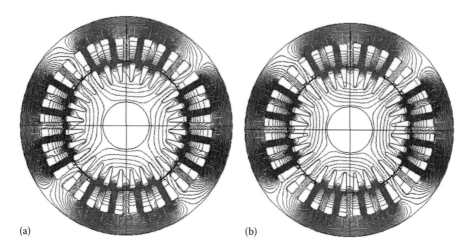

(a)   (b)

FIGURE 6.36
(See color insert.) (a, b) Magnetic flux density lines in the closed-slots (a, left) and open-slots (b, right) machines.
(From Pucci, M. and Serporta, C., *IEEE Trans. Magn.*, 46(2), 2010.)

of closed and open rotor slots with unskewed bars (see Figure 6.37). The test has been done with the machine running at steady-state at no load supplied at fundamental frequency of 1 Hz, at the rated rotor flux linkage of 1 Wb, and with a voltage carrier frequency at 1500 Hz.

The space-vector of the stator inductance $\mathbf{L}_s$ can be defined on the basis of quantities coming from the FEA:

$$\mathbf{L}_s(\vartheta_r) = L_{sD}(\vartheta_r) + jL_{sQ}(\vartheta_r) = \frac{\mathbf{\Psi}_s(\vartheta_r)}{\mathbf{i}_s(\vartheta_r)} = \frac{\psi_{sD}(\vartheta_r) + j\psi_{sQ}(\vartheta_r)}{i_{sD}(\vartheta_r) + ji_{sQ}(\vartheta_r)}$$

$$= \frac{\psi_{sD}(\vartheta_r)i_{sD}(\vartheta_r) + \psi_{sQ}(\vartheta_r)i_{sQ}(\vartheta_r) + j(\psi_{sQ}(\vartheta_r)i_{sD}(\vartheta_r) + \psi_{sD}(\vartheta_r)i_{sQ}(\vartheta_r))}{i_{sD}(\vartheta_r)^2 + i_{sQ}(\vartheta_r)^2} \quad (6.61)$$

FIGURE 6.37
Photograph of the three rotors, skewed (left), unskewed (center), and unskewed with open slots (right). (From Cirrincione, M. et al., *IEEE Trans. Ind. Appl.*, 44, 1683, 2008.)

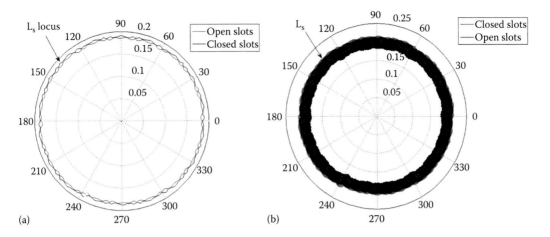

FIGURE 6.38

(a) Locus of the stator inductance space-vector under fundamental supply. (b) **(See color insert.)** Locus of the stator inductance space-vector under fundamental supply and high-frequency carrier. (From Pucci, M. and Serporta, C., *Electric Power Syst. Res.*, (81), 318, 2011.)

Figure 6.38a shows the locus of the stator inductance space-vector, obtained with the two motors, respectively, with closed and open slots, under a pure sinusoidal supply at 1 Hz at no load while Figure 6.38b shows the same locus when the rotating voltage carrier is added to the fundamental supply. It can be clearly observed that the presence of the carrier modifies significantly the stator inductance space-vector locus, introducing additional lobes in it.

As a result from the FEA test, Figure 6.39 shows the relative L_{hexc} under different carrier frequencies. The magnitude of the revolving voltage carrier has been increased linearly with its frequency starting from a value of 20 V at 400 Hz to maintain the stator current signal almost constant. Figure 6.39 plots the relative L_{hexc} versus the carrier frequency in a frequency interval ranging from 400 to 3000 Hz. Results highlight firstly that the relative L_{hexc} of the machine with open rotor slots is always higher than that of the machine with closed rotor slots for any value of the carrier frequency, as expected. Moreover, for both kinds of motors, the relative L_{hexc} increases with the carrier frequency up to a certain value, beyond which, it starts decreasing.

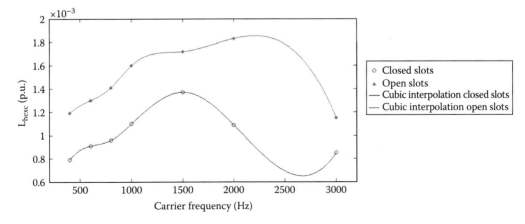

FIGURE 6.39

L_{hexc} versus carrier frequency. (From Pucci, M. and Serporta, C., *IEEE Trans. Magn.*, 46(2), 2010.)

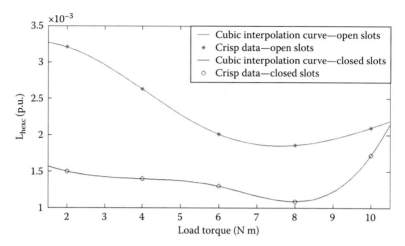

FIGURE 6.40
L_{hexc} versus load torque. (From Pucci, M. and Serporta, C., *Electric Power Syst. Res.*, (81), 318, 2011.)

The behavior of the machine under different load conditions has been analyzed. All the tests have been done at the carrier frequency of 1500 Hz, where the slotting effect in the machine is the maximum (see Figure 6.39). In particular, the machine has been supplied at voltage values increasing in amplitude and frequency with the load torque, so as to make the machine work always at the same rotating speed of 30 rpm (corresponding to the supply frequency of 1 Hz at no load). Figure 6.40 plots the relative L_{hexc} versus the load torque in an interval ranging from 2 to 10 N m (close to the rated one). Results highlight firstly that L_{hexc} of the machine with open rotor slots is always higher than that of the machine with closed rotor slots for any value of the load, as expected. Moreover, for both kinds of motors, the relative L_{hexc} decreases with the load up to a certain value (8 N m in the motors under test), above which the relative L_{hexc} starts increasing. Above a certain load, both kinds of machines start increasing their saliency effects.

Also the behavior of the machine after some modifications of the rotor structure has been studied [54]. Also in this case, the tests have been done at the carrier frequency of 1500 Hz. In particular, two kinds of rotor structure modifications have been taken into consideration: (a) the variation of the rotor bar cross-section area and (b) the variation of the depth of the rotor slot opening. With regard to point (a), the rotor bar shape has been modified so as to maintain the ratio between the quantities a and b (see Figure 6.41) constant. In this way, it can be observed that the bar cross-section area maintains almost constant, according to the variations of a and b from their corresponding values of the prototype a_{prot} and b_{prot}. This has been done to maintain the current density in the bar almost constant. With regard to point (b), the slot depth c (see Figure 6.41) has been increased starting from the value of the machine prototype c_{prot}. Figure 6.42 shows the L_{hexc} versus the ratio a/a_{prot}, while Figure 6.43 shows the L_{hexc} versus the ratio c/c_{prot}.

In synthesis, the following considerations could be done:

1. An optimal carrier frequency range exists where the machine shows at the best its rotor slotting effects. This threshold carrier frequency is higher in the open-slots machine than in the closed-slots machine.

2. Reducing the rotor flux linkage during the operation of the machine, especially in open-slots machine, amplifies the slotting effect.

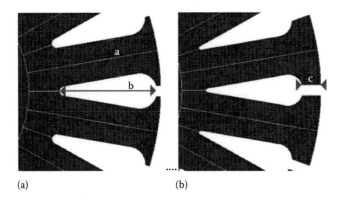

(a)   (b)

FIGURE 6.41
Variations of the bar cross-section area (a) and slot depth (b). (From Pucci, M. and Serporta, C., *Electric Power Syst. Res.*, (81), 318, 2011.)

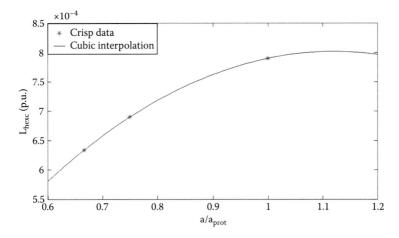

FIGURE 6.42
L_{hexc} versus a/a_{prot}. (From Pucci, M. and Serporta, C., *Electric Power Syst. Res.*, (81), 318, 2011.)

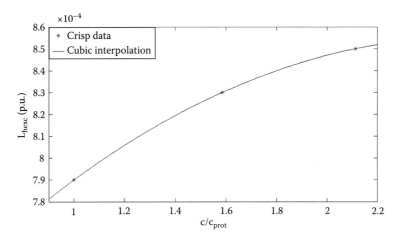

FIGURE 6.43
L_{hexc} versus c/c_{prot}. (From Pucci, M. and Serporta, C., *Electric Power Syst. Res.*, (81), 318, 2011.)

3. Increasing the load torque reduces the slotting effect of the machine up to a certain load above which the slotting effect starts increasing; this effect is more observable in open slots machine than in closed-slots ones.

4. A suitable design criterion for the rotor to amplify the slotting effect is to build the rotor bar thinner.

5. A suitable design criterion for the rotor to amplify the slotting effect is to build the rotor slot deeper.

6.5.1.3 Pulsating Carrier Techniques

Rotating carrier-based techniques permit the entire machine circumferential profile of asymmetries to be properly scanned. An alternative approach is to inject not a rotating but pulsating carrier, variable in time, but fixed in space. The direction in which the carrier should be oriented must be selected so as to maximize the sensitivity of the method versus the anisotropy to be detected.

A first kind of techniques is based on the *balance of quadrature impedances* [55]. This method tries to extract the information of the flux angle, during low- or zero-speed operation. In this case, a high-frequency carrier signal is added to the control input of the PWM, written in this way in field coordinates:

$$\mathbf{u}_s^{(F)} = \left(u_{sx}^{(F)} + u_c \cos(\omega_c t) + j u_{sy}^{(F)} \right) \tag{6.62}$$

The signal (6.62) excites the machine in the direct axis x. The estimated direct axis $\hat{\rho}$ can be different from the real one ρ of the quantity $(\hat{\rho} - \rho)$, being the difference small, since it is assumed that the estimated angle is going to converge toward the real one. Because of the anisotropic impedance of the machine, the high-frequency current due to the corresponding voltage develops a spatial displacement γ with respect to the real flux angle, corresponding to a phase angle $\gamma + (\hat{\rho} - \rho)$ with respect to the injected voltage. It can be observed that, when operated in saturated conditions, the machine offers an impedance $\dot{Z}_c$ at carrier frequency ω_c, function of the circumferential angle α in field coordinates. This impedance presents its maximum value $\dot{Z}_x$ on the direct axis x and its minimum $\dot{Z}_y$ on the quadrature axis y. It should be noted that the impedance $\dot{Z}_c$ is related to the total leakage inductance, and not to the stator or the rotor leakage inductances. The estimated angle $\hat{\rho}$ is thus different from both the stator and rotor flux angle, resulting in a need of correction for correctly performing the field orientation. The tracking of the x-axis is based on the symmetry of the $\dot{Z}_c$ characteristic, that is, $\dot{Z}_c(\alpha) = \dot{Z}_c(-\alpha)$, as shown in Figure 6.44.

Moreover, an orthogonal x-y coordinate system is introduced, with its real axis displaced by $-\pi/4$ with respect to the estimated x-axis and by $-(\pi/4 - \gamma)$ with respect to the real x-axis. Figure 6.45 shows the block diagram of the proposed method. The carrier component of the stator current $\mathbf{i}_c$ as well as the excitation signal u_{cx} is transformed into the aforementioned x-y reference frame. They are then converted in complex vectors characterized in magnitude by their own RMS values and in phase by their proper phase. From these, the complex impedance is obtained as

$$\dot{Z}_c^{(xy)} = \dot{Z}_x + j\dot{Z}_y = \frac{\mathbf{u}_c^{(xy)}}{\mathbf{i}_c^{(xy)}} \tag{6.63}$$

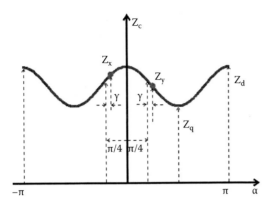

FIGURE 6.44
Impedance at carrier frequency versus circumferential angle in field coordinates.

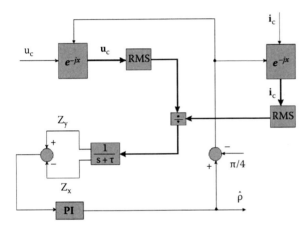

FIGURE 6.45
Block diagram of the field angle estimation based on the impedance measurements in quadrature axis.

$\dot{Z}_x$ and $\dot{Z}_y$ would be equal in condition of proper alignment of the x-y reference frame with the flux, resulting in $\gamma = 0$. Since a positive error angle γ is a cause of an increase in $\dot{Z}_x$ and a decrease in $\dot{Z}_y$, the following error signal can be used:

$$\varepsilon = \dot{Z}_y(\gamma) - \dot{Z}_x(\gamma) \tag{6.64}$$

A PI controller finally processes the error signal ε to retrieve the estimated angle $\hat{\rho}$.

A different approach has been followed by Consoli et al., [56], which is, however, in the framework of the ZST [51]. It proposes an improvement of the sensorless scheme based on the injection of a low-frequency (50/100 Hz) sinusoidal component to the normal stator current reference. This low-frequency component is stationary in space, meaning that it is applied on a constant angular position and generates a sinusoidal mmf interacting with the main rotating field. As a result, a modulation of the saturation level of the machine core which depends on the position of the air-gap flux is obtained. Consequently, a component of the zero-sequence voltage is produced, containing information about the position of the air-gap flux. The main difference of this approach is that low-frequency (0 ÷ 5 Hz) signal demodulation is achieved, thus avoiding any approximation generated by high-frequency signal demodulation.

6.5.1.4 High-Frequency Excitation Techniques

The underlying idea of such techniques is that the switching of a PWM inverter imposes to the machine a set of repetitive transient excitations. Correspondingly, the transient flux components cannot penetrate the rotor sufficiently to create a mutual flux linkage. The resulting flux is the superimposition of separate leakage fluxes, linking, respectively, with the stator and rotor windings, thus contributing to the total leakage flux. In Ref. [50], it has been clearly demonstrated that the fundamental flux linkage and the transient one, caused by the step-like transient excitation, can be treated separately, under the assumption of linearization of the magnetization curve. On this basis, a separate analysis referring only to the transient flux linkage can be made. The transient flux-linkage space-vector $\boldsymbol{\psi}_{\sigma tr}$ can be written as

$$\boldsymbol{\psi}_{\sigma tr} = \mathbf{L}_\sigma \mathbf{i}_{tr} \tag{6.65}$$

where

$\mathbf{i}_{tr}$ is the current component caused by the transient voltage $\mathbf{u}_{tr}$
$\mathbf{L}_\sigma$ is the tensor of the total leakage inductance (see Equation 6.56)

If the stator resistance is neglected, the voltage equation during a switching transient can be written as

$$\mathbf{u}_{tr} = \mathbf{u}_s - \mathbf{u}_{s1} \tag{6.66}$$

where $\mathbf{u}_{s1}$ is the space-vector of the fundamental stator voltage existing at steady-state just before the activation of the switching. This transient voltage will cause a variation of the leakage flux space-vector:

$$\mathbf{u}_{tr} = \frac{d\boldsymbol{\psi}_{\sigma tr}}{dt} = \mathbf{L}_\sigma \frac{d\mathbf{i}_{tr}}{dt} + \mathbf{i}_{tr} \frac{d\mathbf{L}_\sigma}{dt} \tag{6.67}$$

that must be added to the leakage flux $\boldsymbol{\psi}_{\sigma 1}$ caused by the fundamental voltage. As far as the classic model of the IM based on the fundamental mmf is concerned, the leakage inductances of each phase are considered balanced. The presence of various kinds of anisotropies in the machine is a cause of an unbalance of the phase leakage inductances; as a result, each phase can present instantaneously a different value of inductance. It means that the application of the transient voltage causes the variation of the transient current in a spatial direction, which differs from that of the applied voltage. It obviously implies the representation of the leakage inductance as a tensor, instead of a simple scalar, as already mentioned.

To make a proper analysis, the machine is supposed to be saturated on a defined angular position δ with respect to the direct axis in the stationary reference frame (see Figure 6.46a). Moreover, the machine is assumed at no load. If the voltage vector, $\mathbf{u}_s^{(1)}$, is instantaneously activated, the corresponding transient current $\mathbf{i}_{tr}^{(1)}$ can be computed from Equation 6.56, converted in the stationary reference frame considering that $\mathbf{u}_{s1}$ is null because $R_s \cong 0$ and $d\mathbf{L}_\sigma/dt \cong 0$ since δ is constant. To obtain $\mathbf{i}_{tr}^{(1)}$, the global leakage tensor should previously be inverted, giving rise to

$$\mathbf{L}_\sigma^{-1} = \frac{1}{L_{\sigma d} L_{\sigma q}} \begin{bmatrix} \frac{1}{2}(L_{\sigma d} + L_{\sigma q}) - \frac{1}{2}(L_{\sigma d} - L_{\sigma q})\cos(2\delta) & \frac{1}{2}(L_{\sigma d} - L_{\sigma q})\sin(2\delta) \\ \frac{1}{2}(L_{\sigma d} - L_{\sigma q})\sin(2\delta) & \frac{1}{2}(L_{\sigma d} + L_{\sigma q}) + \frac{1}{2}(L_{\sigma d} - L_{\sigma q})\cos(2\delta) \end{bmatrix} \tag{6.68}$$

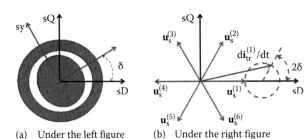

FIGURE 6.46
Effect of the anisotropies on the derivative of the transient current.

(a) Under the left figure (b) Under the right figure

On this basis, the transient current can be written as

$$\frac{d\mathbf{i}_{tr}^{(1)}}{dt} = \frac{1}{2}[(L_{\sigma d} + L_{\sigma q}) - (L_{\sigma d} - L_{\sigma q})\cos(2\delta) + j(L_{\sigma d} - L_{\sigma q})\sin(2\delta)]\mathbf{u}_s^{(1)} \qquad (6.69)$$

Equation 6.69 shows that, if different values of δ are considered, the transient current derivative space-vector is centered on the apex of the voltage vector $\mathbf{u}_s^{(1)}$, describing a circle displaced by an angle 2δ from the direct axis (see Figure 6.46b).

A real IM, besides the saturation effect, presents also the slotting effect. This last effect, more visible in open-slots machines than in closed-slots ones, presents a periodicity of $q_r\vartheta_r$. As a consequence, the saturation and the slotting anisotropies sum their effects, with a resulting vector $d\mathbf{i}_{tr}^{(1)}(\vartheta_r,\delta)/dt$ whose apex describes an epicyclical curve centered on the apex of the voltage vector $\mathbf{u}_s^{(1)}$. The spatial high-frequency component of such a curve is due to the rotor slotting effect, and is displaced by an angle $q_r\vartheta_r$ around the center point of the circle.

If the machine is further loaded, then $d\vartheta_r/dt \neq d\delta/dt$ because of the slip frequency. This is a cause of a modification in time of the $d\mathbf{i}_{tr}^{(1)}(\vartheta_r,\delta)/dt$ trajectory, maintaining, however, the same kind of shape.

If, instead of the $\mathbf{u}_s^{(1)}$ space-vector, the $\mathbf{u}_s^{(3)}$ is applied, then the resulting $d\mathbf{i}_{tr}^{(3)}(\vartheta_r,\delta)/dt$ vector will present an epicyclical curve centered on the apex of the voltage vector $\mathbf{u}_s^{(3)}$ (sB axis of the machine). Finally, the $\mathbf{u}_s^{(5)}$ is applied, then the resulting $d\mathbf{i}_{tr}^{(5)}(\vartheta_r,\delta)/dt$ vector will present an epicyclical curve centered on the apex of the voltage vector $\mathbf{u}_s^{(5)}$ (sC axis of the machine).

6.5.1.4.1 INFORM Technique

The *INFORM* technique has been proposed by Schroedl [20]. Let's consider the stator equation of an anisotropic AC machine written in the stationary reference frame [6]:

$$\mathbf{u}_s = R_\sigma \mathbf{i}_s + \mathbf{L}_\sigma \frac{d\mathbf{i}_s}{dt} + \frac{k_r}{T_r}(j\omega_r T_r - 1)\mathbf{\psi}_r' \qquad (6.70)$$

where
$\mathbf{L}_\sigma$ is the inductance tensor, taking into consideration the saturation-induced anisotropy
$k_r = L_m/L_r$ is the coupling factor of the rotor
$R_\sigma = R_s + k_r^2 R_r$ is the equivalent resistance

The time derivative of the stator current space-vector can be obtained in discrete terms as $\Delta\mathbf{i}_s/\Delta t$, on the basis of the difference $\Delta\mathbf{i}_s$ during the finite time interval Δt, assuming that the stator voltage space-vector $\mathbf{u}_s$ remains constant during the same time interval. The effect of the stator resistance voltage drop and of the back emf in Equation 6.69 can be canceled by applying consecutively two voltage space-vectors in equal direction and opposite orientation. The two subsequent switching configurations in a time interval Δt could then be $\mathbf{u}_s^{(1)} \rightarrow \mathbf{u}_s^{(4)}$, with $\mathbf{u}_s^{(1)} = -\mathbf{u}_s^{(4)}$, or $\mathbf{u}_s^{(2)} \rightarrow \mathbf{u}_s^{(5)}$, with $\mathbf{u}_s^{(2)} = -\mathbf{u}_s^{(5)}$, or finally $\mathbf{u}_s^{(3)} \rightarrow \mathbf{u}_s^{(6)}$ with $\mathbf{u}_s^{(3)} = -\mathbf{u}_s^{(6)}$. The underlying assumption is also that, during the time interval, the fundamental components of both the stator current $\mathbf{i}_s$ and of the rotor flux-linkage $\boldsymbol{\psi}_r'$ space-vectors do not vary significantly. As far as the switching commutation $\mathbf{u}_s^{(1)} \rightarrow \mathbf{u}_s^{(4)}$ is concerned, if each of the two voltage space-vector terms is inserted in Equation 6.70, after making the difference, the following equation can be obtained:

$$\mathbf{u}_s^{(1)} - \mathbf{u}_s^{(4)} = \mathbf{L}_\sigma \left(\frac{\Delta \mathbf{i}_s^{(1)}}{\Delta t} - \frac{\Delta \mathbf{i}_s^{(4)}}{\Delta t} \right) \tag{6.71}$$

Since both $\mathbf{u}_s^{(1)}$ and $\mathbf{u}_s^{(4)}$ lie on the direct axis, it is significant to study principally the real component of the resulting stator current derivative space-vector. This can be deduced from Equation 6.71 after substituting the expression of the inverse of the global leakage inductance tensor in Equation 6.68 and finally taking the real component of the resulting space-vector. As a result, the following equation is deduced:

$$\Re_e(\Delta \mathbf{i}_s^{(4)} - \Delta \mathbf{i}_s^{(1)}) = \left[(L_{\sigma d} + L_{\sigma q}) - (L_{\sigma d} - L_{\sigma q})\cos(2\delta) \right] |\mathbf{u}_s| \Delta t \tag{6.72}$$

Analogous expression of the stator current variation can be obtained for the switching commutations $\mathbf{u}_s^{(2)} \rightarrow \mathbf{u}_s^{(5)}$ and $\mathbf{u}_s^{(3)} \rightarrow \mathbf{u}_s^{(6)}$, taking from the resulting stator current space-vectors only the components lying in the direction of the forcing voltage (sB axis for $\mathbf{u}_s^{(3)}$ and sC axis for $\mathbf{u}_s^{(5)}$). A space-vector of the variation of the stator currents can then be defined in this way:

$$f = \frac{2}{3} \left[\Delta \mathbf{i}_s^{(1)} \Big|_{sA} - \Delta \mathbf{i}_s^{(4)} \Big|_{sA} + \alpha \left(\Delta \mathbf{i}_s^{(5)} \Big|_{sC} - \Delta \mathbf{i}_s^{(2)} \Big|_{sC} \right) + \alpha^2 \left(\Delta \mathbf{i}_s^{(3)} \Big|_{sB} - \Delta \mathbf{i}_s^{(6)} \Big|_{sB} \right) \right] \tag{6.73}$$

The resulting expression of the stator current variation space-vector is

$$f = \frac{1}{2} (L_{\sigma d} - L_{\sigma q}) |\mathbf{u}_s| \Delta t e^{j(2\hat{\delta} + \pi)} \tag{6.74}$$

The space-vector f depends on the difference between the direct and quadrature leakage inductances, null if these inductances coincide; moreover, it can be computed on-line on the basis of the instantaneous measurements of the stator phase currents. Its phase is twice the flux phase, except for the constant phase shift equal to π.

As a result, the estimated flux angle is

$$\hat{\delta} = \frac{1}{2} (\arg(f) - \pi) \tag{6.75}$$

To make this technique work properly, the machine should have closed rotor slots. The closing of the slots, in fact, shields the rotor bars from the high-frequency leakage flux, therefore reducing the effect of the slotting which, in this case, is a disturbance and then should be properly reduced.

6.5.1.4.2 *Instantaneous Rotor Position Measurement*

This technique is based on the idea of exploiting the information linked to the rotor slotting for the retrieval of the rotor position. In this sense, different from *INFORM* method, the rotor slotting is the anisotropy to be tracked and the saturation of the main flux is a disturbance. This technique has been proposed in Refs [21,22] and relies on the instantaneous measurement of the stator voltage, between the inverter terminals and the star point of the machine [23].

As mentioned earlier, the presence of anisotropies in the machine is the cause of an unbalance of the phase leakage inductances. If the rotor slotting effect is accounted for, this effect is visible whenever the number of rotor slots for pole pair is not a multiple of 3. This is a normal design criterion of IMs because machines with the number of rotor slots for pole pair that is a multiple of 3 present a significant torque ripple.

For each phase k of the machine, the dependence of the leakage inductance from the rotor slotting effect can be suitably modeled considering only its fundamental component as [22]

$$L_{\sigma slot}(\vartheta_r) = L_{\sigma 0}\left(1 + \frac{L_{\sigma slot0}}{L_{\sigma 0}}\cos\left(q_r\left(\vartheta_r - (k-1)\frac{2}{3}\pi\right)\right)\right) \tag{6.76}$$

where

$L_{\sigma 0}$ is the average leakage inductance
$L_{\sigma slot0}$ is the amplitude of the slotting leakage inductance variation

At the same time, the saturation caused on the stator and rotor teeth by the magnetizing component of flux density distribution, in proximity of its spatial distribution maximum, causes a variation of each phase leakage inductance. This variation can be modeled, for each phase k, in this way:

$$L_{\sigma sat}(\delta) = L_{\sigma 0}\left(1 + \frac{L_{\sigma sat0}}{L_{\sigma 0}}\cos\left(2\left(\delta - (k-1)\frac{2}{3}\pi + \varphi_{sat}\right)\right)\right) \tag{6.77}$$

where

$L_{\sigma 0}$ is the average leakage inductance
$L_{\sigma sat0}$ is the amplitude of the magnetizing leakage inductance variation
δ is the field angle

Finally, φ_{sat} is the angular displacement between the magnetizing fields with respect to phase sA. Saturation affects, on the one hand, the permeance of the leakage paths as a function of the flux density distribution and, on the other hand, the value of $L_{\sigma slot0}$ itself.

The combined effect of the saturation- and slotting-based anisotropies can be represented taking into consideration a saturation-dependent amplitude modulation factor.

As a result, the total leakage phase inductances can be written, for each phase k, as a function of the rotor angle ϑ_r and of the field angle δ:

$$L_\sigma(\vartheta_r, \delta) = L_{\sigma 0}\left(1 + \frac{L_{\sigma slot0}}{L_{\sigma 0}}\left(1 + k_m\frac{L_{\sigma sat0}}{L_{\sigma 0}}\cos\left(2\left(\delta - (k-1)\frac{2}{3}\pi\right)\right)\right)\cos\left(q_r\left(\vartheta_r - (k-1)\frac{2}{3}\pi\right)\right)\right.$$

$$\left. \times\left(1 + \frac{L_{\sigma sat0}}{L_{\sigma 0}}\cos\left(2\left(\delta - (k-1)\frac{2}{3}\pi\right)\right)\right)\right) \tag{6.78}$$

where $k_m = 0.2,\ldots,0.5$ is a machine-dependent coefficient.

One way to acquire the anisotropy-based signal is to measure the zero-sequence voltage of the stator winding, defined as the average value of the three phase voltages:

$$u_{s0} = \frac{u_{sA} + u_{sB} + u_{sC}}{3} \tag{6.79}$$

Since the back emfs do not present any zero-sequence component,

$$e_{sA} + e_{sB} + e_{sC} = 0 \tag{6.80}$$

with e_{sA}, e_{sB}, and e_{sC} being the back emfs of the three phases, and then it can be written as

$$u_{s0} = L_{\sigma A} \frac{di_{sA}}{dt} + L_{\sigma B} \frac{di_{sB}}{dt} + L_{\sigma C} \frac{di_{sC}}{dt} \tag{6.81}$$

where $L_{\sigma A}$, $L_{\sigma B}$, and $L_{\sigma C}$ are the phase leakage inductances, defined as (6.78).

If the transient generated by the switching of the voltage space-vector $\mathbf{u}_s^{(1)} = -\mathbf{u}_s^{(4)}$ is considered, the zero-sequence voltage can be written as

$$u_{s0}^{(1,4)} = \pm U_d \frac{L_{\sigma A}(L_{\sigma B} + L_{\sigma C}) - 2L_{\sigma B}L_{\sigma C}}{L_{\sigma A}L_{\sigma B} + L_{\sigma B}L_{\sigma C} + L_{\sigma A}L_{\sigma C}} + u_{s0i} \tag{6.82}$$

where

The positive (negative) sign is related to the application of $\mathbf{u}_s^{(1)}(\mathbf{u}_s^{(4)})$

U_d is the DC link voltage

The term u_{s0i} takes into account the effects due to the rotor-induced voltages e_{sA}, e_{sB}, and e_{sC} and can be written as

$$u_{s0i} = 3 \frac{L_{\sigma A}L_{\sigma B}e_{sc} + L_{\sigma A}L_{\sigma C}e_{sB} + L_{\sigma B}L_{\sigma C}e_{sA}}{L_{\sigma A}L_{\sigma B} + L_{\sigma B}L_{\sigma C} + L_{\sigma A}L_{\sigma C}} \tag{6.83}$$

The term u_{s0i} can be neglected at low speed, while it can be properly compensated at high speeds, taking the difference signal $u_{s0A} = u_{s0}^{(1)} - u_{s0}^{(4)}$ or $u_{s0B} = u_{s0}^{(2)} - u_{s0}^{(5)}$ or $u_{s0C} = u_{s0}^{(3)} - u_{s0}^{(6)}$, in dependence of which the switching pattern of the inverter is commanded. Since these three voltages form a symmetric pattern, a complex space-vector depending on the rotor position can be defined as

$$p(\vartheta_r) = \frac{2}{3}[u_{s0A} + \alpha u_{s0B} + \alpha^2 u_{s0C}] = p_{sD}(\vartheta_r) + jp_{sQ}(\vartheta_r) \tag{6.84}$$

The information about the rotor position is contained in the phase of such a signal, while its amplitude does not play any role in it. The position-dependent signal in (6.84) can be further compensated by the saturation disturbing term, giving rise to a signal depending only on the rotor position [22].

6.6 Conclusions on Sensorless Techniques for IM Drives

A definitive judgment on which among the several sensorless techniques proposed in literature is the best is a hard task. Some conclusions could, however, be drawn. All the techniques based on the fundamental mmf model of the IM are preferable in terms of simplicity and computational demand, even if they must adopt complex models and must be integrated with suitable signal processing methodologies for compensating machines and inverter nonlinearities, as long as low-speed operation is the target. Suitable parameter estimation algorithms must be integrated as well to make them properly work in a wide range of working conditions with full torque capability. In general, model-based techniques can operate in a wide speed range, whereas saliency-based methods can be operated only at very low operating speed or at zero speed. On the contrary, saliency-based techniques can properly work at very low and zero speed, even in loaded conditions, where the model-based techniques fail since the machine speed becomes unobservable [3,6]. Moreover, saliency-based techniques permit the drive to be operated in closed-loop position control, whereas model-based techniques can be operated basically only in closed-loop speed control.

Saliency-based techniques can track several saliencies of the motor. Closed-slots machine exhibit basically only the main flux magnetic saturation saliency. Its angular orientation can be detected and tracked, taking properly into consideration its modification with the load. Open-slots machines present further saliencies to be tracked, like the rotor slotting effects. Tracking this saliency gives rise to a high-resolution position signal with high dynamic bandwidth. In this case, however, the main flux saturation should be considered as a disturbance and consequently compensated.

With specific regard to saliency-based sensorless techniques, an extensive theoretical and experimental comparison among the main methods has been proposed in Ref. [58]. The following conclusions have been carried out [57,58]:

1. All the saliency-based techniques could be suitably adopted for rotor position estimation and field orientation, since both slotting effect and saturation-induced saliencies give rise to measurable effects.

2. Almost all the methods present the same limits and restrictions, with specific regard to those methods tracking the rotor slotting saliency.

3. Saturation-induced saliencies in the zero-sequence-voltage methods, independent from the kind of excitation (carrier signal or PWM switching), depend on the main flux for the case of open and semiclosed rotor slot machines, suggesting that they should be suitably used for flux angle estimation (more difficult in closed rotor slot machines).

4. Zero-sequence-voltage carrier-based methods and PWM-switching zero-sequence-voltage-based methods are based on the same physical principles and basically give same kind of results. The differences concern several implementation issues. It is hard to say definitely which is the best.

5. The inverter nonlinearities, as well as other undesired effects like high-frequency phenomena caused by long cables lengths, shielding, and grounding practices, influence all the methods in different ways, giving rise to significantly different performances of the methods themselves with specific regard to robustness and accuracy issues.

6. Significant machine-oriented issues give new prospects to the research, such as the effect of motor design on attainable accuracy and resolution, and the number of harmonics needed to achieve a specified accuracy. These issues influence also the simplicity/complexity of the implementation.

References

1. P. Vas, *Sensorless Vector and Direct Torque Control*, Oxford Science Publication, Oxford, U.K., 1998.
2. K. Rajashekara, A. Kawamura, K. Matsuse, *Sensorless Control of AC Motor Drives*, IEEE Press, Piscataway, NJ, 1996.
3. J. Holtz, Sensorless control of induction motor drives, *Proceedings of the IEEE*, 90(8), 1359–1394, August 2002.
4. R. D. Lorenz, T. A. Lipo, D. W. Novotny, Motion control with induction motors, *Proceedings of the IEEE*, 82(8), 1215–1240, August 1994.
5. W. Leonhard, *Control of Electrical Drives*, Springer-Verlag, Berlin, Germany, 1997.
6. J. Holtz, Sensorless control of induction machines—With or without signal injection? *IEEE Transactions on Industrial Electronics*, 53(1), 7–30, February 2006.
7. R. Joetten, G. Maeder, Control methods for good dynamic performance induction motor drives based on current and voltage measured quantities, *IEEE Transactions on Industry Applications*, IA-19, 356–363, May/June 1983.
8. C. Shauder, Adaptive speed identification for vector control of induction motors without rotational transducers, *IEEE Transactions on Industry Applications*, 28(5), 1054–1061, September/October 1992.
9. H. Tajima, Y. Hori, Speed sensorless field-orientation of the induction machine, *IEEE Transactions on Industry Applications*, 29(1), 175–180, January/February 1993.
10. L. Harnefors, Design and analysis of general rotor-flux-oriented vector control systems, *IEEE Transactions on Industrial Electronics*, 48(2), 383–390, April 2001.
11. G. C. Verghese, S. R. Sanders, Observers for flux estimation in induction machines, *IEEE Transactions on Industrial Electronics*, 35(1), 85–94, February 1988.
12. H. Kubota, K. Matsuse, T. Nakano, DSP-based speed adaptive flux observer of induction motor, *IEEE Transactions on Industry Applications*, 29(2), 344–348, March/April 1993.
13. H. Kubota, I. Sato, Y. Tamura, K. Matsuse, H. Ohta, Y. Hori, Stable operation of adaptive observer based sensorless induction motor drives in regenerating mode at low speeds, *Thirty-Sixth IAS Annual Meeting*, Chicago, IL, Vol. 1, pp. 469–474, October 2001.
14. H. Kubota, K. Matsuse, Y. Hori, Behaviour of sensorless induction motor drives in regenerating mode, *PCC Proceeding*, Nagaoka, Japan, August 3–6, 1997.
15. M. Hinkkanen, J. Luomi, Stabilization of the regenerating mode of full-order flux observer for sensorless induction motors, *IEEE Electrical Machines and Drives Conference, IEMDC 2003*, Madison, WI, Vol. 1, pp. 145–150, June 1–4, 2003.
16. H. Kubota, I. Sato, Y. Tamura, K. Matsuse, H. Otha, Y. Hori, Regenerating-mode low-speed operation of sensorless induction motor drive with adaptive observer, *IEEE Transactions on Industry Applications*, 38(4), 1081–1086, July/August 2002.
17. S. Doki, S. Sangwongwanich, S. Okuma, Implementation of speed-sensor-less field-oriented vector control using adaptive sliding observer, *IEEE IECON 92*, San Diego, CA, pp. 453–458, November 9–13, 1992.
18. M. W. Degner, R. D. Lorenz, Using multiple saliencies for the estimation of flux, position and velocity in AC machines, *IEEE Transactions on Industry Applications*, 34(5), 1097–1104, September/October 1998.
19. J. Cilia, D. M. Asher, K. J. Bradley, Sensorless position detection for vector controlled induction motor drives using an asymmetric outersection cage, *IEEE Transactions on Industry Applications*, 33(5), 1162–1169, September/October 1997.
20. M. Schroedl, Sensorless control of AC machines at low speed and standstill based on the inform method, *Thirty-First IEEE IAS Annual Meeting*, San Diego, CA, pp. 270–277, October 6–10, 1996.
21. J. Jiang, Sensorless field oriented control of induction motors at zero stator frequency, PhD dissertation, Department of Electrical Engineering, Wuppertal University, Wuppertal, Germany, 1999 (in German).

22. J. Holtz, H. Pan, Elimination of saturation effects in sensorless position controlled induction motors, *IEEE Transactions on Industry Applications*, 40(2), 623–631, March/April 2004.

23. J. Holtz, H. Pan, Acquisition of rotor anisotropy signals in sensorless position control systems, *IEEE IAS Annual Meeting*, pp. 1165–1172, October 12–16, 2003.

24. J. Hu, B. Wu, New integration algorithms for estimating motor flux over a wide speed range, *IEEE Transactions on Power Electronics*, 13(5), 969–977, September 1998.

25. K. D. Hurst, T. Hableter, G. Griva, F. Profumo, Zero-speed tacholess IM torque control: Simply a matter of stator voltage integration, *IEEE Transactions on Industry Applications*, 34(4), 790–795, July/August 1998.

26. L. E. Borges de Silva, B. K. Bose, J. O. P. Pinto, Recurrent-neural-network-based implementation of a programmable cascaded low-pass filter used in stator flux synthesis of vector-controlled induction motor drive, *IEEE Transactions on Industrial Electronics*, 46(3), 662–665, June 1999.

27. J. O. P. Pinto, B. K. Bose, L. E. Borges de Silva, A stator-flux-oriented vector-controlled induction motor drive with space-vector PWM and flux-vector synthesis by neural network, *IEEE Transactions on Industry Applications*, 37(5), 1308–1318, September/October 2001.

28. J. Holtz, Q. Juntao, Sensorless vector control of induction motors at very low speed using a nonlinear inverter model and parameter identification, *IEEE Transactions on Industry Applications*, 38(4), 1087–1095, July/August 2002.

29. J. Holtz, Q. Juntao, Drift- and parameter-compensated flux estimator for persistent zero-stator-frequency operation of sensorless-controlled induction motors, *IEEE Transactions on Industry Applications*, 39(4), 1052–1060, July/August 2003.

30. Q. Juntao, Sensorless vector control of induction motors at low and zero speed, Fortschritt-Berichte VDI, Reihe 21, Elektrotechnik, n. 325.

31. B. K. Bose, N. R. Patel, A sensorless stator flux oriented vector controlled induction motor drive with neuro-fuzzy based performance enhancement, *Thirty-Second IEEE IAS Annual Meeting*, New Orleans, LA, October 5–9, 1997.

32. P. L. Jansen, R. D. Lorenz, A physically insightful approach to the design and accuracy assessment of flux observers for field oriented induction machine drives, *IEEE Transactions on Industry Applications*, 30(1), 101–110, January/February 1994.

33. F. Peng, T. Fukao, Robust speed identification for speed-sensorless vector control of induction motors, *IEEE Transactions on Industry Applications*, 30(5), 1234–1240, 1994.

34. P. L. Jansen, R. D. Lorenz, Accuracy limitations of velocity and flux estimation in direct field oriented induction machines, *Proceedings of EPE Conference*, Brighton, U.K., September 13–16, 1993.

35. R. Blasco-Gimenez, G. M. Asher, M. Sumner, K. J. Bradley, Dynamic performance limitations for MRAS based sensorless induction motor drives. I. Stability analysis for the closed loop drive, *Electric Power Applications, IEE Proceedings*, Vol. 143(2), pp. 113–122, March 1996.

36. R. Blasco-Gimenez, G. M. Asher, M. Sumner, K. J. Bradley, Dynamic performance limitations for MRAS based sensorless induction motor drives. II. Online parameter tuning and dynamic performance studies, *Electric Power Applications, IEE Proceedings*, Vol. 143(2), pp. 123–134, March 1996.

37. P. J. Jansen, R. D. Lorenz, Observer-based direct field orientation: Analysis and comparison of alternative methods, *IEEE Transactions on Industry Applications*, 30(4), 945–953, July/August 1994.

38. J. Maes, J. Melkebeek, Speed-sensorless direct torque control of induction motors using an adaptive flux observer, *IEEE Transactions on Industry Applications*, 36(3), 778–785, May/June 2000.

39. C. Lascu, I. Boldea, F. Blaabjerg, A modified direct torque control for induction motor sensorless drive, *IEEE Transactions on Industry Applications*, 36(1), 122–130, 2000.

40. Y.-N. Lin, C.-L. Chen, Adaptive pseudoreduced-order flux observer for speed sensorless field-oriented control of IM, *IEEE Transactions on Industrial Electronics*, 46(5), 1042–1045, October 1999.

41. R. Nilsen, M. P. Kazmierkowski, Reduced-order observer with parameter adaption for fast rotor flux estimation in induction machines, *IEE Proceedings D, Control Theory and Applications*, 136(1), 35–43, January 1989.

42. C.-M. Lee, C.-L. Chen, Observer-based speed estimation method for sensorless vector control of induction motors, *IEE Proceedings D, Control Theory and Applications*, 145(3), 359–363, May 1998.
43. G. Franceschini, M. Pastorelli, F. Profumo, C. Tassoni, A. Vagati, About the gain choice of flux observer in induction servo-motors, *IEEE Industry Applications Society Annual Meeting*, Seattle, WA, pp. 601–606, October 7–12, 1990.
44. M. Cirrincione, M. Pucci, G. Cirrincione, G. Capolino, Sensorless control of induction motors by reduced order observer with MCA EXIN + based adaptive speed estimation, *IEEE Transactions on Industrial Electronics*, 54,(1), 150–166, February 2007, Special Section Neural Network Applications in Power Electronics and Motor Drives (Guest Editor B. K. Bose).
45. Y.-R. Kim, S.-K. Sul, M.-H. Park, Speed sensorless vector control of an induction motor using extended Kalman filter, *IEEE Transactions on Industry Applications*, 30(5), 1225–1233, September/October 1994.
46. J. Holtz, Sensorless position control of induction motors—An emerging technology, *IEEE Transactions on Industrial Electronics*, 45(6), 840–851, 1998.
47. P. L. Jansen, R. D. Lorenz, Transducerless position and velocity estimation in induction and salient AC machines, *IEEE Transactions on Industry Applications*, 31(2), 240–247, March/April 1995.
48. P. L. Jansen, R. D. Lorenz, Transducerless field orientation concepts employing saturation-induced saliencies in induction machines, *IEEE Transactions on Industry Applications*, 32(6), 1380–1393, November/December 1996.
49. N. Teske, G. M. Asher, M. Sumner, K. J. Bradley, Suppression of saturation saliency effects for the sensorless position control of induction motor drives under loaded conditions, *IEEE Transactions on Industrial Electronics*, 47(5), 1142–1149, September/October 2000.
50. N. Teske, G. M. Asher, K. J. Bradley, M. Sumner, Analysis and suppression of inverter clamping saliency in sensorless position controlled of induction motor drives, *IEEE Industry Applications Society Annual Meeting*, Chicago, IL, pp. 2629–2636, 2001.
51. A. Consoli, G. Scarcella, A. Testa, Speed- and current-sensorless field-oriented induction motor drive operating at low stator frequencies, *IEEE Transactions on Industry Applications*, 40(1), 186–193, January/February 2004.
52. A. Consoli, G. Scarcella, G. Bottiglieri, A. Testa, Harmonic analysis of voltage zero-sequence-based encoderless techniques, *IEEE Transactions on Industry Applications*, 42(6), 1548–1557, November/December 2006.
53. M. Pucci, C. Serporta, Finite-element analysis of rotor slotting saliency in induction motors for sensorless control, *IEEE Transactions on Magnetics*, 46(2), 650–659, February 2010.
54. M. Pucci, C. Serporta, Analysis of rotor slotting saliency in induction motor sensorless control, *Electric Power Systems Research*, 81, 318–328, 2011.
55. J.-I. Ha, S.-K. Sul, Sensorless field-oriented control of an induction machine by high-frequency signal injection, *IEEE Transactions on Industry Applications*, 35, 45–51, January/February 1999.
56. A. Consoli, G. Scarcella, G. Bottiglieri, G. Scelba, A. Testa, D. A. Triolo, Low-frequency signal-demodulation-based sensorless technique for induction motor drives at low speed, *IEEE Transactions on Industrial Electronics*, 53(1), 207–215, February 2006.
57. F. Briz, M. W. Degner, A. Diez, R. D. Lorenz, Static and dynamic behavior of saturation-induced saliencies and their effect on carrier-signal-based sensorless AC drives, *IEEE Transactions on Industry Applications*, 38(3), 670–678, May/June 2002.
58. F. Briz, M. W. Degner, P. Garcia, R. D. Lorenz, Comparison of saliency-based sensorless control techniques for AC machines, *IEEE Transactions on Industry Applications*, 40(4), 1107–1115, July/August 2004.

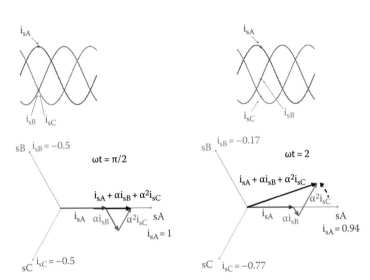

FIGURE 1.4
Graphical description of the space-vector quantity.

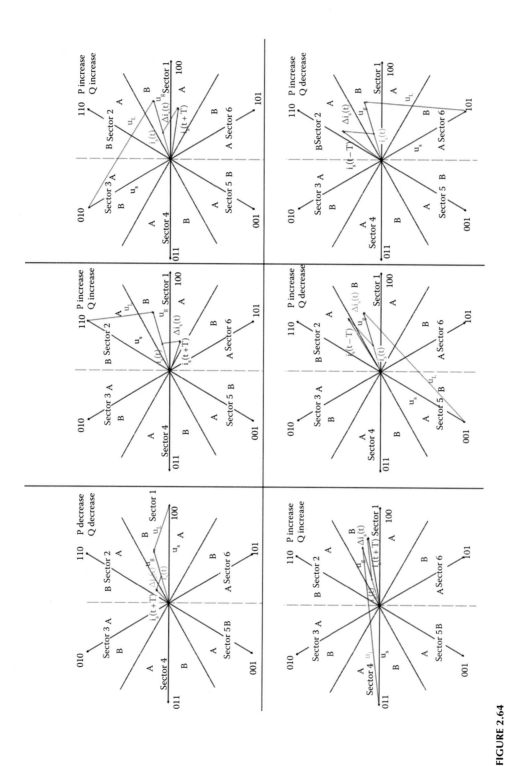

FIGURE 2.64
Effect of the application of any voltage vector on the active and reactive power exchanged with the grid. (From Pucci, M. et al., *Electric Power Syst. Res*, 81(4), 830, April 2011, doi: 10.1016/j.epsr.2010.11.007.)

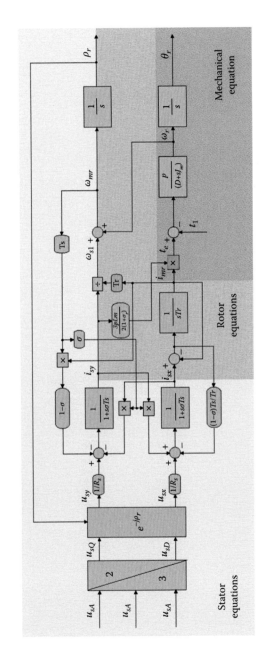

FIGURE 4.4
Block diagram of the dynamical model of the IM in the rotor flux-linkage reference frame.

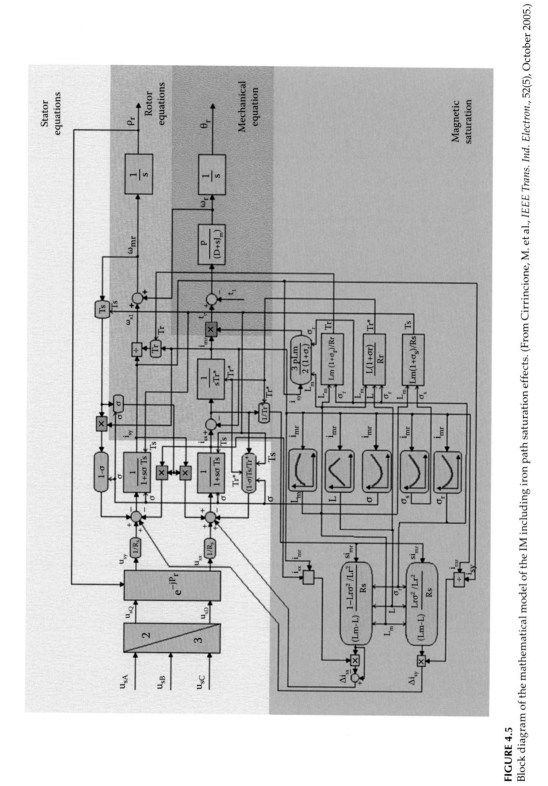

FIGURE 4.5

Block diagram of the mathematical model of the IM including iron path saturation effects. (From Cirrincione, M. et al., *IEEE Trans. Ind. Electron.*, 52(5), October 2005.)

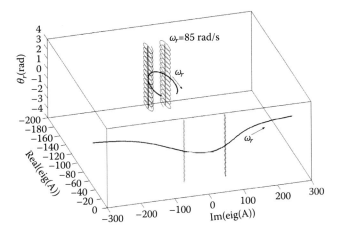

FIGURE 4.23
Eigenvalue locus of matrix **A** obtained with the fundamental mmf model and the model taking into consideration the rotor slotting effects. (From Cirrincione, M. et al., *IEEE Trans. Ind. Appl.*, 44, 1683, 2008.)

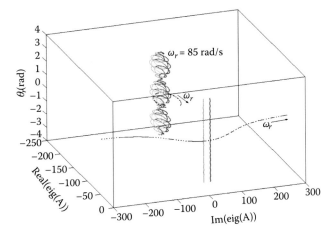

FIGURE 4.24
Eigenvalue locus of matrix **A** obtained with the fundamental mmf model and the model taking into consideration the stator and rotor slotting effects. (From Cirrincione, M. et al., Space-vector state model of induction machines including rotor and stator slotting effects, *IEEE IEMDC '07*, Antalya, Turkey, pp. 673–682, May 3–5, 2007.)

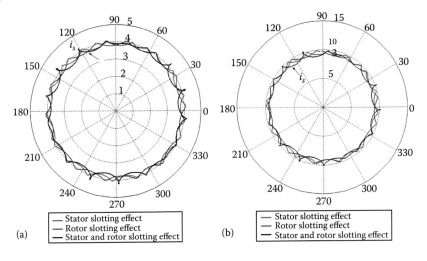

FIGURE 4.33
(a) Locus of the stator current at no load (numerical simulation). (b) Locus of the stator current at 6.37% load slip (numerical simulation). (From Cirrincione, M. et al., Space-vector state model of induction machines including rotor and stator slotting effects, in *IEEE IEMDC '07*, Antalya, Turkey, pp. 673–682, May 3–5, 2007.)

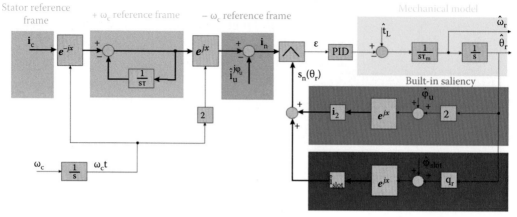

FIGURE 6.33
Block diagram of the rotor position estimator based on PLL for anisotropy tracking.

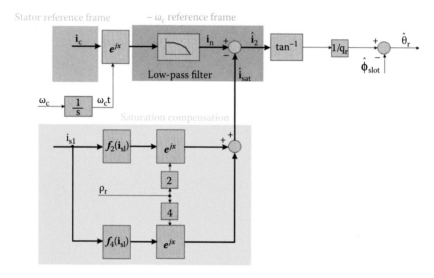

FIGURE 6.34
Block diagram of the rotor position estimator based on rotor slotting anisotropy tracking.

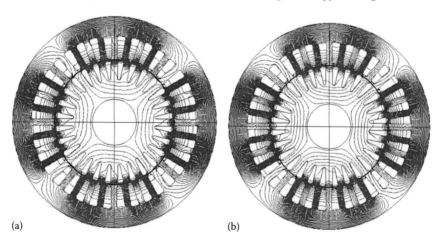

FIGURE 6.36
(a, b) Magnetic flux density lines in the closed-slots (a, left) and open-slots (b, right) machines. (From Pucci, M. and Serporta, C., *IEEE Trans. Magn.*, 46(2), 2010.)

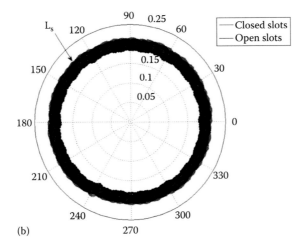

FIGURE 6.38
(b) Locus of the stator inductance space-vector under fundamental supply and high-frequency carrier. (From Pucci, M. and Serporta, C., *Electric Power Syst. Res.*, (81), 318, 2011.)

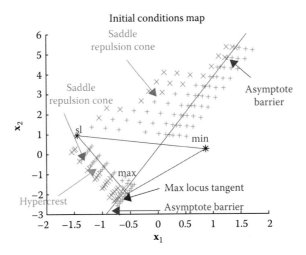

FIGURE 8.18
The initial conditions map for the 2-D TLS problem: the green crosses represent the initial conditions by which a gradient flow algorithm converges, that is, the domain of convergence. The red crosses give divergence. The dark blue triangle has the critical points as its vertices. The dark blue straight line represents the asymptote/barrier z_1. (From Cirrincione, G. and Cirrincione, M., *Neural Based Orthogonal Data Fitting: The EXIN Neural Networks*, Series: Adaptive and Learning Systems for Signal Processing, Communications and Control, Wiley & Sons, New York, 255 pp., November 2010.)

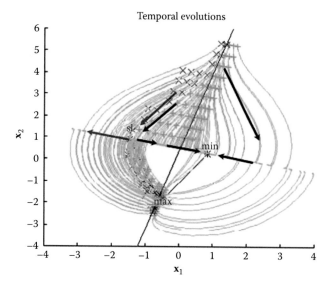

FIGURE 8.19
The temporal evolution of the sequential TLS EXIN neuron for the generic 2-D TLS problem. (From Cirrincione, G. and Cirrincione, M., *Neural Based Orthogonal Data Fitting: The EXIN Neural Networks*, Series: Adaptive and Learning Systems for Signal Processing, Communications and Control, Wiley & Sons, New York, 255 pp., November 2010.)

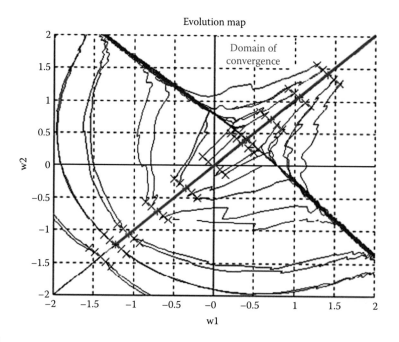

FIGURE 8.21
Evolution map of the sequential TLS EXIN for the first nongeneric TLS benchmark problem. The initial conditions for the converging trajectories are in red, the others are in black. The straight line passing through the red points contains the saddle/solution locus. The dark blue thick straight line is the divergence line. (From Cirrincione, G. and Cirrincione, M., *Neural Based Orthogonal Data Fitting: The EXIN Neural Networks*, Series: Adaptive and Learning Systems for Signal Processing, Communications and Control, Wiley & Sons, New York, 255 pp., November 2010.)

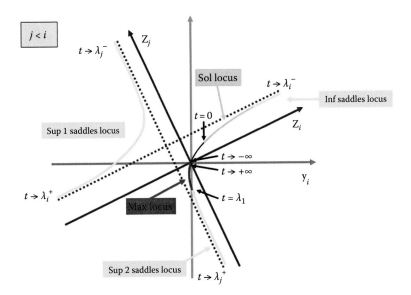

FIGURE 8.23
Critical loci for j < I. (From Cirrincione, G. and Cirrincione, M., *Neural Based Orthogonal Data Fitting: The EXIN Neural Networks*, Series: Adaptive and Learning Systems for Signal Processing, Communications and Control, Wiley & Sons, New York, 255 pp., November 2010.)

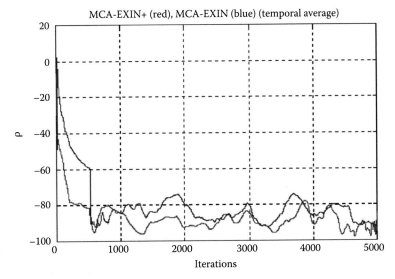

FIGURE 8.26
Index ρ (expressed in dB) for MCA EXIN+ with scheduling (red) and MCA EXIN (blue). (From Cirrincione, G. and Cirrincione, M., *Neural Based Orthogonal Data Fitting: The EXIN Neural Networks*, Series: Adaptive and Learning Systems for Signal Processing, Communications and Control, Wiley & Sons, New York, 255 pp., November 2010.)

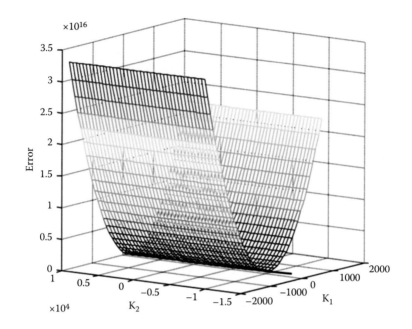

FIGURE 9.49
OLS error function versus K_1, K_2, and estimated K trajectory (black).

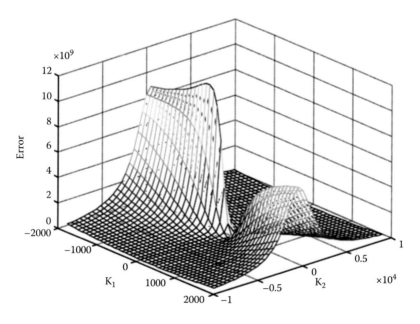

FIGURE 9.50
TLS error function versus K_1, K_2.

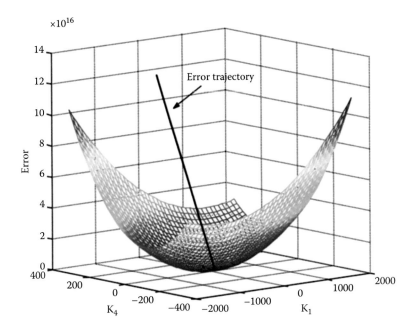

FIGURE 9.51
Error function of the constrained minimization versus K_1, K_4, and estimated K trajectory (black).

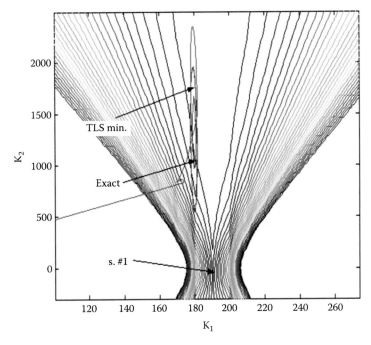

FIGURE 9.54
Contours and critical points of the TLS error in the plane K_1, K_2; the TLS EXIN weight trajectory is also shown. (From Cirrincione, M. et al., *IEEE Trans. Ind. Appl.*, 39(5), 1247, September/October 2003.)

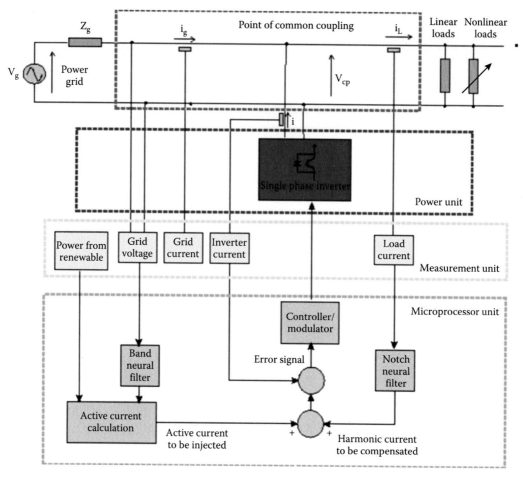

FIGURE 10.1
Block diagram of the neural-enhanced single-phase DG systems with APF capability.

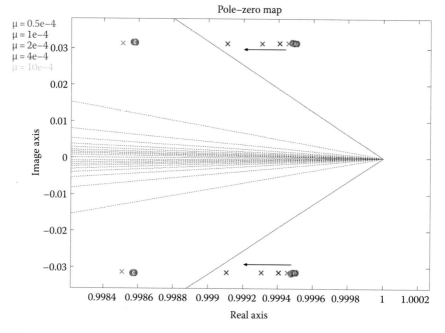

FIGURE 10.13
Position of poles close to the unitary circle for different values of μ. (From Cirrincione, M. et al., *IEEE Trans. Ind. Electron.*, 55(5), 2093, May 2008.)

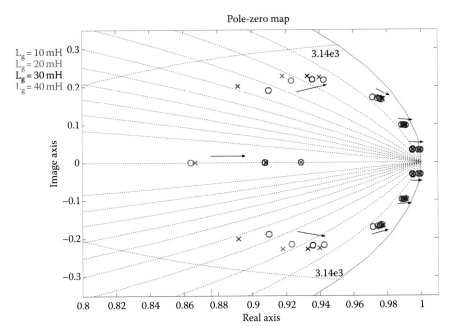

FIGURE 10.14
Position of poles for different values of L_g (L_g increases from left to right). (From Cirrincione, M. et al., *IEEE Trans. Ind. Electron.*, 55(5), 2093, May 2008.)

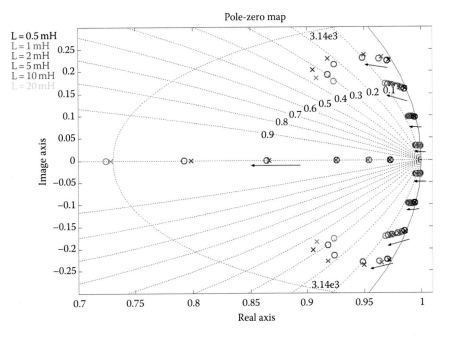

FIGURE 10.15
Position of the poles for different values of L (L increases from right to left). (From Cirrincione, M. et al., *IEEE Trans. Ind. Electron.*, 55(5), 2093, May 2008.)

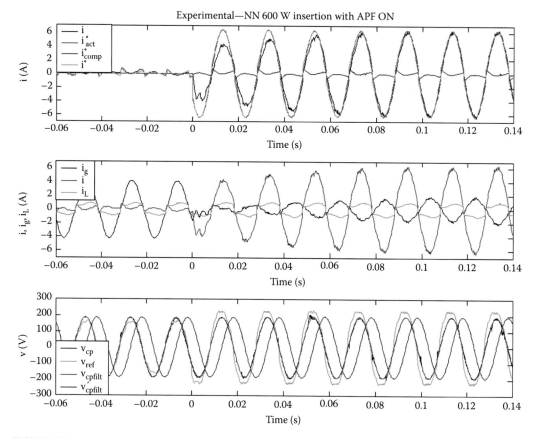

FIGURE 10.21
Grid current, inverter current, load current, and coupling point voltage during the step transient of $P_{ref} = 600\,W$ with APF on with the DG-APF based on neural adaptive filtering (experiment). (From Cirrincione, M. et al., *IEEE Trans. Ind. Electron.*, 55(5), 2093, May 2008.)

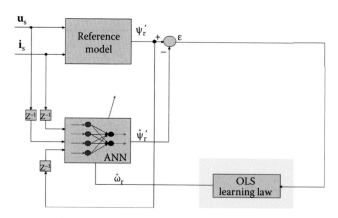

FIGURE 11.6
Block diagram of the OLS MRAS observer.

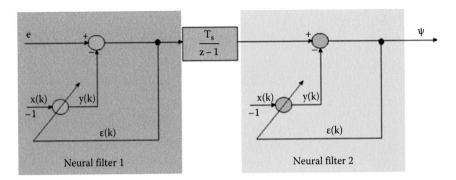

FIGURE 11.16
Block diagram of the neural adaptive integrator. (From Cirrincione, M. et al., *IEEE Trans. Power Electron.*, 19(1), 25, January 2004.)

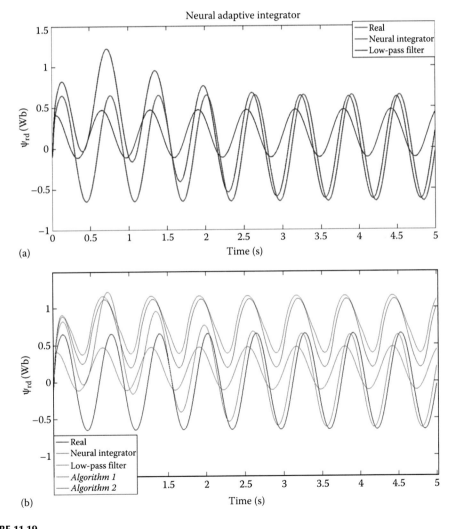

FIGURE 11.19
(a) Transient time waveform of ψ_{rd} component obtained with the neural filter and LPF. (b) Transient time waveform of ψ_{rd} component obtained with the neural filter, LPF, and *algorithms 1* and *2*.

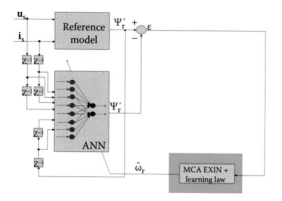

FIGURE 11.32
The block diagram of the MCA EXIN + MRAS speed observer.

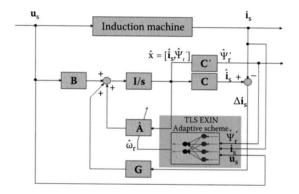

FIGURE 11.40
Block diagram of the TLS EXIN full-order adaptive observer.

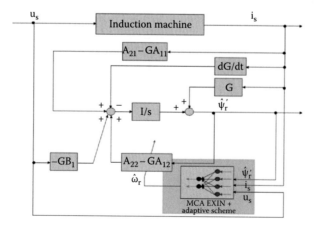

FIGURE 11.61
Block diagram of the MCA EXIN + reduced-order observer.

7

Permanent Magnet Synchronous Motor Drives

Angelo Accetta

Institute of Intelligent Systems for the Automation (ISSIA)
National Research Council (CNR)
Palermo, Italy

7.1 Introduction

Over the last two decades, permanent magnet synchronous motors (PMSMs) have drawn an ever-increasing attention. The idea of substituting the electrical excitation winding of a synchronous machine with the magnets dates back to nineteenth century [1], but the PMSM starts to be interesting for practical applications only with the discovery of rare-earth magnets.

Nowadays, PMSMs are widely used in many applications for the following reasons:

- No rotor windings are present, so the only electrical contacts are the stator ones; in this way, PMSM can be used even in environments with explosion risks.

- The absence of excitation windings reduces electrical losses, entailing a higher motor efficiency and, at the same time, a lower cooling demand.

- Both the torque density and the power density are higher in PMSMs than in externally excited synchronous motors and in induction machines (IMs).

- Since PMSM can achieve higher air-gap flux density, better dynamic performance can be obtained with respect to the externally excited synchronous motors.

Nevertheless, they present some drawbacks because of the presence of the permanent magnets (PMs):

- Being the PM usually made from rare-earth materials, their cost is usually higher than externally excited synchronous motors and in IMs.

- They present a mechanical fragility that makes the PMSM unsuitable in applications with heavy mechanical stress.

- Finally, the PMs may lose their magnetic properties if exposed to high temperature; this limits the overloading capability of the PMSM.

PMSMs are usually divided into two categories, *DC brushless* and *AC brushless* [2]:

- *DC brushless synchronous motors* are PMSMs with a *trapezoidal* air-gap flux distribution.

- *AC brushless synchronous motors* are PMSMs with a *sinusoidal* air-gap flux.

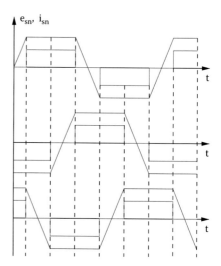

FIGURE 7.1
Back emf and current pulses for control of DC brushless.

Differences between the two categories lie in mechanical characteristics, control methodology, and achievable dynamic performance.

7.1.1 DC Brushless Motors

The PM generates a trapezoidal or quasitrapezoidal magnetic flux. On the stator, armature windings are usually concentrated, and only two windings at a time are commanded to let current flow through them. The primary energy source is generally a DC one, and the controller ensures that the DC flows only through two windings, resulting in a square-waveform current in each armature winding. Each current pulse has a duration equivalent to 120°, and it is centered with the back electromotive force (emf), as shown in Figure 7.1.

Obviously, the control methodology for a DC brushless requires the information of the shaft position to ensure the correct switching sequence for the armature windings. However, it is not required a high precision; the current pulse must be centered about the trapezoidal back emf and each stator winding is not allowed to have a current when the back emf is changing polarity: only the center of the back emf should be detected. Three Hall effect sensors can be therefore used to provide the correct timing pulse for the stator winding switching sequence [3]. The dynamic performance that can be obtained from such motors is strictly dependent on the adopted control methodology; for example, a DC brushless fed by a DC commutator can be suitable for PC application like CD players or hard disk motors, but the same motor fed by a PWM inverter and another control strategy can be used in many servo applications.

7.1.2 AC Brushless Motors

They differ from their DC counterpart in many aspects; the quasisinusoidal distribution of the air-gap flux is obtained both by the design of the rotor magnet and of the armature windings. The latter are sinusoidally distributed over the pole arc [4], while the PMs are usually tapered at the pole edges and shortened in order to occupy a smaller pole arc (120°). AC brushless motors are widely used in automotive applications, especially when high dynamic performance and accurate positioning are the main goals. Vector control is the main control methodology for this kind of motors since it permits obtaining performance comparable with classic DC machine.

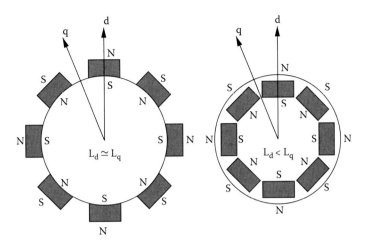

FIGURE 7.2
Magnet placement in a PMSM.

Another criterion for classifying brushless motors is the location of PMs. Actually, they can be placed either on the surface of the rotor, and then the machine will be called *surface mounted*, or buried inside the rotor, and in this case, it is called *interior-mounted* motor; in Figure 7.2, a scheme of the two magnet displacements is shown.

The surface-mounted motor is theoretically an isotropic machine, being the magnet's magnetic permeability almost equal to the corresponding value of the air. However, the speed limit is lower than that of the interior-mounted motor since the magnet is usually glued to the rotor surface.

On the other hand, the interior-mounted motor is an anisotropic machine, and because of the magnet placement, the reluctance of the direct axis, that is, the axis collinear with the direction of the magnetic flux, is greater than that of the quadrature axis. This consideration, as shown in the following, results in a negative reluctance torque. Methods to avoid this effect will be explained in this chapter.

PMs can also be used to improve the efficiency and the power factor of a reluctance motor: this motor is usually called PM-assisted reluctance (PMAR) motor or simply interior permanent magnet (IPM) motor. The synchronous reluctance machines (without PM) are motors with high dynamic performance and have also a good torque-to-mass ratio; nevertheless, their application is limited because of a low power factor. PMs can correct their power factor [5].

Actually, the rotor of an IPM motor has usually several flux barriers per pole, where the PMs are placed, as shown in Figure 7.3.

These flux barriers have different thicknesses to improve the air-gap flux distribution, and the power factor can be increased by properly selecting the PM flux so that the stator flux linkage is in quadrature with the current vector, as shown in Figure 7.4 [6].

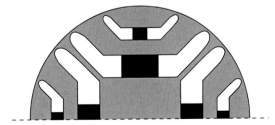

FIGURE 7.3
PMAR motor scheme.

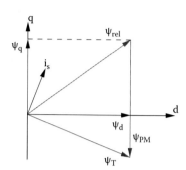

FIGURE 7.4
IPM vector scheme.

Indeed, the IPM and PMAR motors typologies should refer to different machines: if the flux produced by the PMs is greater than the reluctance flux, the machine should be called IPM motor, PMAR motor otherwise [7]. This difference has an influence also in the machine modeling: since the direct axis of the synchronous reference frame is placed along the direction of the main flux (the PM flux for the classic PMSM), IPM motors have the direct axis superimposed to the PM flux, while the PMAR motors have the direct axis coincident with the maximum permeance direction [7].

Nevertheless, IPM motors present, like all synchronous reluctance motors, a higher torque ripple than other PM motors [8,9]. In Ref. [10], it has been shown that rotor skewing is not sufficient to reduce the torque ripple and that the only suitable way is to uniformly distribute the barrier along the rotor, similarly to the stator slot distribution. Also in Ref. [9], two rotor configurations, called "Romeo and Juliet" and "Machaon," has been proposed, whose results show a reduction of the torque ripple.

7.1.3 Permanent Magnets

PMs are a particular kind of magnetic material that, after a magnetization process, can maintain their magnetization level.

The *residual magnetization* B_r and the *coercive field* H_c are the two main parameters used to describe the behavior of the magnets. The residual magnetization is the value of the magnetic field produced by the magnet after the external magnetizing field has been removed. The coercive field is the minimum value of an external opposite magnetic field that can demagnetize the PM. Indeed, the coercive field should be properly divided into *coercive field* H_{cB} and *intrinsic coercive field* H_{cJ}. The intrinsic coercive field H_{cJ} represents the external field that should be applied to the magnet in order to make it lose all its magnetic properties, while the coercive field H_{cB} is the external field that nullifies the PM field. In other words, the intrinsic coercive field is a measure of how "permanent" is the magnet: if, theoretically, a magnet would have $H_{cJ} = \infty$, then that magnet will never lose its magnetic field. Obviously, for every PM, $H_{cB} < H_{cJ}$.

PMs are used to generate a magnetic field across an air-gap. The operating point of a magnet, that is, the resulting value of the magnetic field produced when the magnet is placed in a magnetic circuit, is given by the intersection between its hysteresis curve and the *load line*, given by the geometry of the system. As a consequence, the higher the coercive field, the higher the resulting magnetic field.

Another useful parameter is the *maximum energy product*, equal to the maximum value of the product *BH*, whose value is calculated with the help of the hysteresis curve of a specific PM. For this reason, the maximum energy product corresponds to a specific point on the hysteresis curve. If a PM has its operating point equal to the maximum energy point, then that magnet has the minimum volume.

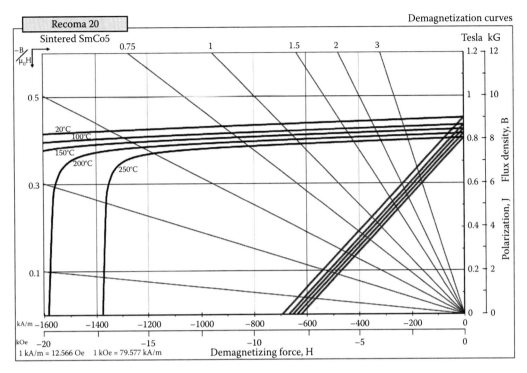

FIGURE 7.5
Hysteresis curve for the Recoma 20 (samarium-cobalt) magnetic material.

Different materials can have very different values of these two parameters; for example, Alnico 5 has a residual magnetization almost equal to $1.22\,\text{Wb}/\text{m}^2$ and a coercive field equal to $49\,\text{kA}/\text{m}$. Electrical steel M-5 has a similar value of residual magnetization ($1.4\,\text{Wb}/\text{m}^2$) but a very smaller value of coercive field ($7\,\text{A}/\text{m}$). Also magnetic ferrites are used to produce PM; they are usually composed by iron oxide and strontium or barium carbonate. They have a residual magnetization lower than the alnico magnets but a higher coercive field. More recent materials used for PMs are the samarium-cobalt, whose hysteresis curve is shown in Figure 7.5, a rare earth made from an alloy of samarium and cobalt that has a great maximum energy product, and the neodymium, whose hysteresis curve is shown in Figure 7.6, an alloy made from neodymium, iron, and boron that has both high residual magnetization and coercive field along with good mechanical properties that make it suitable for the production of motor-designed PMs (Table 7.1).

7.2 Space-Vector Model of Permanent Magnet Synchronous Motors

The most general case for modeling the PMSM is the anisotropic machine; as represented in Figure 7.7, for simplicity, only the case of one pole pair will be considered. It can be proven that the self-inductance of each stator winding must have a $2\theta_r$ periodicity, where $\theta_r = p\theta_m$ is the position of the rotor, measured in electric radians, with p being the pole pair number and θ_m the position of the rotor, measured in mechanical radians.

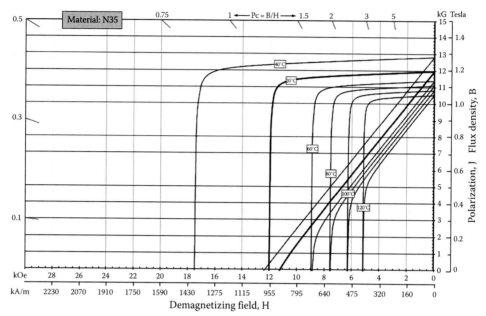

1 kA/m = 12.566 Oe   1 kOe = 79.577 kA/m

FIGURE 7.6
Hysteresis curve for the N35 (neodymium-iron-boron) magnetic material.

TABLE 7.1

Properties of Various Magnetic Materials

Material	B_r (mT)	H_{cB} (kA/m)	H_{cJ} (kA/m)	BH_{max} (kJ/m³)
N35	1195	899	955	275
Alnico G2	720	45	160	12.7
Ferrite c12	400	290	318	32
Recoma 20	900	700	2400	160

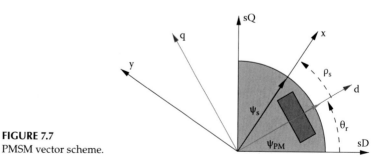

FIGURE 7.7
PMSM vector scheme.

If the rotor direct axis is parallel with the axis of the stator winding named A, then the respective magnetizing inductance L_{mA} attains its minimum value, since the air-gap is the largest in this configuration, equal to

$$L_{mA}(0) = L_{md} = \mu_0 N_s^2 \left(\frac{rl}{g_{\max}} \right) \left(\frac{\pi}{4} \right)$$

(7.1)

where
  μ_0 is the magnetic permeability of the air
  N_s is the number of winding turns
  r is the rotor radius
  l is the rotor length
  $g_{\max}$ is the maximum air-gap
  L_{md} is the magnetizing inductance of the direct axis of the PM reference frame

When $\theta = \pi/2$, the air-gap is minimum, then the magnetizing inductance of the A winding reaches its maximum value:

$$L_{mA}\left(\frac{\pi}{2} \right) = L_{mq} = \mu_0 N_s^2 \left(\frac{rl}{g_{\min}} \right) \left(\frac{\pi}{4} \right)$$

(7.2)

with $g_{\min}$ being the minimum value of the air-gap. Taking into account the leakage inductance L_{sl}, assumed equal for each winding, the self-inductance of the A winding can be expressed as

$$L_{sA}(\theta_r) = L_{sl} + L_{0s} + L_{2s}\cos(2\theta_r)$$

(7.3)

where [3]

$$L_{0s} = \frac{L_{mq} + L_{md}}{2} = \mu_0 N_s^2 rl \left(\frac{1}{g_{\min}} + \frac{1}{g_{\max}} \right) \left(\frac{\pi}{8} \right)$$

$$L_{2s} = \frac{L_{mq} - L_{md}}{2} = \mu_0 N_s^2 rl \left(\frac{1}{g_{\min}} - \frac{1}{g_{\max}} \right) \left(\frac{\pi}{8} \right)$$

(7.4)

Direct- and quadrature-axis inductances L_d and L_q can be therefore expressed as

$$L_d = L_{sl} + L_{md}$$

$$L_q = L_{sl} + L_{mq}$$

(7.5)

As for the self-inductance of the B and C phase windings, they can be easily deduced from (7.3):

$$L_{sB}(\theta_r) = L_{sl} + L_{0s} + L_{2s}\cos\left(2\theta_r + \frac{2\pi}{3} \right)$$

$$L_{sC}(\theta_r) = L_{sl} + L_{0s} + L_{2s}\cos\left(2\theta_r - \frac{2\pi}{3} \right)$$

(7.6)

The mutual inductances between two stator phases can be obtained in a similar way, and their general expression is

$$L_{xy}(\theta_r) = \mu_0 N_x N_y \left(\frac{rl}{g(\theta_r)}\right) \frac{\pi}{4} \cos(\alpha) \tag{7.7}$$

where α is the angle between the axes of the two windings. Taking into account the A and the B windings, when $\theta = 0$, the mutual inductance can be expressed as

$$L_{AB}(0) = \mu_0 N_s^2 \left(\frac{rl}{g_{max}}\right) \frac{\pi}{4} \cos\left(\frac{2\pi}{3}\right) = -\frac{1}{2} L_{md} \tag{7.8}$$

while, if $\theta = \pi/2$,

$$L_{AB}\left(\frac{\pi}{2}\right) = \mu_0 N_s^2 \left(\frac{rl}{g_{min}}\right) \frac{\pi}{4} \cos\left(\frac{2\pi}{3}\right) = -\frac{1}{2} L_{mq} \tag{7.9}$$

Even in this case, it should be remarked that the mutual inductances show a $2\theta_r$ periodicity, with the maximum value attained when the rotor d axis coincides with the bisector of the angle between the two phases, so the following relationships for the three mutual inductances can be written:

$$L_{AB}(\theta_r) = -\frac{1}{2} L_{0s} - L_{2s} \cos\left(2\theta_r - \frac{2\pi}{3}\right)$$

$$L_{AC}(\theta_r) = -\frac{1}{2} L_{0s} - L_{2s} \cos\left(2\theta_r + \frac{2\pi}{3}\right) \tag{7.10}$$

$$L_{BC}(\theta_r) = -\frac{1}{2} L_{0s} - L_{2s} \cos(2\theta_r)$$

A matrix formulation can be used to represent the model of PMSM in a compact form. If R_s is the value of the stator phase resistance, the phase voltages on each stator winding can be obtained as

$$\begin{bmatrix} u_{sA} \\ u_{sB} \\ u_{sC} \end{bmatrix} = R_s \begin{bmatrix} i_{sA} \\ i_{sB} \\ i_{sC} \end{bmatrix} + \frac{d}{dt} \begin{bmatrix} \psi_{sA} \\ \psi_{sB} \\ \psi_{sC} \end{bmatrix} \tag{7.11}$$

where ψ_{sn} ($n = A, B, C$) is the nth phase stator flux linkage. Each flux linkage is generated both by the self-inductance of each phase and the mutual inductance between the three phases. In a matrix form,

$$\begin{bmatrix} \psi_{sA} \\ \psi_{sB} \\ \psi_{sC} \end{bmatrix} = \begin{bmatrix} L_A & L_{AB} & L_{AC} \\ L_{AB} & L_B & L_{BC} \\ L_{AC} & L_{BC} & L_C \end{bmatrix} \begin{bmatrix} i_{sA} \\ i_{sB} \\ i_{sC} \end{bmatrix} + \psi_{PM} \begin{bmatrix} \cos(\theta_r) \\ \cos\left(\theta_r - \frac{2\pi}{3}\right) \\ \cos\left(\theta_r + \frac{2\pi}{3}\right) \end{bmatrix} \tag{7.12}$$

where ψ_{PM} is the magnetic flux produced by the PMs. It is possible to rewrite Equation 7.12 in order to emphasize the inductance terms depending on θ_r:

$$
\begin{bmatrix} \psi_{sA} \\ \psi_{sB} \\ \psi_{sC} \end{bmatrix} = \begin{bmatrix} L_{ls}+L_{0s} & -\dfrac{1}{2}L_{0s} & -\dfrac{1}{2}L_{0s} \\ -\dfrac{1}{2}L_{0s} & L_{ls}+L_{0s} & -\dfrac{1}{2}L_{0s} \\ -\dfrac{1}{2}L_{0s} & -\dfrac{1}{2}L_{0s} & L_{ls}+L_{0s} \end{bmatrix} \begin{bmatrix} i_{sA} \\ i_{sB} \\ i_{sC} \end{bmatrix}
$$

$$
+L_{2s} \begin{bmatrix} \cos(2\theta_r) & \cos\left(2\theta_r-\dfrac{2\pi}{3}\right) & \cos\left(2\theta_r+\dfrac{2\pi}{3}\right) \\ \cos\left(2\theta_r-\dfrac{2\pi}{3}\right) & \cos\left(2\theta_r+\dfrac{2\pi}{3}\right) & \cos(2\theta_r) \\ \cos\left(2\theta_r+\dfrac{2\pi}{3}\right) & \cos(2\theta_r) & \cos\left(2\theta_r-\dfrac{2\pi}{3}\right) \end{bmatrix} \begin{bmatrix} i_{sA} \\ i_{sB} \\ i_{sC} \end{bmatrix} + \psi_{PM}\begin{bmatrix} \cos(\theta_r) \\ \cos\left(\theta_r-\dfrac{2\pi}{3}\right) \\ \cos\left(\theta_r+\dfrac{2\pi}{3}\right) \end{bmatrix}
$$

$$(7.13)$$

or, in a more compact form,

$$\boldsymbol{\psi}_p = \mathbf{L}\mathbf{i}_p - \frac{L_{2s}}{2}\mathbf{T}\mathbf{i}_p + \frac{\psi_{PM}}{2}\mathbf{t} \tag{7.14}$$

where $\mathbf{T}$ is a matrix and $\mathbf{t}$ is a vector defined through the fundamental Euler's formula:

$$
\mathbf{T} = \begin{bmatrix} e^{j2\theta_r} & \alpha^2 e^{j2\theta_r} & \alpha e^{j2\theta_r} \\ \alpha^2 e^{j2\theta_r} & \alpha e^{j2\theta_r} & e^{j2\theta_r} \\ \alpha e^{j2\theta_r} & e^{j2\theta_r} & \alpha^2 e^{j2\theta_r} \end{bmatrix} + \begin{bmatrix} e^{-j2\theta_r} & \alpha e^{-j2\theta_r} & \alpha^2 e^{-j2\theta_r} \\ \alpha e^{-j2\theta_r} & \alpha^2 e^{-j2\theta_r} & e^{-j2\theta_r} \\ \alpha^2 e^{-j2\theta_r} & e^{-j2\theta_r} & \alpha e^{-j2\theta_r} \end{bmatrix}
$$

$$(7.15)$$

$$
\mathbf{t} = \begin{bmatrix} e^{j\theta_r} \\ \alpha^2 e^{j\theta_r} \\ \alpha e^{j\theta_r} \end{bmatrix} + \begin{bmatrix} e^{-j\theta_r} \\ \alpha e^{-j\theta_r} \\ \alpha^2 e^{-j\theta_r} \end{bmatrix}
$$

The space-vector of the stator flux can be obtained by premultiplying the vector $\boldsymbol{\psi}$ by $2/3\cdot[1 \quad \alpha \quad \alpha^2] = 2/3\boldsymbol{\alpha}^T$, thus leading to

$$\boldsymbol{\psi}_s = \frac{2}{3}\boldsymbol{\alpha}^T\boldsymbol{\psi}_p = \frac{2}{3}\begin{bmatrix} 1 & \alpha & \alpha^2 \end{bmatrix}\begin{bmatrix} \psi_{sA} \\ \psi_{sB} \\ \psi_{sC} \end{bmatrix} = \frac{2}{3}\boldsymbol{\alpha}^T\mathbf{L}\mathbf{i}_p - \frac{2}{3}\boldsymbol{\alpha}^T\frac{L_{2s}}{2}\mathbf{T}\mathbf{i}_p + \frac{2}{3}\boldsymbol{\alpha}^T\frac{\psi_{PM}}{2}\mathbf{t} \tag{7.16}$$

Taking into account each term in Equation 7.16,

$$\frac{2}{3}\boldsymbol{\alpha}^T\mathbf{L}\mathbf{i}_p = \frac{2}{3}\begin{bmatrix}1\\\alpha\\\alpha^2\end{bmatrix}^T\begin{bmatrix}L_{ls}+L_{0s} & -\frac{1}{2}L_{0s} & -\frac{1}{2}L_{0s}\\ -\frac{1}{2}L_{0s} & L_{ls}+L_{0s} & -\frac{1}{2}L_{0s}\\ -\frac{1}{2}L_{0s} & -\frac{1}{2}L_{0s} & L_{ls}+L_{0s}\end{bmatrix}\begin{bmatrix}i_{sA}\\i_{sB}\\i_{sC}\end{bmatrix}$$

$$=\frac{2}{3}\begin{bmatrix}(L_{ls}+L_{0s})-\frac{\alpha^2}{2}L_{0s}-\frac{\alpha}{2}L_{0s}\\ \alpha(L_{ls}+L_{0s})-\frac{1}{2}L_{0s}-\frac{\alpha^2}{2}L_{0s}\\ \alpha^2(L_{ls}+L_{0s})-\frac{\alpha}{2}L_{0s}-\frac{1}{2}L_{0s}\end{bmatrix}^T\begin{bmatrix}i_{sA}\\i_{sB}\\i_{sC}\end{bmatrix} = (L_{ls}+L_{0s})\mathbf{i}_s - \frac{\alpha}{2}L_{0s}\mathbf{i}_s - \frac{\alpha^2}{2}L_{0s}\mathbf{i}_s$$

$$=\left(L_{ls}+\frac{3}{2}L_{0s}\right)\mathbf{i}_s$$

$$\frac{2}{3}\frac{L_{2s}}{2}\boldsymbol{\alpha}^T\mathbf{T}\mathbf{i}_p = \frac{2}{3}\frac{L_{2s}}{2}\begin{bmatrix}1\\\alpha\\\alpha^2\end{bmatrix}^T\left\{\begin{bmatrix}e^{j2\theta_r} & \alpha^2 e^{j2\theta_r} & \alpha e^{j2\theta_r}\\ \alpha^2 e^{j2\theta_r} & \alpha e^{j2\theta_r} & e^{j2\theta_r}\\ \alpha e^{j2\theta_r} & e^{j2\theta_r} & \alpha^2 e^{j2\theta_r}\end{bmatrix}+\begin{bmatrix}e^{-j2\theta_r} & \alpha e^{-j2\theta_r} & \alpha^2 e^{-j2\theta_r}\\ \alpha e^{-j2\theta_r} & \alpha^2 e^{-j2\theta_r} & e^{-j2\theta_r}\\ \alpha^2 e^{-j2\theta_r} & e^{-j2\theta_r} & \alpha e^{-j2\theta_r}\end{bmatrix}\right\}\begin{bmatrix}i_{sA}\\i_{sB}\\i_{sC}\end{bmatrix}$$

$$=\frac{2}{3}\frac{L_{2s}}{2}\left\{\begin{bmatrix}3e^{j2\theta_r}\\3\alpha^2 e^{j2\theta_r}\\3\alpha e^{j2\theta_r}\end{bmatrix}^T+\begin{bmatrix}e^{-j2\theta_r}+\alpha^2 e^{-j2\theta_r}+\alpha e^{-j2\theta_r}\\ \alpha e^{-j2\theta_r}+e^{-j2\theta_r}+\alpha^2 e^{-j2\theta_r}\\ \alpha^2 e^{-j2\theta_r}+\alpha e^{-j2\theta_r}+e^{-j2\theta_r}\end{bmatrix}^T\right\}\begin{bmatrix}i_{sA}\\i_{sB}\\i_{sC}\end{bmatrix}$$

$$=3\frac{L_{2s}}{2}\mathbf{i}_s^* e^{j2\theta} + \frac{L_{2s}}{2}\mathbf{i}_s\left[e^{-j2\theta_r}+\alpha^2 e^{-j2\theta_r}+\alpha e^{-j2\theta_r}\right] = \frac{3}{2}L_{2s}\mathbf{i}_s^* e^{j2\theta_r}$$

$$\frac{2}{3}\boldsymbol{\alpha}^T\frac{\psi_{PM}}{2}\mathbf{t} = \frac{2}{3}\frac{\psi_{PM}}{2}\begin{bmatrix}1\\\alpha\\\alpha^2\end{bmatrix}^T\left\{\begin{bmatrix}e^{j\theta_r}\\\alpha^2 e^{j\theta_r}\\\alpha e^{j\theta_r}\end{bmatrix}+\begin{bmatrix}e^{-j\theta_r}\\\alpha e^{-j\theta_r}\\\alpha^2 e^{-j\theta_r}\end{bmatrix}\right\}$$

$$=\frac{2}{3}\frac{\psi_{PM}}{2}\left\{3e^{j\theta_r}+\left[e^{-j\theta_r}+\alpha^2 e^{-j\theta_r}+\alpha e^{-j\theta_r}\right]\right\} = \psi_{PM}e^{j\theta_r}$$

Gathering the earlier terms, the space-vector of the stator flux is written as

$$\boldsymbol{\psi}_s = \left(L_{ls}+\frac{3}{2}L_{0s}\right)\mathbf{i}_s - \frac{3}{2}L_{2s}\mathbf{i}_s^* e^{j2\theta_r} + \psi_{PM}e^{j\theta_r} \tag{7.17}$$

Multiplying Equation 7.17 by $e^{-j\theta_r}$, the space-vector of the stator flux is expressed in the synchronous reference frame, rotating at ω_r, with direct axis lying on the direction of the flux vector of the PMs.

$$\boldsymbol{\psi}_s^{\psi_{PM}} = \boldsymbol{\psi}_s e^{-j\theta_r} = \left(L_{ls} + \frac{3}{2}L_{0s}\right)\mathbf{i}_s e^{-j\theta_r} - \frac{3}{2}L_{2s}\mathbf{i}_s^* e^{j\theta_r} + \psi_{PM}$$

$$= \left(L_{ls} + \frac{3}{2}L_{0s}\right)\mathbf{i}_s^{\psi_{PM}} - \frac{3}{2}L_{2s}\mathbf{i}_s^{*\psi_{PM}} + \psi_{PM} \qquad (7.18)$$

Recalling Equation 7.11, using the space-vectors, the stator voltage $\mathbf{u}_s$ can be obtained as

$$\mathbf{u}_s = R_s\mathbf{i}_s + \frac{d}{dt}\boldsymbol{\psi}_s \qquad (7.19)$$

Using the space-vectors in the synchronous reference frame, Equation 7.19 can be also written as

$$\mathbf{u}_s^{\psi_{PM}}e^{j\theta_r} = R_s\mathbf{i}_s^{\psi_{PM}}e^{j\theta_r} + \frac{d}{dt}\left(\boldsymbol{\psi}_s^{\psi_{PM}}e^{j\theta_r}\right) = R_s\mathbf{i}_s^{\psi_{PM}}e^{j\theta_r} + e^{j\theta_r}\frac{d}{dt}\boldsymbol{\psi}_s^{\psi_{PM}} + j\omega_r\boldsymbol{\psi}_s^{\psi_{PM}}e^{j\theta_r}$$

leading to

$$\mathbf{u}_s^{\psi_{PM}} = R_s\mathbf{i}_s^{\psi_{PM}} + \frac{d}{dt}\boldsymbol{\psi}_s^{\psi_{PM}} + j\omega_r\boldsymbol{\psi}_s^{\psi_{PM}} \qquad (7.20)$$

Decomposing each vector in its direct and quadrature-axis components, we obtain

$$\begin{cases} u_{sd}^{\psi_{PM}} = R_s i_{sd}^{\psi_{PM}} + \left(L_{ls} + \frac{3}{2}L_{0s}\right)\frac{d}{dt}i_{sd}^{\psi_{PM}} - \frac{3}{2}L_{2s}\frac{d}{dt}i_{sd}^{\psi_{PM}} - \omega_r\left(L_{ls} + \frac{3}{2}L_{0s}\right)i_{sq}^{\psi_{PM}} - \frac{3}{2}L_{2s}\omega_r i_{sq}^{\psi_{PM}} \\[3mm] u_{sq}^{\psi_{PM}} = R_s i_{sq}^{\psi_{PM}} + \left(L_{ls} + \frac{3}{2}L_{0s}\right)\frac{d}{dt}i_{sq}^{\psi_{PM}} + \frac{3}{2}L_{2s}\frac{d}{dt}i_{sq}^{\psi_{PM}} + \omega_r\left(L_{ls} + \frac{3}{2}L_{0s}\right)i_{sd}^{\psi_{PM}} - \frac{3}{2}L_{2s}\omega_r i_{sd}^{\psi_{PM}} + \omega_r\psi_{PM} \end{cases}$$
$$(7.21)$$

According to Equation 7.5, it is possible to obtain a more readable form for Equation 7.21, using direct and quadrature-axis inductances, L_d and L_q, defined as

$$\begin{cases} L_d = L_{sl} + \underbrace{\frac{3}{2}L_{0s} - \frac{3}{2}L_{2s}}_{L_{md}} \\[4mm] L_q = L_{sl} + \underbrace{\frac{3}{2}L_{0s} + \frac{3}{2}L_{2s}}_{L_{mq}} \end{cases}$$

obtaining

$$
\begin{cases}
u_{sd}^{\psi_{PM}} = R_s i_{sd}^{\psi_{PM}} + L_d \dfrac{d}{dt} i_{sd}^{\psi_{PM}} - \omega_r L_q i_{sq}^{\psi_{PM}} \\[2ex]
u_{sq}^{\psi_{PM}} = R_s i_{sq}^{\psi_{PM}} + L_q \dfrac{d}{dt} i_{sq}^{\psi_{PM}} + \omega_r L_d i_{sd}^{\psi_{PM}} + \omega_r \psi_{PM}
\end{cases}
\tag{7.22}
$$

The "instantaneous complex power" drawn by the PMSM can be computed considering that no neutral wire exists:

$$
\mathbf{s} = \mathbf{u_s i_s^*} = \frac{2}{3}\left(u_{sA} + \alpha u_{sB} + \alpha^2 u_{sC}\right) \cdot \frac{2}{3}\left(i_{sA} + \alpha^2 i_{sB} + \alpha i_{sC}\right)
$$

$$
= \frac{4}{9}\left[u_{sA}i_{sA} + u_{sB}i_{sB} + u_{sC}i_{sC} + \alpha\left(u_{sA}i_{sC} + u_{sB}i_{sA} + u_{sC}i_{sB}\right) + \alpha^2\left(u_{sA}i_{sB} + u_{sB}i_{sC} + u_{sC}i_{sA}\right)\right]
\tag{7.23}
$$

Then the "instantaneous real power" is given by Equation 1.21, considering that non–power invariant form is used:

$$
p(t) = \frac{3}{2}\operatorname{Re}(\mathbf{s}) = \frac{3}{2}\left(u_{sd}^{\psi_{PM}} i_{sd}^{\psi_{PM}} + u_{sq}^{\psi_{PM}} i_{sq}^{\psi_{PM}}\right)
\tag{7.24}
$$

Given Equation 7.22, Equation 7.24 can be further expanded:

$$
p(t) = \frac{3}{2}\left[i_{sd}^{\psi_{PM}}\left(R_s i_{sd}^{\psi_{PM}} + L_d \frac{d}{dt} i_{sd}^{\psi_{PM}} - \omega_r L_q i_{sq}^{\psi_{PM}}\right) + i_{sq}^{\psi_{PM}}\left(R_s i_{sq}^{\psi_{PM}} + L_q \frac{d}{dt} i_{sq}^{\psi_{PM}} + \omega_r L_d i_{sd}^{\psi_{PM}} + \omega_r \psi_{PM}\right)\right]
$$

$$
= \frac{3}{2}\left[\underbrace{R_s\left(i_{sd}^{\psi_{PM}\,2} + i_{sq}^{\psi_{PM}\,2}\right)}_{P_{je}} + \underbrace{\omega_r(L_d - L_q)i_{sd}^{\psi_{PM}} i_{sq}^{\psi_{PM}} + \omega_r i_{sq}^{\psi_{PM}} \psi_{PM}}_{P_{em}} + \underbrace{L_d i_{sd}^{\psi_{PM}} \frac{d}{dt} i_{sd}^{\psi_{PM}} + L_q i_{sq}^{\psi_{PM}} \frac{d}{dt} i_{sq}^{\psi_{PM}}}_{P_{mag}} \right]
\tag{7.25}
$$

where
  P_{je} represents stator losses, caused by the stator resistance
  P_{em} is the electromagnetic power
  P_{mag} is the magnetic power stored in the stator windings

The electromagnetic torque can be therefore computed by dividing the electromagnetic power P_{em} by the synchronous speed, leading to

$$
t_e = p\frac{P_{em}}{\omega_r} = \frac{3}{2}p\left[\underbrace{(L_d - L_q)i_{sd}^{\psi_{PM}} i_{sq}^{\psi_{PM}}}_{\text{reluctance}} + i_{sq}^{\psi_{PM}} \psi_{PM}\right]
\tag{7.26}
$$

where p is the pole pair number. The first term is the *reluctance torque*; according to Equations 7.1 and 7.2, in the PMSM, $L_d < L_q$; therefore, the reluctance torque is a resistive torque. For this reason, if no flux weakening is mandatory, the direct-axis current is controlled to zero, to nullify the effect of the reluctance torque. More details will be given in Section 7.3.

7.3 Control Strategies of PMSM Drives

Control of PMSM drives is relatively simpler than that of IM drives. To decouple torque and flux controls, a flux model is not strictly required, since the rotor flux is produced by the PM, and thus, it is perfectly known once the rotor position is known. In most cases, the PMSM can work with the maximum possible torque angle, since flux weakening is only required when the speed must exceed the base speed.

In this section, various control strategies will be examined, developed on the basis of the PMSM model described in Section 7.2.

7.3.1 Field-Oriented Control of PMSM Drives

The effectiveness of the field-oriented control (FOC) lies in the possibility of controlling a PMSM as a DC machine, with the same dynamic performance. As known, the DC motor permits a decoupled control of both flux and torque: maintaining constant the field current, the regulation of the torque component of the armature current permits the control of the torque produced by the machine. In the PMSM, the control can be exploited only through the stator windings; given the three-phase stator currents, i_{sA}, i_{sB}, and i_{sC}, the two-phase current components, $i_{sd}^{\psi PM}$ and $i_{sq}^{\psi PM}$, in the PMs reference frame can be obtained. Firstly, the three-phase current is transformed in the stationary reference frame by the $3 \rightarrow 2$ transformation given by Equation 1.6 (non–power invariant form, the two-axis Park transformation):

$$\mathbf{i}_s = \begin{bmatrix} i_{sD} \\ i_{sQ} \end{bmatrix} = \frac{2}{3} \begin{bmatrix} 1 & -\dfrac{1}{2} & -\dfrac{1}{2} \\ 0 & \dfrac{\sqrt{3}}{2} & -\dfrac{\sqrt{3}}{2} \end{bmatrix} \begin{bmatrix} i_{sA} \\ i_{sB} \\ i_{sC} \end{bmatrix} \tag{7.27}$$

Then, the stator current space-vector $\mathbf{i}_s$, expressed in the stationary reference frame, must be converted into the synchronous reference frame, whose direct axis lies in the direction of the PM flux vector (Figure 7.8):

$$\mathbf{i}_s^{\psi PM} = \mathbf{i}_s e^{-j\theta_r} \Rightarrow \begin{bmatrix} i_{sd}^{\psi PM} \\ i_{sq}^{\psi PM} \end{bmatrix} = \begin{bmatrix} \cos\theta_r & -\sin\theta_r \\ \sin\theta_r & \cos\theta_r \end{bmatrix} \begin{bmatrix} i_{sD} \\ i_{sQ} \end{bmatrix} \tag{7.28}$$

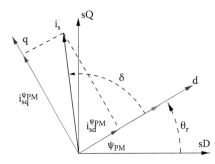

FIGURE 7.8
Vector scheme of the current space-vector in FOC.

By applying the space-vector theory to the PMSM, a DC-equivalent model can be obtained: for a constant value of the direct-axis current component $i_{sd}^{\psi_{PM}}$, the machine flux is constant, and the quadrature-axis current component $i_{sq}^{\psi_{PM}}$ controls the electromagnetic torque, as shown in Equation 7.26:

$$t_e = \frac{3}{2}p\left[(L_d - L_q)i_{sd}^{\psi_{PM}}i_{sq}^{\psi_{PM}} + i_{sq}^{\psi_{PM}}\psi_{PM}\right]$$

If the PMs are mounted on the rotor surface, the machine can be approximately considered isotropic, since the direct and quadrature-axis inductances have practically the same value and no reluctance torque is present: the direct-axis current component has no influence on the torque, which is perfectly controlled by the quadrature-axis current component. In these conditions, the torque equation simplifies to

$$t_e = \frac{3}{2}pi_{sq}^{\psi_{PM}}\psi_{PM} \qquad (7.29)$$

In interior-mounted PM motors, the two-axis reluctances L_d and L_q are different, and the reluctance torque is always present. The value of this torque depends on both $i_{sd}^{\psi_{PM}}$ and $i_{sq}^{\psi_{PM}}$, and its effect can be easily eliminated by imposing $i_{sd}^{\psi_{PM}} = 0$.

Nevertheless, some applications may require a value of $i_{sd}^{\psi_{PM}} < 0$; in fact, as shown in Figure 7.9, if both $i_{sd}^{\psi_{PM}}$ and $i_{sq}^{\psi_{PM}}$ are properly controlled, a higher torque can be achieved. In fact, given the torque angle δ, defined as

$$\delta = \arccos\left(\frac{i_{sd}^{\psi_{PM}}}{i_{sq}^{\psi_{PM}}}\right) \qquad (7.30)$$

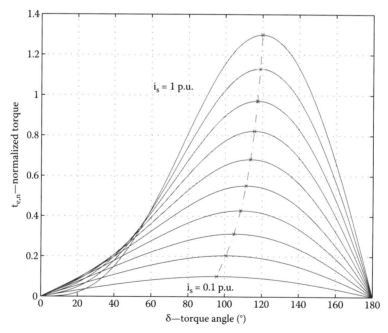

FIGURE 7.9
Normalized torque vs. torque angle for various current values.

when its value is greater than 90°, the reluctance torque becomes positive, since $i_{sd}^{\psi PM} < 0$. The resulting effect on the total torque is an increase, despite the decrease in the $i_{sq}^{\psi PM}$ current component.

It should be also remarked that the value of the torque angle in which the maximum torque is attained varies with the value of stator currents [11]. In Figure 7.9, various torque characteristics for $\left|i_s^{\psi PM}\right|$ varying from 0.1 to 1.0 p.u. are shown.

As for the flux control, it follows from Equation 7.18 that both current components influence the stator flux linkage. In fact, it is possible to infer from Equation 7.18:

$$\psi_s^{\psi PM} = \left(L_{ls} + \frac{3}{2}L_{0s}\right)\mathbf{i}_s^{\psi PM} - \frac{3}{2}L_{2s}\mathbf{i}_s^{*\psi PM} + \psi_{PM} = \left(L_d i_{sd}^{\psi PM} + \psi_{PM}\right) + j\left(L_q i_{sq}^{\psi PM}\right) \qquad (7.31)$$

Thus, the flux control must act on both components, being the stator flux-linkage module equal to

$$\left|\psi_s^{\psi PM}\right| = \sqrt{\left(L_d i_{sd}^{\psi PM} + \psi_{PM}\right)^2 + \left(L_q i_{sq}^{\psi PM}\right)^2} \qquad (7.32)$$

while its phase ϕ is equal to

$$\phi = \tan^{-1}\left[\frac{\left(L_q i_{sq}^{\psi PM}\right)}{\left(L_d i_{sd}^{\psi PM} + \psi_{PM}\right)}\right] \qquad (7.33)$$

7.3.2 Torque-Controlled Drives

Torque-controlled drives are commanded with two input signals: the torque reference t_{eref} and the flux reference $\left|\psi_{sref}^{\psi PM}\right|$; these two signals can be provided by a speed controller or directly imposed as reference signals.

Various control schemes can be used, depending on the PMSM anisotropy, the control variables chosen, and the inverter typology used, that is either the current source inverter (CSI) or the voltage source inverter (VSI). In the following, some solutions will be examined, both for the surface-mounted and interior-mounted PMSM.

7.3.2.1 Surface-Mounted PMSM

Torque control of surface-mounted PMSMs is simple: as a result of the absence of the reluctance torque, the torque depends only on $i_{sq}^{\psi PM}$, while the stator flux linkage depends on both components of the stator current space-vector.

7.3.2.1.1 FOC with Impressed Currents

The electromagnetic torque value and the stator flux-linkage vector amplitude may be written, from Equations 7.29 and 7.32, as

$$t_e = \frac{3}{2}p i_{sq}^{\psi PM}\psi_{PM} \qquad (7.34a)$$

$$\left|\psi_s^{\psi PM}\right| = \sqrt{\left(L_{dq} i_{sd}^{\psi PM} + \psi_{PM}\right)^2 + \left(L_{dq} i_{sq}^{\psi PM}\right)^2} \qquad (7.34b)$$

Given the torque reference, the reference value of the quadrature-axis component of the stator current vector, in the PM reference frame, $i_{sqref}^{\psi_{PM}}$, can be obtained, from Equation 7.34a, as

$$i_{sqref}^{\psi_{PM}} = \frac{2}{3} \frac{t_{eref}}{p\psi_{PM}} \tag{7.35}$$

From Equation 7.35 and the flux reference, the reference value of the direct-axis component of the stator current vector, $i_{sd,ref}$ is computed:

$$i_{sdref}^{\psi_{PM}} = \frac{\sqrt{\left(\psi_{s,ref}^{\psi_{PM}}\right)^2 - \left(L_{dq}(2/3)(t_{e,ref}/p\psi_{PM})\right)^2} - \psi_{PM}}{L_{dq}} \tag{7.36}$$

The torque angle δ and the module of the stator current space-vector $\left|\mathbf{i}_s^{\Psi_{PM}}\right|$ can be easily computed, given its components in the PM reference frame:

$$\left|\mathbf{i}_s^{\psi_{PM}}\right| = \frac{\sqrt{\left(\psi_{s,ref}^{\psi_{PM}}\right)^2 - \psi_{PM}^2 - 2\psi_{PM}L_{dq}i_{sdref}^{\psi_{PM}}}}{L_{dq}}$$

$$\cos\delta = \frac{i_{sdref}^{\psi_{PM}}}{\left|\mathbf{i}_s^{\psi_{PM}}\right|}; \quad \sin\delta = \frac{i_{sqref}^{\psi_{PM}}}{\left|\mathbf{i}_s^{\psi_{PM}}\right|} \tag{7.37}$$

In Figure 7.10, a scheme of the torque control is shown. The polar coordinates of the stator current space-vector with a current-controlled VSI are derived by using the earlier equation [11]. The torque and the flux reference are provided as input signals and then used to compute the reference values of the amplitude and of the angle (torque angle) of the stator current space-vector, expressed in the PM flux reference frame. Adding the torque angle δ_{ref} to the instantaneous rotor position, multiplied by the pole pairs number p, the angle of the stator current space-vector in the stationary reference frame is obtained:

$$\theta_{sref} = \delta_{ref} + \theta_r \tag{7.38}$$

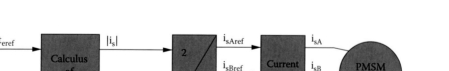

FIGURE 7.10
Current torque control in polar coordinates.

Finally, the reference values of the three-phase currents can be easily obtained, given the angle θ_{sref}:

$$\begin{bmatrix} i_{sAref} \\ i_{sBref} \\ i_{sCref} \end{bmatrix} = \left| \mathbf{i}_s^{\psi PM} \right| \begin{bmatrix} \sin(\theta_{sref}) \\ \sin\left(\theta_{sref} - \dfrac{2\pi}{3}\right) \\ \sin\left(\theta_{sref} - \dfrac{2\pi}{3}\right) \end{bmatrix} \tag{7.39}$$

Equation 7.39 is computed from Equation 1.8 after suitable manipulation. It should be remarked that in Equation 7.39, the amplitude of the stator current space-vector is that computed in the PM reference frame, while it should have been expressed in the stationary reference frame; since the Park transformation is amplitude invariant, both values are equal. These reference currents are provided to the inverter, usually current controlled through hysteresis controllers, as explained in Chapters 2 and 5.

The computation of the module and of the angle of the stator current space-vector can be avoided. In fact, given the components of the space-vector $\mathbf{i}_{sref}^{\psi PM}$ in the PM reference frame, the current references in the stationary reference frame can be obtained through the inverse Park transformation:

$$\mathbf{i}_{sref} = e^{j\theta_r} \mathbf{i}_{sref}^{\psi PM} \Rightarrow \begin{bmatrix} i_{sDref} \\ i_{sQref} \end{bmatrix} = \begin{bmatrix} \cos\theta_r & \sin\theta_r \\ -\sin\theta_r & \cos\theta_r \end{bmatrix} \begin{bmatrix} i_{sdref}^{\psi PM} \\ i_{sqref}^{\psi PM} \end{bmatrix} \tag{7.40}$$

Finally, the two-phase components of $\mathbf{i}_{sref}$ can be transformed in the three-phase references by means of the $2 \rightarrow 3$ transformation, Equation 1.7:

$$\begin{bmatrix} i_{sAref} \\ i_{sBref} \\ i_{sCref} \end{bmatrix} = \begin{bmatrix} 1 & 0 \\ -\dfrac{1}{2} & \dfrac{\sqrt{3}}{2} \\ -\dfrac{1}{2} & -\dfrac{\sqrt{3}}{2} \end{bmatrix} \begin{bmatrix} i_{sDref} \\ i_{sQref} \end{bmatrix} \tag{7.41}$$

These three-phase current references can be fed to a current-controlled VSI to obtain the desired torque and flux in the PMSM, see Figure 7.11.

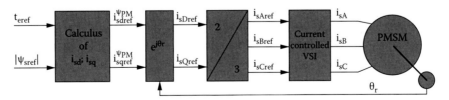

FIGURE 7.11
Current torque control in rectangular coordinates.

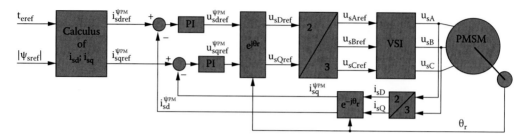

FIGURE 7.12
Voltage torque control in rectangular coordinates.

7.3.2.1.2 FOC with Impressed Voltage

Using a PWM-VSI, current control can be performed in the PM reference frame, where currents components are DC quantities. Thanks to that, simple PI controllers can be used (see Figure 7.12) for the two components of the stator current space-vector to obtain the voltage references $u_{sdref}^{\psi PM}$ and $u_{sqref}^{\psi PM}$. These references are then transformed, using the rotation matrix shown in Equation 7.40:

$$\mathbf{u}_{sref} = e^{j\theta_r} \mathbf{u}_{sref}^{\psi PM} \Rightarrow \begin{bmatrix} u_{sDref} \\ u_{sQref} \end{bmatrix} = \begin{bmatrix} \cos\theta_r & \sin\theta_r \\ -\sin\theta_r & \cos\theta_r \end{bmatrix} \begin{bmatrix} u_{sdref} \\ u_{sqref} \end{bmatrix} \tag{7.42}$$

And then, the reference voltages u_{sAref}, u_{sBref} and u_{sCref} are obtained by Equation 1.7 and fed to the VSI which will synthetize them by the adopted PWM technique:

$$\begin{bmatrix} u_{sAref} \\ u_{sBref} \\ u_{sCref} \end{bmatrix} = \begin{bmatrix} 1 & 0 \\ -\dfrac{1}{2} & \dfrac{\sqrt{3}}{2} \\ -\dfrac{1}{2} & -\dfrac{\sqrt{3}}{2} \end{bmatrix} \begin{bmatrix} u_{sDref} \\ u_{sQref} \end{bmatrix} \tag{7.43}$$

For more details on the voltage control of the inverter, the reader should refer to Chapters 2 and 5.

7.3.2.2 Interior-Mounted PMSM

If a magnetic saliency is present, then resolution of Equations 7.26 and 7.31 is involved, so a real-time computation of $i_{sdref}^{\psi PM} = f\left(t_{eref}, \psi_s^{\psi PM}\right)$ and $i_{sqref}^{\psi PM} = g\left(t_{eref}, \psi_s^{\psi PM}\right)$ is more difficult. To adopt the same approach as in previous paragraph, iterative methods should be used to obtain an off-line resolution and then store the results in a look-up table to be used in real-time control. To take into account parameter variation, further look-up tables should be derived and implemented.

The simplest torque control strategy for the interior-mounted PMSM is based on forcing the torque angle δ to be equal to $\pi/2$, that is, setting the $i_{sdref}^{\psi PM}$ equal to zero. In this way, the stator flux linkage is not controlled, and the torque relies only on $i_{sqref}^{\psi PM}$. Torque control can therefore be exploited similarly to that of surface-mounted PMSM. Obviously, this control

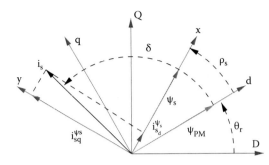

FIGURE 7.13
Vector scheme for interior-mounted PMSM.

method does not permit the drive to obtain an optimal dynamic performance, since the torque does not attain its maximum value; moreover, the maximum speed is limited by the base speed of the machine, since no flux weakening is possible.

Alternatively, FOC for interior-mounted PMSM can be achieved by properly controlling both components of the stator current space-vector, expressed either in the stator flux-linkage synchronous reference frame or in the PM synchronous reference frame [2]. Both these reference frames rotate at synchronous speed, but their respective direct axes create the load angle ρ_s, as shown in Figure 7.13, where the vector diagram of the interior-mounted PMSM is drawn. Vector control in the stator flux-linkage reference frame has the advantage that flux and torque controls are easily decoupled but requires that the instantaneous phase angle ρ_s between the PM flux and the stator flux linkage has to be known. The vector control in the PM flux reference frame requires only the knowledge of PM flux vector position, or the rotor position, but torque and flux control are cross coupled.

Given Equation 7.31, it is possible to rewrite Equation 7.27 in the following form:

$$t_e = \frac{3}{2}p\left[\psi_{sd}^{\psi_{PM}}i_{sq}^{\psi_{PM}} - \psi_{sq}^{\psi_{PM}}i_{sd}^{\psi_{PM}}\right] = \frac{3}{2}p\left|\psi_s^{\psi_s}\right|i_{sq}^{\psi_s} \tag{7.44}$$

where $i_{sq}^{\psi_s}$ is the quadrature-axis component of the stator current space-vector in the stator flux-linkage reference frame. Equation 7.44 and Figure 7.13 show that the electromagnetic torque can be controlled through the $i_{sq}^{\psi_s}$ component, while the stator flux linkage is affected by the $i_{sd}^{\psi_s}$ component.

Figure 7.14 shows the vector control scheme in the PM reference frame. In this reference frame, torque and stator flux linkage are cross coupled, so independent control of these two quantities is not possible. Given the torque reference, both $i_{sdref}^{\psi_{PM}}$ and $i_{sqref}^{\psi_{PM}}$ are computed, according to a predetermined control strategy, usually the maximum torque per ampere (MTPA) control strategy, whereby the dynamic performance of the PMSM drive is optimized. It should be remarked that correspondingly the efficiency of the drive

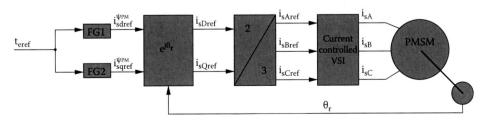

FIGURE 7.14
Current torque control in stator linkage reference frame.

is not optimized. Indeed, although the stator losses are minimized, the core losses are not optimized and affect the overall efficiency of the drive [11].

If i_{sb} is the base value of the stator current, given by

$$i_{sb} = \frac{\psi_{PM}}{L_q - L_d} \tag{7.45}$$

the base value of the electromagnetic torque is equal to

$$t_{eb} = \frac{3}{2} p \psi_{PM} i_{sb} = \frac{3}{2} p \frac{\psi_{PM}^2}{L_q - L_d} \tag{7.46}$$

and this is used to compute the per unit value of the torque:

$$t_{en} = \frac{t_e}{t_{eb}} = \frac{(3/2)p\left[(L_d - L_q)i_{sd}^{\psi_{PM}} i_{sq}^{\psi_{PM}} + i_{sq}^{\psi_{PM}} \psi_{PM}\right]}{(3/2)p(\psi_{PM}^2/L_q - L_d)} = -\frac{(L_q - L_d)^2 i_{sd}^{\psi_{PM}} i_{sq}^{\psi_{PM}}}{\psi_{PM}^2} + \frac{(L_d - L_q)i_{sq}^{\psi_{PM}}}{\psi_{PM}}$$

$$= -i_{sdn}i_{sqn} + i_{sqn} = i_{sqn}(1 - i_{sdn}) \tag{7.47}$$

For a given value of the normalized torque, Equation 7.47 is in the form

$$k = i_{sqn}(1 - i_{sdn}) \Rightarrow i_{sqn} = \frac{k}{(1 - i_{sdn})} \tag{7.48}$$

For each value of k, Equation 7.48 is a hyperbola, and the locus of their points closest to the origin is the locus that satisfies the MTPA control strategy, as shown in Figure 7.15.

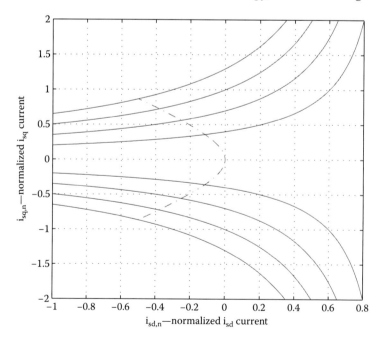

FIGURE 7.15
MTPA locus.

The MTPA criteria ensure minimal ohmic losses, since the same torque value is achieved with the minimum possible value of current. If the PMSM is considered to be surface mounted, then this criteria lead to $i_{sdref}^{\psi PM} = 0$, in order to use the whole current for torque production. According to Equations 7.33 and 7.34, the stator flux-linkage amplitude reference can be computed as

$$|\mathbf{\psi}_{s,ref}| = \sqrt{(\psi_{PM})^2 + \left(L_{dq}\frac{2}{3}\frac{t_{e,ref}}{\psi_{PM}}\right)^2}$$

As already explained in Section 7.3.2, for an interior-mounted PMSM, the torque equation contains also a reluctance term, so forcing $i_{sd}^{\psi PM} = 0$ does not permit obtaining the maximum torque per unit current.

7.3.2.3 Feed-Forward Control

Recalling Equation 7.22 rewritten here for sake of simplicity, it should be remarked that some coupling terms appear in both voltage equations:

$$u_{sd}^{\psi PM} = R_s i_{sd}^{\psi PM} + L_d \frac{d}{dt} i_{sd}^{\psi PM} \underbrace{-\omega_r L_q i_{sq}^{\psi PM}}_{\text{coupling term}}$$

$$u_{sq}^{\psi PM} = R_s i_{sq}^{\psi PM} + L_q \frac{d}{dt} i_{sq}^{\psi PM} \underbrace{+\omega_r L_d i_{sd}^{\psi PM} + \omega_r \psi_{PM}}_{\text{coupling term}}$$

(7.49)

In a current-controlled inverter scheme, these coupling terms are automatically compensated by the control. If a VSI with impressed voltages is used, then a feed-forward compensation term should be used, to decouple flux and torque equivalent circuit. This problem also arises in the vector control of IM drives, as shown in Section 5.3.3.4.

Neglecting the delay introduced by the signal processing, a feed-forward action can be used to effectively decouple the torque and the flux control. This is obtained by adding to the reference voltages $u_{sdref}^{\psi PM}$ and $u_{sqref}^{\psi PM}$ generated by the voltage PI controllers the following terms:

$$\begin{cases} u_{sdref}^{\psi PM}{}^* = u_{sdref}^{\psi PM} + \omega_r L_q i_{sq}^{\psi PM} \\ u_{sqref}^{\psi PM}{}^* = u_{sqref}^{\psi PM} - \omega_r L_d i_{sd}^{\psi PM} - \omega_r \psi_{PM} \end{cases}$$

The decoupling terms can be obtained through the measurement of the stator current and the rotating speed. Not taking into account the feed-forward action may result in worse dynamic performance of the drive and even in unstable behavior of the drive.

7.3.3 Speed-Controlled Drives

Speed vector control for the PMSM can be achieved by adding a speed control loop to the torque control loop (Figure 7.16).

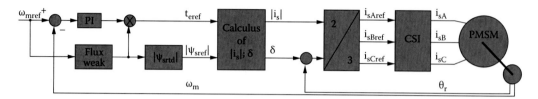

FIGURE 7.16
Current speed control in polar coordinates.

The speed reference ω_{mref} is provided as input signal; the difference between this reference signal and the real speed ω_m of the drive is the speed error $e_\omega = \omega_{mref} - \omega_m$. This error signal is the input of the speed PI controller. The output of the PI speed controller is the torque reference t_{eref}.

The speed reference is also used to compute the flux reference for the PMSM; in fact, if the speed reference is lower than the *base speed*, that is, the speed in which the back emf is equal to the external applied voltage, then the flux reference is equal to the rated flux $\psi_{srat}^{\psi PM}$ of the machine:

$$\omega_{mref} \leq \omega_b \Rightarrow \psi_{sref}^{\psi PM} = \psi_{srat}^{\psi PM} \tag{7.50}$$

On the contrary, if the speed reference is higher than the base speed, then *flux weakening* is required. Imposing a flux reference lower than the rated flux corresponds to imposing a flux-producing current reference $i_{sdref} < 0$, thus producing a flux opposite to that produced by the PMs.

$$\omega_{mref} > \omega_b \Rightarrow \psi_{sref}^{\psi PM} = \frac{\omega_b}{\omega_m} \psi_{srat}^{\psi PM} \tag{7.51}$$

In this way, the back emf is reduced, leading to a potential risk, since the machine can absorb a current higher than the rated value, resulting in a dangerous overheating. For this reason, the torque-producing current reference $i_{sqref}^{\psi PM}$ must be reduced with the same ratio:

$$\omega_{mref} > \omega_b \Rightarrow t_{eref} = \frac{\omega_b}{\omega_m} t_{erat} \tag{7.52}$$

The electromagnetic power is therefore constant and equal to the rated value in the flux weakening region; in fact,

$$P_e = \omega_r t_e = \begin{cases} \omega_r t_{eref} \Leftrightarrow \omega_r \leq p\omega_b \\ \omega_r \dfrac{\omega_b}{p\omega_r} t_{erat} = \dfrac{\omega_b}{p} t_{erat} \Leftrightarrow \omega_r > p\omega_b \end{cases} \tag{7.53}$$

7.3.3.1 Experimental Results

The FOC scheme with impressed voltages has been experimentally verified on a fractional horsepower PMSM, whose test setup is described in Appendix 7.A. The adopted

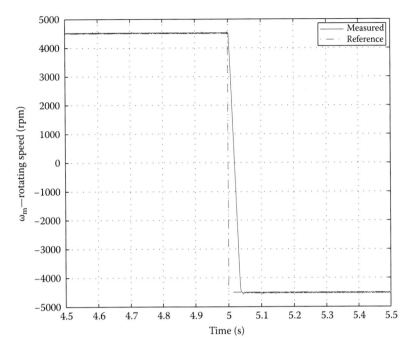

FIGURE 7.17
Speed vector control of a PMSM drive—speed.

control scheme is sketched in Figure 7.12, to which the speed control loop has been added. Figures 7.17 and 7.18 show the speed and current waveforms, respectively, that have been measured from a PMSM drive with vector control in speed control mode (see Appendix).

The drive has been given a speed reversal command at a speed value equal to half the rated speed. As shown in Table 7.2, the PMSM is an interior-mounted PMSM, since $L_d < L_q$, although it has a low saliency ratio. The control has been performed with impressed voltages. The $i_{sdref}^{\psi PM}$ has been set equal to zero, so the torque angle δ has been set equal to 90°. As explained earlier, in this condition, the torque depends only on $i_{sq}^{\psi PM}$, so Figure 7.18 shows also the waveform of the electromagnetic torque. The test has been performed at no load; as a consequence, the $i_{sq}^{\psi PM}$ current component has a very low value at steady-state because of the natural friction load, while its value reaches its maximum rated value during transient to maximize the torque and consequently the dynamic performance.

7.3.4 Direct Torque Control

The direct torque control (DTC) strategy is based on the selection of a proper stator voltage space-vector to directly control the stator flux space-vector and, therefore, the produced electromagnetic torque. DTC is also usually implemented in the control of IM, as shown in Chapter 5. In this section, DTC is devised to PMSM drives.

As seen in Section 7.3.2, FOC can be performed with both impressed currents and voltages, depending on the final control variables: the first case, the three-phase current

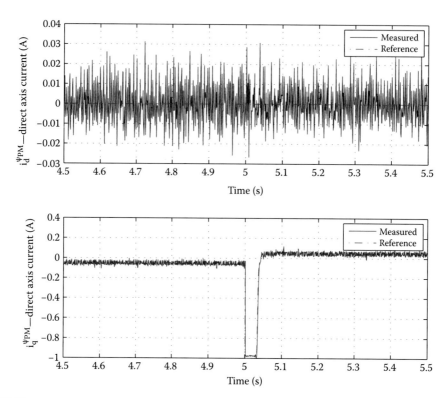

FIGURE 7.18
Speed vector control of a PMSM drive—current components.

TABLE 7.2

Parameters of the Test PMSM Drive

Parameter	Value
Rated power	15.7 W
Rated voltage	36 V
Maximum continuous torque (up to 5000 rpm)	30 mN m
Maximum continuous current	913 mA
Rated speed	6000 rpm
Pole pair	1
Stator resistance	4.3 Ω
Direct-axis inductance	0.359 mH
Quadrature-axis inductance	0.433 mH
Moment of inertia	10×10^{-7} kg · m^2
PM flux	24.35 mWb

references while, in the second, the three-phase voltage references. DTC permits a high dynamic performance to be obtained, without using either hysteresis current comparators or pulsewidth modulators (PWM); indeed, by applying a proper switching patter to the inverter, the stator flux linkage, and consequently the electromagnetic torque, can be controlled. Obviously, the DTC is based on the knowledge of the space-vector of the stator flux linkage, in terms of both its amplitude and phase, and of the electromagnetic torque produced.

7.3.4.1 Electromagnetic Torque Production in the PMSM

Similarly to Equation 5.59, taking into account Equations 7.26 and 7.30, the electromagnetic torque produced by the PMSM can be expressed as

$$t_e = \frac{3}{2}p\left[(L_d - L_q)i_{sd}^{\psi_{PM}}i_{sq}^{\psi_{PM}} + i_{sq}^{\psi_{PM}}\psi_{PM}\right] = \frac{3}{2}p\left[\psi_{sd}^{\psi_{PM}}i_{sq}^{\psi_{PM}} - \psi_{sq}^{\psi_{PM}}i_{sd}^{\psi_{PM}}\right]$$

$$= \frac{3}{2}p\left[\psi_{sd}^{\psi_{PM}}\psi_{sq}^{\psi_{PM}}\left(\frac{L_d - L_q}{L_d L_q}\right) + \frac{\psi_{sq}^{\psi_{PM}}\psi_{PM}}{L_d}\right] \tag{7.54}$$

Neglecting the saliency term, Equation 7.43 can be rewritten as

$$t_e = \frac{3}{2}p\frac{1}{L_{dq}}\left(\boldsymbol{\psi}_s^{\psi_{PM}} \wedge \boldsymbol{\psi}_{PM}^{\psi_{PM}}\right) \tag{7.55}$$

According to Equation 7.55, the electromagnetic torque can be controlled by acting on the amplitude and phase position of the stator flux linkage [12]; Figure 7.13 shows the vector diagram of the PMSM. By keeping the amplitude of stator flux-linkage constant and by quickly varying its angular position, a torque variation is produced. It should be remarked that, unlike the IM, there is not a first-order delay dependence between the stator flux linkage and the PM flux. As a consequence, if the electromagnetic torque should be reduced, the angle between stator flux-linkage vector and PM flux vector must be actively reduced, since it will tend to remain constant. Apart from this difference, the control strategy remains the same as in IM drives. If the stator flux-linkage space-vector is accelerated in the positive direction, keeping its amplitude constant, a fast torque increase occurs. On the contrary, if the stator flux-linkage space-vector is accelerated in the negative direction, the electromagnetic torque is reduced, or even a torque of opposite sign, which is able to brake the machine, can be generated if the angle between the two flux vectors changes its sign.

As for the relationship between the stator flux-linkage space-vector and the inverter configurations, no differences exist with respect to the IM counterpart (see Section 5.4.2).

7.3.4.2 Criteria for the Selection of the Voltage Space-Vectors and Control Strategies

It should be highlighted that, also for the PMSM, the DTC does not permit an intrinsic decoupling, since the application of a nonnull inverter configuration modifies both the amplitude and the angular position of the stator flux-linkage space-vector. The decoupling is obtained only if, for each sampling time interval, the correct inverter configuration is chosen to minimize the error between the references and the actual values of the control variables.

The space-vector plane is divided into six regions, each having an angular amplitude equal to $\pi/3$, centered on the nonnull inverter voltage space-vectors. If the stator flux-linkage vector is inside the sector k, at the nth instant of time, the possible voltage vectors produce the following effects [13]:

- The voltage space-vector $\mathbf{u}_k$ increases the amplitude of the stator flux linkage and keeps almost unchanged the angular position (radial positive configuration).
- The voltage space-vector $\mathbf{u}_{k+1}$ increases the amplitude of the stator flux linkage and rotates it in the counterclockwise direction (direct positive configuration).
- The voltage space-vector $\mathbf{u}_{k+2}$ reduces the amplitude of the stator flux linkage and rotates it in the counterclockwise direction (direct negative configuration).

- The voltage space-vector $\mathbf{u}_{k+3}$ reduces the amplitude of the stator flux linkage and maintains almost unchanged the angular position (radial negative configuration).
- The voltage space-vector $\mathbf{u}_{k-1}$ increases the amplitude of the stator flux linkage and rotates it in the clockwise direction (inverse positive configuration).
- The voltage space-vector $\mathbf{u}_{k-2}$ reduces the amplitude of the stator flux linkage and rotates it in the clockwise direction (inverse negative configuration).

The voltage space-vectors $\mathbf{u}_0$ and $\mathbf{u}_7$ maintain almost unchanged amplitude and the angular position of the stator flux linkage; therefore, unlike the IM, in the PMSM, they do not affect the electromagnetic torque. For this reason, they are not usually employed in the DTC of the PMSM. The following optimal inverter voltage space-vector chosen strategies can be derived:

- *Both electromagnetic torque and stator flux linkage must be increased*: a voltage space-vector with positive tangential and radial component must be chosen, usually $\mathbf{u}_{k+1}$.
- *Stator flux-linkage vector amplitude must be increased, but electromagnetic torque must be reduced*: radial component must be positive, tangential component must be negative, so $\mathbf{u}_{k-1}$ is used.
- *Stator flux-linkage vector amplitude must be reduced, but electromagnetic torque must be increased*: radial component must be negative, while tangential component must be positive, so $\mathbf{u}_{k+2}$ is used.
- *Both electromagnetic torque and stator flux linkage must be reduced*: a voltage space-vector with negative tangential and radial component is used, $\mathbf{u}_{k-2}$.

It should be remarked that the control strategy adopted for the DTC in the PMSM drives is the D strategy of the DTC in the IM drives; therefore, the same considerations made in Section 5.4.3 are also valid here: the torque response is very quick, since only direct and inverse configurations are used to affect it, but as a consequence, the switching frequency may reach high values.

With DTC, it is also possible to control only the i_{sq} component of the stator current space-vector if the stator flux linkage is at the rated value, thus, forcing $i_{sd}^{\psi_{PM}} = 0$ [2]. If the flux current $i_{sd}^{\psi_{PM}}$ is equal to zero, then the stator flux linkage is given by the sum of the PM flux and the flux on the quadrature axis of rotor reference frame:

$$\boldsymbol{\psi}_{s,1}^{\psi_{PM}} = \psi_{PM} + jL_q i_{sq}^{\psi_{PM}} \tag{7.56}$$

If only $i_{sq}^{\psi_{PM}}$ is varied by $\Delta i_{sq}^{\psi_{PM}}$, then $\boldsymbol{\psi}_s^{\psi_{PM}}$ has a variation only on the quadrature axis of the rotor reference frame, as the following:

$$\boldsymbol{\psi}_{s,2}^{\psi_{PM}} = \psi_{PM} + jL_q \left(i_{sq}^{\psi_{PM}} + \Delta i_{sq}^{\psi_{PM}} \right) \tag{7.57}$$

Then, in each sampling time interval, the torque control is performed by selecting the inverter configuration that makes the stator voltage space-vector to be directed along the quadrature axis of the rotor reference frame; the sense of the voltage vector must be positive if the torque must be increased, negative otherwise. Obviously, in this case, the rotor position has to be measured or estimated, but the estimation of the amplitude of stator flux linkage can be avoided.

7.3.4.3 Estimation of the Stator Flux and the Electromagnetic Torque

The estimation of the control variables in DTC of PMSM drives, stator flux linkage and electromagnetic torque, shares some similarities with the estimation of the same variables in IM drives (see Section 5.4.4).

From Equation 7.19, it follows that [2]

$$\mathbf{u}_s = R_s \mathbf{i}_s + \frac{d}{dt}\boldsymbol{\psi}_s \Rightarrow \boldsymbol{\psi}_s = \int (\mathbf{u}_s - R_s \mathbf{i}_s)dt \tag{7.58}$$

Since Equation 7.47 is expressed in the stator reference frame, the components of $\boldsymbol{\psi}_s$ are

$$\begin{cases} \psi_{sD} = \int (u_{sD} - R_s i_{sD})\,dt \\ \psi_{sQ} = \int (u_{sQ} - R_s i_{sQ})\,dt \end{cases} \tag{7.59}$$

As it is known, only one parameter is required in Equation 7.59, the stator resistance R_s. However, this parameter is not perfectly known and varies with the temperature. At high speed, the ohmic voltage drop on the stator resistance is negligible, with respect to the stator voltage, and so does the parameter variation. On the contrary, at low speed, the ohmic drop and the stator voltage are comparable, and therefore, an error on the value of R_s heavily affects the flux estimation (see the sensitivity analysis in Chapter 9).

An open-loop integration method is not therefore the best way to integrate Equation 7.59. This is one of the reasons why a closed-loop scheme should be used, which is notoriously less affected by parameter variation. Moreover, apart from this problem, open-loop integration has some problems (DC drift, initial value) so closed-loop integration must be adopted to cope with these problems. A first-order low-pass filter (LPF) with a low cutoff frequency can be used, whose expression in the Laplace domain s is

$$G(s) = \frac{T}{1+sT} \tag{7.60}$$

where T is a time constant, chosen large enough to approximate an ideal integrator (see Figure 7.19). To further improve the flux estimation, a thermal model of the stator resistance should be used. Inaccurate flux estimation at frequencies below $1/T$ can be avoided using

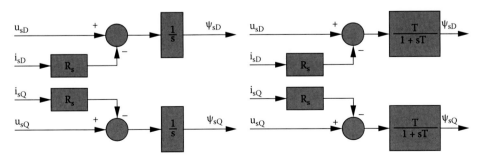

FIGURE 7.19
Open-loop and closed-loop integration schemes.

the reference value of the stator flux linkage instead of the estimated value; the following Equation 7.61 implements this strategy with a smooth transition below the cutoff frequency:

$$\boldsymbol{\psi}_s = \frac{T(\mathbf{u}_s - R_s\mathbf{i}_s) + \boldsymbol{\psi}_{sref}}{1 + sT} \tag{7.61}$$

In PMSM drives, the reference value of the stator flux linkage can be assumed equal to

$$\boldsymbol{\psi}_{sref} = \psi_{PM}e^{j\theta_r} \tag{7.62}$$

For the adoption of the flux estimator represented by Equation 7.61, rotor position is necessary; this solution can provide good results if a surface-mounted PMSM is used, but at very low rotating speed, other solutions are needed.

Another method for estimating the stator flux linkage can be obtained in the stator linkage reference frame. Equation 7.19 gives the vector voltage equation in the stationary reference frame that yields

$$\mathbf{u}_s = R_s\mathbf{i}_s + \frac{d}{dt}\boldsymbol{\psi}_s \Rightarrow \mathbf{u}_s^{\psi_s}e^{-j\omega_{ms}t} = R_s\mathbf{i}_s^{\psi_s}e^{-j\omega_{ms}t} + \frac{d}{dt}\left(\boldsymbol{\psi}_s^{\psi_s}e^{-j\omega_{ms}t}\right)$$

$$\Rightarrow \mathbf{u}_s^{\psi_s} = R_s\mathbf{i}_s^{\psi_s} + \frac{d}{dt}\boldsymbol{\psi}_s^{\psi_s} + j\omega_{ms}\boldsymbol{\psi}_s^{\psi_s} \tag{7.63}$$

Taking the real part of Equation 7.63, the time derivative of the stator flux-linkage vector amplitude can be computed, considering that $\left|\boldsymbol{\psi}_s^{\psi_s}\right| = \psi_{sx}^{\psi_s}$:

$$\frac{d}{dt}\left|\boldsymbol{\psi}_s^{\psi_s}\right| = u_{sd}^{\psi_s} - R_s i_{sd}^{\psi_s} \tag{7.64}$$

while its rotating speed, ω_{ms}, is obtained from the imaginary part of Equation 7.63:

$$\omega_{ms} = \frac{u_{sq}^{\psi_s} - R_s i_{sq}^{\psi_s}}{\left|\boldsymbol{\psi}_s^{\psi_s}\right|} \tag{7.65}$$

As shown in Figure 7.20, the angle ρ_s, needed for the coordinate transformation in the stator linkage reference frame, is obtained through the integration of the rotating speed ω_{ms}. Only one coordinate transformation is used in this scheme.

All the flux estimators presented above need the knowledge of the stator voltages, which can be retrieved by using the DC link voltage (see Chapter 5 for further details). DC link voltage U_d measurement requires cheap sensors, since it is normally slow varying. Three-phase voltages can be then computed through Equations 5.70a through 5.70c, written in the following for convenience:

$$u_{sA} = \frac{U_d}{3}(2S_a - S_b - S_c)$$

$$u_{sB} = \frac{U_d}{3}(2S_b - S_a - S_c) \tag{5.70a,b,c}$$

$$u_{sC} = \frac{U_d}{3}(2S_c - S_b - S_a)$$

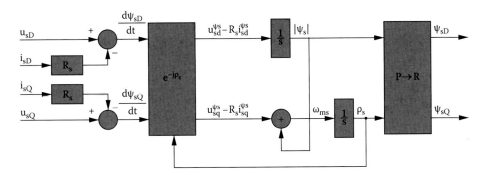

FIGURE 7.20
Flux estimation in the stator linkage reference frame.

It should be however remarked that, although some of the flux estimators above described permit obtaining an accurate estimation of the angle ρ_s of the stator flux-linkage vector, this is not strictly required in DTC. Indeed, since the application of the inverter voltage space-vector is limited to six different vectors, the precision required for the flux vector position is equal to $\pi/3$. That is why it is possible to avoid using inverse trigonometric functions except for the estimator in Figure 7.20, where the angle ρ_s is needed for the coordinate transformation. In general, the sector in which the stator flux-linkage vector lies can be determined with the flowchart in Figure 5.36.

Finally, a good torque estimation is required to obtain a proper DTC; electromagnetic torque can be computed, without needing the rotor position information, transforming Equation 7.43 in the stationary reference frame, which yields

$$t_e = \frac{3}{2} p \left[\psi_{sD} i_{sQ} - \psi_{sQ} i_{sD} \right] \tag{7.66}$$

The components of the stator flux linkage can be obtained by one of the earlier explained estimators; if the rotor position is also known, it is possible to adopt Equation 7.26 which is in the PM reference frame, where direct and quadrature-axis inductances are present:

$$t_e = \frac{3}{2} \left[(L_d - L_q) i_{sd}^{\psi_{PM}} i_{sd}^{\psi_{PM}} + i_{sd}^{\psi_{PM}} \psi_{PM} \right] \tag{7.67}$$

Saturation effects on these inductances should also be considered. A third method to estimate the electromagnetic torque is to store in a look-up table the characteristics:

$$t_e = f(\theta_r, \mathbf{i}_s)$$

With this method, the L_d and L_q inductances are not required, but a self-commissioning phase is necessary, where the electromagnetic torque is obtained as the rate of change of the machine coenergy. More details on this method are given in Ref. [2].

7.3.4.4 The Direct Torque Control Scheme

In Figure 7.21 a block diagram of the DTC for PMSM is shown. This control method has only one control input, the electromagnetic torque reference t_{eref}; its difference with the

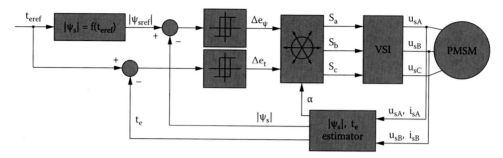

FIGURE 7.21
DTC scheme for PMSM.

estimated torque produced in the PM motor is the torque error e_t, which is fed to a hysteresis controller, whose output Δe_t is −1 or 1. The torque error is also used to compute the flux reference $|\psi_{sref}|$: two different approaches can be used, maximum torque per unit current or maximum torque per unit flux [2].

The stator flux-linkage reference is compared with the estimated amplitude, and the computed error e_ψ is given as input to another hysteresis comparator. Depending on the values of Δe_t and Δe_ψ, the optimal inverter configuration is selected and then fed to the VSI. In this scheme, two current sensors and one voltage sensor are needed: the rotor position is not to be acquired.

7.4 Sensorless Control of PMSM Drives

The main advantage of the adoption of the sensorless control techniques is the possibility to control the speed and/or the position of electrical machines without any speed/position sensor. Some of these techniques have been already presented for IM drives in Chapter 6. The removal of the encoder, or other speed/position sensors, permits a cost reduction, a size reduction, and an increased reliability for the whole drive, and these are the three main reasons why the sensorless control has been so deeply investigated in these years. In particular, the application of sensorless control is a key issue for the PMSM. Indeed, given the higher torque density, with respect to other electrical machines, the size of the encoder can be a significant part of the whole drive, especially for low-power PMSMs.

Two main categories of sensorless control can be described:

- *Anisotropy-based sensorless techniques,* which exploit the magnetic anisotropy of the PMSM to obtain an electrical signal containing the information for estimating the spatial position of the investigated anisotropy.

- *Model-based sensorless techniques,* which are based on the dynamic model of the PMSM. The position information is obtained through the emf estimation, or the deviation between the measured and estimated currents, or by solving some model-derived linear equations.

In the following section, the main techniques are examined, with their pros and cons.

7.4.1 Anisotropy-Based Sensorless Techniques

The rotor position information can be retrieved by properly tracking the magnetic anisotropy of the machine. By injecting a high-frequency signal into the stator windings, a high-frequency leakage flux is created. A voltage signal injection is generally used; nevertheless, also a current signal injection may be employed [14]. Voltage carrier signal can be easily injected if a VSI is used, while a current carrier signal provides larger voltage signals due to the increase of the voltage magnitude resulting from the high frequency of the injected signal: that is why a lower frequency is used with current carrier signals, which however leads up to a larger torque ripple [15].

As a matter of fact, PMSMs do not present a sinusoidal air-gap flux distribution; even surface-mounted PMSMs exhibit a nonnull magnetic saliency ratio, since their pole arc is less than 180° [11]. As a result, a magnetic anisotropy linked with the rotor position can be tracked both in surface- and interior-mounted PMSMs: the higher the saliency ratio, the easier the extraction of the rotor position information.

The various techniques based on the saliency tracking can be classified as follows:

1. Techniques based on the injection of a carrier signal
    a. Injection of a rotating carrier signal
    b. Injection of an alternating (or pulsating) carrier signal
2. Techniques based on the selection of proper inverter switching pattern

These techniques can work also at low and zero speed, but they also tend to fail if the rotating speed exceeds a threshold value or if the saliency ratio is poor. Actually, if the "position signal" information entailed by the injection of the carrier signal has a little amplitude, the position information may be corrupted and the estimator may show an unstable behavior. In general, the position signal depends on the position estimation error, and the position signal can be therefore attenuated if its frequency exceeds the bandwidth of the estimator during the transient.

In the following, the carrier signal injection technique will be examined and some state-of-the-art control schemes will be proposed.

7.4.1.1 Rotating Carrier Signal Injection

These sensorless techniques are based on the application of a rotating voltage carrier signal superimposed to the fundamental supply stator voltage. If ω_c is the frequency of the voltage carrier signal, the carrier signal $\mathbf{u}_c$ can be described in the stationary reference frame as [16,17]

$$\mathbf{u}_c = u_c e^{j\omega_c t} \tag{7.68}$$

The frequency ω_c should be significantly higher than the fundamental frequency ω_r. Since at a high frequency the ohmic drop on the stator resistance R_s is negligible compared to the flux variation, Equation 7.19 can be rewritten at the carrier frequency as

$$\mathbf{u}_c = \frac{d}{dt}\mathbf{\psi}_c \tag{7.69}$$

By exploiting Equation 7.68, it follows that

$$\mathbf{\psi}_c = \int \mathbf{u}_c \, dt = -j\frac{u_c}{\omega_c} e^{j\theta_c} \tag{7.70}$$

where

$$\omega_c = \frac{d\theta_c}{dt} \tag{7.71}$$

With the help of Equation 7.17, the flux at the carrier frequency can be expressed in matrix form as

$$
\begin{bmatrix} \psi_{cD} \\ \psi_{cQ} \end{bmatrix} = \begin{bmatrix} L_{sl} + \dfrac{3}{2}L_{0s} - \dfrac{3}{2}L_{2s}\cos(2\theta_r) & -\dfrac{3}{2}L_{2s}\sin(2\theta_r) \\ -\dfrac{3}{2}L_{2s}\sin(2\theta_r) & L_{sl} + \dfrac{3}{2}L_{0s} + \dfrac{3}{2}L_{2s}\cos(2\theta_r) \end{bmatrix} \begin{bmatrix} i_{cD} \\ i_{cQ} \end{bmatrix}
$$

$$
= \begin{bmatrix} L - \Delta L\cos(2\theta_r) & -\Delta L\sin(2\theta_r) \\ -\Delta L\sin(2\theta_r) & L + \Delta L\cos(2\theta_r) \end{bmatrix} \begin{bmatrix} i_{cD} \\ i_{cQ} \end{bmatrix} \tag{7.72}
$$

One simple way of solving for the currents i_{cD} and i_{cQ} in Equation 7.72 takes into account Equation 7.18, expressed in the PM reference frame:

$$\mathbf{\psi}_c^{\psi PM} = L\mathbf{i}_c^{\psi PM} - \Delta L\mathbf{i}_c^{*\psi PM} = -j\frac{u_c}{\omega_c} e^{j(\theta_c - \theta_r)} \tag{7.73}$$

Decomposing Equation 7.73 into the real and imaginary parts yields

$$
\begin{cases} \psi_{cd}^{\psi PM} = L i_{cd}^{\psi PM} - \Delta L i_{cd}^{\psi PM} = \dfrac{u_c}{\omega_c}\sin(\theta_c - \theta_r) \\[4mm] \psi_{cq}^{\psi PM} = L i_{cq}^{\psi PM} + \Delta L i_{cq}^{\psi PM} = -\dfrac{u_c}{\omega_c}\cos(\theta_c - \theta_r) \end{cases} \tag{7.74}
$$

The direct- and quadrature-axis components of the carrier current space-vector $\mathbf{i}_c^{\psi PM}$ in the PM reference frame can be easily obtained from Equation 7.74:

$$
\begin{cases} i_{cd}^{\psi PM} = \dfrac{u_c}{\omega_c(L - \Delta L)}\sin(\theta_c - \theta_r) \\[4mm] i_{cq}^{\psi PM} = -\dfrac{u_c}{\omega_c(L + \Delta L)}\cos(\theta_c - \theta_r) \end{cases} \tag{7.75}
$$

The components of the $\mathbf{i}_c$ vector in the stationary reference frame are therefore

$$
\begin{cases} i_{cD} = i_{cd}^{\psi PM}\cos\theta_r - i_{cq}^{\psi PM}\sin\theta_r \\[2mm] i_{cQ} = i_{cd}^{\psi PM}\sin\theta_r + i_{cq}^{\psi PM}\cos\theta_r \end{cases} \tag{7.76}
$$

Substituting Equation 7.75 into 7.76 yields

$$\begin{cases} i_{cD} = \dfrac{u_c}{\omega_c}\left[\dfrac{1}{(L-\Delta L)}\sin(\theta_c-\theta_r)\cos\theta_r + \dfrac{1}{(L+\Delta L)}\cos(\theta_c-\theta_r)\sin\theta_r \right] \\[3mm] i_{cQ} = \dfrac{u_c}{\omega_c}\left[\dfrac{1}{(L-\Delta L)}\sin(\theta_c-\theta_r)\sin\theta_r - \dfrac{1}{(L+\Delta L)}\cos(\theta_c-\theta_r)\cos\theta_r \right] \end{cases}$$

$$\Rightarrow \begin{cases} i_{cD} = \dfrac{u_c}{\omega_c(L^2-\Delta L^2)}[(L+\Delta L)\sin(\theta_c-\theta_r)\cos\theta_r + (L-\Delta L)\cos(\theta_c-\theta_r)\sin\theta_r] \\[3mm] i_{cQ} = \dfrac{u_c}{\omega_c(L^2-\Delta L^2)}[(L+\Delta L)\sin(\theta_c-\theta_r)\sin\theta_r - (L-\Delta L)\cos(\theta_c-\theta_r)\cos\theta_r] \end{cases}$$

And finally,

$$\begin{cases} i_{cD} = \dfrac{u_c}{\omega_c(L^2-\Delta L^2)}[L\sin(\theta_c)+\Delta L\sin(\theta_c-2\theta_r)] \\[3mm] i_{cQ} = \dfrac{u_c}{\omega_c(L^2-\Delta L^2)}[-L\cos(\theta_c)+\Delta L\cos(\theta_c-2\theta_r)] \end{cases} \tag{7.77}$$

Equation 7.77 can be rewritten as follows:

$$\mathbf{i}_c = \underbrace{\frac{u_c L}{\omega_c(L^2-\Delta L^2)}e^{j\left(\omega_c t-\frac{\pi}{2}\right)}}_{\mathbf{i}_{cp}} + \underbrace{\frac{u_c \Delta L}{\omega_c(L^2-\Delta L^2)}e^{-j\left[(\omega_c-2\omega_r)t+\frac{\pi}{2}\right]}}_{\mathbf{i}_{cn}} \tag{7.78}$$

Equation 7.78 states that the $\mathbf{i}_c$ vector has two components, $\mathbf{i}_{cp}$ and $\mathbf{i}_{cn}$, rotating at different speeds and in opposite direction [17]. Figure 7.22 shows both these components: the rotating speed of the drive, in mechanical radians, is equal to $\omega_m = 500\,\text{rpm}$, and the machine has three pole pairs; the carrier frequency is $\omega_c = 400\,\text{Hz}$.

It is noteworthy that the rotating speed information is contained only in the $\mathbf{i}_{cn}$ component of the carrier current space-vector. The demodulation of the speed signal can be achieved in various ways. In the following, one typical demodulation technique is explained, which is the basis of many others.

In Figure 7.23, a first demodulation scheme is shown [18], adopting a PLL to observe the rotor position. After measuring the three-phase currents, the stator current space-vector $\mathbf{i}_s$ is obtained, then it is filtered by a band-pass filter, whose central frequency is ω_c: in this way, the carrier current space-vector $\mathbf{i}_c$ is isolated; $\mathbf{i}_c$ is vector multiplied with a reference vector $\mathbf{i}_{cref}$, and the amplitude of the resulting vector ε is extracted:

$$\varepsilon = \left| \mathbf{i}_c \wedge \mathbf{i}_{cref} \right| = \left| \mathbf{i}_c \wedge e^{j(2\hat{\theta}_r-\theta_c)} \right| = i_{cD}\sin(2\hat{\theta}_r-\theta_c) - i_{cQ}\cos(2\hat{\theta}_r-\theta_c) \tag{7.79}$$

where $\hat{\theta}_r$ is the estimated position. Equation 7.79 can be rewritten as

$$\varepsilon_{lpf} = \frac{u_c L}{\omega_c(L^2-\Delta L^2)}\cos[2(\hat{\theta}_r-\theta_c)] + \frac{u_c \Delta L}{\omega_c(L^2-\Delta L^2)}\cos[2(\theta_r-\hat{\theta}_r)] \tag{7.80}$$

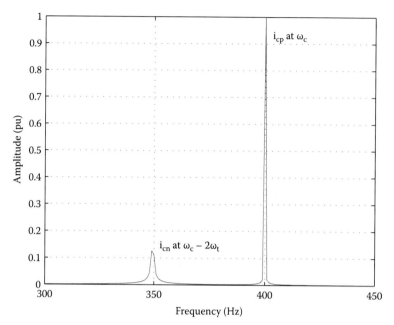

FIGURE 7.22
Harmonics generated by the rotating voltage carrier injection.

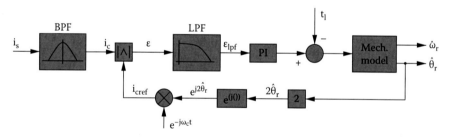

FIGURE 7.23
Demodulation scheme for rotating carrier signal injection.

The signal ε is then filtered by an LPF to eliminate the first term of Equation 7.80, as shown in Figure 7.23. The second term contains then the position error, which is fed to a PI controller to give the electromagnetic torque reference. If the mechanical model is used, the estimated rotating speed and the position are obtained and fed back in the PLL.

This rotating carrier signal injection sensorless scheme permits the drive to work even at low and zero speed but presents some drawbacks [18]. Firstly, it has a poor signal-to-noise ratio, as well as a parameter dependence on the error demodulation, making this method unsuitable for low-saliency machines. Finally, the estimation fails if the rotating speed is too high since it attenuates the carrier signal as it is often the case in injection techniques.

Figure 7.24 shows the simulation results of a speed reversal test performed on an interior-mounted PMSM. Speed vector control with impressed voltages has been used. The described sensorless scheme is shown in Figure 7.12 to which a speed control loop has been added. Figures 7.17 and 7.18 show the speed and current waveforms, respectively, that have been measured from a PMSM drive with FOC in speed control mode.

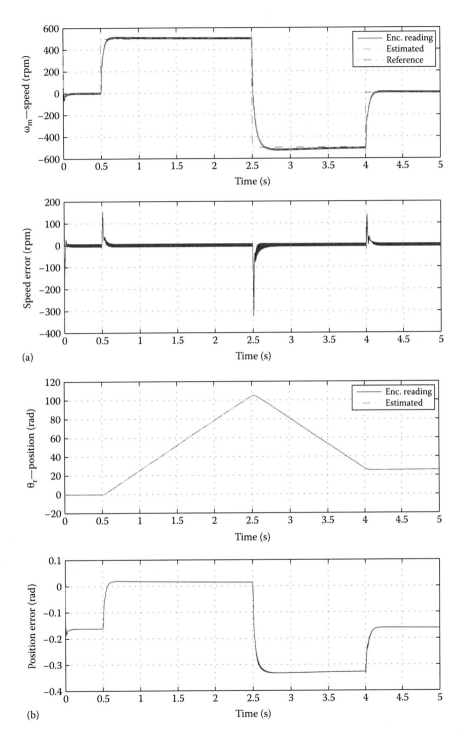

FIGURE 7.24
Results for speed control of rotating carrier sensorless technique. (a) Rotor speed and (b) rotor position.

As shown in Figure 7.24a, the estimator is able to track the rotating speed of the drive: the steady-state speed error has a null mean value. When the speed reference changes, a transient speed estimation error occurs, governed by the bandwidth of the observer. As a consequence, the position error can be either increased or decreased.

7.4.1.2 Alternating Carrier Signal Injection

A pulsating carrier voltage signal is obtained by injecting in the estimated PM reference frame a signal at the frequency ω_c:

$$\mathbf{u}_c^{\hat{\psi}_{PM}} = u_c \cos(\omega_c t) \tag{7.81}$$

This is then a signal acting as a pulsating signal along the direct axis of the PM reference frame. In the actual PM reference frame, Equation 7.81 becomes

$$\mathbf{u}_c^{\psi_{PM}} = u_c \cos(\omega_c t) e^{j(\hat{\theta}_r - \theta_r)} \tag{7.82}$$

Resolution of the simplified differential equation of the PMSM, as shown in the earlier paragraph, yields the following solution [19]:

$$\mathbf{i}_c^{\hat{\psi}_{PM}} = \left[\frac{u_c L}{\omega_c(L^2 - \Delta L^2)} + \frac{u_c \Delta L}{\omega_c(L^2 - \Delta L^2)} e^{j2(\theta_r - \hat{\theta}_r)} \right] \sin(\omega_c t) \tag{7.83}$$

where the carrier current space-vector is expressed in the estimated PM reference frame. This carrier current has one component which is not useful for the position estimation and another term containing the information about the position error. Expressing Equation 7.83 in the stationary reference frame yields

$$\mathbf{i}_c = e^{j\hat{\theta}_r} \mathbf{i}_c^{\hat{\psi}_{PM}} = e^{j\hat{\theta}_r} \left[\frac{u_c L}{\omega_c(L^2 - \Delta L^2)} + \frac{u_c \Delta L}{\omega_c(L^2 - \Delta L^2)} e^{j2(\theta_r - \hat{\theta}_r)} \right] \sin(\omega_c t)$$

$$= \left[\frac{u_c L}{\omega_c(L^2 - \Delta L^2)} e^{j\hat{\theta}_r} + \frac{u_c \Delta L}{\omega_c(L^2 - \Delta L^2)} e^{j2\theta_r - \hat{\theta}_r} \right] \sin(\omega_c t)$$

$$= \left[\frac{u_c L}{\omega_c(L^2 - \Delta L^2)} e^{j\hat{\theta}_r} + \frac{u_c \Delta L}{\omega_c(L^2 - \Delta L^2)} e^{j2\theta_r - \hat{\theta}_r} \right] \frac{e^{j\omega_c t} - e^{-j\omega_c t}}{2j}$$

$$= \frac{1}{2j} \left[\frac{u_c L}{\omega_c(L^2 - \Delta L^2)} e^{j(\omega_c t + \hat{\theta}_r)} - \frac{u_c L}{\omega_c(L^2 - \Delta L^2)} e^{-j(\omega_c t - \hat{\theta}_r)} \right.$$

$$\left. + \frac{u_c \Delta L}{\omega_c(L^2 - \Delta L^2)} e^{j(\omega_c + 2\theta_r - \hat{\theta}_r)} - \frac{u_c \Delta L}{\omega_c(L^2 - \Delta L^2)} e^{-j(\omega_c - 2\theta_r + \hat{\theta}_r)} \right] \tag{7.84}$$

By decomposing Equation 7.83 into its real and imaginary parts, it is apparent that the quadrature-axis component, $i_{cq}^{\hat{\psi}_{PM}}$, depends on the sinus of the position error, for small values of it can be treated as the position error itself. As a result, an algorithm can be devised using $i_{cq}^{\hat{\psi}_{PM}}$ component as an error signal, as shown in Figure 7.26.

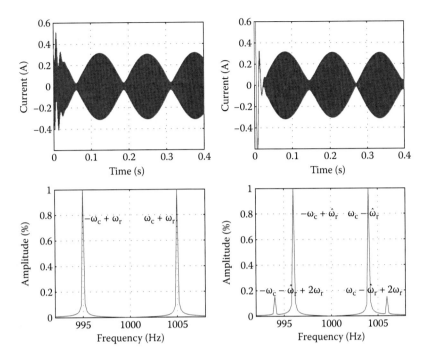

FIGURE 7.25
Desynchronization effects for pulsating carrier signal injection. (From Accetta, A. et al., *IEEE Trans. Ind. Electron.*, 59(3), 1355, March 2012.)

Figure 7.25 shows the time waveform and the frequency content (obtained with the FFT) of the carrier current both in synchronized and desynchronized conditions. In synchronized conditions, when $\hat{\omega}_r = \omega_r$, the four terms in Equation 7.84 are overlapped, while in desynchronized conditions, all terms appear.

The sensorless scheme shown in Figure 7.26 [20] uses the stator current space-vector, expressed in the estimated PM reference frame, given the computed estimated position $\hat{\theta}_r$. The resulting space-vector $i_s^{\hat{\psi}_{PM}}$ is then band-pass filtered to extract the carrier space-vector $i_c^{\hat{\psi}_{PM}}$; according to Equation 7.83, the digital filter should be centered about the carrier frequency. Finally, the position estimation error is obtained from the $i_{cq}^{\hat{\psi}_{PM}}$, multiplied for the sign function of the $i_{cd}^{\hat{\psi}_{PM}}$ current.

Figure 7.27 shows the experimental results of a speed reversal at no load performed on the PMSM drive described in Appendix 7.A. Speed vector control with impressed voltages has been used. The adopted control scheme is described in Figure 7.12 to which a speed

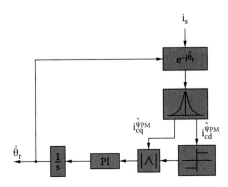

FIGURE 7.26
Demodulation scheme for pulsating carrier signal injection. (From Accetta, A. et al., *IEEE Trans. Ind. Electron.*, 59(3), 1355, March 2012.)

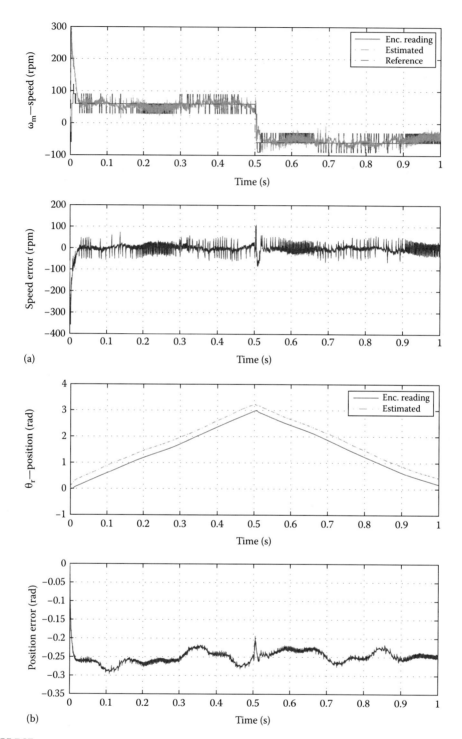

FIGURE 7.27
Experimental results for pulsating voltage carrier sensorless technique. (a) Rotor speed, (b) rotor position, and (c) current components. (From Accetta, A. et al., *IEEE Trans. Ind. Electron.*, 59(3), 1355, March 2012.)

control loop has been added. The speed reference is initially set to 60 rpm and suddenly changes to −60 rpm at 0.5 s. The whole test has a duration equal to 1 s. The injected voltage carrier signal has a frequency equal to 1 kHz and an amplitude equal to 5 V. After an initial convergence transient due to the fact that all the digital filters are starting from the zero state, the estimator is able to correctly track the rotating speed of the drive (see Figure 7.27a). The estimated position, shown in Figure 7.27b, has a constant bias of 0.25 rad because of the initial speed convergence transient after which the position error remains constant.

7.4.1.2.1 Experimental Application of Pulsating Voltage Carrier Estimator with the Adaptive Linear Neural Filter*

The pulsating carrier voltage carrier sensorless technique can be improved by using an adaptive linear neural filter (ADALINE), described in Chapter 8. The strategy to be employed with such digital filter is to change the bandwidth according to the working conditions of the drive; during the transient, the rotor position information may be attenuated by the band-pass filter, so a wider bandwidth should be used. On the contrary, in steady-state, a narrower bandwidth should be used to increase the signal-to-noise ratio and obtain a more accurate estimation.

Figure 7.28 shows the experimental results obtained performing a speed reversal, from 60 to −60 rpm at load, on the PMSM drive described in the Appendix to this chapter. The load torque

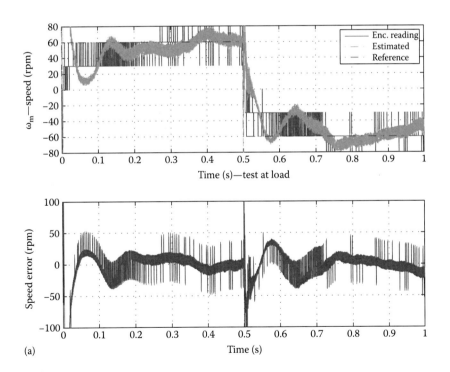

FIGURE 7.28
Low-speed load test results for adaptive pulsating voltage carrier sensorless technique. (a) Rotor speed, (b) rotor position, and (c) current components. (From Accetta, A. et al., *IEEE Trans. Ind. Electron.*, 59(3), 1355, March 2012.)

(continued)

---

* See Ref. [21].

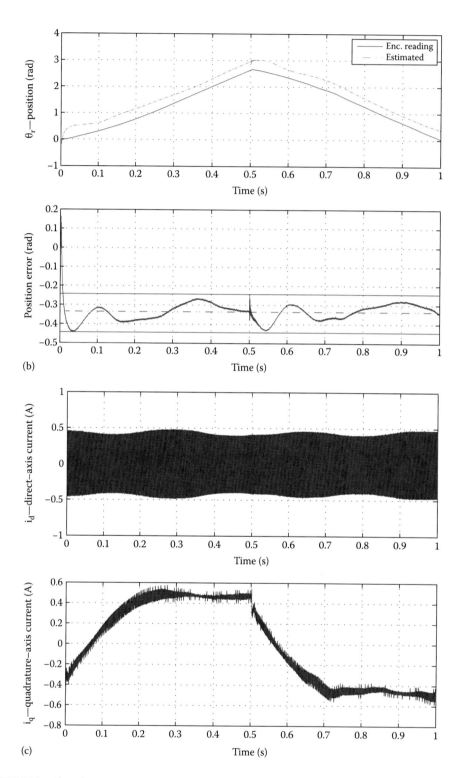

FIGURE 7.28 (continued)

applied is linearly varying with the rotating speed, since a friction load has been used. During the speed steady-state, it reaches the value of 18.4 mN m, almost equal to half the rated torque.

The speed estimation error is almost zero, except for the initial speed convergence transient. The position error, as seen in Figure 7.28b, remains almost constant, apart from the bias generated by the speed convergence transient.

Figures 7.29 and 7.30 show a high-speed reversal at no load, performed with both adaptive and not adaptive sensorless scheme. It should be remarked that, while the adaptive scheme is able to control the drive even in a speed reversal at 1/2 of the rated speed, the not adaptive scheme fails at 600 rpm, 1/10 of the rated speed.

7.4.1.3 INFORM Sensorless Technique

The indirect flux detection by on-line reactance measurement (INFORM) technique has been already described in Chapter 6 for the IM (see Ref. [20] in Chapter 6). The same technique can be conveniently used for the position estimation in the PMSM drive; hence, it is briefly summarized here.

Remembering Equations 7.17 and 7.19, the stator voltage space-vector equations in the stator reference frame can be expressed as

$$\mathbf{u}_s = R_s \mathbf{i}_s + \frac{d}{dt}\boldsymbol{\psi}_s$$

$$\boldsymbol{\psi}_s = \left(L_{ls} + \frac{3}{2}L_{0s}\right)\mathbf{i}_s - \frac{3}{2}L_{2s}\mathbf{i}_s^* e^{j2\theta_r} + \psi_{PM}e^{j\theta_r}$$

or, in an equivalent form,

$$\mathbf{u}_s = R_s \mathbf{i}_s + \mathbf{L}(2\theta_r)\frac{d}{dt}\mathbf{i}_s + \omega_r \Delta\mathbf{L}(2\theta_r)\mathbf{i}_s + \omega_r \boldsymbol{\psi}_{PM}(\theta_r) \qquad (7.85)$$

Assuming a short time interval Δt, the voltage space-vector can be considered constant; by applying two subsequent inverter voltage configurations, in order to obtain two inverter voltage space-vectors having the same direction but opposite orientation, both the ohmic drop on the stator resistance and the back emf terms can be neglected. If the two inverter voltage space-vectors are, for example, $\mathbf{u}_s^{(1)}$ and $\mathbf{u}_s^{(4)}$, then Equation 7.85 can be rewritten, substituting the continuous time derivatives with their discrete-time counterpart, as

$$\mathbf{u}_s^{(1)} - \mathbf{u}_s^{(4)} = \mathbf{L}(2\theta_r)\left[\frac{\Delta\mathbf{i}_s^{(1)}}{\Delta t} - \frac{\Delta\mathbf{i}_s^{(4)}}{\Delta t}\right] \qquad (7.86)$$

Taking into account the inverse of the $\mathbf{L}(2\theta_r)$ matrix, and extracting the real part of the vector solution, since both $\mathbf{u}_s^{(1)}$ and $\mathbf{u}_s^{(4)}$ lie on the direct axis of the stationary reference frame, the following solution can be found:

$$\Re\left[\Delta\mathbf{i}_s^{(1)} - \Delta\mathbf{i}_s^{(4)}\right] = \Re\left\{\Delta t\mathbf{L}^{-1}(2\theta_r)\left[\Delta\mathbf{i}_s^{(1)} - \Delta\mathbf{i}_s^{(4)}\right]\right\} = \frac{2\Delta t}{L^2 - \Delta L^2}[L + \Delta L\cos(2\theta_r)]|\mathbf{u}_s| \qquad (7.87)$$

The same procedure can then be applied for the other two pairs of opposite inverter voltage space-vectors, extracting only the component of the vector solution that lies on the

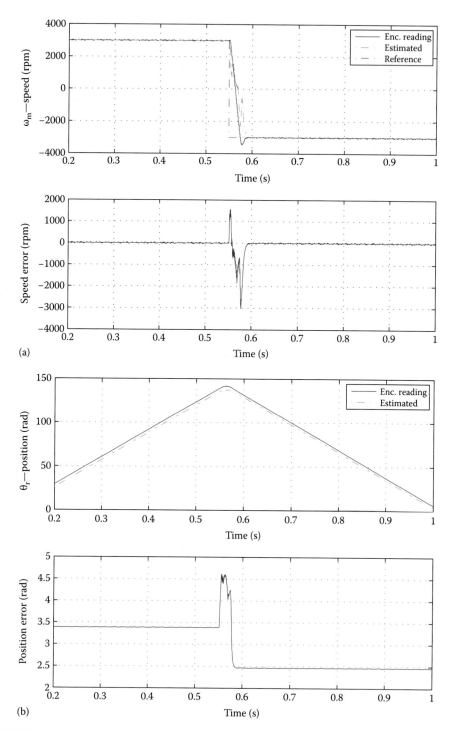

FIGURE 7.29
High-speed load test results for adaptive pulsating voltage carrier sensorless technique. (a) Rotor speed and (b) rotor position. (From Accetta, A. et al., *IEEE Trans. Ind. Electron.*, 59(3), 1355, March 2012.)

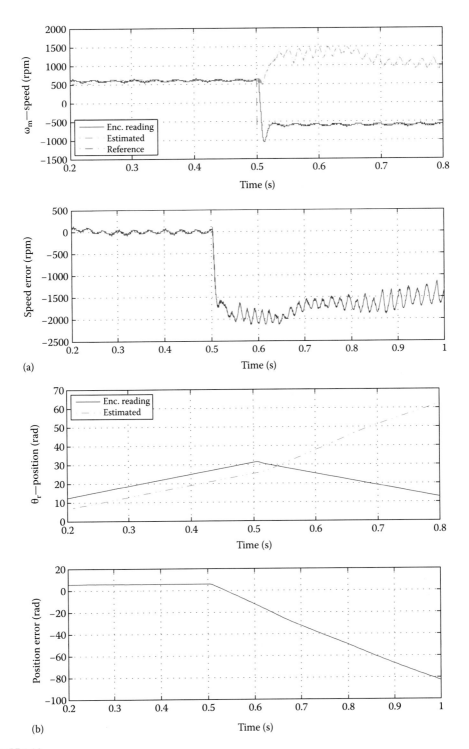

FIGURE 7.30
High-speed load test results for not adaptive pulsating voltage carrier sensorless technique. (a) Rotor speed and (b) rotor position. (From Accetta, A. et al., *IEEE Trans. Ind. Electron.*, 59(3), 1355, March 2012.)

direction of the forcing inverter voltage. Similarly to the IM, a space-vector of the stator current variation can be computed:

$$f = \frac{1}{2} \frac{\Delta t}{L^2 - \Delta L^2} \Delta L |\mathbf{u}_s| e^{j(2\hat{\theta}_r + \pi)} \tag{7.88}$$

from which the estimated rotor position can be obtained.

It should be however noted that the INFORM sensorless method requires the measurement of the current response to the application of the chosen inverter test vectors. Furthermore, the application of the test vector should be integrated with the PWM for normal operation. In order to reduce the algorithm complexity and the hardware requirements, hence the cost of the drive, an optimized INFORM measurement sequence should be chosen and the current response be obtained through the measurement of the DC link current [22].

7.4.2 Model-Based Sensorless Techniques

Many sensorless techniques based on the fundamental excitation of the PMSM exist in literature. Actually, the PMSM dynamic model is well known and simple compared to the model of the IM, so suitable sensorless techniques can be derived. These techniques can be divided into the following:

- *Open-loop estimators*: Usually, the stator flux linkage is estimated, both in amplitude and phase. From the instantaneous value of its phase, it is possible to compute the synchronous rotating speed of the drive. The current error between measured and estimated values can be also used to correct the speed estimation. Also, the back emf can be used to obtain the position estimation, since it is strictly related to the rotor position.

- *Observer-based estimators*: An extended Kalman filter or a Luenberger observer can be used in high-performance PMSM drives. The main idea is the possibility to observe unmeasurable states, that is, speed and position, through measurable quantities, usually stator voltages and currents. Some schemes can avoid the voltages measurement, and the consequent filtering, and use the reference three-phase voltages.

Fundamental excitation–based sensorless techniques for the PMSM present some difficulties that have been already dealt with in Section 6.4 for the IM sensorless control and, briefly, in Section 7.3.4, regarding the DTC of the PMSM.

7.4.2.1 Open-Loop Estimators

Differently from the IM, in a PMSM, the stator flux linkage has the same rotating speed of the rotor, in electric radians. This can be easily exploited to obtain an estimation of the rotating speed, since the stator flux-linkage space-vector can be, in theory, computed from the stator voltages.

7.4.2.1.1 emf-Based Estimators

A sensorless scheme based on the estimation of the back emf for the PMSM drive has been presented for the first time in Ref. [23], and it has been since then analyzed and improved.

Given the stator voltages and current measurements, the back emf space-vector $\mathbf{e}_s$ can be obtained as follows:

$$\mathbf{e}_s = \mathbf{u}_s - R_s \mathbf{i}_s \tag{7.89}$$

Integrating Equation 7.88, the stator flux linkage is then obtained:

$$\boldsymbol{\psi}_s = \int \mathbf{e}_s \, dt = \int \mathbf{u}_s - R_s \mathbf{i}_s \, dt \qquad (7.90)$$

From Equation 7.90, the stator flux-linkage angle ρ_s can be computed through a trigonometric inverse function or by the knowledge of $\sin(\rho_s)$ and $\cos(\rho_s)$ as well as the stator flux-linkage space-vector amplitude.

Drift in the integration process is a common source of error between real and estimated rotor position. Ideally, in steady-state, the stator flux-linkage space-vector describes a circle, centered into the origin of the stationary reference frame; drift error shifts this circle, entailing a big position estimation error (130% at 0.4 p.u. drift [23]). Drift compensation can be performed with the method explained in Section 6.4.1.

It should be remarked that, at zero or close to zero speed, the back emf is too small to produce reliable position estimation. For this reason, this kind of sensorless technique tends to fail at low speed. As a consequence, the PMSM drive is not capable of self-starting, and also position control is not possible. This problem can be solved by providing a self-starting open-loop strategy that controls the drive until the rotating speed exceeds the threshold level for obtaining a good position estimation. Alternatively, hybrid estimators, based both on the back emf and the signal injection, can be devised; however, the position estimations computed by the two algorithms must be properly managed [24,25].

Another issue with the back emf-based sensorless techniques is that they are only suitable for surface-mounted PMSM; indeed, in interior-mounted PMSMs, the saliency modifies the machine model, and the position information is present not only in the back emf but also in the saliency term: some corrections are therefore to be made. Nevertheless, these correction terms can worsen the estimation and even cause instability in the algorithm.

Recently, an extended emf model for the sensorless control of interior-mounted PMSM has been proposed [26]. The extended emf is derived from the model of the IM-PMSM and takes into account both the position information contained in the back emf and in the saliency voltage:

$$\mathbf{e}_{ext} = \left[(L_d - L_q) \left(\omega_r i_{sd}^{\psi_{PM}} - \frac{d}{dt} i_{sq}^{\psi_{PM}} \right) + \omega_r K_E \right] [-\sin(\theta_r) + j\cos(\theta_r)] \qquad (7.91)$$

Remark that Equation 7.91 is valid for all the brushless synchronous machines. By forcing $(L_d - L_q) = 0$, the extended emf becomes equal to that of a surface-mounted PMSM, while if $K_E = 0$, the back emf of a synchronous reluctance motor is obtained [26].

7.4.2.2 Observer-Based Estimators

Sensorless PMSM drive control can also be obtained with an observer, in which rotating speed and rotor position are considered as unmeasurable states. This observer permits estimating both quantities with measured stator currents and voltages.

7.4.2.2.1 Extended Kalman Filter

An extended Kalman filter can be chosen for sensorless control; it is a recursive optimum state estimator that can be used especially when the measured quantities are supposed to be affected by noise. The state estimation is performed in two phases. In the first, a rough prediction is made on the basis of the mathematical model chosen for the drive; in the second, the estimation is corrected using a feedback scheme. In Ref. [2], details on how the Kalman filter can be designed for the sensorless control are given.

7.4.2.2.2 TLS EXIN Neuron–Based Speed Estimator

In Chapter 8, the TLS EXIN neuron has been deeply described; here, only its application to the sensorless control of PMSM is covered. This estimator arises from the application of the neural networks to the fundamental excitation–based model of the PMSM. The main idea is to consider the rotating speed of the PMSM drive as the solution of an overdetermined linear system, based upon the PMSM model, and therefore obtain its estimation as a total least squares (TLS) solution of this system.

The PMSM stator voltage space-vector equation in the stationary reference frame, can be written as

$$\mathbf{u}_s = R_s \mathbf{i}_s + \mathbf{L}(2\theta_r)\frac{d}{dt}\mathbf{i}_s + \omega_r \Delta\mathbf{L}(2\theta_r)\mathbf{i}_s + \omega_r \boldsymbol{\psi}_{PM}(\theta_r) \tag{7.92}$$

where

$$\mathbf{L}(2\theta_r) = \begin{bmatrix} L - \Delta L\cos(2\theta_r) & -\Delta L\sin(2\theta_r) \\ -\Delta L\sin(2\theta_r) & L + \Delta L\cos(2\theta_r) \end{bmatrix}$$

$$\Delta\mathbf{L}(2\theta_r) = \begin{bmatrix} \Delta L\sin(2\theta_r) & -\Delta L\cos(2\theta_r) \\ -\Delta L\cos(2\theta_r) & -\Delta L\sin(2\theta_r) \end{bmatrix} \tag{7.93}$$

Splitting Equation 7.84 into direct- and quadrature-axis components, the two following scalar equations can be found:

$$u_{sD} = R_s i_{sD} + L\frac{di_{sD}}{dt} - \Delta L\cos(2\theta_r)\frac{di_{sD}}{dt} - \Delta L\sin(2\theta_r)\frac{di_{sQ}}{dt} + 2\omega_r \Delta L\sin(2\theta_r)i_{sD} + $$

$$- 2\omega_r \Delta L\cos(2\theta_r)i_{sQ} - \omega_r \psi_{PM}\sin(\theta_r)$$

$$u_{sQ} = R_s i_{sQ} + L\frac{di_{sQ}}{dt} + \Delta L\cos(2\theta_r)\frac{di_{sQ}}{dt} - \Delta L\sin(2\theta_r)\frac{di_{sD}}{dt} - $$

$$- 2\omega_r \Delta L\sin(2\theta_r)i_{sQ} - 2\omega_r \Delta L\cos(2\theta_r)i_{sD} + \omega_r \psi_{PM}\cos(\theta_r) \tag{7.94}$$

Equation 7.94 can be rearranged, to put it in the form of a linear system, in the following way:

$$u_{sD} - R_s i_{sD} - L\frac{di_{sD}}{dt} + \Delta L\cos(2\theta_r)\frac{di_{sD}}{dt} + \Delta L\sin(2\theta_r)\frac{di_{sQ}}{dt}$$

$$= \omega_r[2\Delta L\sin(2\theta_r)i_{sD} - 2\Delta L\cos(2\theta_r)i_{sQ} - \psi_{PM}\sin(\theta_r)]$$

$$u_{sQ} - R_s i_{sQ} - L\frac{di_{sQ}}{dt} - \Delta L\cos(2\theta_r)\frac{di_{sQ}}{dt} + \Delta L\sin(2\theta_r)\frac{di_{sD}}{dt} \tag{7.95}$$

$$= \omega_r[-2\Delta L\sin(2\theta_r)i_{sQ} - 2\Delta L\cos(2\theta_r)i_{sD} + \psi_{PM}\cos(\theta_r)]$$

Taking into account a discrete-time representation, choosing T_{sp} as sampling time, Equation 7.95 can be modified according to the following relationship between continuous and discrete time:

$$s = \frac{1 - z^{-1}}{T_{sp}} \qquad (7.96)$$

where
s is the Laplace variable
z the discrete-time variable, eventually obtaining

$$\underbrace{T_s \mathbf{u}_s(k) - T_{sp} R_s \mathbf{i}_s(k) + \mathbf{L}(2\theta_r(k))[\mathbf{i}_s(k) - \mathbf{i}_s(k-1)]}_{\mathbf{S}} = \omega_r(k) \underbrace{T_{sp}[\Delta \mathbf{L}(2\theta_r(k))\mathbf{i}_s(k) + \boldsymbol{\psi}_{PM}(\theta_r(k))]}_{\mathbf{q}} \qquad (7.97)$$

where k is the actual time instant. The **S** vector is also called "data vector" and the **q** vector is called "observation vector." Equation 7.97 represents a TLS problem, where the solution $\hat{\omega}_r(k)$ is the estimation of the rotating speed of the drive. It should be noted that the solution $\hat{\omega}_r(k)$ at the kth instant of time depends only on the stator current space-vector $\mathbf{i}_s(k)$ and $\mathbf{i}_s(k-1)$, and on the estimated position $\theta_r(k)$. As a consequence, the TLS problem is perfectly known at the kth instant of time.

The TLS solution is the most viable way to compute the solution of Equation 7.97. In fact, it should be emphasized that both stator voltages and currents are required and that these measurements are affected by noise; therefore, noise is contained in both data and observation vectors. Ordinary, least squares methods consider only the observation vector affected by noise; thus, it is not suitable for this application (see Chapter 8 for further details).

The TLS EXIN neuron presented in the next chapter can solve the linear system described by Equation 7.97 that can be rewritten as a matrix linear equation:

$$\mathbf{S}\hat{\omega}_r \approx \mathbf{q}$$

and find the TLS solution $\hat{\omega}_r$.

Figure 7.31 shows the block scheme of this sensorless technique. Both three-phase voltages and currents must be acquired to obtain the data and observation vectors in Equation 7.96. From these measured quantities, the current and voltage space-vector in the stationary reference frame are obtained and then fed to the TLS EXIN algorithm to solve the correspondent TLS problem. The TLS estimated speed is then integrated to obtain the estimated position. The speed control scheme has already been presented in Section 7.3.3.

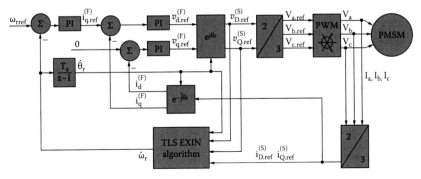

FIGURE 7.31
TLS EXIN sensorless scheme.

The proposed TLS EXIN–based speed observer has been tested experimentally on the fractional power PMSM drive, whose experimental rig is described in Appendix 7.A.

7.4.2.2.2.1 Test 1: Speed Reversal at High Speed with No Load and with Load   The drive has been initially given a speed reversal from 1500 to –1500 rpm with no load torque applied on the shaft. Figure 7.32a shows the reference, measured and estimated speeds, as well as

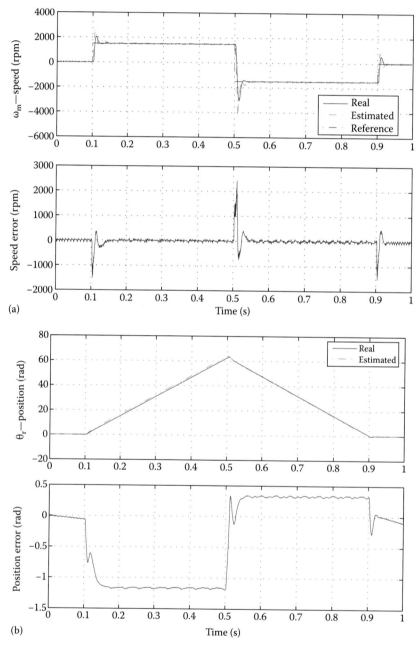

FIGURE 7.32
TLS EXIN sensorless technique—test 1 results in no load. (a) Rotor speed and (b) rotor position.

the speed estimation error, as obtained during this test. Figure 7.32b shows the corresponding waveform of the measured and estimated position as well as the position estimation error. It can be observed that the estimated speed properly tracks the measured one and its reference. This is confirmed by the speed estimation error waveform, which is almost null in average at speed steady-state, getting high values only during the fast transient and in particular during the overshoot.

Also, the estimated position correctly tracks the measured one, with the position estimation error being almost constant at speed steady-state and varying during the transient. This confirms that estimation errors mainly occur during the fast transient.

The drive has been further given the same speed reversal from 1500 to −1500 rpm, but with a load torque. The machine has been in this case loaded with a friction type load, whose amplitude is almost linearly increasing with the speed. Figure 7.33a shows the reference and measured and estimated speeds as well as the speed estimation error, while Figure 7.33b shows the corresponding waveform of the measured and estimated position as well as the position estimation error. Even in loaded conditions, results show a proper behavior of the observer at speed steady-state with a significant estimation error in transient.

7.4.2.2.2.2 Test 2: Triangular Speed Reference with No Load and with Load   The drive has been given a speed triangular waveform of peak value 1500 rpm and frequency 1 Hz, respectively, with no load and with load. This test has been done to verify the correct behavior of the observer in case of continuously varying speed references. Figure 7.34a shows the reference, measured and estimated speeds, as well as the speed estimation error. Figure 7.34b shows the corresponding waveform of the measured and estimated position as well as the position estimation error. It can be observed that the estimated speed properly tracks the measured one and its reference during the entire test, with almost null estimation errors increasing only during each zero crossing, as expected. This is confirmed also by the position waveforms, showing an estimated position distant from the measured one of an almost constant quantity (position estimation error quasi constant), confirming that the distance between the two curves is due to the errors occurring during the initial transient. Figure 7.35a and b show the same waveforms obtained in loaded conditions. Even in this case, the load is a friction type, whose amplitude is almost linearly increasing with the speed. These waveforms show the correct behavior of the speed estimator also in loaded conditions. The main difference with the no-load curves is that speed estimation error during the zero crossing is higher, as expected.

7.4.2.2.2.3 Test 3: Speed Step Reference at Low Speed with No Load   To verify the proper behavior of the speed observer at very low speed, the drive has been given a speed step reference of 100 rpm with no load. Figure 7.36a shows the reference, measured and estimated speeds, as well as the speed estimation error. Figure 7.36b shows the corresponding waveform of the measured and estimated position as well as the position estimation error. It could be observed that the estimated speed correctly tracks the measured one and its reference, even if after a slightly longer transient (0.1 s in this case) caused by the reduced bandwidth of the observer at lower speeds. The average speed estimation error is very low at steady-state, equal in percentage to 1.87%, while during the transient, this error is much bigger. It is confirmed by the position waveforms, showing a linear variation of the position where the distance between the two curves of the estimated and measured position remains almost constant.

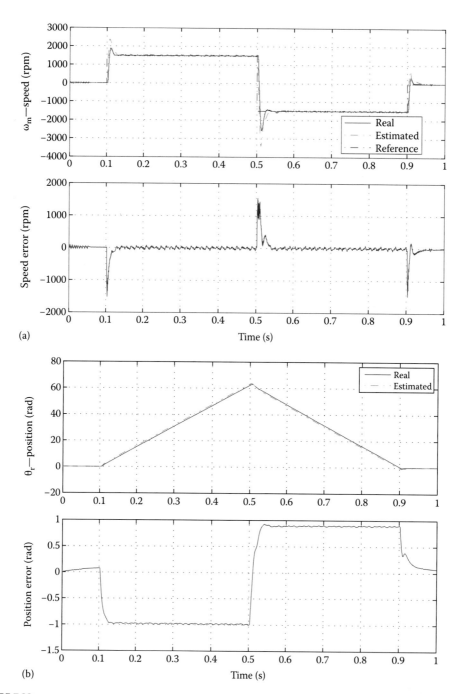

FIGURE 7.33
TLS EXIN sensorless technique—test 1 results in load. (a) Rotor speed and (b) rotor position.

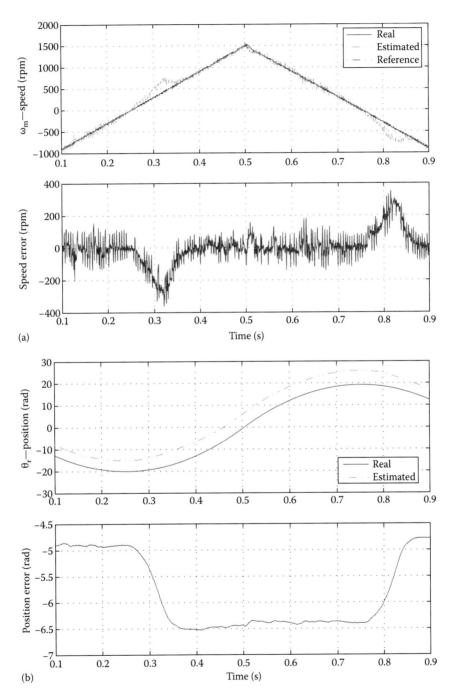

FIGURE 7.34
TLS EXIN sensorless technique—test 2 results in no load. (a) Rotor speed and (b) rotor position.

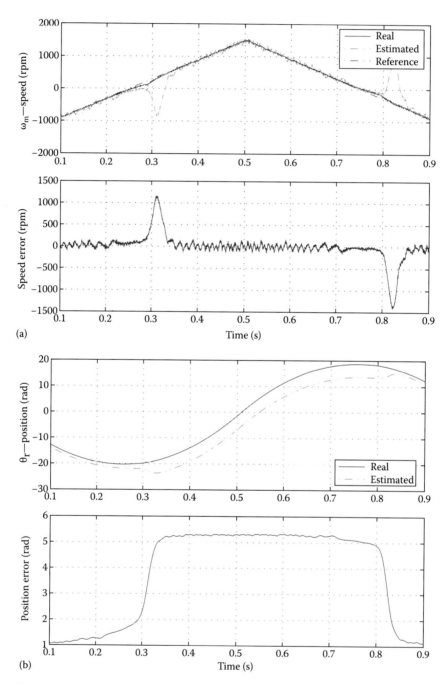

FIGURE 7.35
TLS EXIN sensorless technique—test 2 results in load. (a) Rotor speed and (b) rotor position.

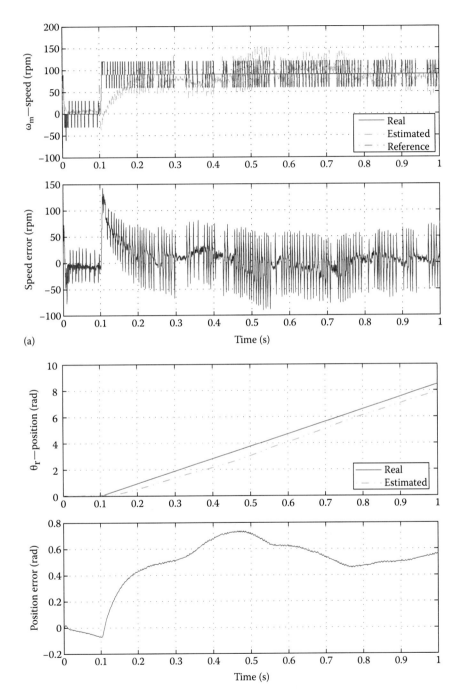

FIGURE 7.36
TLS EXIN sensorless technique—test 3 results. (a) Rotor speed and (b) rotor position.

FIGURE 7.37
Photograph of the fractional power PMSM drive.

Appendix: Experimental Test Setup

The employed test setup is based on a PMSM drive consisting of

- A three-phase PMSM model MBE.300.E500, whose parameters are shown in Table 7.2
- A development system model Technosoft MSK28335 with a DSC motion controller model TMS320F28335
- A three-phase VSI model PM50 power module board with rated voltage 36 V, rated current 2.1 A, DC link voltage in a range 12–36 V, working at PWM frequencies up to 25 kHz

The photographs of the adopted test setup is shown in Figure 7.37.

The PMSM has a saliency ratio equal to 0.83 and is equipped with a 500 pulse per round quadrature incremental encoder. The efficiency of the PMSM at rated load torque is about 81%.

References

1. J. F. Gieras, M. Wing, *Permanent Magnet Motor Technology: Design and Applications*, Marcel Dekker Inc., New York, 2002.
2. P. Vas, *Sensorless Vector and Direct Torque Control*, Oxford University Press, Oxford, U.K., 1998.
3. D. W. Novotny, T. A. Lipo, *Vector Control and Dynamics of AC Drives*, Oxford University Press, Oxford, U.K., 1997.
4. T. J. E. Miller, *Brushless Permanent-Magnet and Reluctance Motor Drives*, Oxford University Press, Oxford, U.K., 1989.
5. A. Fratta, A. Vagati, F. Villata, Permanent magnet assisted synchronous reluctance drive for constant-power application: Drive power limit, *Proceedings of Intelligent Motion European Conference (PCIM)*, Nürnberg, Germany, pp. 196–203, April 1992.

6. Bianchi, N., S. Bolognani, Interior PM synchronous motor for high performance applications, *Proceedings of the Power Conversion Conference, 2002. PCC Osaka 2002*, Osaka, Vol. 1, pp. 148–153, 2002.

7. E. Armando, P. Guglielmi, G. Pellegrino, M. Pastorelli, A. Vagati, Accurate modeling and performance analysis of IPM-PMASR motors, *IEEE Transactions on Industry Applications*, 45(1), 123–130, January–February 2009.

8. A. Fratta, G. P. Troglia, A. Vagati, F. Villata, Evaluation of torque ripple in high performance synchronous reluctance machines, *Conference Records of IEEE IAS Annual Meeting*, Toronto, ON, Canada, Vol. I, pp. 163–170, October 1993.

9. N. Bianchi, S. Bolognani, D. Bon, M. Dai Pre, Rotor flux-barrier design for torque ripple reduction in synchronous reluctance and PM-assisted synchronous reluctance motors, *IEEE Transactions on Industry Applications*, 45(3), 921–928, May–June 2009.

10. A. Vagati, M. Pastorelli, G. Franceschini, S. C. Petrache, Design of low-torque-ripple synchronous reluctance motors, *IEEE Transactions on Industry Applications*, 34(4), 758–765, July–August 1998.

11. R. Krishnan, *Permanent Magnet Synchronous and Brushless DC Motor Drives*, CRC Press, Boca Raton, FL, 2010.

12. S. Dun, F. Weizhong, H. Yikang, Study on the direct torque control of permanent magnet synchronous motor drives, *Proceedings of the Fifth International Conference on Electrical Machines and Systems (ICEMS)*, Shenyang, China, pp. 571–574, 2001.

13. G. Buja, D. Casadei, J. Lalu, G. Serra, Il controllo diretto di coppia negli azionamenti elettrici con motore asincrono, Specialistic Course ANAE-UCIREV, Milano, Italy, November 7–8, 1996.

14. D. Raca, P. Garcia, D. D. Reigosa, F. Briz, R. D. Lorenz, Carrier-signal selection for sensorless control of PM synchronous machines at zero and very low speeds, *IEEE Transactions on Industry Applications*, 46(1), 167–178, January–February 2010.

15. O. Benjak, D. Gerling, Review of position estimation methods for PMSM drives without a position sensor, part III: Methods based on saliency and signal injection, *International Conference on Electrical Machines and Systems (ICEMS), 2010*, Incheon, Korea, pp. 873–878, October 10–13, 2010.

16. P. L. Jansen, R. D. Lorenz, Transducerless position and velocity estimation in induction and salient AC machines, *IEEE Transactions on Industry Applications*, 31(2), 240–247, March–April 1995.

17. W. Limei, G. Qingding, R. D. Lorenz, Sensorless control of permanent magnet synchronous motor, *The Third International Power Electronics and Motion Control Conference, 2000. Proceedings. IPEMC 2000*, Beijing, China, Vol. 1, pp. 186–190, 2000.

18. M. Linke, R. Kennel, J. Holtz, Sensorless position control of permanent magnet synchronous machines without limitation at zero speed, *IEEE 2002 28th Annual Conference of the IECON 02 [Industrial Electronics Society]*, Seville, Spain, Vol. 1, pp. 674–679, November 5–8, 2002.

19. H. Kim, R. D. Lorenz, Carrier signal injection based sensorless control methods for IPM synchronous machine drives, *Industry Applications Conference, 2004. 39th IAS Annual Meeting. Conference Record of the 2004 IEEE*, Seattle, USA, Vol. 2, pp. 977–984, October 3–7, 2004.

20. O. C. Ferreira, R. Kennel, Encoderless control of industrial servo drives, *12th International Power Electronics and Motion Control Conference, 2006. EPE-PEMC 2006*, pp. 1962–1967, August 30–September 1, 2006.

21. A. Accetta, M. Cirrincione, M. Pucci, G. Vitale, Sensorless control of PMSM fractional horsepower drives by signal injection and neural adaptive-band filtering, Special section fractional horsepower electric drives, *IEEE Transactions on Industrial Electronics*, 59(3), 1355–1366, March 2012.

22. E. Robeischl, M. Schroedl, Optimized INFORM measurement sequence for sensorless PM synchronous motor drives with respect to minimum current distortion, *IEEE Transactions on Industry Applications*, 40(2), 591–598, March–April 2004.

23. R. Wu, G. R. Slemon, A permanent magnet motor drive without a shaft sensor, *IEEE Transactions on Industry Applications*, 27(5), 1005–1011, September–October 1991.

24. L. Harnefors, H. P. Nee, A general algorithm for speed and position estimation of AC motors, *IEEE Transactions on Industrial Electronics*, 47(1), 77–83, February 2000.

25. M. Schrodl, M. Hofer, W. Staffler, Sensorless control of PM synchronous motors in the whole speed range including standstill using a combined INFORM/EMF model, *12th International Power Electronics and Motion Control Conference, 2006. EPE-PEMC 2006*, Portoroz, Slovenia, Vol., pp. 1943–1949, August 30–September 1, 2006.
26. Z. Chen, M. Tomita, S. Doki, S. Okuma, An extended electromotive force model for sensorless control of interior permanent-magnet synchronous motors, *IEEE Transactions on Industrial Electronics*, 50(2), 288–295, April 2003.

Part III

Neural-Based Orthogonal Regression

8

Neural-Based Orthogonal Regression

Giansalvo Cirrincione, PhD

Associate Professor
University of Picardie Jules Verne
Amiens, France

8.1 Introduction: ADALINE and Least Squares Problems

This chapter introduces the well-known concept of linear neural network, in particular, the one made up of one neuron. In literature, it is called with the acronym ADALINE, which historically has two meanings; firstly introduced as ADAptive Linear Neuron, it changed its name when neural networks were less studied because of the attack of the famous book of Minsky and Papert [1] and got its name "ADAptive LINear Element" [2]. With respect to the general neural network model, the ADALINE, in its simplest form, consists only of the adaptive linear combiner, and its output is only the result of this summation. No output activation function is present, or, if preferred, the output function is the identity function. Figure 8.1 shows this simple general ADALINE model.

The output y can then be described by

$$b_j = \sum_{i=1}^{n} a_{ji} x_i = \mathbf{a}_j^T \mathbf{x} \tag{8.1a}$$

where
  a_{ji} is the ith component of the $\mathbf{a}_j$ vector given as input at the jth instant of time
  x_i is the ith component of the x weight vector
  b_j is the corresponding output at the jth instant of time

If the vector $\mathbf{b} = [b_1 \ldots b_j \ldots b_m]^T \in \Re^m$ is considered as well as the matrix $A \in \Re^{m \times n}$ composed of m row vectors $\mathbf{a}_j^T$, then finding the weight vector $\mathbf{x}$ (the problem of *linear parameter estimation*) is generally equivalent to solving the following overdetermined set of linear equations, on the basis of the training set (TS) made up of the expanded matrix $[A;\mathbf{b}]$:

$$A\mathbf{x} \approx \mathbf{b} \tag{8.1b}$$

Generally, $A \in \Re^{m \times n}$ is called *data matrix*, and $\mathbf{b} \in \Re^m$ is called *observation vector*. According to the classical ordinary least squares (OLS) approach, errors are implicitly assumed to be confined to the observation vector. This assumption is however unrealistic. Actually, also the data matrix is affected by noise, like sampling errors, human errors, modeling errors, and measurement errors. In Refs [3,4], some methods are presented to estimate the influence of

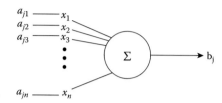

FIGURE 8.1
Linear neuron network.

these errors on the OLS solution. The total least squares (TLS) method is a technique devised to make up for these errors. The TLS problem has been presented for the first time in Ref. [5], where it is solved by using the singular value decomposition (SVD), as proposed in Ref. [4] and more completely in Ref. [36]. This estimation method stems historically from statistics literature, where it is called *orthogonal regression* or errors-in-variables (EIV) regression.* As a matter of fact, the problem of the regression straight line has been considered since last century [6]. The main contributions are in Refs [6–10]. About 30 years ago, this technique has been extended to multivariable cases and later to multidimensional cases (where several observation vectors b are treated), as in Refs [11] and [12]. A complete analysis of the TLS problem can be found in Ref. [13], where the algorithm of Ref. [5] is generalized to the nongeneric case (no, generic TLS), where the initial algorithm failed to find a solution. According to the data least squares (DLS) approach, errors are assumed to be confined only to the data matrix [14]. The DLS case is particularly suitable for certain deconvolution problems, like in system identification or channel equalization [14].

8.2 Approaches to the Linear Regression

8.2.1 OLS Problem

The least-square solution is the one minimizing

$$\min_{b' \in \Re^m} \|\mathbf{b} - \mathbf{b}'\|_2 \text{ with the constraint } \mathbf{b}' \in R(A) \tag{8.2}$$

where $R(A)$ is the column space of A. When a minimum of b′ is found, every vector satisfying

$$A\mathbf{x}' = \mathbf{b}' \tag{8.3}$$

is then called the OLS solution. It corresponds to the point minimizing the following cost function:

$$E_{OLS}(x) = (A\mathbf{x} - \mathbf{b})^T (A\mathbf{x} - \mathbf{b}) \tag{8.4}$$

8.2.2 DLS Problem

The least-square solution is the one minimizing

$$\min_{A'' \in \Re^{mxn}} \|A - A''\|_F \text{ with the constraint } \mathbf{b} \in R(A'') \tag{8.5}$$

---

* In EIV models, the true values of observed variables satisfy unknown but exact linear relationships.

where $\|\ldots\|_F$ is the Frobenius norm. Once a minimum A'' is found, every $\mathbf{x}''$ satisfying

$$A''\mathbf{x}'' = \mathbf{b}' \tag{8.6}$$

is called the DLS solution. In [15, p. 120], it is proved that it corresponds to the minimization of the cost function:

$$E_{DLS}(\mathbf{x}) = \frac{(A\mathbf{x} - \mathbf{b})^T (A\mathbf{x} - \mathbf{b})}{\mathbf{x}^T \mathbf{x}} \tag{8.7}$$

8.2.3 TLS Problem

The least-square solution is the one minimizing

$$\min_{[\hat{A};\hat{\mathbf{b}}] \in \mathfrak{R}^{mx(n+1)}} \left\| [A;\mathbf{b}] - [\hat{A};\hat{\mathbf{b}}] \right\|_F \text{ with the constraint } \hat{\mathbf{b}} \in R(\hat{A}) \tag{8.8}$$

When a minimum of $[\hat{A};\hat{\mathbf{b}}]$ is found, every $\hat{x}$ satisfying

$$\hat{A}\hat{\mathbf{x}} = \hat{\mathbf{b}} \tag{8.9}$$

is called TLS solution. It minimizes the sum of orthogonal squared distances (weighted residues squared sum):

$$E_{TLS}(\mathbf{x}) = \frac{(A\mathbf{x} - \mathbf{b})^T (A\mathbf{x} - \mathbf{b})}{1 + \mathbf{x}^T \mathbf{x}} = \frac{\left\| [A;\mathbf{b}] \left[\mathbf{x}^T; -1 \right] \right\|_2^2}{\left\| \left[\mathbf{x}^T; -1 \right] \right\|_2^2} \tag{8.10}$$

which corresponds to the Raleigh quotient (RQ) of $[A;\mathbf{b}]^T[A;\mathbf{b}]$:

$$E_{MCA}(\mathbf{x}) = \frac{\mathbf{u}^T [A;\mathbf{b}]^T [A;\mathbf{b}]\mathbf{u}}{\mathbf{u}^T \mathbf{u}} \tag{8.11}$$

with the constraint $u_{n+1} = -1$ (u_{n+1} is the last component of the vector $\mathbf{u}$), which represents a hyperplane in the space u, called *TLS hyperplane*. Equation 8.11 shows that the solution can be found by using the minor component analysis (MCA), followed by a normalization of the result.

It can be proven that the TLS solution is parallel to the right singular vector ($\in \mathfrak{R}^{n+1}$) corresponding to the smallest singular value of $[A;\mathbf{b}]$.

8.3 Minor Component Analysis and the MCA EXIN Neuron

The MCA is a technique extracting the second-order statistics of the input signal and plays an ever increasingly important role in data analysis and signal processing.

The *minor components* are the eigenvectors corresponding to the smallest eigenvalues of the data autocorrelation matrix [16].

8.3.1 Some MCA Applications

MCA has several applications, especially in adaptive signal processing. It has been applied to frequency estimation [17,18], bearing [19], beamforming [20], moving target detection [21], and clutter cancelation [22]. It has also been applied to TLS algorithms for parameter estimation [23,24]; Reference [16] shows the relationship between the MCA and the function approximation with the TLS criterion, which enlarges the application domain of the MCA to several engineering areas as well as to cognitive science, for instance, computer vision.

8.3.2 Neural Approach

Several neural networks are able to solve the MCA problem. The only nonlinear neural network is the Hopfield network presented in Mathew et al. [17,18], which, however, has some serious drawbacks (see Cirrincione [15], p. 120). All other existing methods use only *one linear neuron*.

Let a linear neuron be considered with input vector $\mathbf{x}(t) = [x_1(t),\dots, x_N(t)]^T$ and with output $y(t)$:

$$y(t) = \sum_{i=1}^{N} w_i(t)\, x_i(t) = \mathbf{w}^T(t)\mathbf{x}(t) \tag{8.12}$$

where $\mathbf{w}(t) = [w_1(t),\dots, w_N(t)]^T$ is the weight vector. The RQ of the autocorrelation matrix of $\mathbf{x}(t)$ is the Liapounov function of these neurons, employing a gradient descent method for minimizing this function. The most important linear neurons are OJA, OJAn, OJA+ [16], LUO [29,41], FENG, FENG1 [32], and MCA EXIN [25,26,37–41]. This terminology has been introduced in Refs [25,26] and is currently used in literature. In Cirrincione et al. [25], the following propositions are proved by analyzing the RQ properties and the gradient flows of the corresponding ordinary differential equations (ODE) of the earlier neurons:

1. The ODE of LUO, OJAn, and MCA EXIN are equivalent, since they differ only in the Riemann metrics. This implies a similar stability analysis, with the exception of the critical points.

2. The RQ critical points are *singular*, since the Hessian matrix is not invertible in these points. Because of this, the RQ *is not a Morse function** in every open space of the domain containing a critical point.

3. As a result, the phase portrait of the gradient flows is only made up of straight lines along the direction of the RQ eigenvectors, that is, the critical points are not isolated.

---

* A function $f: U \to \mathfrak{R}$, where U is an open subset of $\mathfrak{R}^n$, is a *Morse function* if all of its critical points $\mathbf{x}_0 \in U$ are not singular.

4. In case of a nonsingular critical point, the local stability properties of the gradient flow around this point do not change with the Riemann metrics. Nevertheless, in the case of a singular critical point, the local phase portrait of the gradient flow around the RQ critical point can change substantially with the Riemann metrics [27,28].

8.4 MCA EXIN Neuron

The learning law of the MCA EXIN neuron derives from discretization of the sequential version of the exact gradient flow of the RQ:

$$\mathbf{w}(t+1) = \mathbf{w}(t) - \frac{\alpha(t)y(t)}{\mathbf{w}^T(t)\mathbf{w}(t)}\left[\mathbf{x}(t) - \frac{y(t)\mathbf{w}(t)}{\mathbf{w}^T(t)\mathbf{w}(t)}\right] \tag{8.13}$$

where $\alpha(t)$ is the learning rate.

8.4.1 Convergence during the First Transient Phase

This analysis deals with the ODE MCA EXIN and is therefore limited by the ODE assumptions: it is a first approximation theory of the time behavior of the neuron, particularly until the minimum is reached for the first time. In Cirrincione et al. [25,26], the following theorem is proved:

Theorem 8.1 (Convergence of the MCA EXIN)

Let R be the n × n autocorrelation matrix input data, with eigenvalues $0 \le \lambda_N \le \lambda_{N-1} \le \cdots \le \lambda_1$ *and correspondingly orthonormal eigenvectors* $\mathbf{z}_N, \mathbf{z}_{N-1},\ldots, \mathbf{z}_1$. *If* $\mathbf{w}(0)$ *satisfies* $\mathbf{w}^T(0)$ $\mathbf{z}_N \ne 0$ *and* λ_n *is single, then, in the limits of validity of the ODE approximation, for MCA EXIN, it holds that*

$$\left\|\mathbf{w}(t)\right\|_2^2 \approx \left\|\mathbf{w}(0)\right\|_2^2 \quad t > 0 \tag{8.14}$$

$$\mathbf{w}(t) \rightarrow \pm\left\|\mathbf{w}(0)\right\|_2 \mathbf{z}_N \quad \vee \quad \mathbf{w}(t) \rightarrow \infty \tag{8.15}$$

The phase portrait within the limits of the ODE approximation is explained in the following remarks [29,30] (also valid for the other neurons).

- The RQ has a critical direction which is a global minimum and a critical direction which is a global maximum. The other critical directions are saddles.
- The critical direction is a global minimum in the direction of any eigenvector with a larger eigenvalue and a maximum in the direction of any eigenvector with a smaller eigenvalue.
- The critical direction is a global minimum in the direction of any linear combination of eigenvectors with larger eigenvalues and a maximum in the direction of any linear combination of eigenvectors with smaller eigenvalues.

- Every saddle direction has an infinity of cones of attraction, each with centers on this direction and containing the directions of the linear combinations of the eigenvectors with bigger eigenvalues than the saddle eigenvalue. It has also an infinity of cones of repulsion, each with axes orthogonal to the axes of the cones of attraction, with centers on the saddle direction and containing the directions (dimensions of escape) of the linear combinations of the eigenvectors with smaller eigenvalues than the saddle eigenvalue.
- The hyperplane through the origin and perpendicular to the minimum direction is made up of points which are maxima along the minimum direction (hypercrest along the minimum direction). The hyperplane through the origin and perpendicular to the maximum direction is made up of points which are minima along the maximum direction (hypervalley along the maximum direction). The hyperplane through the origin and perpendicular to a saddle direction is made up of two kinds of points: points given by the intersection with the saddle cone of repulsion, which are minima along the saddle direction (hypervalley along the saddle direction), and points given by the intersection with the saddle cone of attraction, that is, all the other points, which are maxima along the saddle direction (hypercrest along the saddle direction).

Figure 8.2 shows a 3-D portrait of the RQ as a function of the MCA EXIN ODE.

If the weight vector modulus is less than unity, the weight vector of the MCA EXIN neuron reaches the minor component direction more quickly than OJAn, which is quicker that LUO. The contrary is true if the weight vector modulus is greater than unity. If the weight vector lies on the unity sphere, the three neurons have the same speed. Moreover, their time constants are inversely proportional to $\lambda_{N-1} - \lambda_N$: eigenvalues well far away from each other result in a quicker response. OJA and OJAn have the same time constant.

It seems that OJA and LUO can give better results than MCA EXIN if large initial conditions are chosen. Unfortunately, this is not a good choice because of the flatness of the RQ landscape and the fact that, for the learning laws of all the neurons different from the MCA EXIN

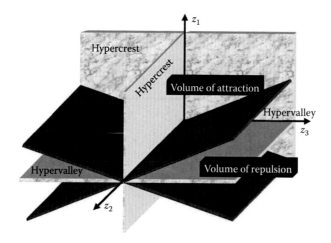

FIGURE 8.2
MCA EXIN ODE stability analysis and phase portrait (3-D case). (From Cirrincione, G. et al., *IEEE Trans. Neural Netw.*, 13(1), 160, January 2002; Cirrincione, G. and Cirrincione, M., *Neural Based Orthogonal Data Fitting: The EXIN Neural Networks*, Series: Adaptive and Learning Systems for Signal Processing, Communications and Control, Wiley & Sons, New York, 255pp., November 2010.)

neuron, it is difficult to stop the algorithm (see the divergence analysis in Ref. [31]). From this analysis, the best choice for the MCA EXIN neuron would be *null initial conditions*. However, this is not possible, which can be easily verified. Moreover, too low values of the weight vector may lead up to strong oscillation (see Equation 8.13). In the following, it will be shown that the MCA neuron can be provided with a particular scheduling (DLS scheduling) which allows it to start from infinitesimal initial conditions, to keep low weight values for some time, and to follow a stable trajectory in the phase portrait: this version is called MCA EXIN+.

8.4.2 Dynamic Behavior of the MCA Neuron

The analysis of the temporal behavior of all MCA neurons can be carried out by using the stochastic discrete laws, since the mere use of the ODE approximation fails to reveal some important features of these neurons.

The weight vectors OJAn, LUO, and MCA EXIN have constant modulus during their temporal evolution according to the ODE approximation. Nevertheless, in Refs [25,26], it is shown that the modulus value always increases. With respect to other neurons, MCA EXIN has the largest increment for weight moduli less than unity: this accounts for its large transient oscillations and the fact that the MC direction (the global minimum of the RQ) is reached more rapidly with null or very low initial conditions. Moreover, the beginning of the divergence, as shown in Ref. [31], implies large moduli: in this case, MCA EXIN has the smallest modulus increment, which means that its divergence is the slowest. In a nutshell, the dynamic behavior of all MCA neurons, with the exception of FENG, FENG1, and OJA+, is as follows (see Figure 8.3 for the 3-D case):

1. An initial transient phase.
2. There are fluctuations around the locus of constant modulus, but with an increasing bias; the fluctuations are functions of the learning rate.
3. Ability to arrive in the direction desired.
4. By increasing the modulus, the fluctuations push the weight to another critical point and so on, until ∞.

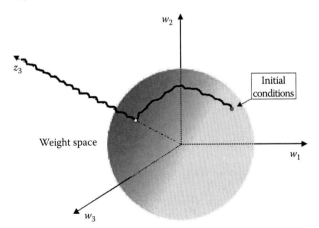

FIGURE 8.3
Dynamic behavior of the weights for MCA EXIN, LUO, OJAn, and OJA (for particular initial conditions): 3-D case. (From Cirrincione, G. et al., *IEEE Trans. Neural Netw.*, 13(1), 160, January 2002; Cirrincione, G. and Cirrincione, M., *Neural Based Orthogonal Data Fitting: The EXIN Neural Netw.*, Series: Adaptive and Learning Systems for Signal Processing, Communications and Control, Wiley & Sons, New York, 255pp., November 2010.)

In Refs [25,26,31], the existence of a *"sudden divergence"* at a finite time is proved for some linear neurons. For LUO, this time depends on the extension of the eigenvalue spectrum of the autocorrelation matrix R. For OJA and OJA+, it depends on the inverse of the smallest eigenvalue of R, which means that noisy data can worsen the divergence phenomenon. Figure 8.4 shows an example: OJA has the worst performance, MCA EXIN diverges very slowly, and OJA+ converges. Figure 8.5 shows the deviation from the MC direction for

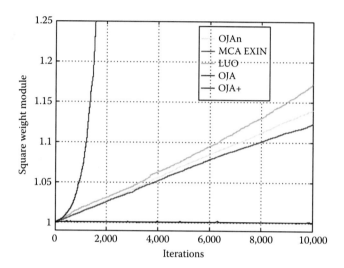

FIGURE 8.4
The divergences (represented by the evolution of the squared weight modulus) of the MCA linear neurons with initial conditions equal to the solution. (From Cirrincione, G. et al., *IEEE Trans. Neural Netw.*, 13(1), 160, January 2002; Cirrincione, G. and Cirrincione, M., *Neural Based Orthogonal Data Fitting: The EXIN Neural Networks*, Series: Adaptive and Learning Systems for Signal Processing, Communications and Control, Wiley & Sons, New York, 255pp., November 2010.)

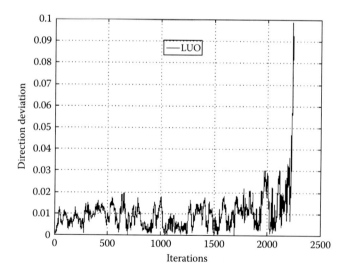

FIGURE 8.5
The deviation from the MC direction for LUO, measured as the squared sine of the angle between the weight vector and the MC direction. (From Cirrincione, G. et al., *IEEE Trans. Neural Netw.*, 13(1), 160, January 2002.)

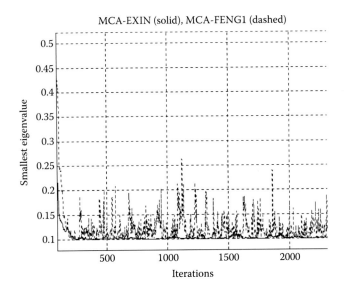

FIGURE 8.6
Smallest eigenvalue computation by MCA EXIN (lower curve) and FENG for a 3-D well-conditioned autocorrelation matrix. (From Cirrincione, G. et al., *IEEE Trans. Neural Netw.*, 13(1), 160, January 2002; Cirrincione, G. and Cirrincione, M., *Neural Based Orthogonal Data Fitting: The EXIN Neural Networks*, Series: Adaptive and Learning Systems for Signal Processing, Communications and Control, Wiley & Sons, New York, 255pp., November 2010.)

LUO. The effects of the weight increments orthogonally to the MC direction and the beginning of the sudden divergence are apparent.

Because of divergence, it is difficult to come up with a suitable stop criterion. In Refs [25,26], some techniques are developed to solve this problem for MCA EXIN as well as an analysis of FENG and FENG1, whose behaviors correspond to an anti-Hebbian law with resulting large oscillations around the solution, unlike MCA EXIN which has the smallest variance around the solution. Figure 8.6 shows an example of this.

8.4.3 Dynamic Stability and Learning Rate

In Refs [25,26], an original technique is shown for analyzing the fluctuations of the weight vector $\Psi = \mathbf{w}$ as a function of the learning rate α. With this regard, the time analysis of the weight vector subspace is presented in Refs [25,26] and Figure 8.7 shows an example. Particularly, the following facts have been proved:

- There exists a new form of divergence, *the instability divergence.*
- A small value of α results in a low learning time.
- It is difficult to find a good value of α for avoiding the instability divergence.
- Both the transient phase and the accuracy of the solution depend on the choice of α.

LUO, OJAn, and MCA EXIN are iterative algorithms with high variance and low bias. However, OJAn and, above all, LUO, require many more iterations to converge than MCA EXIN because they have more fluctuations around the MC direction and cannot be stopped earlier by a stop criterion that requires the flatness of the weight time evolution. On the contrary, OJA, OJA+, and FENG are algorithms with low variance and high bias. However, FENG has larger fluctuations

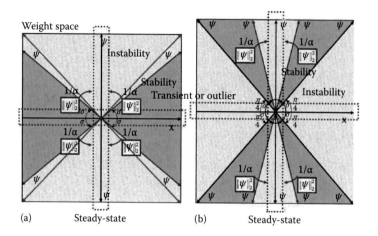

FIGURE 8.7

Stability subspaces of the weight vector $\Psi = \mathbf{w}$ with respect to the input vector (2-D case) for OJA and OJA+ (a) and MCA EXIN (b). Arrows indicate the time evolution of Ψ. (From Cirrincione, G. et al., *IEEE Trans. Neural Netw.*, 13(1), 160, January 2002; Cirrincione, G. and Cirrincione, M., *Neural Based Orthogonal Data Fitting: The EXIN Neural Networks*, Series: Adaptive and Learning Systems for Signal Processing, Communications and Control, Wiley & Sons, New York, 255pp., November 2010.)

and is unreliable for near-singular matrices. Only in the case of MCA EXIN, the choice of low initial conditions results in a large variance in the first transient phase of the weights but not at the expense of the final accuracy, since, as seen in Refs [25,26], fluctuations go down to zero during the divergence phase.

8.4.4 Numerical Considerations

8.4.4.1 Computational Cost

The MCA learning laws are iterative algorithms with a different computational cost per iteration, shown in Table 8.1, showing that OJA has the lowest cost. All costs depend on the dimensionality n of the data: for high-dimensional data, all learning laws have the same cost with the exception of OJA which has as much as 33% fewer flops per iteration.

8.4.4.2 Quantization Errors

Quantization errors can worsen the solution of the gradient-based algorithms with respect to the performance achievable in infinite precision. These errors accumulate in time

TABLE 8.1

Cost per Iteration for the MCA Neurons

	Flops per Iteration
OJA+	$8n + 3$
LUO	$8n + 1$
EXIN	$8n + 1$
OJAn	$8n$
FENG	$8n - 1$
OJA	$6n$

Source: From Cirrincione, G. et al., *IEEE Trans. Neural Netw.*, 13(1), 160, January 2002.

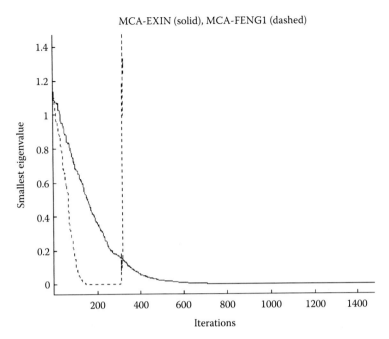

FIGURE 8.8
Numerical sudden divergence of FENG in case of a singular autocorrelation matrix. (From Cirrincione, G. et al., *IEEE Trans. Neural Netw.*, 13(1), 160, January 2002; Cirrincione, G. and Cirrincione, M., *Neural Based Orthogonal Data Fitting: The EXIN Neural Networks*, Series: Adaptive and Learning Systems for Signal Processing, Communications and Control, Wiley & Sons, New York, 255pp., November 2010.)

without bound, leading in the long term (tens of millions of iterations) to an eventual overflow [32]. This kind of divergence is here called *numerical divergence.*

The degradation of the solution is proportional to the spread of the eigenvalue spectrum of the input autocorrelation matrix, which is relevant for near-singular matrices. An example with FENG is shown in Figure 8.8 showing the computation of the smallest eigenvalue of a singular matrix whose eigenvalues are 0, 1, and 1.5. Notice the finite-time divergence as a result of numerical problems.

8.4.5 Acceleration Techniques

The MCA learning laws are instantaneous adaptive gradient algorithms and then work sequentially. Nonetheless, batch techniques can further accelerate the MCA. If incoming inputs are collected in blocks and are fed to the neuron, which changes its weights only after the whole block presentation, all methods typical of the batch learning can be used. Despite this, it is not practical to use Newton and quasi-Newton techniques because the RQ Hessian matrix at the minimum is singular and then the inverse does not exist. On the other hand, the conjugate gradient approach can face this problem [33, pp. 256–259] [34] and can be used for accelerating MCA EXIN in block mode [15].

8.4.6 Simulations

The following simulations use, as data, a zero mean Gaussian random vector x(t) generated by an autocorrelation matrix R whose spectrum is chosen in advance. The goal of this approach is the analysis of the behavior of the MCA laws with respect to the

TABLE 8.2

Total Cost of the MCA Learning Laws for
Autocorrelation Matrices of Increasing Size

	κ_2	EXIN	LUO	FENG	OJA
3	20	42726	37466	41133	20888
5	40	65966	59131	60345	42942
7	60	183033	640074	742346	div
8	70	205968	706075	830000	div
9	80	228681	965544	a	div
10	90	252002	1061098	div	div
15	140	366255	div	div	div
25	240	2003755	div	div	div
50	490	3725868	div	div	div
75	740	4488496	div	div	div
100	990	6258001	div	div	div

Source: From Cirrincione, G. et al., *IEEE Trans. Neural Netw.*,
13(1), 160, January 2002.
Note: div, divergence; [a]Inaccurate result.

dimensionality n of data and the conditioning of R. In the first group of simulations, the components of the initial weight vector are chosen randomly in [0, 1]. λ_n is always chosen equal to 1. The other eigenvalues are given by the law $\lambda_i = n - i$, then the condition number $\kappa_2(R) = \lambda_1/\lambda_n$ increases with n, but R always remains a well-conditioned matrix. Table 8.2 shows, for four MCA laws, the best results,* in terms of total flops cost obtained for each value of n. With the exception of EXIN, all other laws diverge for low values of n: from OJA which diverges for only n = 7 to LUO which diverges for n = 10. This problem can be explained by the choice of the initial conditions: by increasing the number of components, the initial weight modulus increases and quickly becomes greater than 1. About OJA, there is sudden divergence. About FENG, there is instability divergence. Indeed fluctuations only depend on the modulus and the data inputs. The modulus is large and remains so because of the too large oscillations which prevent the weights from approaching the MC direction. As a consequence, this generates more and more instability until the finite-time divergence. Obviously, for increasing n, the divergence is anticipated. About LUO, two explanations are possible: sudden and instability divergence. Nevertheless, in these experiments, the divergence is due to instability, because it is accompanied by very large oscillations and certainly occurs before the sudden divergence. Figures 8.9 through 8.11 show some of these experiments for, respectively, n = 5, 11, and 100. Figure 8.9 confirms that FENG has instability divergence (see peaks before divergence). Figure 8.11 shows the good results obtained with MCA EXIN.

In order to avoid starting with too high initial conditions, a second group of experiments has been done where the initial weight vector is the same as in the first experiment, but after dividing its norm by 10. Table 8.3 shows the results about the divergence of all MCA laws with the exception of MCA EXIN, which always converges. Obviously, the divergence occurs later than in the first group of experiments because the weights are lower. Figures 8.12 and 8.13 show some results. A deeper discussion of these results can be found in Refs [25,26].

---

* For each value of n, several experiments have been done by changing the learning rate (initial and final values, monotonic decreasing law) and only the best results have been reported.

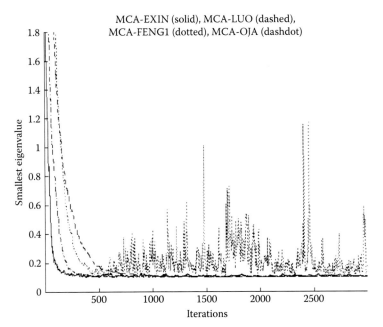

FIGURE 8.9
Computation of the smallest eigenvalue of a 5 × 5 autocorrelation matrix for the first choice of initial conditions. (From Cirrincione, G. et al., *IEEE Trans. Neural Networks*, 13(1), 160, January 2002; Cirrincione, G. and Cirrincione, M., *Neural Based Orthogonal Data Fitting: The EXIN Neural Netw.*, Series: Adaptive and Learning Systems for Signal Processing, Communications and Control, Wiley & Sons, New York, 255pp., November 2010.)

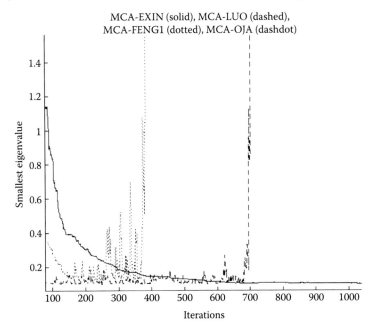

FIGURE 8.10
Computation of the smallest eigenvalue of an 11 × 11 autocorrelation matrix for the first choice of initial conditions. (From Cirrincione, G. et al., *IEEE Trans. Neural Netw.*, 13(1), 160, January 2002; Cirrincione, G. and Cirrincione, M., *Neural Based Orthogonal Data Fitting: The EXIN Neural Networks*, Series: Adaptive and Learning Systems for Signal Processing, Communications and Control, Wiley & Sons, New York, 255pp., November 2010.)

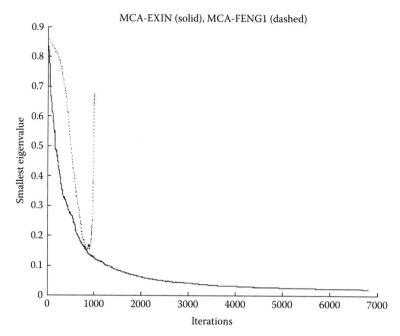

FIGURE 8.11

Computation of the smallest eigenvalue of a 100×100 autocorrelation matrix for the first choice of initial conditions. (From Cirrincione, G. et al., *IEEE Trans. Neural Netw.*, 13(1), 160, January 2002.)

TABLE 8.3

Divergence of the MCA Learning Laws for
Autocorrelation Matrices of Increasing Size

dim	LUO	FENG	OJA	OJAn	OJA+
3	conv.	conv.	conv.	conv.	conv.
7	conv.	conv.	conv.	conv.	conv.
9	conv.	413	conv.	conv.	conv.
10	conv.	460	conv.	conv.	conv.
15	conv.	531	conv.	conv.	conv.
18	conv.	371	conv.	conv.	conv.
20	conv.	700	180	conv.	65
25	conv.	27	27	conv.	26
30	conv.	27	27	1848	17
40	conv.	400	17	975	12
50	conv.	370	6	670	12
60	conv.	540	3	550	14
70	conv.	260	7	520	12
80	545	220	5	400	6
90	8	7	8	355	8
100	8	8	8	250	5

Source: From Cirrincione, G. et al., *IEEE Trans. Neural Netw.*, 13(1), 160, January 2002.

Note: The numbers show at which iteration the divergence occurs; conv., convergence.

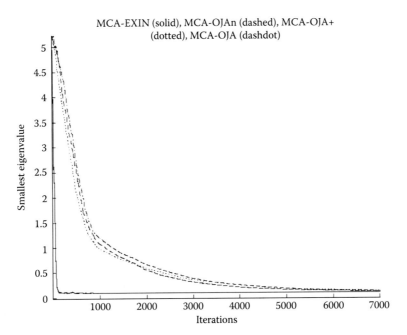

FIGURE 8.12
Computation of the smallest eigenvalue of a 7×7 autocorrelation matrix for the first choice of initial conditions. (From Cirrincione, G. et al., *IEEE Trans. Neural Netw.*, 13(1), 160, January 2002.)

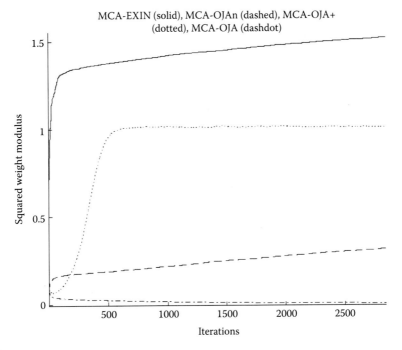

FIGURE 8.13
Plot of the squared weight moduli in the case of a 10×10 autocorrelation matrix for the first choice of initial conditions. (From Cirrincione, G. et al., *IEEE Trans. Neural Netw.*, 13(1), 160, January 2002.)

8.4.7 Conclusions and Prospects for the MCA Neuron

The MCA is becoming ever increasingly important not only in signal processing but also in data analysis (orthogonal regression, TLS): important applications in electrical engineering systems are the topic of this book, but others do exist, as in computer vision for the estimation of the parameters in the essential matrix (structure from motion) [15].

MCA EXIN is by far the best MCA law. It has the best convergence in the MC direction and the slowest divergence, and has neither sudden divergence nor instability divergence problems. It works properly in higher dimension spaces as well and has been used throughout this book in real applications. It is robust to outliers because of the presence of the squared modulus weight vector in the denominator of the weight increment law. Nevertheless, there exists a variant (NMCA EXIN [35]) which is robust to outliers since it implements the M estimators theory [36]. MCA EXIN can be easily stopped; it is a high-variance/small-bias algorithm with a good dynamic performance thanks to inertia due to high-modulus weight vectors. It has the same computational burden as the other MCA laws, except for OJA.

If the sign of the weight vector increment is changed in Equation 8.13, the PCA EXIN learning law is obtained [15] which performs the principal components analysis (PCA) of the data.

A variant of the MCA EXIN (with a different learning rate) which does not diverge is presented in Refs [37,38] and uses a new theory for analyzing the convergence. In Refs [39–41], the MCA EXIN neuron is used for implementing a nonlinear Volterra filter.

8.5 TLS EXIN Neuron

The complete theory of the TLS EXIN neuron can be found in Refs [15,26].

In Section 8.2.1, the cost function $E_{TLS}(\mathbf{x})$ has been introduced in (8.10), which is to be minimized to find the TLS solution and rewritten in the following for convenience:

$$E_{TLS}(\mathbf{x}) = \frac{(A\mathbf{x} - \mathbf{b})^T (A\mathbf{x} - \mathbf{b})}{1 + \mathbf{x}^T \mathbf{x}} = \frac{\left\| [A;\mathbf{b}] [\mathbf{x}^T;-1] \right\|_2^2}{\left\| [\mathbf{x}^T;-1] \right\|_2^2} \tag{8.10}$$

It is apparent then that this function is the RQ of $[A;\mathbf{b}]^T[A;\mathbf{b}]$ constrained to the TLS hyperplane $x_{n+1} = -1$, which means that the TLS solution is parallel to the right singular value $\in \Re^{n+1}$ corresponding to the minimum singular value of $[A;\mathbf{b}]$. Define

$$\xi_i = \left[\mathbf{a}_i^T; b_i \right]^T \quad \text{and} \quad y_i = \Psi^T \xi_i \tag{8.16}$$

with $\mathbf{a}_i^T$ being the ith row of A and y_i being the output of the MCA linear neuron of weight vector $\Psi \in \Re^{n+1}$ and input $\xi_i \in \Re^{n+1}$. Its learning law minimizes the RQ of the autocorrelation matrix of the input data R which is *equivalent* to $[A;\mathbf{b}]^T[A;\mathbf{b}]/m$ where m is the number of rows. Indeed,

$$R = R\left[\xi(t)\xi^T(t) \right] = \begin{bmatrix} \bar{R} & \mathbf{r} \\ \mathbf{r}^T & \Gamma \end{bmatrix} \approx \frac{1}{m} \begin{bmatrix} A^T A & A^T \mathbf{b} \\ \mathbf{b}^T A & \mathbf{b}^T \mathbf{b} \end{bmatrix} = \frac{[A;\mathbf{b}]^T [A;\mathbf{b}]}{m} \tag{8.17}$$

Thus, to find the TLS solution, the MCA solution must be normalized to have the last component of the solution equal to –1. This way of finding the TLS solution poses two fundamental problems:

1. An additional operation of division must be made because of the normalization of the output.
2. All the problems on the behavior of the MCA neurons, cited in Section 8.4, are valid here too.

If the weight vector is constrained to lie on the TLS hyperplane (i.e., losing a degree of freedom), it can be argued that all the convergence problems, which are the consequence of the degeneracy property of the Rayleigh quotient, are no more valid because on the TLS hyperplane the only critical points are the intersections with the critical straight lines, and therefore, they are isolated (it implies the global asymptotic stability of the solution on the hyperplane plus the possible solution ∞). This reasoning, illustrated in Figure 8.14 for the 3-D space $\Psi \in \mathfrak{R}^3$, derives from Section 8.5. Moreover, the MCA landscape changes in the following way:

- The MCA attraction and repulsion volumes become volumes of the same kind for each saddle point. The intersections of the hypervalleys and hypercrests with the hyperplane TLS result in $(n - 1)$-dimensional planes which are, respectively, constituted of either the minima or the maxima with respect to the direction of their orthogonal vector. In Figure 8.14, the TLS plane is divided into two zones: one attraction zone and one repulsion zone (partially shown in the figure). The straight

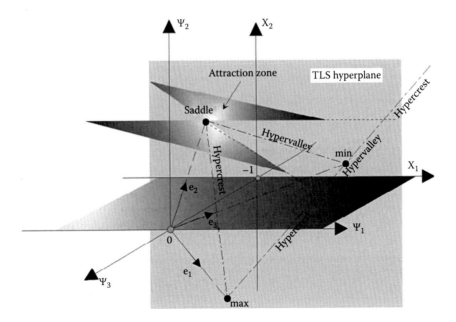

FIGURE 8.14
Cost landscape stability analysis for MCA EXIN and TLS EXIN (TLS hyperplane in light gray) in the $\mathfrak{R}^3$ space. (From Cirrincione, G. and Cirrincione, M., *Neural Based Orthogonal Data Fitting: The EXIN Neural Networks*, Series: Adaptive and Learning Systems for Signal Processing, Communications and Control, Wiley & Sons, New York, 255pp., November 2010.)

line passing in the saddle point and the maximum is a *crest* along the maximal direction. The straight line passing in the saddle point and the minimum is a *valley* along the minimal direction. The straight line passing in the minimum and the maximum is a crest in the saddle direction, except at the intersection with the repulsion zone which is a valley in the saddle direction.

- For MCA, the hypercrest along the minimal direction is not a barrier (limit of the attraction area) for the minimum. Indeed, if initial conditions are chosen for the learning law inside one of the half-spaces delimited by the hypercrest, the weight vector always converges toward the minimal direction in the orientation of the half-space containing the initial conditions. This reasoning is not valid any longer in the TLS case, since the crest is a barrier. In this case, there is only a minimum point and it is then compulsory to choose the initial conditions in the half-space containing this point. This crest is not the only TLS barrier: it is necessary to consider it with the saddle volumes.

Replacing in Equation 8.13 $\Psi(t) = \mathbf{w}(t)$ with $[\mathbf{x}^T(t); -1]^T$ and ξ_i with $\left[\mathbf{a}_i^T; b_i\right]^T$, i being the index of the $[A;\mathbf{b}]$ row presented as input to the neuron at the time t, and taking only the first n components yields

$$\mathbf{x}(t+1) = \mathbf{x}(t) - \alpha(t)\gamma(t)\mathbf{\alpha}_i + [\alpha(t)\gamma^2(t)]\mathbf{x}(t) \tag{8.18}$$

where

$$\gamma(t) = \frac{\delta(t)}{1 + \mathbf{x}(t)^T\mathbf{x}(t)} \tag{8.19}$$

and

$$\delta(t) = \mathbf{x}^T(t)\mathbf{\alpha}_i - b_i \tag{8.20}$$

This is the *TLS EXIN learning law*. The TLS EXIN neuron is a linear unit with n inputs (vector $\mathbf{a}_i$), n weights (vector $\mathbf{x}$), one output (scalar $y_i = \mathbf{x}^T\mathbf{a}_i$), and one training error (scalar $\delta(t)$). With this typology, the training is considered as *supervised*, being b_i the target. The quantity in brackets is positive: it implies that the second term in Equation 8.18 is a *reverse ridge regression*.

TLS EXIN can work either sequentially or in batch and block modes. Several methods for accelerating the TLS EXIN convergence speed have been implemented (in mode block and batch).

- To improve conjugate gradient (CG) methods, the scaled conjugate gradient (SCG) [54] has been employed. It combines the CG with the trust region model of the Levenberg–Marquardt. SCG does not need an a priori parameter chosen by the user.
- The Broyden–Fletcher–Goldfarb–Shanno (BFGS) algorithm [42] can be used, since Newton or quasi-Newton can be employed, since the n × n Hessian matrix of $E_{TLS}(\mathbf{x})$ is proven to be positive definite and therefore is not singular [15,26].

8.5.1 Stability Analysis (Geometrical Approach)

Let the singular value decomposition (SVD) of the matrix $[A;\mathbf{b}]$ be

$$[A;B] = U\Sigma V^T \tag{8.21}$$

where

$$\Sigma = diag(\sigma_1, \ldots, \sigma_{n+1}) \in \Re^{mx(n+1)}, \quad \sigma_1 \geq \cdots \geq \sigma_{n+1} \geq 0 \tag{8.22}$$

$$V = [\mathbf{v}_1, \ldots, \mathbf{v}_{n+1}], \quad \mathbf{v}_i \in \Re^{n+1}$$

$$U = [\mathbf{u}_1, \ldots, \mathbf{u}_m], \quad \mathbf{u}_i \in \Re^m \tag{8.23}$$

And let the SVD of the matrix A be

$$A = U'\Sigma'V'^T \tag{8.24}$$

where

$$\Sigma' = diag(\sigma_1', \ldots, \sigma_{n+1}') \in \Re^{mxn}, \quad \sigma_1' \geq \cdots \geq \sigma_n' \geq 0 \tag{8.25}$$

$$V' = [\mathbf{v}_1', \ldots, \mathbf{v}_n'], \quad \mathbf{v}_i' \in \Re^n$$

$$U' = [\mathbf{u}_1', \ldots, \mathbf{u}_m'], \quad \mathbf{u}_i \in \Re^m \tag{8.26}$$

Then it follows

$$A^T A = V'\Lambda'V'^T \tag{8.27}$$

where $\Lambda = diag(\lambda_1, \ldots, \lambda_n) = \Sigma'^T\Sigma$, with $\lambda_i = \sigma_i'^2$.

The error function (8.10) can be considered as the ratio between two definite positive quadratic forms, which leads up to a geometrical analysis of the family of equilevel hypersurfaces:

$$E_{TLS}(\mathbf{x}) = \gamma \tag{8.28}$$

where γ plays the role of the family parameter.

If the unknown vector $\mathbf{x}$ is substituted by $\mathbf{y} + \mathbf{x}_c(\gamma)$ (translation), where

$$\mathbf{x}_c(\gamma) = \left[A^T A - \gamma I_n\right]^{-1} A^T \mathbf{b} \tag{8.29}$$

coincides with the center of the hypersurface (hyperconic) with level γ, followed by a rotation $\mathbf{z} = V'^T\mathbf{y}$, then the hyperconic with level γ is represented by

$$\mathbf{z}^T\left[\Lambda - \gamma I_n\right]\mathbf{z} = g(\gamma) \tag{8.30}$$

where

$$g(\gamma) = \mathbf{b}^T A V'\left[\Lambda - \gamma I_n\right]^{-1} V'A^T\mathbf{b} - \mathbf{b}^T\mathbf{b} + \gamma \tag{8.31}$$

If the vector $\mathbf{q} \equiv [q_1, q_2, \ldots, q_n]^T = V'A^T\mathbf{b}$ is introduced, then Equation 8.31 becomes

$$g(\gamma) = \sum_{i=1}^{n} \frac{q_i^2}{\lambda_i - \gamma} - \mathbf{b}^T\mathbf{b} = \mathbf{q}^T(\Lambda - \gamma I_n)^{-1}\mathbf{q} + \gamma - \mathbf{b}^T\mathbf{b} \qquad (8.32)$$

The components q_i are called *convergence keys* because of their fundamental role in the analysis of the TLS convergence domain.

The zeros of Equation 8.32 coincide with the squared singular values of the matrix $[A;\mathbf{b}]$ and with the values of $E_{TLS}(\mathbf{x})$ at its critical points (Corollary 92 of Ref. [26]). It can be therefore concluded that Equation 8.32 is a version of the secular TLS equation [3,11] and that the smallest zero of $g(\gamma)$, that is, $\gamma_{min} = \sigma_{n+1}^2$, gives the level of the TLS solution. There exists one and only one value γ_{min} of γ in the interval $(0, \lambda_n)$ such that $g(\gamma_{min}) = 0$. As a consequence, the equilevel hyperconics of Equation 8.30 are $(n - 1)$-dimensional hyperellipsoids $\forall \gamma \in (\gamma_{min}, \lambda)$. For $\gamma = \gamma_{min}$, the hyperconic reduces into one point $\mathbf{z} = \mathbf{y} = \mathbf{0}$. This point corresponds to the unique minimum of $E_{TLS}(\mathbf{x})$, and its position is given by

$$\hat{\mathbf{x}} = \mathbf{x}_c(\gamma_{min}) = \left[A^TA - \gamma I_n\right]^{-1}A^T\mathbf{b} = \left(A^TA - \sigma_{n+1}^2 I\right)^{-1}A^T\mathbf{b} \qquad (8.33)$$

that is the TLS solution [23].

The asymptotes of $g(\gamma)$ are given by the squared SV $\sigma_i'^2 = \lambda_i$ of the data matrix A. They are between the zeros σ_i^2, complying with the interlace theorem, as shown in Figure 8.15.

The cost function TLS EXIN of an n-dimensional unknown $\mathbf{x}$ has $n + 1$ critical points: one minimum, $n - 1$ saddle points, and one maximum.

Figure 8.16 shows an example of $E_{TLS}(\mathbf{x})$ for n = 2 (one minimum, one saddle point, and one maximum). The equilevel curves are better described in Figure 8.17 that also shows the critical points locus and one section of the cost function in the direction of the straight line passing along the minimum and the saddle point, which is a valley along the maximal direction. There appears also the *barrier* given by the saddle point in this section: a gradient descent method with initial conditions on the left of the saddle point cannot reach the minimum.

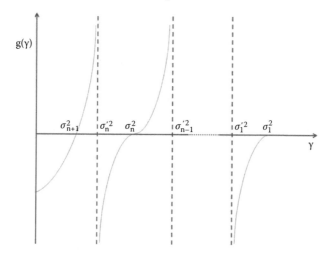

FIGURE 8.15
The $g(\gamma)$ as a function of the $E_{TLS}(\mathbf{x})$ level. (From Cirrincione, G. and Cirrincione, M., *Neural Based Orthogonal Data Fitting: The EXIN Neural Networks*, Series: Adaptive and Learning Systems for Signal Processing, Communications and Control, Wiley & Sons, New York, 255pp., November 2010.)

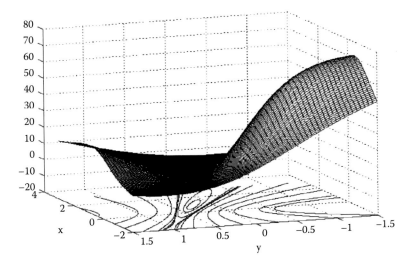

FIGURE 8.16
TLS cost landscape and corresponding equilevel curves for an overdetermined system of the two unknowns x and y. (From Cirrincione, G. and Cirrincione, M., *Neural Based Orthogonal Data Fitting: The EXIN Neural Networks*, Series Adaptive and Learning Systems for Signal Processing, Communications and Control, Wiley & Sons, New York, 255pp., November 2010.)

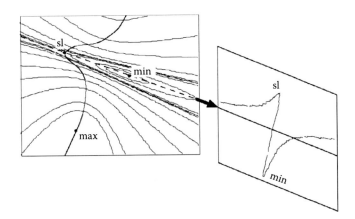

FIGURE 8.17
TLS equilevel curves for the benchmark overdetermined system with the corresponding critical points and the barrier of convergence, together with the section along the straight line passing through the minimum and the saddle. (From Cirrincione, G. and Cirrincione, M., *Neural Based Orthogonal Data Fitting: The EXIN Neural Networks*, Series: Adaptive and Learning Systems for Signal Processing, Communications and Control, Wiley & Sons, New York, 255pp., November 2010.)

8.5.2 Convergence Domain

The TLS domain of convergence is affected by the composed action of the following frontiers:

1. *The barriers (asymptotes)*: The TLS barrier convergence follows the **z** directions asymptotically: that is the direction of the eigenvalues v'_j for $j \neq n$. Furthermore, in every plane $z_i z_j$ with $j < i$, there is a unique barrier for the $\mathbf{v}'_i$ direction, parallel to $\mathbf{v}'_j$. The origin always lies between the minimum and the barrier.

2. *The saddle cone projections (volumes) in the TLS hyperplane*: These repulsive cone projections border that part of the frontier between the saddle and the corresponding barrier.

3. *The maximum locus tangent*: If the maximum is close to the vertical asymptote, it approximately corresponds to this asymptote. With respect to a vertical asymptote corresponding to a far away eigenvalue λ_i ($\lambda_i \ll \lambda_1$), its effect is not negligible.

4. *The saddle-maximum hypercrest projection*: In the TLS hyperplane, the hyperplane representing the maxima in the minimum direction projects onto an $(n-1)$-dimensional plane which corresponds to a straight line in every plane $z_i z_j$.

The origin of the TLS hyperplane is always between the saddle-maximum hypercrest projection and the TLS solution (Theorem 113 in Ref. [25]), as a consequence of which the fundamental theorem of the TLS EXIN (Theorem 114 in Ref. [25]) can be deduced.

Theorem 8.2 (Origin and TLS Domain of Convergence)

The TLS origin belongs to the TLS domain of convergence.

This theorem shows the existence of universal initial conditions: with null initial conditions, the convergence is always ensured.

As an example, Figures 8.18 and 8.19 show the domain of convergence of the generic TLS benchmark problem shown in Figure 8.16. Figure 8.18 shows that the domain of convergence for a certain number of initial conditions for the sequential TLS EXIN neuron (it is also valid for every gradient flow of the TLS energy): the initial conditions for which the neuron converges are green. The dark blue straight line represents the asymptote/barrier z_1. The bottom part of the frontier is given by the asymptote because the maximum is close to it. For increasing x_2, the frontier is given by the action of the saddle-maximum line and

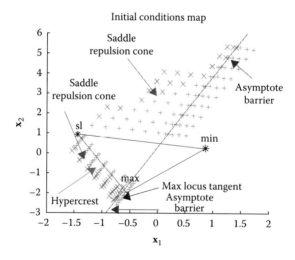

FIGURE 8.18
(See color insert.) The initial conditions map for the 2-D TLS problem: the green crosses represent the initial conditions by which a gradient flow algorithm converges, that is, the domain of convergence. The red crosses give divergence. The dark blue triangle has the critical points as its vertices. The dark blue straight line represents the asymptote/barrier z_1. (From Cirrincione, G. and Cirrincione, M., *Neural Based Orthogonal Data Fitting: The EXIN Neural Networks*, Series: Adaptive and Learning Systems for Signal Processing, Communications and Control, Wiley & Sons, New York, 255pp., November 2010.)

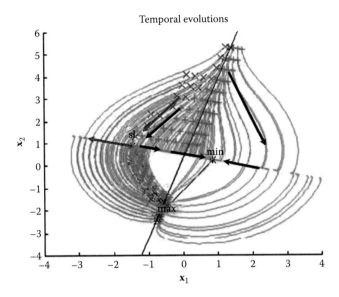

FIGURE 8.19
(See color insert.) The temporal evolution of the sequential TLS EXIN neuron for the generic 2-D TLS problem. (From Cirrincione, G. and Cirrincione, M., *Neural Based Orthogonal Data Fitting: The EXIN Neural Networks*, Series: Adaptive and Learning Systems for Signal Processing, Communications and Control, Wiley & Sons, New York, 255pp., November 2010.)

the repulsion area of the saddle. The latter is responsible for the upper part to the top part, given by the asymptote. Figure 8.19 shows the corresponding weight vector evolution by means of the blue lines. All the trajectories beginning from the green points go to the straight line through the minimum and the saddle (TLS projection of the saddle-minimum hypervalley) on the right side of the saddle and then converge to the minimum. All the trajectories beginning from the red points go to the same straight line on the left side of the saddle and then diverge in the direction of the straight line.

8.5.3 Nongeneric TLS Problem

The TLS problem (8.8) has no solutions if

$$\sigma_p > \sigma_{p+1} = \cdots = \sigma_{n+1}, \quad p \leq n \quad \text{and} \quad v_{n+1,i} = 0, \quad i = p+1, \ldots, n+1 \tag{8.34}$$

This is the nongeneric TLS problem. In this case, the existence conditions are not satisfied. These problems occur when A is *rank deficient* ($\sigma'_p \approx \sigma_{p+1}$ large). This last situation can be determined by examining the smallest σ_i for which $v_{n+1,i} \neq 0$. If this singular value is large, the user can turn down the problem and declare it as noncompatible with a linear modeling. Alternatively, the problem can be made generic by adding more equations. This is the case if the model is EIV and the observation errors are i.i.d (statistically independent and identically distributed, i.e., with the same variance). In the case of $\sigma'_p \approx 0$, it is possible to cancel the dependence between the columns of A by eliminating some columns of A so that the resulting submatrix is full rank. The TLS method is then applied to this subproblem (*subset selection* [43–45]). Although an exact nongeneric TLS problem is rare, practically close to nongeneric situations do occur. In these cases, the generic TLS solution can be still computed, but it becomes unstable and very sensitive to data errors as $\sigma'_p - \sigma_{p+1}$ approaches zero [5].

If $\sigma_n > \sigma_{n+1}$ and $v_{n+1,n+1} = 0$, then [13]

$$\sigma_{n+1} > \sigma'_n; \quad v_{n+1,n+1} = \begin{bmatrix} \pm \mathbf{v}'_n \\ 0 \end{bmatrix} \tag{8.35}$$

The nongeneric TLS finds

$$\min_{[\hat{A};\hat{b}] \in \mathfrak{R}^{mx(n+1)}} \left\| [A;\mathbf{b}] - [\hat{A};\hat{\mathbf{b}}] \right\|_F \quad \text{with the constraint } \hat{\mathbf{b}} \in R(\hat{A})$$

and $\hspace{11cm}$ (8.36)

$$\begin{bmatrix} \hat{\mathbf{x}} \\ -1 \end{bmatrix} \perp \mathbf{v}_j, \quad j = p+1, \dots, n+1 \quad \text{(provided that } v_{n+1,p} \neq 0 \text{ and } v_{n+1,j} = 0)$$

Once the minimum $[\hat{A};\hat{\mathbf{b}}]$ is found, every $\hat{\mathbf{x}}$ verifying

$$\hat{A}\hat{\mathbf{x}} = \hat{\mathbf{b}} \tag{8.37}$$

is called a nongeneric TLS solution.

The nongeneric TLS algorithm [13] must firstly identify the close to nongeneric or the nongeneric situations before applying the corresponding formula: the computation of the numerical rank of a noisy matrix is necessary. It can be proven that TLS EXIN solves automatically this problem without changing the learning law [26]. From a geometric point of view (8.34), it means that the last $n - p + 1$ critical points (the smallest) go to infinity, that is, the corresponding zeros in $g(\gamma)$ coincide with the first $n - p + 1$ asymptotes. The saddle associated with σ_p^2 is the first critical point not coincident with an asymptote.

This saddle is the nongeneric TLS solution. If $\sigma_{p+1} \to \sigma'_p$ (highly conflicting equations), the domain of convergence worsens because of the approaching of the saddle to the corresponding asymptote (good equations have faraway saddles). Furthermore, the motor gap of the saddles is smaller, thus worsening the convergence speed of the iterative methods, and the maximum and the origin are much farther from the saddle/solution.

For $p = n$, the lowest critical point, associated with σ_{n+1}, goes to infinity in the direction of $\mathbf{v}'_n$. Equation 8.35 expresses the fact that the $(n + 1)$-dimensional vector $\mathbf{v}_{n+1}$ is parallel to the TLS hyperplane, just in the direction of $\mathbf{v}'_n$. Hence, the solution is contained in the hyperplane (*nongeneric TLS subspace*) through the origin and normal to the $A^T A$ eigenvector $\mathbf{v}'_n$.

The nongeneric TLS subspace has dimension $n - 1$, being n the dimension of the TLS hyperplane. In this subspace, the lower saddle, that is, the critical point corresponding to σ_n, loses its dimension of escape, and therefore, it represents a minimum. The most important consequence is the possibility to exactly repeat all the TLS stability analysis on this lower dimensional subspace. Figure 8.20 shows that, for every plane $z_n z_i$, the saddle/minimum is always closer to the asymptote than the other inferior ($t < \lambda_i$) saddles. This implies that the origin is always between the saddle/solution and the maximum. Hence, the choice of initial conditions for every TLS gradient flow, just like TLS EXIN, assures the convergence. In addition, in every plane $z_n z_i$, there exists a divergence straight line which is between the saddle/minimum and the maximum [15]. The convergence analysis for $p < n$ is similar and can be found in Ref. [15] too.

Figure 8.21 shows an example of the TLS EXIN 2-D phase diagram for different initial conditions: The red (black) initial conditions give convergent (divergent) trajectories.

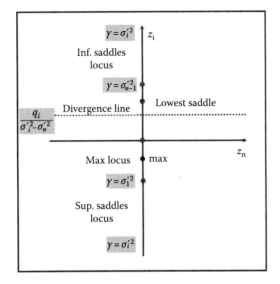

FIGURE 8.20
Solution/center locus in the plane $z_n z_i$ for the nongeneric TLS. The dashed line represents the divergence line. (From Cirrincione, G. and Cirrincione, M., *Neural Based Orthogonal Data Fitting: The EXIN Neural Networks*, Series: Adaptive and Learning Systems for Signal Processing, Communications and Control, Wiley & Sons, New York, 255pp., November 2010.)

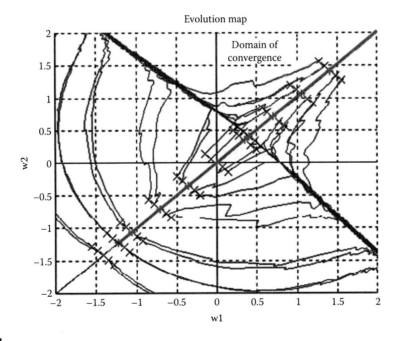

FIGURE 8.21
(See color insert.) Evolution map of the sequential TLS EXIN for the first nongeneric TLS benchmark problem. The initial conditions for the converging trajectories are in red, the others are in black. The straight line passing through the red points contains the saddle/solution locus. The dark blue thick straight line is the divergence line. (From Cirrincione, G. and Cirrincione, M., *Neural Based Orthogonal Data Fitting: The EXIN Neural Networks*, Series: Adaptive and Learning Systems for Signal Processing, Communications and Control, Wiley & Sons, New York, 255pp., November 2010.)

The black straight line passing through the red points contains the saddle/solution locus and coincides with the axis z_i. The domain of convergence is given by the half line with origin the maximum $(-1.2247, -1.2247)$ and containing the saddle/solution. All the nonconverging trajectories tend to the divergence straight line (dark blue thick line), which is the asymptote z_2, and then diverge in the $\mathbf{v}_2'$ direction.

8.6 Generalization of Linear Least Squares Problems

The generalized TLS problem (GeTLS, [15,26,31,46,47]) deals with the case in which the errors Δa_{ij} in the data matrix A are i.i.d. with zero average and same variance σ_a^2, and the errors Δb_i are also i.i.d. with zero average and same variance σ_b^2. This formulation is justified by the fact that $\mathbf{b}$ and A represent different physical quantities and, as a consequence, are often measured by methods of different accuracy. Solving GeTLS implies finding the vector $\bar{\mathbf{x}}$ by minimizing

$$\zeta \|\Delta A\|_F^2 + (1-\zeta)\|\Delta \mathbf{b}\|_F^2 \quad \text{with } 0 \leq \zeta \leq 1 \tag{8.38}$$

where

$$\frac{\zeta}{1-\zeta} = \frac{\sigma_a^2}{\sigma_b^2} \tag{8.39}$$

Formulation (8.38) comprises the three problems: OLS, TLS, and DLS (Section 8.2). Indeed, $\zeta = 0$ results in the OLS formulation because $\sigma_a^2 = 0$, whereas $\zeta = 0.5$ results in the TLS formulation because $\sigma_a^2 = \sigma_b^2 = \sigma_v^2$; if $\zeta = 1$, the DLS formulation is obtained because $\sigma_b^2 = 0$. Equation 8.39 defines the problem for intermediate values of ζ: in an experiment, variances σ_a^2 and σ_b^2 are estimated under the spherical Gaussian assumption in order to compute the corresponding ζ by Equation 8.39.

The optimal solution of the problem (8.38) is given by minimizing the following error function:

$$E_{GeTLS}(\mathbf{x}) = \frac{1}{2} \frac{(A\mathbf{x} - \mathbf{b})^T (A\mathbf{x} - \mathbf{b})}{(1-\zeta) + \zeta \mathbf{x}^T \mathbf{x}} = \frac{\mathbf{u}^T R \mathbf{u}}{\mathbf{u}^T D \mathbf{u}} \tag{8.40}$$

where $\mathbf{u} = \left[\mathbf{x}^T; -1\right]^T$, $R = [A; \mathbf{b}]^T [A; \mathbf{b}]$, and $D = diag\left(\overbrace{2\zeta, \ldots, 2\zeta}^{n \ \text{times}}, 2(1-\zeta)\right)$.

For $\zeta = 0$, 0.5 and 1, this error function is equal, respectively, to the OLS, TLS, and DLS. In Equation 8.40, it is apparent that the difference with the weighted TLS [48–50], which needs only a scalar parameter, is going to infinity to obtain the DLS solution. The same problem can be found in the generalization [51], which uses only one parameter. Hence, the need of a finite parameter in the numerical algorithms limits the accuracy. This does not happen in the GeTLS formulation because DLS is represented by a finite parameter.

For $0 < \zeta < 1$, the error Equation 8.40 is the ratio of positive definite quadratic forms and represents the generalized Rayleigh quotient of the symmetric positive definite pencil (R,D) [52]. Its critical points solve the generalized eigenvalue problem $R\mathbf{u} = \gamma D\mathbf{u}$, that is,

$$\begin{bmatrix} A^T A & A^T \mathbf{b} \\ \mathbf{b}^T A & \mathbf{b}^T \mathbf{b} \end{bmatrix} \mathbf{u} = 2\gamma \begin{bmatrix} \zeta I_n & 0 \\ 0^T & 1-\zeta \end{bmatrix} \mathbf{u} \tag{8.41}$$

The last equation in Equation 8.41 yields the GeTLS EXIN secular equation:

$$\mathbf{b}^T A \mathbf{x}_{crit} + 2\gamma(1-\zeta) - \mathbf{b}^T \mathbf{b} = 0 \Rightarrow \tag{8.42}$$

$$\mathbf{q}^T(\Lambda - 2\gamma\zeta I_n)^{-1}\mathbf{q} + 2\gamma(1-\zeta) - \mathbf{b}^T \mathbf{b} = g(\gamma,\zeta) = 0 \tag{8.43}$$

where

$g(\gamma, \zeta)$ is the generalization of $g(\gamma)$ [15] ($g(\gamma) = g(\gamma, 0.5)$) and can be computed in the same way as $g(\gamma)$ (compare with Equation 8.32)

$\mathbf{x}_{crit}$ is the value of x corresponding to γ, which is one of the critical values for $E_{GeTLS}(\zeta, \mathbf{x})$

The first n equations in (2.41) yield the GeTLS EXIN critical value corresponding to γ:

$$\mathbf{x}_{crit} = \left[A^T A - 2\gamma\zeta I_n \right]^{-1} A^T \mathbf{b} \tag{8.44}$$

which is the generalization of Equation 8.29. Both Equations 8.43 and 8.44 are the bases of the GeTLS EXIN theory.

In conclusion, the GeTLS EXIN approach can be analyzed as a generalized eigenvalue problem (GEVD, generalized eigenvalue decomposition) because of the form of the error function Equation 8.40. This interpretation is important both from the theoretical point of view (new theorems and inequalities have been deduced) and the numerical point of view (other algorithms can be used, like the Wilkinson's algorithm [52]).

In Ref. [26], the GEVD approach is used in order to study the OLS and DLS spectra:

OLS	There are n infinite eigenvalues whose associated eigenvectors do not intersect the TLS hyperplane and one zero eigenvalue associated with the eigenvector which is the minimum of E_{GeTLS} and intersects the TLS hyperplane in the OLS solution
DLS	There is one infinite eigenvalue, whose associated eigenvector intersects the TLS hyperplane in the origin of the axes and represents the maximum of E_{GeTLS}

8.7 GeMCA EXIN Neuron

In the case $0 < \zeta < 1$, R can be diagonalized by a D-orthogonal transformation [52]. Indeed,

$$D = Z^T Z \tag{8.45}$$

where

$$Z = \begin{bmatrix} \sqrt{2\zeta} I_n & 0 \\ 0^T & \sqrt{2(1-\zeta)} \end{bmatrix} \tag{8.46}$$

Define matrix K as

$$K = Z^{-T}RZ^{-1} = \frac{1}{2}\begin{bmatrix} \dfrac{A^TA}{\zeta} & \dfrac{A^T\mathbf{b}}{\sqrt{\zeta(1-\zeta)}} \\[2ex] \dfrac{\mathbf{b}^TA}{\sqrt{\zeta(1-\zeta)}} & \dfrac{\mathbf{b}^T\mathbf{b}}{1-\zeta} \end{bmatrix} = Z^{-T}[A;\mathbf{b}]^T[A;\mathbf{b}]Z^{-1}$$

$$= \left\| [A;\mathbf{b}]Z^{-1} \right\|_2^2 = \left\| \dfrac{A}{\sqrt{2\zeta}}; \dfrac{\mathbf{b}}{\sqrt{2(1-\zeta)}} \right\|_2^2 \qquad (8.47)$$

and the corresponding eigendecomposition is given by

$$V^TKV = diag(\alpha_1,\ldots,\alpha_{n+1}) = D_\alpha \qquad (8.48)$$

The eigenvector matrix Y is defined as

$$Y = [\mathbf{y}_1,\ldots,\mathbf{y}_{n+1}] = Z^{-1}V \qquad (8.49)$$

and is D-orthogonal, that is,

$$Y^TDY = 1 \qquad (8.50)$$

Hence, $(\alpha_i, \mathbf{y}_i)$ is an eigenpair of the symmetric positive definite pencil (R,D).

From an MCA point of view, the aforementioned theory can be reformulated as a Rayleigh quotient:

$$E_{GeTLS}(x) = \frac{1}{2}\frac{(A\mathbf{x}-\mathbf{b})^T(A\mathbf{x}-\mathbf{b})}{(1-\zeta)+\zeta\mathbf{x}^T\mathbf{x}} = \frac{\mathbf{u}^TR\mathbf{u}}{\mathbf{u}^TD\mathbf{u}} = \frac{\mathbf{u}^TR\mathbf{u}}{\mathbf{u}^TZ^TZ\mathbf{u}} = \frac{\mathbf{v}^TZ^{-T}RZ\mathbf{v}}{\mathbf{v}^T\mathbf{v}} = \frac{\mathbf{v}^TK\mathbf{v}}{\mathbf{v}^T\mathbf{v}} = E_{MCA}(\mathbf{v}) \quad (8.51)$$

where

$$\mathbf{u} = Z^{-1}\mathbf{v} \qquad (8.52)$$

The critical vectors $\mathbf{v}^c$ of E_{MCA} correspond to the columns of matrix V defined in Equation 8.48. Hence, Equation 8.52 corresponds to Equation 8.49. From a numerical point of view, it means that an MCA algorithm, like the MCA EXIN algorithm (neuron) (basically a gradient flow of the Rayleigh quotient), can replace the GeTLS EXIN algorithm by using matrix K as autocorrelation matrix. The estimated minor component must be scaled by using Equation 8.52 and normalized by constraining the last component equal to −1. Choosing K as autocorrelation matrix implies that the MCA EXIN neuron is fed by input vectors m_i defined as

$$\mathbf{m}_i = \left[\dfrac{\mathbf{a}_i^T}{\sqrt{2\zeta}}; \dfrac{b_i}{\sqrt{2(1-\zeta)}} \right]^T \qquad (8.53)$$

being $\mathbf{a}_i$ the column vector representing the ith row of matrix A. The MCA EXIN neuron whose input is preprocessed by means of Equation 8.53 is called GeMCA EXIN neuron.

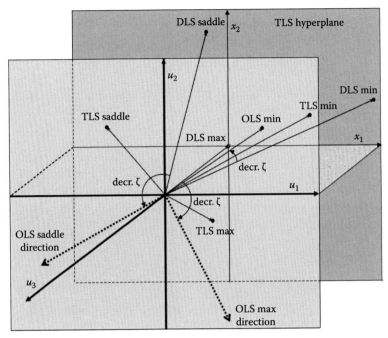

FIGURE 8.22
GeMCA EXIN and GeTLS EXIN critical directions and points. (From Cirrincione, G. and Cirrincione, M., *Neural Based Orthogonal Data Fitting: The EXIN Neural Networks*, Series: Adaptive and Learning Systems for Signal Processing, Communications and Control, Wiley & Sons, New York, 255pp., November 2010.)

8.7.1 Qualitative Analysis of the Critical Points of the GeMCA EXIN Error Function

Figure 8.22 resumes the analysis made in Ref. [26]. For visualization purposes, the GeTLS problem is 2-D ($n = 2$). The components of vector $\mathbf{u}$ are u_1, u_2, and u_3. The plane normal to u_3 has been drawn, together with the TLS hyperplane, whose equation is $u_3 = -1$, which contains the axes x_1 and x_2 for the representation of the GeTLS solution $\mathbf{x}$. All vectors point to the TLS hyperplane (the black dots), except the $\mathbf{u}$ axes and the dashed directions. All critical directions have been estimated by using Equation 8.51 and then scaled by means of Equation 8.52. The results for the three particular cases, namely, OLS, TLS, and DLS, are considered. All minima, that is, the GeTLS solutions, are shown. For DLS, the maximum is parallel to $\mathbf{u}_3 = (0, 0, 1)^T$ and corresponds to an infinite eigenvalue. For decreasing (decr.), the critical vector corresponding to a saddle or to a maximum tends to move toward a direction parallel to the TLS hyperplane (front plane in the figure) approaching the OLS case. For OLS, only the minimum vector intersects the TLS hyperplane and so yields the corresponding OLS solution. The other critical directions correspond to infinite eigenvalues.

8.7.2 Analysis of the Error Function GeTLS (Geometrical Approach)

The analysis of the geometry of E_{GeTLS} is the same as E_{TLS}, as already anticipated in Section 8.4. In particular, Equation 8.44 becomes, by using decomposition (8.27),

$$\mathbf{x}_c(\gamma, \zeta) = V'\left(\Lambda - 2\gamma\zeta I_n\right)^{-1}\mathbf{q} \qquad (8.54)$$

There is also a family of hyperconics (generalization of Equation 8.30):

$$\mathbf{z}^T \left[\Lambda - 2\gamma\zeta I_n\right]\mathbf{z} = g(\gamma,\zeta) \tag{8.55}$$

which can be recast as

$$\mathbf{z}^T \left[\Lambda - 2\gamma\zeta I_n\right]\mathbf{z} = g(\gamma,\zeta) \rightarrow \sum_{i=1}^{n} \frac{z_i^2}{k_i} = 1 \tag{8.56}$$

where
$k_i = g(\gamma, \zeta)/\lambda_i - 2\lambda\zeta$
z_i is the ith component of vector $\mathbf{z}$ and, also, the coordinate along the direction of the eigenvector $\mathbf{v}_i'$ associated with the eigenvalue $\lambda_i = \sigma_i^2$ because of the rotation $\mathbf{z} = V'^T\mathbf{y}$

This family has been analyzed in Ref. [15]. The main results are the following:

- The asymptotes of $g(\gamma, \zeta)$ with respect to γ are given by the values $\gamma_i = \left(\sigma_i'^2/2\zeta\right)(i=1,\ldots,n)$. They are located among the $g(\gamma, \zeta)$ zeros and vice versa (interlacing theorem [53] for $\zeta = 0.5$).
- The GeTLS error function of an nth-dimensional unknown has $n + 1$ critical points: one minimum, $n - 1$ saddle points, and one maximum.
- The zeros of $g(\gamma, \zeta)$ are coincident with the heights of $E_{GeTLS}(\mathbf{x})$ at its critical points. In particular, the GeTLS solution, for each ζ, corresponds to the height γ_{min} and is given by

$$\mathbf{x}_{GeTLSsol} = \mathbf{x}_c(\gamma_{min},\zeta) = \left[A^T A - 2\gamma_{min}\zeta I_n\right]^{-1} A^T \mathbf{b} \tag{8.57}$$

- The zeros of $g(\gamma, 0.5)$ are coincident with the squares of the singular values of $[A;\mathbf{b}]$.

8.7.3 Critical Loci: Center Trajectories

Equation 8.54 represents the locus of the equilevel hypersurface centers in the $\mathbf{x}$ reference frame; in the $\mathbf{z}$ reference frame, after the application of the V' rotation matrix around the origin, this equation becomes

$$\mathbf{z}_c(\gamma,\zeta) = \left[\Lambda - 2\gamma\zeta I_n\right]^{-1} \mathbf{q} \tag{8.58}$$

That is, for each component ($i = 1,\ldots, n$),

$$z_{ci} = \frac{q_i}{\lambda_i - 2\lambda\zeta} \tag{8.59}$$

In every plane $z_n z_i$, this locus is an equilateral hyperbola translated with respect to the origin:

$$z_i = \frac{q_i z_j}{(\lambda_i - \lambda_j)z_j + q_j} \tag{8.60}$$

All of these loci cross in the origin **z** which coincides with the origin **x**. Its asymptotes are given by

$$z_j = -\frac{q_j}{\Delta\lambda} \quad \text{and} \quad z_i = \frac{q_i}{\Delta\lambda} \quad \text{where } \Delta\lambda = \lambda_i - \lambda_j \qquad (8.61)$$

The asymptote of the hyperbolas parallel to z_i (the component in the direction of the eigenvector $\mathbf{v}_i'$ associated with $\lambda_i = \sigma_i'^2$) is defined as "z_i asymptote" in Ref. [15].

Under the assumption of distinct singular values of A, these loci tend to degenerate into two orthogonal straight lines for $q_j \to 0$ or $q_i \to 0$.

The global locus (8.59) is parameterized with the product $\gamma\zeta$. The parameter $t = 2\gamma\zeta$ is used. With no a priori hypothesis, the effects of γ and ζ cannot be separated. The hyperbolas cross in the origin for $t = \infty$ and $t = -\infty$.

The following interpretations can be done about the locus:

1. If $\zeta = constant$, the hyperbolas, which refer to couples of coordinates, represent the locus of the corresponding coordinates of the centers of the equilevel hypersurfaces for the energy cost of the GeTLS problem at hand.

2. If $\gamma = constant$, the hyperbolas express the way the variability of ζ moves the centers of the equilevel hypersurfaces, just giving an idea of the difficulties of the different GeTLS problems.

3. If $\forall\zeta$, the zeros of $g(\gamma, \zeta)$ are computed, the hyperbola is divided into loci (critical loci), representing the critical points.

Among these interpretations, the third is the most interesting from the point of view of the study of the domain of convergence and for the introduction of the principle of the GeTLS scheduling. The critical loci (see Figure 8.23) are

- The *solution locus* composed of the points having $t \leq \lambda_n$ and given $\forall\zeta$ by $2\zeta\gamma_{min}$ (γ_{min} is the smallest zero of $g(\gamma, \zeta)$). This locus extends itself to infinity only if one plane coordinate is z_n; if it is not the case, it represents the part of the hyperbola branch just until λ_n.

- The *saddle locus* composed of the points having $\lambda_j \leq t \leq \lambda_{j-1}$ and given $\forall\zeta$ by $2\zeta\gamma_{saddle}$ (γ_{saddle} is the zero of $g(\gamma, \zeta)$ in the corresponding interval); it represents this saddle for every GeTLS problem. It is represented by an entire branch of the hyperbola only in the plane $z_j z_{j-1}$; in all other planes, it represents the part of the branch between $t \leq \lambda_j$ and $t \leq \lambda_{j-1}$.

- The *maximum locus* composed of the points having $t \geq \lambda_1$ and given $\forall\zeta$ by $2\zeta\gamma_{max}$ (γ_{max} is the largest zero of $g(\gamma, \zeta)$). This locus extends itself to infinity only if one plane coordinate is z_1; if it is not the case, it represents the part of the hyperbola branch from λ_1 to ∞, corresponding to the origin. This point is attained only by the DLS maximum because, in this case, the zero of $g(\gamma, 1)$ is at infinity: Equation 8.43 shows that $g(\gamma, 1)$ is monotonically increasing for $t \geq \lambda_1$ and tends to $-\mathbf{b}^T\mathbf{b}$ for $t \to \infty$.

The position in the curve for $t = 0$ (origin of the parametric curve) coincides with the OLS solution and is the only center for the OLS equilevel hypersurfaces. The hyperbolas can be described as a function of this point [26].

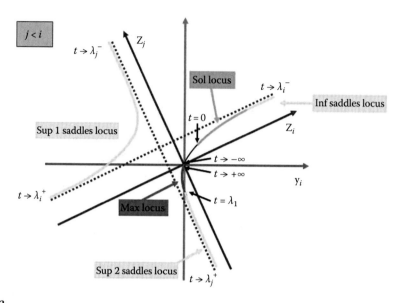

FIGURE 8.23
(See color insert.) Critical loci for j < I. (From Cirrincione, G. and Cirrincione, M., *Neural Based Orthogonal Data Fitting: The EXIN Neural Networks*, Series: Adaptive and Learning Systems for Signal Processing, Communications and Control, Wiley & Sons, New York, 255 pp., November 2010.)

The value of the parameter t of a GeTLS solution (position in the solution locus with respect to the origin, which corresponds to the OLS solution) is proportional to the value of ζ. In particular, it means that $t_{DLS} \geq t_{TLS} \geq t_{OLS} = 0$. In every saddle locus and in the maximum locus, the critical points have a position t on the corresponding locus proportional to the parameter ζ of the GeTLS problem.

In Refs [25,26], it is proven that the hyperbola branch positions in the plane $z_i z_j$ depend on the sign of the corresponding q_i and q_j. Hence, knowledge of the OLS solution indicates the quadrant containing the solution locus for every plane $z_i z_j$.

8.8 GeTLS EXIN Neuron

The optimal solution of the problem (8.38) is obtained by minimizing the cost function Equation 8.40, which can be expressed as

$$E_{GeTLS\ EXIN}(\mathbf{x}) = \sum_{i=1}^{m} E^{(i)}(\mathbf{x}) \tag{8.62}$$

where

$$E^{(i)}(\mathbf{x}) = \frac{1}{2} \frac{\left(\mathbf{a}_i^T \mathbf{x} - b_i\right)^2}{(1-\zeta) + \zeta \mathbf{x}^T \mathbf{x}} = \frac{1}{2} \frac{\delta^2}{(1-\zeta) + \zeta \mathbf{x}^T \mathbf{x}} \tag{8.63}$$

It follows

$$\frac{dE^{(i)}(\mathbf{x})}{d\mathbf{x}} = \frac{\delta \mathbf{a}_i}{(1-\zeta)+\zeta\mathbf{x}^T\mathbf{x}} - \frac{\delta^2\zeta\mathbf{x}}{\left[(1-\zeta)+\zeta\mathbf{x}^T\mathbf{x}\right]^2} \tag{8.64}$$

And the corresponding steepest descent discrete-time formula is

$$\mathbf{x}(t+1) = \mathbf{x}(t) - \alpha(t)\gamma(t)\boldsymbol{\alpha}_i + \left[\zeta\alpha(t)\gamma^2(t)\right]\mathbf{x}(t) \tag{8.65}$$

where

$$\gamma(t) = \frac{\delta(t)}{(1-\zeta)+\zeta\mathbf{x}(t)^T\mathbf{x}(t)} \tag{8.66}$$

Equation 8.65 represents the GeTLS EXIN learning law. Its analysis is similar to the TLS EXIN analysis.

8.8.1 GeTLS Domain of Convergence

The GeTLS domain of convergence is similar to the TLS domain of convergence, that is, it is bordered by the same barriers/asymptotes and the same maximum locus tangent [15].

For $\zeta = 0$ (OLS), the domain of convergence is the whole x space. To increase ζ, the domain of convergence becomes smaller and smaller. It has been proved [13] that the origin of axes of the solution space x always belongs to the domain of convergence for $\zeta \in [0, 1]$, that is, the null initial conditions always guarantee the solution. In the DLS case, the origin is not a regular point of the cost function Equation 8.7 and then it can be used as initial conditions. The shape of the domain depends on several factors: for instance, the position of the saddle points and the asymptotes of $\mathbf{z}$ influence the position of the frontier (convergence barrier). In the following, some considerations are given which are derived by the analysis of the critical loci.

The origin of the x space, which coincides with the origin of the z space, has some very important features, as a consequence of the existence of the hyperbolas:

1. In every parametric hyperbola, the origin corresponds to the point for $t = -\infty$; recalling that there are no zeros between this point and the GeTLS solution, it follows that the origin is in the same branch of the solution locus, and no convergence barriers (caused by saddles) exist between the origin and this locus.
2. In every parametric hyperbola, the origin also corresponds to the point for $t = +\infty$; it follows that the origin is the superior limit of the maximum locus.

From these considerations, it follows the fundamental result that the origin of the TLS hyperplane is the best choice for the initial conditions of the GeTLS EXIN neuron if no a priori information is given.

The maximum is always nearer to the minimum (GeTLS solution) than every saddle in every plane $z_n z_i$. In every plane $z_i z_j$, the maximum is nearer to the minimum than all saddles corresponding to $t > \lambda_j$. The origin is always positioned between the maximum and the solution locus, and therefore, the remarks about the maximum position are, a fortiori, even more valid for the origin. From the point of view of the convergence, the group of saddles for $t < \lambda_j$ causes no problems if the origin is chosen as initial condition because the weight vector, if the learning rate is not too large, approximately follows the solution locus (it coincides with the locus of the centers of the equilevel hyperellipsoids). Resuming, a scheduling of the parameter ζ, defined as a continuous function from 0 to 1, combined with not too high a learning rate, accelerates the DLS EXIN neuron and guarantees its convergence in the absence of a priori knowledge about the initial conditions, that is, the null initial conditions are the universal choice for the convergence.

8.8.2 Scheduling

From the point of view of the neural approaches and, more generally, of the gradient flows, the DLS problem ($\zeta = 1$) is the most difficult (for $b \neq 0$) for the following reasons:

- The domain of convergence is smaller.
- The cost function Equation 8.7 is the flattest.
- The DLS solution is the farthest GeTLS solution from the origin.
- The DLS error cost is not defined for null initial conditions, and every gradient flow algorithm diverges for this choice of initial conditions; hence, there is not a universal choice which guarantees the convergence as for the other GeTLS EXIN neurons.

In order to solve these problems, a novel technique of parametric programming, called *scheduling* [26], is proposed. It is a GeTLS EXIN neuron parameterized by a nonconstant ζ with a predefined temporal evolution. The GeTLS EXIN weight vector follows the solution locus toward the DLS solution if its parameter ζ is made variable and increasing from 0 to 1 according to a predefined law (scheduling). The first iteration, for $\zeta = 0$, is in the direction of the OLS solution (minimum): therefore, in a case where both the null initial conditions and the values of the weights at the first iteration are in the domain of convergence. This new position is also in the DLS domain because it is in the direction of the solution locus. Changing slowly ζ makes the weight vector follow the hyperbola branch containing the solution locus for every x plane (see Figure 8.24). Furthermore, the GeTLS neuron, for not too high an initial learning rate, begins with very low initial conditions from the first iteration. This accelerates the learning law.

The following example shows a DLS hyperbolic scheduling:

$$\zeta(i) = 1 - \frac{1}{i} \tag{8.67}$$

where $i \geq 1$ is the iteration number. The initial conditions are null. Figure 8.25 shows the weight trajectory in the solution space. The weights approach the solution locus (parameterized by ζ) just before the TLS solution because the hyperbolic scheduling only yields a single learning step for $\zeta = 0$. Note the *attractive* effect of the solution locus. The transient and the accuracy are very good.

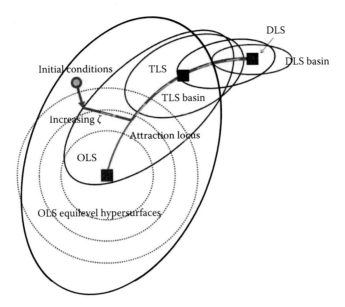

FIGURE 8.24
DLS scheduling. (From Cirrincione, G. and Cirrincione, M., *Neural Based Orthogonal Data Fitting: The EXIN Neural Networks*, Series: Adaptive and Learning Systems for Signal Processing, Communications and Control, Wiley & Sons, New York, 255 pp., November 2010.)

Weight phase diagram and the benchmark solution locus

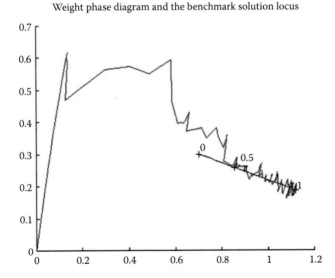

FIGURE 8.25
Trajectory of the weights of DLS scheduling EXIN (hyperbolic scheduling and null initial conditions). The solution locus is indicated together with the OLS, TLS, and DLS solutions. (From Cirrincione, G. and Cirrincione, M., *Neural Based Orthogonal Data Fitting: The EXIN Neural Networks*, Series: Adaptive and Learning Systems for Signal Processing, Communications and Control, Wiley & Sons, New York, 255pp., November 2010.)

The DLS scheduling cannot have null initial conditions when the observation vector (**b**) is null; indeed, the first step (OLS $\zeta = 0$) has $x^T A^T A x$ as energy function, which is null for null initial conditions. The problem of the lack of universal initial conditions can be circumvented by inserting one or more OLS learning steps in the scheduling in order to enter the domain of convergence. If the initial conditions are very small, one OLS step is enough to assure the convergence.

8.8.3 Accelerated MCA EXIN Neuron (MCA EXIN+)

The DLS scheduling EXIN can be used in order to improve the MCA EXIN neuron. Indeed, the DLS energy cost for **b** = **0** becomes (see Equation 8.7)

$$E_{DLS}(\mathbf{x}) = \frac{1}{2} \frac{\mathbf{x}^T A^T A \mathbf{x}}{\mathbf{x}^T \mathbf{x}} = \frac{1}{2} r\left(\mathbf{x}, A^T A\right) \quad \forall \mathbf{x} \in \mathfrak{R}^n - \{\mathbf{0}\} \tag{8.68}$$

that is, the Rayleigh quotient of $A^T A$. Recalling that the neuron is fed with the rows of the matrix A and that the autocorrelation matrix R of the input data is equivalent to $A^T A / m$, it follows that the DLS neural problem is equivalent to an MCA neural problem with the same inputs. This equivalence is only possible for DLS EXIN (GeTLS EXIN with $\zeta = 1$) and MCA EXIN because both neurons use the *exact* error gradient in their learning law.

The DLS EXIN neuron fed by the rows of a matrix A and with null target is equivalent to the MCA EXIN neuron with the same input. Both neurons find the minor component of $A^T A$, that is, the right singular vector associated with the smallest singular value of A. The drawback of the equivalence is the fact that, instead of MCA EXIN which always converges, DLS EXIN is not guaranteed to converge. This equivalence allows using the DLS scheduling for improving MCA EXIN.

Definition 8.1 (MCA EXIN+)

MCA EXIN+ is a linear neuron with a DLS scheduling learning law, the same inputs of the MCA EXIN, and null target [26].

Large fluctuations of the weights imply that the learning law increases the estimation error, and when this increase is too large, it will make the weight vector deviate drastically from the normal learning, which may result in divergence or an increased learning time. This is a serious problem for MCA EXIN when the initial conditions are infinitesimal. On the other hand, MCA EXIN converges faster for smaller initial conditions. These observations justify the MCA EXIN+ improvements with respect to MCA EXIN:

- *A smoother dynamics* because the weight path in every plane $z_i z_j$ remains near the hyperbola branch containing the solution locus
- *A faster convergence* because of the smaller DLS scheduling fluctuations which reduce the settling time
- *A better accuracy* because of the small deviations from the solution

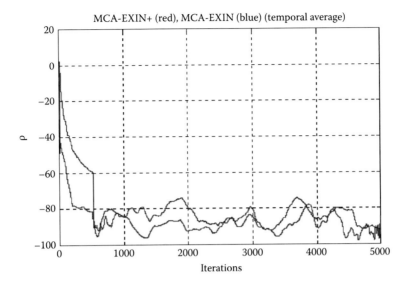

FIGURE 8.26
(See color insert.) Index ρ (expressed in dB) for MCA EXIN+ with scheduling (red) and MCA EXIN (blue). (From Cirrincione, G. and Cirrincione, M., *Neural Based Orthogonal Data Fitting: The EXIN Neural Networks*, Series: Adaptive and Learning Systems for Signal Processing, Communications and Control, Wiley & Sons, New York, 255 pp., November 2010.)

The next simulation deals with a 2-D problem for comparing MCA EXIN+ and MCA EXIN. The scheduling is linear. Figure 8.26 shows, for both neurons, the plot of an index parameter (expressed in decibels), which measures the accuracy and is defined as

$$\rho = \frac{\left\| \mathbf{w}_i(t) - \mathbf{w}_i^* \right\|_2^2}{n} \tag{8.69}$$

where
    $\mathbf{w}_i(t)$ is the n dimensional neuron weight vector
    $\mathbf{w}_i^*$ is the n dimensional desired weight vector

The figure stresses well the faster convergence and better accuracy of MCA EXIN+.
    It follows a flowchart for the computer implementation of the MCA EXIN+ algorithm [26].

1. *Goal*: to find the minimum eigenvector x of the matrix A.
2. *Inputs*:
   a. $\eta(t)$: learning rate, decreasing to zero.
   b. $\mathbf{x}(0)$: initial conditions (better as small as possible, but non null).
   c. $\zeta(t)$: GeTLS parameter, increasing from 0 to 1.
   d. $\mathbf{a}(t)$: the row of A which is input at instant t.
   e. ε: stop threshold.
   f. t_{max}: maximum number of iterations.

3. *Algorithm*:

    a. For each t

        i. Compute

$$\mathbf{x}(t+1) = \mathbf{x}(t) - \eta(t)\gamma(t)\mathbf{a}(t) + \left[\zeta(t)\eta(t)\gamma^2(t)\right]\mathbf{a}(t)$$

        where

$$\gamma(t) = \frac{\mathbf{a}^T(x)\mathbf{x}(t)}{1 - \zeta(t) + \zeta(t)\mathbf{x}^T(t)\mathbf{x}(t)}$$

        ii. Compute

$$E_{GeTLS}(t+1) = \frac{1}{2} \frac{\left(A\mathbf{x}(t)\right)^T\left(A\mathbf{x}(t)\right)}{1 - \zeta(t) + \zeta(t)\mathbf{x}^T(t)\mathbf{x}(t)}$$

$$= \frac{\gamma(t)}{2} \frac{\left(A\mathbf{x}(t)\right)^T\left(A\mathbf{x}(t)\right)}{\mathbf{a}^T(x)\mathbf{x}(t)}$$

        iii. If $\|E_{GeTLS}(t+1) - E_{GeTLS}(t)\|_2 < \varepsilon$ for a certain number of iterations, then STOP.

        iv. If $t > t_{max}$, then STOP.

        v. Update $\eta(t)$ and $\zeta(t)$.

Note that $\zeta(t)$ must be equal to 1 well before the algorithm is stopped.

References

1. M. Minsky, S. Papert, *Perceptrons*, MIT Press, Cambridge, MA, 1969.
2. B. Widrow, S. D. Stearns, *Adaptive Signal Processing*, Signal Processing Series, Prentice-Hall, Englewood Cliffs, NJ, 1985.
3. S. D. Hodges, P. G. Moore, Data uncertainties and least square regression, *Applied Statistics*, 21, 185–195, 1972.
4. G. W. Stewart, Sensitivity coefficients for the effects of errors in the independent variables in a line regression, Technical Report TR-571, Department of Computer Science, University of Maryland, College Park, MD, 1977.
5. G. H. Golub, C. F. Van Loan, An analysis of the total least squares problem, *SIAM Journal of Numerical Analysis*, 17, 883–893, 1980.
6. R. J. Adcock, A problem in least squares, *The Analyst*, 5, 53–54, 1878.
7. K. Pearson, On lines and planes of closest fit to points in space, *Philosophical Magazine*, 2, 559–572, 1901.
8. T. C. Koopmans, *Linear Regression Analysis of Economic Time Series*, De Erver F. Bohn, N.V., Haarlem, the Netherlands, 1937.
9. A. Madansky, The fitting of straight lines when both variables are subject to error, *Journal of the American Statistical Association*, 54, 173–205, 1959.

10. D. York, Least squares fitting of a straight line, *Canadian Journal of Physics*, 44, 1079–1086, 1966.
11. P. Sprent, *Models in Regression and Related Topics*, Methuen, London, U.K., 1969.
12. L. J. Gleser, Estimation in a multivariate "errors in variables" regression model: Large sample results, *Annals of Statistics*, 9, 24–44, 1981.
13. S. Van Huffel, J. Vandewalle, *The Total Least Squares Problems: Computational Aspects and Analysis. Frontiers in Applied Mathematics*, SIAM, Philadelphia, PA, 1991.
14. R. D. Degroat, E. Dowling, The data least squares problem and channel equalization, *IEEE Transactions on Signal Processing*, 41(1), 407–411, January 1993.
15. G. Cirrincione, *A Neural Approach to the Structure from Motion Problem*, PhD thesis, LIS INPG, Grenoble, France, December 1998.
16. L. Xu, E. Oja, C. Suen, Modified Hebbian learning for curve and surface fitting, *Neural Networks*, 5, 441–457, 1992.
17. G. Mathew, V. Reddy, Development and analysis of a neural network approach to Pisarenko's harmonic retrieval method, *IEEE Transactions on Signal Processing*, 42, 663–667, 1994.
18. G. Mathew, V. Reddy, Orthogonal eigensubspace estimation using neural networks, *IEEE Transactions on Signal Processing*, 42, 1803–1811, 1994.
19. R. Schmidt, Multiple emitter location and signal parameter estimation, *IEEE Transactions on Antennas and Propagation*, 34, 276–280, 1986.
20. J. W. Griffiths, Adaptive array processing, a tutorial, *IEE Proceedings*, 130(pt.F), 3–10, 1983.
21. R. Klemm, Adaptive airborne MTI: An auxiliary channel approach, *IEE Proceedings*, 134(pt.F), 269–276, 1987.
22. S. Barbarossa, E. Daddio, G. Galati, Comparison of optimum and linear prediction technique for clutter cancellation, *IEE Proceedings*, 134(pt.F), 277–282, 1987.
23. K. Gao, M. O. Ahmad, M. N. Swamy, Learning algorithm for total least-squares adaptive signal processing, *Electronic Letters*, 28(4), 430–432, February 1992.
24. K. Gao, M. O. Ahmad, M. N. Swamy, A constrained anti-Hebbian learning algorithm for total least-squares estimation with applications to adaptive FIR and IIR filtering, *IEEE Transactions on Circuits and Systems-II: Analog and Digital Processing*, 41(11), 718–729, November 1994.
25. G. Cirrincione, M. Cirrincione, J. Hérault, S. Van Huffel, The MCA EXIN neuron for the minor component analysis, *IEEE Transactions on Neural Networks*, 13(1), 160–187, January 2002.
26. G. Cirrincione, M. Cirrincione, *Neural Based Orthogonal Data Fitting: The EXIN Neural Networks*, Series: Adaptive and Learning Systems for Signal Processing, Communications and Control, Wiley & Sons, New York, 255pp., November 2010.
27. U. Helmke, J. B. Moore, *Optimization and Dynamical Systems*, Springer-Verlag, London, U.K., 1994.
28. D. Shafer, Gradient vector fields near degenerate singularities, In Z. Nitecki, C. Robinson, eds., *Global Theory of Dynamical Systems*, Lecture Notes in Mathematics, Springer-Verlag, Berlin, Germany, Vol. 819, pp. 410–417, 1980.
29. F. Luo, R. Unbehauen, A. Cichocki, A minor component analysis algorithm, *Neural Networks*, 10(2), 291–297, 1997.
30. D. Z. Feng, Z. Bao, L. C. Jiao, Total least mean squares algorithm, *IEEE Transactions on Signal Processing*, 46, 2122–2130, August 1998.
31. G. Cirrincione, G. Ganesan, K. V. S. Hari, S. Van Huffel, Direct and neural techniques for the data least squares problem, *MTNS 2000*, Perpignan, France, p. 10, June 2000.
32. J. M. Cioffi, Limited-precision effects in adaptive filtering, *IEEE Transactions on Circuits and Systems*, 34(7), 821–833, July 1987.
33. M. Hestenes, *Conjugate Direction Methods in Optimization*, Springer-Verlag, New York, 1980.
34. X. Yang, T. K. Sarkar, E. Arvas, A survey of conjugate gradient algorithms for solution of extreme eigen-problems of a symmetric matrix, *IEEE Transactions on Acoustics, Speech and Signal Processing*, 37(10), 1550–1556, October 1989.
35. G. Cirrincione, M. Cirrincione, Robust total least squares by the nonlinear MCA EXIN neuron, *Proceedings of IEEE International Joint Symposia on Intelligence and Systems*, Rockville, MD, pp. 295–300, May 1998.

36. P. Huber, *Robust Statistics*, Wiley, New York, 1981.
37. C. Han, X. Kong, R. Wei, A stable total least square adaptive algorithm for the nonlinear volterra filter volterra filter, *Seventh International Conference on Information Fusion*, Stockholm, Sweden, June–July 2004.
38. C. Han, X. Kong, R. Wei, Modified gradient algorithm for total least square filtering, *Neurocomputing*, 70, 568–576, 2006.
39. D. Peng, Z. Yi, X. Xiang, A modified MCA EXIN algorithm and its convergence analysis, *Advances in Neural Networks, Second International Symposium on Neural Networks, Chongqing, China, May 30–June 1, 2005, Proceedings, Part II*, Vol. 3496/2005 of Lecture Notes in Computer Science, Springer, Berlin, Germany, pp. 1028–1033, May 02, 2005.
40. D. Peng, Z. Yi, Convergence analysis of an effective MCA learning algorithm, *ICNN&B'05. International Conference on Neural Networks and Brain*, Beijing, China, 2005, Vol. 3, pp. 2003–2008, October 13–15, 2005.
41. D. Peng, Z. Yi, W. Luo, Convergence analysis of a simple minor component analysis algorithm, *Neural Networks*, 20(7), 842–850, 2007.
42. W. H. Press, S. A. Teukolsky, W. T. Wetterling, B. P. Flannery, *Numerical Recipes in C: The Art of Scientific Computing*, 2nd edn., Cambridge University Press, Cambridge, U.K., 1992.
43. R. R. Hocking, The analysis and selection of variables in linear regression, *Biometrics*, 32, 1–49, 1976.
44. R. R. Hocking, Developments in linear regression methodology 1959–1982, *Technometrics*, 25, 219–230, 1983.
45. S. Van Huffel, J. Vandewalle, Subset selection using the total least squares approach in collinearity problems with errors in the variables, *Linear Algebra and Its Applications*, 88/89, 695–714, 1987.
46. A. Premoli, M. L. Rastello, G. Cirrincione, A new approach to total least squares techniques for metrological applications. In P. Ciarlini, M. G. Cox, F. Pavese, D. Richter, eds., *Advanced Mathematical Tools in Metrology II Series on Advances in Mathematics for Applied Sciences*, World Scientific, Singapore, Vol. 40, pp. 206–215, 1996.
47. G. Cirrincione, M. Cirrincione, S. Van Huffel, The GeTLS EXIN neuron for linear regression, *Proceedings of the IEEE-INNS-ENNS International Joint Conference on Neural Networks, IJCNN*, Como, Italy, 2000, Vol. 6, pp. 285–289, July 24–27, 2000.
48. B. D. Rao, Unified treatment of LS, TLS and truncated SVD methods using a weighted TLS framework. In S. Van Huffel, ed., *Recent Advances in Total Least Squares Techniques and Errors-in-Variables Modeling*, SIAM, Philadelphia, PA, pp. 11–20, 1997.
49. C. C. Paige, Z. Strakos, Unifying least squares, total least squares and data least squares problems. In S. Van Huffel, P. Lemmerling, eds., *Proceedings of the Third International Workshop on TLS and Errors-in-Variables Modelling*, Kluwer Academic, Dordrecht, the Netherlands, pp. 35–44, 2001.
50. C. C. Paige, Z. Strakos, Scaled total least squares fundamentals, *Numerische Mathematik*, 91, 117–146, 2002.
51. A. Cichocki, R. Unbehauen, *Neural Networks for Optimization and Signal Processing*, John Wiley & Sons, New York, 1993.
52. G. W. Stewart, *Matrix Algorithms, Volume II: Eigensystems*, SIAM, Philadelphia, PA, 2001.
53. R. C. Thompson, Principal submatrices ix: Interlacing inequalities for singular values of submatrices, *Linear Algebra and Its Applications*, 5, 1–12, 1972.
54. M. F. Moller, A scaled conjugate gradient algorithm for fast supervised learning, *Neural Networks*, 6, 525–533, 1993.

Part IV

Selected Applications

9

Least-Squares and Neural Identification of Electrical Machines

9.1 Parameter Estimation of Induction Machines (IMs)

The instantaneous knowledge of the electrical parameters of an IM, either supplied by the power grid or in the framework of a variable speed electric drive, is extremely important. It is proved by the huge amount of scientific papers published in the last 20 years in the field of on-line parameter estimation of IMs [1–80].

The instantaneous knowledge of the electrical parameters of an IM is useful in the power system analysis. It is well known that, in the analysis of such systems, the knowledge of the dependence of the loads from the power grid voltage and frequency is crucial for retrieving the information about the behavior of the load themselves, especially in faulty conditions. To this aim, IMs have been studied for their potential contribution to the short-circuit currents, whose dependence from the power grid voltage and frequency is not straightforward. The on-line estimation of the electrical parameters of the machine could be helpful for the computation of the contribution of the motor to the short-circuit currents.

Another case, in which the on-line estimation of the IM parameters is important, with particular reference to the estimation of the stator and rotor resistances, is that of motors directly supplied by the power grid, subject to frequent start-ups and braking phases. The stator and rotor resistance estimation gives useful information about the variations of the temperature of windings, which is useful for deciding whether and when the subsequent start-up should take place in security conditions for avoiding potential damages to the machine.

Furthermore, with reference to an IM variable speed drive, the instantaneous knowledge of the electrical parameters of the IM is crucial, since it significantly affects the performance and the stability of the control system. In particular, as far as the rotor-flux-oriented control is concerned, the estimation of the amplitude and phase of the rotor flux space-vector is usually made either with the "current model" (see Section 5.3.3.2), whose implementation requires the correct knowledge of the rotor time constant, or with the "voltage model" (see Section 5.3.3.1), whose implementation requires the correct knowledge of the stator resistance. With reference to the direct torque control (DTC), the estimation of the amplitude and phase (or just the sector) of the stator flux space-vector is usually made by a "voltage model." Also in this case, the correct knowledge of the stator resistance is necessary.

The electrical parameters of the machine, in rated working conditions, are usually measured with off-line methodologies at standstill (no-load and locked-rotor tests) [4,5]. During normal working conditions, however, these parameters can change. Particularly, the values of the stator and rotor resistances vary because of the heating or cooling of the machine, while the stator and rotor inductances as well as the leakage coefficients can vary with the magnetization level of the machine core (saturation of the main flux). However, while the variation of the resistances is quite slow, since it is determined by the thermal time constants of the machine, the variation of the magnetic parameters is very fast; moreover, the magnetic parameters vary whenever the machine works at speeds higher than the rated one, since in these working conditions the magnetic flux is reduced (field-weakening) with consequent variation of all of the magnetic parameters. The rotor time constant can thus vary because of both the heating/cooling effects and the modification of the machine magnetization. In normal working conditions, the parameters could have percent variations even as much as 50% [1,2,5,6].

If the parameters of the machine vary while the corresponding values assigned to the flux models are kept constant, the so-called detuning of the flux model occurs. The detuning of the flux model significantly modifies the performance of the electric drive and can even lead up to its instability. This requires the necessity for developing algorithms for the on-line estimation of the electrical parameters to be used for the on-line adaptation of the corresponding flux model with regard to the variations of the machine parameters.

9.2 Sensitivity of the Flux Model to Parameter Variations

In the following, a sensitivity analysis of the flux models to the variations of the electrical parameters of the machine is presented [6,8]. In particular, with regard to the "voltage" and "current" flux models usually adopted in field-oriented control (FOC) and DTC schemes, the variations of the following variables will be analyzed with respect to the load:

1. The steady-state ratio between the amplitudes of the estimated and real rotor (stator) flux linkages
2. The steady-state difference between the phases of the estimated and real rotor (stator) flux linkages
3. The steady-state ratio between the estimated and real electromagnetic torque

9.2.1 Sensitivity of the Current Flux Model

The "current flux model" of the IM will be considered, written in the rotor-flux-oriented reference frame, where the $\wedge$ symbol will indicate all the variables and parameters estimated by the flux model, while the corresponding real variables of the machine are represented without $\wedge$. For simplicity's sake, the apex denoting the reference frame in which equations are written is not expressed in the following. They are supposed to be written in the rotor-flux-oriented reference frame.

The current flux model can be written in this form:

$$\begin{cases} \hat{T}_r \dfrac{d\left|\hat{\mathbf{i}}_{\mathbf{mr}}\right|}{dt} + \left|\hat{\mathbf{i}}_{\mathbf{mr}}\right| = \hat{i}_{sx} \\[3mm] \omega_{mr} = \omega_r + \dfrac{\hat{i}_{sy}}{\hat{T}_r \left|\hat{\mathbf{i}}_{\mathbf{mr}}\right|} \end{cases} \qquad (9.1)$$

with $\hat{\boldsymbol{\psi}}_{\mathbf{r}} = \hat{L}_m \hat{\mathbf{i}}_{\mathbf{mr}}$. For the symbols, see the list at the end of the chapter.

The same equations, written for the machine, are as follows:

$$\begin{cases} T_r \dfrac{d\,|\mathbf{i_{mr}}|}{dt} + |\mathbf{i_{mr}}| = i_{sx} \\[3mm] \omega_{mr} = \omega_r + \dfrac{i_{sy}}{T_r\,|\mathbf{i_{mr}}|} \end{cases} \tag{9.2}$$

with $\psi_r = L_m \mathbf{i_{mr}}$.

Since the slip pulsation ω_{sl} (in electrical angles) depends only on the load, the following equality is valid:

$$\omega_{sl} = \frac{i_{sy}}{T_r i_{sx}} = \frac{\hat{i}_{sy}}{\hat{T}_r \hat{i}_{sx}} \tag{9.3}$$

Combining Equation 9.3 with Equations 9.1 and 9.2, the following steady-state space-vector relationship between the estimated and real rotor flux linkages can be found:

$$\frac{\boldsymbol{\psi_r}}{\hat{\boldsymbol{\psi_r}}} = \frac{L_m\,(1 + j\omega_{sl}\hat{T}_r)}{\hat{L}_m\,(1 + j\omega_{sl}T_r)} \tag{9.4}$$

From this vector equation, the following two scalar equations can be deduced, valid, respectively, for the amplitude and for the phase angle of the rotor flux space-vector:

$$\frac{|\boldsymbol{\psi_r}|}{|\hat{\boldsymbol{\psi_r}}|} = \frac{L_m}{\hat{L}_m}\frac{\sqrt{1 + (\omega_{sl}\hat{T}_r)^2}}{\sqrt{1 + (\omega_{sl}T_r)^2}} \tag{9.5}$$

$$\delta_{pr} = \rho_r - \hat{\rho}_r = \arctan\left(\frac{\omega_{sl}(\hat{T}_r - T_r)}{1 + \omega_{sl}^2 T_r \hat{T}_r}\right) = \arctan\left(\left(1 - \frac{T_r}{\hat{T}_r}\right)\frac{\hat{i}_{sx}\hat{i}_{sy}}{\hat{i}_{sx}^2 + (T_r/\hat{T}_r)\hat{i}_{sy}^2}\right) \tag{9.6}$$

where δ_{pr} is the error in the estimation of the rotor flux-linkage phase position. Figure 9.1 shows the stator current space-vector and its decomposition on the $x - y$ reference frame aligned with the real rotor flux as well as its decomposition on the $\hat{x} - \hat{y}$ reference frame aligned with the estimated rotor flux. It sketches the situation when heating of the windings (and resulting increase of their resistances) or flux reduction of the machine occurs, as explained in the following.

Since $\beta_i = \arctan(i_{sy}/i_{sx}) = \arctan(\omega_{sl}T_r)$ and $\hat{\beta}_i = \arctan(\hat{i}_{sy}/\hat{i}_{sx}) = \arctan(\omega_{sl}\hat{T}_r)$, the following conclusions can be drawn:

- In case of machine heating, the steady-state value of the rotor flux-linkage amplitude tends to be larger than that of the model; therefore, $R_r > \hat{R}_r$, then $T_r < \hat{T}_r$, and $\beta_i < \hat{\beta}_i$: this means that the reference frame aligned with the estimated rotor flux lags the one aligned with the real rotor flux and the stator current components are wrongly estimated: $i_{sx} > \hat{i}_{sx}$ and $i_{sy} < \hat{i}_{sy}$.

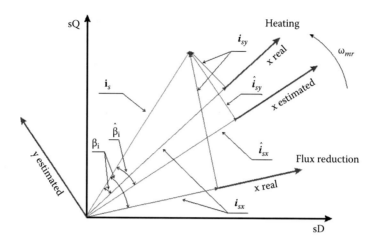

FIGURE 9.1
Decomposition of the stator current space-vector in case of detuning of the flux model.

- In case the machine works in the linear part of the magnetic characteristic (field-weakening phase), the steady-state value of the rotor flux-linkage amplitude tends to be lower than that of the model; therefore, $L_r > \hat{L}_r$, then $T_r > \hat{T}_r$ and $\beta_i > \hat{\beta}_i$, resulting in the wrong estimation of the stator current components since, in this case, the estimated reference frame leads the real one: $i_{sx} < \hat{i}_{sx}$ and $i_{sy} > \hat{i}_{sy}$.

The same conclusions can be drawn when studying the functions described by Equations 9.4 and 9.5. Figure 9.2 shows the steady-state curves $|\psi_r|/|\hat{\psi}_r|$ versus $\hat{R}_r/R_r$, drawn for a fixed value of $\hat{i}_{sy}/\hat{i}_{sx}$. The parameters of the machine under study are shown in Table 9.1. All of the curves intersect in the point $\hat{R}_r/R_r = 1$ and $|\psi_r|/|\hat{\psi}_r| = 1$, in correspondence to which the flux model is correctly tuned. When $\hat{R}_r/R_r < 1$, then the machine is in condition of heating, and when $|\psi_r|/|\hat{\psi}_r| > 1$, the machine rotor flux is higher than that in the flux model. The opposite occurs when the machine temperature decreases and the real flux in the machine is lower than that in the flux model. The higher the difference of $\hat{R}_r/R_r$ from 1, the higher the

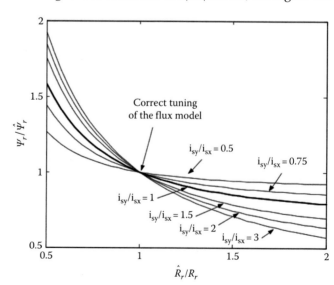

FIGURE 9.2
$|\psi_r|/|\hat{\psi}_r|$ versus $\hat{R}_r/R_r$ parameterized with $\hat{i}_{sy}/\hat{i}_{sx}$.

TABLE 9.1

K-Parameters of the IM and Results of the Estimation

	OLS	True	Err. %
K_1	188.6	189.0088	−0.2136
K_2	1267	946.4266	33.8906
K_{31}	127.1	127.5	−0.3765
K_4	32.8	32.8711	−0.1550
K_5	243.2	243.9244	−0.2913

Source: Cirrincione, M. and Pucci, M., Experimental verification of a technique for the real-time identification of induction motors based on the recursive least-squares, *International Workshop on Advanced Motion Control (IEEE AMC'02)*, Maribor, Slovenia, 2002.

difference between the real and the estimated fluxes. The higher the ratio $\hat{i}_{sy}/\hat{i}_{sx}$, that is, the higher the load, the higher the difference between the real and the estimated fluxes.

Figure 9.3 shows the $|\psi_r|/|\hat{\psi}_r|$ versus the slip pulsation ω_{sl} for fixed values of the ratio $\hat{R}_r/R_r$. It can be observed that, if the load is null ($\omega_{sl} = 0$), the flux model, even if detuned, is able to correctly estimate the amplitude of the rotor flux linkage. The higher the load torque, the higher the difference between the estimated and the real fluxes. This means that, as far as the current model is concerned, the detuning of the model is apparent at load. Figure 9.4, showing the surface $|\psi_r|/|\hat{\psi}_r|$ versus $\hat{R}_r/R_r$ and ω_{sl}, summarizes the results of Figures 9.2 and 9.3.

Figure 9.5 shows the set of curves providing $\hat{\rho}_r - \rho_r$ versus $\hat{R}_r/R_r$, for fixed values of the ratio $\hat{i}_{sy}/\hat{i}_{sx}$. All the curves intersect in the point $\hat{R}_r/R_r = 1$ and $\delta_{pr} = 0$, in correspondence to which the flux model is correctly tuned. Figure 9.6 shows the set of curves providing δ_{pr} versus ω_{sl}, for fixed values of $\hat{T}_r/T_r$. This curve shows that, if the load is null ($\omega_{sl} = 0$), the flux model, even if detuned, is able to correctly estimate the phase of the rotor flux-linkage space-vector. To a load torque increase corresponds firstly an increase of the flux angle error, followed by a reduction of it.

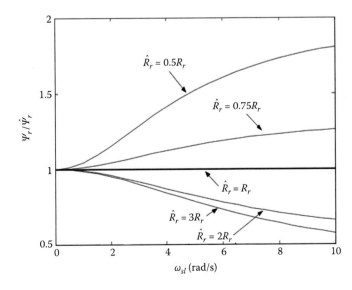

FIGURE 9.3
$|\psi_r|/|\hat{\psi}_r|$ versus ω_{sl} parameterized with $\hat{R}_r/R_r$.

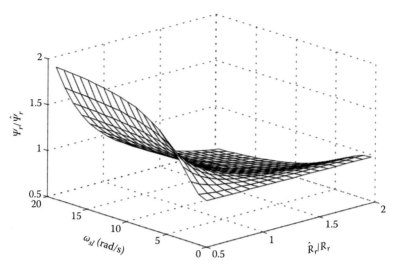

FIGURE 9.4
$|\psi_r|/|\hat{\psi}_r|$ versus $\hat{R}_r/R_r$ and ω_{sl}.

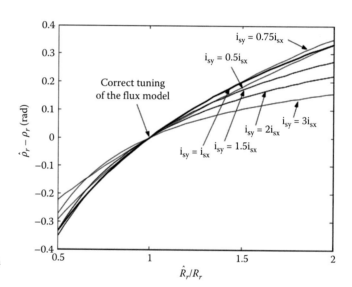

FIGURE 9.5
$\hat{\rho}_r - \rho_r$ versus ω_{sl} parameterized with $\hat{R}_r/R_r$.

Finally, Figures 9.7 through 9.10 show the same kind of curves, related to sensitivity of the current flux model to the variations of the three-phase magnetizing inductance and the rotor leakage factor.

The relationship between the estimated $\hat{t}_e$ and real t_e electromagnetic torque can be written, by using Equations 9.4 and 9.5 as follows:

$$\frac{t_e}{\hat{t}_e} = \frac{L_m^2}{L_r} \frac{\hat{L}_r}{\hat{L}_m^2} \frac{T_r}{\hat{T}_r} \frac{1+(\omega_{sl}\hat{T}_r)^2}{1+(\omega_{sl}T_r)^2} \tag{9.7}$$

On the basis of Equation 9.7, the set of curves drawn in Figure 9.11 have been found, showing the ratio $t_e/\hat{t}_e$ versus $\hat{R}_r/R_r$, for fixed values of the ratio $\hat{i}_{sy}/\hat{i}_{sx}$. This figure has been

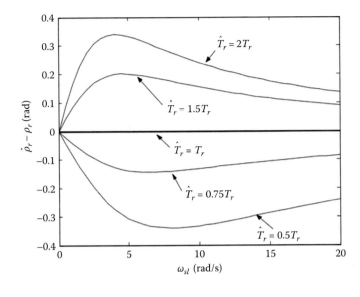

FIGURE 9.6
$\hat{\rho}_r - \rho_r$ versus $\hat{R}_r/R_r$ and ω_{sl}.

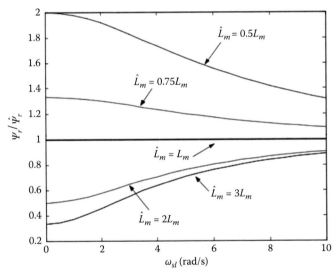

FIGURE 9.7
$|\psi_r|/|\hat{\psi}_r|$ versus ω_{sl} parameterized with $L_m/\hat{L}_m$.

obtained under the hypothesis of linearity from the point of view of the magnetization of the machine. All of the curves intersect in the point $\hat{R}_r/R_r = 1$ and $t_e/\hat{t}_e = 1$, in correspondence to which the flux model is correctly tuned. It can be further observed that there are working conditions in which, even if the flux model is detuned, the dynamic performance of the drive is better than that achievable when it is tuned. More precisely, if $\hat{R}_r/R_r < 1$ (heating) and $\hat{i}_{sy}/\hat{i}_{sx}$ is high (big load torque or low value of the magnetizing current), the real electromagnetic torque is much bigger than the estimated one. Finally, the relationship between the electromagnetic torque and the estimated current components $\hat{i}_{sx}$ and $\hat{i}_{sy}$ is given by the following:

$$t_e = \frac{3}{2} p \frac{L_m^2}{L_r} \frac{T_r}{\hat{T}_r} \hat{i}_{sx} \hat{i}_{sy} \frac{\hat{i}_{sx}^2 + \hat{i}_{sy}^2}{\hat{i}_{sx}^2 + (T_r/\hat{T}_r)^2 \hat{i}_{sy}^2} \tag{9.8}$$

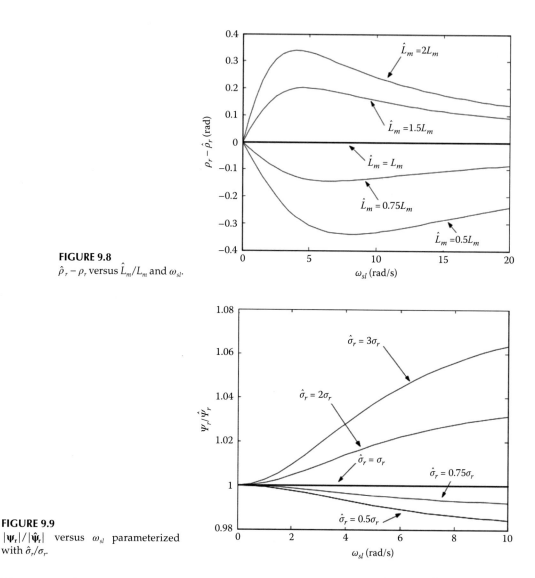

FIGURE 9.8
$\hat{\rho}_r - \rho_r$ versus $\hat{L}_m/L_m$ and ω_{sl}.

FIGURE 9.9
$|\boldsymbol{\psi_r}|/|\hat{\boldsymbol{\psi_r}}|$ versus ω_{sl} parameterized with $\hat{\sigma}_r/\sigma_r$.

Figure 9.12 shows the relationship between the real torque t_e and $\hat{i}_{sy}$, once $\hat{i}_{sx}$ is fixed, for different values of the ratio $\hat{R}_r/R_r$. It highlights that the relationship between t_e and $\hat{i}_{sy}$ is almost linear only for $\hat{R}_r/R_r = 1$, that is, in conditions of correct field orientation. For values of $\hat{R}_r/R_r$ far different from 1, this relationship becomes nonlinear.

9.2.2 Sensitivity of the Voltage Flux Model

As in the case of the current flux model, it is possible to make an analysis of the sensitivity of the "voltage flux model" versus the variation of the parameters of the machine. Starting from Equation 5.19, it is possible to retrieve the relationship between the space-vectors of the estimated and real rotor flux linkages:

$$\frac{\hat{\boldsymbol{\psi_r}}}{\boldsymbol{\psi_r}} = \frac{L_m \hat{L}_r}{\hat{L}_m L_r}\left[1 + \frac{L_r^2}{R_r L_m^2}\left(\frac{R_r}{L_r} + j\omega_{sl}\right)\left(\left(\sigma L_s - \hat{\sigma}\hat{L}_s\right) - j\frac{(R_s - \hat{R}_s)}{(\omega_{sl} + \omega_r)}\right)\right] \tag{9.9}$$

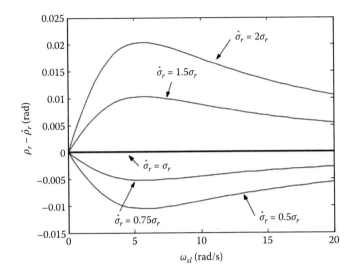

FIGURE 9.10

$\hat{\rho}_r - \rho_r$ versus $\hat{\sigma}_r/\sigma_r$ and ω_{sl}.

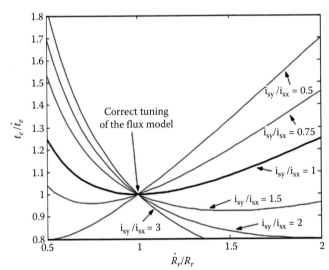

FIGURE 9.11

$t_e/\hat{t}_e$ versus $\hat{R}_r/R_r$ parameterized with $\hat{i}_{sy}/\hat{i}_{sx}$.

From the vector Equation 9.9, the following scalar equations can be deduced:

$$\frac{|\hat{\boldsymbol{\psi}}_r|}{|\boldsymbol{\psi}_r|} = \frac{L_m \hat{L}_r}{\hat{L}_m L_r} \sqrt{\left(1 + \frac{L_r}{L_m^2}(\sigma L_s - \hat{\sigma}\hat{L}_s) + \frac{\omega_{sl} L_r^2}{R_r L_m^2} \frac{(R_s - \hat{R}_s)}{(\omega_{sl} + \omega_r)}\right)^2 + \left(-\frac{L_r}{L_m^2} \frac{(R_s - \hat{R}_s)}{(\omega_{sl} + \omega_r)} + \frac{\omega_{sl} L_r^2}{R_r L_m^2}(\sigma L_s - \hat{\sigma}\hat{L}_s)\right)^2}$$

(9.10)

and

$$\delta_{pr} = \hat{\rho}_r - \rho_r = \arctan\left(\frac{-\dfrac{L_r}{L_m^2}\dfrac{(R_s - \hat{R}_s)}{(\omega_{sl} + \omega_r)} + \dfrac{\omega_{sl} L_r^2}{R_r L_m^2}(\sigma L_s - \hat{\sigma}\hat{L}_s)}{1 + \dfrac{L_r}{L_m^2}(\sigma L_s - \hat{\sigma}\hat{L}_s) + \dfrac{\omega_{sl} L_r^2}{R_r L_m^2}\dfrac{(R_s - \hat{R}_s)}{(\omega_{sl} + \omega_r)}}\right)$$

(9.11)

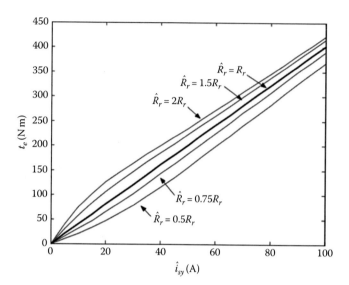

FIGURE 9.12

t_e versus $\hat{i}_{sy}$ parameterized with $\hat{R}_r/R_r$.

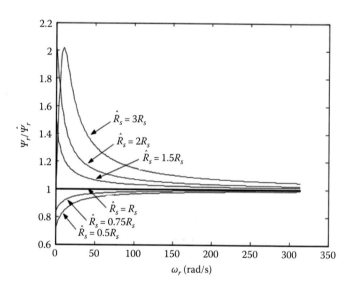

FIGURE 9.13

$|\boldsymbol{\psi}_r|/|\hat{\boldsymbol{\psi}}_r|$ versus ω_r parameterized with $\hat{R}_s/R_s$.

Figure 9.13 shows the set of curves $|\boldsymbol{\psi}_r|/|\hat{\boldsymbol{\psi}}_r|$ versus the rotor speed ω_r for fixed values of the ratio $\hat{R}_s/R_s$, drawn for a unique value of slip pulsation equal to 10 rad/s. This figure clearly shows that, if $\hat{R}_s/R_s < 1$ (heating of the machine), $|\boldsymbol{\psi}_r|/|\hat{\boldsymbol{\psi}}_r| < 1$, that is, the real flux is lower than the estimated one. Vice versa, if the machine cools down. The higher the difference $\hat{R}_s/R_s$ from 1, the higher difference between the estimated and the real fluxes. Furthermore, the highest sensitivity to any variation of the stator resistance can be observed at low speeds, in correspondence to which the ohmic drop on the stator resistance plays a significant role. At increasing rotor speeds, the sensitivity of the flux model versus stator resistance variations reduces significantly. Figure 9.14 shows the set of curves $\rho_r - \hat{\rho}_r$ versus the rotor speed ω_r for fixed values of the ratio $\hat{R}_s/R_s$, drawn for a unique value of slip pulsation equal to 10 rad/s. Also in this case, the error in the estimation of the rotor flux angle is very high at low rotating speeds, and it reduces significantly at increasing speeds.

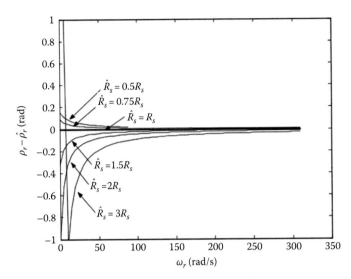

FIGURE 9.14
$\rho_r - \hat{\rho}_r$ versus ω_r parameterized with $\hat{R}_s/R_s$.

Similarly, a sensitivity analysis of the voltage model of the IM estimating the stator flux linkage, instead of the rotor one, can be performed. Such a flux model, which can be suitably adopted both in stator-flux-oriented vector-controlled and DTC drives, is described by Equation 5.46. Also in this case, it is possible to retrieve the steady-state relationship between the estimated and real stator flux-linkage space-vectors:

$$\frac{\hat{\boldsymbol{\psi}}_s}{\boldsymbol{\psi}_s} = 1 - j\frac{1}{L_s}\left(\frac{1+jT_r\omega_{sl}}{1+j\sigma T_r\omega_{sl}}\right)\frac{R_s - \hat{R}_s}{\omega_r + \omega_{sl}} \tag{9.12}$$

From the vector Equation 9.12, the following scalar equations can be deduced:

$$\frac{|\hat{\boldsymbol{\psi}}_s|}{|\boldsymbol{\psi}_s|} = \sqrt{\left(1 - \frac{(\sigma T_r\omega_{sl} - T_r\omega_{sl})(R_s - \hat{R}_s)}{L_s(\omega_r + \omega_{sl})(1+\sigma^2 T_r^2\omega_{sl}^2)}\right)^2 + \left(\frac{(1+\sigma T_r^2\omega_{sl}^2)(R_s - \hat{R}_s)}{L_s(\omega_r + \omega_{sl})(1+\sigma^2 T_r^2\omega_{sl}^2)}\right)^2} \tag{9.13}$$

and

$$\delta_{ps} = \hat{\rho}_s - \rho_s = \arctan\left(\frac{-\dfrac{(1+\sigma T_r^2\omega_{sl}^2)(R_s - \hat{R}_s)}{L_s(\omega_r + \omega_{sl})(1+\sigma^2 T_r^2\omega_{sl}^2)}}{1 - \dfrac{(\sigma T_r\omega_{sl} - T_r\omega_{sl})(R_s - \hat{R}_s)}{L_s(\omega_r + \omega_{sl})(1+\sigma^2 T_r^2\omega_{sl}^2)}}\right) \tag{9.14}$$

From Equations 9.13 and 9.14, the set of curves describing the sensitivity analysis of the voltage flux model can be drawn.

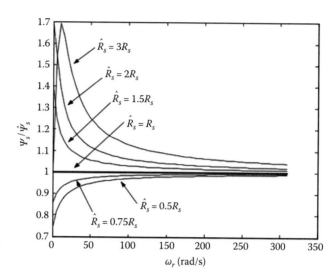

FIGURE 9.15

$|\boldsymbol{\psi}_s|/|\hat{\boldsymbol{\psi}}_s|$ versus ω_r parameterized with $\hat{R}_s/R_s$.

Figure 9.15 shows the set of curves $|\boldsymbol{\psi}_s|/|\hat{\boldsymbol{\psi}}_s|$ versus the rotor speed ω_r for fixed values of the ratio $\hat{R}_s/R_s$, drawn for a unique value of slip pulsation equal to 10 rad/s. This figure clearly shows that, if $\hat{R}_s/R_s < 1$ (heating of the machine), $|\boldsymbol{\psi}_s|/|\hat{\boldsymbol{\psi}}_s| < 1$, that is, the real flux is lower than the estimated one. Vice versa, if the machine cools down. The higher the difference $\hat{R}_s/R_s$ from 1, the higher difference between the estimated and the real fluxes. Furthermore, the highest sensitivity to any variation of the stator resistance can be observed at low speeds, in correspondence to which the ohmic drop on the stator resistance plays a significant role. At increasing rotor speeds, the sensitivity of the flux model versus stator resistance variations reduces significantly.

Figure 9.16 shows the set of curves $\rho_s - \hat{\rho}_s$ versus the rotor speed ω_r for fixed values of the ratio $\hat{R}_s/R_s$, drawn for a unique value of slip pulsation equal to 10 rad/s. Also in this case, the error in the estimation of the rotor flux angle is very high at low rotating speeds, reducing significantly at increasing speeds.

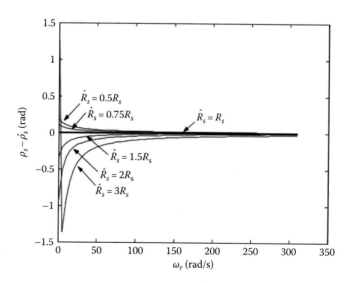

FIGURE 9.16

$\rho_s - \hat{\rho}_s$ versus ω_r parameterized with $\hat{R}_s/R_s$.

9.3 Experimental Analysis of the Effects of Flux Model Detuning on the Control Performance

In this section, the experimental verification of the sensitivity analysis in Section 9.2 is proposed. With this regard, the machine under test is a 22 kW IM whose parameters are shown in Table 4.2. The adopted control scheme is a rotor-flux-oriented control with impressed voltages (see Figure 5.16). The adopted flux model is the current flux model in the rotor-flux-oriented reference frame (see Figure 5.11b).

The drive has been given a t = 0.1 s and a constant rotor flux amplitude of 1.5 Wb, corresponding to a working point on the knee of the magnetization curve (see Figure 4.14). Subsequently, at t = 0.6 s, the drive has been given a step speed reference of 10 Hz. The test has been performed twice, respectively, with the rotor resistance of the flux model equal to that of the motor (flux model correctly tuned) and with the rotor resistance of the flux model half of that of the motor (detuned flux model). Since the machine under test presents a wound rotor, its real resistance has been modified by adding in series a resistor whose resistance has a value equal to that of the winding itself (the rotor resistance doubles).

Figure 9.17 shows the curves of the rotor speed, the rotor flux amplitude, and the i_{sy} current component during this test. The same test has been repeated with a rotor flux reference equal to 0.6 Wb, corresponding to a working point in the linear part of the magnetic characteristic. Figure 9.18 shows the same waveforms, obtained during this test. The analysis of Figures 9.17 and 9.18 highlights that, if the flux model is not correctly tuned, when $\hat{R}_r/R_r < 1$ and the machine works in the saturation part of the magnetic characteristic, the

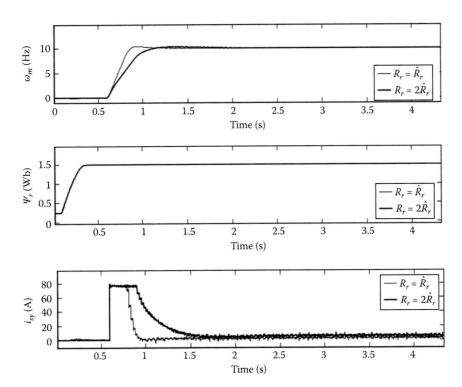

FIGURE 9.17
Rotor speed, rotor flux, and i_{sy} with flux amplitude = 1.5 Wb.

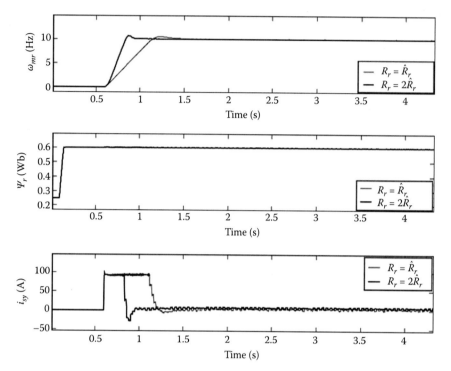

FIGURE 9.18
Rotor speed, rotor flux, and i_{sy} with flux amplitude = 0.6 Wb.

dynamic performance of the drive with detuned flux model is worse than the corresponding with the flux model correctly tuned. On the contrary, when $\hat{R}_r/R_r < 1$ and the machine works in the linear part of the magnetic characteristic, the dynamic performance of the drive with detuned flux model is better than the corresponding with the flux model correctly tuned, as shown in Figure 9.11.

9.4 Methods for the On-Line Tracking of the Machine Parameter Variations

To properly cope with the negative effects on the flux models caused by the variation of the electrical parameters of the IM drive during on-line operation, several approaches have been devised. A simple flux (either rotor or stator) model can be used, supplemented with an algorithm for the on-line estimation of some IM parameters, typically the rotor time constant for current flux models and the stator resistance for voltage ones. Alternatively, the machine flux can be estimated with an observer whose estimation is robust to parameter variations. Finally, an observer could be used, estimating simultaneously the flux linkage and one (or several) electrical parameter(s) of the machine.

Figure 9.19 shows a classification of the main methodologies that can be adopted for the on-line tracking of the variations of the electrical parameters of the IM.

The first known papers in the literature have been dedicated to the formulation of identification techniques using several steady-state measurements [4]. Then, other methods have been proposed on the basis of nonlinear formulation of the machine equations. The drawback of

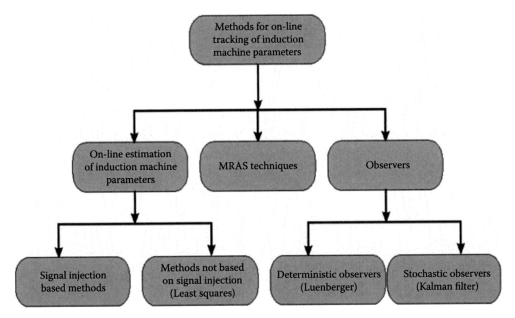

FIGURE 9.19
Classification of methodologies for on-line tracking of electrical parameter variations.

these methods is well known, since the amount of necessary computations avoids them to be implemented on-line even with a powerful digital processor. Only in the last few years, from the 1980s onward, have new methods been developed, mainly for control purposes, such as recursive least squares (RLS) methods [20,84–87], extended Kalman filter (EKF) [25,81], and model reference adaptive system (MRAS) [2,51,82,83]. Their target was that of tracking the fluctuations of the machine parameters and keeping the system as it was initially tuned for both static and dynamic performance. However, both EKF and MRAS methods are more cumbersome from a numerical point of view with respect to RLS methods which, additionally, have a more reliable theory as for convergence and stability issues of the algorithms. In particular, in Refs [84–92], a least squares (LS) approach has been developed that estimates the parameters on the basis of stator currents and voltages and the rotational speed. This method, however, has always proven difficult as auxiliary variables have been introduced in order to retrieve the physical parameters, and unfortunately, not all parameters are easy to detect. Only in Ref. [86], a more complete method has been developed to overcome this problem, but with more computation burden and without giving any proof of convergence. In Refs [93,94], emphasis has been therefore laid on the theoretical analysis of LS methods applied to the identification of an IM, which is the topic of the rest of this chapter.

9.5 On-Line Estimation of the IM Parameters with the Ordinary Least Squares Method

This paragraph shows essentially how the LS methods can be applied to the estimation of the parameters of an IM, by using different kinds of LSs. Particularly it shows how the estimation of the electrical parameters of the motor can be performed in an unconstrained way, indirectly taking into consideration the constraints which inevitably arise when the well-known stator and rotor voltage equations are rearranged so as to allow the application of the LS method. The assumptions under which this technique is valid are also presented as well as the identifiability criteria both for transient and sinusoidal steady-state. Moreover, some practical issues about the choice of the most suitable reference frame are also developed on the basis of numerical considerations. In the end, simulations under various load conditions are shown, and the results are discussed. All simulations have been carried out taking into consideration an IM directly connected to the utility grid: it is apparent that the same considerations can be extended also to the case of a motor in an AC drive.

9.5.1 Space-Vector Voltage Equations in the General Reference Frame

It is well known that the IM model can be described by the following stator and rotor space-vector voltage equations in the general reference frame, which rotates at a general speed ω_g (in electrical angles per second) as described in Chapter 4:

$$\mathbf{u}_s^g = R_s \mathbf{i}_s^g + \frac{d\psi_s^g}{dt} + j\omega_g \psi_s^g \tag{9.15a}$$

$$\mathbf{u}_r^g = R_r \mathbf{i}_r^g + \frac{d\psi_r^g}{dt} + j(\omega_g - \omega_r)\psi_r^g \tag{9.15b}$$

$$\psi_s^g = L_s \mathbf{i}_s^g + L_m \mathbf{i}_r^g \tag{9.15c}$$

$$\psi_r^g = L_r \mathbf{i}_r^g + L_m \mathbf{i}_s^g \tag{9.15d}$$

For the symbols, see the list at the end of the chapter.

In the previous equations, the different parameters of the formulas could be constant or time and space variant depending on preliminary assumptions. With the important assumption that [86,93,94]

$$\frac{d\omega_r}{dt} \approx 0 \tag{9.16}$$

that is, under the assumption that the motor be at standstill, in slow transients, or in sinusoidal steady-state, and expressing the direct and quadrature components by eliminating the space-vector of the rotor currents, which are not measurable quantities, the following two simplified scalar equations result:

$$\frac{d^2 i_{sx}^g}{dt^2} + \left(\frac{1}{\sigma T_s} + \frac{\beta_0}{\sigma}\right)\frac{di_{sx}^g}{dt} - (2\omega_g - \omega_r)\frac{di_{sy}^g}{dt} - \omega_g(\omega_g - \omega_r)i_{sx}^g + \frac{\beta_0}{\sigma T_s}i_{sx}^g - \frac{\omega_g - \omega_r}{\sigma T_s}i_{sy}^g - \frac{\beta_0}{\sigma}\omega_g i_{sy}^g =$$

$$= \frac{1}{\sigma L_s}\frac{du_{sx}^g}{dt} - \frac{\omega_g - \omega_r}{\sigma L_s}u_{sy}^g + \frac{\beta_0}{\sigma L_s}u_{sx}^g \tag{9.17a}$$

$$\frac{d^2 i_{sy}^g}{dt^2} + \left(\frac{1}{\sigma T_s} + \frac{\beta_0}{\sigma}\right)\frac{di_{sy}^g}{dt} + (2\omega_g - \omega_r)\frac{di_{sx}^g}{dt} - \omega_g(\omega_g - \omega_r)i_{sy}^g + \frac{\beta_0}{\sigma T_s}i_{sy}^g + \frac{\omega_g - \omega_r}{\sigma T_s}i_{sx}^g + \frac{\beta_0}{\sigma}\omega_g i_{sx}^g =$$

$$= \frac{1}{\sigma L_s}\frac{du_{sy}^g}{dt} + \frac{\omega_g - \omega_r}{\sigma L_s}u_{sx}^g + \frac{\beta_0}{\sigma L_s}u_{sy}^g \tag{9.17b}$$

The following parameters, called K-parameters, are then defined, using the notations of [86]:

$$K_1 = \frac{1}{\sigma T_s} + \frac{\beta_0}{\sigma} \tag{9.18a}$$

$$K_2 = \frac{\beta_0}{\sigma T_s} \tag{9.18b}$$

$$K_{31} = \frac{1}{\sigma T_s} \tag{9.18c}$$

$$K_{32} = \frac{\beta_0}{\sigma} \tag{9.18d}$$

$$K_4 = \frac{1}{\sigma L_s} \tag{9.18e}$$

$$K_5 = \frac{\beta_0}{\sigma L_s} \tag{9.18f}$$

Between these six parameters, a linear and a quadratic relationship exist, as shown in the following:

$$K_1 = K_{31} + K_{32} \tag{9.19a}$$

$$K_2 K_4 = K_{31} K_5 \tag{9.19b}$$

It may be useful to substitute one of the two relationships for a combination of them, that is,

$$K_1 = \frac{K_2 K_4}{K_5} + K_{32} \tag{9.19c}$$

From the K-parameters, not all the five electrical parameters (R_s, R_r, L_s, L_r, L_m can be retrieved as no rotor measurements are available [86,93,94]: in fact, the K-parameters determine only four independent parameters, that is, R_s, R_r, σ, and $\beta_0 = 1/T_r$, in the following way:

$$T_r = \frac{K_4}{K_5} \tag{9.20a}$$

$$\sigma = \frac{K_5}{K_4 K_{32}} = \frac{K_5}{K_4(K_1 - K_{31})} \tag{9.20b}$$

$$L_s = \frac{K_{32}}{K_5} = \frac{K_1 - K_{31}}{K_5} \tag{9.20c}$$

$$R_s = \frac{K_2}{K_5} = \frac{K_{31}}{K_4} = \frac{K_1 K_5 - K_{32}}{K_4 K_5} \tag{9.20d}$$

As recalled in Ref. [86], L_m, L_r, and R_r cannot be obtained independently as rotor flux linkages are unknown: thus machines with identical rotor time constant and identical ratio L_m^2/L_r have the same input/output (voltage/current) equations.

According to the reference frame that is chosen, both the equations and the relationships can be simplified, in the sense that the number of the relationships and the K-parameters can be reduced, as discussed in the following.

In the stationary reference frame fixed to the stator, the Equations 9.17a and 9.17b becomes

$$\frac{d^2 i_{sD}}{dt^2} + \left(\frac{1}{\sigma T_s} + \frac{\beta_0}{\sigma} \right) \frac{d i_{sD}}{dt} + \omega_r \frac{d i_{sQ}}{dt} + \frac{\beta_0}{\sigma T_s} i_{sD} + \frac{\omega_r}{\sigma T_s} i_{sQ} =$$

$$= \frac{1}{\sigma L_s} \frac{d u_{sD}}{dt} + \frac{\omega_r}{\sigma L_s} u_{sQ} + \frac{\beta_0}{\sigma L_s} u_{sD} \tag{9.21a}$$

$$\frac{d^2 i_{sQ}}{dt^2} + \left(\frac{1}{\sigma T_s} + \frac{\beta_0}{\sigma} \right) \frac{d i_{sQ}}{dt} - \omega_r \frac{d i_{sD}}{dt} + \frac{\beta_0}{\sigma T_s} i_{sQ} - \frac{\omega_r}{\sigma T_s} i_{sD} =$$

$$= \frac{1}{\sigma L_s} \frac{d u_{sQ}}{dt} - \frac{\omega_r}{\sigma L_s} u_{sD} + \frac{\beta_0}{\sigma L_s} u_{sQ} \tag{9.21b}$$

The parameters to be estimated are then five, as the K_{32} is not present; the corresponding term in Equations 9.17a and 9.17b is null. As a result, only relationship (9.19b) holds. Also Equations 9.20a through 9.20d holds, but obviously, only the second members where K_{32} is not present can be used.

Conversely, in the rotating reference frame fixed to the rotor, the Equations 9.17a and b becomes

$$\frac{d^2 i_{sd}}{dt^2} + \left(\frac{1}{\sigma T_s} + \frac{\beta_0}{\sigma} \right) \frac{d i_{sd}}{dt} - \omega_r \frac{d i_{sq}}{dt} + \frac{\beta_0}{\sigma T_s} i_{sd} - \frac{\beta_0}{\sigma} \omega_r i_{sq}$$

$$= \frac{1}{\sigma L_s} \frac{d u_{sd}}{dt} + \frac{\beta_0}{\sigma L_s} u_{sd} \tag{9.22a}$$

$$\frac{d^2 i_{sq}}{dt^2} + \left(\frac{1}{\sigma T_s} + \frac{\beta_0}{\sigma} \right) \frac{d i_{sq}}{dt} + \omega_r \frac{d i_{sd}}{dt} + \frac{\beta_0}{\sigma T_s} i_{sq} + \frac{\beta_0}{\sigma} \omega_r i_{sd}$$

$$= \frac{1}{\sigma L_s} \frac{d u_{sq}}{dt} + \frac{\beta_0}{\sigma L_s} u_{sq} \tag{9.22b}$$

Also in this case, the parameters to be estimated are then five, as the K_{31} is not present; the corresponding term in Equations 9.17a and 9.17b is null. As a result, only relationship (9.19c) holds. Also Equations 9.20a through 9.20d holds, but obviously, only the second members where K_{31} is not present can be used.

In any case, the coefficients of the above differential equations are time varying. If the rotor speed is known, these equations can be expressed in matrix form as follows, where the coefficients of the matrix are all known:

$$\begin{pmatrix} \dfrac{di_{sx}^g}{dt} & i_{sx}^g & -(\omega_g - \omega_r)i_{sy}^g & -\omega_g i_{sy}^g & -\left(\dfrac{du_{sx}^g}{dt} - (\omega_g - \omega_r)u_{sy}^g \right) & -u_{sx}^g \\ \dfrac{di_{sy}^g}{dt} & i_{sy}^g & (\omega_g - \omega_r)i_{sx}^g & \omega_g i_{sx}^g & -\left(\dfrac{du_{sy}^g}{dt} + (\omega_g - \omega_r)u_{sx}^g \right) & -u_{sy}^g \end{pmatrix} \begin{pmatrix} K_1 \\ K_2 \\ K_{31} \\ K_{32} \\ K_4 \\ K_5 \end{pmatrix}$$

$$= \begin{pmatrix} -\dfrac{d^2 i_{sx}^g}{dt^2} + (2\omega_g - \omega_r)\dfrac{di_{sy}^g}{dt} + \omega_g(\omega_g - \omega_r)i_{sx}^g \\ -\dfrac{d^2 i_{sy}^g}{dt^2} - (2\omega_g - \omega_r)\dfrac{di_{sx}^g}{dt} + \omega_g(\omega_g - \omega_r)i_{sy}^g \end{pmatrix} \tag{9.23}$$

This matrix equation can be solved for by using the different LS methods presented in Chapter 8, in particular, the ordinary least squares (OLS) and the total least squares (TLS) methods.

It should be remarked that the previous equations have been obtained under the assumption that the rotor speed varies slowly. From this standpoint, it would be more correct to use it in steady-state or slowly varying transient or at standstill, but even with less slowly varying rotor speed, these equations can be used as shown in the following. The drawback of the previous equation is also, as pointed out in Refs [86,88], that differentiators are required as well as voltage sensors, which requires that the derivatives be reconstructed by filtered differentiation, while the use of voltage sensors requires analog filtering.

It is to remark that expressing the Equation 9.17 in the reference frames fixed to the rotor and to the stator, the following matrix equations are determined:

$$\begin{pmatrix} \dfrac{di_{sd}}{dt} & i_{sd} & -\omega_r i_{sq} & -\dfrac{du_{sd}}{dt} & -u_{sd} \\ \dfrac{di_{sq}}{dt} & i_{sq} & \omega_r i_{sd} & -\dfrac{du_{sq}}{dt} & -u_{sq} \end{pmatrix} \begin{pmatrix} K_1 \\ K_2 \\ K_{32} \\ K_4 \\ K_5 \end{pmatrix} = \begin{pmatrix} -\dfrac{d^2 i_{sd}}{dt^2} + \omega_r \dfrac{di_{sq}}{dt} \\ -\dfrac{d^2 i_{sq}}{dt^2} - \omega_r \dfrac{di_{sd}}{dt} \end{pmatrix} \tag{9.24}$$

$$\begin{pmatrix} \dfrac{di_{sD}}{dt} & i_{sD} & \omega_r i_{sQ} & -\left(\dfrac{du_{sD}}{dt} + \omega_r u_{sQ} \right) & -u_{sD} \\ \dfrac{di_{sQ}}{dt} & i_{sQ} & -\omega_r i_{sD} & -\left(\dfrac{du_{sQ}}{dt} - \omega_r u_{sD} \right) & -u_{sQ} \end{pmatrix} \begin{pmatrix} K_1 \\ K_2 \\ K_{31} \\ K_4 \\ K_5 \end{pmatrix} = \begin{pmatrix} -\dfrac{d^2 i_{sD}}{dt^2} - \omega_r \dfrac{di_{sQ}}{dt} \\ -\dfrac{d^2 i_{sQ}}{dt^2} + \omega_r \dfrac{di_{sD}}{dt} \end{pmatrix} \tag{9.25}$$

The use of Equations 9.23 through 9.25 gives rise to three different approaches for parameter estimation. The use of Equation 9.23 only has been used in Ref. [90], Equation 9.24 in Ref. [91], and Equation 9.25 in Ref. [88]. Equation 9.24 has also been used in Ref. [89]. However, all of these approaches consider the LS resolution of these equations as an unconstrained optimization problem in the search of the minimum two-norm solution, while the presence of one or two constraints should be accounted for. As a matter of fact, [86] also considers an RLS with an algorithm for updating the K_2 parameter by directly using the constraint, without giving any proof about the convergence of the algorithm. In this chapter, it is shown that the solutions can be retrieved by OLS in an unconstrained way, taking indirectly into consideration the constraint: in this way, no constrained minimization convergence is necessary as the well-known results of LS methodology can be used.

It is apparent from the matrix Equation 9.23 (and obviously from Equations 9.24 and 9.25) that the two scalar equations that make it up are linearly independent. This means that in transient conditions or in nonsinusoidal steady-state, the data matrix is full rank, and the LS solution can be obtained. However, in sinusoidal steady-state, the data matrix has rank 2, and, thus, only two parameters can be computed. This means that the estimation of all the parameters cannot be made in steady-state.

One remark has to be made on the choice of the set of equations. In a synchronous reference frame (for instance, either in the rotor flux-linkage reference frame or in the stator flux-linkage reference frame or in the magnetizing flux-linkage reference frame), all signals, at least in the steady-state, are constant or vary slower than those in the stationary reference frame fixed to the stator or to the rotor reference frame. Since faster varying signals have wealthier frequency information, either the stator or the rotor reference frame should be used for the purpose of parameter estimation. Indeed constant or slow-varying signals can give rise to an ill-conditioned problem as, in transient state, the data matrix can happen not to be full rank with resulting numerical problems. The choice of the stator (or rotor) reference frame is therefore caused by numerical reasons, as explained more fully in Section 9.5.2.

9.5.2 Estimation of the Magnetizing Curve

A direct consequence of the retrieval of the four electrical parameters given by Equation 9.20 is the possibility to estimate the magnetizing curve of the machine, that is, the curve of the rotor flux-linkage amplitude versus the rotor magnetizing current amplitude $|\psi_r| = f(|\mathbf{i}_{mr}|)$. For this purpose, some tests should be made at different magnitudes of the supply voltage and at the same supply frequency in order to make the machine work under different magnetizing excitations. At each supply voltage after a speed transient from zero to steady-state speed (start-up test), the electrical parameters rotor time constant T_r, stator inductance L_s, stator resistance R_s, and global leakage factor σ are estimated. Either an unconstrained or constrained (see Section 9.6.1.2.1) LS method can be used. At the same time, the rotor magnetizing current $|\mathbf{i}_{mr}|$ can be computed by means of the well-known flux model based on the rotor equations of the IM in the rotor flux reference frame [1–3] (see Section 5.3.3.2). It should be remarked that, because the start-up tests are made at no-load, the steady-state value of the estimated $|\mathbf{i}_{mr}|$ does not suffer of any inaccurate guess in the T_r parameter needed by the flux estimator. In fact, the ratio between the true and the estimated rotor flux linkage is given by Equation 9.5, rewritten as follows in terms of rotor magnetizing currents [6,8]:

$$\frac{|\mathbf{i}_{mr}|}{|\hat{\mathbf{i}}_{mr}|} = \frac{\sqrt{1+(\omega_{sl}\hat{T}_r)^2}}{\sqrt{1+(\omega_{sl}T_r)^2}} \tag{9.26}$$

where
  the "^" stands for estimated variables
  ω_{sl} is the slip pulsation, depending on the load torque

From Equations 9.5 or 9.26 and from Figure 9.9, it is apparent that the magnetizing current estimation is independent from the parameter $\hat{T}_r$ used in the flux model at no-load (slip frequency is 0), as in the case under test.

Afterward, from L_s and σ, L_m is computed for each operating point as explained in Section 9.6.1.2.1. Finally, at the end of the tests at different voltage levels, the magnetization curve is estimated.

However, it should be noticed that only the global leakage factor can be estimated and not the stator or rotor leakage factors individually; consequently, the estimation of the static magnetizing inductance has been made under the usual assumption that $\sigma_r = 2\sigma_s$ [5]. Under this assumption, σ_r and σ_s are computed from σ by the following [5]:

$$\sigma_s = \sqrt{\frac{1}{2(1-\sigma)} + \frac{1}{16} - \frac{3}{4}} \tag{9.27a}$$

$$\sigma_r = \sqrt{\frac{2}{1-\sigma} + \frac{1}{4} - \frac{3}{2}} \tag{9.27b}$$

Therefore, under each magnetic excitation, the total magnetizing inductance L_m is computed from the estimated values of L_s and σ as follows:

$$L_m = \frac{L_s}{1+\sigma_s} = \frac{L_s}{1+\sqrt{\dfrac{1}{2(1-\sigma)} + \dfrac{1}{16} - \dfrac{3}{4}}} \tag{9.28}$$

Finally, the magnetization curve of the machine, which gives the nonlinear relationship between the rotor magnetizing current and the rotor magnetic flux $|\psi_r| = f(|i_{mr}|)$, is computed as follows:

$$\left|\psi_r(i_{mr})\right| = L_m(i_{mr})\left|i_{mr}\right| \tag{9.29}$$

9.5.3 Ordinary Least Squares Identification

Any of the Equations 9.23 through 9.25 can be written in the form of Chapter 8:

$$\mathbf{Ak} \approx \mathbf{b} \tag{9.30}$$

where $\mathbf{k}$, the column vector of the unknown parameters, is in this case the K-parameters. At each instant of time, a couple of equations are processed. Because of the error in modeling, as a result of the neglecting of the terms depending on $d\omega_r/dt$, the noisy measurements of the signals and all of the nonlinearities not accounted for by the linear dynamic model of the machine make the two terms of the equation only approximately equal [86]. This would then cause the matrix Equation 9.30 to have errors both in the observation vector as well as in the data matrix. The most suitable method of solution would then be using TLS rather than OLS [95–97]. In this paragraph, only the OLS estimation is presented, while that based on the TLS is presented subsequently.

For the implementation of the LS technique, it is necessary to use both analog and digital filters: this makes the errors in the data matrix less critical than those in the observation vector, which are more affected by error modeling (neglecting the terms depending on $d\omega_r/dt$). This justifies then the use of OLS; however, more accurate results can be obtained with the TLS as shown later for comparison: see, however, also [98,99].

The algorithm used here is the same RLS algorithm presented in Ref. [86]. However, a deeper analysis of the error estimate is developed, based on the classical linear regression theory.

In the following, the stationary reference frame will be used for parameter estimation: this choice will be made later on the basis of numerical analysis considerations.

9.5.4 RLS Algorithm

The target is to minimize the two-norm of the residual of Equation 9.30 (see also (8.2)), that is, $\|Ak - b\|_2$, with $k \in \mathfrak{R}^n$, $A \in \mathfrak{R}^{2m \times n}$, $b \in \mathfrak{R}^{2m}$, $\mathfrak{R}$ being the set of real numbers, m the number of time instants in the interval time of observation (in this case, there are two observations picked out at each instant of time, so the total number of rows of the A matrix is $2m$), and n the number of unknown variables (five K-parameters in this case). The unique minimum solution, in case the data matrix is full rank, is given by the following:

$$\mathbf{k}' = (\mathbf{A}^T\mathbf{A})^{-1}\mathbf{A}^T\mathbf{b} = \mathbf{P}^{-1}\mathbf{r} \tag{9.31a}$$

$$\mathbf{P} = \mathbf{A}^T\mathbf{A} \tag{9.31b}$$

$$\mathbf{r} = \mathbf{A}^T\mathbf{b} \tag{9.31c}$$

where $\mathbf{P}$ is the autocorrelation matrix of the rows of the data matrix, while $\mathbf{r}$ is the cross-correlation vector between the rows of the data matrix and the observation vector $\mathbf{b}$. A recursive solution can be easily obtained for each instant of time. The $\mathbf{A}$ matrix can be partitioned into two submatrices of n columns. The first is the $\mathbf{A}$ matrix after $m - 1$ instants of time, called $\tilde{\mathbf{A}}$ and therefore composed of the first $2m - 2$ rows, while the other is composed of the two last rows of the $\mathbf{A}$ matrix. The same can be done for the observation vector $\mathbf{b}$, which can be partitioned into $\tilde{\mathbf{b}}$, of $2m - 2$ components, and $\breve{\mathbf{b}}$ of 2 components. That is,

$$\mathbf{A} = \begin{bmatrix} \tilde{\mathbf{A}} \\ \breve{\mathbf{A}} \end{bmatrix} \begin{matrix} 2m-2 \\ 2 \end{matrix} \tag{9.32a}$$

$$\mathbf{b} = \begin{bmatrix} \tilde{\mathbf{b}} \\ \breve{\mathbf{b}} \end{bmatrix} \begin{matrix} 2m-2 \\ 2 \end{matrix} \tag{9.32b}$$

Therefore, after m instants of time, the matrix $\mathbf{P}$ and the vector $\mathbf{r}$ can be computed as follows:

$$\mathbf{P}(m) = \mathbf{A}^T\mathbf{A} = \tilde{\mathbf{A}}^T\tilde{\mathbf{A}} + \breve{\mathbf{A}}^T\breve{\mathbf{A}} = \mathbf{P}(m-1) + \breve{\mathbf{A}}^T\breve{\mathbf{A}} \tag{9.33a}$$

$$\mathbf{r}(m) = \mathbf{A}^T\mathbf{b} = \tilde{\mathbf{A}}^T\tilde{\mathbf{b}} + \breve{\mathbf{A}}^T\breve{\mathbf{b}} = \mathbf{r}(m-1) + \breve{\mathbf{A}}^T\breve{\mathbf{b}} \tag{9.33b}$$

Then, after m instants of time, the OLS solution $\mathbf{k}'$ is given by the following:

$$\mathbf{k}'(m) = \mathbf{P}^{-1}(m)\mathbf{r}(m) \tag{9.34}$$

As $\mathbf{P}$ is in general a matrix of order $n \times n$, in this case, the inversion of a 5×5 matrix is required at each instant of time. However, as suggested by Stephen and Bodson [86], the strategy used consisted in updating $\mathbf{P}$ and $\mathbf{r}$ at each instant and the solution $\mathbf{k}'$ at a lower rate, thus avoiding the computational burden of the inversion and permitting the continuous monitoring of the process for parameter estimation. Moreover, since the covariance matrix of **k**, depending on the inverse of **P**, is of interest for giving a measure of the reliability of the estimation, the computation of the inverse of **P** is also of help.

By the inspection of Equations 9.24 and 9.25, it is easy to recognize that the second column of the **A** matrix has values which are pretty lower than those of the other columns. This means that the estimation of K_2 is critical, which is also confirmed in the works [87,89].

As mentioned earlier, two ways can be followed now: either a constrained LS minimization or a unconstrained one. The constrained minimization has been followed by Stephen and Bodson [86], but no proof about the convergence of the algorithm has been shown. More will be said in the following paragraph. As explained in the next few lines, the indirect use of the constraint allows an unconstrained minimization and therefore the employment of the sound theory of OLS. In Ref. [88], an LS constrained minimization has been suggested, but too many constraints are present due to the high order of the differential equations derived from Equation 9.15: from the discussion shown earlier, this minimization process would require only one constraint if the stator or rotor reference frame is chosen, which simplifies the computation.

The unconstrained minimization has been used by Boussak and Capolino [90]. However, the mere computation of the K-parameters, for the numerical issue mentioned earlier, causes errors if the additional constraint is not accounted for. As a matter of fact, it is not the computation of the K-parameters which is important, but that of the electrical parameters. If the constraint is used, these parameters are given explicitly from Equation 9.20 without making use of the K_2. Therefore, the K-parameters can be computed with an unconstrained minimization, but then Equation 9.20 without using K_2 is to be used at the same time. Finally, the use of an unconstrained method permits the adoption of the reliable theory of OLS, so no convergence problem exists. The choice of the reference frame most suitable for the parameter estimation is not trivial. One consideration has to be made. With regard to the estimation of all parameters, the worst case would be the steady-state, as the rank of the data matrix is only 2. This is because, in the frequency domain, all the information is given by the fundamental harmonic. It is true, however, that inverter-fed IMs are seldom fed by a single frequency voltage source because of the presence of time harmonics of the voltage source inverter (VSI), but in the identification scheme used in this work, all signals are filtered in such a way that any contribution of higher order time harmonics is almost irrelevant for the computation. In transient condition, it would be important to have an idea of how far the data matrix is from singularity, in other terms, how far the transient is from the steady-state. A natural measure of this distance would be the condition number of the data matrix, which is defined as follows:

Let $\mathbf{A} \in \mathfrak{R}^{m \times n}$ have full rank, then the condition number of A is as follows:

$$\kappa(\mathbf{A}) = \left\| \mathbf{A} \right\|_2 \left\| \mathbf{A}^\dagger \right\|_2$$

where
   $\|\mathbf{A}\|_2$ is the two-norm of the matrix **A**
   $\mathbf{A}^\dagger$ is the pseudoinverse of **A**

It should be remarked that the quantity $1/\kappa(\mathbf{A})$ represents, for a nonsingular matrix **A**, the relative distance of **A** from the set of all singular matrices of order $m \times n$. Matrices with small (respectively, large) condition numbers are said to be well-conditioned (respectively, ill-conditioned). Numerical results, show that the condition number of the data matrix **A** obtained in the reference frame fixed to the stator is lower than the one obtained in the reference frame fixed to the rotor [95]. This could be explained by considering that in the frequency domain the span of harmonics over the fundamental is larger in the stator reference

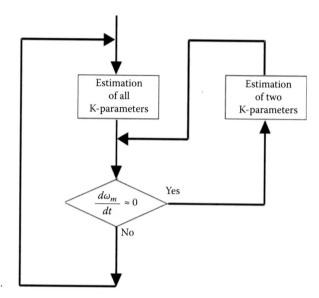

FIGURE 9.20
Block diagram of the selection algorithm.

frame than in the rotor reference frame: this is essentially due to the fact that in the rotor reference frame the frequencies of the fundamental and other harmonics are lower than the corresponding ones in the stator reference frame, so the harmonics obtained in the rotor reference frame are closer to one another and to the fundamental one than the ones obtained in the stator reference frame. This situation of "closeness" to the fundamental gives a measure of how far the transient is from the steady-state, that is, how far the data matrix is from singularity to solve the problem, a selection algorithm chooses the number of K-parameters, and thus of electrical parameters, which are to be calculated on the basis of the rate of change of the speed. During the first speed transient of the drive, the algorithm calculates the five K-parameters and consequently the four electrical parameters, while, if the motor is in speed steady-state, the algorithm calculates only two K-parameters and the only electrical parameter that is selected on the basis of the requirements of the control system (typically T_r for FOC and R_s for DTC). It should be noted that, without this kind of selection algorithm, it is not possible to track the variation of any parameter of the motor in speed steady-state. Figure 9.20 shows the block diagram of the selection algorithm, derived to estimate at speed transient all the five K-parameters, and at speed steady-state only two K-parameters. The two K-parameters to be estimated at steady-state are chosen on the basis of the adopted flux model. If a current model is adopted, then the K-parameters depending T_r are chosen, while if the voltage model is adopted, then those depending on the stator resistance R_s are chosen. Therefore, the tuning of the decoupling circuit, which depends on L_s and σ, can be consequently updated at each speed transient of the drive, while the flux model, which depends on T_r or R_s, is updated at each instant.

9.5.5 Signal Processing System

Special care should be taken for processing the machine signals. The estimation algorithm needs the signals of the stator voltages and currents, their derivatives, up to the second order for the current and the first order for the voltage, and the machine speed.

Since the motor can be supplied both by the electric grid and by a VSI, low-pass analog filters for stator voltage and current signals are needed to avoid aliasing phenomena. The presence of filters, however, causes distortion and time delays of the processed signals which, therefore, at each time instant, should be synchronized with one another in order to respect the dynamic equation of the IM. Figure 9.21 shows the complete scheme used for processing all the signals needed by the identification algorithm. It is composed of the following [93,99,100]:

- Four analog low-pass antialiasing filters (B(s) block in Figure 9.21) that filter the stator voltage and current signals from the voltage and current sensors in the drive
- Four digital low-pass filters (F(z) block in Figure 9.21) reducing high-order harmonics and the noise of the stator voltage and current signals that can be amplified by the following differentiator filters
- Six digital differentiator filters (D(z) block in Figure 9.21) that allow to obtain the derivatives of the stator voltages (up to the first order) and currents (up to the second order) of the drive

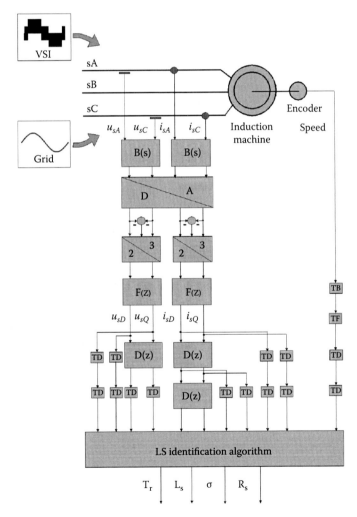

FIGURE 9.21
Block diagram of the signal processing system. (From Cirrincione, M. et al., *IEEE Trans. Ind. Electron.*, 52(5), 1391, October 2005.)

Since, in the application under study, the analog-to-digital converter (ADC) of the voltage and current signals has a sampling time $T_{sp} = 100\,\mu s$ (sampling frequency $f_s = 10\,kHz$), it is necessary to cut off all harmonics of these signals from 5 kHz upward to avoid aliasing phenomena. Bessel low-pass analog filters have been chosen for this task because they can be easily designed with an almost exactly linear phase characteristic.

Since the derivatives of voltage and current signals are necessary, digital finite impulse response (FIR) differentiator filters (blocks D(z) in Figure 9.21) have been designed. FIR filters have been chosen for their characteristic to have an exactly linear phase diagram. These filters implement the transfer function $D(j\omega) = j\omega$ in the ω-domain. Finally, digital low-pass FIR filters have been chosen (blocks F(z) in Figure 9.21). Since all of the signals processed by the estimation algorithm must be synchronized, whenever a stator voltage or current signal is processed by the differentiator filter, the other signals, which are not differentiated and are used in the identification algorithm, must be delayed in time with the group delay of the differentiator filter. With reference to the experimental implementation of this methodology, an electronic card with six analog low-pass fourth-order Bessel filters has been built while all digital filters (low-pass FIR filters and differentiator FIR filters) along with the LS algorithm have been implemented by software on a digital signal processor (DSP).

In particular, Figure 9.22 shows the frequency response of the Bessel filter board which has been built together with the frequency response of the ideal filter. Figures 9.23 and 9.24 show the frequency response of the digital low-pass FIR filter and of the differentiator FIR filter which have been implemented on the DSP.

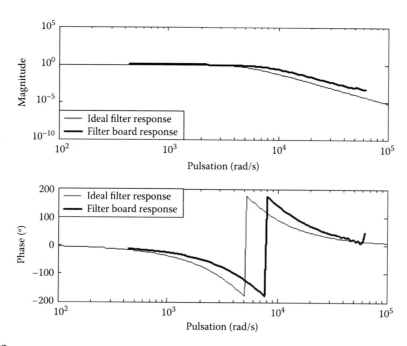

FIGURE 9.22

Frequency response of the low-pass analog Bessel filter. (From Cirrincione, M. et al., *IEEE Trans. Ind. Electron.*, 52(5), 1391, October 2005.)

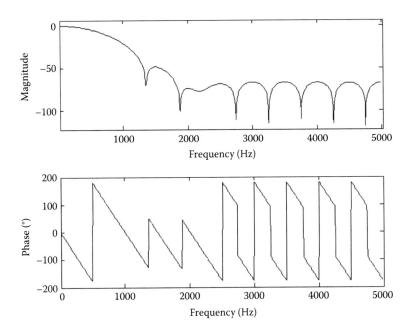

FIGURE 9.23
Frequency response of the low-pass FIR filter. (From Cirrincione, M. et al., *IEEE Trans. Ind. Electron.*, 52(5), 1391, October 2005.)

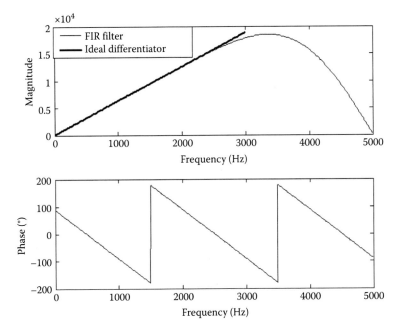

FIGURE 9.24
Frequency response of the differentiator FIR filter. (From Cirrincione, M. et al., *IEEE Trans. Ind. Electron.*, 52(5), 1391, October 2005.)

9.5.6 Description of the Test Setup for the Experimental Application

The test setup that has been built for the experimental verification of the parameter estimation algorithm consists of the following [101]:

- One three-phase IM with rated values shown in Table 4.1
- One electronic power converter (three-phase diode rectifier and VSI composed of three IGBT modules without any control system) of rated power 7.5 kVA
- One electronic card with voltage sensors (model LEM LV 25-P) and current sensors (model LEM LA 55-P) for monitoring the instantaneous values of the stator phase voltages and currents
- One voltage sensor (model LEM CV3-1000) for monitoring the instantaneous value of the DC link voltage
- One electronic card with analog fourth-order low-pass Bessel filters and cutoff frequency of 800 Hz
- One incremental encoder (model RS 256-499, 2500 pulses per round)
- One dSPACE card (model DS1103) with a floating-point DSP

The VSI is driven by an asynchronous space-vector pulsewidth modulation (SV-PWM) technique (switching frequency f_{PWM} = 5 kHz) implemented by software on the dSPACE card, and the DC link voltage sensor permits taking into account the instantaneous value of the DC link voltage for the modulation. Figure 9.25 shows the electric scheme of the adopted test setup, while Figure 9.26 shows its photograph.

It should be noted that the current sensors employed in test setup present a percent accuracy of 0.65% with a linearity percent error less than 0.15% and with a − 1 dB bandwidth of 200 kHz, while the voltage sensors present a percent accuracy of 0.9% with a linearity

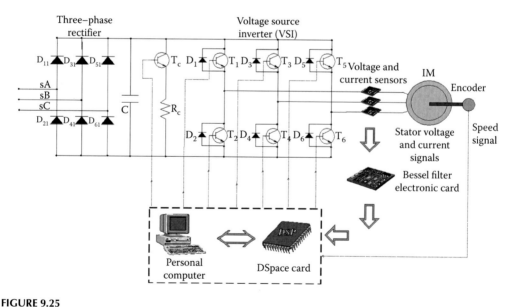

FIGURE 9.25
Electric scheme diagram of the experimental test setup. (From Cirrincione, M. et al., *IEEE Trans. Ind. Appl.*, 39(5), 1247, September/October 2003.)

FIGURE 9.26
Photograph of the test setup. (From Cirrincione, M. et al., *IEEE Trans. Ind. Appl.*, 39(5), 1247, September/October 2003.)

percent error less than 0.2% and with a response time of 400 μs. Moreover, the six analog signals have been acquired with six 16 bit ADC, multiplexed in groups of three, with 1 μs sampling time for each channel, and therefore the associated percent quantization error, evaluated as $100/2^{N+1}$ with N number of bits of the ADC, is $7.6 \ 10^{-4}$%. On the basis of the previous text, the quantization percent error is negligible in comparison with the transducers' error: in particular, the global percent error associated with the acquisition of the current signals is 0.80076%, and the corresponding one associated with the voltage signals is 1.10076%.

9.5.7 Simulation and Experimental Results

The capability of the RLS algorithm to estimate, by exploiting a speed transient of the machine, all the electrical parameters of the IM (rotor time constant, stator resistance, stator inductance, global leakage factor) have been verified both numerically and experimentally by using the previously described test setup [94,101]. The model of the IM for the simulation has the same parameters of the real motor, as shown in Table 4.1. Simulations have been performed in MATLAB®–Simulink® environment. In particular, the following test has been performed. The motor, both in simulation and in the experimental application, has been supplied by the VSI, driven by means of the SV-PWM, to which a reference sinusoidal voltage of 220 V and 50 Hz has been provided. In both tests, the entire speed transient from zero speed to steady-state speed has been exploited to estimate all four electrical parameters of the IM.

Figure 9.27 shows the rotor speed and the i_{sD} and i_{sQ} stator currents during the start-up of the motor with no load in the simulated test. Figure 9.28 shows that the corresponding estimation of the parameters converges to the expected value after some iterations.

Figure 9.29 shows the rotor speed and the i_{sD} and i_{sQ} stator currents during the start-up of the motor with no load in the experimental test. Figure 9.30 shows that the corresponding estimation of the parameters converges to the expected value after some iterations. The values of the electrical parameters of the machine assumed to be real are those measured with the usual no-load and locked-rotor tests. In both the simulated and experimental test, the K-parameters have been all set to one at the beginning of the estimation, but it has been verified that the final value of the estimated parameters is quite independent from the initial one.

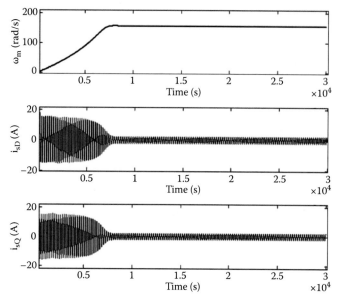

FIGURE 9.27
Rotor speed, i_{sD}, and i_{sQ} waveforms (simulation results). (From Cirrincione, M. and Pucci, M., Experimental verification of a technique for the real-time identification of induction motors based on the recursive least-squares, *International Workshop on Advanced Motion Control (IEEE AMC'02)*, Maribor, Slovenia, 2002.)

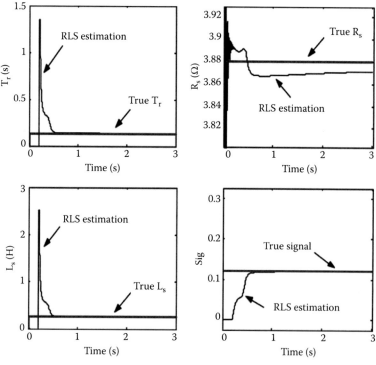

FIGURE 9.28
Real and estimated electrical parameters of the motor (simulation results). (From Cirrincione, M. and Pucci, M., Experimental verification of a technique for the real-time identification of induction motors based on the recursive least-squares, *International Workshop on Advanced Motion Control (IEEE AMC'02)*, Maribor, Slovenia, 2002.)

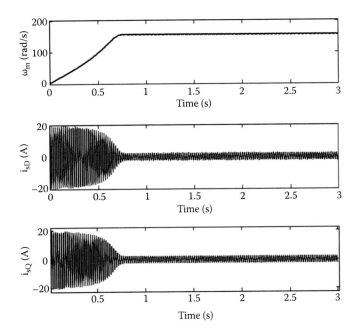

FIGURE 9.29
Rotor speed, i_{sD}, and i_{sQ} waveforms (experimental results). (From Cirrincione, M. and Pucci, M., Experimental verification of a technique for the real-time identification of induction motors based on the recursive least-squares, *International Workshop on Advanced Motion Control (IEEE AMC'02)*, Maribor, Slovenia, 2002.)

Table 9.1 shows the true and the estimated values of the K-parameters of the IM under test as well as the estimation errors in the simulation verification. It can be observed that except K_2, which is not used for computing the electrical parameters of the machine, the percent estimation error on all K-parameters is always below 0.3%.

In general, this methodology presents the following advantages:

- It is simpler and less computationally cumbersome than the other algorithms in literature like EKF, ELO, and MRAS.
- It allows the four electrical parameters to be estimated during a speed transient of the machine and one electrical parameter in sinusoidal steady-state.
- It can be applied with few modifications for the off-line identification of the machine at standstill to be used in the self-commissioning of the drive.
- It does not require any a priori knowledge of the electrical parameters of the motor or its plate data.
- It can be applied both to FOC and DTC drive with slight modifications: this is useful in industrial applications as it cuts down the development time.
- It permits computing the variation of the parameters due to different magnetic excitations: this feature is useful when the drive should work in the field-weakening region.
- It can work at a much lower adapting rate than that of the control system and the flux model, and thereby the computational burden can be spread over more sampling time intervals of the control system.

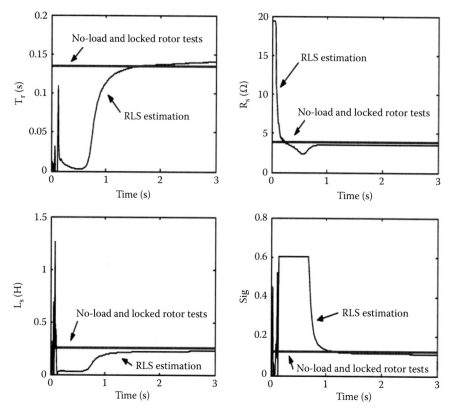

FIGURE 9.30
Real and estimated electrical parameters of the motor (experimental results). (From Cirrincione, M. and Pucci, M., Experimental verification of a technique for the real-time identification of induction motors based on the recursive least-squares, *International Workshop on Advanced Motion Control* (*IEEE AMC'02*), Maribor, Slovenia, 2002.)

On the contrary, the method presents the following drawbacks:

- It requires a signal processing system that must be accurately designed and built.
- It requires differentiator filters that amplify high-frequency noise.
- It requires high-quality (high-bandwidth) stator voltage sensors that are much more expensive than current sensors.

9.6 Constrained Minimization for Parameter Estimation of IMs in Saturated and Unsaturated Conditions

The matrix Equation 9.21 together with the constraint Equation 9.19 can be written in the following form [100,102]:

$$\begin{cases} \mathbf{Ak} \approx \mathbf{b} \\ f(\mathbf{k}) = 0 \end{cases} \tag{9.35}$$

where $f(.)$ is the constraint function Equation 9.19. The first equation in Equation 9.35 can be solved for the K-parameters both in steady-state and transient in real time by using an OLS method, because the main cause of error, the modeling error, is present in the observation vector, as explained in the earlier paragraph.

By the inspection of the Equation 9.21 or the like in the different reference frames (Equations 9.17 or 9.22), it is easy to recognize that the second column of the matrix has values which are lower than those of the other columns. This means two things:

1. A gradient descent method is unsuitable, as it is too slow to converge.

2. If a constrained minimization is used, it is intuitive to realize that the difference between the true value, that is, the solution satisfying the constraints, and the two-norm solution, that is, the unconstrained minimum, can be quite apart from each other along the direction of minimum gradient (K_2).

This makes the estimation of K_2 critical, which is also confirmed in Refs [86,90] (K_2 problem). Two paths can be then followed to overcome this difficulty:

1. The first employs an unconstrained minimization, which in a way, takes into account the constraint. This is the method followed by Stephen and Bodson [86], but no proof about the convergence of the algorithm has been shown. Also in Ref. [88], an LS constrained minimization has been suggested, but too many constraints are present, due to the high order of the differential equations derived from Equation 9.21, which makes the method unsuitable for real-world applications.

2. The second method is a constrained analytical minimization adopted for overcoming the K_2 problem. This method is fully explained in Ref. [102] and is summarized in the following.

9.6.1 Constrained Minimization: Analytical Solution

It is known that in parameter estimation, if the stator reference frame is chosen (the other reference frames can be chosen also, but the stationary reference frame is advisable for the previously mentioned numerical considerations), an equation as $\mathbf{Ak} = \mathbf{b}$ appears, where $\mathbf{A}$ is an m × 5 matrix (data matrix), $\mathbf{k}$ is the vector of the five unknowns (K-parameters), and $\mathbf{b}$ is the m × 1 vector (observation vector). The $\mathbf{k}$ vector is computed by an OLS algorithm, which means that the following function error should be minimized:

$$\mathbf{E} = \|\mathbf{Ak} - \mathbf{b}\|_2^2 = (\mathbf{b} - \mathbf{Ak})^T(\mathbf{b} - \mathbf{Ak})$$

$$= \mathbf{b}^T\mathbf{b} - 2\mathbf{P}^T\mathbf{k} + \mathbf{k}^T\mathbf{Rk} \tag{9.36}$$

where
$\mathbf{R} = \mathbf{A}^T\mathbf{A}$ is the autocorrelation matrix (5 × 5)
$\mathbf{p}^T = \mathbf{b}^T\mathbf{A}$ is the mutual correlation vector ($\mathbf{p} \in R^{5\times1}$)

However, between the k_i (i = 1,..., 5) components, the following constraint exists:

$$k_2k_4 = k_3k_5 \tag{9.37}$$

This means that a constrained LS minimization should be performed. To perform this goal, two paths can be followed. One reduces the equations into a canonical form, which permits a geometrical insight into the problem and then uses a Lagrangian optimization technique in this new reference framework, fully explained in Ref. [100] and the other explained in Ref. [102], which is simpler to deduce but does not convey the same geometrical meaning.

9.6.1.1 First Constrained Minimization Method

To perform the constrained LS minimization, Equation 9.35 should be transformed into a canonical form with a suitable reference frame and then accordingly the constraint should be modified. As explained in the following, three translations and one rotation are necessary to achieve this goal [103].

1. Translation
   The expression in Equation 9.36 can be easily simplified into

$$z_1 = -2\mathbf{P}^T\mathbf{k} + \mathbf{k}^T\mathbf{R}\mathbf{k} \tag{9.38}$$

   where

$$z_1 = E - \mathbf{b}^T\mathbf{b}$$

2. Rotation
   Let $\mathbf{y} = \mathbf{V}^T\mathbf{k}$ and $\mathbf{V}$ the matrix whose columns are the normalized eigenvectors of $\mathbf{R}$. Then, by applying this rotation, it follows that

$$z_1 = -2\mathbf{P}^T\mathbf{V}\mathbf{y} + \mathbf{y}^T\Lambda_R\mathbf{y} \tag{9.39}$$

   where $\Lambda_R = \mathbf{V}^T\mathbf{R}\mathbf{V}$ is a diagonal matrix formed by the eigenvalues of $\mathbf{R}$, which are all real.

3. Translation
   Let now translate $\mathbf{y}$ by a vector $\mathbf{h}$, that is,

$$\mathbf{y} = \hat{\mathbf{y}} + \mathbf{h} \tag{9.40}$$

The purpose is to determine the value of $\mathbf{h}$ so that no first-order term appears in (9.39). The method of the squares completion can be adopted. Indeed by substituting (9.40) into (9.39) and eliminating the first-order terms, it results

$$\mathbf{h} = \Lambda_R^{-1}\mathbf{V}^T\mathbf{P} \tag{9.41}$$

then (9.39) can be written as

$$z_1 = -2\mathbf{P}^T\mathbf{V}\mathbf{h} + \mathbf{h}^T\Lambda_R\mathbf{h} + \hat{\mathbf{y}}^T\Lambda_R\hat{\mathbf{y}} \tag{9.42}$$

4. Translation
   From preceding text, it results

$$\hat{z} = \hat{\mathbf{y}}^{\mathrm{T}} \Lambda_R \hat{\mathbf{y}} \tag{9.43}$$

because

$$z_1 + 2\mathbf{P}^{\mathrm{T}}\mathbf{V}\mathbf{h} - \mathbf{h}^{\mathrm{T}}\Lambda_R\mathbf{h} = \hat{\mathbf{y}}^{\mathrm{T}}\Lambda_R\hat{\mathbf{y}}$$

Now the constraint must be rewritten considering that $\mathbf{k} = \mathbf{V}\mathbf{y} = \mathbf{V}(\hat{\mathbf{y}} + \mathbf{h})$. Let $\mathbf{e}_i^{\mathrm{T}} = (0 \cdots 1 \cdots 0)$ be the unit vector whose components are all null, save for the ith component which is one.

Then

$$k_i = \mathbf{e}_i^{\mathrm{T}}\mathbf{V}(\hat{\mathbf{y}} + \mathbf{h}) \quad \text{with } i = 1,\dots,5.$$

The constraint can be then so rewritten as

$$(\hat{\mathbf{y}} + \mathbf{h})^T \mathbf{V}^{\mathrm{T}}\mathbf{e}_2\mathbf{e}_4^{\mathrm{T}}\mathbf{V}(\hat{\mathbf{y}} + \mathbf{h}) = (\hat{\mathbf{y}} + \mathbf{h})^T \mathbf{V}^{\mathrm{T}}\mathbf{e}_3\mathbf{e}_5^{\mathrm{T}}\mathbf{V}(\hat{\mathbf{y}} + \mathbf{h})$$

which means that the constraint is as follows:

$$(\hat{\mathbf{y}}^{\mathrm{T}}\mathbf{v}_2 + \mathbf{h}^{\mathrm{T}}\mathbf{v}_2)(\hat{\mathbf{y}}^{\mathrm{T}}\mathbf{v}_4 + \mathbf{h}^{\mathrm{T}}\mathbf{v}_4) = (\hat{\mathbf{y}}^{\mathrm{T}}\mathbf{v}_3 + \mathbf{h}^{\mathrm{T}}\mathbf{v}_2)(\hat{\mathbf{y}}^{\mathrm{T}}\mathbf{v}_5 + \mathbf{h}^{\mathrm{T}}\mathbf{v}_5) \tag{9.44}$$

where $\mathbf{v}_i$ is the ith column of $\mathbf{V}^{\mathrm{T}}$ ($i = 1,\dots,5$).
   Let now f_i and f be as follows:

$$f_i = (\hat{\mathbf{y}}^{\mathrm{T}}\mathbf{v}_i + \mathbf{h}^{\mathrm{T}}\mathbf{v}_i) \quad (i = 1,\dots,5) \tag{9.45}$$

$$f = f_2 f_4 - f_3 f_5 \tag{9.46}$$

The gradient of f ($grad_{\hat{y}} f$, from now on abbreviated as $grad\, f$) can be written as

$$grad\, f = f_4(grad\, f_2) + f_2(grad\, f_4) - f_5(grad\, f_3) - f_3(grad\, f_5) \tag{9.47}$$

But $grad\, f_i = \mathbf{v}_i$ and $grad\, \hat{z} = 2\Lambda_R\hat{\mathbf{y}}$.
   Let α be the Lagrangian multiplier and $C(\hat{y}, \alpha)$ the total cost function to be minimized, defined as follows:

$$C(\hat{y}, \alpha) = \hat{y}^T \Lambda_R \hat{y} + \alpha f \tag{9.48}$$

so

$$grad\ C = 2\Lambda_R \hat{\mathbf{y}} + \alpha\ grad\ f \tag{9.49}$$

Equation 9.47 can be developed in the following way:

$$grad\ f = (W + W^T)\hat{\mathbf{y}} + (W + W^T)\mathbf{h} \tag{9.50}$$

where

$$W = \mathbf{v}_2\mathbf{v}_4^T - \mathbf{v}_3\mathbf{v}_5^T \tag{9.51}$$

Then to find the minimum of the cost function, it is necessary that *grad C* = 0, that is,

$$\begin{cases} 2\Lambda_R\hat{\mathbf{y}} + \alpha(\mathbf{W} + \mathbf{W}^T)\hat{\mathbf{y}} + \alpha(\mathbf{W} + \mathbf{W}^T)\mathbf{h} = 0 \\ f_2 f_4 = f_3 f_5 \end{cases} \tag{9.52}$$

With this regard, the constraint (9.41) can be rewritten as follows:

$$\frac{1}{2}\hat{\mathbf{y}}^T(\mathbf{W} + \mathbf{W}^T)\hat{\mathbf{y}} + \mathbf{h}^T(\mathbf{W} + \mathbf{W}^T)\hat{\mathbf{y}} + \frac{1}{2}\mathbf{h}^T(\mathbf{W} + \mathbf{W}^T)\mathbf{h} = 0 \tag{9.53}$$

Equations 9.32 and 9.33 can therefore be written as follows:

$$\begin{cases} 2\Lambda_R\hat{\mathbf{y}} + \alpha(\mathbf{W} + \mathbf{W}^T)\hat{\mathbf{y}} + \alpha(\mathbf{W} + \mathbf{W}^T)\mathbf{h} = 0 & (9.54\,a) \\ \frac{1}{2}\hat{\mathbf{y}}^T(\mathbf{W} + \mathbf{W}^T)\hat{\mathbf{y}} + \mathbf{h}^T(\mathbf{W} + \mathbf{W}^T)\hat{\mathbf{y}} + \frac{1}{2}\mathbf{h}^T(\mathbf{W} + \mathbf{W}^T)\mathbf{h} = 0 & (9.54\,b) \end{cases}$$

From (9.43a), it results that

$$\hat{\mathbf{y}} = -\alpha\,[2\Lambda_R + \alpha(\mathbf{W} + \mathbf{W}^T)]^{-1}\,(\mathbf{W} + \mathbf{W}^T)\mathbf{h} \tag{9.55}$$

By substituting (9.55) into (9.54b), a nonlinear scalar equation with the scalar unknown α, which can be easily solved for by a numerical technique (e.g., the Newton–Raphson method, the bisection method, or even with the old "regula falsi," a variation of the secant method [103], p. 338).

9.6.1.1.1 *First Constrained Minimization Method: Simulation and Experimental Results*

The capability of this method to estimate, by exploiting a speed transient of the machine, all the electrical parameters of the IM (rotor time constant, stator resistance, stator inductance, and global leakage factor) has been verified in simulation and experimentally [102], by using the test setup described earlier. The model of the IM for the simulation has the same parameters of the real motor, as shown in Table 4.1. Simulations have been performed in MATLAB®–Simulink® environment. In particular, the following test has been performed.

The motor, both in simulation and in the experimental application, has been supplied by the VSI, driven by means of the SV-PWM, to which a reference sinusoidal voltage of 220 V and 50 Hz has been provided. The entire speed transient from zero speed to steady-state speed has been exploited to estimate all four electrical parameters of the IM.

Figure 9.31 shows the rotor speed and the i_{sD} and i_{sQ} stator currents during the start-up of the motor with no load in the simulated test. Figure 9.32 shows the corresponding estimation of the parameters, obtained with the constrained analytical solution.

Figure 9.33 shows the rotor speed and the i_{sD} and i_{sQ} stator currents during the start-up of the motor with no load in the experimental test. Figure 9.34 shows the corresponding estimation of the parameters, obtained with the constrained analytical solution.

The values of the electrical parameters of the real machine are measured with the usual no-load and locked-rotor tests and have been considered as the true values.

Tables 9.2 and 9.3 show the true and estimated values of the K-parameters of the IM under test, obtained with the constrained analytical solution in the simulation and experimental tests, respectively. The graphs show that the estimated electrical parameters converge quickly and smoothly to true ones. Moreover, the tables highlight the correct estimation of all the K-parameters, including K_2.

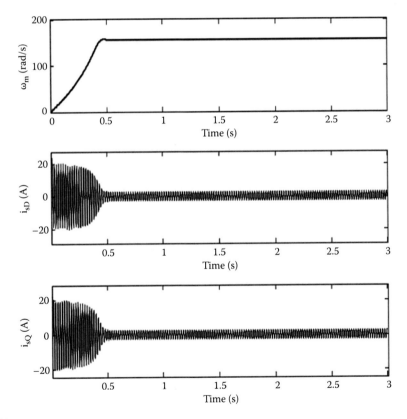

FIGURE 9.31
Rotor speed and i_{sD}, i_{sQ} waveforms (simulation results). (From Cirrincione, M. et al., Capolino, Constrained least-squares method for the estimation of the electrical parameters of an induction motor, *COMPEL (The International Journal for Computation and Mathematics in Electrical and Electronic Engineering), Special Issue: Selected papers from the International Conference on Electrical Machines (ICEM) 2002*, Bruges, Belgium, Vol. 22(4), pp. 1089–1101, 2003.)

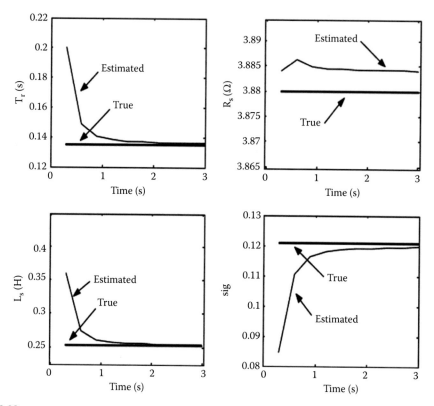

FIGURE 9.32
Real and estimated electrical parameters of the motor (simulation results). (From Cirrincione, M. et al., Constrained least-squares method for the estimation of the electrical parameters of an induction motor, *COMPEL* (*The International Journal for Computation and Mathematics in Electrical and Electronic Engineering*), *Special Issue: Selected papers from the International Conference on Electrical Machines (ICEM) 2002, Bruges, Belgium, Vol. 22(4), pp. 1089–1101, 2003.*)

9.6.1.2 Second Constrained Minimization Method

This method is in a way a simplified version of the previous method [101]. Let α be the Lagrangian multiplier and E_c the total cost function to be minimized, then from Equation 9.35, it follows that

$$E_c = \mathbf{b}^T \mathbf{b} - 2\mathbf{p}^T \mathbf{k} + \mathbf{k}^T \mathbf{R} \mathbf{k} + \alpha (k_2 k_4 - k_3 k_5) \tag{9.56}$$

so the gradient of this cost function is given as follows:

$$grad(\mathbf{E}_c) = -2\mathbf{p} + 2\mathbf{R}\mathbf{k} + \begin{pmatrix} 0 \\ \alpha k_4 \\ -\alpha k_5 \\ \alpha k_2 \\ -\alpha k_3 \end{pmatrix} = 0 \tag{9.57}$$

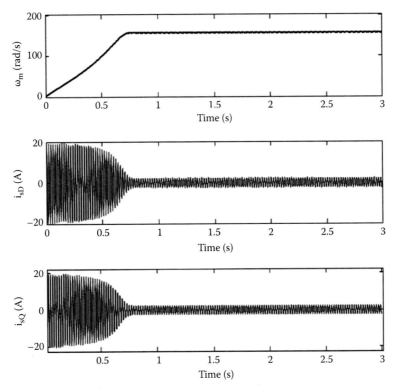

FIGURE 9.33
Rotor speed, i_{sD}, and i_{sQ} waveforms (experimental results). (From Cirrincione, M. et al., Constrained least-squares method for the estimation of the electrical parameters of an induction motor, *COMPEL* (*The International Journal for Computation and Mathematics in Electrical and Electronic Engineering*), *Special Issue: Selected papers from the International Conference on Electrical Machines (ICEM) 2002, Bruges, Belgium,* Vol. 22(4), pp. 1089–1101, 2003.)

If the following matrix **E** is introduced,

$$\mathbf{E} = \begin{pmatrix} 0 & 0 & 0 & 0 & 0 \\ 0 & 0 & 0 & 1 & 0 \\ 0 & 0 & 0 & 0 & 1 \\ 0 & 1 & 0 & 0 & 0 \\ 0 & 0 & 1 & 0 & 0 \end{pmatrix} \tag{9.58}$$

then the Equation 9.46 can be written as follows:

$$-2\mathbf{p} + 2\mathbf{R}\mathbf{k} + \alpha\mathbf{E}\mathbf{k} = 0 \tag{9.59}$$

which implies

$$(2\mathbf{R} + \alpha\mathbf{E})\mathbf{k} = 2\mathbf{p}$$

Or if the inverse of the matrix exists, then

$$\mathbf{k} = (2\mathbf{R} + \alpha\mathbf{E})^{-1}2\mathbf{p} \tag{9.60}$$

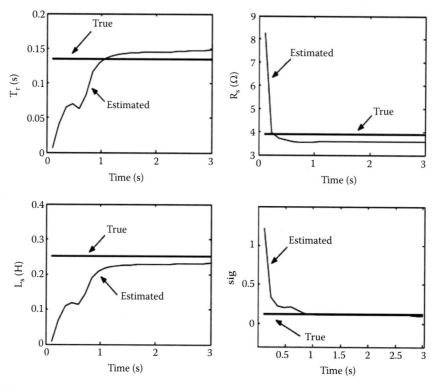

FIGURE 9.34
Real and estimated electrical parameters of the motor (experimental results). (From Cirrincione, M. et al., Constrained least-squares method for the estimation of the electrical parameters of an induction motor, *COMPEL* (*The International Journal for Computation and Mathematics in Electrical and Electronic Engineering*), *Special Issue: Selected papers from the International Conference on Electrical Machines (ICEM) 2002*, Bruges, Belgium, Vol. 22(4), pp. 1089–1101, 2003.)

TABLE 9.2

Steady-State Estimated K-Parameters
(Simulation Results)

	True	Estim.	Err. %
K_1	189.0088	189.4606	0.2390
K_2	946.4266	941.3659	−0.5347
K_{31}	127.5398	128.0074	0.3666
K_4	32.8711	32.9568	0.2609
K_5	243.9244	242.3646	−0.6395

Source: From Cirrincione, M. et al., Constrained least-squares method for the estimation of the electrical parameters of an induction motor, *COMPEL* (*The International Journal for Computation and Mathematics in Electrical and Electronic Engineering*), *Special Issue: Selected papers from the International Conference on Electrical Machines (ICEM) 2002*, Bruges, Belgium, Vol. 22(4), pp. 1089–1101, 2003.

TABLE 9.3

Steady-State Estimated K-Parameters (Experimental Results)

	True	Estim.	Err. %
K_1	189.0088	212.2766	12.3104
K_2	946.4266	995.9520	5.2329
K_{31}	127.5398	147.4474	15.6089
K_4	32.8711	41.2711	25.5544
K_5	243.9244	278.7709	14.2858

Source: Cirrincione, M. et al., Constrained least-squares method for the estimation of the electrical parameters of an induction motor, *COMPEL* (*The International Journal for Computation and Mathematics in Electrical and Electronic Engineering*), Special Issue: *Selected papers from the International Conference on Electrical Machines (ICEM) 2002*, Bruges, Belgium, Vol. 22(4), pp. 1089–1101, 2003.

If the value **k** given in (9.60) is substituted into (9.56), a scalar equation in the unknown α is obtained, which can be easily solved for with a nonlinear numerical method (see Section 9.6.1.1). Then the **K** vector is obtained with (9.60).

As for the error estimate, the same considerations made in respect with the unconstrained minimization can be made. If the unconstrained minimization is used, then the formula (42) in Ref. [86] can be applied, that is,

$$\delta k_i = \sqrt{\overline{\mathbf{Re}^*(\mathbf{A}^T\mathbf{A})_{ii}}} \tag{9.61}$$

where

δk_i indicates the amount by which the ith component of **k** could vary without causing more than a doubling of the residual error (parametric error index)

$\mathbf{Re}^*$ is defined as the value obtained by the residual error when **k** is obtained as the solution of the unconstrained LS method

$()_{ii}$ indicates the iith (diagonal) element of the matrix inside the brackets, that is, the minimum residual error

Thanks to formula (9.60), a similar formula can be obtained from (9.61) for the constrained case, that is,

$$\delta k_i^c = \sqrt{\overline{\mathbf{Re}_c^*(\mathbf{A}^T\mathbf{A}+\alpha\mathbf{E})_{ii}}} \tag{9.62}$$

where

δk_i^c indicates the amount by which the ith component of **k** could vary without causing more than a doubling of the residual error (parametric error index)

$\mathbf{Re}_c^*$ is defined as the value obtained by $\mathbf{E}_c$ when **k** is obtained by (9.60), that is, the minimum residual error with a constraint

From the definition of the matrix $\mathbf{E}$, it results

$$\delta k_i^c = \sqrt{\mathbf{Re}_c^*(\mathbf{A}^T\mathbf{A} + \alpha\mathbf{E})_{ii}} = \sqrt{\mathbf{Re}_c^*(\mathbf{A}^T\mathbf{A})_{ii}} \qquad (9.63)$$

and since $\mathbf{Re}_c^* < \mathbf{Re}^*$, it results that $\delta k_i^c < \delta k_i$. This means that with the constrained minimization the parametric error index diminishes. A large parametric error index indicates less accuracy in the results [86]. So with constrained minimization, the accuracy of the results is more than in the case of the unconstrained minimization. In any case, it should be remarked that with this OLS approach the errors are assumed to be confined mainly in the observation vector, assumption in a way acceptable because there exist second derivatives in the observation vector. If also the uncertainty in the data matrix is to be accounted for, then a constrained TLS [95–97] technique should be used.

9.6.1.2.1 Second Constrained Minimization Method: Simulation and Experimental Results

The proposed methodology has been tested in simulation and experimentally to retrieve the electrical parameters of an IM under both linear and magnetically saturated working conditions [101], by using the dynamical models presented in Chapter 4 (Equations 4.23, 4.63, 4.89, 4.88, 4.90, and 4.91). In particular, the nonlinear relationships between L_m, L, σ, σ_s, σ_r, and the rotor magnetizing current as obtained experimentally in Ref. [5] on the real IM will be used (see Figure 4.5).

9.6.1.2.1.1 Simulation Results At first, a comparison has been made between the estimation of the electrical and magnetic parameters obtained with the proposed algorithm (constrained minimization) and the classical OLS algorithm, like the one explained in Section 9.5.4 and Refs [94,102,105].

In particular, various tests have been performed for verifying the capability of the parameter estimation algorithm to work correctly under both slow- and fast-speed transients and under different steady-state magnetization levels. Different steady-state magnetization levels in fact correspond to different values of the magnetic parameters of the machine. Since the equations employed for the identification model come from the mathematical model of the IM that does not take into consideration the effects of magnetic saturation of the iron path, the identification algorithm can compute only the steady-state values of the parameters themselves, that is, the values corresponding to the steady-state magnetization of the machine, as highlighted earlier and in Ref. [105].

For this purpose, a set of start-up tests has been done under different voltage levels at the frequency of 50 Hz, to make the machine work under different steady-state magnetization excitations, that is, with different values of $|\mathbf{i}_{mr}|$ and $|\mathbf{\psi}_r|$. Correspondingly, both slow- and fast-speed transients have been created: low voltage/frequency ratios correspond to slow-speed transients, while high voltage/frequency ratios correspond to fast-speed transients. At the end of each test, the four electrical parameters L_s, σ, T_r, and R_s have been retrieved with both methods.

Figure 9.35 shows the rotor speed and the stator i_{sA} current waveforms during a start-up test under a 155 V and 50 Hz supply. Figure 9.36 shows the corresponding curves of the estimated electrical parameters in comparison with the real ones, obtained with the constrained minimization. It shows that, as obviously expected, the magnetic parameters of the machine (T_r, L_s, and σ) vary during the speed transient of the machine from zero to speed steady-state because of the magnetization of the machine. It should be noted that the adopted identification model of the IM does not take into consideration the saturation

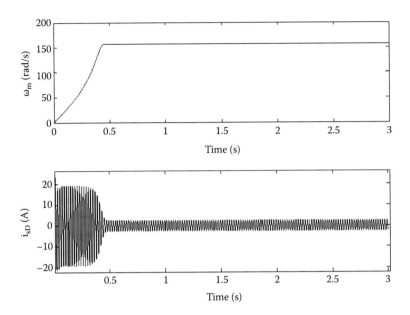

FIGURE 9.35
Rotor speed and stator current at start-ups under 155 V, 50 Hz supply (simulation). (From Cirrincione, M. et al., *IEEE Trans. Ind. Electron.*, 52(5), 1391, October 2005.)

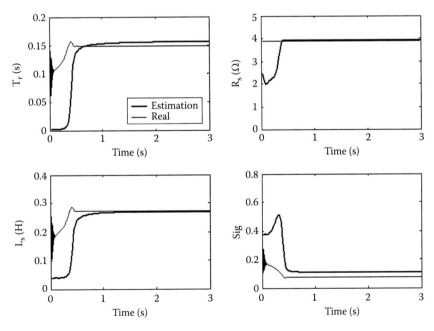

FIGURE 9.36
Estimated and real electrical parameters of the machine under a 155 V, 50 Hz supply (simulation). (From Cirrincione, M. et al., *IEEE Trans. Ind. Electron.*, 52(5), 1391, October 2005.)

TABLE 9.4

Percent Estimation Errors Obtained with the RLS

u_s (V)	Err. % T_r	Err. % R_s	Err. % L_s	Err. % σ
30	4.42	−2.33	3.13	−6.11
60	0.87	−0.67	−0.66	−4.08
90	3.70	0.18	2.85	−4.83
120	6.14	0.73	4.85	−6.86
150	7.56	1.18	6.20	−8.18
180	6.27	1.61	4.94	−7.28
210	5.26	2.03	4.09	−6.45
240	6.32	2.53	5.32	−7.02
270	4.28	2.85	4.20	−4.89
300	2.00	3.10	3.33	−2.54

Source: Cirrincione, M. et al., *IEEE Trans. Ind. Electron.*, 52(5), 1391, October 2005.

effects of the iron path of the machine, so the constrained minimization algorithm is able to retrieve only the steady-state values of the parameters, while no correct estimation is possible during the speed transient.

Tables 9.4 and 9.5 show the relative percent error of both estimations with reference to the true values of the parameters reported in the look-up tables of the model described in Figure 4.5. From these tables and Figure 9.37, it is apparent that the results obtained with the constrained minimization method are superior to those obtained with the classical unconstrained LS method.

The variation of the estimation error according to the load torque has been also investigated. Figure 9.38 shows the percent estimation error of every K-parameter under different load conditions ranging from no load to rated load. It can be observed that both K_1 and K_4 have a low percent error, which is almost independent from load conditions while K_{31} and K_5 depend heavily from them and moreover have higher percent errors in respect with the former parameters. An explanation can be given by considering that the columns

TABLE 9.5

Percent Estimation Errors Obtained with the Constrained Minimization

u_s (V)	Err. % T_r	Err. % R_s	Err. % L_s	Err. % σ
30	−7.66	−2.82	−7.82	5.81
60	−0.801	−0.78	−2.14	−2.27
90	2.94	0.21	2.06	−4.11
120	3.47	0.39	2.50	−4.33
150	3.52	0.45	2.79	−4.36
180	2.84	0.52	2.63	−4.00
210	2.33	0.63	2.04	−3.58
240	2.80	0.8	2.99	−3.76
270	1.62	1.00	2.55	−2.52
300	0.05	1.30	2.50	−0.84

Source: Cirrincione, M. et al., *IEEE Trans. Ind. Electron.*, 52(5), 1391, October 2005.

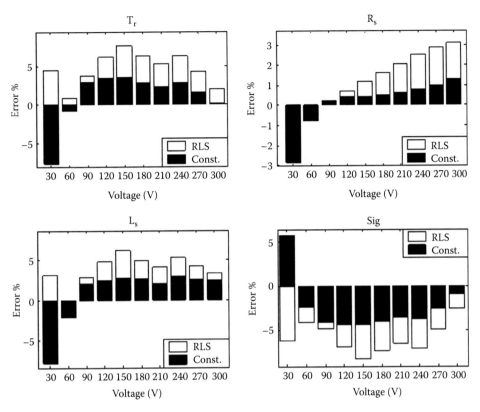

FIGURE 9.37
Percent estimation errors on the electrical parameters (simulation). (From Cirrincione, M. et al., *IEEE Trans. Ind. Electron.*, 52(5), 1391, October 2005.)

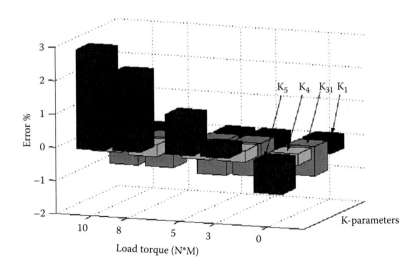

FIGURE 9.38
Percent estimation errors on the K-parameters according to the variation of the load (simulation). (From Cirrincione, M. et al., *IEEE Trans. Ind. Electron.*, 52(5), 1391, October 2005.)

of the data matrix corresponding to K_1 and K_4 have a magnitude which is much higher than those corresponding to K_{31} and K_5. Moreover, the errors in K_5 are higher than those in K_{31} as the corresponding column of the data matrix of this last parameter is dependent on the load: the higher the load, the higher the stator current, and the higher the value of the corresponding column in the data matrix, the lower the error of the K_{31} parameter with respect to the error on the K_5 parameter.

9.6.1.2.1.2 Experimental Results   The presented methodology has been verified experimentally on the test setup described in Section 9.5.6. Table 4.1 shows the nameplate data and the electrical parameters of the employed IM, obtained with the no-load and locked-rotor tests. In particular, as explained in Section 9.5.2, a set of start-up tests has been done under different voltage levels at the frequency of 50 Hz with no load, to make the machine work under different steady-state magnetization excitation, that is, with different values of $|i_{mr}|$ and $|\psi_r|$. Figures 9.39 and 9.40 show the rotor speed and the stator i_{sD} current waveforms, during four start-up tests, respectively, under a 55, 105, 155, and 220 V and 50 Hz supply. Figures 9.41 through 9.44 show the corresponding curves of the estimated electrical parameters in comparison with those measured with the usual no-load and locked-rotor tests. It should be noted that, in dependence on the steady-state magnetization level of the machine, the estimated electrical parameters, after convergence, can be either closer or not to the values of the parameters measured with the no-load and locked-rotor tests. For example, the steady-state estimation of L_s obtained under 220 V, 50 Hz supply is closer to the L_s measured with the no-load test than the estimation of σ at the same supply conditions to the σ measured with the locked-rotor test. This is easily explained by the fact that in computing σ with the

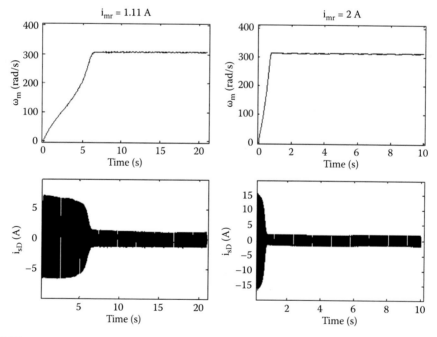

FIGURE 9.39

Rotor speed and stator current during start-up tests under, respectively, 55 and 105 V, 50 Hz supply (experiment). (From Cirrincione, M. et al., *IEEE Trans. Ind. Electron.*, 52(5), 1391, October 2005.)

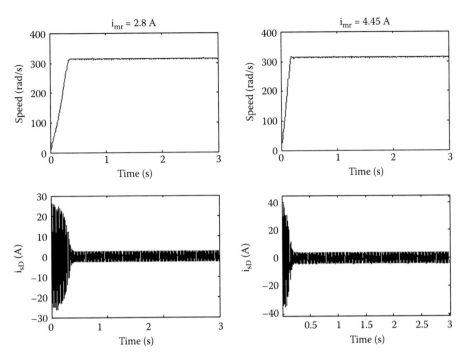

FIGURE 9.40
Rotor speed and stator current during start-up tests under, respectively, 155 and 220 V, 50 Hz supply (experiment). (From Cirrincione, M. et al., *IEEE Trans. Ind. Electron.*, 52(5), 1391, October 2005.)

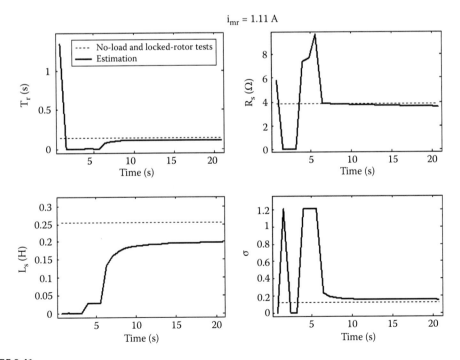

FIGURE 9.41
Estimated and no-load and locked-rotor tests electrical parameters of the machine under a 55 V, 50 Hz supply (experiment). (From Cirrincione, M. et al., *IEEE Trans. Ind. Electron.*, 52(5), 1391, October 2005.)

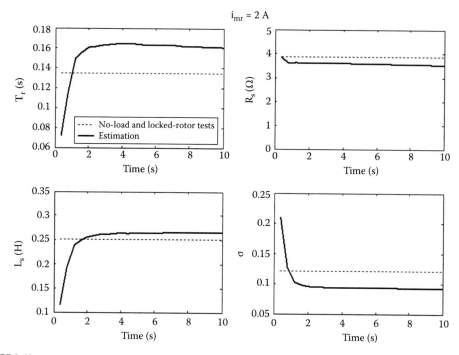

FIGURE 9.42

Estimated and no-load and locked-rotor tests electrical parameters of the machine under a 105 V, 50 Hz supply (experiment). (From Cirrincione, M. et al., *IEEE Trans. Ind. Electron.*, 52(5), 1391, October 2005.)

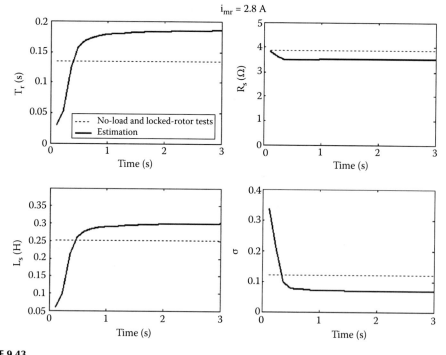

FIGURE 9.43

Estimated and no-load and locked-rotor tests electrical parameters of the machine under a 155 V, 50 Hz supply (experiment). (From Cirrincione, M. et al., *IEEE Trans. Ind. Electron.*, 52(5), 1391, October 2005.)

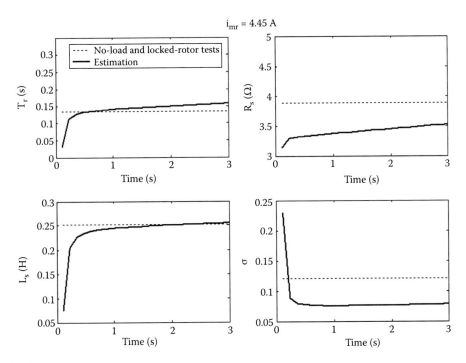

FIGURE 9.44
Estimated and no-load and locked-rotor tests electrical parameters of the machine under a 220 V, 50 Hz supply (experiment). (From Cirrincione, M. et al., *IEEE Trans. Ind. Electron.*, 52(5), 1391, October 2005.)

locked-rotor test, the supply voltage is reduced to ensure rated current, with resulting unsaturated working conditions, while for the estimation of L_s, the no-load test is employed, which is carried out under a condition similar to this experiment, which is also at no load. Several more tests have been made under different voltage levels at the frequency of 50 Hz by employing Equations 4.90 through 4.92 so that the curves in Figures 9.45 through 9.48 have been obtained, which represent, respectively, the variation of L_s, L_m, σ, σ_r, σ_s, T_r, and $|\psi_r|$ as a function of $|i_{mr}|$. Each set of experimental data of each electrical parameter has been then interpolated with a polynomial curve. In particular, the L_s, L_m, T_r, and the magnetization curves have been interpolated with a third-order polynomial, while the σ, σ_r, and σ_s curves have been interpolated with a fifth-order polynomial, as shown in the figures.

It can be concluded that the employment of the constrained minimization algorithm permits a better overall estimation of the K-parameters in comparison with that obtainable with an unconstrained classical LS method, as shown also theoretically. The proposed methodology in general offers the following advantages:

- The possibility to work with the motor supplied either by the sinusoidal voltage waveform from the electric grid or the more general voltage waveform generated by a converter
- No need of an a priori knowledge of the electrical parameters or all the nameplate data of the machine (except for the rated voltage and frequency)
- The simultaneous estimation of the four electrical parameters

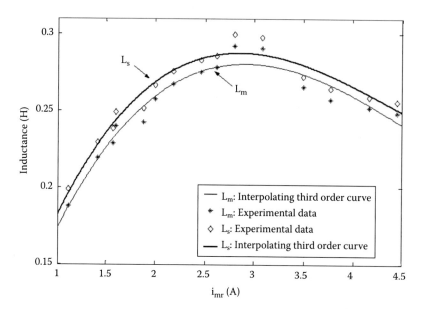

FIGURE 9.45

Estimated L_s and L_m versus $|i_{mr}|$: experimental results and corresponding interpolating curves (experiment). (From Cirrincione, M. et al., *IEEE Trans. Ind. Electron.*, 52(5), 1391, October 2005.)

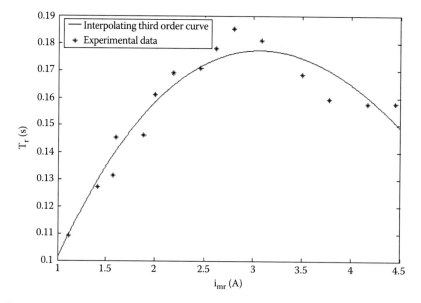

FIGURE 9.46

Estimated T_r versus $|i_{mr}|$: experimental results and corresponding interpolating curve (experiment). (From Cirrincione, M. et al., *IEEE Trans. Ind. Electron.*, 52(5), 1391, October 2005.)

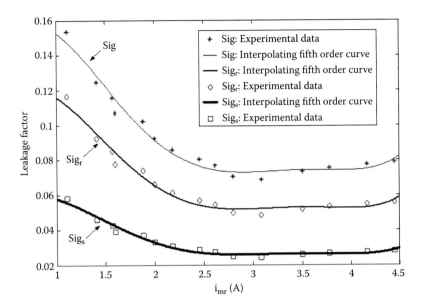

FIGURE 9.47
Estimated σ, σ_r, and σ_s versus $|\mathbf{i}_{mr}|$: experimental results and corresponding interpolating curves (experiment). (From Cirrincione, M. et al., *IEEE Trans. Ind. Electron.*, 52(5), 1391, October 2005.)

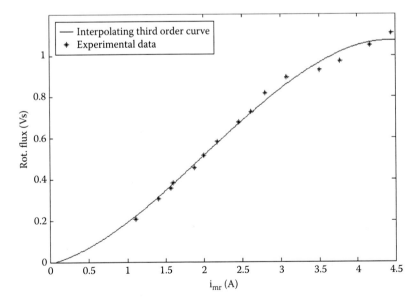

FIGURE 9.48
Estimated magnetization curve: experimental results and corresponding interpolating curve (experiment). (From Cirrincione, M. et al., *IEEE Trans. Ind. Electron.*, 52(5), 1391, October 2005.)

The next paragraph deals with use of the TLS (TLS function) to increase the robustness of the estimated parameters in respect to unavoidable errors present both in the observation vector and in the data matrix.

9.7 Parameter Estimation of an IM with the Total Least Squares Method

The purpose of this paragraph is to show theoretically and experimentally that the OLS method mentioned earlier for retrieving the parameters of an IM can be inadequate if signals are affected by noise. For example, fast commutations of power devices produce a huge rate of change of the current with consequent radiated fields. Moreover, the parasitic capacitance between the motor and the ground cause common-mode high-frequency currents conducted disturbances. These interferences can couple easily with the signal line, and thus measurement errors are always present; therefore, the corresponding uncertainty of the elements of the data matrix used in LS methods is to be taken into consideration. From this standpoint, the TLS presented in Chapter 8 should be used, since it also considers errors in the data matrix. In literature, only Ref. [88] uses the TLS for parameter estimation of AC machines, but without proper constraints. This results in an involved algorithm, too high an order of the differential equations, and also a cumbersome signal processing system, unsuitable for real-world applications. Also in Refs [106–108], the TLS method, together with maximum likelihood methods, has been proposed for a similar problem (parameter estimation of the transformer and the synchronous machine), but in the frequency domain, it is not applied to the on-line identification for AC drives. In this paragraph, it is shown, both in simulation and experimentally, that the presence of progressively increasing noise both in the data matrix and in the observation vector deteriorates the accuracy of the results of the OLS method [99]. It is also proved that the TLS method is more robust. In particular, the TLS EXIN neuron will be used to compute the value of the parameters, as it is the only neuron that can work on-line for this purpose. Moreover, its theoretical behavior is completely known, particularly its convergence and tracking ability [97]. To refine the obtained accuracy of the results, an optimization technique explicitly taking into account the constraint is also presented.

In solving Equation 9.35, TLS EXIN neuron is used, that is, the following error function (see (8.10)),

$$E_{TLS}(\mathbf{k}) = \frac{(\mathbf{Ak} - \mathbf{b})^T (\mathbf{Ak} - \mathbf{b})}{1 + \mathbf{k}^T \mathbf{k}} = \frac{\left\| [\mathbf{A}; \mathbf{b}][\mathbf{k}^T; -1]^T \right\|_2^2}{\left\| [\mathbf{k}^T; -1]^T \right\|_2^2} \tag{9.64}$$

and the corresponding TLS EXIN neuron learning law (see (8.18) and (8.19))

$$\mathbf{k}(t+1) = \mathbf{k}(t) - \alpha(t)\gamma(t)a_i + [\alpha(t)\gamma^2(t)]\mathbf{k}(t) \tag{9.65}$$

where

$$\gamma(t) = \frac{\delta(t)}{1 + \mathbf{k}^T(t)\mathbf{k}(t)} \tag{9.66}$$

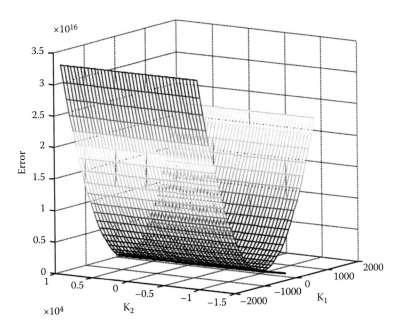

FIGURE 9.49
(See color insert.) OLS error function versus K_1, K_2, and estimated K trajectory (black).

This unconstrained minimization using the simple gradient descent algorithm, as recalled earlier, fails in computing K_2, because the particular structure of matrix **A** implies a very low value of the second column of the matrix $\mathbf{R} = \mathbf{A}^T\mathbf{A}$ making this problem ill-conditioned (K_2 problem): geometrically, this results in a flat error surface along the K_2 direction, and numerical scaling does not help in solving this problem. Figure 9.49 shows the OLS error function versus K_1, K_2 (all the other K-parameters are assumed equal to the correct ones), and the corresponding estimated K trajectory (black). The flatness of the OLS error surface in the K_2 direction is easily observable. Moreover, it can further be observed that the estimated K trajectory remains trapped in the direction of K_2, which makes the estimation of this parameter very difficult. Correspondingly, Figure 9.50 shows the TLS error function versus K_1, K_2 (all the other K-parameters are assumed equal to the correct ones). The more complex shape of the TLS error surface with respect to the OLS one can easily be noticed.

A constrained minimization could then be devised also in this case to approach the true solution. Of course, also in this case, in transient conditions, all four electrical parameters can be retrieved since the data matrix is full rank, but in sinusoidal steady-state, only two K-parameters can be computed since the data matrix has rank 2. Consequently, only one electrical parameter can be obtained.

In any case, if no constraints are taken into account, the TLS solution becomes close-to-nongeneric and less accurate (above all for K_2) when the noise is present in the data. In this case (strong noise in the data matrix and the observation vector), the Equation 9.35 becomes such that the assumption of linearity is no longer valid. Thus a nonlinear minimization method which explicitly takes into account the constraint in Equation 9.35 should be used to refine the TLS estimation, in the same fashion as the algorithms presented in Section 9.6.1.

Now consider Equations 9.36 and 9.37 and let $\mathbf{P}^T = (p_1\, p_2\, p_3\, p_4\, p_5)$ and $\tilde{\mathbf{k}} = (k_1\, k_3\, k_4\, k_5)$ be the reduced **k** vector, where the k_2 component has been omitted. Similarly, let all vectors with

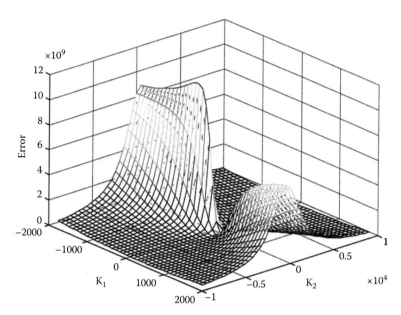

FIGURE 9.50
(See color insert.) TLS error function versus K_1, K_2.

the tilde ~ be the reduced vectors where the second component has been omitted. Thus, the following formula can be written:

$$\mathbf{P}^\mathsf{T}\mathbf{k} = \tilde{\mathbf{P}}^\mathsf{T}\tilde{\mathbf{k}} + p_2 \frac{k_3 k_5}{k_4} \tag{9.67}$$

The $\mathbf{k}^\mathsf{T}\mathbf{R}\mathbf{k}$ quadratic form can be reformulated by using the reduced vectors. At first, the reduced matrix $\tilde{\mathbf{R}}$ can be defined as the submatrix obtained from $\mathbf{R}$ by removing the second row and the second column. Then, by letting $\mathbf{r}_2^\mathsf{T} = (r_{12}\ r_{22}\ r_{32}\ r_{42}\ r_{52})$ be the second column of $\mathbf{R}$ and $\tilde{\mathbf{r}}_2^\mathsf{T} = (r_{12}\ r_{32}\ r_{42}\ r_{52})$ be the corresponding reduced vector, and remembering that $\mathbf{R}^T = \mathbf{R}$, the quadratic form can be rewritten as follows:

$$\mathbf{k}^\mathsf{T}\mathbf{R}\mathbf{k} = \tilde{\mathbf{k}}^\mathsf{T}\tilde{\mathbf{R}}\tilde{\mathbf{k}} + 2\frac{k_3 k_5}{k_4}\mathbf{r}_2^\mathsf{T}\mathbf{k} =$$

$$= \tilde{\mathbf{k}}^\mathsf{T}\tilde{\mathbf{R}}\tilde{\mathbf{k}} + 2\frac{k_3 k_5}{k_4}\tilde{\mathbf{r}}_2^\mathsf{T}\tilde{\mathbf{k}} + 2r_{22}\left(\frac{k_3 k_5}{k_4}\right)^2 \tag{9.68}$$

In the end, the error function to be minimized is as follows:

$$E = \mathbf{b}^\mathsf{T}\mathbf{b} - 2\tilde{\mathbf{P}}^\mathsf{T}\tilde{\mathbf{k}} + \tilde{\mathbf{k}}^\mathsf{T}\tilde{\mathbf{R}}\tilde{\mathbf{k}} +$$

$$- 2p_2\frac{k_3 k_5}{k_4} + 2\frac{k_3 k_5}{k_4}\tilde{\mathbf{r}}_2^\mathsf{T}\tilde{\mathbf{k}} + 2r_{22}\left(\frac{k_3 k_5}{k_4}\right)^2, \tag{9.69}$$

which depends solely on k_1, k_3, k_4, and k_5.

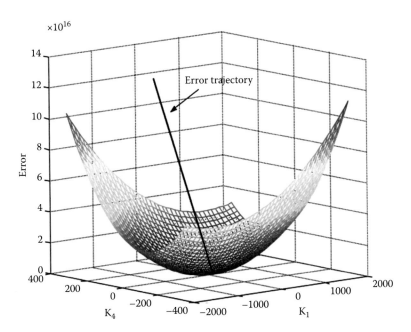

FIGURE 9.51
(See color insert.) Error function of the constrained minimization versus K_1, K_4, and estimated K trajectory (black).

The nonlinear part of Equation 9.69 can be neglected compared to the linear one because it depends on the size of the second column of the data matrix, which is the smallest studying this particular case. Hence, the asymptotic properties of E depend on $\tilde{R}$.

The data autocorrelation matrix R is positive definite or semidefinite if A is full rank or not, respectively. In this case, all possible principal submatrices of R (e.g., $\tilde{R}$) have this property [108] as well. As a consequence, the error E given by Equation 9.69 is always convex, and therefore only one critical point exists. This guarantees the convergence of the minimization method for any choice of the initial conditions. In the case of $\tilde{R}$, for the theorem of Eckart–Young–Mirsky [108], the smallest eigenvalue is greater, or in the worst case equal to that of R, thus implying a better conditioning of the problem than the unconstrained one. Hence, in the constrained algorithm, not only the convergence is guaranteed but also the estimate is more accurate. Figure 9.51 shows the error function of the constrained minimization algorithm (see Equation 9.69) versus K_1, K_4 (all the other K-parameters are assumed equal to the correct ones), and the corresponding estimated K trajectory (black). It can be noticed that, after eliminating the parameter K_2 from the error function, its shape is almost a parable, as expected. As a result, the error trajectory easily converges to its minimum.

9.7.1 Simulation and Experimental Results

The LS method has been verified numerically in simulation and applied experimentally [100] on the test setup described in Section 9.5.6. Simulations have been performed in MATLAB and Simulink. The parameters of the IM used in the simulation are listed in Table 4.1. In the experiments, the motor has been supplied by an asynchronous SV-PWM driven VSI using a voltage/frequency open-loop control. The pulsewidth modulation as well as the open-loop scalar control algorithm has been implemented in software on the DSP of the dSPACE 1103 board employing the MATLAB–Simulink–Real Time Workshop®-Real Time Interface® software. Virtual instruments have been used for controlling the

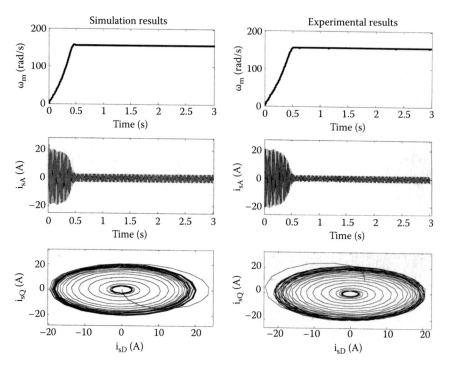

FIGURE 9.52

Rotor speed, i_{sD}, and i_{sQ} waveforms (simulation and experimental results). (From Cirrincione, M. et al., *IEEE Trans. Ind. Appl.*, 39(5), 1247, September/October 2003.)

drive and for monitoring on-line all the electrical and mechanical signals of the motor, for example, the rotor speed, the DC link voltage, and the stator voltages/currents. The LS algorithm has been implemented in software on the DSP board. In all shown tests, the motor, both in simulation and experimental, has been supplied by the VSI with sinusoidal reference voltage of 220 V at 50 Hz. Figure 9.52 shows, both in simulation and experimentally, the rotor speed, the i_{sA} stator current of the phase A, and the stator current locus $i_{sD} - i_{sQ}$ during one of these tests with no load.

The implementation of this method during these tests both in simulation and on the test setup under usual operating conditions, that is, without noisy perturbations, has given the results shown in Table 9.6. The OLS and the TLS methods have been implemented on the DSP. The initial values of the K-parameters have always been set to 0, to ensure convergence as explained in Theorem 8.2. From these tables, it is apparent that the TLS solution is from a vector point of view closer to the true value than OLS: K_2 is computed with good accuracy, while the other parameters are practically the same, even if there is a slight deterioration in the experimental test in the computation of K_{31}. With this regard it should be remarked that, in the experimental tests, the true values of the parameters are those measured with the usual no-load and locked-rotor tests, which are retrieved under operating conditions that differ from those of the test: therefore the machine parameters computed by the LS method can be slightly different from those obtained with the traditional methods.

A deeper insight into these results is given in the following with the off-line analysis of the data.

In the first place, the error surfaces, obtained with the simulation data, by employing the OLS, the TLS, and also the constrained OLS are summarized in Figure 9.53. For obvious

TABLE 9.6

Estimated K-Parameters (Simulation and Experimental Results, Unconstrained OLS and TLS)

	OLS $\times 10^3$	TLS $\times 10^3$	OLS Err. %	TLS Err. %
Simulation results				
K_1 [s^{-1}]	0.1886	0.1888	0.21	0.11
K_2 [s^{-2}]	1.267	0.9364	33.8	1.05
K_{31} [s^{-1}]	0.1271	0.1280	0.37	0.34
K_4 [H^{-1}]	0.0328	0.0329	0.15	0.09
K_5 [s^{-1} H^{-1}]	0.2432	0.2413	0.29	1.06
Experimental results				
K_1 [s^{-1}]	0.1887	0.1794	0.14	5.09
K_2 [s^{-2}]	1.373	0.9480	45.11	0.16
K_{31} [s^{-1}]	0.1229	0.1172	3.60	8.09
K_4 [H^{-1}]	0.0328	0.0324	0.16	1.53
K_5 [s^{-1} H^{-1}]	0.2580	0.2470	5.78	1.17

Source: Cirrincione, M. et al., *IEEE Trans. Ind. Appl.*, 39(5), 1247, September/October 2003.

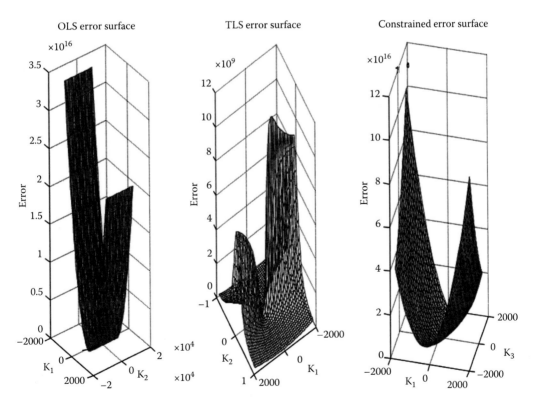

FIGURE 9.53
3-D error surfaces for OLS, TLS, and constrained OLS. (From Cirrincione, M. et al., *IEEE Trans. Ind. Appl.*, 39(5), 1247, September/October 2003.)

visualization reasons, the three dimension (3-D) plots are drawn by varying just two K-parameters and letting the remaining ones be at the values given by Table 4.1. From these 3-D plots, the K_2 problem is apparent in the first of these graphs, as the error surface is flat along the K_2 direction see also Figure 9.49; the same is also true from the second graph, where the more complex TLS surface is shown. In the third graph, the constrained OLS surface is drawn, in which the K_2 variable is eliminated by using the constraint; thus the solution of the problem does not present any flatness.

The flatness in the OLS error cost along the K_2 direction is due to the very small size of the singular value of **A** corresponding to K_2, which implies high two-norm condition numbers [108]: the relative condition number, relating the perturbation in **A** to the perturbation in the solution, is given by 6578 for the simulation data and by 8820 for the experimental data; the relative condition number, relating the perturbation in **b** to the perturbation in the solution, is given by 98.4 for the simulation data and by 116 for the experimental data. The computation of the other singular values shows that they are grouped together and this confirms that only K_2 is difficult to be estimated. Furthermore, also scaling **A** does not change the condition numbers, which also shows that the solution is more influenced by perturbations in **A** than in **b**. Thus these considerations justify the use of TLS techniques which, unlike OLS, take into account the noise in **A**. The presence of the flatness along the K_2 direction in the TLS error function, however, suggests a "nongenericity" in the TLS problem, which is confirmed by the analysis of the eigenvalues of the data autocorrelation matrix **R**. For both simulation and experimental examples, the eigenvector corresponding to K_2 (the MC vector of **A**) tends (partially) to be parallel to the TLS hyperplane in the direction of the true value of K_2. Hence, the problem is close-to-nongeneric because of K_2. The application of the TLS EXIN neuron automatically implements the approximate (because the problem is not exactly nongeneric) constraint that the weights have to remain orthogonal to the K_2 solution [109]. For the case of the experimental data, Figure 9.54 shows the contour plots for the TLS error in the plane $K_1 - K_2$ for values ranging from the lowest saddle point (s #1) to the minimum point (TLS min); these two critical points, together with the solution and the temporal evolution of the weights of the TLS EXIN neuron, are also shown. It can be seen that neither the TLS minimum (generic TLS solution) nor the TLS lowest saddle #1 (nongeneric TLS solution) corresponds to the true solution, which is somehow in-between (close-to-nongeneric TLS solution). Note the particular shape of the contours that confirms the nongenericity around the K_2 direction.

Also by using off-line OLS direct methods [108], that is, the Cholesky decomposition of the normal equations, the Householder factorization, and the SVD technique, no improvement has been achieved in the OLS solution. On the contrary, TLS EXIN (for null initial conditions) has always given a better estimation of K_2, because it automatically solves the close-to-nongeneric TLS problem. The constraint (9.37) is well satisfied by the K-parameters estimated by TLS. This suggests a possible correspondence between the TLS nongeneric constraint and this constraint. Indeed, constraint (9.37) depends on the choice of the K-parameters. This choice also influences the Equation 9.36, which in turn implies the nongeneric constraint.

Afterward, to check the robustness of TLS against noise, a uniformly distributed noise between −5% and +5% of the rated voltage and current has been given to each acquired signal so as to have noisy elements both in the data matrix and in the observation vector. The electrical drive has been submitted to the same tests as described earlier but, to take into account the statistical effect of the random noise, each test has

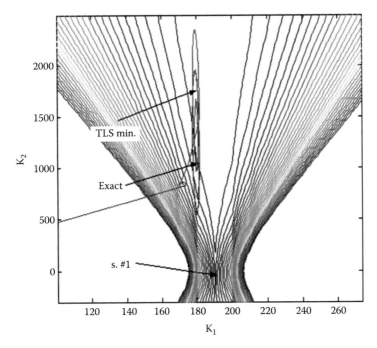

FIGURE 9.54
(See color insert.) Contours and critical points of the TLS error in the plane K_1, K_2; the TLS EXIN weight trajectory is also shown. (From Cirrincione, M. et al., *IEEE Trans. Ind. Appl.*, 39(5), 1247, September/October 2003.)

been repeated 50 times: in this way, the statistical average of all the estimation has been computed. To speed convergence, a BFGS method, which requires blocks of data, has been implemented on the DSP. The size of these blocks has been kept low not to overload the DSP.

Figures 9.55 and 9.56 show the true parameters of the machine and the waveforms of the average parameter estimation, computed by the OLS and TLS, respectively, in simulation and in the experimentation, when the signals are corrupted by the noise. The shift in time between the OLS and TLS curves is due to the different length of the block for each of them. The electrical parameters are updated every 0.3 s.

Table 9.7 shows the percentage error of the K-parameters at the end of the estimating process both with OLS and TLS as well as their global errors; this global error is computed as the two-norm, divided by the two-norm of the true K-parameters vector, of the difference between the solution vector obtained with each method and the true K-parameters vector. These results show on the one hand that the OLS solution deteriorates in comparison with the TLS one in the presence of the noise and on the other hand that the TLS solution goes away from the solution, especially in the estimation of K_2, in comparison with the no-noise case shown on Table 9.6. This result is then refined by the use of the constrained algorithm which further approaches the true values as shown in Table 9.7 with substantial lower errors than those obtained with TLS and OLS. It should be remarked that the constrained algorithm is to be used in real-time applications after the TLS algorithm converged when the signals are very noisy. The use of the OLS constrained algorithm gives good results only after the convergence has been reached, that is, when the initial conditions are in the neighborhood of the solution. Tests made in the same noisy environment with initial

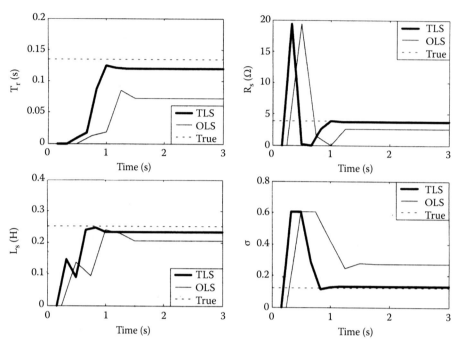

FIGURE 9.55
Real and estimated electrical parameters of the motor with noise (simulation results). (From Cirrincione, M. et al., *IEEE Trans. Ind. Appl.*, 39(5), 1247, September/October 2003.)

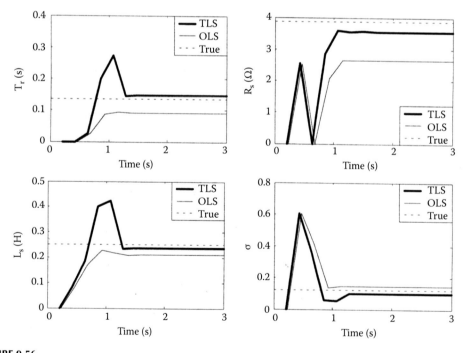

FIGURE 9.56
Real and estimated electrical parameters of the motor with noise (experimental results). (From Cirrincione, M. et al., *IEEE Trans. Ind. Appl.*, 39(5), 1247, September/October 2003.)

TABLE 9.7

Estimation Relative Errors on K-Parameters

	Simulation Results			Experimental Results		
	TLS	**OLS**	**Const.**	**TLS**	**OLS**	**Const.**
e K_1%	0.53	49.20	0	17.46	14.3	12.2
e K_2%	8.14	69.23	5.40	241.9	1394	0.32
e K_{31}%	0	62.99	0.80	24.41	27.6	17.32
e K_4%	0	43.75	0	34.38	3.10	28.12
e K_5%	1.65	8.60	6.20	8.23	37.0	7.41
E. TLS	0.07			2.28		
E. OLS	0.66			13.13		
E. const.	0.05			0.04		

Source: Cirrincione, M. et al., *IEEE Trans. Ind. Appl.*, 39(5), 1247, September/October 2003.

conditions far from the solution have led to worse results than those obtained with the unconstrained TLS.

It should be emphasized that the problem of the estimation of the parameters of the IM is to be placed in the context of a close-to-nongeneric problem and that the TLS EXIN neuron learning law accurately estimates all parameters by implicitly satisfying the nonlinear constraint. Moreover, the TLS EXIN is a robust technique as to noise, unlike OLS. On the other hand, for progressively increasing noise, obviously also the TLS solution can deteriorate, and in this case, a nonlinear constrained minimization algorithm for the refinement of the solution should be used as well.

9.8 Application of the RLS-Based Parameter Estimation to Flux Model Adaptation in FOC and DTC IM Drives

This section shows the application of the LS-based on-line parameter estimation algorithm to the flux model adaptation in high-performance FOC and DTC IM drives [111–113]. In this case, for simplicity of implementation, the RLS algorithm has been used (see Section 9.5.4). The adopted FOC scheme is based on voltage control with rotor flux orientation (Figure 5.16), while the adopted DTC scheme is the classic switching-table-based DTC (Figure 5.37). The parameters of the IM under test are shown in Table 4.1. In particular, some simulation results are shown highlighting how the LS technique, integrated with the selection algorithm in Figure 9.20, is able to make the control system adaptive versus machine parameters variations. In both control schemes, the control strategy is made adaptive by computing in real-time values of some electrical parameters of the machine, thus allowing the flux models in each scheme to be correctly tuned at each instant. In particular, in both control schemes, all the electrical parameter (T_r, L_s, R_s, and σ) are computed during each speed transient, while in speed steady-state, only the parameter that is the most critical from the control point of view is chosen for estimation, as the others are assumed constant, equal to the values estimated during the last speed transient.

The flux model adopted in the FOC scheme is the current model based on the rotor equations in the rotor flux reference frame (Figure 5.11b), and, therefore, the estimation of the rotor flux requires the knowledge of the T_r, to the variation of which the flux model is sensible as shown in Section 9.2.1. The value of T_r needed by the flux model is usually set to that corresponding to the rated temperature and the rated magnetization level of the machine. The parameter T_r, however, changes slowly with the temperature of the rotor and more quickly with the state of magnetization of the machine (field-weakening). Thus a variation of the temperature and the magnetic saturation of the machine causes errors in the flux model (detuning) and can be critical from the point of view of both the stability and the performance of the control scheme. For this reason, the most critical parameter in the FOC scheme is T_r and then it is chosen as the unique parameter to be estimated in speed steady-state.

The flux model adopted in the DTC scheme is the voltage model, which is based on the stator equations in the stationary reference frame, requiring the knowledge of the stator resistance R_s. The value of R_s needed by the flux model is usually set to that corresponding to the rated temperature of the machine while the stator resistance of the machine changes with the temperature of the stator winding. Thus a variation of the temperature of the machine causes even in a DTC drive a detuning of the flux model, which reduces the performance of the drive and can be critical from the point of view of the stability. For this reason, the most critical parameter in the DTC scheme is R_s and then it is chosen as the unique parameter to be estimated in speed steady-state. Simulations have been performed making use of the MATLAB–Simulink software of the Mathworks®. Some tests have been made to check the parameter estimation algorithm both in a very fast speed transient and in speed steady-state. The same kind of tests have been carried out in both drives.

In the first test, a speed reference step of 100 rad/s and a rotor flux reference step of 0.8 Wb have been given to the drive. Under these conditions, the rate of change of the speed is high, and the assumption of slow transient does not exactly hold. Figure 9.57a and b shows the waveforms of the rotor speed, the stator voltage of phase sA, and the stator current of phase sA, respectively, for the FOC and DTC drives. Figure 9.58a and b shows the estimated electrical parameters during the speed transient and the real ones in both control techniques. It can be noticed that the RLS algorithm is able to correctly estimate all the four electrical parameters of the machine at the end of the speed transient, despite of the different stator voltage and current waveforms.

In the second test, the capability of the algorithm to track on-line the variation of the most critical electrical parameter in speed steady-state has been tested. In particular, at the constant speed of 100 rad/s and constant load torque of 10 N m, a step variation of the most critical parameter, respectively, an increase as much as twice of the T_r of the motor in the FOC drive and of the R_s of the motor in the DTC drive, has been imposed. Figure 9.59a and b shows respectively the imposed variation of the true T_r compared with the estimated one in the FOC drive and the imposed variation of the true R_s compared with the estimated one in the DTC drive. It can be seen that the algorithm is able to track the variation of the parameter, even if the rotor speed is constant. Figure 9.60a and b shows, respectively, the true rotor flux compared with the estimated one in the FOC drive and the true stator flux compared with the estimated one in the DTC drive. It can be noticed that, thanks to the adaptation algorithm, the real magnetic flux is different from the estimated one until the estimated parameter becomes close to the real one in both schemes.

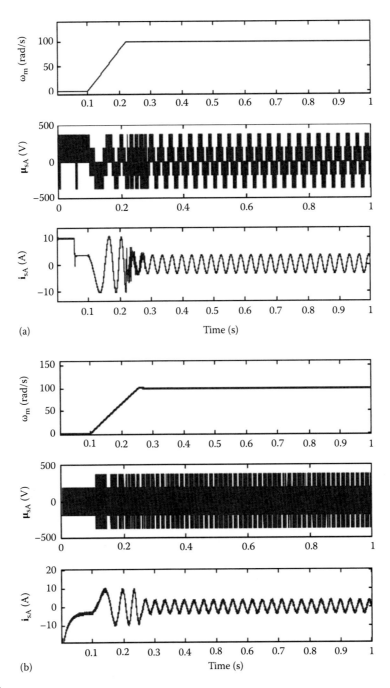

FIGURE 9.57
(a) Rotor speed, stator voltage, stator current (FOC). (b) Rotor speed, stator voltage, stator current (DTC). (From Cirrincione, M. and Pucci, M., A direct-torque-control of an AC drive based on a recursive-least-squares (RLS) method, *Proceedings of the IEEE International Symposium on Diagnostics for Electric Machines, Power Electronics and Drives (IEEE SDEMPED'01)*, Grado, Italy, 2001.)

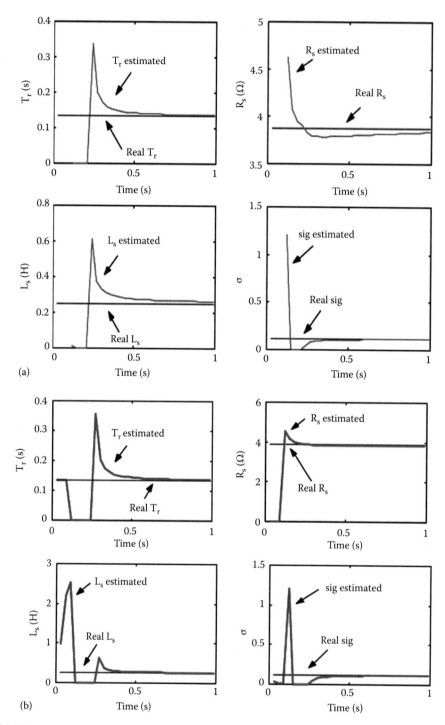

FIGURE 9.58
(a) Estimated and real parameters in a speed transient (FOC). (b) Estimated and real parameters in a speed transient (DTC). (From Cirrincione, M. and Pucci, M., A direct-torque-control of an AC drive based on a recursive-least-squares (RLS) method, *Proceedings of the IEEE International Symposium on Diagnostics for Electric Machines, Power Electronics and Drives (IEEE SDEMPED'01)*, Grado, Italy, 2001.)

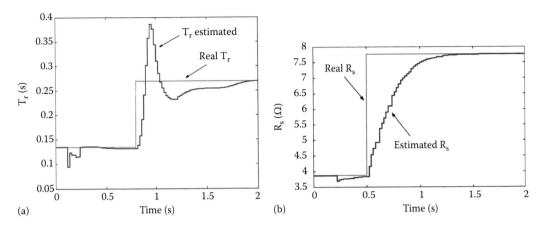

FIGURE 9.59
(a) Tracking of the T_r in speed steady-state (FOC). (b) Tracking of the R_s in speed steady-state (DTC). (From Cirrincione, M. and Pucci, M., A direct-torque-control of an AC drive based on a recursive-least-squares (RLS) method, *Proceedings of the IEEE International Symposium on Diagnostics for Electric Machines, Power Electronics and Drives (IEEE SDEMPED'01)*, Grado, Italy, 2001.)

9.9 Estimation of the IM Parameters at Standstill

Off-line parameter estimation techniques for IMs [4,5,10,13,19,68,113] are essential for the self-commissioning of the corresponding drives, consisting in the proper tuning of the controllers, of the decoupling circuits, if present, and of the flux models. Traditionally the no-load and locked-rotor tests are used, but the accuracy they provide is often insufficient for the previous applications; moreover, these tests are sometimes hard to perform both because the motor is usually coupled to the mechanical load, and then it is not always possible to lock the rotor, and because they are difficult to automate [4]. Identification methods of the IM in standstill have been therefore developed, so machine does not produce any torque, and the locking of the rotor is undesirable. These techniques have been proposed both for the frequency domain and the time domain. The frequency domain requires a sinusoidal supply with variable frequency in order to obtain a frequency characteristic from which the electrical parameters of the motor can be inferred. In the time domain, several techniques can be used, like the MRAS, EKF, LS, the two-frequency method, etc.

A possible approach for the time domain identification of IMs at standstill is using LS techniques [114]. Starting from Equation 9.25, under the hypothesis of working at standstill, that is, $\omega_r = 0$ and $\dfrac{d\omega_r}{dt} = 0$, the equation simplifies as follows:

$$\begin{pmatrix} \dfrac{di_{sD}}{dt} & i_{sD} & -\dfrac{du_{sD}}{dt} & -u_{sD} \\ \dfrac{di_{sQ}}{dt} & i_{sQ} & -\dfrac{du_{sQ}}{dt} & -u_{sQ} \end{pmatrix} \begin{pmatrix} K_1 \\ K_2 \\ K_4 \\ K_5 \end{pmatrix} = \begin{pmatrix} -\dfrac{d^2 i_{sD}}{dt^2} \\ -\dfrac{d^2 i_{sQ}}{dt^2} \end{pmatrix} \qquad (9.70)$$

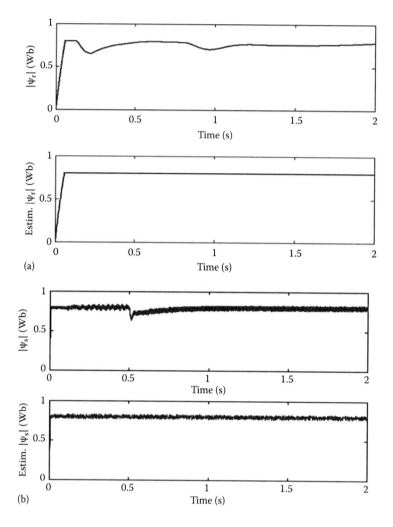

FIGURE 9.60
(a) Estimated and real rotor flux (FOC). (b) Estimated and real stator flux (DTC). (From Cirrincione, M. and Pucci, M., A direct-torque-control of an AC drive based on a recursive-least-squares (RLS) method, *Proceedings of the IEEE International Symposium on Diagnostics for Electric Machines, Power Electronics and Drives (IEEE SDEMPED'01)*, Grado, Italy, 2001.)

The following remarks are to be made on the aforementioned equations:

- No modeling error exists differently from the on-line counterpart as explained in Ref. [114], and, therefore, the equation is almost exactly linear (the error is only due to the measurement errors and the filtering).
- K_{31} is not present any longer.
- No constraint exists.

To solve this equation, an LS method can be used that determines the electrical parameters of the motor on the basis of the stator voltage and current signals and their derivatives.

Moreover, because of the absence of any modeling error, a more accurate result can be achieved than that of the on-line counterpart. In Ref. [114], a classic OLS algorithm has been used to solve this problem. From the K-parameters, as it is known, the four electrical parameters must be retrieved in this way:

$$T_r = \frac{K_4}{K_5} \tag{9.71a}$$

$$R_s = \frac{K_2}{K_5} \tag{9.71b}$$

$$L_s = \frac{K_5 K_1 - K_2 K_4}{K_5^2} \tag{9.71c}$$

$$\sigma = \left(\frac{K_5}{K_1 K_5 - K_4 K_2} \right) \left(\frac{K_5}{K_4} \right) \tag{9.71d}$$

All four parameters are therefore computed in an unconstrained way. By the inspection of Equation 9.70, a supply waveform with at least two different harmonics would be enough for retrieving all the four K-parameters. A special care has to be taken in the selection of the suitable voltage waveform for supplying the motor, which could be easily synthesized by the VSI. This waveform has been selected following the criteria that no electromagnetic torque should be produced at standstill and consequently any mechanical stress on the rotor. For obtaining this, a zero voltage reference has been given to phase sB of the motor whilst two transient voltage waveforms varying from zero to a constant value have been given to phases sA and sC as represented in Figure 9.61 [114].

In particular, no step voltage input can be given as its derivatives are not functions, nor an input signal whose first derivative is a threshold function can be given as the data matrix is not full rank. Thus only an input signal whose first derivative has an interval in which it is linear can be given for retrieving all parameters.

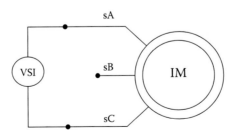

FIGURE 9.61
Schematics of the supply. (From Cirrincione, M. et al., A least-squares based methodology for estimating the electrical parameters of induction machine at standstill, *Proceedings of the IEEE International Symposium on Industrial Electronics (ISIE'02)*, L'Aquila, Italy, 2002.)

In particular, these three voltage references have been given to the three phases of the motor:

$$u_{Aref} = \begin{cases} \lambda t^2 & \text{for } t < 1s \\ \lambda(2 - e^{\alpha(1-t)}) & \text{for } t > 1s \end{cases}$$

$$u_{Bref} = 0 \tag{9.72}$$

$$u_{Cref} = \begin{cases} -\lambda t^2 & \text{for } t < 1s \\ -\lambda(2 - e^{\alpha(1-t)}) & \text{for } t > 1s \end{cases}$$

where $\alpha = 3$ and λ is selected by means of an automatized procedure so to supply the motor with a stator current of amplitude close to the magnetizing current of the motor in normal operating conditions. A drawback is that, with this kind of supply waveform, the electrical parameters of the motor are estimated generating a magnetization of the machine, which is somewhat different from that in normal operating condition. In particular, with the previously described supply, a magnetomotive force (mmf) that is fixed in space and with amplitude variable in time is generated, while, in normal operating conditions, an mmf that is rotating in space and with constant amplitude is generated. The parameters of the machine under test are shown in Table 4.1.

Figure 9.62 shows the stator u_{sD}, u_{sQ} voltage signals, filtered both analogically by means of the low-pass Bessel filter card and digitally by software in the dSPACE card, obtained on the basis of the three voltage references of Equation 9.72. Figure 9.63 shows the corresponding stator i_{sD}, i_{sQ} current signals. Figure 9.64 shows the electrical parameters estimated by the OLS algorithm and the real ones, which have been obtained with the usual no-load and locked-rotor tests. It shows that the estimation process of all the electrical parameters of the machine requires less than 1 s.

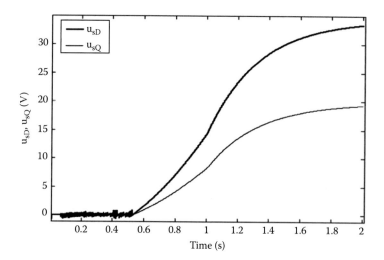

FIGURE 9.62
u_{sD} and u_{sQ} filtered stator voltages. (From Cirrincione, M. et al., A least-squares based methodology for estimating the electrical parameters of induction machine at standstill, *Proceedings of the IEEE International Symposium on Industrial Electronics (ISIE'02)*, L'Aquila, Italy, 2002.)

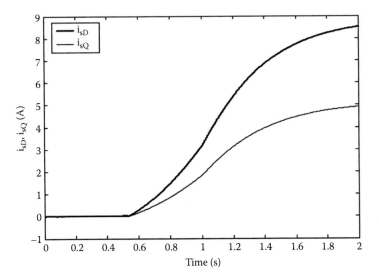

FIGURE 9.63

i_{sD} and i_{sQ} filtered stator currents. (From Cirrincione, M. et al., A least-squares based methodology for estimating the electrical parameters of induction machine at standstill, *Proceedings of the IEEE International Symposium on Industrial Electronics (ISIE'02)*, L'Aquila, Italy, 2002.)

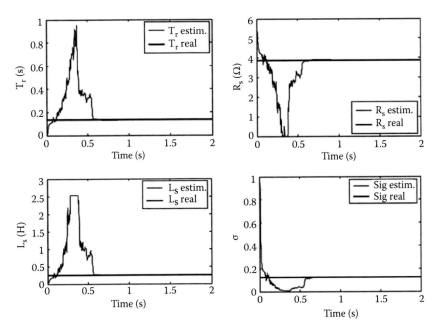

FIGURE 9.64

Estimated electrical parameters and real ones. (From Cirrincione, M. et al., A least-squares based methodology for estimating the electrical parameters of induction machine at standstill, *Proceedings of the IEEE International Symposium on Industrial Electronics (ISIE'02)*, L'Aquila, Italy, 2002.)

TABLE 9.8

Estimated and True Electrical Parameters

	OLS	True	Err. %
T_r (s)	0.1341	0.1348	−0.4649
R_s (Ω)	3.88	3.88	−1.6699 10^{-4}
L_s (H)	0.2518	0.2520	−0.0957
σ	0.1208	0.1207	0.1048

Source: Cirrincione, M. et al., A least-squares based methodology for estimating the electrical parameters of induction machine at standstill, *Proceedings of the IEEE International Symposium on Industrial Electronics (ISIE'02)*, L'Aquila, Italy, 2002.

TABLE 9.9

Estimated and True K-Parameters

	OLS	True	Err. %
K_1	189.2196	189.0088	0.1115
K_2	950.7603	946.4266	0.4579
K_4	32.8681	32.8711	−0.0090
K_5	245.0417	243.9244	0.4581

Source: Cirrincione, M. et al., A least-squares based methodology for estimating the electrical parameters of induction machine at standstill, *Proceedings of the IEEE International Symposium on Industrial Electronics (ISIE'02)*, L'Aquila, Italy, 2002.

Tables 9.8 and 9.9 show the true and estimated values of the electrical parameters and of the K-parameters of the IM under test as well as the estimation errors. It should be remarked that, due to the previously explained reasons, the off-line estimation by means of the LS methodology is more accurate than its on-line counterpart.

List of Symbols

$u_{sA},\, u_{sB},\, u_{sC}$	stator phase voltages
$i_{sA},\, i_{sB},\, i_{sC}$	stator phase currents
$\mathbf{u}_s = u_{sD} + ju_{sQ}$	space-vector of the stator voltages in the stator reference frame
$\mathbf{u}_s^g = u_{sx}^g + ju_{sy}^g$	space-vector of the stator voltages in a generic rotating reference frame
$\mathbf{i}_s = i_{sD} + ji_{sQ}$	space-vector of the stator currents in the stator reference frame
$\mathbf{i}_s^g = i_{sx}^g + ji_{sy}^g$	space-vector of the stator currents in a generic rotating reference frame
$\mathbf{i}_r' = i_{rd} + ji_{rq}$	space-vector of the rotor currents in the stator reference frame
$\mathbf{i}_r^g = i_{rx}^g + ji_{ry}^g$	space-vector of the rotor currents in a generic rotating reference frame
$\mathbf{\psi}_r' = \psi_{rd} + j\psi_{rq}$	space-vector of the rotor flux linkages in the stator reference frame
$\mathbf{\psi}_s = \psi_{sD} + j\psi_{sQ}$	space-vector of the stator flux linkages in the stator reference frame
$\mathbf{i}_{mr} = i_{mrD} + ji_{mrQ}$	space-vector of the rotor magnetizing current in the stator reference frame

$\mathbf{i}_{ms} = i_{msD} + ji_{msQ}$	space-vector of the stator magnetizing current in the stator reference frame
$\mathbf{i}_{mm} = i_{mmD} + ji_{mmQ}$	space-vector of the magnetizing current in the stator reference frame
ρ_r	phase angle of the rotor flux-linkage space-vector with respect to the sD axis
ρ_s	phase angle of the stator flux-linkage space-vector with respect to the sD axis
ρ_m	phase angle of the magnetizing flux-linkage space-vector with respect to the sD axis
ϑ_r	angular position of the rotor with respect to the sD axis
L_s	stator inductance
L'_s	stator transient inductance
L_r	rotor inductance
L_m	total static magnetizing inductance
$L_{s\sigma}$	stator leakage inductance
$L_{r\sigma}$	rotor leakage inductance
R_s	resistance of a stator phase winding
R_r	resistance of a rotor phase winding
T_s	stator time constant
T_r	rotor time constant
$\beta_0 = R_r/L_r$	inverse of the rotor time constant T_r
T'_s	stator transient time constant
T'_r	rotor transient time constant
$T_{r\sigma}$	rotor leakage time constant
$\sigma = 1 - L_m^2/(L_s L_r)$	total leakage factor
σ_r	rotor leakage factor
σ_s	stator leakage factor
p	number of pole pairs
ω_{mr}	angular speed of the rotor flux space-vector
ω_{ms}	angular speed of the stator flux space-vector
ω_{mm}	angular speed of the magnetizing flux space-vector
ω_{sl}	angular slip speed
ω_r	angular rotor speed (in electrical angles per second)
t_e	electromagnetic torque

References

1. W. Leonhard, *Control of Electrical Drives*, Springer-Verlag, Berlin, Germany, 1997.
2. P. Vas, *Sensorless Vector and Direct Torque Control*, Oxford Science Publications, Oxford, U.K., 1998.
3. P. Vas, *Vector Control of AC Machines*, Oxford Science Publications, Oxford, U.K., 1990.
4. P. Vas, *Parameter Estimation and Condition Monitoring*, Cambridge University Press, Cambridge, U.K., 1996.
5. H. Klaassen, Selbsteinstellende, Feldorientierte regelung einer asynchronmaschine und geberlose drehzahlregelung, Dissertation, Technische Universität Braunschweig, Braunschweig, Germany, 1999.

6. N. P. Quang, J.-A. Dittrich, *Praxis der feldorientierten Drehstrom-antriebsregelungen*, Expert-Verlag, 1999.

7. J. Holtz, Sensorless control of induction motor drives, *Proceedings of the IEEE*, 90(8), 1359–1394, August 2002.

8. P. L. Jansen, R. D. Lorenz, A physically insightful approach to the design and accuracy assessment of flux observers for field oriented induction machine drives, *IEEE Transactions on Industry Applications*, 30(1), 101–110, January/February 1994.

9. P. J. Chrzan, H. Klaassen, Parameter identification of vector-controlled induction machines, *Electrical Engineering*, 79(1), 39–46, 1996.

10. S. I. Moon, A. Keyhani, Estimation of induction machine parameters from standstill time-domain data, *IEEE Transactions on Industry Applications*, 30(6), 1606–1615, November/December 1994.

11. A. Consoli, L. Fortuna, A. Gallo, Induction motor identification by a microcomputer-based structure, *IEEE Transactions on Industry Applications*, 34, 422–428, November 1987.

12. C. Wang, D. W. Novotny, T. A. Lipo, An automated rotor time constant measurement system for indirect field-oriented drives, *IEEE Transactions on Industry Applications*, 24, 151–159, January/February 1988.

13. M. Bertoluzzo, G. S. Buja, R. Menis, Self-commissioning of RFO IM drives: One-test identification of the magnetization characteristic of the motor, *IEEE Transactions on Industry Applications*, 37, 1801–1806, November/December 2001.

14. E. Levi, M. Sokola, S. N. Vukosavic, A method for magnetizing curve identification in rotor flux oriented induction machines, *IEEE Transactions on Energy Conversion*, 15, 157–162, June 2000.

15. E. Levi, Method of magnetizing curve identification in vector controlled induction machines, *European Transactions on Electrical Power*, 2(5), 309–314, 1992.

16. A. Ganji, P. Guillaume, R. Pintelon, P. Lataire, Induction motor dynamic and static inductance identification using a broadband excitation technique, *IEEE Transactions on Energy Conversion*, 13, 15–20, March 1998.

17. D. E. Borgard, G. Olsson, R. D. Lorenz, Accuracy issues for parameter estimation of field oriented induction machine drives, *IEEE Transactions on Industry Applications*, 31, 795–801, July/August 1995.

18. A. Gastli, Identification of induction motor equivalent circuit parameters using the single-phase test, *IEEE Transactions on Energy Conversion*, 14, 51–56, March 1999.

19. T. Kudor, K. Ishihara, H. Naitoh, Self-commissioning for vector controlled induction motors, in *Power Electronics Technology and Applications II*, F. C. Lee, (Ed.), IEEE Press, New York, pp. 509–516, 1997.

20. L. A. de S. Ribeiro, C. B. Jacobina, A. M. N. Lima, A. C. Oliviera, Real-time estimation of the electric parameters of an induction machine using sinusoidal PWM voltage waveforms, *IEEE Transactions on Industry Applications*, 36, 743–754, May/June 2000.

21. J. K. Seok, S. K. Sul, Induction motor parameter tuning for high-performance drives, *IEEE Transactions on Industry Applications*, 37, 35–41, January/February 2001.

22. T. Matsuo, T. A. Lipo, A rotor parameter identification scheme for vector controlled induction motor drives, *IEEE Transactions on Industry Applications*, 21, 624–632, May/June 1985.

23. H. Chai, P. P. Acarnley, Induction motor parameter estimation algorithm using spectral analysis, *IEE Proceedings of the Electric Power Applications, Part B*, 139(3), 165–174, 1992.

24. H. Sugimoto, S. Tamai, Secondary resistance identification of an induction motor—Applied model reference adaptive system and its characteristics, *IEEE Transactions on Industry Applications*, 23, 296–303, March/April 1987.

25. L. C. Zai, C. L. DeMarco, T. A. Lipo, An extended Kalman filter approach to rotor time constant measurement in PWM induction motor drives, *IEEE Transactions on Industry Applications*, 28, 96–104, January/February 1992.

26. J. W. Finch, D. J. Atkinson, P. P. Acarnley, Full-order estimator for induction motor states and parameters, *IEE Proceedings on Electric Power Applications*, 145(3), 169–179, 1998.

27. D. J. Atkinson, J. W. Finch, P. P. Acarnley, Estimation of rotor resistance in induction motors, *IEE Proceedings on Electric Power Applications*, 143(1), 87–94, 1996.

28. L. J. Garces, Parameter adaption for the speed-controlled static AC drive with a squirrel-cage induction motor, *IEEE Transactions on Industry Applications*, IA–16, 173–178, March/April 1980.

29. T. M. Rowan, R. J. Kerkman, D. Leggate, A simple on-line adaption for indirect field orientation of an induction machine, *IEEE Transactions on Industry Applications*, 27, 720–727, July/August 1991.

30. S. N. Vukosavic, M. R. Stojic, On-line tuning of the rotor time constant for vector-controlled induction motor in position control applications, *IEEE Transactions on Industry Electronics*, 40, 130–138, February 1993.

31. R. D. Lorenz, D. B. Lawson, A simplified approach to continuous, on-line tuning of field-oriented induction machine drives, *IEEE Transactions on Industry Applications*, 26, 420–424, May/June 1990.

32. R. Lessmeier, W. Schumacher, W. Leonhard, Microprocessor-controlled AC-servo drives with synchronous or induction motors: Which is preferable? *IEEE Transactions on Industry Applications*, 22, 812–819, September/October 1986.

33. N. R. Klaes, On line tuning of the rotor resistance in an inverter fed induction machine with direct-self-control, *European Transactions on Electrical Power*, 4(1), 5–11, 1994.

34. K. Ohnishi, Y. Ueda, K. Miyachi, Model reference adaptive system against rotor resistance variation in induction motor drive, *IEEE Transactions on Industry Applications*, 33, 217–223, August 1986.

35. R. J. Wai, D. C. Liu, F. J. Lin, Rotor time-constant estimation approaches based on energy function and sliding mode for induction motor drive, *Electric Power Systems Research*, 52, 229–239, 1999.

36. S. K. Sul, A novel technique of rotor resistance estimation considering variation of mutual inductance, *IEEE Transactions on Industry Applications*, 25, 578–587, July/August 1989.

37. C. C. Chan and H. Wang, An effective method for rotor resistance identification for high-performance induction motor vector control, *IEEE Transactions on Industry Electronics*, 37(6), 477–482, 1990.

38. H. Toliyat, M. S. Arefeen, K. M. Rahman, M. Ehsani, Rotor time constant updating scheme for a rotor flux oriented induction motor drive, *IEEE Transactions on Power Electronics*, 14, 850–857, September 1999.

39. H. T. Yang, K. Y. Huang, C. L. Huang, An artificial neural network based identification and control approach for the field-oriented induction motor, *Electric Power Systems Research*, 30, 35–45, 1994.

40. F. Zidani, M. S. Nait-Said, M. E. H. Benbouzid, D. Diallo, R. Abdessemed, A fuzzy rotor resistance updating scheme for an IFOC induction motor drive, *IEEE Power Engineering Review*, 21(11), 47–50, November 2001.

41. E. Bim, Fuzzy optimization for rotor constant identification of an indirect FOC induction motor drive, *IEEE Transactions on Industry Electronics*, 48, 1293–1295, December 2001.

42. E. Cerruto, A. Consoli, A. Raciti, A. Testa, Fuzzy adaptive vector control of induction motor drives, *IEEE Transactions on Power Electronics*, 12(6), 1028–1040, November 1997.

43. H. Kubota, K. Matsuse, Speed sensorless field-oriented control of induction motor with rotor resistance adaptation, *IEEE Transactions on Industry Electronics*, 30, 1219–1224, September/October 1994.

44. J. Faiz, M. B. B. Sharifian, Different techniques for real time estimation of an induction motor rotor resistance in sensorless direct torque control for electric vehicle, *IEEE Transactions on Energy Conversion*, 16, 104–109, March 2001.

45. K. Akatsu, A. Kawamura, Online rotor resistance estimation using the transient state under the speed sensorless control of induction motor, *IEEE Transactions on Power Electronics*, 15, 553–560, May 2000.

46. H. Kubota, D. Yoshihara, K. Matsuse, Rotor resistance adaptation for sensorless vector-controlled induction machines, *Electrical Engineering in Japan*, 125(2), 65–72, 1998.

47. R. Krishnan, A. S. Bharadwaj, A review of parameter sensitivity and adaptation in indirect vector controlled induction motor drive systems, *IEEE Transactions on Power Electronics*, 6, 695–703, October 1991.

48. L. Umanand, S. R. Bhat, Online estimation of stator resistance of an induction motor for speed control applications, *IEE Proceedings of Electric Power Applications*, 142(2), 97–103, 1995.

49. M. Tsuji, S. Chen, K. Izumi, E. Yamada, A sensorless vector control system for induction motors using q-axis flux with stator resistance identification, *IEEE Transactions on Industry Electronics*, 48, 185–194, February 2001.

50. R. Marino, S. Peresada, P. Tomei, On-line stator and rotor resistance estimation for induction motors, *IEEE Transactions on Control Systems Technology*, 8(3), 570–579, May 2000.

51. V. Vasic, S. Vukosavic, Robust MRAS-based algorithm for stator resistance and rotor speed identification, *IEEE Power Engineering Review*, 21(11), 39–41, November 2001.

52. K. Shinohara, T. Nagano, H. Arima, W. Z. W. Mustafa, Online tuning method of stator and rotor resistances in both motoring and re-generating operations for vector controlled induction machines, *Electrical Engineering in Japan*, 135(1), 56–64, 2001.

53. E. D. Mitronikas, A. N. Safacas, E. C. Tatakis, A new stator resistance tuning method for stator-flux-oriented vector-controlled induction motor drive, *IEEE Transactions on Industry Electronics*, 48, 1148–1157, December 2001.

54. L. A. Cabrera, E. Elbuluk, I. Husain, Tuning the stator resistance of induction motors using artificial neural network, *IEEE Transactions on Power Electronics*, 12, 779–787, September 1997.

55. B. K. Bose, N. R. Patel, Quazi-fuzzy estimation of stator resistance of induction motor, *IEEE Transactions on Power Electronics*, 13, 401–409, May 1998.

56. S. Mir, M. E. Elbuluk, D. S. Zinger, PI and fuzzy estimators for tuning the stator resistance in direct torque control of induction machines, *IEEE Transactions on Power Electronics*, 13, 279–287, March 1998.

57. T. G. Habetler, F. Profumo, G. Griva, M. Pastorelli, A. Bettini, Stator resistance tuning in a stator-flux field-oriented drive using an instantaneous hybrid flux estimator, *IEEE Transactions on Power Electronics*, 13, 125–133, January 1998.

58. M. Wang, E. Levi, M. Jovanovic, Compensation of parameter variation effects in sensorless indirect vector controlled induction machines using model based approach, *Electric Power Components and Systems*, 27(9), 1001–1027, 1999.

59. O. Touhami, R. Ibtiouen, S. Moulahoum, S. Mekhtoub, Analysis and compensation of parameter variation effect in vector controlled induction machine, *Archives of Electrical Engineering*, 50(2), 165–182, 2001.

60. H. A. Toliyat, E. Levi, M. Raina, A review of RFO induction motor parameter estimation techniques, *IEEE Transactions on Energy Conversion*, 18(2), 271–283, 2003.

61. H. A. Toliyat, E. Levi, M. Raina, A review of RFO induction motor parameter estimation techniques, *IEEE Power Engineering Review*, 22(7), 52, 2002.

62. M. S. Zaky, M. M. Khater, S. S. Shokralla, H. A. Yasin, Wide-speed-range estimation with online parameter identification schemes of sensorless induction motor drives, *IEEE Transactions on Power Electronics*, 56(5), 1699–1707, 2009.

63. M. Rashed, P. F. A. MacConnell, A. F. Stronach, Nonlinear adaptive state-feedback speed control of a voltage-fed induction motor with varying parameters, *IEEE Transactions on Industry Electronics*, 42(3), 723–732, 2006.

64. D. J. Atkinson, P. P. Acarnley, J. W. Finch, Observers for induction motor state and parameter estimation, *IEEE Transactions on Industry Electronics*, 27(6), 1119–1127, 1991.

65. S. Bolognani, L. Peretti, M. Zigliotto, Parameter sensitivity analysis of an improved open-loop speed estimate for induction motor drives, *IEEE Transactions on Power Electronics*, 23(4), 2127–2135, 2008.

66. D. P. Marcetic, S. N. Vukosavic, Speed-sensorless AC drives with the rotor time constant parameter update, *IEEE Transactions on Industry Electronics*, 54(5), 2618–2625, 2007.

67. C. Attaianese, G. Tomasso, A. Damiano, I. Marongiu, A. Perfetto, A novel approach to speed and parameters estimation in induction motor drives, *IEEE Transactions on Energy Conversion*, 14(4), 939–945, 1999.

68. P. Castaldi, A. Tilli, Parameter estimation of induction motor at standstill with magnetic flux monitoring, *IEEE Transactions on Control Systems Technology*, 13(3), 386–400, 2005.

69. J. L. Zamora, A. Garcia-Cerrada, Online estimation of the stator parameters in an induction motor using only voltage and current measurements, *IEEE Transactions on Industry Electronics*, 36(3), 805–816, 2000.

70. A. Ukil, R. Bloch, A. Andenna, Estimation of induction motor operating power factor from measured current and manufacturer data, *IEEE Transactions on Energy Conversion*, 26(2), 699–706, 2011.

71. E. Laroche, M. Boutayeb, Identification of the induction motor in sinusoidal mode, *IEEE Transactions on Energy Conversion*, 25(1), 11–19, 2010.

72. G. Kang, J. Kim, K. Nam, Parameter estimation scheme for low-speed linear induction motors having different leakage inductances, *IEEE Transactions on Industry Electronics*, 50(4), 708–716, 2003.

73. E. Laroche, E. Sedda, C. Durieu, Methodological insights for online estimation of induction motor parameters, *IEEE Transactions on Control Systems Technology*, 16(5), 1021–1028, 2008.

74. J. A. de Kock, F. S. van der Merwe, H. J. Vermeulen, Induction motor parameter estimation through an output error technique, *IEEE Transactions on Energy Conversion*, 9(1), 69–76, 1994.

75. Y.-S. Kwon, J.-H. Lee, S.-H. Moon, B.-K. Kwon, C.-H. Choi, J.-K. Seok, Standstill parameter identification of vector-controlled induction motors using the frequency characteristics of rotor bars, *IEEE Transactions on Industry Electronics*, 45(5), 1610–1618, 2009.

76. G. Kenné, R. S. Simo, F. Lamnabhi-Lagarrigue, A. Arzandé, J. C. Vannier, An online simplified rotor resistance estimator for induction motors, *IEEE Transactions on Control Systems Technology*, 18(5), 1188–1194, 2010.

77. T. Du, P. Vas, F. Stronach, Design and application of extended observers for joint state and parameter estimation in high-performance AC drives, *IEE Proceedings on Electric Power Applications*, 142(2), 71–78, 1995.

78. D. S. Raptis, A. G. Kladas, J. A. Tegopoulos, Accurate induction motor estimator based on magnetic field analysis, *IEEE Transactions on Magnetics*, 44(6), 1574–1577, 2008.

79. S. Rao, M. Buss, V. Utkin, Simultaneous state and parameter estimation in induction motors using first- and second-order sliding modes, *IEEE Transactions on Industry Electronics*, 56(9), 3369–3376, 2009.

80. J. Pedra, L. Sainz, Parameter estimation of squirrel-cage induction motors without torque measurements, *IEE Proceedings on Electric Power Applications*, 153(2), 263–270, 2006.

81. J. Holtz, H. Pan, Elimination of saturation effects in sensorless position controlled induction motors, *IEEE Transactions on Industry Electronics*, 40(2), 623–631, March/April 2004.

82. J. Holtz, H. Pan, Acquisition of rotor anisotropy signals in sensorless position control systems, *IEEE Industry Applications Society Annual Meeting*, Salt Lake City, UT, 2003.

83. W. Leonhard, G. Heinemann, Self-tuning field oriented control of an induction motor drive, *International Power Electronics Conference*, Tokyo, Japan, 1990.

84. C. Cecati, N. Rotondale, On-line identification of electrical parameters of the induction motor using RLS estimation, *Proceedings of the 24th Annual Conference of the IEEE Industry Electronics Society (IECON'98)*, Aachen, Germany, Vol. 4, pp. 2263–2268, August 31–September 4, 1998.

85. M. Velez-Reyes, K. Minami, G. C. Verghese, Recursive speed and parameter estimation for induction machines, *IEEE Industry Applications Society Annual Meeting*, San Diego, CA, 1989.

86. J. Stephan, M. Bodson, Real-time estimation of the parameters and fluxes of induction motors, *IEEE Transactions on Industry Electronics*, 30(3), 746–759, May/June 1994.

87. L. A. de S. Ribeiro, C. B. Jacobina, A. M. N. Lima, The influence of the slip and the speed in the parameter estimation of induction machines, *Power Electronics Specialists Conference (PESC'97)*, St. Louis, MO, June 1997.

88. C. Moons, B. De Moor, Parameter identification of induction motor drives, *Automatica*, 31(8), 1137, 1995.

89. M. Velez-Reyes, Speed and parameter estimation for induction motors, MSc Electrical Engineering dissertation, Massachusetts Institute of Technology (MIT), Boston, MA, May 1988.

90. M. Boussak, G. A. Capolino, Recursive least-squares rotor time-constant identification for vector-controlled induction machine, *Electric Machines and Power Systems*, 20(2), 137–147, 1992.

91. A. Bellini, A. De Carli, M. La Cava, Parameter identification for induction motor simulation, *Automatica*, 12, 383–386, 1976.

92. C. B. Jacobina, J. E. C. Filho, A. M. N. Lima, Estimating the parameters of induction machines at standstill, *IEEE Transactions on Energy Conversion*, 17(1), 85–89, March 2002.

93. L. Salvatore, S. Stasi, Application of EKF to parameter and state estimation of PMSM drive, *IEE Proceedings B Electric Power Applications*, 139(3), 155–164, 1992.

94. M. Pucci, Novel numerical techniques for the identification of induction motors for the control of AC drives: Simulations and experimental implementations (in Italian), PhD thesis, University of Palermo, Palermo, Italy, January 2001.

95. M. Cirrincione, M. Pucci, Identification of an induction motor with the least-squares method, Electrical Engineering Research Report, No. 10, pp. 22–30, December 2000.

96. G. Cirrincione, M. Cirrincione, J. Hérault, S. Van Huffel, The MCA EXIN neuron for the minor component analysis: Fundamentals and comparisons, *IEEE Transactions on Neural Networks*, 13(1), 160–187, January 2002.

97. G. Cirrincione, M. Cirrincione, Linear system identification by using the TLS EXIN neuron, *Neurocomputing*, 28(1–3), 53–74, October 1999.

98. G. Cirrincione, M. Cirrincione, *Neural-Based Orthogonal Data Fitting: The EXIN Neural Networks*, John Wiley & Sons, Inc., Hoboken, NJ, 2010.

99. M. Cirrincione, M. Pucci, G. Cirrincione, Estimation of the electrical parameters of an induction motor with the TLS EXIN neuron, *Proceedings of the 22nd International Conference on Distributed Computing Systems (ICDCS'02)*, Aruba, Dutch Caribbean, 2002.

100. M. Cirrincione, M. Pucci, G. Cirrincione, G. A. Capolino, A new experimental application of least-squares techniques for the estimation of the induction motor parameters, *IEEE Transactions on Industry Applications*, 39(5), 1247–1256, September/October 2003.

101. M. Cirrincione, M. Pucci, G. Cirrincione, G. Capolino, Constrained minimization for parameter estimation of induction motors in saturated and unsaturated conditions, *IEEE Transactions on Industrial Electronics*, 52(5), 1391–1402, October 2005.

102. M. Cirrincione, M. Pucci, Experimental verification of a technique for the real-time identification of induction motors based on the recursive least-squares, *International Workshop on Advanced Motion Control (IEEE AMC'02)*, Maribor, Slovenia, 2002.

103. M. Cirrincione, M. Pucci, G. Cirrincione, G. A. Capolino, Constrained least-squares method for the estimation of the electrical parameters of an induction motor, *COMPEL (The International Journal for Computation and Mathematics in Electrical and Electronic Engineering), Special Issue: Selected Papers from the International Conference on Electrical Machines (ICEM) 2002*, Bruges, Belgium, Vol. 22(4), pp. 1089–1101, 2003.

104. A. Ralston, P. Rabinowits, *A First Course in Numerical Analysis*, 2nd edn., International Student Edition, McGraw-Hill, New York, 1978.

105. M. Pucci, M. Cirrincione, Estimation of the electrical parameters of an induction motor in saturated and unsaturated conditions by use of the least-squares method, *IEEE International Aegean Conference on Electrical Machines and Power Electronics (IEEE ACEMP '01)*, Kusadasi, Turkey, 2001.

106. P. Guillame, R. Pintelon, J. Schoukens, A weighted total least squares estimator for multivariable systems with nearly maximum likelihood properties, *IEEE Transactions on Instrumentation and Measurement*, 47(4), 818–822, August 1998.

107. J. Verbeeck, R. Pintelon, P. Lataire, Identification of synchronous machine parameters using a multiple input multiple output approach, *IEEE Transactions on Energy Conversion*, 14(14), 909–917, December 1999.

108. N. Dedene, R. Pintelon, P. Lataire, Estimation of a global synchronous machine model using a multiple-input multiple-output estimator, *IEEE Transactions on Energy Conversion*, 18(1), 11–16, February 2003.

109. L. N. Trefethen, D. Bau, *Numerical Linear Algebra*, Society for Industrial and Applied Mathematics (SIAM), Philadelphia, PA, 1997.

110. S. Van Huffel, J. Vandewalle, Analysis of the total least squares problem and its use in parameter estimation, PhD thesis, Society for Industrial and Applied Mathematics (SIAM), Leuven, Belgium, 1987.

111. M. Cirrincione, M. Pucci, A rotor-flux-oriented vector control of an AC drive based on a recursive least-squares (RLS) method, Electrical Engineering Research Report, No. 9, pp. 11–22, July 2000.

112. M. Pucci, C. Serporta, A recursive least-squares method for real-time parameter estimation in adjustable-speed induction motor drives, Electrical Engineering Research Report, No. 10, pp. 31–37, December 2000.

113. M. Cirrincione, M. Pucci, A direct-torque-control of an AC drive based on a recursive-least-squares (RLS) method, *Proceedings of the IEEE International Symposium on Diagnostics for Electric Machines, Power Electronics and Drives (IEEE SDEMPED'01)*, Grado, Italy, 2001.

114. M. Cirrincione, M. Pucci, G. Vitale, A least-squares based methodology for estimating the electrical parameters of induction machine at standstill, *Proceedings of the IEEE International Symposium on Industrial Electronics (ISIE'02)*, L'Aquila, Italy, 2002.

10

Neural-Enhanced Single-Phase DG Systems with APF Capability

This chapter deals with the use of neural network (NN) to improve the performance of a distributed generation (DG) system with active power filtering (APF) capability. These two areas of power electronics have developed with recent advances in control system technology. As shown in Chapters 2 and 3, the same power circuit topology can be utilized for power control and active filtering as well. These functions can be performed by the same circuit with suitable NN control.

In particular, the inverter function is performed by a DG unit that, connected in parallel with the power grid, injects into the grid a current with phase and frequency equal to the corresponding ones of the grid voltage and with amplitude depending on the power available from the renewable sources; on the other hand, the APF unit injects system harmonic currents equal in amplitude to those of the loads but with opposite phase, thus keeping the line current almost sinusoidal. The role of NN consists in these applications in both the detection of the grid voltage fundamental and the computation of the load harmonic compensation current.

Two sections are present in this chapter. The former deals with a single-phase unit and the latter with a three-phase one.

10.1 Introduction

This chapter presents and discusses the results obtained experimentally on a 130 V RMS, 50 Hz single-phase power distribution network, which in itself contains the distorted currents arising from both the utility and a nonlinear load. Several concepts explained in the previous part of this book are utilized, such as the use of the current control inverter (Chapter 2), the shunt active power filter (Chapter 3), and the neural adaptive filter (Chapter 8). These concepts are exploited to set up an experimental test rig to verify the proposed methodology. Moreover, some issues such as the global stability analysis of the proposed control approach are discussed and shown in the discrete domain.

The literature proposes some examples of converters for DG also from renewable sources [1–4] and APFs [5–10]; the integration of both functions is presented in Refs [11–13] in which, starting from a DG unit connected to the grid, the additional operation of an active filter is implemented.

The correct DG connection to the grid depends strongly on the correct detection of the fundamental harmonic frequency and phase of the voltage at the coupling point. In principle, grid voltage should be sinusoidal, but, as explained in Section 3.1, difficulties due to the grid impedance and nonlinear loads arise. Moreover, the presence of the DG itself worsens the situation, increasing the coupling point voltage distortion.

As for the grid fundamental voltage extraction, two main strategies exist: the phase-locked loop (PLL) [14,15] and the zero crossing detection (ZCD) [13–15].

In Section 3.4.3.3, the control of APF based on instantaneous active-reactive (p-q) power theory has been explained; in literature, some solutions based on NN are proposed. Some of these are back propagation network (BPN) multilayer perceptrons like in Refs [16,17] that are trained off-line to learn main harmonic characteristics, then they are utilized on-line to obtain the switching pattern of the inverter. In general, for all BPN-based techniques, several well-known problems have to be accounted for, such as the choice of the number of neurons, the selection of the learning parameters, a meaningful training set and test set, the convergence time if it converges, the computational time, and so on. An alternative way to estimate the active component of the fundamental load current, obtained on-line by subtracting the compensating current component from the total current is proposed in Ref. [18], where a linear NN (ADALINE), trained by a least squares algorithm, is used.

This last approach is similar to the one used in this chapter, however, as shown in Section 7.6 of this book, the NN used for the system described in this chapter is less complex and less computationally demanding, since it is equivalent simply to a second-order digital filter.

The system described in this chapter utilizes twice the same neural adaptive filter based on linear neurons (ADALINEs); it is employed to compute the fundamental component of the grid voltage waveform for the grid connection and the overall harmonic component of the load current for the current harmonic compensation respectively. In the first configuration, the network allows the fundamental grid frequency voltage to be detected, suppressing other components, and is called "band" configuration. In the second, all current harmonic components are detected, and the fundamental is eliminated; it is called "notch" configuration. It should be noted that this approach does not need any a priori training of the NN, which adapts itself on-line [19–21].

Other interesting issues, discussed in the following, are the design criteria of the NN for the particular application, the control of the DG unit to deliver the available power from renewable sources, and the use of a parallel active filter (PAF) by multiresonant current control with resonant frequencies at the grid fundamental and its third, fifth, and seventh harmonics. Such a controller permits controlling in a decoupled way and with zero steady-state error the grid fundamental component of the inverter current, responsible for the power transfer to the grid, and its harmonics, responsible for the APF capability.

10.2 General Operating Principle

Figure 10.1 shows a block diagram of the neural-enhanced single-phase DG systems with APF capability; four main blocks are recognizable: the grid with its point of common coupling (PCC), the power unit that contains the inverter, the measurement units with all parameters to be acquired, and finally the microprocessor units that include the NN-based filters.

The grid is modeled by a sinusoidal generator with a series impedance. Various loads that could be both linear and nonlinear are connected. The PCC contains the inverter terminals connected to the power line and the current and voltage transducers. In particular, the grid current i_g, the inverter current i, the load current i_L, and the voltage v_{cp} are measured at the PCC.

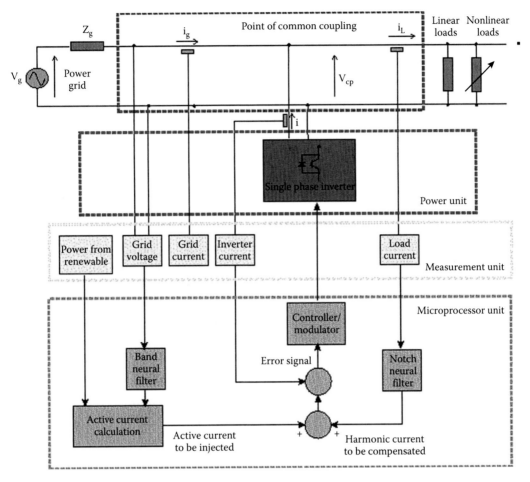

FIGURE 10.1
(See color insert.) Block diagram of the neural-enhanced single-phase DG systems with APF capability.

The power unit contains the single-phase H-bridge inverter with its interconnecting inductance. The measurement unit contains all transducers to allow the main parameters to be converted for a suitable scaled voltage to acquire. Finally, the microprocessor unit receives information by transducers; it calculates both the active current to be injected and the harmonic current to be compensated on the basis of the NN-based filter performance and outputs the signal to drive the inverter power devices.

10.3 ADALINE Design Criteria

The representation of the NN is sketched in Figure 10.2.

As described in Section 7.6, the network presents two inputs and two outputs. The first input is the signal to be processed; the second is a sinusoidal sequence with reference pulsation ω_0. The two outputs give the notch and the band behavior.

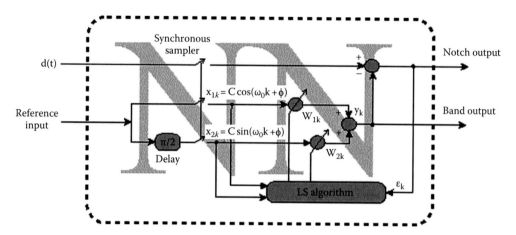

FIGURE 10.2
Schematic representation of the NN.

It should be noted that the neuron weights are adapted by a least mean squares (LMS) on-line training algorithm because of its low complexity, low computational demand, and high-speed convergence.

Moreover, this does not cause a significant increase of the computational demand and complexity of the filters, complying with the target of this work to develop a simple DG control system.

The reference input is delayed by $\pi/2$ to build the second sinusoidal reference, and the frequency of this reference input corresponds to the frequency of the primary input signal d_k that should be canceled or let pass where k is the current sampled time instant.

Update of the weights, using the LMS, is given by

$$\begin{cases} w_1(k+1) = w_1(k) + 2\mu\varepsilon(k)x_1(k) \\ w_2(k+1) = w_2(k) + 2\mu\varepsilon(k)x_2(k) \end{cases} \tag{10.1}$$

where
$w_i(k)$ is the weight of the ith neuron at the kth time sample
μ is the learning rate
$\varepsilon_i(k)$ is the difference between the primary input signal $d(k)$ and the band filter output $y(k)$; $\varepsilon(k)$ is also the notch filter output

The sampled reference inputs are given by

$$\begin{cases} x_1(k) = C\cos(k\omega_0 + \phi) \\ x_2(k) = C\sin(k\omega_0 + \phi) \end{cases} \tag{10.2}$$

where C is the amplitude of the sinusoidal sequence with reference frequency ω_0.

The network transfer function in z-domain has the following form:

$$\begin{cases} H(z) = \dfrac{z^2 - 2z\cos\omega_0 + 1}{z^2 - 2(1-\mu C^2)z\cos\omega_0 + 1 - 2\mu C^2} & \textit{notch filter transfer function} \\[4mm] K(z) = \dfrac{2\mu C^2(z\cos\omega_0 - 1)}{z^2 - 2(1-\mu C^2)z\cos\omega_0 + 1 - 2\mu C^2} & \textit{band filter transfer function} \end{cases} \tag{10.3}$$

The internal parameter μ has to be set to obtain a good trade-off between the bandwidth and the convergence speed; this is crucial for the network performance as a filter and for the overall stability of the system. As a matter of fact, a slow convergence, corresponding to a lower value of μ and a narrower band, introduces a delay that, in a feedback action, could be unacceptable.

In particular, the constraints for the harmonic voltage fundamental frequency extraction are as follows: the reference sequence pulsation has to be fixed to the nominal value of the grid pulsation and the bandwidth has to contain the slight variation permitted by standard to maintain that the filter looked even in the presence of grid frequency variation. On the other hand, the second harmonic and interharmonics are suppressed being out of the bandwidth of the filter.

In addition, filter stability considerations impose the upper limit of μ on the basis of the maximum eigenvalue of the autocorrelation matrix λ_{max} then: $1/\lambda_{max} > \mu > 0$.

The optimal choice of μ is different for notch and band operations, and is discussed in the following subsections.

10.3.1 Notch Operation

In this configuration, the NN filter is able to suppress a single frequency (or a very narrow band frequency) of the input signal. The frequency outside the band of the filter remains unchanged both in amplitude and phase. This function is used to eliminate the fundamental load current component and, consequently, to detect all current harmonic components.

If the stability of the filter is ensured, the following considerations for choosing the values of the parameter μ hold:

1. A narrow notch permits the compensation even of the low frequency load current harmonics; on the other hand, even if the grid frequency presents a slight shift with respect to the nominal value, in the range allowed by standards, it has to be eliminated as well.

2. A narrow notch permits the filter to have better phase characteristics around the notch frequency, with phase characteristics equal to 0 (zero delay time of the filter at all harmonics) and at frequencies very close to the notch one. In this way, there is no phase distortion of the load harmonic compensation current even at low frequency harmonics.

Figure 10.3 shows the Bode diagram of the notch filter for different values of the parameter μ. It should be noted that constraints are satisfied for $\mu = 8 \times 10^{-4}$ providing a bandwidth equal to 2 Hz (see Figure 10.3a) and exhibits a linear phase in the region where there are harmonics to be compensated (see Figure 10.3b). Two wrong choices are also plotted. It should be noted that for $\mu = 1 \times 10^{-4}$, the bandwidth is too narrow, and, for $\mu = 20 \times 10^{-4}$, the phase is different from zero in the range of current harmonics.

10.3.2 Band Operation

The band operation function is used to detect the fundamental grid frequency voltage by suppressing other harmonics and interharmonics coming from the grid.

Taking into consideration, as an example, the standard IEC 61727 [22], the frequency range for normal grid operation is 50 ± 1 Hz. As a consequence, the fundamental frequency is expected in the range from 49 to 51 Hz.

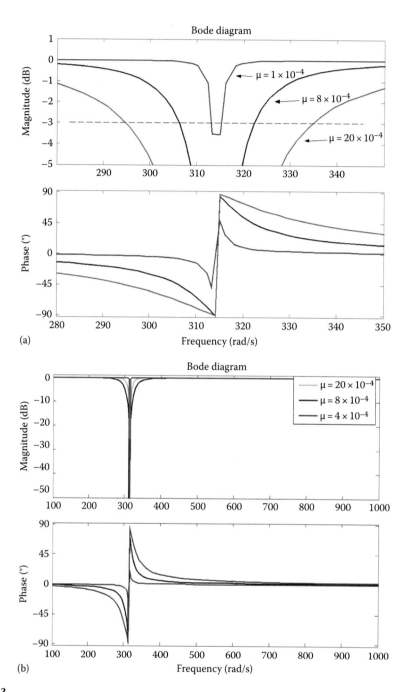

FIGURE 10.3
(a) Frequency response of the neural adaptive filter in notch configuration for different values of μ (zoom).
(b) Frequency response of the neural adaptive filter in notch configuration for different values of μ.

With regard to the band neural filter, μ can be chosen for a slightly wider band compared to the allowed grid fundamental frequencies for two reasons:

1. A slightly wider band of the filter permits it to converge quickly, which is very important for a fast connection of the DG to the grid.
2. A slightly wider band of the filter permits the DG to be connected properly to the grid even in the presence of small variations of the grid frequency. With this regard, choosing a wider band permits the filter to have better phase characteristics around the band frequency (close to 0), which is particularly important when the grid connection is to be made in the presence of a modified grid frequency. In this way, a grid frequency tracking system is not strictly needed (see the PLL in Ref. [15]).

The plot of the Bode diagram of the NN filter in band configuration is shown in Figure 10.4a and 10.4b. It should be noted that by adopting $\mu = 20 \times 10^{-4}$, a bandwidth from about 43–57 Hz is obtained. Moreover, the Bode diagram strongly decreases outside the filter bandwidth so as to suppress other harmonics. As example, it can be observed that the third harmonic has an attenuation greater than 20 dB.

10.3.3 MATLAB®–Simulink® Implementation

In Figure 10.5, the Simulink® implementation of the NN-based filter is shown. The on-line adaptation of two weights is implemented on the basis of (10.1). Firstly, the input sequence $x(k)$ is multiplied by the error $\varepsilon(k)$, and a gain equal to 2μ allows the term $(2\mu \, \varepsilon_k x_k)$ to be generated. This term is summed, inside the dotted square, to the last sampled weight (obtained by a unitary delay), and the new weight is obtained.

The sum of the weights is the band signal, and by subtracting it from the original signal, the notch signal is obtained.

10.3.4 Comparison with Traditional Digital Filters

The performance of the NN notch adaptive filter in the z-domain, described by Equation 10.3, is compared in this section with traditional digital notch filters of the same order, typically Butterworth and Chebyshev ones. For a sampling frequency of 15 kHz and $\mu = 2 \times 10^{-4}$, the NN gives a 6 Hz band centered on the notch at 50 Hz; the corresponding Chebyshev and Butterworth filters of the first order (with two poles and zeros like the NN filter in the z-domain) have been designed. With regard to the Chebyshev filter, the peak-to-peak ripple in the passband has been set to 0.5 dB, but it can be seen that its variation does not affect in a significant way the final transfer function. The resulting Butterworth filter Bode diagram practically matches the NN adaptive filter, while the Chebyshev one is quite different. Figure 10.6 shows the Bode diagrams for the three filters in a frequency range around the notch. The Chebyshev filter presents the worst ratio between filter attenuation at the notch and amplitude of the frequency band. The diagrams of the NN and the classical Butterworth filters are superimposed, as expected.

It could be claimed that the NN adaptive filter implements a Butterworth filter. However, the NN filter is clearly dependant on the notch frequency; this permits the filter coefficients to be varied in accordance with a variation of the notch frequency.

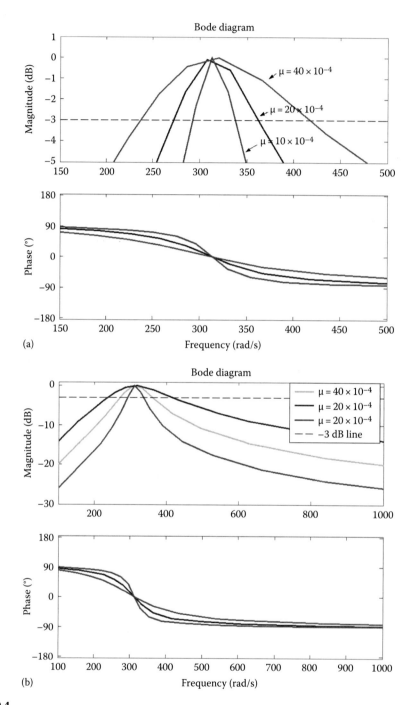

FIGURE 10.4

(a) Frequency response of the neural adaptive filter in band configuration for different values of μ (zoom).

(b) Frequency response of the neural adaptive filter in band configuration for different values of μ.

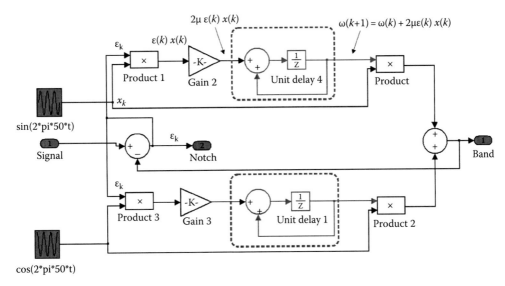

FIGURE 10.5
Block diagram of the NN-based filter Simulink® implementation.

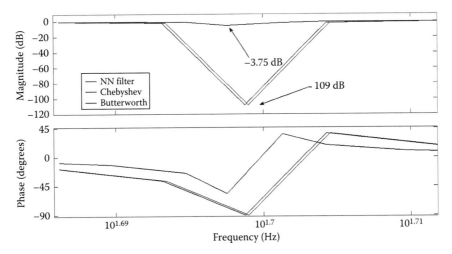

FIGURE 10.6
Bode diagrams of the NN adaptive filter, of the Butterworth filter, and of the Chebyshev filter. (From Cirrincione, M. et al., *IEEE Trans. Ind. Electron.*, 56(8), 3128, August 2009.)

More precisely, it can be said that NN behaves as an adaptive Butterworth filter. In the considered application, this kind of adaptation makes the NN filter robust to any possible shift of the grid frequency.

10.3.5 NN Band Filter versus PLL: Theoretical Comparison

A PLL solution has been implemented and compared with the proposed NN band filter to lock the grid fundamental frequency since PLL is the most classic solution adopted for grid connection of DGs. Figure 10.7 shows the block diagram of the PLL system in the continuous Laplace domain.

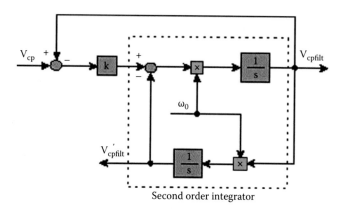

FIGURE 10.7
Block diagram of the PLL system in the continuous domain. (From Cirrincione, M. et al., *IEEE Trans. Ind. Electron.*, 55(5), 2093, May 2008.)

The corresponding transfer function is

$$H_{PLL}(s) = \frac{V_{cpfilt}(s)}{V_{cp}(s)} = \frac{k\omega_0 s}{s^2 + k\omega_0 s + \omega_0^2} \tag{10.4}$$

where
$V_{cpfilt}(s)$ and $V_{cp}(s)$ are the PLL output voltage and the voltage on PCC in the Laplace domain
ω_0 represents the resonance (band) frequency
k is a parameter that influences the width of the band and, consequently, the speed
   convergence of the PLL: a lower k corresponds to a narrower width of the band and to
   a slower convergence speed

In this case, k has been set to 0.1 on the basis of a trade-off between these aspects.
   In Figure 10.8, the frequency response of the PLL for different values of k is plotted.
   For digital implementation of the PLL, a second-order integrator discretization [15] has
been chosen, which is a modified Euler integrator obtained by approximating the integrator $1/s$ in the continuous domain with the discrete filter in the discrete z-domain, with T_{sp}
the sampling time of the control system:

$$\frac{T_{sp}}{2} \frac{3z-1}{z^2-z} \tag{10.5}$$

This choice permits a higher discrete integration accuracy with respect to the classic forward
Euler integrator. The transfer function of the PLL in the discrete domain z becomes

$$H_{PLL}(z) = \frac{V_{cpfilt}(z)}{V_{cp}(z)} = \frac{2kT_{sp}(3z^3 - 4z^2 + z)}{4z^4 - 8z^3 + \left(4 + 6kT_{sp}\omega_0 + 9\omega_0^2 T_{sp}^2\right)z^2 - \left(8kT_{sp}\omega_0 + 6\omega_0^2 T_{sp}^2\right)z + 2kT_{sp}\omega_0 + \omega_0^2 T_{sp}^2} \tag{10.6}$$

The PLL is thus a fourth-order digital band filter. Comparing Equation 10.6 with Equation
10.3, it could be noted that the neural adaptive band filter is a second-order filter, while the

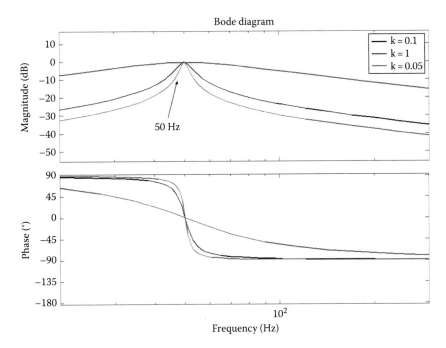

FIGURE 10.8
Frequency response of the PLL for different values of k. (From Cirrincione, M. et al., *IEEE Trans. Ind. Electron.*, 55(5), 2093, May 2008.)

PLL is a fourth-order one. This implies a lower complexity and computational demand for the NN filter. Moreover, the neural adaptive band filter has a faster convergence and a better connection capability than the PLL, when the grid frequency slightly changes with respect to the rated one.

10.4 Building the Current Reference

The desired current to be supplied by the grid is sinusoidal; as a consequence, all harmonic components different from fundamental ones have to be suppressed, and the reference i^*_{comp} will contain these terms with opposite phase. This function is performed by a notch neural filter (see Figure 10.1) that eliminates from the load current the fundamental current component.

On the basis of the power given by the renewable source, P_{ref}, the active current reference is calculated, and is added to the reference to supply active power to the grid. This reference component i^*_{act} is obtained as

$$i^*_{act} = \frac{2P_{ref}}{v^2_{cpfilt} + v'^2_{cpfilt}} v_{cpfilt} \tag{10.7}$$

where v_{cpfilt} is the output of the neural band adaptive filter; it contains only the grid voltage fundamental component. A fictitious voltage, v'_{cpfilt}, which is in quadrature with v_{cpfilt}, is obtained by shifting this last signal in time by a discrete time delay z^{-75} (75 samples at 15 kHz sampling frequency of the control system correspond to 5 ms, which is a quarter of period at 50 Hz). The denominator of Equation 10.7 is the square of the amplitude of the coupling point voltage. With such a scheme, the reference active current is generated to produce an amount of power equal to P_{ref} and is a sinusoid in steady-state, exactly in phase with the fundamental of the coupling point voltage v_{cp}: the DG therefore generates only active power with no reactive power exchanged with the grid. The availability of the grid harmonic voltage is crucial when a DG unit is connected close to a nonlinear current load. As a matter of fact, the coupling point voltage v_{cp} can be influenced by the voltage drop on the grid impedance of the load current i_L becoming distorted too; moreover, the DG can inject current harmonics into the same grid impedance, worsening the shape of the grid voltage.

10.5 Multiresonant Current Controller

The classical PI compensator expressed in s-domain as

$$H(s) = K_P + \frac{K_I}{s} \tag{10.8}$$

when used in an AC system, as in the case of the presented application, introduces a steady-state error both in amplitude and phase. To avoid this, a synchronous frame regulator that acts on a DC signal is necessary (see VOC in Chapter 2). As shown in Refs [19,23], a control network with the same DC control response of (10.8) but centered around the AC control frequency ω_0 has the form

$$G_r(s) = K_P + \frac{K_I s}{s^2 + \omega_0^2} \tag{10.9}$$

Equation 10.9 expresses a proportional gain summed to a resonant transfer function having a resonant frequency equal to ω_0 and an ideally infinite gain at the same frequency.

In the proposed system, different sinusoidal components of a variable frequency have to be controlled in a decoupled way. The first of is tuned at the grid frequency, and the others, at its third, fifth, and seventh harmonics. The controller tuned at the fundamental is used for the active power generation toward the grid, while the three harmonic controllers are used either for the load current harmonic compensation (APF capability set on) or to actively control to zero the first three odd harmonics of the injected current (APF capability set off).

For what was explained earlier, the resulting controller is a multiresonant controller composed of four resonant controllers; its transfer function controller in the continuous Laplace domain s is as follows:

$$G_{MR}(s) = \sum_{n=1,3,5,7} \left(K_{1n} + \frac{s K_{2n}}{s^2 + (n\omega_0)^2} \right) \tag{10.10}$$

where
  ω_0 is the resonant pulsation of the controller at the fundamental, tuned at the grid frequency
  K_{1n} and K_{2n} are the gain terms for the different harmonics of order n

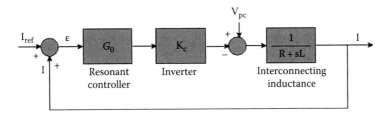

FIGURE 10.9
Block diagram of current control loop.

These terms can be calculated on the basis of the closed-loop transfer function of the current control loop. As a matter of fact, the current control loop can be schematically drawn as shown in Figure 10.9:

where
K_C is the pulsewidth modulated (PWM) inverter assuming its delay as negligible
L is the interconnecting grid inductance with its parasitic resistance R

The current open-loop transfer function is

$$G_O(s) = K_C \left[\sum_{n=1,3,5,7} \left(K_{1n} + \frac{sK_{2n}}{s^2 + (n\omega_0)^2} \right) \right] \left(\frac{1}{R+sL} \right) \tag{10.11}$$

which can be expressed as

$$G_O(s) = \left[\left(\frac{\sum_{m=0}^{8} c_m s^m}{\prod_{n=1,3,5,7} (s^2 + (n\omega_0)^2)} \right) \right] \left(\frac{1}{R+sL} \right) \tag{10.12}$$

Finally, the characteristic polynomial of the closed-loop transfer function has the form

$$G_O(s) = \left(\sum_{m=0}^{8} c_m s^m + \left(\prod_{n=1,3,5,7} (s^2 + (n\omega_0)^2) \right) (R_L + sL) \right) \tag{10.13}$$

It corresponds to a ninth-order s polynomial whose coefficients can be calculated by equating $G_0(s)$ with a Naslin polynomial of the same order [1]:

$$P_N(s) = a_0 \left(1 + \sum_{n=0}^{7} \frac{s^{n+1}}{\alpha^n \omega^{n+1}} \right) \tag{10.14}$$

The equivalent multiresonant controller expressed in the discrete domain z is

$$G_R(z) = \sum_{n=1,3,5,7} \left(\frac{(K_{1n} + K_{2n}T_{sp})z^2 - \cos(n\omega_0 T_{sp})(2K_{1n} + K_{2n}T_{sp})z + K_{1n}}{z^2 - 2\cos(n\omega_0 T_{sp})z + 1} \right) \tag{10.15}$$

where T_{sp} is the sampling time of the control system.

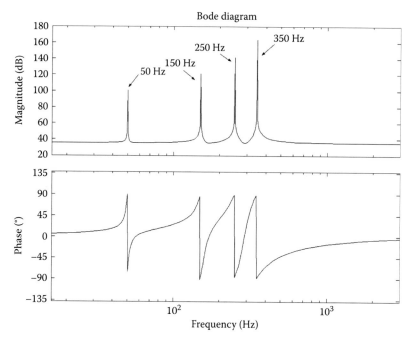

FIGURE 10.10

Frequency response of the designed multiresonant controller. (From Cirrincione, M. et al., *IEEE Trans. Ind. Electron.*, 55(5), 2093, May 2008.)

Figure 10.10 shows the frequency response of the designed multiresonant controller.

The output of the current controller is the voltage reference of the inverter, which is fed to the PWM block. In this case, a bipolar control of the voltage source inverter (VSI) is performed, and, therefore, the duty cycle d of a leg can be computed on the basis of the DC link voltage u_d and the reference voltage v_{ref} as

$$d = \frac{v_{ref}}{2u_d} + \frac{1}{2} \qquad (10.16)$$

10.6 Stability Issues

The stability analysis focusses on the pole position of the transfer function of the system, that is, the inverter/load current ratio. In particular, these poles could be subjected to variations due to the main system parameters. Among these, particular care has to be taken for the learning factor μ of the notch filter H(z); the filtering inductance L and the grid inductance L_g. This last has been considered because it cannot be chosen by the designer.

If all the inductances of the system are considered constant, which is a reasonable hypothesis, the system is linear. The voltage equations of the system in the Laplace domain are the following:

$$I(s) = \frac{1}{R + sL}(V_{inv}(s) - V_{cp}(s)) \qquad (10.17)$$

$$I_g(s) = \frac{1}{R_g + sL_g}(V_g(s) - V_{cp}(s)) \tag{10.18}$$

$$V_{cp}(s) = R_L(I(s) + I_g(s) - I_L(s)) \tag{10.19}$$

In the discrete domain z, the reference current can be written as

$$I^*(z) = H(z)I_L(z) + K(z)V_{cp}(z)$$

and the reference voltage v_{ref} provided to the PWM is

$$v_{ref}(z) = G_R(z)[I^*(z) - I(z)]$$

where $G_R(z)$ is defined by Equation 10.15. Finally, the inverter voltage $v_{inv}(z)$ is supposed to be equal to the reference $v_{ref}(z)$ provided to the PWM, delayed of one sampling time of the control system.

Figure 10.11 shows the block diagram of the entire system in the hybrid continuous/discrete domain. Actually the grid, the interconnecting inductance, and the load impedances are considered in the s-domain. The load current, the inverter current, and the voltage in the PCC are sampled and then they are processed in the z-domain. After processing, the inverter voltage returns in the s-domain by means of the zero-holder (ZOH) block.

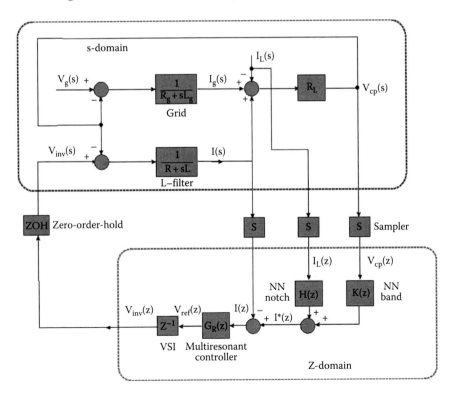

FIGURE 10.11
Block diagram of the entire system in the mixed continuous/discrete domain.

TABLE 10.1

Poles and Zeros of w(z)

	Zeros		Poles
z_1	−0.8366	$p_{1,2}$	0.1838 ± 0.7558i
$z_{2,3}$	0.9234 ± 0.2151i	$p_{3,4}$	0.9179 ± 0.2279i
$z_{4,5}$	0.9741 ± 0.1684i	$p_{5,6}$	0.9757 ± 0.1689i
z_6	0.8642	p_7	0.8667
$z_{7,8}$	0.9897 ± 0.0979i	$p_{8,9}$	0.9900 ± 0.0982i
$z_{9,10}$	0.9949 ± 0.0323i	$p_{10,11}$	0.9949 ± 0.0320i
$z_{11,12}$	0.9995 ± 0.0315i	$p_{12,13}$	0.9986 ± 0.0316i
$z_{13,14}$	0.9986 ± 0.0316i	$p_{14,15}$	0.9993 ± 0.0314i

Source: Cirrincione, M. et al., *IEEE Trans. Ind. Electron.*, 55(5), 2093, May 2008.

If Equations 10.17 through 10.19 are converted into the z-domain, the input-output transfer function $w(z) = I(z)/I_L(z)$ can be deduced. w(z) presents 15 poles and 14 zeros, as shown in Table 10.1.

All the poles of w(z), obtained with the rated parameters of the system given in Table 10.2, are inside the unit circle in the z-domain, and, therefore, the system is stable. Figure 10.12 shows the pole-zero map of $w(z)$.

Figure 10.13 shows the pole position for a given variation of μ. It can be observed that, for decreasing values of μ, the poles approach the unitary circle, getting closer to the position of the zeros, which lie exactly on the unit circle. This is reasonable since, for decreasing values of μ, the filter becomes closer to the ideal one. For increasing values of μ, the poles move far away from the unit circle. However, there is a upper limit for the values that can be given to μ as described in Section 10.3.1.

Figure 10.14 shows how the position of the poles varies for different values of L_g. It can be observed that for increasing values of L_g, the poles approach the unit circle and go outside of it for a value of L_g that is about as much as 10 times the rated value 1. This is an interesting result since, because of the grid protections operation in faulty conditions, the equivalent grid inductance can increase with potential instability of the DG-APF system.

Finally, Figure 10.15 shows the position of the poles for different values of L. It can be observed that for increasing values of L of the poles, the damping factor of the system reduces. On the contrary, a strong reduction of L leads the poles close to the unit circle and,

TABLE 10.2

Parameters of the Electrical Grid

L_g [mH]	20
R_g [Ω]	1
L [mH]	4
R [Ω]	0.2
V_g rms [V]	130
U_d [V]	250

Source: Cirrincione, M. et al., *IEEE Trans. Ind. Electron.*, 55(5), 2093, May 2008.

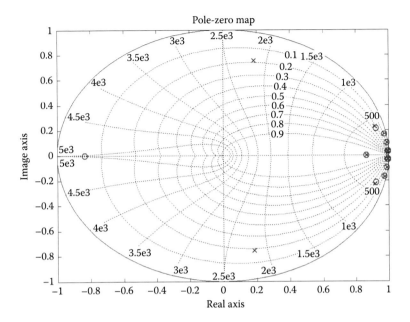

FIGURE 10.12
Pole-zero map of w(z). (From Cirrincione, M. et al., *IEEE Trans. Ind. Electron.*, 55(5), 2093, May 2008.)

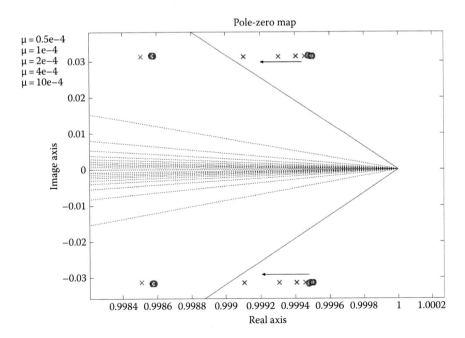

FIGURE 10.13
(See color insert.) Position of poles close to the unitary circle for different values of μ. (From Cirrincione, M. et al., *IEEE Trans. Ind. Electron.*, 55(5), 2093, May 2008.)

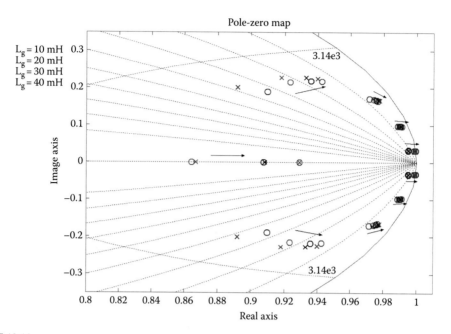

FIGURE 10.14
(See color insert.) Position of poles for different values of L_g (L_g increases from left to right). (From Cirrincione, M. et al., *IEEE Trans. Ind. Electron.*, 55(5), 2093, May 2008.)

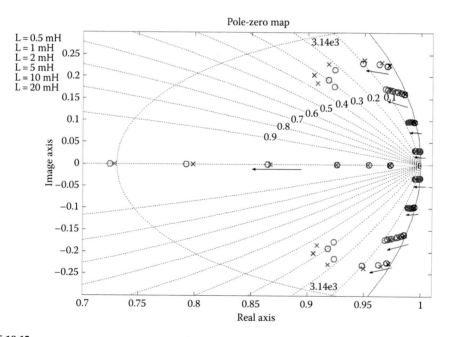

FIGURE 10.15
(See color insert.) Position of the poles for different values of L (L increases from right to left). (From Cirrincione, M. et al., *IEEE Trans. Ind. Electron.*, 55(5), 2093, May 2008.)

for very low values of L, can lead them outside the circle with consequent instability of the system. It should be recalled that each position of the poles has been computed here by considering a correct tuning of the multiresonant current controller for the corresponding value of L. Usually, once the L value is fixed by the designer, the permitted variation is due to aging and fabrication tolerance. Obviously, a completely different pole-zero pattern is to be expected if the controller is assumed to be tuned once for a precise value of L and then a variation of L is imposed to the system.

10.7 Test Rig

The proposed DG unit with APF capability has been tested on a single-phase power grid. The grid is represented by a sinusoidal voltage generator v_g while the grid inductance and resistance are respectively L_g and R_g; it supplies a linear resistive load with resistance equal to 50 Ω and a nonlinear current load, obtained with a diode bridge supplying a highly inductive load.

A renewable source (photovoltaic [PV], eolic, fuel cells, etc.) supplies with a DC voltage of $U_d = 250$ V a single-phase inverter. It is connected to the grid by a filtering inductance L (with parasitic resistance R).

The DG-APF needs, to properly work, the acquisition of the inverter current i, the load current i_L, and the voltage of the coupling point v_{cp}. These signals are acquired by a programmable hardware, a digital signal processing (DSP) in the case under study, which implements both the load current harmonic compensation and the grid connection features.

Figure 10.16 shows the electric scheme of the DG unit with APF capability, as implemented for the experimental verification of the proposed neural control scheme.

10.8 Experimental Results

10.8.1 APF Insertion

The single-phase sinusoidal electrical grid at 130 V rms, 50 Hz, sketched in Figure 10.16 with parameters of Table 10.2, has been considered for the experimental assessment of the methodology.

The nonlinear load current is obtained with a diode bridge supplying a highly inductive load, so the required current is almost a square wave. In addition, an auxiliary nonlinear load can be connected as an alternative to the previous one. This auxiliary load is obtained with a saturable transformer supplying a diode plus a resistance load. The current waveform required by it is a complex waveform with both odd and even harmonics.

The single-phase inverter is connected to the grid by a filtering inductance, which smoothes the inverter current waveform. The inverter employs four IGBT modules type Semikron SMK 50 GB 123 in an H-bridge connection with their proper drivers. Three current sensors type LEM LA 55-P are used to measure respectively the grid, the inverter, and the load current. A voltage sensor type LEM CV3-1000 has been used to measure the coupling point voltage. A sampling frequency of 15 kHz has been adopted both for the simulation and the experimental tests.

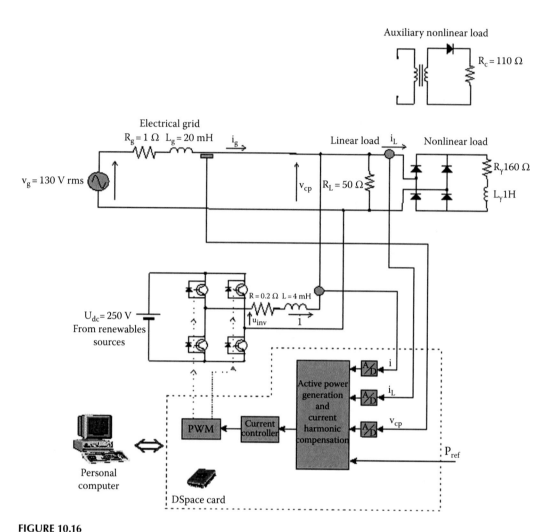

FIGURE 10.16
Electrical scheme of the DG unit with active filtering capability. (From Cirrincione, M. et al., *IEEE Trans. Ind. Electron.*, 55(5), 2093, May 2008.)

Figure 10.17 shows on the left the steady-state time waveform of the current required by the nonlinear current load and its harmonic content obtained with the FFT (fast Fourier transform). It is a quasi-square waveform, presenting only odd harmonics with amplitude decreasing with inverse proportionality with the frequency, whose %THD (total harmonic distortion) computed up to the 20th harmonics is 25.2%. On the right side of Figure 10.17, there is the steady-state time waveform of the current required by the auxiliary nonlinear current load and its harmonic content. It can be noted that also even harmonics are present, and the corresponding %THD is equal to 43%.

Two kinds of transients have been studied in the proposed system. The first consists in the sudden insertion of the active filtering capability (APF), maintaining an active power reference equal to 0. It corresponds to the operation of a pure shunt power APF.

The second transient occurs when the APF is working and additional power is available from the renewable source. In this case, the generation of active power to supply the grid is required in addition to APF performance.

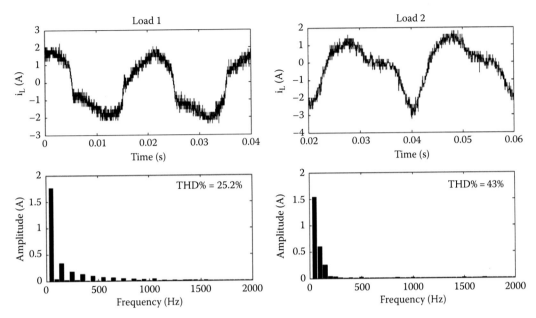

FIGURE 10.17
Measured current waveform required by the nonlinear load and its spectrum (left) and by the auxiliary nonlinear load (right). (From Cirrincione, M. et al., *IEEE Trans. Ind. Electron.*, 56(8), 3128, August 2009.)

Both of these transient conditions have been reproduced, in particular in the second, a step reference of active power of 600 W when the DG was already working in active filtering mode.

The test conditions are particularly challenging for the nonlinear load and for step variation of reference signal corresponding to the sudden start of the APF and to the sudden power availability from renewable sources.

Figure 10.18 shows the effects of the sudden insertion of the APF on the grid, with null active power reference, obtained experimentally with the DG-APF based on neural adaptive filtering. The APF is operated at t = 0; before this time, the grid current is equal to that required by the nonlinear load. In particular, Figure 10.18 shows the inverter current waveforms (reference and measured), the load and grid current and finally the coupling point voltage (measured and filtered by the NN band filter) and its fictitious quadrature component, and the reference voltage provided to the PWM. It can be observed that, after the APF insertion, the inverter current rapidly reaches the reference (in few cycles of the fundamental), while the load grid current, initially equal to the load one, becomes close to a sinusoid. At the same time, the coupling point voltage becomes slightly more noisy after the APF insertion, as expected, due to the current injected by the inverter, and the NN filter correctly extracts its fundamental. Figure 10.19 shows the steady-state waveform of the inverter current and its spectrum, obtained with the FFT, at the end of the transient shown in Figure 10.18. In APF mode, the inverter generates a current with a harmonic content that presents a high third harmonic of the grid fundamental (see Figure 10.19) controlled by the corresponding resonant controller plus some undesired multiples of it. Figure 10.20 shows the steady-state grid current and its spectrum, obtained with the FFT, at the end of the transient in Figure 10.18. It shows that the grid current, after the insertion of the APF, is basically a sinusoid: the third, fifth, and seventh harmonics are correctly eliminated by the multiresonant controller driven by the neural notch filter, and the first meaningful

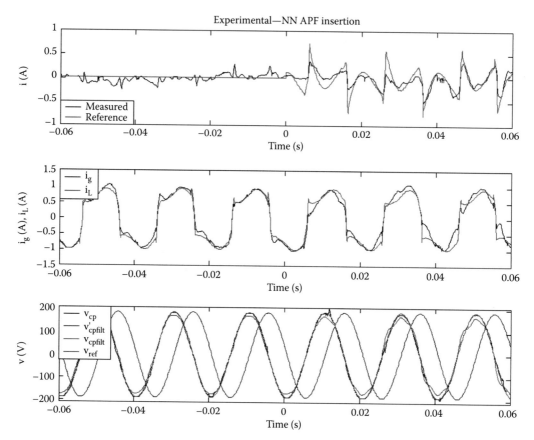

FIGURE 10.18

Measured grid current, inverter current, load current, and coupling point voltage during the transient of the insertion of APF with the DG-APF based on neural adaptive filtering. (From Cirrincione, M. et al., *IEEE Trans. Ind. Electron.*, 55(5), 2093, May 2008.)

harmonic is the ninth, which, even if not controlled, has an amplitude that is less than 4% of the fundamental. Furthermore, the %THD of the grid current is 4.51%.

10.8.2 Power Reference Insertion

In this test, during the APF operation, a step power reference is given, as it happens when energy is suddenly available from a renewable source. A step of 600 W has been chosen. Both linear (R_l) and nonlinear loads are connected to the power grid.

The effects of this step power reference, given at t = 0, are shown in Figure 10.21. In particular, the following parameters are included: the inverter current waveforms (reference and measured), the load and grid current and finally the coupling point voltage (measured and filtered by the NN band filter) and its fictitious quadrature component, and finally the reference voltage provided to the PWM. The top diagram contains the measured current (i), the reference active current due to power step (i_{act}^*), the reference current due to the APF operation (i_{comp}^*), and the sum of the last two terms ($i^* = i_{act}^* + i_{comp}^*$). After the power step, the active current reference component rises; it is in phase with the filtered coupling point voltage because of the action of the NN-based filter. The inverter current rapidly converges to the reference one in about three cycles of the grid fundamental and without any overshoot.

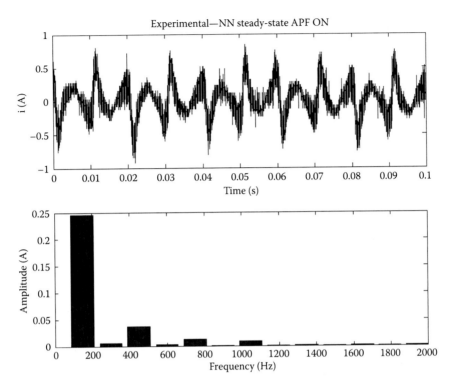

FIGURE 10.19
Measured steady-state inverter current with APF on with the DG-APF based on neural adaptive filtering. (From Cirrincione, M. et al., *IEEE Trans. Ind. Electron.*, 55(5), 2093, May 2008.)

The middle diagram contains the grid current (i_g), the load current (i_L), and again the inverter current (i). After the power step, the grid current changes its sign because the amount of power generated by the DG is higher than the sum of that required both by the linear and nonlinear loads. The inverter and grid currents present correspondingly in steady-state opposite signs.

Finally, the bottom diagram shows the voltage and, in particular, voltage in the PCC (V_{cp}), the corresponding filtered one (V_{cpfilt}) and delayed of $\pi/4$ (V'_{cpfilt}), and the reference voltage (V_{ref}). The reference voltage provided to the PWM is lower than the coupling point voltage before the power step and then, when it operates as DG, it becomes higher.

Figure 10.22 shows the steady-state time waveform of the inverter current and its spectrum obtained with the FFT at the end of the transient in Figure 10.21. The aim of this picture is to show that the inverter current is basically a sinusoid at 50 Hz, apart from the third, fifth, and seventh harmonics that are exactly equal to the corresponding harmonic components of the load current (before AFP operation the current was that shown in Figure 10.17), but now they are generated by the inverter, because the APF capability is activated. If the APF capability were set off, the multiresonant controller would have zero references for these harmonics, which would therefore be actively controlled to zero. The resulting %THD is equal to 4.74%.

10.8.3 Load Variation

The aim of this section is to verify the proposed system behavior under sudden variations of the nonlinear load, in order to test the behavior of the neural adaptive filter. Two tests

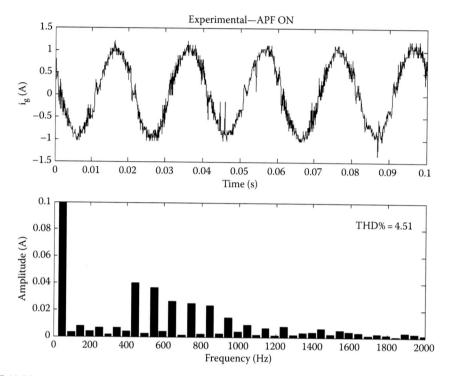

FIGURE 10.20
Measured steady-state grid current with APF on with the DG-APF based on neural adaptive filtering. (From Cirrincione, M. et al., *IEEE Trans. Ind. Electron.*, 55(5), 2093, May 2008.)

have been done in numerical simulation. The first is a step variation of the load from 100% to 200% (where 100% corresponds to the load in Figure 10.16), the second is a step variation of the load from 0% to 100%. Figure 10.23 shows on the left the effects of 100%–200% transition of load and on the right the effects of 0%–100% transition of load. In the top diagrams, the reference and measured inverter current are shown; in the middle, the grid current; and in the bottom, the load current.

It can be observed that, for the 100%–200% load transition, the load system rapidly converges to the new system configuration in almost one period of the grid frequency. For the 0%–100% load transition, the transient is longer (about six periods of the grid frequency). This is caused by the convergence of both the multiresonant controller and the notch adaptive filter. It should be noted that this last test is done in more difficult working conditions, since initially the inverter is connected to the grid without exchanging any power or compensating any load harmonic currents.

10.8.4 NN Filter versus PLL

Finally, in order to compare correctly the grid connection capability of the NN band adaptive filter with that of the PLL, some additional tests in particularly challenging conditions have been performed. The numerical tests have been done in three different working conditions.

The first test has the aim to compare the tracking speed of the grid voltage (transient response), and results are shown in Figure 10.24 upper plot that contains the coupling

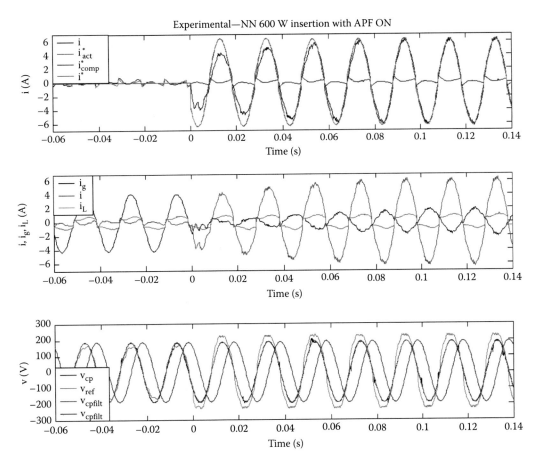

FIGURE 10.21
(See color insert.) Grid current, inverter current, load current, and coupling point voltage during the step transient of P_{ref} = 600 W with APF on with the DG-APF based on neural adaptive filtering (experiment). (From Cirrincione, M. et al., *IEEE Trans. Ind. Electron.*, 55(5), 2093, May 2008.)

point sinusoidal voltage and its tracking by the output of both the NN band filter and the PLL. It shows that the NN band filter exhibits a faster convergence toward the grid voltage than the PLL.

The second test shows the behavior of NN filter and PLL when the voltage is distorted (33% of the third harmonic is present) at rated frequency of 50 Hz. The corresponding results are in the middle diagram of Figure 10.24. It shows that both the NN filter and the PLL work properly, extracting correctly (in magnitude and in phase) the fundamental component of v_{cp}.

In the third test, the grid voltage v_{cp} is distorted (33% of the third harmonic is present), but, in addition, frequency variation (45 Hz) occurs with the NN filter and the PLL tuned at the grid rated frequency (50 Hz); it represents a very challenging situation, and is shown in the bottom plot of Figure 10.24. The results show that, in presence of a grid frequency shift from the fundamental, the NN band filter has a better behavior of the estimated fundamental grid voltage than the PLL. The PLL presents a higher phase shift and amplitude error than the NN filter. Neither the NN filter nor the PLL have a frequency tracking method which would further increase the complexity and computational demand.

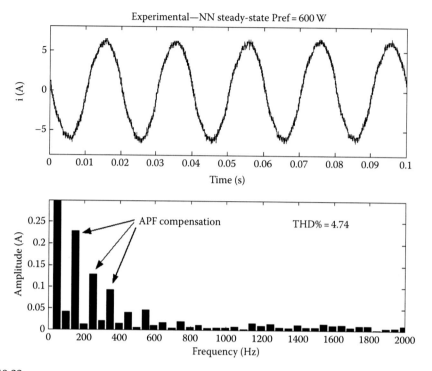

FIGURE 10.22

Steady-state inverter current with $P_{ref} = 600\,W$ with APF on with the DG-APF based on neural adaptive filtering (experiment). (From Cirrincione, M. et al., *IEEE Trans. Ind. Electron.*, 55(5), 2093, May 2008.)

10.8.5 NN Filter versus p-q Theory

In this case, an auxiliary nonlinear load is used as an alternative to the previously described one. This auxiliary load is obtained with a saturable transformer supplying a diode plus a resistance load. The current waveform required by it is a complex waveform with both odd and even harmonics. The p-q theory–based algorithm has been implemented experimentally on the dSPACE DS1103 board which also hosts the NN adaptive filter–based control. A sampling frequency of 15 kHz has been adopted for the entire control system in both cases.

The performed tests consist in the insertion test of APF with the nonlinear load and with the auxiliary nonlinear load using firstly the p-q theory–based algorithm and secondly the NN adaptive filter–based control.

Figure 10.25 shows the reference current (i^{*}_{comp}), the measured inverter current (i), the load current (i_L), and the grid current (i_g) during the APF insertion, obtained with the NN APF control under nonlinear load. As it can be seen, when at the beginning the APF is in parallel with the grid but no current harmonic compensation occurs, the corresponding inverter current is controlled to zero and the grid current coincides with the load one. After about 40 ms, the APF is operated and begins to correctly compensate load current harmonics, maintaining the grid current almost sinusoidal, as expected.

In Figure 10.26, the same test is performed by connecting the auxiliary nonlinear load instead of the nonlinear load used in the previous test. Also in this case, the APF works correctly, compensating properly the grid harmonic currents. The NN adaptive filter very

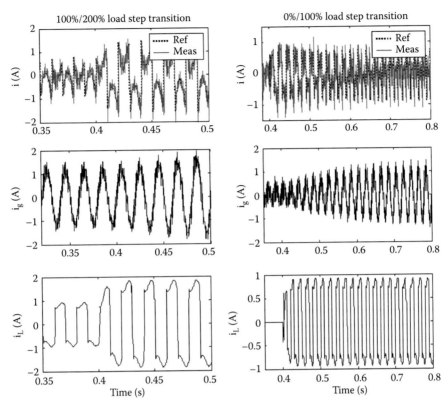

FIGURE 10.23

i, i_g, and i_L under two load step variations, 0%–100% and 100%–200% (simulation). (From Cirrincione, M. et al., *IEEE Trans. Ind. Electron.*, 55(5), 2093, May 2008.)

quickly adapts itself, giving the correct load harmonics estimation. Correspondingly, the multiresonant controller quickly controls the inverter reference current to its reference.

Figures 10.27 and 10.28 show the same tests performed under the same nonlinear loads but obtained with the classic p-q theory control. The comparison shows no appreciable difference in the performance in the time domain analysis; it means that both the NN adaptive filter–based control and the p-q theory–based control correctly work, permitting a full compensation of the grid current harmonics.

10.8.6 Compliance with International Standards

The results obtained with the NN-based DG-APF have been compared to the prescribed limits of the American and European International Standards. In particular, the IEEE standard 1547-2003 "IEEE Standard for interconnecting distributed resources with electric power systems" [24], the European standard CEI/EN61727 "Characteristics of the utility interface" [22], and the CEI/EN61000-3-2 "Limits for harmonic current emissions (equipment input current ≤ 16 A per phase)" [25] have been taken into account.

The IEEE standard 1547-2003 "IEEE Standard for interconnecting distributed resources with electric power systems" claims that the output of any renewable source–based system at the PCC should comply with clause 10 of IEEE standard 519-1992 in which the total harmonic current distortion (TDD: total demand distortion, harmonic current distortion in % of

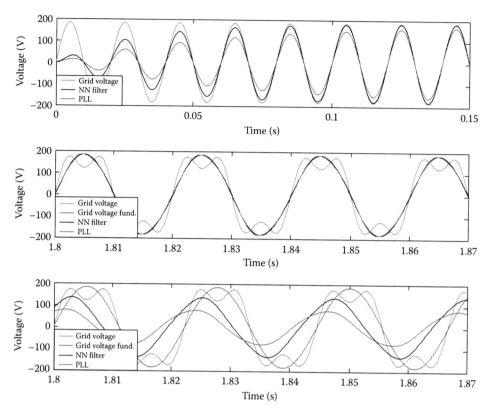

FIGURE 10.24
Steady-state grid voltage, NN band filter, and PLL output under different working conditions (simulation). (From Cirrincione, M. et al., *IEEE Trans. Ind. Electron.*, 55(5), 2093, May 2008.)

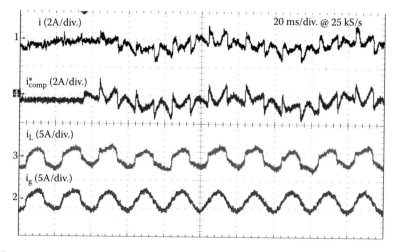

FIGURE 10.25
i^*_{comp}, i, i_L, and i_g during APF insertion with NN control under nonlinear load (experiment). (From Cirrincione, M. et al., *IEEE Trans. Ind. Electron.*, 56(8), 3128, August 2009.)

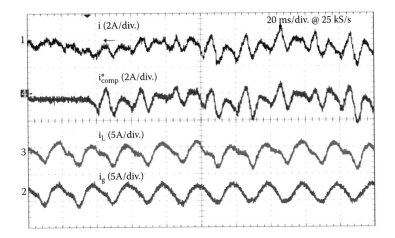

FIGURE 10.26
i^*_{comp}, i, i_L, and i_g during APF insertion with NN control under auxiliary nonlinear load (experiment). (From Cirrincione, M. et al., *IEEE Trans. Ind. Electron.*, 56(8), 3128, August 2009.)

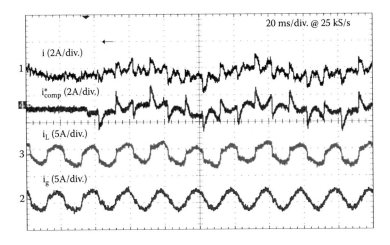

FIGURE 10.27
i^*_{comp}, i, i_L, and i_g during APF insertion with p-q theory–based control under nonlinear load (experiment). (From Cirrincione, M. et al., *IEEE Trans. Ind. Electron.*, 56(8), 3128, August 2009.)

maximum demand load current [15 or 30 min demand]) shall be less than 5% of the fundamental frequency current at full system output. The limits are specified as a percentage of the fundamental frequency current at full system output. It permits considering the whole effect of the DG and the nonlinear load (EPS, electrical power system), as seen from the PCC (red dotted line in Figure 10.29). Under this hypothesis, the harmonic content of i_g is to be analyzed.

The European standard CEI/EN61727 "Characteristics of the utility interface" addresses the interface requirements between PV systems and the utility, and provides technical recommendations. It has been written for PV systems, but its philosophy can be extended to any renewable source–based DG. About power quality and in particular harmonics, the standard claims that low levels of current and voltage harmonics are desirable. Suggested design targets harmonic limits are only in terms of current THD (limit 5%), while no limit

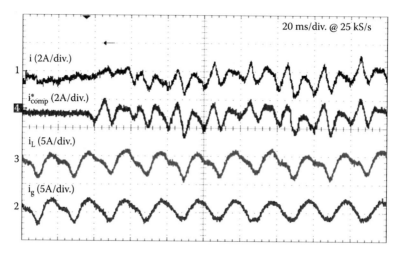

FIGURE 10.28
i^*_{comp}, i, i_L, and i_g during APF insertion p-q theory–based control under auxiliary nonload (experiment). (From Cirrincione, M. et al., *IEEE Trans. Ind. Electron.*, 56(8), 3128, August 2009.)

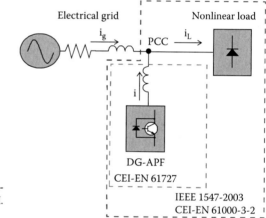

FIGURE 10.29
Schematics of the DG-APF with reference to international standard compliance. (From Cirrincione, M. et al., *IEEE Trans. Ind. Electron.*, 55(5), 2093, May 2008.)

on each single harmonic is provided. Referring to this standard, the only DG-APF has to be considered (blue dotted line in Figure 10.29).

Finally, the also the standard CEI/EN61000-3-2 "Limits for harmonic current emissions (equipment input current ≤ 16 A per phase)" [25] has been considered. Even if the standard has been devised for loads, in the case under study, the DG-APF plus the nonlinear load have considered an EPS seen from the PCC. As a matter of fact, the nonlinear load under study, considered alone, would be highly beyond the limits. Absolute values of amplitude for each harmonic up to 40th are specified. Since this standard refers to the maximum current of 16 A, the same value has been adopted as reference value also in the IEEE standard to define the customer's average maximum demand load current.

With regard to the %THD limits, [24] and [25] impose a limit of 5% on the grid current i_g, while [30] imposes a limit of 5% on inverter current i. Figures 10.20 and 10.22 show that in both cases, the THD limits are respected. It should be borne in mind, however, that the injected current i in Figure 10.22 includes also the third, fifth, and seventh harmonics that

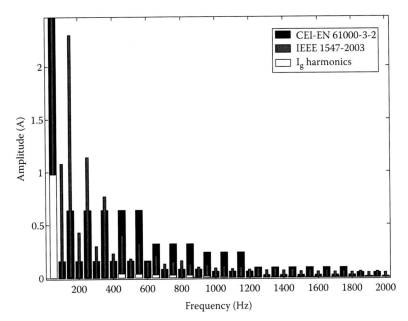

FIGURE 10.30
Harmonic content of grid current and limits prescribed by the international standards. (From Cirrincione, M. et al., *IEEE Trans. Ind. Electron.*, 55(5), 2093, May 2008.)

are generated for load current harmonic compensation. If no load harmonic compensation were performed, the %THD would be even lower.

With regard to the harmonic-by-harmonic analysis, Figure 10.30 shows the limits prescribed by the standards [29,31] as well as the harmonic content of the current i_g. It can be observed that the proposed system complies with both the American and European standards.

10.9 APF Connection Procedure

The connection of the APF to the power grid is a critical issue. To deal with this procedure, the experimental rig described in Section 10.7 is equipped with one hardware (a switch) and two software commands (selected by the user on the virtual instrument on the PC). For the sake of simplicity, these commands are not illustrated in Figure 10.16. The hardware command acts directly on the IGBT driver signals and has the task to enable/disable the devices driving. It is necessary, since the inverter is driven with a bipolar modulation, to avoid that, in absence of modulation, two devices are always actively commanded in on state. The first software command corresponds to the standby/run selection (output of the modulator respectively 0 or 1). The second corresponds to the enabling/disabling of the APF feature. The parallel connection of the APF is therefore performed with these steps:

1. Connection of the nonlinear load to the power grid
2. DC link voltage increase to the rated value
3. Commutation of the hardware command from disable to enable mode (physical connection of the inverter with the grid without any control)

4. Commutation of the software command 1 from standby to run (APF in parallel with the grid but current compensation not still enabled)

5. Commutation of the software command 2 from disable to enable (APF current compensation working)

References

1. M. P. Kazmierkowski, R. Krishnan, F. Blaabjerg, *Control in Power Electronics*, Academic Press, London, U.K., 2002.
2. F. Blaabjerg, F. Lov, R. Teodorescu, Z. Chen, Power electronics in renewable energy systems, *Proceedings of the 12th International Power Electronics and Motion Control Conference (EPE-PEMC'06)*, Portorož, Slovenia, pp. 1–17, August 2006.
3. F. Blaabjerg, A. Consoli, J. A. Ferreira, J. J. D. van Wyk, The future of electronic power processing and conversion, *IEEE Transactions on Power Electronics*, 20(3), 715–720, May 2005.
4. F. Blaabjerg, R. Teodorescu, M. Liserre, A. V. Timbus, Overview of control and grid synchronization for distributed power generation systems, *IEEE Transactions on Industrial Electronics*, 53(5), 1398–1409, October 2006.
5. S. A. Gonzalez, R. Garcia-Retegui, M. Benedetti, Harmonic computation technique suitable for active power filters, *IEEE Transactions on Industrial Electronics*, 54(5), 2791–2796, October 2007.
6. K. M. Cho, W. S. Oh, Y. T. Kim, H. J. Kim, A new switching strategy for pulse width modulation (PWM) power converters, *IEEE Transactions on Industrial Electronics*, 54(1), 330–337, February 2007.
7. D. O. Abdeslam, P. Wira, J. Merckle, D. Flieller, Y.-A. Chapuis, A unified artificial neural network architecture for active power filters, *IEEE Transactions on Industrial Electronics*, 54(1), 61–76, February 2007.
8. J. R. Vazquez, P. Salmeron, Active power filter control using neural network technologies, *IEE Proceedings on Electric Power Applications*, 150(2), 139–145, March 2003.
9. A. Zouidi, F. Fnaiech, K. Al-Haddad, Neural network controlled three-phase three-wire shunt active power filter, *IEEE International Symposium on Industrial Electronics (ISIE 2006)*, Montreal, Quebec, Canada, Vol. 1, pp. 5–10, July 2006.
10. A. Elmitwally, S. Abdelkader, M. El-Kateb, Neural network controlled three-phase four-wire shunt active power filter, *IEE Proceedings on Generation, Transmission and Distribution*, 147(2), 87–92, March 2000.
11. W. Tsai-Fu, S. Chih-Lung, H. Chien-Chang, C. Jeiyang, 1/spl phi/ 3W grid-connection PV power inverter with partial active power filter, *IEEE Transactions on Aerospace and Electronic Systems*, 39(2), 635–646, April 2003.
12. W. Tsai-Fu, S. Chih-Lung, N. Hung-Shou, L. Guang-Feng, A 1/spl phi/3W inverter with grid connection and active power filtering based on nonlinear programming and fast-zero-phase detection algorithm, *IEEE Transactions on Power Electronics*, 20(1), 218–226, January 2005.
13. T.-F. Wu, H.-S. Nei, C.-L. Shen, G.-F. Li, A single-phase two-wire grid-connection PV inverter with active power filtering and nonlinear inductance consideration, *Proceedings of the 19th Annual IEEE Applied Power Electronics Conference and Exposition (APEC 2004)*, Anaheim, CA, Vol. 3, pp. 1566–1571, 2004.
14. G. Hsieh, J. C. Hung, Phase-locked loop techniques: A survey, *IEEE Transactions on Industrial Electronics*, 43(6), 609–615, December 1996 (invited paper).
15. M. Ciobotaru, R. Teodorescu, F. Blaabjerg, A new single-phase PLL structure based on second order generalized integrator, *Proceedings of the 37th IEEE Power Electronics Specialists Conference (PESC'06)*, Jeju, South Korea, pp. 1–6, June 18–22, 2006.

16. M. Rukonuzzaman, M. Nakaoka, Single-phase shunt active power filter with adaptive neural network method for determining compensating current, *Proceedings of the 27th Annual Conference of the IEEE Industrial Electronics Society (IECON'01)*, Denver, CO, Vol. 3, pp. 2032–2037, November 29–December 2, 2001.

17. C. Sharmeela, M. R. Mohan., G. Uma, Line harmonics reduction using neural based controller for shunt active filters, *Conference on Convergent Technologies for Asia-Pacific Region (TENCON 2003)*, Bangalore, India, Vol. 4, pp. 1554–1557, October 15–17, 2003.

18. D. Gao, X. Sun, A shunt active power filter with control method based on neural network, *Proceedings of the PowerCon 2000 International Conference on Power System Technology*, Perth, WA, Vol. 3, pp. 1619–1624, December 4–7, 2000.

19. M. Cirrincione, M. Pucci, G. Vitale, A single-phase DG generation unit with shunt active power filter capability by adaptive neural filtering, *IEEE Transactions on Industrial Electronics*, 55(5), 2093–2110, May 2008.

20. M. Cirrincione, M. Pucci, G. Vitale, A single-phase shunt active power filter for current harmonic compensation by adaptive neural filtering, *European Power Electronics and Drives Journal*, 19(1), 40–49, March 2009.

21. M. Cirrincione, M. Pucci, G. Vitale, A. Miraoui, Current harmonic compensation by a single-phase shunt active power filter controlled by adaptive neural filtering, *IEEE Transactions on Industrial Electronics*, 56(8), 3128–3143, August 2009.

22. International Standard IEC 61727, Characteristics of the utility interface, 1997.

23. D. N. Zmood, D. G. Holmes, G. H. Bode, Frequency-domain analysis of three-phase linear current regulators, *IEEE Transactions on Industry Applications*, 37(2), 601–610, March/April 2001.

24. IEEE standard 1547-2003 IEEE standard for interconnecting distributed resources with electric power systems, Institute of Electrical and Electronics Engineers, New York, 2003.

25. International Standard IEC 61000-3-2, Limits for harmonic current emissions (equipment input current $\leq 16\,A$ per phase), 1998.

11

Neural Sensorless Control of AC Drives

11.1 NN-Based Sensorless Control

Sensorless control of AC motor drives is a huge topic, which has been faced up to by plenty of researchers [1–6]. Among the various techniques proposed by literature, many of which have been described in Chapter 6, a possible approach is the adoption of *artificial intelligence* (AI)–based techniques. As a result, an electrical drive can assume an intelligent behavior, meaning that it can embed some features of "learning," "self-organizing," or "self-adapting" [7–12]. Besides other advantages, the integration of AI in electrical adjustable speed drives can frequently lead to a reduction of the development times and can avoid mathematical difficulties in system development, since AI techniques in general do not require the mathematical model of system. With specific regard to automatic control aspects, generally speaking, the correct knowledge of the system (plant) model to be controlled is usually needed, typically expressed in terms of algebraic or differential equations, providing the system input–output relationship. These models can frequently be complex, rely on unrealistic assumptions and contain parameters that are measurable with difficulty, or can vary during the operation of the system. There are cases in which the mathematical model of the system is not determinable at all. These problems could be overcome by adopting intelligent control techniques that can be implemented even in the absence of the plant mathematical model and can be significantly less sensitive versus parameter variations. All the earlier considerations make AI an interesting tool for improving drive control.

Currently, only two large manufactures, Hitachi and Yaskawa, have incorporated AI features in control of their industrial drives [7]. In particular, Hitachi drive model J300 is a sensorless vector-controlled induction machine (IM) drive, adopting fuzzy logic control for the calculation of the optimal acceleration and deceleration times as well as for the control of the stator currents. Yaskawa drive model VS-616G5, called "true flux vector inverter," adopts a neural-based control. In particular, it contains a "flux observer with intelligent neurocontrol" and permits the drive to work in simple scalar (V/f) mode, vector control with speed feedback, and sensorless vector control.

AI techniques have been divided mainly into two categories: *hard computing* and *soft computing*. To the first category belong the so-called expert systems (ES), which are based on the principles of certainty, rigour, and rigidity, being strictly linked to the rigidity of the Boolean logic. To the second category belong the so-called *fuzzy logic systems* (FLS) [13–18], the *artificial neural networks* (ANN) [19–29], the *genetic algorithms* (GA) [30], and the hybrid systems [7], in particular the neurofuzzy (fuzzy-neural) and the genetic-assisted neural and fuzzy systems. All soft-computing techniques have been conceived for suitably dealing with uncertainty of knowledge and data, typical of real-world applications. While both ES and FLSs are rule-based techniques, aiming to reproduce the behavior of the human brain, ANNs tend to reproduce a biological architecture. A different approach is that of GAs, also known as evolutionary computation, which are based on the principles of genetics (Darwin's theory or fittest theory of the evolution). GAs solve a problem of optimization on the basis of an evolutionary process providing the best (fittest) solution (survivor). Among the different AI techniques, ANNs seem to have maximum impact on power electronics and electrical drives [11,12,31,32,33].

With specific regard to sensorless techniques for IM drives, literature reports several approaches. Reference [34] proposes a simple open-loop speed estimator, based on Equation 6.20, where the speed estimator is integrated with a model reference adaptive system (MRAS) technique estimating on-line the stator resistance of the machine. In particular, the adaptive model of the MRAS is a *multilayer perceptron* (MLP) trained on-line by a *back propagation* network (BPN) algorithm. Reference [35] describes a neural-based MRAS observer. This MRAS scheme adopts an NN rotor flux observer to entirely replace the conventional voltage model and not the current model as in other NN sensorless schemes to improve the drive performance at low and zero speeds. An MLP estimates the rotor flux from present and past samples of the terminal voltages and currents, without any open-loop integration and with less sensitivity versus motor parameter variations. As claimed by the authors, the training data for the NN are obtained from experimental measurements, giving a more accurate model that includes all drive nonlinearities. In Ref. [36], the ANNs are used to correct the estimated rotor speed provided by a speed observer. The internal signals of the speed observer system are thus used to correct the observer's errors at steady states and during transients. Even in this case, an MLP trained on-line by a BPN algorithm has been adopted. It has been sufficient to use a double-layer ANN with fewer neurons in the hidden layer and one in the output layer. For the determination of the neuron number in the hidden layer of the feed-forward network, an evolutionary programming method has been used. The developed system, based on the speed observer, is stable and robust as claimed by the authors. In Ref. [37], an MRAS observer is proposed, where the reference and adaptive models are the conventional voltage and current models of the IM. An NN having in input the rotor fluxes, estimated, respectively, with the reference and adaptive models and the delayed estimated speed, is used to estimate the current value of the speed. The NN has therefore partially a recurrent structure of Jordan's sequential network and can be trained by conventional BPN algorithms. In this case, the NN is trained on-line and does not need any a priori training. Orlowska-Kowalska et al. [38] explore the goodness of an adaptive sliding-mode neurofuzzy speed controller (ASNFC) devised for a two-mass sensorless vector-controlled IM drive. The ANN weights of this controller are trained on-line by a modified gradient descent algorithm, according to the error between the estimated speed of the IM and the reference model output. Moreover, a priori off-line learning of neurofuzzy speed controller (NFC) is not needed. The speed has been estimated by an adaptive observer. A more classic neural approach has been followed by Heredia et al., [39], which proposes the adoption of a neural network (NN) to estimate the speed. The idea consists in an association of some inputs (stator currents, voltage, and frequency) with some outputs (speed and torque). In this case, for each set of inputs, there is a set of outputs. To accomplish this operation, the ANN must be trained initially. It is not necessary to carry out this phase on-line and to give the net all the possible input–output combinations since it has the capacity to generalize results starting from a limited set of inputs–outputs. Once the phase of training has been accomplished, the net should be able to estimate the speed for any set of inputs. The final structure of the adopted ANN is a feed-forward MLP with three layers: the first composed of 2 neurons, the second by 10 neurons to reach the objective of the stipulated error, and the third by 2 neurons to give the estimated speed and torque outputs. An approach similar to that proposed by Karanayil et al., [34] has been followed by Campbell and

Summer [40], which proposes a classic MRAS observer based on the flux error minimization, integrated with a neural estimator of the stator resistance. The estimation of the stator resistance is used on-line to adapt the reference model. A very interesting neural-based sensorless MRAS observer has been proposed by Ben-Brahim and Kurosawa [41], Elloumi et al., [42], Ben-Brahim [43], and Ben-Brahim et al., [44]. Since they have directly inspired the LS-based MRAS observers described in the following, it will be treated separately in more detail.

11.2 BPN-Based MRAS Speed Observer

One of the most interesting approaches to neural-based sensorless control of high-performance IM drives has been proposed in Refs [41–44]. This set of articles present a neural-based MRAS speed observer. Among the various MRAS solutions discussed in Section 6.4.6, this observer is in the framework of MRASs based on rotor flux error minimization [45]. Like in Ref. [45], the MRAS is composed of a *reference* and an *adaptive model*, both used for simultaneously estimating the rotor flux-linkage space-vector amplitude and phase. In this case, however, a two-layered "real-time" NN has been adopted as the adaptive model. This network is able to learn on-line during drive operation and, therefore, does not require any off-line training phase, different from traditional supervised ANNs. The on-line learning of the ANN is governed by the reference model, which the ANN tries to track with its on-line training. As a result of the training process, the rotational speed of the drive is estimated.

As far as the *reference model* is concerned, it is described by the stator voltage equations in stator reference frame, as in Ref. [45], rewritten here for the sake of simplicity:

$$\begin{cases} \dfrac{d\psi_{rd}}{dt} = \dfrac{L_r}{L_m}\left(u_{sD} - R_s i_{sD} - \sigma L_s \dfrac{di_{sD}}{dt}\right) \\ \dfrac{d\psi_{rq}}{dt} = \dfrac{L_r}{L_m}\left(u_{sQ} - R_s i_{sQ} - \sigma L_s \dfrac{di_{sQ}}{dt}\right) \end{cases} \tag{11.1}$$

For the symbols, see the list at the end of the chapter. The block diagram of such a voltage model is sketched in Figure 6.23.

As far as the *adaptive model* is concerned, a neural model of the system has been suitably created, starting from the current model of the IM, rewritten here for the sake of simplicity:

$$\begin{cases} \dfrac{d\hat{\psi}_{rd}}{dt} = \dfrac{1}{T_r}\left(L_m i_{sD} - \hat{\psi}_{rd} - \omega_r T_r \hat{\psi}_{rq}\right) \\ \dfrac{d\hat{\psi}_{rq}}{dt} = \dfrac{1}{T_r}\left(L_m i_{sQ} - \hat{\psi}_{rq} + \omega_r T_r \hat{\psi}_{rd}\right) \end{cases} \tag{11.2}$$

To be rearranged as an ANN, the current model in Equation 11.2 should be firstly converted from the continuous to the discrete domain. If the simple forward Euler method is adopted, the following finite difference equations are obtained:

$$\begin{cases} \hat{\psi}_{rd}(k) = w_1 \hat{\psi}_{rd}(k-1) - w_2 \hat{\psi}_{rq}(k-1) + w_3 i_{sD}(k-1) \\ \hat{\psi}_{rq}(k) = w_1 \hat{\psi}_{rq}(k-1) + w_2 \hat{\psi}_{rd}(k-1) + w_3 i_{sQ}(k-1) \end{cases} \tag{11.3}$$

where

$\wedge$ marks the variables estimated with the adaptive model

k is the current time sample

A linear NN can reproduce these equations, where w_1, w_2, and w_3 are the weights of the NNs defined as $w_1 = 1 - T_{sp}/T_r$, $w_2 = \omega_r T_{sp}$, and $w_3 = T_{sp} L_m/T_r$, T_{sp} is the sampling time of the control system, L_m is the three-phase magnetizing inductance of the motor, T_r is rotor time constant, and ω_r is the rotor speed in electrical angles. Equation 11.3 describes a four inputs–two outputs linear ANN, whose inputs are $i_{sD}(k - 1)$, $i_{sQ}(k - 1)$, $\hat{\psi}_{rd}(k - 1)$, $\hat{\psi}_{rq}(k - 1)$, whose outputs are $\hat{\psi}_{rd}(k)$, $\hat{\psi}_{rq}(k)$, and whose weights are w_1, w_2, and w_3 [41–44]. As a matter of fact, this linear ANN is not a feed-forward network but a recurrent one, that is the delayed outputs are fed back to the input. In this case, the *adaptive model* is said to be used as a "simulator" [46].

In theory, all the three weights w_1, w_2, and w_3 could be trained on-line; practically, however, the weights w_1 and w_3 depend only on known machine parameters and can be therefore kept constant. In this way, only w_2 is trained on-line, since it contains the information about the rotor speed. Figure 11.1 shows the block diagram of the BPN MRAS observer.

11.2.1 On-Line Training of the BPN MRAS Observer

As recalled earlier, the linear NN constituting the adaptive model is trained on-line, i.e., it does not need any off-line training given on the basis of previously acquired data. This is a key issue for the good behavior of the observer. To perform the on-line training, the ANN outputs $\hat{\psi}_{rd}(k)$, $\hat{\psi}_{rq}(k)$ are compared with their target values $\psi_{rd}(k)$, $\psi_{rq}(k)$, provided simultaneously by the reference model. Figure 11.2 shows the block diagram of the training process.

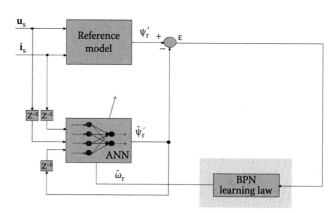

FIGURE 11.1
Block diagram of the BPN MRAS observer.

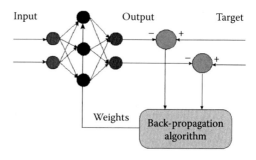

FIGURE 11.2
Block diagram of the training process.

The error between the estimated fluxes, provided, respectively, from the reference and adaptive models, is back propagated to adjust the neural weights (in particular w_2, depending on the rotor speed) to minimize the error itself. In practice, any mismatch between the speed estimated by the neural model and the real motor speed generates an error between the two estimated fluxes. This error between $\hat{\psi}_{rd}(k)$, $\hat{\psi}_{rq}(k)$ and $\psi_{rd}(k)$, $\psi_{rq}(k)$ is used to adjust w_2, according to a BPN learning law. Passing from the complex notation to the vectorial one, if $\boldsymbol{\varepsilon}(k) = \boldsymbol{\psi}_r'(k) - \hat{\boldsymbol{\psi}}_r'(k)$ (the space-vector error), w_2 is adjusted to minimize to squared-error energy function $E = 1/2\,\boldsymbol{\varepsilon}^2(k)$, as follows:

$$\Delta w_2(k) \propto -\frac{dE}{dw_2} = -\frac{dE}{d\hat{\boldsymbol{\psi}}_r'(k)} \frac{d\hat{\boldsymbol{\psi}}_r'(k)}{dw_2} = -\boldsymbol{\delta}(k)\mathbf{J}\,\hat{\boldsymbol{\psi}}_r'(k-1) \tag{11.4}$$

where $\boldsymbol{\delta}(k) = \dfrac{dE}{d\hat{\boldsymbol{\psi}}_r'(k)} = \left(\boldsymbol{\psi}_r'(k) - \hat{\boldsymbol{\psi}}_r'(k) \right)^T$.

The weight variation $\Delta w_2(k)$ is therefore given from a gradient descent error minimization:

$$\Delta w_2 = \left(\boldsymbol{\psi}_r'(k) - \hat{\boldsymbol{\psi}}_r'(k) \right)^T \mathbf{J}\,\hat{\boldsymbol{\psi}}_r'(k-1) \tag{11.5}$$

with $\mathbf{J} = \begin{bmatrix} 0 & -1 \\ 1 & 0 \end{bmatrix}$.

Finally, the adaptation law of w_2 is

$$w_2(k) = w_2(k-1) + \eta \Delta w_2(k) \tag{11.6}$$

The estimated speed can be computed therefore as

$$\hat{\omega}_r(k) = \hat{\omega}_r(k-1) + \eta \Delta w_2(k) + \alpha \Delta w_2(k-1) \tag{11.7}$$

where
  η is the learning rate
  α (called momentum) takes into consideration the effects of the past weight changes on
    the current one

The dynamics of the convergence of the BPN-trained ANN depend on the parameters η and α.

11.2.2 Implementation Issues of the BPN MRAS Observer

To increase the performance of the BPN MRAS observer, in Refs [41–44], some further issues have been accounted for. In particular, as far as the adaptive model is concerned, the simplified model in Equation 11.3 has been used only for the derivation of the BPN algorithm. To ensure a more accurate estimation of the rotor flux, a better discrete adaptive model has been used, as in the following.
If the adaptive model in the continuous domain is written in the form

$$\frac{d\hat{\boldsymbol{\psi}}_r'}{dt} = \mathbf{A}\hat{\boldsymbol{\psi}}_r' + B\,\mathbf{i}_s \tag{11.8}$$

the corresponding accurate discrete model can be written as

$$\hat{\psi}_r'(k+1) = e^{\mathbf{A}T_{sp}} \, \hat{\psi}_r'(k) + \frac{e^{\mathbf{A}T_{sp}} - \mathbf{I}}{\mathbf{A}} B\,\mathbf{i}_s(k) \tag{11.9}$$

Writing $\vartheta_r = \omega_r T_{sp}$ and approximating $e^{\mathbf{A}T_{sp}} \cong \mathbf{I} + \mathbf{A}T_{sp}$, it can be written as

$$\hat{\psi}_r'(k+1) = e^{-T_{sp}/T_r}(\cos(\vartheta_r)\mathbf{I} + \sin(\vartheta_r)\mathbf{J})\,\hat{\psi}_r'(k) + B T_{sp}\,\mathbf{i}_s(k) \tag{11.10}$$

The experimental implementation of Equation 11.10 can be particularly demanding from the computational point of view, which is a drawback of this approach. For this reason, look-up tables can be used for their implementation. This drawback can be overcome by adopting another kind of discretization method, the so-called modified Euler, as shown in the following.

Finally, in Refs [41–44], the problem of the open-loop integration of the flux in the reference model has been solved adopting a simple first-order low-pass filter (LPF) with low cutoff frequency. This is the simplest solution but also worse as far as the low-speed behavior of the drive is concerned (see Chapter 6).

11.2.3 Experimental Results with the BPN MRAS Observer

The BPN MRAS observer in Refs [41–44] has been tested experimentally on a 2.2 kW FOC IM drive. In particular, Appendices 11.A and 11.B describe, respectively, the adopted control scheme and the test setup on which all tests have been performed. Figure 11.3 shows the reference, the measured, and the estimated speeds as well as the speed estimation

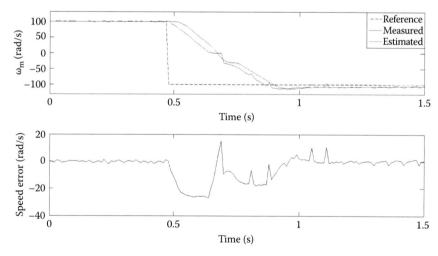

FIGURE 11.3
Reference, measured, and estimated speeds during a speed reversal 100→−100 rad/s with the BPN MRAS observer.

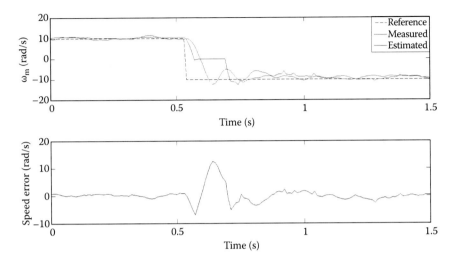

FIGURE 11.4
Reference, measured, and estimated speeds during a speed reversal 10→−10 rad/s with the BPN MRAS observer.

error during a speed reversal from 100 to −100 rad/s at no load. It shows that the estimated speed properly tracks both the measured one and its reference, with an instantaneous speed estimation error, which is slightly lower than 30 rad/s. This peak value of the estimation error, which is not very low, is mainly due to fact that the estimated speed must be low-pass filtered before being fed back to the speed controller. Reducing the value of the learning rate η below a certain value can solve this problem but at the expense of a reduction of the convergence time of the algorithm. A trade-off between these two requirements must be found for the selection of the learning rate η. The need of low-pass filtering the ANN estimation is due to the noisy nature of the estimation itself, whose main reason is the fact that the ANN is used in "simulation" mode or, alternatively, that the ANN is of recurrent type. The BPN MRAS observer has been tested also in the low-speed region. Figure 11.4 shows the reference, the measured, and the estimated speeds as well as the speed estimation error during a speed reversal from 10 to −10 rad/s at no load. It shows that the estimated speed properly tracks both the measured one and its reference at steady-state, with an instantaneous speed estimation error around 10 rad/s. Moreover, in transient, during the zero crossing of the speed, there is a time interval (about 0.15 s) in which the measured speed remains equal to 0 before starting the convergence to the reference. As expected, this causes a significant reduction of the bandwidth of the speed control loop. This behavior at low speed is mainly due to the mismatch between the observer and the machine parameters and to the adoption of an LPF as an integrator in the reference model. This last issue causes also a lack of field orientation, with resulting reduction of the drive dynamic performance. The BPN MRAS observer has been finally tested at zero-speed operation at no load. This is a particularly challenging working condition because the machine speed is not observable. Figure 11.5 shows the reference, the measured, and the estimated speeds during this test in a time interval of 60 s. It can be observed that, in average, both the estimated and measured speeds are controlled to 0. There are, however, time intervals in which both the estimated and measured speeds are quite different from 0, getting values near to 25 rad/s in transient.

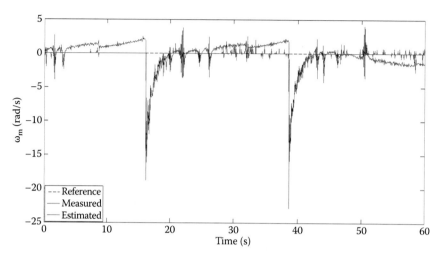

FIGURE 11.5
Reference, measured, and estimated speeds during zero-speed operation with the BPN MRAS observer.

11.3 LS-Based MRAS Speed Observer

The NNs equations (11.3) represent a linear relationship between the inputs and outputs of the system, describing a classic linear NN, well known as ADALINE (ADAptive LINEar Neural Network). Because of the linear nature of the NN, the BPN algorithm adopted in Refs [41–44] for its on-line training does not result the best solution, since it requires a heuristic choice of the learning rate and the momentum. It is in fact questionable to use a nonlinear method like the BPN algorithm which could cause local minima, paralysis of the NN, need of two heuristically chosen parameters, initialization problems, and convergence problems. In Ref. [44], this linearity problem has been recognized, but the minimization has been performed with a gradient descent algorithm, dependent also on the momentum, which is not strictly necessary. Because of the linearity of the network, it is therefore preferable to utilize on-line training algorithms more suitable for linear networks.

Equation 11.3 can be written in the following matrix form, considering the fact that the ANN can be used as a predictor and not as a simulator:

$$\begin{bmatrix} \psi_{rq}(k-1) \\ -\psi_{rd}(k-1) \end{bmatrix} w_2 = \begin{bmatrix} \hat{\psi}_{rd}(k) - w_1\psi_{rd}(k-1) - w_3 i_{sD}(k-1) \\ \hat{\psi}_{rq}(k) - w_1\psi_{rq}(k-1) - w_3 i_{sQ}(k-1) \end{bmatrix} \tag{11.11}$$

which is a classic matrix equation of the type $\mathbf{Ax} \approx \mathbf{b}$, where $\mathbf{A}$ is called "data matrix," $\mathbf{b}$ is called "observation vector," and $\mathbf{x}$ is the vector consisting only of the unknown scalar w_2.

Least squares (LS) techniques reveal in this case, the best solution to be adopted for on-line training of the ADALINE and thus to solve in recursive form Equation 11.11. Furthermore, the corpus of theory developed in this framework permits to theoretically justify the choice of any network parameter, with the guarantee of the stability of the convergence (see Chapter 8 for details). It is not the case of the BPN algorithm. A first attempt to improve the BPN MRAS observer [41–44] has been therefore to substitute

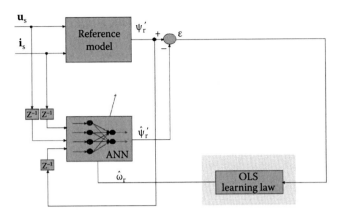

FIGURE 11.6
(See color insert.) Block diagram of the OLS MRAS observer.

the BPN training algorithm with a more suitable ordinary least squares (OLS) one [47]. At the same time, after observing that the adaptive model in [41–44] had been used in "simulation mode," which means that its outputs are fed back recursively, the modified adaptive model has been used in "prediction mode," with resulting no feedbacks, no need of filtering the estimated signal, and higher accuracy both in transient and steady-state operation. Figure 11.6 shows the block diagram of the adopted OLS MRAS observer [47].

11.3.1 Experimental Results with the OLS MRAS Observer

The OLS MRAS observer [47] has been experimentally tested on both a field-oriented control (FOC) and a direct torque control (DTC) IM drive. Appendices 11.A and 11.B describe, respectively, the adopted control scheme and the test setup on which all the tests have been performed. Figures 11.7 and 11.8 show the reference, estimated, and measured speeds as well as the speed estimation error, obtained, respectively, with the FOC and the ST-DTC sensorless drives, with a series of speed step references of the type 130→100→70→30 rad/s. It can be observed that, with both control schemes, the estimated speed tracks properly its reference and the measured one. The steady-state accuracy of the estimation is good, with almost null average estimation error down to 30 rad/s. Even the transient accuracy is quite good, with peak estimation errors rarely exceeding 10 rad/s. It should be noticed that the estimated speed waveform obtained in DTC drive is slightly more noisy than that obtained with FOC, which is an expectable result due to the inherent bang-bang nature of DTC control. Figures 11.9 and 11.11 show the reference, estimated, and measured speeds as well as the speed estimation error, obtained, respectively, with the FOC and the ST-DTC sensorless drives, with a series of speed step references, including a speed reversal in field-weakening region of the type $0 \rightarrow 200 \rightarrow -200 \rightarrow 0$ rad/s. It could be noted that the OLS MRAS observer works properly with both control techniques in field-weakening region. Even during the speed reversal, the peak estimation error does not exceed 20 rad/s. Figures 11.10 and 11.12 show the corresponding waveforms of rotor flux amplitude and i_{sy} current component, as far as FOC drive is concerned, and stator flux amplitude and electromagnetic torque, as far as DTC drive is concerned. The DTC sensorless drive exhibits a higher torque ripple with respect to the FOC drive, as expected.

Figures 11.13 and 11.14 show the steady-state mean speed estimation errors and standard deviations, obtained, respectively, with the FOC and the DTC OLS with MRAS sensorless drives, for different values of reference speeds at no load and at rated load. They show

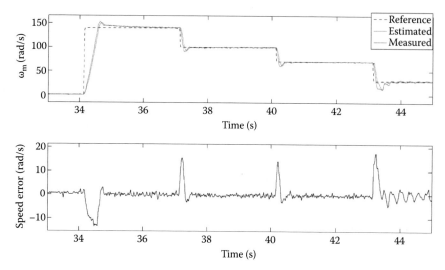

FIGURE 11.7
Reference, measured, and estimated speeds with OLS MRAS observer in FOC drive. (From Cirrincione, M. and Pucci, M., An MRAS based speed estimation method with a linear neuron for high performance induction motor drives and its experimentation, *IEEE International Electric Machines and Drives Conference (IEMDC'2003)*, Madison, WI, 2003.)

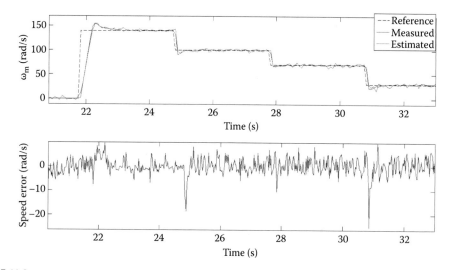

FIGURE 11.8
Reference, measured, and estimated speeds with OLS MRAS observer in DTC drive. (From Cirrincione, M. and Pucci, M., An MRAS based speed estimation method with a linear neuron for high performance induction motor drives and its experimentation, *IEEE International Electric Machines and Drives Conference (IEMDC'2003)*, Madison, WI, 2003.)

that the OLS MRAS observer integrated in both control techniques guarantees a very low speed estimation error down to 20 rad/s. Below this speed, in both cases, the speed estimation error rapidly increases up to 80% at 10 rad/s. As expected, in loaded conditions, the speed estimation error is higher especially at low speed. For the earlier-cited reasons, the standard deviation with DTC is always higher than that with FOC.

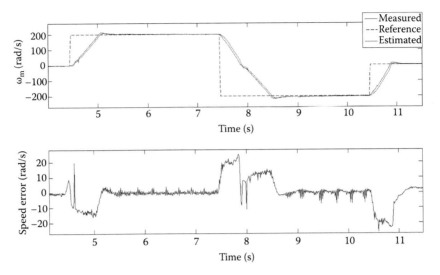

FIGURE 11.9
Reference, measured, and estimated speeds with OLS MRAS observer in FOC drive. (From Cirrincione, M. and Pucci, M., An MRAS based speed estimation method with a linear neuron for high performance induction motor drives and its experimentation, *IEEE International Electric Machines and Drives Conference (IEMDC'2003)*, Madison, WI, 2003.)

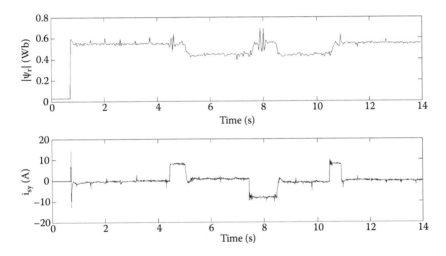

FIGURE 11.10
Rotor flux linkage and i_{sy} current component with OLS MRAS observer in FOC drive. (From Cirrincione, M. and Pucci, M., An MRAS based speed estimation method with a linear neuron for high performance induction motor drives and its experimentation, *IEEE International Electric Machines and Drives Conference (IEMDC'2003)*, Madison, WI, 2003.)

11.3.2 TLS EXIN MRAS Observer

The matrix Equation 11.11 can be solved with any LS technique, either with OLS if the errors are present only in the observation vector or with data ordinary least squares (DLS) if the errors are present only in the data matrix or with total least squares (TLS) if the errors are present both in the data matrix and in the observation vector. In this respect, Equation 11.11 shows that the matrix **A** is composed of the d-q-axis components of the

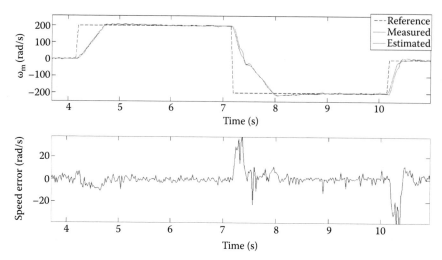

FIGURE 11.11
Reference, measured, and estimated speeds with OLS MRAS observer in DTC drive. (From Cirrincione, M. and Pucci, M., An MRAS based speed estimation method with a linear neuron for high performance inductio motor drives and its experimentation, *IEEE International Electric Machines and Drives Conference (IEMDC'2003)*, Madison, WI, 2003.)

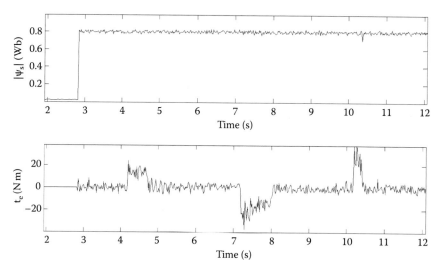

FIGURE 11.12
Stator flux linkage and electromagnetic torque with OLS MRAS observer in DTC drive. (From Cirrincione, M. and Pucci, M., An MRAS based speed estimation method with a linear neuron for high performance induction motor drives and its experimentation, *IEEE International Electric Machines and Drives Conference (IEMDC'2003)*, Madison, WI, 2003.)

rotor flux linkage which can be affected by errors and noise resulting from open-loop integration of the model reference or measurements, and the same can be said for the observation vector **b** which is also composed of the d-q-axis components of the rotor flux linkage and the d-q-axis components of the stator current space-vector; this can be therefore described as a TLS problem, rather than an OLS problem. Any LS technique different

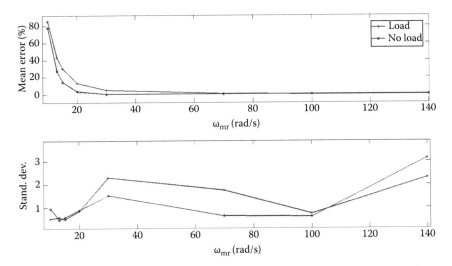

FIGURE 11.13
Steady-state mean estimation error and standard deviation with OLS MRAS observer in FOC drive. (From Cirrincione, M. and Pucci, M., An MRAS based speed estimation method with a linear neuron for high performance induction motor drives and its experimentation, *IEEE International Electric Machines and Drives Conference (IEMDC'2003)*, Madison, WI, 2003.)

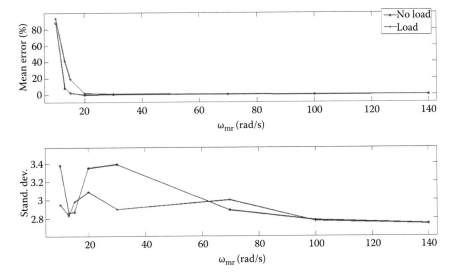

FIGURE 11.14
Steady-state mean estimation error and standard deviation with OLS MRAS observer in DTC drive. (From Cirrincione, M. and Pucci, M., An MRAS based speed estimation method with a linear neuron for high performance induction motor drives and its experimentation, *IEEE International Electric Machines and Drives Conference (IEMDC'2003)*, Madison, WI, 2003.)

from TLS would be therefore inadequate [48,49]. The TLS EXIN neuron has been then applied to Equation 11.11 for the speed estimation [50] since it is the best algorithm for solving the TLS problem in a recursive way. The block diagram of the TLS EXIN MRAS observer, as proposed in Ref. [50], is the same as the one sketched in Figure 11.6, being the difference only in the ADALINE training algorithm (TLS instead of OLS).

11.3.2.1 Neural Adaptive Integrator

To solve the problem of open-loop flux integration in the reference model, present also in the MRAS systems mentioned earlier, the so-called neural adaptive integrator has been developed [51]. In particular, a linear NN, an ADALINE, has been used for the integration of a signal to eliminate the DC component, which results in having a pure integrator unaffected by the DC drift and the initial conditions. This integrator uses two neural adaptive filters [52], each of which operates with two basic processes forming a feedback loop: an adaptive process to adjust only one parameter and a filtering process where an error signal is computed and then fed back to actuate the adaptive process, and therefore, eliminating the DC component.

The simplicity of this linear neuron (just one weight is adapted) accounts for the use of this kind of NNs in this application, whereas traditional fixed notch filters would be unfeasible because of their complexity and computational burden.

Figure 11.15 shows the block diagram of a neural filter or adaptive noise canceler, as called in Ref. [52]. The input signal is the signal affected from noise, and it can been considered as $s + n_0$ where s is the signal and n_0 is the noise, uncorrelated with the signal. This input signal $s + n_0$ is the so-called "primary input" to the neural filter. Let a second noise n_1 be received by the neural filter and let this noise be also uncorrelated with s but in a way correlated with n_0. This second noise is the so-called "reference input" to the neuron. The noise n_1 is given as input to the neuron, and the output y is then subtracted to the primary input so as to obtain the system output $z = s + n_0 - y$, which is also the error ε between the primary input and y. The reference input is processed by the linear neuron by an LS algorithm in order to minimize the total power output $E[\varepsilon^2]$ where $E[.]$ is the expectation function. This goal is reached by feeding back the output signal z to the linear neuron. If this minimization is reached, then also $E[(n_0 - y)^2]$ is minimized as the signal power $E[s^2]$ is not affected. This is the same as approximating s with ε in an LS sense. As a consequence, the minimization of the total power output implies the maximization of the output signal-to-noise ratio. It must be emphasized that only an approximate correlation between the noises n_1 and n_0 is necessary and nothing more.

In case of the problem at hand, the noise n_0 is the DC component. Thus a notch filter with a notch at zero frequency can be achieved if a neuron with only one bias weight is used, that is a neuron whose input is a constant, for example -1.

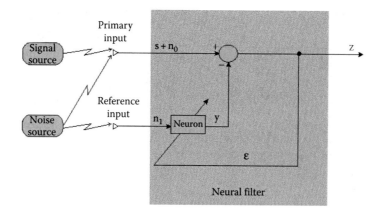

FIGURE 11.15
Block diagram of the neural adaptive filter. (From Cirrincione, M. et al., *IEEE Trans. Power Electron.*, 19(1), 25, January 2004.)

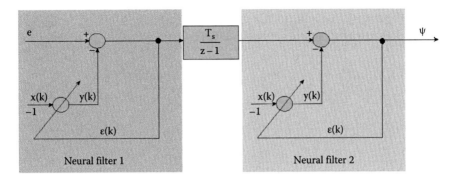

FIGURE 11.16
(See color insert.) Block diagram of the neural adaptive integrator. (From Cirrincione, M. et al., *IEEE Trans. Power Electron.*, 19(1), 25, January 2004.)

The learning law of the neural adaptive filter is then the following:

$$y(k+1) = y(k) + 2\tau(d(k) - y(k)) \tag{11.12}$$

where
k is the actual time instant, $d(k)$ is either e (in the first notch filter) or the output of the integrator (in the second notch filter)
τ is the learning rate

This single weight neuron is able to remove not only a constant bias, but also a slowly varying drift in the primary input [52]. Figure 11.16 shows the adaptive integrator with two identical neural notch filters before and after the pure integrator. It should be remarked that two neural filters must be used in the neural-based integrator: the neural filter 1 eliminates the DC component of the signal to be processed and the neural filter 2 eliminates the DC drift appearing at the output of the integrator because of the initial conditions and of the filtering error of the neural filter 1 during its adaptation.

By converting the equation of the neural adaptive filter from the discrete domain into the continuous one, the global transfer function of the neural adaptive filter is given by

$$G_{filt}(s) = \frac{s}{s + 2\tau/T_{sp}} \tag{11.13}$$

where s is the Laplace variable. Figure 11.17, which shows the frequency response of the adaptive notch filter with two different values of the learning factor τ (respectively, 2×10^{-4} and 1×10^{-5}), highlights that the lower the value of τ, the smaller the frequency amplitude of the notch. By converting the equation of the whole neural integrator from the discrete domain into the continuous one, the global transfer function of the neural adaptive integrator is given by

$$G_{int}(s) = \frac{s}{s^2 + 4\tau/T_{sp}s + 4\tau^2/T_{sp}^2} \tag{11.14}$$

Figure 11.18 shows the frequency response of an ideal integrator, an LPF-based integrator (cutoff frequency = 15 rad/s), and the neural adaptive integrator with two different values of τ, respectively, 2×10^{-4} and 1×10^{-5}. This figure shows that the adaptive neural integrator with $\tau = 2 \times 10^{-4}$ outperforms the LPF, both in its magnitude and phase characteristics, in

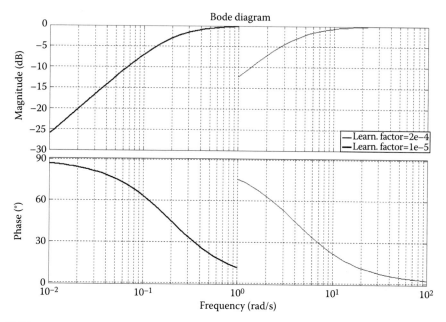

FIGURE 11.17

Frequency response of the adaptive notch filter with two values of the learning factor. (From Cirrincione, M. et al., *IEEE Trans. Ind. Electron.*, 54(1), 127, February 2007.)

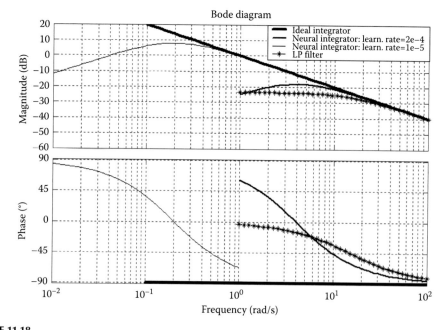

FIGURE 11.18

Frequency response of the adaptive integrator with two values of the learning factor, of the LP integrator, and of the ideal integrator. (From Cirrincione, M. et al., *IEEE Trans. Ind. Electron.*, 54(1), 127, February 2007.)

the neighborhood of a reference speed of about 10 rad/s in electrical angles (corresponding to 5 rad/s in mechanical ones).

11.3.2.1.1 Experimental Results with the Neural Adaptive Integrator   In Ref. [51], the neural adaptive integrator sketched in Figure 11.16 has been proposed and implemented experimentally on a rotor flux–oriented IM drive. Firstly, some simulation tests have been done in MATLAB®–Simulink® environment.

Figure 11.19a and b shows the transient time waveforms of the ψ_{rd} component, as obtained with the neural integrator, the LPF, and *algorithms 1* and *2* (see Section 6.4.1), with a step 2% DC signal superimposed to the voltage signal on phase sA. The integration methods have all been tested in open-loop flux integration, meaning that the flux feedback control has not been activated. They clearly show that the neural adaptive filter exhibits the best behavior as far as DC drift canceling is concerned. As a matter of fact, the neural filter permits the DC bias to be completely canceled; this is the case neither with the LPF, presenting

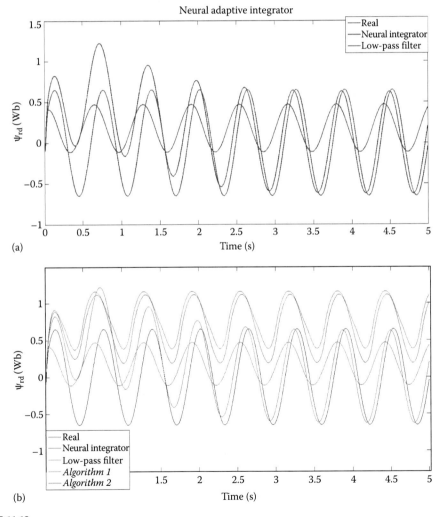

FIGURE 11.19
(See color insert.) (a) Transient time waveform of ψ_{rd} component obtained with the neural filter and LPF. (b) Transient time waveform of ψ_{rd} component obtained with the neural filter, LPF, and *algorithms 1* and *2*.

a huge error in flux estimation (both in magnitude and phase), nor with *algorithms 1* and *2*, presenting a large DC bias on the estimated flux (being *algorithm 2* better than *algorithm 1*).

Figure 11.20 shows the experimental steady-state locus of the rotor flux-linkage space-vector obtained with the neural adaptive integrator. Experimental results obtained with the neural adaptive integrator have been further compared experimentally with the *LPF* and the *algorithms 1, 2,* and *3* described in Section 6.4.1. All the integration methods have been tested experimentally on the FOC IM drive described in Appendices 11.A and 11.B. They have all been tested in open-loop flux integration, meaning that the flux feedback control has not been activated.

Figure 11.21 shows the steady-state D, Q rotor flux components versus time, and the rotor flux loci, obtained at the pulsation of 60 rad/s. In the loci diagrams, the shift of the center of

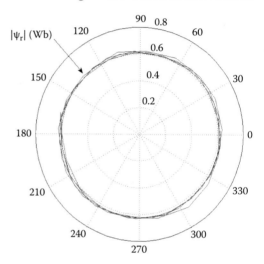

FIGURE 11.20
Steady-state locus of the rotor flux linkage with the neural adaptive integrator.

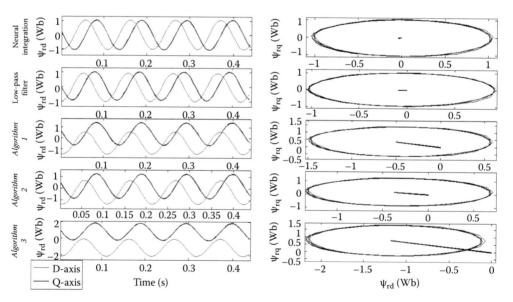

FIGURE 11.21
D, Q rotor flux components versus time and rotor flux locus estimated with the five integrators in open-loop estimation. (From Cirrincione, M. et al., *IEEE Trans. Power Electron.*, 19(1), 25, January 2004.)

the locus from the origin is shown in bold line to prove how the bias affects the flux estimation. With this last regard, the neural integrator is far better than all the others, as it presents a very low bias, then there are in order the LPF, the *algorithms 2* and *1*, and the *algorithm 3*. At this frequency, the LPF still works correctly as its cutoff frequency (15 rad/s) is well apart from the operating conditions. However, it does not outperform the neural integrator.

11.3.2.2 Experimental Results with the TLS EXIN MRAS Observer

The TLS EXIN MRAS observer has been tested experimentally on both FOC and DTC IM drives. The adopted FOC [50] and DTC [53] schemes are fully described in Appendix 11.A, while the corresponding test setup has been described in Appendix 11.B. In Ref. [50], the reference model of the MRAS has been integrated with the *algorithm 2* for the on-line stator resistance estimation algorithm, as described in Section 6.4.3.

11.3.2.2.1 TLS EXIN MRAS Observer with FOC Drive The TLS EXIN MRAS observer with the neural integrator and with the R_s estimation algorithm has been tested in some challenging working conditions. Firstly, a speed reversal $-10 \rightarrow 10$ rad/s at no load has been given to verify the dynamic performance of the drive at low speed. Figure 11.22 shows the waveforms of the estimated, measured, and reference speeds: it is apparent that the speed response is very quick and that after few oscillations it converges to the reference value.

Secondly, the load torque rejection capability has been verified both at medium and low speeds. Figures 11.23 and 11.24 show the reference, the estimated, and the measured speeds when a sudden load torque insertion and uninsertion of 11 N m is applied at the working speeds, respectively, at 70 and 10 rad/s. It could be noted that, at medium speeds, the drive response occurs immediately when the torque steps are given, no matter what their directions are. Moreover, even during the speed transient caused by the torque, the estimated speed follows the real one very well. As far as the low-speed test is concerned, it can be observed that the drive speed response follows the reference speed in spite of the torque perturbation even at this very critical speed: there is only a small transient during which the estimated speed slightly differs from the real one. This time interval, in which the measured speed is different from the estimated one and is also negative for some instants, is due to a nonperfect tuning of the speed observer at this speed.

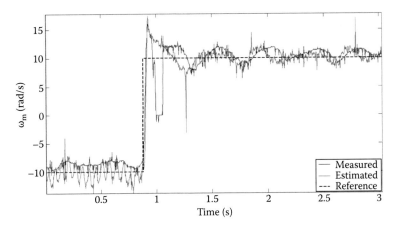

FIGURE 11.22
Speed reversal from −10 to 10 rad/s at no load with the TLS MRAS observer. (From Cirrincione, M. et al., *IEEE Trans. Ind. Appl.*, 1, 140, June/July 2004.)

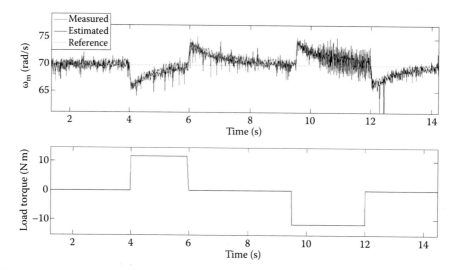

FIGURE 11.23
Speed response of the drive to a sudden torque insertion at 70 rad/s with the TLS MRAS observer. (From Cirrincione, M. et al., *IEEE Trans. Ind. Appl.*, 1, 140, June/July 2004.)

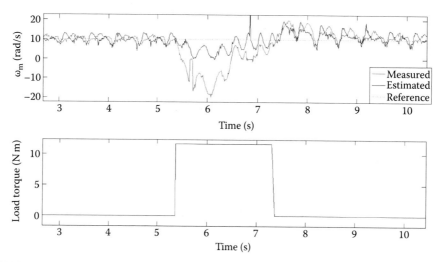

FIGURE 11.24
Speed response of the drive to a sudden torque insertion at 10 rad/s with the TLS MRAS observer. (From Cirrincione, M. et al., *IEEE Trans. Ind. Appl.*, 1, 140, June/July 2004.)

Thirdly, the accuracy of the speed estimation at low speed has been verified. Figure 11.25 shows the reference, the estimated, and the measured speeds as well as the speed estimation error and the estimated stator resistance when the machine is running at steady-state at the speed of 6 rad/s, respectively, at no load and with load. It could be observed that the TLS MRAS observer with the neural integrator and with R_s estimation permits the drive to work at 6 rad/s with a mean estimation error of 0.83% at no load and 23.59% at rated load. Table 11.1 summarizes the results obtained with the TLS MRAS observer and compares them with those obtained with the classic MRAS observer [45] and with the OLS MRAS

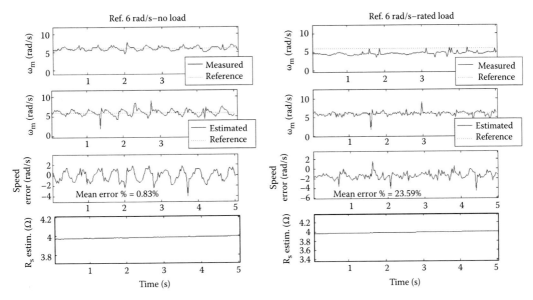

FIGURE 11.25

Estimated speed, measured speed, speed error, and estimated R_s at the reference speeds of 6 rad/s at no load and at load with the TLS MRAS observer. (From Cirrincione, M. et al., *IEEE Trans. Ind. Appl.*, 1, 140, June/July 2004.)

TABLE 11.1

Comparison of Accuracy in Speed Estimation

	TLS and Neural Integrator		OLS and LPF		Classic MRAS Observer	
	Rated Load	No Load	Rated Load	No Load	Rated Load	No Load
Reference Speed (rad/s)	Mean Estimation Error (%)	Mean Estimation Error (%)	Mean Estimation Error (%)	Mean Estimation Error (%)	Mean Estimation Error (%)	Mean Estimation Error (%)
10	0.72	7	90.25	5.49	12.26	21.89
9	6.21	4.19	99.28	12.40	17.29	23.44
8	19.55	1.03	100.25	99.09	27	24.31
7	53.55	32.7	/	/	/	/
6	48.6	38.71	/	/	/	/

Source: From Cirrincione, M. et al., *IEEE Trans. Ind. Appl.*, 1, 140, June/July 2004.

observer [47]. It could be noted that the TLS MRAS observer offers the best behavior, followed by the OLS MRAS observer, and finally by the classic MRAS observer.

Fourthly, the drive has been operated at the rated rotor flux linkage at zero speed. The test has been performed at no load and at the load of 3.5 N m (about 30% of the rated load). Figure 11.26 shows the waveforms of the reference, measured, and estimated speeds for a time interval of about 60 s. It shows that, after the magnetization of the machine, the drive can work properly at zero speed and at no load, even without any signal injection: this is mainly due to the fact that the adaptive model is used as a predictor and is therefore more stable than when used as a simulator, that is, with feedback loops of the estimated rotor flux linkage. In particular, the estimated speed has slight oscillations around 0 rad/s while the measured speed is always 0. Figure 11.27 shows the waveforms of the reference, measured, and estimated speeds at the load torque of 3.5 N m for a time interval of about 60 s. It shows that in average the

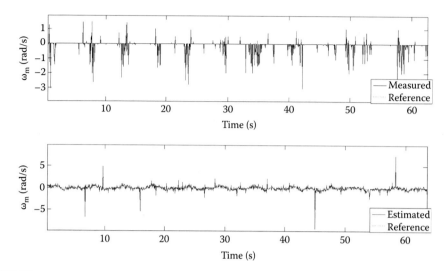

FIGURE 11.26
Speed response of the drive with TLS, neural integrator, and R_s estimation at the reference speed of 0 rad/s at no load with the TLS MRAS observer. (From Cirrincione, M. et al., *IEEE Trans. Ind. Appl.*, 1, 140, June/July 2004.)

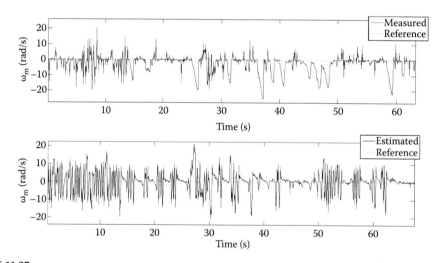

FIGURE 11.27
Speed response of the drive with TLS, neural integrator, and R_s estimation at the reference speed of 0 rad/s at a constant load of 3.5 N m with the TLS MRAS observer. (From Cirrincione, M. et al., *IEEE Trans. Ind. Appl.*, 1, 140, June/July 2004.)

machine is able to stay at zero speed with a medium/small load, even if there are very short time intervals in which the machine moves, but without ever exceeding the speed of 10 rad/s during transients. Above a 3.5 N m load torque, the drive exhibits an unstable behavior.

11.3.2.2.2 TLS EXIN MRAS Observer with DTC Drive   The TLS-based MRAS speed observer with the neural integrator has been tested on both the classic ST-DTC and the DTC-SVM (Space-Vector Modulation). Firstly, the four-quadrant capability of the sensorless drive has been verified by giving as input commands speed steps with a speed reversal from 100

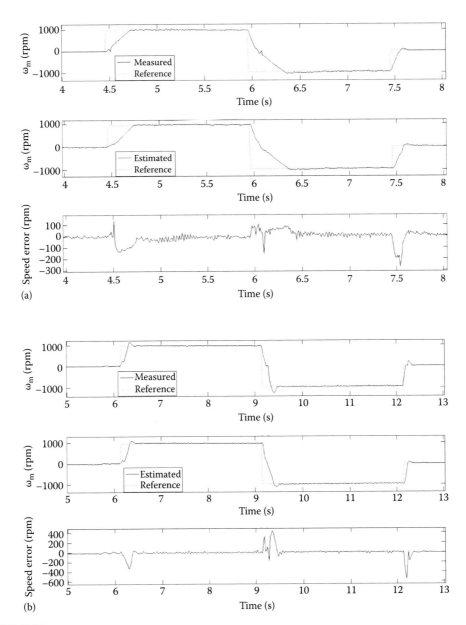

FIGURE 11.28
Classic DTC (a) and DTC-SVM (b)—estimated and measured speeds and speed error during a speed reversal from 954.9 to –954.9 rpm at no load—TLS with adaptive integrator. (From Cirrincione, M. and Pucci, M., *Automatica*, 41(11), 1843, November 2005.)

to –100 rad/s (from 954.9 to –954.9 rpm) at no load. Figure 11.28a and b shows the estimated and measured speed and the speed error, while Figure 11.29a and b shows the stator flux-linkage amplitude and the electromagnetic torque during this test. They show that the speed reversal is accomplished in less than 1 s, respectively, in 0.39 s with the classic DTC and 0.3 s with the DTC-SVM, and the corresponding achieved 3 dB bandwidth of the speed loops at the reference speed of 100 rad/s is, respectively, 14.6 and 16.9 rad/s.

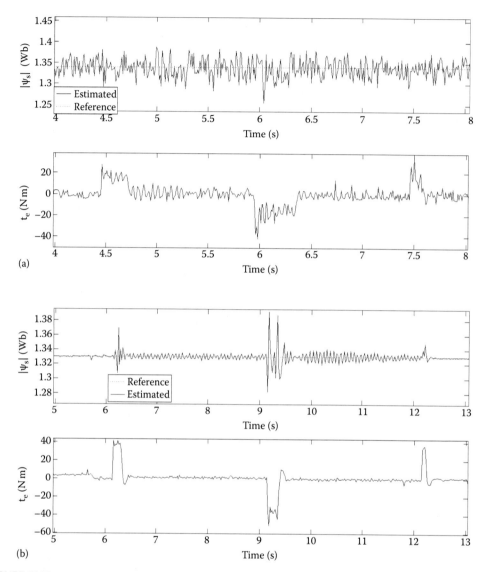

FIGURE 11.29
Classic DTC (a) and DTC-SVM (b)—stator flux-linkage amplitude and electromagnetic torque during a speed reversal from 954.9 to –954.9 rpm at no load—TLS with adaptive integrator. (From Cirrincione, M. and Pucci, M., *Automatica*, 41(11), 1843, November 2005.)

Secondly, to verify the dynamic performance of the drive at low speed, a set of speed step references of low value have been given the drive starting from zero speed and with the machine already magnetized. Figure 11.30 shows the reference, measured, and estimated speeds obtained during four tests with the speed step references of values 5, 10, 15, and 20 rad/s (47.7, 95.5, 143.2, and 191 rpm). It shows a very good behavior of the drive, in terms of both transient and steady-state accuracy even at the lowest speed of 5 rad/s.

Thirdly, the drive has been operated at the rated stator flux linkage at zero speed. The test has been performed only at no load, since zero-speed operation with load does not work, leading the drive into instability. Figure 11.31a and b shows the waveforms of the

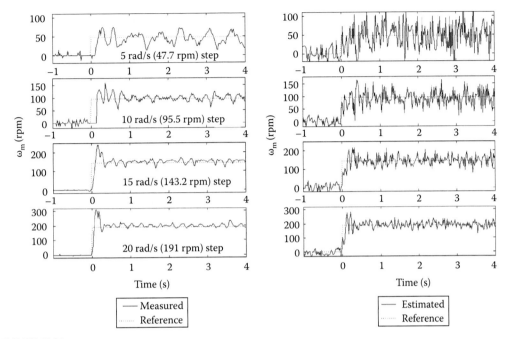

FIGURE 11.30
Classic DTC—reference, estimated, and measured speeds with low-speed step—TLS MRAS observer. (From Cirrincione, M. and Pucci, M., *Automatica*, 41(11), 1843, November 2005.)

reference, measured, and estimated speeds, as obtained, respectively, with the ST-DTC and the DTC-SVM. These figures show that, after the magnetization of the machine, the drive can work properly at zero speed and at no load. In particular, the estimated speed has slight oscillations around 0 rad/s (lower with the classic DTC than with DTC-SVM) and the measured speed is always 0.

11.3.3 Modified Euler Neural Adaptive Model

Equation 11.2 can be written in the following manner too:

$$\dot{\boldsymbol{\psi}}_r' = \mathbf{A}_x \boldsymbol{\psi}_r' + B_x \mathbf{i}_s \tag{11.15}$$

where

$$\boldsymbol{\psi}_r' = \begin{bmatrix} \psi_{rd} \\ \psi_{rq} \end{bmatrix}$$

$$\mathbf{A}_x = \begin{bmatrix} -1 & -\omega_r T_r \\ \omega_r T_r & -1 \end{bmatrix}$$

$$B_x = L_m$$

$$\mathbf{i}_s = \begin{bmatrix} i_{sD} \\ i_{sQ} \end{bmatrix}$$

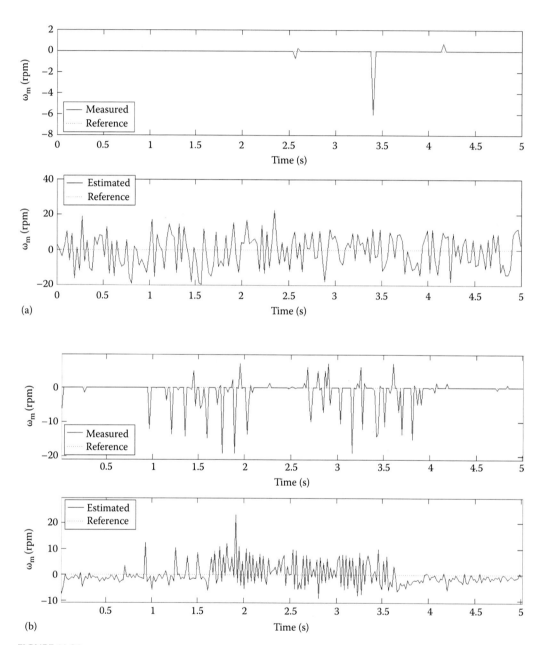

FIGURE 11.31
Classic DTC (a) and DTC-SVM (b)—estimated and measured speeds at zero speed with no load—TLS with adaptive integrator. (From Cirrincione, M. and Pucci, M., *Automatica*, 41(11), 1843, November 2005.)

Its corresponding discrete model is therefore given by

$$\hat{\boldsymbol{\psi}}_r'(k) = e^{\mathbf{A}_x T_{sp}} \hat{\boldsymbol{\psi}}_r'(k-1) + \left[e^{\mathbf{A}_x T_{sp}} - \mathbf{I} \right] \mathbf{A}_x^{-1} B_x \mathbf{i}_s (k-1) \tag{11.16}$$

$e^{\mathbf{A}_x T_{sp}}$ is generally computed by truncating its power series expansion, that is,

$$e^{\mathbf{A}_x T_{sp}} = \mathbf{I} + \frac{\mathbf{A}_x T_{sp}}{1!} + \frac{\mathbf{A}_x^2 T_{sp}^2}{2!} + \cdots + \frac{\mathbf{A}_x^n T_{sp}^n}{n!} \tag{11.17}$$

If n = 1, the simple forward Euler method is obtained, which gives the finite difference equations in Equation 11.13.

An integration method more efficient than that used in Equation 11.3 is the so-called modified Euler integration, which also takes into consideration the values of the variables in two previous time steps [54–57]. From (11.16), the following discrete time equations can be obtained:

$$\begin{cases} \hat{\psi}_{rd}(k) = w_{1n}\hat{\psi}_{rd}(k-1) - w_{2n}\hat{\psi}_{rq}(k-1) + w_{3n}i_{sD}(k-1) \\ \qquad + w_{4n}\hat{\psi}_{rd}(k-2) + w_{5n}\hat{\psi}_{rq}(k-2) - w_{6n}i_{sD}(k-2) \\ \hat{\psi}_{rq}(k) = w_{1n}\hat{\psi}_{rq}(k-1) + w_{2n}\hat{\psi}_{rd}(k-1) + w_{3n}i_{sQ}(k-1) \\ \qquad + w_{4n}\hat{\psi}_{rq}(k-2) - w_{5n}\hat{\psi}_{rq}(k-2) - w_{6n}i_{sQ}(k-2) \end{cases} \tag{11.18}$$

Also in this case, an NN can reproduce these equations, where w_{1n}, w_{2n}, w_{3n}, w_{4n}, w_{5n}, and w_{6n} are the weights of the NNs defined as $w_{1n} = 1 - 3T_{sp}/(2T_r)$, $w_{2n} = 3\omega_r T_{sp}/2$, $w_{3n} = 3T_{sp}L_m/(2T_r)$, $w_{4n} = T_{sp}/(2T_r)$, $w_{5n} = \omega_r T_{sp}/2$, and $w_{6n} = T_{sp}L_m/(2T_r)$. Equation 11.18 describes an eight inputs–two outputs linear ANN, whose inputs are $i_{sD}(k-1)$, $i_{sQ}(k-1)$, $i_{sD}(k-2)$, $i_{sQ}(k-2)$, $\hat{\psi}_{rd}(k-1)$, $\hat{\psi}_{rq}(k-1)$, $\hat{\psi}_{rd}(k-2)$, $\hat{\psi}_{rq}(k-2)$, whose outputs are $\hat{\psi}_{rd}(k)$, $\hat{\psi}_{rq}(k)$, and whose weights are six: w_{1n}, w_{2n}, w_{3n}, w_{4n}, w_{5n}, and w_{6n}.

Rearranging Equation 11.18, the following matrix equation is obtained in prediction mode:

$$\begin{bmatrix} -3/2 T_{sp}\psi_{rq}(k-1) + 1/2 T_{sp}\psi_{rq}(k-2) \\ 3/2 T_{sp}\psi_{rd}(k-1) - 1/2 T_{sp}\psi_{rd}(k-2) \end{bmatrix} \omega_r(k-1)$$

$$= \begin{bmatrix} \hat{\psi}_{rd}(k) - w_{1n}\psi_{rd}(k-1) - w_{3n}i_{sD}(k-1) - w_{4n}\psi_{rd}(k-2) + w_{6n}i_{sD}(k-2) \\ \hat{\psi}_{rq}(k) - w_{1n}\psi_{rq}(k-1) - w_{3n}i_{sQ}(k-1) - w_{4n}\psi_{rq}(k-2) + w_{6n}i_{sQ}(k-2) \end{bmatrix} \tag{11.19}$$

This matrix equation can be solved by any LS technique. Even in this case, since the two matrices are affected by errors especially in the estimation of the rotor flux, a TLS technique should be used instead of OLS. The TLS EXIN neuron, or its improvement called minor component analysis (MCA) EXIN + neuron, should then been employed for on-line applications.

Figure 11.32 shows the block diagram of the corresponding MRAS speed observer.

11.3.3.1 Simulation Mode and Prediction Mode in MRAS Observers: Modified Euler against Simple Euler

Some considerations fully justify the use of the adaptive model in prediction mode with the modified Euler integration. It can be shown that, even in simulation mode, the modified Euler integration trained by the TLS method gives better performances in comparison

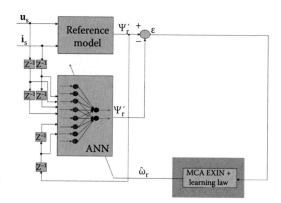

FIGURE 11.32
(See color insert.) The block diagram of the MCA EXIN + MRAS speed observer.

with the results obtainable by using the simple Euler integration method, trained either by the TLS or by the BPN. It can be further shown that, as far as stability is concerned, the use of the simple Euler method in prediction mode gives better results than those obtained with the same Euler integration trained by a BPN in simulation mode.

When used in simulation mode, the process output, that is, the rotor flux linkage, is delayed and then used as an input. In case the simple Euler integration method were used, then the transfer function X(z) of the flux model in the z-domain is

$$X(z) = Z\left(\hat{\mathbf{\Psi}}_r'(k)\right)/Z\left(\mathbf{i}_S(k-1)\right) = \frac{w_3}{(jw_2 + w_1)z^{-1} - 1} = \frac{w_3 z}{(jw_2 + w_1) - z} \qquad (11.20)$$

which has one pole $z_1 = w_1 + jw_2$ and one zero at the origin of the z-domain. For stability reasons, the poles of the transfer function must lie within the unit circle in the z-domain. There is therefore a critical value of the rotor speed that causes instability of the system. More precisely, the following relationships must be satisfied:

$$-\frac{1}{T_{sp}}\sqrt{\frac{T_{sp}}{T_r}\left(2 - \left(\frac{T_{sp}}{T_r}\right)\right)} < \omega_r < \frac{1}{T_{sp}}\sqrt{\frac{T_{sp}}{T_r}\left(2 - \left(\frac{T_{sp}}{T_r}\right)\right)} \qquad (11.21)$$

and

$$T_{sp} < \frac{2T_r}{1 + T_r^2\omega_r^2} \qquad (11.22)$$

Relationship (11.21) shows that the drive goes into instability for increasing values of the rotor speed, while relationship (11.22) shows that there is an upper limit of stability of sampling time T_{sp} if the motor runs at a defined angular speed.

For instance, for the motor at hand whose rated speed is 314 electrical rad/s and $T_r = 0.134\,\text{s}$, this upper limit for the sampling time is of 0.15 ms. Conversely, if a sampling time of 0.1 ms is employed, which is the case under study, the highest limit of the speed is of 385 electrical rad/s (Figure 11.33, upper graph), which implies that the speed can be increased to as much as 18% of the rated speed and not over this limit, with resulting difficulties in using the drive in the field-weakening region.

To overcome this difficulty, the adaptive model should be used in prediction mode, that is, the delayed outputs of the reference model are used as inputs to the adaptive model. In this case, no feedback exists and no stability problems occur.

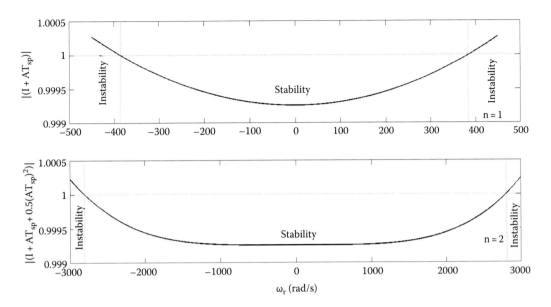

FIGURE 11.33
Amplitude of the poles in simple Euler integration with the approximated exponential function with n = 1 and n = 2. (From Cirrincione, M. and Pucci, M., *IEEE Trans. Ind. Electron.*, 52(2), 532, April 2005.)

The simple Euler method was obtained by using n = 1 in (11.17). Better stability results can be obtained if n = 2 is chosen in (11.17). Then the speed stability limit increases as shown in Figure 11.33, bottom graph. This approximation has been at least used in Refs [41–44] to avoid the stability problems in simulation mode. It should be emphasized that this last method implies at least the on-line computation of the square of the $\mathbf{A_x}T_{sp}$ matrix, which makes this method too cumbersome for on-line applications.

Better results, at the expense of a slight increase of computation in comparison with the simple Euler method, can be obtained with the modified Euler method [57].

As for the integration of the equations (11.2) in the discrete domain, the pure integrator $1/s$ in the continuous domain s has been replaced by the following discrete filter in the z-domain $\dfrac{T_{sp}\left(3/2z-1/2\right)}{z(z-1)}$. This is obtained by the Z transform of the following discrete equation $y(k)=\underbrace{y(k-1)+T_{sp}x(k-1)}_{simple\ Euler}+\underbrace{T_{sp}/2(x(k-1)-x(k-2))}_{additional\ term}$, where $x(k)$ is the integrator input at the current time sample k and $y(k)$ is the corresponding integrator output. This formula is the sum of simple Euler integrator and an additional term taking into consideration the values of the integrand variables in two previous time steps; it ensures a correct integration of the state equations and, thus, a correct flux estimation with the adopted value of T_{sp} different from the simple forward Euler integrator $T_{sp}/z-1$.

In this case, a similar analysis of stability shows that two poles of the transfer function vary with speed, but the resulting speed stability limit is much higher than that obtained with the simple Euler method, thus allowing the exploitation of the field-weakening region (Figure 11.34).

Figure 11.35 shows the reference, the real, and the estimated speeds, obtained in numerical simulation, with the MRAS BPN observer where the adaptive model is the current mode discretized with the simple forward Euler method and used in simulation mode, during two speed step references, respectively, at 100 and 200 rad/s. It clearly shows that,

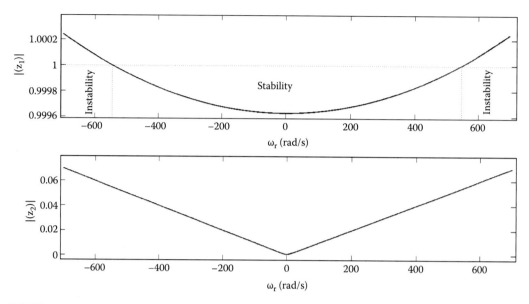

FIGURE 11.34
Amplitude of the poles with the modified Euler integration. (From Cirrincione, M. and Pucci, M., *IEEE Trans. Ind. Electron.*, 52(2), 532, April 2005.)

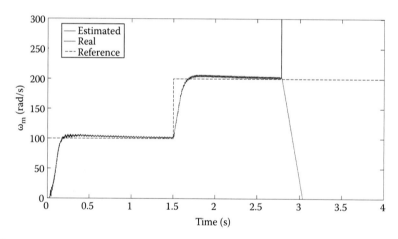

FIGURE 11.35
Estimated and measured speeds in field-weakening, BPN MRAS observer with the discrete Euler current model. (From Cirrincione, M. and Pucci, M., *IEEE Trans. Ind. Electron.*, 52(2), 532, April 2005.)

when the speed reaches about 200 rad/s, the drive becomes unstable, as expected. The same test has also been done, in numerical simulation by employing the MRAS BPN observer, with the approximated adaptive model of Ben-Brahim et al., [44] in simulation mode. Figure 11.36 shows the reference, the real, and the estimated speeds during two speed step references, respectively, at 100 and 200 rad/s, obtained with a specific value of the learning rate $\eta = 0.003$ of the BPN NN. This graph clearly shows that with this value of η, the drive at 200 rad/s tends to approach to instability with a large estimation error and with enormous oscillations of the estimated speed.

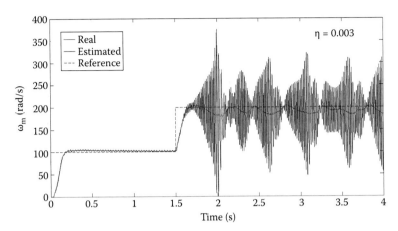

FIGURE 11.36
Estimated and measured speeds in field-weakening, BPN MRAS observer with the approximate current model.
(From Cirrincione, M. and Pucci, M., *IEEE Trans. Ind. Electron.*, 52(2), 532, April 2005.)

11.3.4 MCA EXIN + MRAS Observer

The so-called MCA EXIN + MRAS observer is the latest upgrade of the series of LS-based
MRAS observers [55,56]. In this observer, the modified Euler neural adaptive model of
Section 11.3.3 has been adopted. As far as the ADALINE training algorithm is concerned,
the MCA EXIN + has been adopted, which is a further improvement of the TLS EXIN (see
Chapter 8). A further improvement has been added, which permits the lowest working
speed to be reduced. The neural adaptive integrator used in the reference model has been
improved in its low-frequency behavior by adopting a learning factor varying according
to the reference speed of the drive.

Looking at Figure 11.18, it can be observed that, if a speed below 5 rad/s is required,
a neural integrator with $\tau = 1 \times 10^{-5}$ offers a better behavior, since it approximates the
ideal integrator well at much lower frequencies. Nevertheless, a neural integrator with
$\tau = 1 \times 10^{-5}$ cannot be suitably employed in the whole speed range of the drive, since the
adaptation time of the filter increases when the value of τ decreases, as clearly shown in
Figure 11.37, which shows the difference between the adaptation times obtained with
the two values of τ at the operating speed of 2 rad/s, when a 2% (of the rated voltage) DC
signal is superimposed to the voltage signal on phase sA. This figure clearly shows that
a lower value of τ permits a better flux estimation, but at the expense of a high filtering
adaptation time. For this reason, the use of low values of τ can bring about stability problems in the flux control loop, especially when a speed transient is required. In this regard,
the problem has been solved as follows: at reference speeds above 10 rad/s (in electrical
angles) and during each speed transient, the value of τ has been set to 2×10^{-4}, while in
speed steady-state at references from 10 down to 4 rad/s, the value of τ has been varied
linearly from 2×10^{-4} to 1×10^{-5} and then kept to this last value for lower reference speeds,
as shown in Figure 11.38.

It should be remarked that reducing the value of τ corresponds to moving the poles
of the neural filter toward the origin, which however does not affect the accuracy of
the integrator. This is not the case for the LPF integrator, where the amplitude of the
pole cannot be reduced too much, since the lower the amplitude of the pole, the higher
the drift at the LPF output caused by a DC drift at its input. Actually, if the final value
theorem is applied to both filters assuming a DC drift of amplitude E_{dr} at the input, the

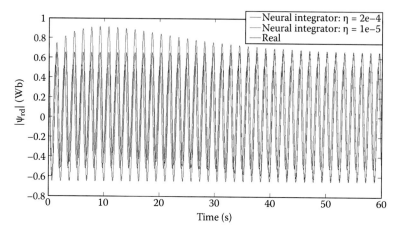

FIGURE 11.37
Rotor flux D-axis component obtained with a 2% (of the rated voltage) DC signal superimposed to the voltage signal on phase sA, with two values of τ (simulation). (From Cirrincione, M. et al., *IEEE Trans. Ind. Electron.*, 54(1), 127, February 2007.)

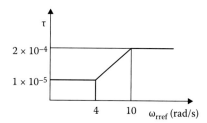

FIGURE 11.38
Variation of the learning factor according to the reference speed of the machine. (From Cirrincione, M. et al., *IEEE Trans. Ind. Electron.*, 54(1), 127, February 2007.)

output is a constant for the LPF and zero for the neural integrator, thanks to the presence of a zero in the origin:

LP filter:

$$\lim_{s\to 0} s\, G_{LP}(s)\frac{E_{dr}}{s} = \lim_{s\to 0} s\frac{1}{s+\alpha_p}\frac{E_{dr}}{s} = \frac{E_{dr}}{\alpha_p}$$

Neural integrator:

$$\lim_{s\to 0} s\, G_{int}(s)\frac{E_{dr}}{s} = \lim_{s\to 0} s\frac{s}{s^2+4\tau/T_{sp}s+4\tau^2/T_{sp}^2}\frac{E_{dr}}{s} = 0 \qquad (11.23)$$

where α_p is the pole of the LPF.

The MCA EXIN + MRAS observer has been tested experimentally on an FOC IM drive (see Appendices 11.A and 11.B). Figure 11.39 shows the reference, measured, and estimated speeds obtained, giving the TLS MRAS observer a square-waveform reference of amplitude 7 rad/s and pulsation 0.2 rad/s. These last figures show the capability of the observer

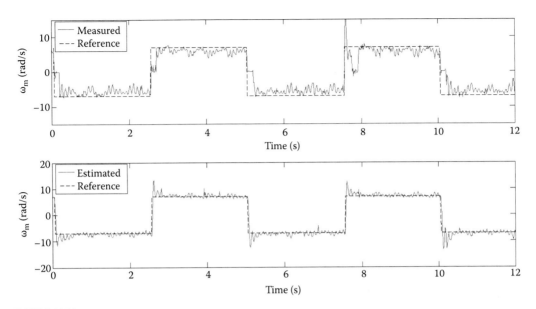

FIGURE 11.39
Reference, measured, and estimated speeds in a square-waveform reference of amplitude 7 rad/s and pulsation 0.2 rad/s with the MCA EXIN + MRAS observer. (From Cirrincione, M. et al., *IEEE Trans. Ind. Electron.*, 54(1), 127, February 2007.)

to follow a square-waveform reference of very low amplitude and high frequency that is a very challenging working condition.

11.4 TLS EXIN Full-Order Luenberger Adaptive Observer

Starting from the structure of the full-order Luenberger adaptive observer described in Section 6.4.7, a neural-based version of it has been developed in Refs [55] and [58]. The state equations of the IM have been discretized and rearranged so as to be represented by a linear NN (ADALINE). Even in this case, the TLS EXIN neuron has been adopted to train on-line the ADALINE in order to properly estimate the rotor speed. The NN observer has been further integrated with the on-line stator resistance estimation algorithm (*algorithm 3*) [59] described in Section 6.4.3.

11.4.1 State-Space Model of the IM

If the stator current and the rotor flux-linkage space-vectors are chosen as state variables, the state equations of the IM in the stationary reference frame can be written as [59]

$$\frac{d}{dt}\begin{bmatrix} \mathbf{i}_s \\ \boldsymbol{\psi}_r' \end{bmatrix} = \frac{d\mathbf{x}}{dt} = \begin{bmatrix} \mathbf{A}_{11} & \mathbf{A}_{12} \\ \mathbf{A}_{21} & \mathbf{A}_{22} \end{bmatrix} \begin{bmatrix} \mathbf{i}_s \\ \boldsymbol{\psi}_r' \end{bmatrix} + \begin{bmatrix} \mathbf{B}_1 \\ 0 \end{bmatrix} \mathbf{u}_s = \mathbf{A}\mathbf{x} + \mathbf{B}\mathbf{u}_s \qquad (11.24a)$$

$$\mathbf{i}_s = \mathbf{C}\mathbf{x} \qquad (11.24b)$$

where

$$\mathbf{A_{11}} = -\left\{ R_s/(\sigma L_s) + (1-\sigma)/\sigma T_r \right\} \mathbf{I} = a_{11}\mathbf{I} \tag{11.25a}$$

$$\mathbf{A_{12}} = L_m/(\sigma L_s L_r) \left\{ (1/T_r)\mathbf{I} - \omega_r \mathbf{J} \right\} = a_{12}\left\{ (1/T_r)\mathbf{I} - \omega_r \mathbf{J} \right\} \tag{11.25b}$$

$$\mathbf{A_{21}} = \left\{ L_m/T_r \right\}\mathbf{I} = a_{21}\mathbf{I} \tag{11.25c}$$

$$\mathbf{A_{22}} = -(1/T_r)\mathbf{I} + \omega_r \mathbf{J} = a_{22}\left\{ (1/T_r)\mathbf{I} - \omega_r \mathbf{J} \right\} \tag{11.25d}$$

$$\mathbf{B_1} = 1/(\sigma L_s)\mathbf{I} = b\mathbf{I} \tag{11.25e}$$

with

$$\mathbf{i_s} = \begin{bmatrix} i_{sD} & i_{sQ} \end{bmatrix}^T, \quad \mathbf{u_s} = \begin{bmatrix} u_{sD} & u_{sQ} \end{bmatrix}^T, \quad \mathbf{\Psi_r'} = \begin{bmatrix} \Psi_{rd} & \Psi_{rq} \end{bmatrix}^T, \quad \mathbf{C'} = \begin{bmatrix} 0 & \mathbf{I} \end{bmatrix},$$

$$\mathbf{C} = \begin{bmatrix} \mathbf{I} & 0 \end{bmatrix}, \quad \mathbf{I} = \begin{bmatrix} 1 & 0 \\ 0 & 1 \end{bmatrix}, \quad \mathbf{J} = \begin{bmatrix} 0 & -1 \\ 1 & 0 \end{bmatrix}.$$

In the earlier state representation, $\mathbf{x} = \begin{bmatrix} \mathbf{i_s}, \mathbf{\psi_r'} \end{bmatrix}$ is the state vector, composed of the stator current and rotor flux-linkage direct and quadrature components in the stationary reference frame, $\mathbf{u_s}$ is the input vector composed of the stator voltage direct and quadrature components in the stationary reference frame, $\mathbf{A}$ is the state matrix (4 × 4 matrix) depending on the rotor speed ω_r, $\mathbf{B}$ is the input matrix, and finally $\mathbf{C}$ is the output matrix.

11.4.2 Adaptive Speed Observer

The full-order Luenberger state observer can be obtained from Equations 11.24 if a correction term is added containing the difference between the actual and estimated states. In particular, since the only measurable state variables are the stator currents, the correction term involves only the error vector on the stator current $\mathbf{e}_{rr} = (\mathbf{i_s} - \hat{\mathbf{i}}_s)$, as in the following:

$$\frac{d\hat{\mathbf{x}}}{dt} = \hat{\mathbf{A}}\mathbf{x} + \mathbf{B}\mathbf{u_s} + \mathbf{G}(\hat{\mathbf{i}}_s - \mathbf{i_s}) \tag{11.26}$$

where $\wedge$ means the estimated values and $\mathbf{G}$ is the observer gain matrix which is designed so that the observer mentioned earlier is stable. For the proper choice of the observer gain matrix, see Section 6.4.7.

11.4.3 TLS-Based Speed Estimation

The TLS-based speed observer derives from a modification of Equation 11.24, in the sense that it exploits the first two scalar equations to estimate the rotor speed, as shown in the following in discrete form for digital implementation [58].

The first two scalar equations of the matrix equation (11.24) can be written as

$$\begin{cases} \dfrac{di_{sD}}{dt} = a_{11}i_{sD} + a_{12}p_r\hat{\psi}_{rd} + a_{12}\hat{\omega}_r\hat{\psi}_{rq} + bu_{sD} \\ \dfrac{di_{sQ}}{dt} = a_{11}i_{sQ} + a_{12}p_r\hat{\psi}_{rq} - a_{12}\hat{\omega}_r\hat{\psi}_{rd} + bu_{sQ} \end{cases} \qquad (11.27)$$

where the current components are measured variables and the rotor flux and speed are estimated ones. Moving from the continuous domain to the discrete one, and approximating the continuous derivative with the discrete filter $(1 - z^{-1})/T_{sp}\, z^{-1}$, where T_{sp} is the sampling time of the control systems, the following equations can be deduced:

$$\begin{cases} \dfrac{i_{sD}(k)-i_{sD}(k-1)}{T_{sp}} = a_{11}i_{sD}(k-1) + a_{12}p_r\hat{\psi}_{rd}(k-1) + a_{12}\hat{\omega}_r(k-1)\hat{\psi}_{rq}(k-1) + b\hat{u}_{sD}(k-1) \\ \dfrac{i_{sQ}(k)-i_{sQ}(k-1)}{T_{sp}} = a_{11}i_{sQ}(k-1) + a_{12}p_r\hat{\psi}_{rq}(k-1) - a_{12}\hat{\omega}_r(k-1)\hat{\psi}_{rd}(k-1) + b\hat{u}_{sQ}(k-1) \end{cases} \qquad (11.28)$$

where k is the current time sample. From (11.28), the following matrix equation can be easily written:

$$\begin{bmatrix} a_{12}T_{sp}\,\hat{\psi}_{rq}(k-1) \\ -a_{12}T_{sp}\hat{\psi}_{rd}(k-1) \end{bmatrix} \hat{\omega}_r(k-1)$$

$$= \begin{bmatrix} i_{sD}(k)-i_{sD}(k-1) - a_{11}T_{sp}i_{sD}(k-1) - a_{12}p_rT_{sp}\,\hat{\psi}_{rd}(k-1) - bT_{sp}\hat{u}_{sD}(k-1) \\ i_{sQ}(k)-i_{sQ}(k-1) - a_{11}T_{sp}i_{sQ}(k-1) - a_{12}p_rT_{sp}\,\hat{\psi}_{rq}(k-1) - bT_{sp}\hat{u}_{sQ}(k-1) \end{bmatrix} \qquad (11.29)$$

where $p_r = 1/T_r$.

A further improvement has been made in Ref. [55], where all the four scalar equations in Equation 8.7 have been exploited in order to retrieve the speed estimation, as shown in the following:

$$\begin{bmatrix} a_{12}T_{sp}\hat{\psi}_{rq}(k-1) \\ -a_{12}T_{sp}\hat{\psi}_{rd}(k-1) \\ -a_{22}T_{sp}\hat{\psi}_{rq}(k-1) \\ a_{22}T_{sp}\hat{\psi}_{rd}(k-1) \end{bmatrix} \hat{\omega}_r(k-1)$$

$$= \begin{bmatrix} i_{sD}(k)-i_{sD}(k-1) - a_{11}T_{sp}i_{sD}(k-1) - a_{12}p_rT_{sp}\hat{\psi}_{rd}(k-1) - bT_{sp}\hat{u}_{sD}(k-1) \\ i_{sQ}(k)-i_{sQ}(k-1) - a_{11}T_{sp}i_{sQ}(k-1) - a_{12}p_rT_{sp}\hat{\psi}_{rq}(k-1) - bT_{sp}\hat{u}_{sQ}(k-1) \\ -\hat{\psi}_{rd}(k) + \hat{\psi}_{rd}(k-1) + a_{21}T_{sp}i_{sd}(k-1) + a_{22}\,T_{sp}/T_r\,\hat{\psi}_{rd}(k-1) \\ -\hat{\psi}_{rq}(k) + \hat{\psi}_{rq}(k-1) + a_{21}T_{sp}i_{sq}(k-1) + a_{22}\,T_{sp}/T_r\,\hat{\psi}_{rq}(k-1) \end{bmatrix} \qquad (11.30)$$

Both Equations 11.29 and 11.30 are classical matrix equations of the type $\mathbf{Ax} \approx \mathbf{b}$, where $\mathbf{A}$ is called "data matrix," $\mathbf{b}$ is called "observation vector," and $\mathbf{x}$ is the unknown vector, equal to $\hat{\omega}_r$. LS techniques reveal, even in this case, the best solution to be adopted for on-line training of the ADALINE and, therefore, to solve in recursive form equations (11.29) and (11.30).

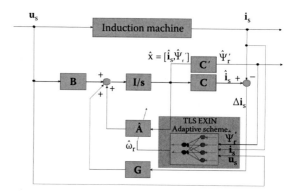

FIGURE 11.40
(See color insert.) Block diagram of the TLS EXIN full-order adaptive observer.

The matrix **A** is composed of the d-q-axis components of the rotor flux linkage which can be affected by errors and noise measurements, and the same can be said for the observation vector **b** which is also composed of the d-q-axis components of the rotor flux linkage and the d-q-axis components of the stator current space-vector; the problem under hand is therefore a TLS problem rather than an LS problem. Any LS technique different from TLS reveals inadequate [48,49].

Figure 11.40 shows the block diagram of the TLS EXIN full-order adaptive observer. It should be remarked that the computation of the rotor speed by means of the TLS estimator is performed through the minimization of the residual of the matrix equation (11.29) or (11.30). The residual is strongly dependent on the rotor flux estimation error, while all its other terms are dependent on the measured values of the electrical variables (i_s and u_s), and are also affected by measurement errors. The TLS inherently gives the best solution for the rotor speed in spite of these uncertainties.

It should be remarked that, to compute the rotor speed, all the four scalar equations of Equation 11.30 are not strictly required. There are basically three possibilities:

1. To solve the first two scalar equations of Equation 11.24, which corresponds to the discretization of a combination of the stator and rotor equations of the machine
2. To solve the last two equations of Equation 11.24, which corresponds to a discretization of the rotor equations of the machine
3. To solve simultaneously all of the four scalar equations as in Equation 11.24

The third possibility reveals the best because, even if it is the most cumbersome from the computational point of view, it offers the highest robustness to any parameter uncertainty in the data matrix and the observation vector. In real applications such uncertainties could be caused, for example, by a nonperfect measurement of the magnetizing inductance. Figure 11.41 shows the real and estimated speeds as well as the real and estimated amplitudes of the rotor flux linkage, obtained in numerical simulation, at the reference speed of 1 rad/s when the parameter a_{12} of the observer has been set equal to 1.5 times the corresponding one in the machine. The test has been done twice, respectively, adopting Equations 11.29 and 11.30. The figure clearly shows that the estimated values of the variables correctly follow their references, since they are controlled in feedback, while their real values are closer to their references when all the four scalar equation are employed than when only the first two scalar equations are adopted. Thus, the employment of all scalar equations of the observer to compute the rotor speed offers a higher robustness to parameter uncertainty of the observer, as expected.

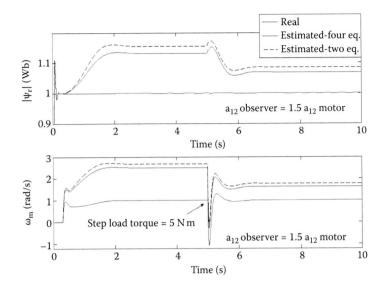

FIGURE 11.41
Rotor flux linkage and rotor speed of the machine at 1 rad/s when all four or only two equations of the full-order observer are used by the TLS algorithm, and the a_{12} parameter of the observer is detuned. (From Cirrincione, M. et al., *IEEE Trans. Ind. Electron.*, 54(1), 127, February 2007.)

11.4.4 Stability Issues of the TLS EXIN Full-Order Adaptive Observer

In the following, the demonstration of the stability of the TLS EXIN full-order adaptive observer is provided. The demonstration refers to the most general case when Equation 11.30 is adopted.
In the TLS adaptive speed observer, three relations are to be considered:

1. The real motor equations, that is, Equation 11.24a, which are rewritten here in a slightly different way:

$$\dot{\mathbf{x}} = \begin{pmatrix} a_{11}\mathbf{I} \\ a_{21}\mathbf{I} \end{pmatrix} \mathbf{i}_s + \frac{1}{T_r} \begin{pmatrix} a_{12}\mathbf{I} \\ a_{22}\mathbf{I} \end{pmatrix} \mathbf{\psi}_r - \omega_r \begin{pmatrix} a_{12}\mathbf{I} \\ a_{22}\mathbf{I} \end{pmatrix} \mathbf{\psi}_r + \mathbf{B}\mathbf{u}_s \quad \text{where } \mathbf{x} = \begin{pmatrix} \mathbf{i}_s \\ \mathbf{\psi}_r \end{pmatrix} \quad (11.31)$$

2. The full-order observer equations, which estimate the stator current and the rotor flux (Equation 11.26), are rewritten as follows:

$$\dot{\hat{\mathbf{x}}} = \begin{pmatrix} a_{11}\mathbf{I} \\ a_{21}\mathbf{I} \end{pmatrix} \hat{\mathbf{i}}_s + \frac{1}{T_r} \begin{pmatrix} a_{12}\mathbf{I} \\ a_{22}\mathbf{I} \end{pmatrix} \hat{\mathbf{\psi}}_r - \hat{\omega}_r \begin{pmatrix} a_{12}\mathbf{I} \\ a_{22}\mathbf{I} \end{pmatrix} \hat{\mathbf{\psi}}_r + \mathbf{B}\mathbf{u}_s + \mathbf{G}(\hat{\mathbf{i}}_s - \mathbf{i}_s) \quad \text{where } \hat{\mathbf{x}} = \begin{pmatrix} \hat{\mathbf{i}}_s \\ \hat{\mathbf{\psi}}_r \end{pmatrix} \quad (11.32)$$

where $\wedge$ stands for the estimated values. In particular, $\hat{\omega}_r$ is the speed estimated with the TLS speed observer through the equations in the following.
3. The TLS speed observer (Equation 11.30) which is written in continuous form as follows:

$$\dot{\tilde{\mathbf{x}}} \cong \begin{pmatrix} a_{11}\mathbf{I} \\ a_{21}\mathbf{I} \end{pmatrix} \mathbf{i}_s + \frac{1}{T_r} \begin{pmatrix} a_{12}\mathbf{I} \\ a_{22}\mathbf{I} \end{pmatrix} \hat{\mathbf{\psi}}_r - \hat{\omega}_r \begin{pmatrix} a_{12}\mathbf{I} \\ a_{22}\mathbf{I} \end{pmatrix} \hat{\mathbf{\psi}}_r + \mathbf{B}\mathbf{u}_s \quad \text{where } \tilde{\mathbf{x}} = \begin{pmatrix} \mathbf{i}_s \\ \hat{\mathbf{\psi}}_r \end{pmatrix} \quad (11.33)$$

where

$$\mathbf{J} = \begin{bmatrix} 0 & -1 \\ 1 & 0 \end{bmatrix}$$

$$\mathbf{I} = \begin{bmatrix} 1 & 0 \\ 0 & 1 \end{bmatrix}$$

By defining $\mathbf{e} = \mathbf{x} - \hat{\mathbf{x}}$, the following equation can be obtained if (11.32) is subtracted from (11.31):

$$\dot{\mathbf{e}} = (\mathbf{A} - \mathbf{GC})\mathbf{e} - \Delta\mathbf{A}\hat{\mathbf{x}} \tag{11.34}$$

where $\Delta\mathbf{A} = \hat{\mathbf{A}} - \mathbf{A} = \begin{bmatrix} \mathbf{O}_2 & -(\hat{\omega}_r - \omega_r)\mathbf{J}a_{12} \\ \mathbf{O}_2 & (\hat{\omega}_r - \omega_r)\mathbf{J} \end{bmatrix}$

where $\mathbf{O}_2 = \begin{bmatrix} 0 & 0 \\ 0 & 0 \end{bmatrix}$

and $\mathbf{C}$ has been previously defined.

If the matrix

$$\tilde{\mathbf{C}} = \begin{bmatrix} \mathbf{O}_2 & \mathbf{O}_2 \\ \mathbf{O}_2 & \mathbf{I} \end{bmatrix}$$

is defined, then the difference between (11.31) and (11.33) is given by

$$\tilde{\mathbf{C}}\dot{\mathbf{e}} \cong \mathbf{A}\tilde{\mathbf{C}}\mathbf{e} - \Delta\mathbf{A}\hat{\mathbf{x}} \tag{11.35}$$

Now it is important to remark that relation (11.35) comes directly from the adoption of an LS speed observer (11.33) and is therefore peculiar of this method.

If (11.34) and (11.35) are subtracted, then the following relation is given:

$$(\mathbf{I}_4 - \tilde{\mathbf{C}})\dot{\mathbf{e}} \cong (\mathbf{A} - \mathbf{GC} - \mathbf{A}\tilde{\mathbf{C}})\mathbf{e} \tag{11.36}$$

where $\mathbf{I}_4 = \begin{bmatrix} 1 & 0 & 0 & 0 \\ 0 & 1 & 0 & 0 \\ 0 & 0 & 1 & 0 \\ 0 & 0 & 0 & 1 \end{bmatrix}$

This is the key formula for the stability of the observer.

If the following vector is defined as follows,

$$\mathbf{e} = \begin{pmatrix} \mathbf{e}_a \\ \overline{\mathbf{e}_b} \end{pmatrix} = \begin{pmatrix} e_1 \\ e_2 \\ e_3 \\ e_4 \end{pmatrix} \tag{11.37}$$

then from (11.36), the following relations hold if, without loss of generality, $\mathbf{G} = \mathbf{0}$:

$$\dot{\mathbf{e}}_\mathbf{a} \cong (a_{11}\mathbf{I})\,\mathbf{e}_\mathbf{a} \tag{11.38a}$$

$$\dot{\mathbf{e}}_\mathbf{b} \cong (a_{21}\mathbf{I})\,\mathbf{e}_\mathbf{a} \tag{11.38b}$$

From (11.38a), it immediately results that $\mathbf{e}_\mathbf{a} \to \mathbf{0}$, that is, $\hat{\mathbf{i}}_\mathbf{s} \to \mathbf{i}_\mathbf{s}$ if $t \to \infty$, since a_{11} is less than 0. From (11.38b), it results that $\mathbf{e}_\mathbf{b} \to const$ if $t \to \infty$. In the following, it will be proved by the Lyapunov theorem that this constant is 0.

Let the Lyapunov function be

$$v = \mathbf{e}^T\mathbf{e} + (\hat{\omega}_r - \omega_r)^2/\mu \tag{11.39}$$

which is null for $\mathbf{e} = \mathbf{0}$ and $\hat{\omega}_r = \omega_r$, and μ is a positive constant.

Then by differentiation, it results that

$$\frac{dv}{dt} = \left(\frac{d\mathbf{e}^T}{dt}\right)\mathbf{e} + \mathbf{e}^T\left(\frac{d\mathbf{e}}{dt}\right) + \frac{1}{\mu}\frac{d}{dt}(\hat{\omega}_r - \omega_r)^2 = \frac{d\tilde{v}}{dt} + \frac{1}{\mu}\frac{d}{dt}(\hat{\omega}_r - \omega_r)^2 \tag{11.40}$$

where $\tilde{v} = \mathbf{e}^T\mathbf{e}$. Then by recalling (11.37), the following relation exists:

$$\frac{d\tilde{v}}{dt} = \frac{d\mathbf{e}^T}{dt}\mathbf{e} + \mathbf{e}^T\left(\frac{d\mathbf{e}}{dt}\right) = \left(\frac{d\mathbf{e}_a^T}{dt}, \frac{d\mathbf{e}_b^T}{dt}\right)\binom{\mathbf{e}_a}{\mathbf{e}_b} + \left(\mathbf{e}_a^T, \mathbf{e}_b^T\right)\binom{d\mathbf{e}_a/dt}{d\mathbf{e}_b/dt}$$

that is, by using (11.38) and recalling that $(a_{11}\mathbf{I}) = \mathbf{A}_{11}$ and $(a_{21}\mathbf{I}) = \mathbf{A}_{21}$, it results that

$$\frac{d\tilde{v}}{dt} = \frac{d\mathbf{e}_a^T}{dt}\mathbf{e}_a + \frac{d\mathbf{e}_b^T}{dt}\mathbf{e}_b + \mathbf{e}_a^T\left(\frac{d\mathbf{e}_a}{dt}\right) + \mathbf{e}_b^T\left(\frac{d\mathbf{e}_b}{dt}\right) =$$

$$= \mathbf{e}_a^T\mathbf{A}_{11}^T\mathbf{e}_a + \mathbf{e}_a^T\mathbf{A}_{11}\mathbf{e}_a + \mathbf{e}_a^T\mathbf{A}_{21}\mathbf{e}_b + \mathbf{e}_b^T\mathbf{A}_{21}\mathbf{e}_a =$$

$$= \mathbf{e}_a^T\left(\mathbf{A}_{11}^T + \mathbf{A}_{11}\right)\mathbf{e}_a + \left(\mathbf{e}_a^T\mathbf{e}_b + \mathbf{e}_b^T\mathbf{e}_a\right)a_{21} = a_{11}\mathbf{e}_a^T\mathbf{e}_a + 2\mathbf{e}_a^T\mathbf{e}_b a_{21} \tag{11.41}$$

This means that, since $\mathbf{e}_\mathbf{a} \to \mathbf{0}$ if $t \to \infty$, then $\dfrac{d\tilde{v}}{dt} \to 0$ and finally that

$$\frac{dv}{dt} \cong \frac{1}{\mu}\frac{d}{dt}(\hat{\omega}_r - \omega_r)^2 = 2\frac{(\hat{\omega}_r - \omega_r)}{\mu}\frac{d\hat{\omega}_r}{dt} \tag{11.42}$$

In case an LS method were used for computing the speed, either an OLS or TLS, a quadratic error function is used, which is null if $\hat{\omega}_r = \omega_r$. The minimum value of this error function can be reached with any gradient method. This is the method used here, and it can be approximated asymptotically with the following ordinary differential equation (ODE):

$$\frac{d\hat{\omega}_r}{dt} = -\alpha\frac{dE_{LS}}{d\hat{\omega}_r} \tag{11.43}$$

where
  α is a positive constant
  E_{LS} is the error function of the chosen LS method

If the OLS method for the speed estimation is used, then this error function is a parabola whose minimum value, which is null in case of data not corrupted by noise, is reached for $\hat{\omega}_r = \omega_r$. This means that

$$\left(\hat{\omega}_r - \omega_r\right)\frac{dE_{OLS}}{d\hat{\omega}_r} > 0 \quad \forall \hat{\omega}_r \neq \omega_r \tag{11.44}$$

then (11.43) yields

$$\left(\hat{\omega}_r - \omega_r\right)\frac{d\hat{\omega}_r}{dt} = -\alpha\left(\hat{\omega}_r - \omega_r\right)\frac{dE_{OLS}}{d\hat{\omega}_r} < 0 \tag{11.45}$$

which means obviously, from (11.42), that for $t \to \infty$,

$$\frac{dv}{dt} \cong \frac{1}{\mu}\frac{d}{dt}\left(\hat{\omega}_r - \omega_r\right)^2 = 2\frac{\left(\hat{\omega}_r - \omega_r\right)}{\mu}\frac{d\hat{\omega}_r}{dt} < 0 \tag{11.46}$$

which implies the stability of the OLS observer.
  In case a TLS error were used, then

$$\left(\hat{\omega}_r - \omega_r\right)\frac{dE_{TLS}}{d\hat{\omega}_r} > 0 \quad \forall \hat{\omega}_r \neq \omega_r \wedge \hat{\omega}_r < \left|\hat{\omega}_{r\,max}\right| \tag{11.47}$$

where $\hat{\omega}_{r\,max}$ is the value at which E_{TLS} reaches its unique maximum. Remember in fact that

$$E_{TLS} = \frac{E_{OLS}}{1 + \hat{\omega}_r^2}$$

Thus, whenever the initial estimated speed verifies $\hat{\omega}_r < \left|\hat{\omega}_{r\,max}\right|$, then a relation similar to (11.46) holds, to which the same considerations can be made, and this proves the stability of the TLS observer. For further details on the TLS domain of convergence, see Ref. [48].

11.4.5 Experimental Results with the TLS EXIN Full-Order Luenberger Adaptive Observer

The TLS EXIN full-order Luenberger adaptive observer has been tested experimentally on an FOC IM drive. The adopted FOC scheme is fully described in Appendix 11.A, while the corresponding test setup has been described in Appendix 11.B. The observer has been integrated with the *algorithm 3* for the on-line stator resistance estimation described in Section 6.4.3. It has been tested in several challenging working conditions, as shown in the following.

11.4.5.1 Dynamic Performance

The dynamic performance of the drive has been verified by giving the following set of speed step variations $0 \rightarrow 100 \rightarrow -100 \rightarrow 0$ rad/s at no load. Figure 11.42 shows the estimated speed, measured speed, and the speed estimation error obtained with the TLS EXIN full-order Luenberger adaptive observer. Figure 11.43 shows the corresponding rotor flux-linkage amplitude and electromagnetic torque. Figure 11.44, which draws a zoom of the speed and torque waveforms during the speed reversal obtained with both the NN and the classic full-order Luenberger adaptive observer, shows that the speed reversal with both the observers is accomplished in 0.29 s, and the torque response is practically instantaneous. These figures show also that the TLS observer outperforms the classic one in terms of maximum instantaneous speed estimation error in speed reversal. In particular, during the zero crossing in the speed reversal, the estimation error with the TLS observer is quite less than that obtained with the classic one. On the other hand, the two observers are equivalent during the steady-state as expected, since they both work at high speeds. By employing the TLS-based observer, the instantaneous speed estimation error is maximum during the first speed transient, when the adaptive observer is not yet correctly tuned due to the mismatch between the parameters of the observer and the real ones of the machine, while afterward it is almost negligible also during the speed reversal. This is not the case of the classic adaptive observer, presenting an instantaneous speed estimation error during the speed reversal of almost 20 rad/s.

Afterward, the drive has been given a start-up transient with a step speed reference of 5 rad/s (47.7 rpm) at no load, with the machine previously magnetized. Figure 11.45, which plots the estimated, measured, and reference speed waveforms during this test, shows a

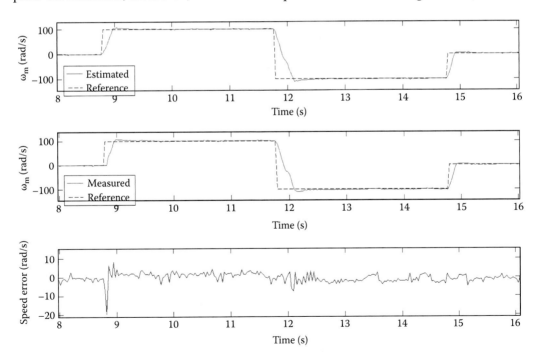

FIGURE 11.42
Reference, estimated, and measured speeds and speed estimation error during a set of speed steps with a reversal from 100 (955 rpm) to −100 rad/s with the TLS EXIN full-order Luenberger adaptive observer. (From Cirrincione, M. et al., *IEEE Trans. Ind. Appl.*, 42(1), 89, January/February 2006.)

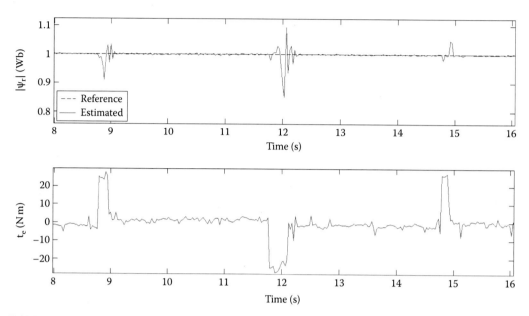

FIGURE 11.43
Rotor flux-linkage amplitude and electromagnetic torque during a set of speed steps with a reversal from 100 (955 rpm) to −100 rad/s (−955 rpm) with the TLS EXIN full-order Luenberger adaptive observer. (From Cirrincione, M. et al., *IEEE Trans. Ind. Appl.*, 42(1), 89, January/February 2006.)

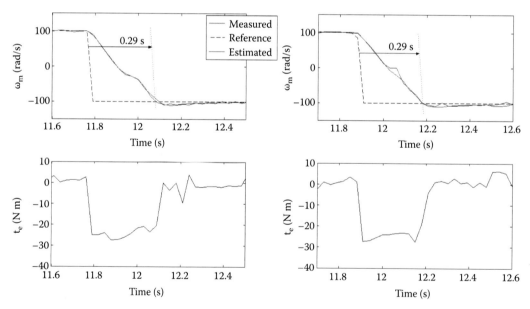

FIGURE 11.44
Reference, estimated, and measured speeds and electromagnetic torque during the speed reversal from 100 (955 rpm) to −100 rad/s (−955 rpm) with both the TLS EXIN and the classic full-order Luenberger adaptive observers. (From Cirrincione, M. et al., *IEEE Trans. Ind. Appl.*, 42(1), 89, January/February 2006.)

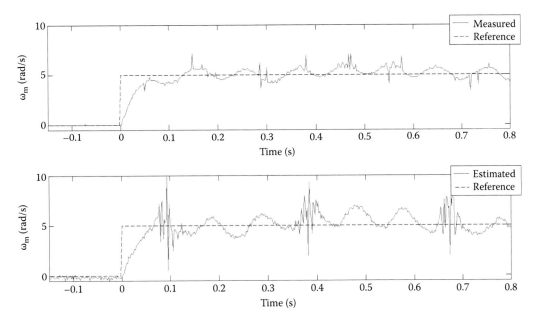

FIGURE 11.45
Reference, estimated, and measured speeds during a speed step reference of 5 rad/s (47.7 rpm) at no load with the TLS EXIN full-order Luenberger adaptive observer. (From Cirrincione, M. et al., *IEEE Trans. Ind. Appl.*, 42(1), 89, January/February 2006.)

good match of the estimated speed with the measured one also during the start-up transient at low speed. Moreover, there is practically no time delay between the speed step command and the movement of the rotor.

Finally, the dynamic performance of the proposed observer has been tested by giving it a square-wave speed reference with the lowest possible amplitude and the highest possible pulsation. Figure 11.46 shows the reference, measured, and estimated speeds obtained, giving the drive a square-waveform speed reference of amplitude 2 rad/s and pulsation 0.3 rad/s. It shows the capability of the observer to follow a square-waveform reference of very low amplitude and high frequency. It should be noted that, during the zero crossing of the speed, there is a time interval during which the real speed of the machine remains equal to 0.

11.4.5.2 Accuracy in the Low-Speed Ranges

The drive has been given a constant speed reference of 5 rad/s (47.7 rpm), with no load. Figure 11.47 shows the reference, the measured, and the estimated speeds and the estimation error with the TLS EXIN full-order Luenberger adaptive observer. Figure 11.48 shows the corresponding waveforms of the measured and estimated i_{sD} as well as the estimated values of R_s and R_r. Figure 11.47 clearly shows that the TLS observer has a slight ripple in the estimated speed with a relative speed error of 3.4%. Figure 11.48 shows a good agreement between the measured and the estimated i_{sD}. Even the values of R_s and R_r are properly estimated.

11.4.5.3 Accuracy at Very Low Speed (below 2 rad/s)

The drive has been given a constant speed reference of 0.5 rad/s at no load. Figure 11.49 shows the reference, the measured, and the estimated speeds and the speed estimation during this test. Figure 11.50 shows the corresponding waveforms of the measured and

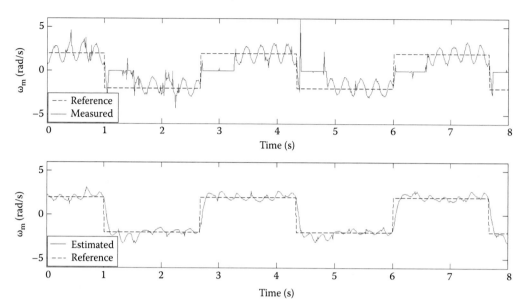

FIGURE 11.46
Reference, measured, and estimated speeds in a square-waveform reference of amplitude 2 rad/s and pulsation 0.3 rad/s with the TLS EXIN full-order Luenberger adaptive observer. (From Cirrincione, M. et al., *IEEE Trans. Ind. Electron.*, 54(1), 127, February 2007.)

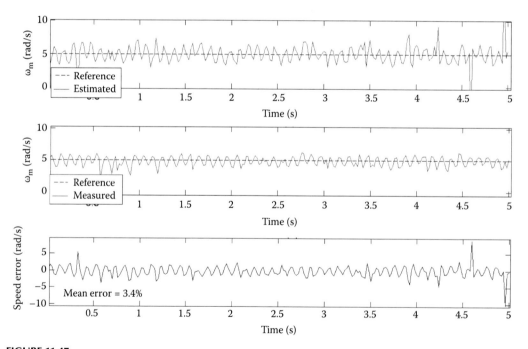

FIGURE 11.47
Reference, estimated, and measured speeds and speed estimation error during a constant reference of 5 rad/s (47.7 rpm) at no load with the TLS EXIN full-order Luenberger adaptive observer. (From Cirrincione, M. et al., *IEEE Trans. Ind. Appl.*, 42(1), 89, January/February 2006.)

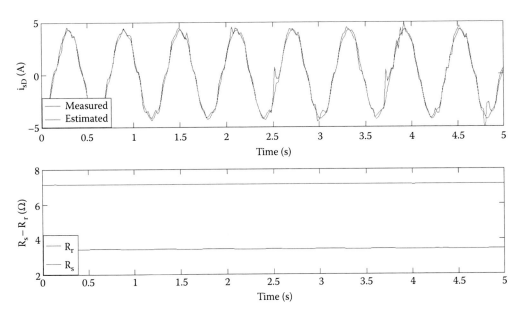

FIGURE 11.48
Measured and estimated i_{sD} and estimated R_s and R_r during a constant reference of 5 rad/s (47.7 rpm) at no load with the TLS EXIN full-order Luenberger adaptive observer. (From Cirrincione, M. et al., *IEEE Trans. Ind. Appl.*, 42(1), 89, January/February 2006.)

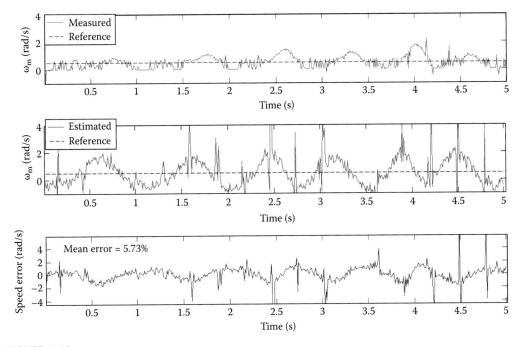

FIGURE 11.49
Reference, estimated, and measured speeds and speed estimation error during a constant reference of 0.5 rad/s (4.77 rpm) with the TLS EXIN full-order Luenberger adaptive observer. (From Cirrincione, M. et al., *IEEE Trans. Ind. Appl.*, 42(1), 89, January/February 2006.)

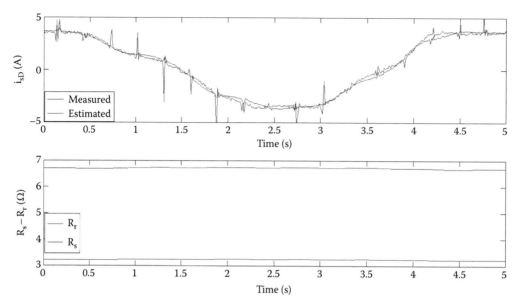

FIGURE 11.50
Measured and estimated i_{sD} and estimated R_s and R_r during a constant reference of 0.5 rad/s (4.77 rpm) with the TLS EXIN full-order Luenberger adaptive observer. (From Cirrincione, M. et al., *IEEE Trans. Ind. Appl.*, 42(1), 89, January/February 2006.)

TABLE 11.2

Mean Estimation Error % at the Speeds
of 1 rad/s (9.55 rpm) and 0.5 rad/s (4.77 rpm)
at No Load and Rated Load

	Mean Error %	
Speed (rad/s)	No Load	Rated Load
1	8	22.97
0.5	5.73	24.6

Source: From Cirrincione, M. et al., *IEEE Trans. Ind. Appl.*, 42(1), 89, January/February 2006.

estimated i_{sD} current components and of the estimated R_s and R_r. These figures show that the measured speed of the motor follows its reference correctly, even if with some small oscillations; at the same time, thanks to the on-line estimation of R_s and R_r, the estimated direct component of the stator current fits the measured one well at both reference speeds.

These figures clearly show that the TLS-based observer with the R_s and R_r on-line estimators and with the inverter nonlinearity compensation permits the drive to work in the very-low-speed region, with the average estimation error shown in Table 11.2. This performance is unattainable by the classic full-order Luenberger adaptive observer.

11.4.5.4 Robustness to Load Perturbations

The robustness of the speed response of the TLS observer to a sudden load torque variation has been verified. The drive, operated at the constant speed of 20 rad/s (191 rpm), has been given two subsequent load torque steps, in particular 0→11→0 N m. Figure 11.51 shows the waveforms of the electromagnetic torque of the drive and the load torque created by the

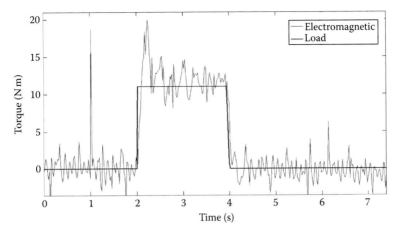

FIGURE 11.51
Electromagnetic torque and load torque during a set of load torque steps of 11 N m with the TLS EXIN full-order Luenberger adaptive observer. (From Cirrincione, M. et al., *IEEE Trans. Ind. Appl.*, 42(1), 89, January/February 2006.)

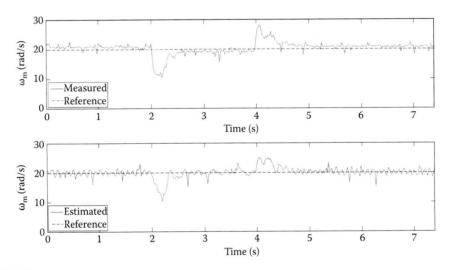

FIGURE 11.52
Reference, estimated, and measured speeds during a set of load torque steps of 11 N m with the TLS EXIN full-order Luenberger adaptive observer. (From Cirrincione, M. et al., *IEEE Trans. Ind. Appl.*, 42(1), 89, January/February 2006.)

torque-controlled DC machine. Figure 11.52 shows the corresponding reference, measured, and estimated speeds during this test. These figures clearly show that the drive response occurs immediately when the torque steps are given. Moreover, even during the speed transient caused by the torque step, the estimated speed follows the real one very well.

11.4.5.5 Regenerative Mode at Very Low Speed

The drive has been operated at very low speed with a negative load torque (regenerative mode). The classic full-order Luenberger adaptive observer, operated with $\mathbf{G} = \mathbf{0}$, has an unstable behavior in these working conditions as clearly shown in Figure 11.53, which

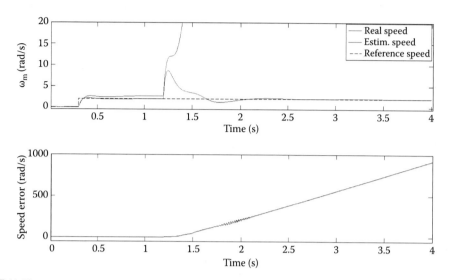

FIGURE 11.53
Reference, estimated, and real speeds and speed estimation error during a constant speed reference of 2 rad/s and a load torque step of −10 N m (regenerative mode) with the full-order Luenberger adaptive observer. (From Cirrincione, M. et al., *IEEE Trans. Ind. Appl.*, 42(1), 89, January/February 2006.)

shows the reference, real, and estimated speed waveforms, obtained in numerical simulations, when the drive is given a constant speed reference of 2 rad/s and then a negative load torque step of −10 N m.

In contrast to this, the TLS observer has proved to have a stable behavior in such conditions. Figure 11.54 shows the reference, the measured, and the estimated speeds and the

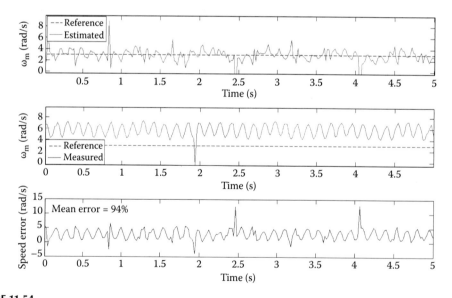

FIGURE 11.54
Reference, estimated, and measured speeds and speed estimation error during a constant speed reference of 3 rad/s with a negative load torque of −10 N m (regenerative mode) with the TLS EXIN full-order Luenberger adaptive observer. (From Cirrincione, M. et al., *IEEE Trans. Ind. Appl.*, 42(1), 89, January/February 2006.)

speed estimation error obtained in an experimental test with the TLS-based observer, when the drive has been given a constant speed reference, respectively, of 3 rad/s (28.64 rpm) and a constant negative load torque of −10 N m. This figure clearly shows that the TLS observer works properly even at very low speed and with negative torque, whereas the classic adaptive observer presents its well-known instability problems [60–63]. However, there exists a big average percent speed error (94%) in this speed range. Such bias error is due to the detuning of the TLS observer in this test due to parameter mismatch.

11.4.5.6 Field-Weakening Operation

The drive has been operated in the field-weakening zone, above its rated speed of 157 rad/s (1500 rpm). To test the speed accuracy both in speed transient and in steady-state, a start-up with a speed step reference of 200 rad/s (1910 rpm) at no load has been given with the machine previously magnetized. Figure 11.55 plots the reference, measured, and the estimated speeds as well as the rotor flux linkage and the i_{sy} current component; it shows a very good match between the estimated and measured speeds during the whole speed transient, both below and above the rated speed when the control system commands the rotor flux reduction. Also in steady-state, the accuracy of the speed estimation is very good with a negligible percent error.

11.4.5.7 Zero-Speed Operation

The drive has been operated at the rated rotor flux linkage at zero speed. The test has been performed at no load and at the load of 5 N m (about 50% of the rated load). Figures 11.56 and 11.57 show the waveforms of the reference, measured, and estimated speeds and speed estimation error for a time interval of 60 s, obtained, respectively, with the TLS and the classic full-order observers. They show that with the TLS observer, after

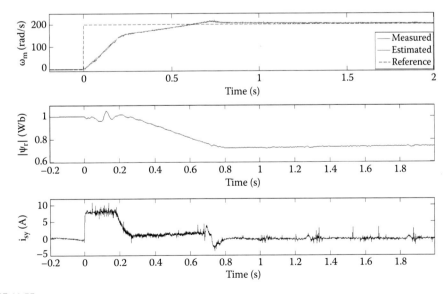

FIGURE 11.55
Reference, estimated, and measured speeds during a speed step reference of 200 rad/s (1910 rpm) in the field-weakening zone with the TLS EXIN full-order Luenberger adaptive observer. (From Cirrincione, M. et al., *IEEE Trans. Ind. Appl.*, 42(1), 89, January/February 2006.)

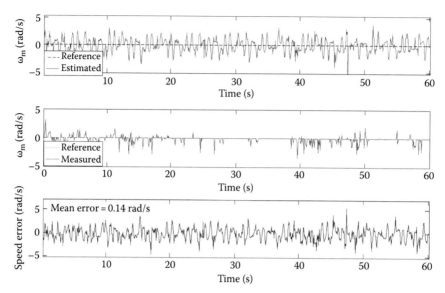

FIGURE 11.56
Reference, estimated, and measured speeds and speed estimation error during zero-speed operation at no load with the TLS EXIN full-order Luenberger adaptive observer. (From Cirrincione, M. et al., *IEEE Trans. Ind. Appl.*, 42(1), 89, January/February 2006.)

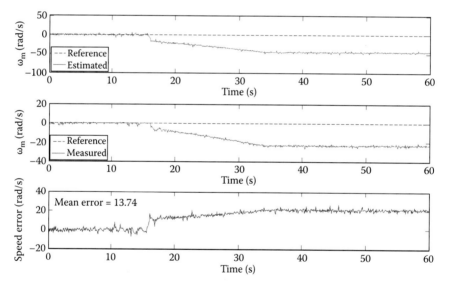

FIGURE 11.57
Reference, estimated, and measured speeds and speed estimation error during zero-speed operation at no load with the full-order Luenberger adaptive observer. (From Cirrincione, M. et al., *IEEE Trans. Ind. Appl.*, 42(1), 89, January/February 2006.)

the magnetization of the machine, the drive can work properly at zero speed and at no load, even without any signal injection.

This is mainly due to the fact that the TLS algorithm estimating the rotor speed uses the measured stator current signals and not the estimated ones, as in the classic observer. In particular, the estimated speed has slight oscillations around 0 rad/s and the measured

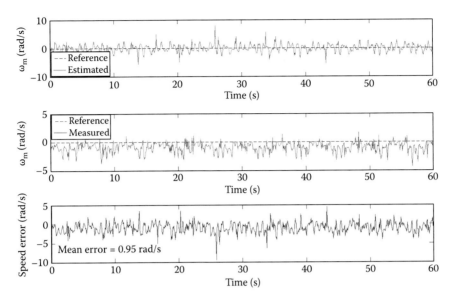

FIGURE 11.58
Reference, estimated, and measured speeds and speed estimation error during zero-speed operation with a
5 N m load torque with the TLS EXIN full-order Luenberger adaptive observer.

speed is always 0. In contrast to this, the classic observer at almost 15 s after the magnetization of the machine has an unstable behavior with the machine running at 45 rad/s with a mean speed estimation error of 13.74 rad/s. The drive has exhibited this kind of unstable behavior after repeating such tests more times.

Figure 11.58 shows the waveforms of the reference, measured, and estimated speeds and speed estimation error at the load torque of 5 N m for a time interval of 60 s obtained with the TLS observer. It shows that, on average, the machine can stay at zero speed even with half the rated load. On the contrary, the classic full-order observer was proved to go to instability at zero speed even with light load torque.

11.4.6 Experimental Comparative Tests

The behavior of five speed observers has been experimentally compared in two working conditions: a speed reversal 100→−100 rad/s and in zero-speed operation at no load. The speed observers under test are the following:

1. The TLS EXIN full-order Luenberger adaptive observer
2. The full-order Luenberger adaptive observer
3. The TLS EXIN MRAS observer
4. The BPN MRAS observer
5. The open-loop speed estimator (scheme in Figure 6.18)

Figure 11.59 shows the reference, measured, and estimated speeds and the instantaneous speed estimation error during the speed reversal. This figure shows that the speed reversal is accomplished with all five observers in less than 0.5 s, with the TLS adaptive observer,

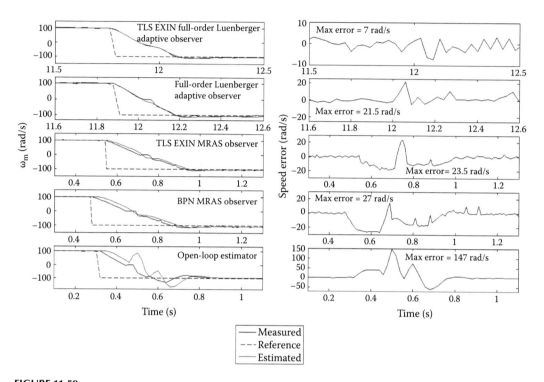

FIGURE 11.59

Reference, measured, and estimated speeds and speed estimation error with all five speed observers during a speed reversal from 100 to −100 rad/s. (From Cirrincione, M. et al., *IEEE Trans. Ind. Electron.*, 54(1), 127, February 2007.)

the classic adaptive observer, and the TLS MRAS observer overcoming in reversal time the BPN MRAS observer and the open-loop estimator. The TLS MRAS observer is slightly better than the BPN MRAS in terms of improved speed bandwidth because of the necessary digital filtering of the estimated speed, which is not necessary either in the TLS MRAS observer, whose adaptive model is employed in prediction mode, or in the TLS adaptive observer. In particular, the TLS MRAS observer can exploit a 3 dB bandwidth which is almost 16% higher than the one of the BPN MRAS observers.

With reference to the estimation accuracy in speed transient conditions, this figure shows that the best results are achieved by the TLS adaptive observer which presents a maximum instantaneous estimation error of 7 rad/s, then there is the classic adaptive observer which presents a maximum instantaneous estimation error of 21.5 rad/s, then the TLS MRAS observer which presents a maximum instantaneous estimation error of 23.5 rad/s, then the BPN MRAS observer which presents a maximum instantaneous estimation error of 27 rad/s, and finally the open-loop estimator which presents a maximum instantaneous estimation error of 147 rad/s.

With regard to the open-loop estimator, these very high values of the instantaneous speed estimation error are somehow typical of all open-loop speed estimators, but they can be certainly explained also by considering that the slip pulsation is computed using the time derivative of the quadrature component of stator current in the stator flux–oriented reference frame (see Equation 6.19), which tends to amplify any high-frequency noise present

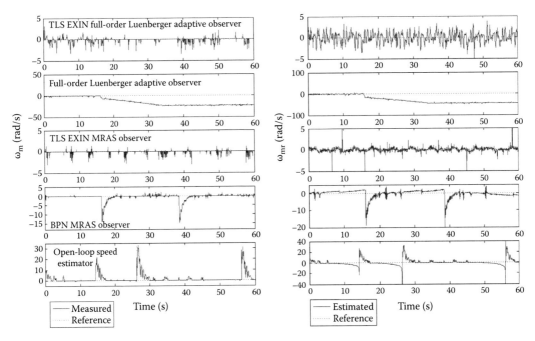

FIGURE 11.60

Reference, measured, and estimated speeds and speed estimation error with all five speed observers during zero-speed operation at no load. (From Cirrincione, M. et al., *IEEE Trans. Ind. Electron.*, 54(1), 127, February 2007.)

in measurements. This kind of phenomenon is present in none of the other four speed observers, as they are all closed-loop speed observers.

All five observers have been compared during a test at the rated rotor flux linkage at zero speed at no load. Figure 11.60 shows the waveforms of the reference, measured, and estimated speeds during the time interval of 60 s, obtained with all five observers. This figure shows that the best results are achieved by the TLS adaptive observer and by TLS MRAS observer, which can work correctly at zero speed at no load; then there is the BPN MRAS observer and the open-loop estimator. The former, because of the employment of the adaptive model in simulation mode, has a worse behavior than the TLS MRAS observer. Finally, there is the classic adaptive observer which has an unstable behavior with the machine running at 45 rad/s after some seconds. The correct behavior of the TLS MRAS observer and the TLS adaptive observer at zero speed are caused, first, by the employment of the adaptive model in prediction mode, and second, by the fact that the TLS algorithm estimates the rotor speed by using the measured stator current signals and not the estimated ones, like in the classic adaptive observer.

Table 11.3 summarizes and compares the results of the five tested speed observers with regard to the main issues of sensorless algorithms (score 1 corresponds to the best and 5 to the worst). It shows that the proposed TLS adaptive observer outperforms all the other observers in all the specified issues. The proposed TLS MRAS observer outperforms the classic adaptive observer, the BPN MRAS observer, and the open-loop estimator in all issues, except for the accuracy in speed transient, where the classic adaptive observer has a better behavior.

TABLE 11.3

Comparative Analysis of the Five Speed Observers

	Accuracy in Speed Transient	Minimum Speed	Zero-Speed Operation	Field-Weakening Behavior	Regeneration Mode at Low Speed
TLS MRAS observer	3	2	2	1	1
TLS adaptive observer	1	1	1	1	1
Classic adaptive observer	2	3	5	1	No
BPN MRAS observer	4	5	3	No	1
Open-loop estimator	5	4	4	1	1

Source: From Cirrincione, M. et al., *IEEE Trans. Ind. Electron.*, 54(1), 127, February 2007.
Note: Score 1 corresponds to the best and 5 to the worst.

11.5 MCA EXIN + Reduced-Order Observer

The reduced-order observer, as described in Section 6.4.9, permits estimating the rotor flux-linkage components of the IM, exploiting an observer of reduced dynamic order (2 instead of 4). The consequent main advantage is obviously the reduction of the complexity and computational demand required for its implementation. Starting from the structure of the reduced-order observer, a neural-based version of it has been developed [64]. The state equations of the IM have been discretized and rearranged so as to be represented by a linear NN (ADALINE). Also in this case, the MCA EXIN + neuron has been adopted to train the ADALINE on-line in order to properly estimate the rotor speed.

11.5.1 Reduced-Order Observer Equations

Starting from the space-state representation of the IM model, the matrix equations of the reduced-order flux observer, with a *voltage error* used for corrective feedback, can be deduced [64]:

$$\frac{d}{dt}\hat{\boldsymbol{\psi}}'_r = \mathbf{A}_{22}\hat{\boldsymbol{\psi}}'_r + \mathbf{A}_{21}\mathbf{i}_s + \mathbf{G}\left(\frac{d}{dt}\mathbf{i}_s - \mathbf{A}_{12}\hat{\boldsymbol{\psi}}'_r - \mathbf{A}_{11}\mathbf{i}_s - \mathbf{B}_1\mathbf{u}_s\right) =$$

$$= \left(\mathbf{A}_{22} - \mathbf{G}\mathbf{A}_{12}\right)\hat{\boldsymbol{\psi}}'_r + \left(\mathbf{A}_{21} - \mathbf{G}\mathbf{A}_{11}\right)\mathbf{i}_s - \mathbf{G}\mathbf{B}_1\mathbf{u}_s + \mathbf{G}\frac{d}{dt}\mathbf{i}_s \qquad (11.48)$$

where

$$\mathbf{A}_{11} = -\left\{R_s/(\sigma L_s) + (1-\sigma)/(\sigma T_r)\right\}\mathbf{I} = a_{11}\mathbf{I} \qquad (11.49a)$$

$$\mathbf{A}_{12} = L_m/(\sigma L_s L_r)\left\{(1/T_r)\mathbf{I} - \omega_r\mathbf{J}\right\} = a_{12}\left\{(1/T_r)\mathbf{I} - \omega_r\mathbf{J}\right\} \qquad (11.49b)$$

$$\mathbf{A}_{21} = \left\{L_m/T_r\right\}\mathbf{I} = a_{21}\mathbf{I} \qquad (11.49c)$$

$$\mathbf{A}_{22} = -(1/T_r)\mathbf{I} + \omega_r\mathbf{J} = a_{22}\left\{(1/T_r)\mathbf{I} - \omega_r\mathbf{J}\right\} \qquad (11.49d)$$

$$\mathbf{B}_1 = 1/(\sigma L_s)\mathbf{I} = b\mathbf{I} \qquad (11.49e)$$

where all space-vectors are in the stator reference frame: $\mathbf{i}_s = [i_{sD} \quad i_{sQ}]^T$ stator current vector, $\mathbf{u}_s = [u_{sD} \quad u_{sQ}]^T$ stator voltage vector, $\hat{\boldsymbol{\psi}}'_r = \begin{bmatrix} \hat{\psi}_{rd} & \hat{\psi}_{rq} \end{bmatrix}^T$ rotor flux vector, $\mathbf{I} = \begin{bmatrix} 1 & 0 \\ 0 & 1 \end{bmatrix}$, $\mathbf{J} = \begin{bmatrix} 0 & -1 \\ 1 & 0 \end{bmatrix}$, ω_r is the rotor speed, and $\mathbf{G}$ is the observer gain matrix.

11.5.2 MCA EXIN + Based Speed Estimation

The MCA EXIN + reduced-order observer derives from a modification of Equation 11.24, in the sense that it exploits the two first scalar equations to estimate the rotor speed, as shown in the following [58].

The first two scalar equations of the matrix equation (11.24) can be written as

$$\begin{cases} \dfrac{di_{sD}}{dt} = a_{11}i_{sD} + a_{12}p_r\hat{\psi}_{rd} + a_{12}\hat{\omega}_r\hat{\psi}_{rq} + bu_{sD} \\[3mm] \dfrac{di_{sQ}}{dt} = a_{11}i_{sQ} + a_{12}p_r\hat{\psi}_{rq} - a_{12}\hat{\omega}_r\hat{\psi}_{rd} + bu_{sQ} \end{cases} \qquad (11.50)$$

where the current components are measured variables and the rotor flux and speed are estimated ones. By converting from the continuous domain into the discrete one, and approximating the continuous derivative with the discrete filter $(1 - z^{-1})/T_{sp} z^{-1}$, where T_{sp} is the sampling time of the control systems, the following equations can be deduced:

$$\begin{cases} \dfrac{i_{sD}(k) - i_{sD}(k-1)}{T_{sp}} = a_{11}i_{sD}(k-1) + a_{12}p_r\hat{\psi}_{rd}(k-1) + a_{12}\hat{\omega}_r(k-1)\hat{\psi}_{rq}(k-1) + b\hat{u}_{sD}(k-1) \\[3mm] \dfrac{i_{sQ}(k) - i_{sQ}(k-1)}{T_{sp}} = a_{11}i_{sQ}(k-1) + a_{12}p_r\hat{\psi}_{rq}(k-1) - a_{12}\hat{\omega}_r(k-1)\hat{\psi}_{rd}(k-1) + b\hat{u}_{sQ}(k-1) \end{cases} \qquad (11.51)$$

where k is the current time sample. From (11.51), the following matrix equation can be easily deduced:

$$\begin{bmatrix} a_{12}T_{sp}\hat{\psi}_{rq}(k-1) \\ -a_{12}T_{sp}\hat{\psi}_{rd}(k-1) \end{bmatrix} \hat{\omega}_r(k-1)$$

$$= \begin{bmatrix} i_{sD}(k) - i_{sD}(k-1) - a_{11}T_{sp}i_{sD}(k-1) - a_{12}p_rT_{sp}\hat{\psi}_{rd}(k-1) - bT_{sp}\hat{u}_{sD}(k-1) \\ i_{sQ}(k) - i_{sQ}(k-1) - a_{11}T_{sp}i_{sQ}(k-1) - a_{12}p_rT_{sp}\hat{\psi}_{rq}(k-1) - bT_{sp}\hat{u}_{sQ}(k-1) \end{bmatrix} \qquad (11.52)$$

where $p_r = 1/T_r$.

Equation 11.52 is a classical matrix equation of the type $\mathbf{Ax} \approx \mathbf{b}$ as seen in Chapter 8, where the $\mathbf{A}$ is the "data matrix," $\mathbf{b}$ is the "observation vector," and $\mathbf{x}$ is the unknown vector, equal to $\hat{\omega}_r$, being the unknown scalar. LS techniques reveal, even in this case, the best solution to be adopted for on-line training of the ADALINE and, thus, to solve in recursive form equation (11.52). The matrix $\mathbf{A}$ is composed of the d-q-axis components of the rotor flux linkage which can be affected by errors and noise resulting measurements, and the same can be said for the observation vector $\mathbf{b}$ which is also composed of the d-q-axis components of the rotor flux linkage and the d-q-axis components of the stator current space-vector. The problem under hand is thus a TLS problem rather than an OLS problem. Consequently, any LS technique different from TLS would be inadequate [48,49].

Figure 11.61 shows the block diagram of the MCA EXIN + reduced-order adaptive observer. It should be noted that the computation of the rotor speed by means of the

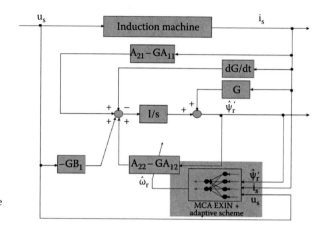

FIGURE 11.61
(See color insert.) Block diagram of the MCA EXIN + reduced-order observer.

MCA EXIN + estimator is performed through the minimization of the residual of the matrix equation (11.52). The residual is strongly dependent on the rotor flux estimation error, while all its other terms are dependent on the measured values of the electrical variables (i_s and u_s), and they are also affected by measurement errors. The TLS inherently gives the best solution for the rotor speed in spite of these uncertainties.

11.5.3 Proposed Choice of the Observer Gain Matrix

The poles of the reduced-order observer are the couple of eigenvalues $\alpha \pm j\beta$ of the matrix $(\mathbf{A}_{22} - \mathbf{GA}_{12})$, where

$$\alpha = -p_r - \frac{L_m}{\sigma L_s L_r} p_r g_{re} - \frac{L_m}{\sigma L_s L_r} \omega_r g_{im}$$

$$\beta = \omega_r + \frac{L_m}{\sigma L_s L_r} \omega_r g_{re} - \frac{L_m}{\sigma L_s L_r} p_r g_{im}$$

and

$$\mathbf{G} = \mathbf{G}_{re} + \mathbf{G}_{im} = g_{re}\mathbf{I} + g_{im}\mathbf{J}$$

The choice of the gain matrix $\mathbf{G}$ proposed in Ref. [65], called *FPP* (fixed poles position), fixes the poles position, in spite of the rotor speed. The *FPP* choice reveals the best for sensorless control for the following explained reasons.

The *FPP* gain matrix choice permits the position of the poles of the observer to be fixed on the negative part of the real semiaxis at distance R from the origin, according to the variation of the rotor speed, to ensure the stability of the observer itself. The proposed gain choice is obtained by imposing $\alpha = -R$ and $\beta = 0$ and gives

$$\mathbf{G} = \mathbf{G}_{re} + \mathbf{G}_{im} = \frac{\sigma L_s L_r}{L_m}\left(-1 + \frac{Rp_r}{p_r^2 + \omega_r^2}\right)\mathbf{I} + \frac{\sigma L_s L_r}{L_m}\frac{R\omega_r}{p_r^2 + \omega_r^2}\mathbf{J} \tag{11.53}$$

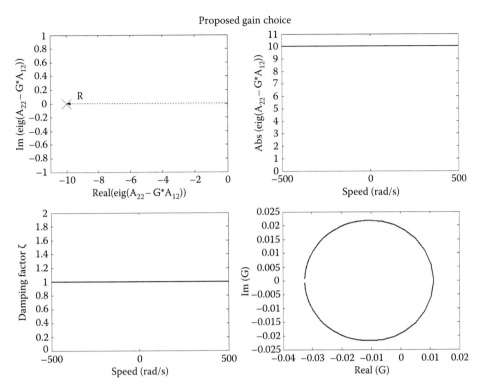

Proposed gain choice

FIGURE 11.62
Pole locus, amplitude versus speed, ζ versus speed, and gain locus with the proposed gain matrix choice. (From Cirrincione, M. et al., *IEEE Trans. Ind. Appl.*, 54(1), 150, February 2007.)

where $p_r = 1/T_r$. Correspondingly, the time derivative of the gain matrix to be used in the observer scheme is

$$\frac{d\mathbf{G}}{dt} = \frac{d\mathbf{G}_{re}}{dt} + \frac{d\mathbf{G}_{im}}{dt} = -2\frac{\sigma L_s L_r}{L_m} \frac{R\,p_r\,\omega_r}{\left(p_r^2 + \omega_r^2\right)^2} \frac{d\omega_r}{dt}\mathbf{I} + \frac{\sigma L_s L_r}{L_m} R\left(\frac{1}{p_r^2 + \omega_r^2} - \frac{2\omega_r^2}{\left(p_r^2 + \omega_r^2\right)^2}\right)\frac{d\omega_r}{dt}\mathbf{J} \quad (11.54)$$

Figure 11.62 shows the observer pole locus, the amplitude of poles versus rotor speed, the damping factor ζ versus rotor speed, and gain locus ($\mathbf{G}_{im}$ versus $\mathbf{G}_{re}$) as obtained with the *FPP* gain matrix choice. It shows that this solution permits keeping the dynamic of the flux estimation constant, because the amplitude of the poles is the constant *R* and the damping factor ζ is always equal to 1. This last feature is particularly important for sensorless control in a high speed range. Most of the choices of the matrix gain result in low damping factor at high rotor speed, which can easily lead up to instability phenomena. Actually, higher values of the damping factor result in low sensitivity to estimated speed perturbations or parameter variations.

11.5.4 Computational Complexity

From the computational point of view, the MCA EXIN + reduced-order observer has been compared here with some NN-based observers, in particular with the TLS EXIN full-order observer, with the TLS EXIN MRAS observer with adaptive neural integrator, and with the classic full-order observer. This comparison has been made on the basis of the number of floating operations (flops) needed by each algorithm for each iteration. The comparison is

TABLE 11.4

Complexity of the Proposed Observer Compared with Others in Literature

		State Equation Integration	Inverter Nonlinearity Compensation	R_s/R_r Estimation	Speed Estimation	Total Flops
MCA EXIN + reduced-order observer ($G \neq 0$)	Without dG/dt	50	25 + 3 IF-THEN	—	45 + 1 IF-THEN	120 + 4 IF-THEN
	With dG/dt	77	25 + 3 IF-THEN	—	45 + 1 IF-THEN	147 + 4 IF-THEN
TLS EXIN full-order observer ($G = 0$)	52	25 + 3 IF-THEN	9	40	126 + 3 IF-THEN	
Classic full-order observer ($G = 0$)	52	—	9	15	76	
TLS EXIN MRAS observer with neural integrator	42	—	11 + 3 IF-THEN	34	87 + 3 IF-THEN	

Source: From Cirrincione, M. et al., *IEEE Trans. Ind. Appl.*, 54(1), 150, February 2007.

shown in Table 11.4. If the correction term dG/dt is not adopted in the proposed observer, the most demanding observer is the TLS EXIN full-order observer which requires 126 flops + 3 IF-THEN instructions, and then there is the MCA EXIN + reduced-order speed observer with 120 flops + 4 IF-THEN instructions. If the correction term dG/dt is adopted, the MCA EXIN + reduced order observer is the most complex, with 147 flops + 4 IF-THEN instructions. Then follows TLS EXIN MRAS observer with adaptive neural integrator which requires 87 flops + 3 IF-THEN instructions, and finally the classic full-order observer requiring 76 flops.

11.5.5 Experimental Results with the MCA EXIN + Reduced-Order Adaptive Observer

The MCA EXIN + reduced-order adaptive observer has been tested experimentally on an FOC IM drive. The adopted FOC scheme is fully described in Appendix 11.A, while the corresponding test setup has been described in Appendix 11.B. It has been tested in several challenging working conditions, as shown in the following.

11.5.5.1 Dynamic Performance

The dynamic performance of the observer has been tested at very low speeds. The drive has been given a set of speed step references at very low speed, ranging from 3 (28.65 rpm) to 6 rad/s (57.29 rpm). Figure 11.63 shows the reference, estimated, and measured speeds during this test, and Table 11.5 shows the 3 dB bandwidth of the speed loop versus the reference speed of the drive. Both Figure 11.63 and Table 11.5 show a good dynamic behavior of the drive with a 3 dB bandwidth decreasing from 69.3 rad/s at 6 rad/s to 12.3 rad/s at 3 rad/s. Same conclusions could be deduced by analyzing Figure 11.64 that shows the reference, estimated, and measured speeds during a set of speed reversal of the types $3 \rightarrow -3$ rad/s, $4 \rightarrow -4$ rad/s, $5 \rightarrow -5$ rad/s, and $6 \rightarrow -6$ rad/s. These last figures show that the drive is able to perform a speed reversal also at very low speeds. However, it should be noted that the lower the speed reference, the higher the time needed for the speed reversal, as expected, because of the reduction of the speed bandwidth of the observer at decreasing speed references; this phenomenon is typical of all observers.

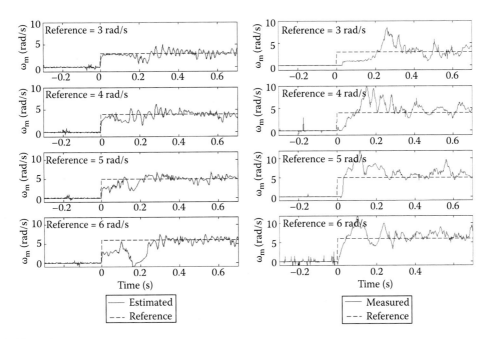

FIGURE 11.63
Reference, estimated, and measured speeds during a set of speed step references with the MCA EXIN + reduced-order observer. (From Cirrincione, M. et al., *IEEE Trans. Ind. Appl.*, 54(1), 150, February 2007.)

TABLE 11.5

Bandwidth of the Speed Loop versus the Reference Speed

Speed (rad/s)	3	4	5	6
3 dB bandwidth	12.3	28.3	80.04	69.38

Source: From Cirrincione, M. et al., *Automatika*, 46(1–2), 59, 2005.

11.5.5.2 Accuracy at Low Speed

The drive has been operated at a constant very low speed (3 rad/s corresponding to 28.65 rpm), at no load, and at rated load. Figure 11.65 shows the reference, estimated, and measured speeds during these tests. It shows that the steady-state speed estimation error is very low, equal to 2.45% at no load and to 7.67% with rated load.

11.5.5.3 Zero-Speed Operation

To test the zero-speed operation capability of the observer, the drive has been operated for 60 s fully magnetized at zero speed with no load. Figure 11.66, which shows the reference, estimated, and measured speeds and position during this test shows the zero-speed capability of this observer. The same kind of test has been performed at the constant load torque of 5 N m. Figure 11.67 shows the reference, estimated, and measured speeds and position during this test and highlights that the measured speed is in average close to 0 and the rotor has an undesired angular movement of 2 rad, achieved in 60 s with a constant applied load torque of 5 N m. This is the ultimate working condition at zero speed. Above 5 N m load torque, the rotor begins to move and instability occurs.

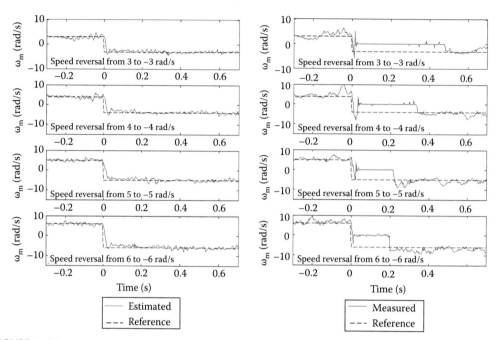

FIGURE 11.64
Reference, estimated, and measured speeds during a set of speed reversal with the MCA EXIN + reduced-order observer. (From Cirrincione, M. et al., *IEEE Trans. Ind. Appl.*, 54(1), 150, February 2007.)

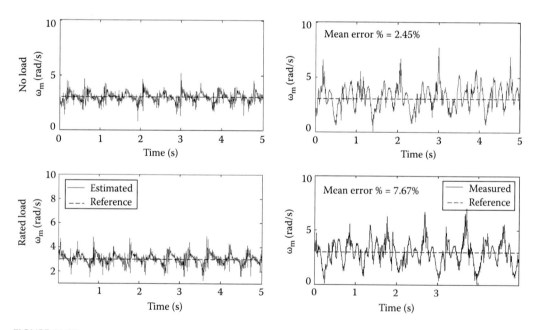

FIGURE 11.65
Reference, estimated, and measured speeds during a constant speed of 3 rad/s at no load and rated load with the MCA EXIN + reduced-order observer. (From Cirrincione, M. et al., *IEEE Trans. Ind. Appl.*, 54(1), 150, February 2007.)

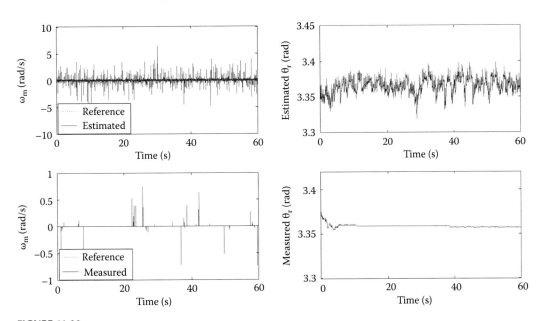

FIGURE 11.66
Reference, estimated, and measured speeds and position at zero speed at no load with the MCA EXIN + reduced-order observer. (From Cirrincione, M. et al., *IEEE Trans. Ind. Appl.*, 54(1), 150, February 2007.)

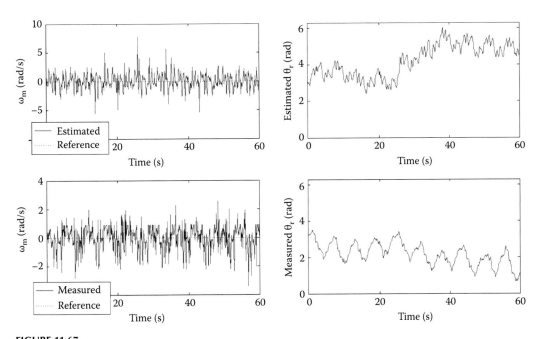

FIGURE 11.67
Reference, estimated, and measured speeds and position at zero speed with 5 N m load torque with the MCA EXIN + reduced-order observer. (From Cirrincione, M. et al., *IEEE Trans. Ind. Appl.*, 54(1), 150, February 2007.)

Appendix A: Implemented Control Schemes

The earlier-described NN-based speed observers have been tested on two kinds of control schemes, FOC and DTC. All the implemented sensorless techniques have been integrated with a suitable methodology for the compensation of the dead times needed to drive the VSI power devices as well as with the methodology for the compensation of the nonlinearity of the VSI characteristics described in Section 6.4.2. In the following, the description of the main characteristics of the adopted schemes is given.

A.1 Sensorless Field-Oriented-Controlled IM Drive

Figure 11.68 shows the block diagram of the adopted sensorless rotor flux–oriented IM drive with impressed voltages. Current control is performed here in the field-oriented reference frame, and a decoupling circuit is present (see Figure 5.18). A control of both the angular speed of the drive ω_m and the amplitude of its rotor flux linkage ψ_r is adopted.

On the direct axis, three controllers are present. The voltage u_0, equal to the product between the absolute value of the estimated machine speed ω_m and the amplitude of the rotor flux linkage ψ_r, is kept constant by an integral (I) controller. The output of this controller is the reference value of the rotor flux-linkage amplitude ψ_{rref}. The rotor flux-linkage amplitude ψ_r is closed-loop controlled by a PI controller, whose output is the reference value of the direct component of the stator current in the field-oriented reference frame $i_{sxref}^{\psi_r}$. The rotor flux amplitude ψ_r is estimated by the block "NN speed observer," which implements the earlier-described neural-based speed observers. It provides also the angle ρ_r, expressing the angular position of the rotor flux-linkage space-vector, necessary for performing the field orientation by the coordinate rotations. The direct component of the stator current $i_{sx}^{\psi_r}$ is closed-loop controlled by a PI controller, whose output is the reference value of the direct component of stator voltage in the field-oriented reference frame.

On the quadrature axis, two controllers are present as well. The estimated motor speed $\hat{\omega}_m$ is closed-loop controlled to its reference ω_{mref} by a PI controller. The output of this controller is the reference value of the quadrature component of the stator current in the field-oriented reference frame $i_{syref}^{\psi_r}$. The quadrature component of the stator current $i_{sy}^{\psi_r}$ is closed-loop controlled by a PI controller, whose output is the reference value of the quadrature component of stator voltage in the field-oriented reference frame. The block $e^{j\rho_r}$ performs a vector rotation from the field-oriented to the stationary reference frame on the basis of the instantaneous knowledge of the rotor flux-linkage angle ρ_r provided by the flux model.

A.2 Sensorless Direct Torque–Controlled IM Drive

The earlier-described NN-based sensorless techniques have been implemented also on DTC schemes. In particular, both the classic switching table–based DTC [66] and the DTC-SVM [67,68] have been adopted.

Figure 11.69 shows the block diagram of DTC IM drive. In this case, the closed-loop control of the estimated machine speed $\hat{\omega}_m$ and the stator flux ψ_s is performed. The machine-estimated speed $\hat{\omega}_m$ is compared with its reference ω_{mref} and the tracking error is processed by a PI controller. The output of the speed controller is the torque reference t_{ref} which is compared with the estimated torque t_e, computed by the flux and torque model, being the tracking error processed by a hysteresis comparator, whose output is one of the inputs of the block "inverter optimal switching table." The stator flux reference ψ_{sref} is compared

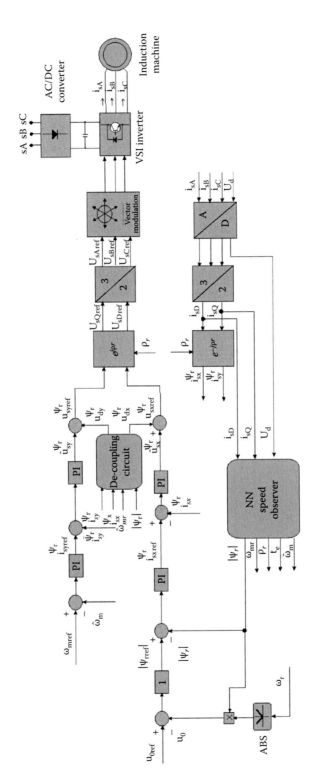

FIGURE 11.68
Block diagram of the NN sensorless FOC drive.

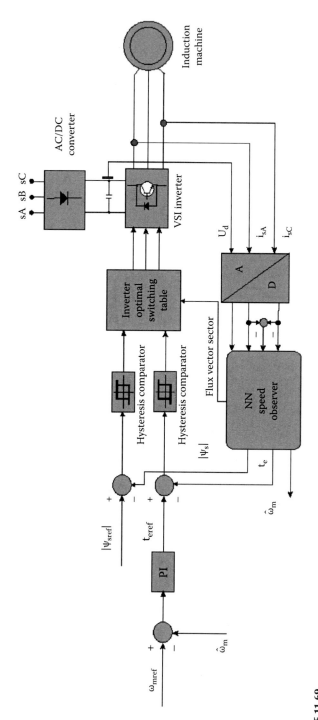

FIGURE 11.69
Block diagram of the NN sensorless ST-DTC drive.

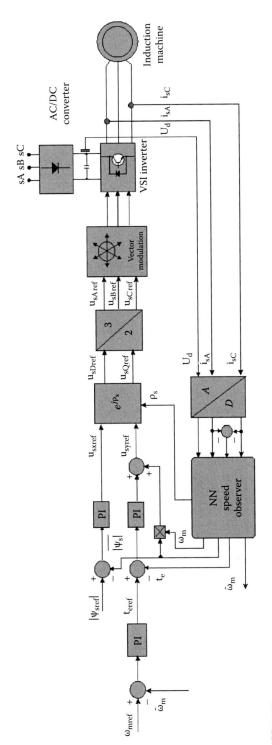

FIGURE 11.70
The block diagram of the NN sensorless DTC-SVM drive.

with the estimated one ψ_s, estimated even in this case by "NN speed observer," and the tracking error is processed by a hysteresis comparator, whose output is the other input of the block "inverter optimal switching table." The flux and torque model is that represented in Figure 5.35. The block "inverter optimal switching table" selects, at each sampling time of the control system, the optimal configuration of the inverter on the basis of the angular position of the stator flux-linkage space-vector. The algorithm adopted for the quadrant identification is that shown in Figure 5.36. The control strategy is usually the D, as defined in Table 5.1, since it permits the best dynamic response. Figure 11.70 shows the block diagram of the DTC-SVM IM drive.

In this scheme, a closed-loop control of both the stator flux-linkage amplitude and the rotor speed is performed. Speed control is achieved by employing a PI controller for processing the speed error resulting from the comparison between the reference and the estimated speeds. The output of the speed controller is the reference torque, which is compared with the estimated one, being the tracking error processed by a PI controller. The output of the torque controller is further added to the decoupling term in Equation 5.76, giving then the quadrature-axis reference voltage $u_{sy}^{\psi_s}$. On the direct axis, the stator flux reference is compared with the estimated flux, being the tracking error processed by a PI controller. The stator flux amplitude and the electromagnetic torque are estimated by the "NN speed observer." The output of the flux controller is the direct-axis reference voltage $u_{sx}^{\psi_s}$. The reference voltages are then transformed from the stator flux linkage to the stationary reference frame by a means of a vector rotation on the basis of the knowledge of the stator flux angle ρ_s. The stator voltages are finally synthesized by a PWM VSI.

Appendix B: Description of the Test Setup

The employed test setup consists of [69] the following:

- A three-phase IM with parameters shown in Table 11.6
- A frequency converter which consists of a three-phase diode rectifier and a 7.5 kVA, three-phase VSI
- A DC machine for loading the IM with parameters shown in Table 11.7
- An electronic AC–DC converter (three-phase diode rectifier and a full-bridge DC–DC converter) for supplying the DC machine of rated power 4 kVA
- A dSPACE card (DS1103) with a PowerPC 604e at 400 MHz and a floating-point DSP TMS320F240
- One electronic card with three voltage sensors (model LEM LV 25-P) and three current sensors (model LEM LA 55-P) for monitoring the instantaneous values of the stator phase voltages and currents
- One voltage sensor (Model LEM CV3-1000) for monitoring the instantaneous value of the DC link voltage
- One electronic card with analog fourth-order low-pass Bessel filters and cutoff frequency of 800 Hz for filtering the stator voltages
- One incremental encoder (model RS 256-499, 2500 pulses per round)

TABLE 11.6

Parameters of the IM

Rated power P_{rated} [kW]	2.2
Rated voltage U_{rated} [V]	220
Rated frequency f_{rated} [Hz]	50
Rated speed [rad/s]	149.75
Pole pairs	2
Stator resistance R_s [Ω]	3.88
Stator inductance L_s [mH]	252
Rotor resistance R_r [Ω]	1.87
Rotor inductance L_r [mH]	252
Three-phase magnetizing inductance L_m [mH]	236
Moment of inertia J [kg·m²]	0.0266

TABLE 11.7

Parameters of the DC Machine

Rated power P_{rated} [kW]	1.5
Rated voltage U_{rated} [V]	300
Rated current I_{rated} [A]	5
Rated speed [rad/s]	150
Rated excitation voltage u_{exrat} [V]	300
Rated excitation current i_{exrat} [A]	0.33

Figure 11.71 shows a photograph of the test setup. Figure 11.72 shows the electric scheme of the test setup. The IM is supplied by a three-phase diode rectifier with a cascaded voltage source inverter (VSI). For control purposes, the DC link voltage and the three stator currents are measured by LEM sensors and acquired by the A/D channels of the dSPACE board. The VSI is driven by a space-vector pulsewidth modulation (SV-PWM) technique (see Section 2.2.1.4 for details), with switching frequency $f_{PWM} = 5\,kHz$. The instantaneous values of the stator phase voltages have been reconstructed from the measurement of the

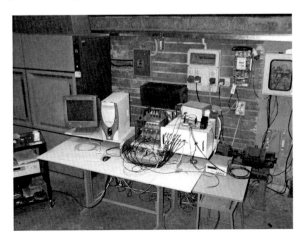

FIGURE 11.71
Photograph of the test setup. (From Cirrincione, M. et al., *IEEE Trans. Ind. Electron.*, 54(1), 127, February 2007.)

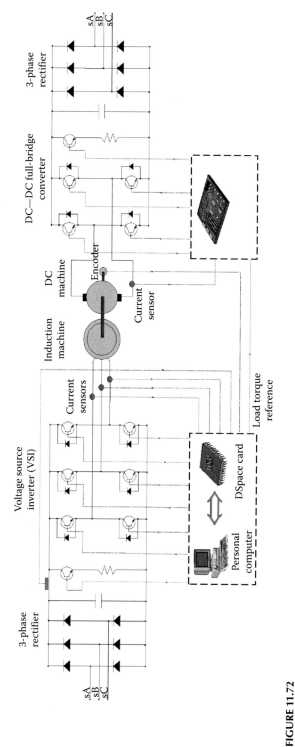

FIGURE 11.72
Electric scheme of the test setup. (From Cirrincione, M. et al., *IEEE Trans. Ind. Appl.*, 1, 140, June/July 2004.)

DC link voltage U_d and from the knowledge of the switching status of the inverter, as $\mathbf{u}_s = \dfrac{2}{3} U_d \left[S_a + a S_b + a^2 S_c \right]$. The speed measurement from the encoder has been used only for comparison reasons with the observer estimations, since only the computed speed has been fed back to the speed controller. The IM drive has been loaded by a controlled DC drive. The DC machine is supplied by a three-phase rectifier with a DC–DC full-bridge converter. A current control is embedded in the DC drive. A torque signal, set by the user, is provided from the dSPACE board to the DC–DC drive in order to command the DC machine torque.

List of Symbols

$u_{sA},\, u_{sB},\, u_{sC}$	stator phase voltages
$i_{sA},\, i_{sB},\, i_{sC}$	stator phase currents
$\mathbf{u}_s = u_{sD} + j u_{sQ}$	space-vector of the stator voltages in the stator reference frame
$\mathbf{u}_s = u_{sx} + j u_{sy}$	space-vector of the stator voltages in a generic rotating reference frame
u_{com}	common-mode voltage
$\mathbf{i}_s = i_{sD} + j i_{sQ}$	space-vector of the stator currents in the stator reference frame
$\mathbf{i}_s = i_{sx} + j i_{sy}$	space-vector of the stator currents in a generic rotating reference frame
$\mathbf{i}'_r = i_{rd} + j i_{rq}$	space-vector of the rotor currents in the stator reference frame
$\mathbf{i}_r = i_{rx} + j i_{ry}$	space-vector of the rotor currents in a generic rotating reference frame
$\boldsymbol{\psi}'_r = \psi_{rd} + j \psi_{rq}$	space-vector of the rotor flux linkages in the stator reference frame
$\boldsymbol{\psi}_s = \psi_{sD} + j \psi_{sQ}$	space-vector of the stator flux linkages in the stator reference frame
$\psi_{s\beta A},\, \psi_{s\beta B},\, \psi_{s\beta C}$	stator phase virtual fluxes
$\mathbf{i}_{mr} = i_{mrD} + j i_{mrQ}$	space-vector of the rotor magnetizing current in the stator reference frame
$\mathbf{i}_{ms} = i_{msD} + j i_{msQ}$	space-vector of the stator magnetizing current in the stator reference frame
$\mathbf{i}_{mm} = i_{mmD} + j i_{mmQ}$	space-vector of the magnetizing current in the stator reference frame
ρ_r	phase angle of the rotor flux-linkage space-vector with respect to the sD axis
ρ_s	phase angle of the stator flux-linkage space-vector with respect to the sD axis
ρ_m	phase angle of the magnetizing flux-linkage space-vector with respect to the sD axis
ϑ_r	angular position of the rotor with respect to the sD axis
L_s	stator inductance
L'_s	stator transient inductance
L_r	rotor inductance
L_m	total static magnetizing inductance
$L_{s\sigma}$	stator leakage inductance
$L_{r\sigma}$	rotor leakage inductance
R_s	resistance of a stator phase winding
R_r	resistance of a rotor phase winding
T_s	stator time constant
T_r	rotor time constant

T'_s	stator transient time constant
T'_r	rotor transient time constant
$T_{r\sigma}$	rotor leakage time constant
$\sigma = 1 - L_m^2/(L_s L_r)$	total leakage factor
σ_r	rotor leakage factor
σ_s	stator leakage factor
p	number of pole pairs
ω_{mr}	angular speed of the rotor flux space-vector
ω_{ms}	angular speed of the stator flux space-vector
ω_{mm}	angular speed of the magnetizing flux space-vector
ω_{sl}	angular slip speed
ω_r	angular rotor speed (in electrical angles per second)
t_e	electromagnetic torque
T_{sp}	sampling time of the control system
U_d	DC link voltage
S_a, S_b, S_c	command signals of the VSI legs

All quantities with *ref* in pedex are reference quantities.

The symbols ψ'_r, ψ_s, ψ_m in apex mean the reference frame in which the variables are expressed.

References

1. P. Vas, *Sensorless Vector and Direct Torque Control*, Oxford Science Publication, Oxford, U.K., 1998.
2. K. Rajashekara, A. Kawamura, K. Matsuse, *Sensorless Control of AC Motor Drives*, IEEE Press, New York, 1996.
3. J. Holtz, Sensorless control of induction motor drives, *Proceedings of the IEEE*, 90(8), 1359–1394, August 2002.
4. R. D. Lorenz, T. A. Lipo, D. W. Novotny, Motion control with induction motors, *Proceedings of the IEEE*, 82(8), 1215–1240, August 1994.
5. W. Leonhard, *Control of Electrical Drives*, Springer-Verlag, Berlin, Germany, 1997.
6. J. Holtz, Sensorless control of induction machines—With or without signal injection? *IEEE Transactions on Industrial Electronics*, 53(1), 7–30, February 2006.
7. P. Vas, *Artificial-Intelligence-Based Electrical Machines and Drives*, Oxford Science Publications, Oxford, U.K., 1999.
8. P. Vas, W. Drury, A. F. Stronach, Recent developments in artificial intelligence based drives—A review, *Proceedings of the International Power Conversion Conference*, 32, 59–70, 1996.
9. B. K. Bose, Fuzzy logic and neural networks in power electronics and drives, *IEEE Industry Applications Magazine*, 6(3), 57–63, 2000.
10. B. K. Bose, Power electronics-an emerging technology, *IEEE Transactions on Industrial Electronics*, 36(3), 403–412, 1989.
11. B. K. Bose, Expert system, fuzzy logic, and neural network applications in power electronics and motion control, *Proceedings of the IEEE*, 82(8), 1303–1323, 1994.
12. B. K. Bose, Neural network applications in power electronics and motor drives—An introduction and perspective, *IEEE Transactions on Industrial Electronics*, 54(1), 14–33, 2007.
13. L. Zadeh, Fuzzy sets, *Information Control*, 8, 338–353, 1965.
14. L. Zadeh, Similarity relations and fuzzy ordering, *Information Sciences*, 3, 177–206, 1971.
15. L. Zadeh, Outline of a new approach to the analysis of complex systems and decision processes, *IEEE Transactions on Systems, Man, and Cybernetics*, 3(1), 28–44, January 1973.

16. E. H. Mamdani, Application of fuzzy algorithms for simple dynamic plant, *Proceedings of the IEEE*, 121(12), 1585–1588, 1974.
17. M. Sugeno, G. T. Kang, Structure identification of fuzzy model, *Fuzzy Sets and Systems*, 28, 15–33, 1988.
18. D. Driankov, H. Hellendoorn, M. Reinfrank, *An Introduction to Fuzzy Control*, Springer-Verlag, Berlin, Germany, 1996.
19. W. S. McCulloch, W. Pitts, A logical calculus of the ideas immanent in nervous activity, *Bulletin of Mathematical Biophysics*, 5, 115–133, 1943.
20. F. Rosenblatt, The perceptron: A probabilistic model for information storage and organization in the brain, *Psychological Review*, 65(6), 386–404, 1958.
21. K. Levemberg, A method for the solution of certain problems in least squares, *The Quarterly of Applied Mathematics*, 2, 164–168, 1944.
22. A. N. Kolgomorov, On the representation of continuous functions of several variables by the superposition of continuous functions of one variable and addition, *Doklady Akademiia Nauk SSSR*, 114(5), 953–956, 1957.
23. R. Jacobs, Increased rates of convergence through learning rate adaptation, *Neural Networks*, 1(4), 295–307, 1988.
24. T. Kohonen, An introduction to neural computing, *Neural Networks*, 1(1), 3–16, 1988.
25. T. Kohonen, The self-organizing map, *Proceedings of IEEE*, 78(9), 1464–1480, 1990.
26. T. Kohonen, *Self-Organizing and Associative Memory*, Springer-Verlag, New York, 1988.
27. B. Kosko, *Neural Networks for Signal Processing*, Prentice Hall, New York, 1991.
28. S. Haykin, *Neural Networks*, Macmillan Publishers, New York, 1994.
29. C. M. Bishop, *Neural Networks for Pattern Recognition*, Oxford University Press, Oxford, U.K., 1995.
30. D. E. Goldberg, *Genetic Algorithm in Search, Optimization and Machine Learning*, Addison-Wesley Publishing Company, Boston, MA, 1989.
31. S. K. Mondal, J. O. P. Pinto, B. K. Bose, A neural-network-based space-vector PWM controller for a three-level voltage-fed inverter induction motor drive, *IEEE Transactions on Industry Applications*, 38(3), 660–669, 2002.
32. J. O. P. Pinto, B. K. Bose, L. E. B. Da Silva, M. P. Kazmierkowski, A neural-network-based space-vector PWM controller for voltage-fed inverter induction motor drive, *IEEE Transactions on Industry Applications*, 36(6), 1628–1636, 2000.
33. M.-H. Kim, M. G. Simoes, B. K. Bose, Neural network-based estimation of power electronic waveforms, *IEEE Transactions on Power Electronics*, 11(2), 383–389, 1996.
34. B. Karanayil, M. F. Rahman, C. Grantham, Online stator and rotor resistance estimation scheme using artificial neural networks for vector controlled speed sensorless induction motor drive, *IEEE Transactions on Industry Applications*, 54(1), 167–176, 2007.
35. S. M. Gadoue, D. Giaouris, J. W. Finch, Sensorless control of induction motor drives at very low and zero speeds using neural network flux observers, *IEEE Transactions on Industry Applications*, 56(8), 3029–3039, 2009.
36. M. Wlas, Z. Krzeminski, J. Guzinski, H. Abu-Rub, H. A. Toliyat, Artificial-neural-network-based sensorless nonlinear control of induction motors, *IEEE Transactions on Energy Conversion*, 20(3), 520–528, 2005.
37. S.-H. Kim, T.-S. Park, J.-Y. Yoo, G.-T. Park, Speed-sensorless vector control of an induction motor using neural network speed estimation, *IEEE Transactions on Industrial Electronics*, 48(3), 609–614, 2001.
38. T. Orlowska-Kowalska, M. Dybkowski, K. Szabat, Adaptive sliding-mode neuro-fuzzy control of the two-mass induction motor drive without mechanical sensors, *IEEE Transactions on Industrial Electronics*, 57(2), 553–564, 2010.
39. J. R. Heredia, F. P. Hidalgo, J. L. D. Paz, Sensorless control of induction motors by artificial neural networks, *IEEE Transactions on Industrial Electronics*, 48(5), 1038–1040, 2001.
40. J. Campbell, M. Sumner, Practical sensorless induction motor drive employing an artificial neural network for online parameter adaptation, *IEE Proceedings on Electric Power Applications*, 149(4), 255–260, 2002.

41. L. Ben-Brahim, R. Kurosawa, Identification of induction motor speed using neural networks, *IEEE Power Conversion Conference*, Yokohama, Japan, 1993.

42. M. Elloumi, L. Ben-Brahim, M. Al-Hamadi, Survey of speed sensorless controls for IM drives, *Proceedings of the 24th Annual Conference of the IEEE Industrial Electronics Society*, Aachen, Germany, 1998.

43. L. Ben-Brahim, Motor speed identification via neural networks, *IEEE Industry Applications Magazine*, 1(1), 28–32, 1995.

44. L. Ben-Brahim, S. Tadakuma, A. Akdag, Speed control of induction motor without rotational transducers, *IEEE Transactions on Industry Applications*, 35(4), 844–850, July/August 1999.

45. C. Shauder, Adaptive speed identification for vector control of induction motors without rotational transducers, *IEEE Transactions on Industry Applications*, 28(5), 1054–1061, September/October 1992.

46. O. Sorensen, Neural network in control application, PhD thesis, Department of Control Engineering, Aalborg University, Aalborg, Denmark, 1994.

47. M. Cirrincione, M. Pucci, An MRAS based speed estimation method with a linear neuron for high performance induction motor drives and its experimentation, *IEEE International Electric Machines and Drives Conference (IEMDC'2003)*, Madison, WI, 2003.

48. G. Cirrincione, M. Cirrincione, J. Hérault, S. Van Huffel, The MCA EXIN neuron for the minor component analysis: Fundamentals and comparisons, *IEEE Transactions on Neural Networks*, 13(1), 160–187, January 2002.

49. G. Cirrincione, M. Cirrincione, Linear system identification by using the TLS EXIN neuron, *Neurocomputing*, 28(1–3), 53–74, October 1999.

50. M. Cirrincione, M. Pucci, G. Cirrincione, G. A. Capolino, A new TLS based MRAS speed estimation with adaptive integration for high performance induction motor drives, *IEEE Transactions on Industry Applications*, 1, 140–151, June/July 2004.

51. M. Cirrincione, M. Pucci, G. Cirrincione, G. A. Capolino, A new adaptive integration methodology for estimating flux in induction machine drives, *IEEE Transactions on Power Electronics*, 19(1), 25–34, January 2004.

52. B. Widrow, S. D. Stearns, *Adaptive Signal Processing*, Prentice Hall, Englewood Cliffs, NJ, 1985.

53. M. Cirrincione, M. Pucci, Sensorless direct torque control of an induction motor by a TLS-based MRAS observer with adaptive integration, *Automatica (Elsevier)*, Adaptive and Intelligent Control, 41(11), 1843–1854, November 2005.

54. M. Cirrincione, M. Pucci, An MRAS speed sensorless high performance induction motor drive with a predictive adaptive model, *IEEE Transactions on Industrial Electronics*, 52(2), 532–551, April 2005.

55. M. Cirrincione, M. Pucci, G. Cirrincione, G. Capolino, Sensorless control of induction machines by a new neural algorithm: The TLS EXIN neuron, *IEEE Transactions on Industrial Electronics*, 54(1), 127–149, February 2007. Special section neural network applications in power electronics and motor drives (Guest editor: B.K. Bose).

56. M. Cirrincione, M. Pucci, G. Cirrincione, G. Capolino, An MRAS sensorless technique based on the MCA EXIN + Neuron for high performance induction motor drives, *Automatika (Korema)*, 46(1–2), 59–72, 2005.

57. J. H. Matheus, K. D. Fink, *Numerical Methods Using Matlab*, 4th edn., Prentice-Hall Publishers Inc., Upper Saddle River, NJ, 2004.

58. M. Cirrincione, M. Pucci, G. Cirrincione, G. Capolino, An adaptive speed observer based on a new total least-squares neuron for induction machine drives, *IEEE Transactions on Industry Applications*, 42(1), 89–104, January/February 2006.

59. H. Kubota, K. Matsuse, T. Nakano, DSP-based speed adaptive flux observer of induction motor, *IEEE Transactions on Industry Applications*, 29(2), 344–348, March/April 1993.

60. H. Kubota, I. Sato, Y. Tamura, K. Matsuse, H. Ohta, Y. Hori, Stable operation of adaptive observer based sensorless induction motor drives in regenerating mode at low speeds, *The 36th IEEE Industry Applications Society Annual Meeting (IAS'2001)*, Chicago, IL, October 2001.

61. H. Kubota, K. Matsuse, Y. Hori, Behaviour of sensorless induction motor drives in regenerating mode, *Proceedings of the IEEE Power Conversion Conference (PCC'97)*, Nagaoka, Japan, 1997.

62. M. Hinkkanen, J. Luomi, Stabilization of the regenerating mode of full-order flux observer for sensorless induction motors, *IEEE Electrical Machines and Drives Conference IEMDC 2003*, Madison, WI, Vol. 1, pp. 145–150, June 1–4, 2003.

63. H. Kubota, I. Sato, Y. Tamura, K. Matsuse, H. Otha, Y. Hori, Regenerating-mode low-speed operation of sensorless induction motor drive with adaptive observer, *IEEE Transactions on Industry Applications*, 38(4), 1081–1086, July/August 2002.

64. M. Cirrincione, M. Pucci, G. Cirrincione, G. Capolino, Sensorless control of induction motors by reduced order observer with MCA EXIN + based adaptive speed estimation, *IEEE Transactions on Industrial Electronics*, 54(1), 150–166, February 2007. Special section neural network applications in power electronics and motor drives (Guest editor: B.K. Bose).

65. H. Tajima, Y. Hori, Speed sensorless field-orientation of the induction machine, *IEEE Transactions on Industry Applications*, 29(1), 175–180, January/February 1993.

66. I. Takahashi, T. Noguchi, A new quick-response and high efficiency control strategy of an induction machine, *IEEE Transactions on Industry Applications*, IA-22, 820–827, September/October 1986.

67. C. Lascu, I. Boldea, F. Blaabjerg, A modified direct torque control for induction motor sensorless drive, *IEEE Transactions on Industry Applications*, 36(1), 122–130, 2000.

68. T. G. Habetler, F. Profumo, M. Pastorelli, L. M. Tolbert, Direct torque control of induction machines using space vector modulation, *IEEE Transactions on Industry Applications*, 28(5), 1045–1053, September/October 1992.

69. M. Pucci, Novel numerical techniques for the identification of induction motors for the control of AC drives: Simulations and experimental implementations (in Italian), PhD thesis, University of Palermo (Italy), Palermo, Italy, January 2001.

Index

A

AC brushless motors, *see* Brushless motors, PMSMs
Active power filters (APFs)
  classification, 106
  conductors, 105
  description, 105
  hybrid active filters, *see* Hybrid active filters
  operating issues, parallel and series filters
    PPF, basic configuration, 107
    scheme, PPF, 107
    SPF, basic configuration, 108
    suppression, 107
  PAFs *vs.* SAFs
    applicative issues, 123
    basic operating principle, 122
    compensation properties, 123
    nonlinear load type, 122
    operating condition, 123
  "power conditioning", 106
  SAF
description, 117
    harmonic current loads, 117–119
    p-q theory, 121–122
    voltage harmonic loads, 119–121
  shunt active filters, 108–117
  three-phase, 107
Adaptive linear neural (ADALINE) filter, 357
Adaptive sliding-mode neurofuzzy speed
    controller (ASNFC), 532
Anisotropy-based sensorless techniques,
    PMSMs
  alternating carrier signal injection
    demodulation scheme, 355
    desynchronization effects, 355
    frequency, 354
    pulsating voltage carrier estimator,
      357–359
    quadrature-axis component, 354
    speed reference, 357
    speed reversal, 355, 356
    speed vector control, 355
    stationary reference frame, 354
  classification, 349
  description, 349
  INFORM, 359–362
  position signal, 349
    rotating carrier signal injection
      demodulation scheme, 351, 352
      description, 349
      direct-and quadrature-axis components,
        350–351
      drawbacks, 352
      frequency, 349–350
      harmonics generation, 351, 352
      matrix formulation, 350
      speed reference changes, 354
      speed reversal test, 352, 353
ANNs, *see* Artificial neural networks (ANNs)
APFs, *see* Active power filters (APFs)
Artificial neural networks (ANNs)
  biological architecture, 532
  BPN-trained, 535
  feed-forward MLP, 532
  off-line training phase, 533
  rigidity, Boolean logic, 531
  simulation mode, 537
ASNFC, *see* Adaptive sliding-mode neurofuzzy
    speed controller (ASNFC)
"Asymmetrical regular sampling", 29

B

Back propagation network (BPN), 498, 532;
    see also BPN-based MRAS speed
    observer
BPN, *see* Back propagation network (BPN)
BPN-based MRAS speed observer
  block diagram, 534
  Euler method, 533–534
  experimental results
    speed reversals, 10r-10 rad/s, 537
    speed reversals, 100r-100 rad/s, 536
    zero-speed operation, 537, 538
  implementation issues
    discrete adaptive model, 535–536
    modified Euler, 536
  on-line training
    block diagram, 534
    errors, 535
    gradient descent error
      minimization, 535
  reference and adaptive model, 533
  "simulator", 534
Brushless motors, PMSMs

AC
    description, 320
    IPM motor, 321–322
    PMAR motor, 321–322
    surface and interior-mounted, 321
DC
    back emf and current pulses, 320
    control methodology, 320
    description, 320

C

Clarke transformation, 115, 121
Closed-loop control, VSIs
    control strategy, 51, 52
    current control
        virtual flux-oriented control, 71–72
        VOC, 66–71
        VOC and VF-OC, 72
    field orientation
        rectangular current error boundary, 59
        rotor-flux space-vector, 60
        switching frequency, 60
    hysteresis current control
        controller, 53
        impressed currents, 53
        reference current, 54
        waveforms, reference and measured
            stator currents, 53
    inverter, current feedback, 51
    power control
        DPC, 72–73
        DPC and VF-DPC, 78–79
        DPC, DPC-EMC 1 and DPC-EMC 2, 84–91
        DPC EMC, 79–82
        DPC-EMC 1 and DPC-EMC 2, 83–84
        switching table DPC, 73–77
        %THD, 91–92
        virtual flux, 78
    predictive current control, 58–59
    real-time optimization, 57–58
    six-pulse to active rectifier
        average power, 64
        characteristic waveforms, 63
        electric scheme, three-phase uncontrolled
            six-pulse rectifier, 62, 63
        mean value, 63
        motor-generator, 61
        rms value, current, 64
        static rectifiers, 61
        three-phase mercury arc rectifier, 62
        three-phase uncontrolled six-pulse
            rectifier, 63

        time domain current waveform, 64, 65
        VOC, 66
    space-vector current control
        stationary resonant controller, 57
        stationary vector controller, 55–56
        synchronous vector controller, 56–57
    suboscillation current control
        controller, 54
        maximum slope, inverter current, 55
        proportional integral (PI) controller, 54
        slope current error and carrier
            slope, 55
    trajectory tracking control, 60–61
Coercive field, 322–323
Common-mode voltage (CMV), 24
Constrained minimization, IMs parameter
        estimation
    description, 451
    function error, 451
    Lagrangian optimization technique, 452
    method, first, 452–456
    method, second, 456–470
"Current" flux model
    advantages, 207
    description, 172
    disadvantages, 208
    rotor-flux-oriented reference frame,
        173–174, 205, 207
    space-vector rotor equation, 173
    stationary reference frame, 205
    types, 206
Current voltage model (CVM), 294

D

DC brushless motors, *see* Brushless motors,
    PMSMs
Direct power control (DPC)
    vs. DPC-EMC 1 and DPC-EMC 2
        CMV and spectrum, 89–90
        electrical scheme, DG generation unit, 86
        P_{ref}, P, Q_{ref}, Q, i_{sA}, i_{sB}, and i_{sC}, 87–88
        ST-DPC techniques, 87
        test setup, 84, 87
    electromagnetic compatible
        CMV variation, 79
        DPC-EMC 1, 80–82
        DPC-EMC 2 technique, 82
    simulation results, DPC-EMC 1
        and DPC-EMC 2
        steady-state inverter current time
            waveforms, 83, 84
        VF-DPC-EMC 2, 83

switching table
  active and reactive power demand, 73
  application, voltage vector, 73, 76
  control scheme, 77
  experimental application, 75, 77
  VF DPC, 72, 78
Direct self-control (DSC)
  characteristics, 254
  zero states, 253
Direct torque control (DTC), IM
  classic DTC and DTC EMC
    common-mode voltage waveform,
      247–250
    rotor speed and electromagnetic torque,
      246–247
  DSC, 253–254
  DTC EMC
    common-mode voltage, 242
    common-mode voltage spectrum, 246
    switching strategy, 242–246
  DTC-SVM, 249–251
  DTC-SVM drive, 251–253
  electromagnetic torque generation
    reciprocal position, stator and rotor flux
      linkage, 232
    stator and rotor flux-linkage
      space-vectors, 230
  vs. FOC, 254–256
  scheme
    flux and torque model, 238, 242
    "inverter optimal switching table",
      240, 242
    stator flux and electromagnetic torque,
      237–240
    stator flux-linkage space-vector and inverter
      configurations, 232–234
    voltage space-vectors, 234–237
Discontinuous PWM (DPWM)
  generalized DPWM (GDPWM)
    technique, 36
  maximum magnitude test, 36
  "minimum switching losses PWM", 35
  modulation waveforms, 35
  switching patterns, inverter phase, 34
  zero-sequence signal, 35
DPC, *see* Direct power control (DPC)
DPWM, *see* Discontinuous PWM (DPWM)
Dynamic and steady-state models, IM
  machine space-vector quantities, definition,
    135–140
  mathematical dynamic model, 154–158
  phase equations, 141–142
  slotting effects, *see* Slotting effects, IM model

space-vector equations
  generalized reference frame, 144–154
  rotor reference frame, 143–144
  stator reference frame, 142–143
space-vector model, 162–165
steady-state space-vector model, 158–162

E

EKF, *see* Extended Kalman filter (EKF)
ELO, *see* Extended Luenberger observer
    (ELO)
Estimator, defined, 275
Euler's formula, 327
Extended Kalman filter (EKF)
  description, 296
  estimation, 297
  state estimation, 363–364
  state vector equation, 297
  vector equation, 298
Extended Luenberger observer (ELO), 275

F

Feedback linearization control (FLC), 189
Feed-forward artificial neural network
    (FFANN), 266, 268
FFANN, *see* Feed-forward artificial neural
    network (FFANN)
FFT, *see* Fourier transform (FFT)
Field-oriented control (FOC), IM
  magnetizing flux-oriented control
    decoupling circuit, 230
    electromagnetic torque, 226
    impressed currents, 227, 228
    rotor space-vector equation, 226
    scalar equations, 227
    space-vector and stator current, 226
    stator leakage inductance, 229
    undesired coupling, 227
    voltage generator, 229
  principle
    DC motor, 200
    decoupled control, 201
    DFOC and IFOC, 201
    electromagnetic torque, 201
    slip angle, 202
  rotor flux-linkage acquisition
    current flux models, 206–208
    drawbacks, direct measurement, 203
    flux models, 204
    flux-oriented control, impressed currents,
      208–213

impressed voltages and rotor flux-
oriented control, 213–216
rotor flux-oriented drive, impressed
voltages, 216–218
voltage flux models, 204–206
rotor flux-oriented control
expression, electromagnetic torque, 202
rotor flux-linkage space vector and stator
current, 202
space-vector equation, 202
stator flux-oriented control
decoupling circuit, 220, 221
electromagnetic torque, 218
impressed currents, 221, 222
impressed voltages, 223, 224
rotor space-vector equation, 220
slip pulsation, 221
stator flux-linkage acquisition, 223–225
stator magnetizing current
space-vector, 219
voltage generator, 221
First constrained minimization method, IMs
parameter estimation
defined, total cost function, 453–454
Newton-Raphson method, 454
simulation and experimental results
capability, 454
electrical parameters, IM, 455
real and electrical parameters, 455, 456, 458
rotor speed and i_{sD}, i_{sQ} waveforms,
455, 457
steady state, K-parameters, 455, 458, 459
translations and rotation, 452–453
FLC, *see* Feedback linearization control (FLC)
FOLO, *see* Full-order Luenberger observer (FOLO)
Fourier transform (FFT), 181, 182
Fryze method
calculation, 117
description, 116
phase voltages, 116
Full-order Luenberger adaptive observer
adaptive speed observer
advantages, 287
experimental time waveform, 287–289
gain matrix, 285, 286
i_{sD} and i_{sQ} current components, 289
Lyapunov stability theorem, 287
observer poles, 286
speed estimation error, time waveform,
289, 290
speed tuning signal, 287
state-space model, IM, 284
Full-order Luenberger observer (FOLO), 274, 275

G

GeMCA EXIN neurons
central trajectories, 406–408
defined, 404
defined, matrix K and eigenvector
matrix Y, 404
D-orthogonal transformation, 403
error function analysis, GeTLS, 405–406
qualitative analysis, error function, 405
GeTLS EXIN neurons
domain, convergence
description, 409
existence, hyperbolas, 409
null initial conditions, 410
equations, 408–409
MCAEXIN+, 412–414
scheduling
DLS hyperbolic, 410, 411
neural approach, 410
OLS learning steps, 412
trajectory, weight, 410, 411
steepest descent discrete-time
formula, 409

H

Harmonic sources
current
circuit scheme, 97, 98
current load, 97, 98
description, 97
spectrum, 97, 98
voltage
description, 97
spectrum, 97, 99
voltage and current, voltage load, 97, 99
voltage load and circuital scheme,
97, 99
Hybrid active filters
"active impedance", 124–125
combination PAF, SPF, 126
current harmonic loads, combination
PAF with SAF, 126, 127
PPF with SPF, 127, 128
description, 123
distribution factor, 125–126
PAF fundamental voltage reduction, scheme
1 to 4, 127–130
parallel combination
PAF with PPF, 123, 124
SAF with SPF, 126, 127
SAF combination, PPF, 123, 124

SAF fundamental current reduction, scheme
    1 to 4, 129–130
series combination
    PAF with PPF, 126
    SAF with SPF, 123, 124
voltage harmonic loads, combination
    SAF with PAF, 127
    SPF with PPF, 127, 128
Hysteresis current control, 52–54

I

Indirect flux detection by on-line reactance
        measurement (INFORM) technique
    forcing inverter voltage, 359, 362
    high-frequency excitation techniques,
        311–312
    rate of change, stator current, 263
    stationary reference frame, 359
    stator reference frame, 359
    test vector, application, 362
Induction machines (IMs)
    control techniques
        classification, 190
        scalar control, *see* Scalar control, IMs
        vector, 189
    drives, sensorless control, *see* Sensorless
        control, IM drives
    DTC
        classic DTC and DTC EMC, 246–249
        DSC, 253–254
        DTC EMC, 242–246
        DTC-SVM, 249–251
        DTC-SVM drive, 251–253
        electromagnetic torque generation,
            230–232
        FOC *vs.* DTC, 254–256
        scheme, 240–242
        stator flux and electromagnetic torque,
            237–240
        stator flux-linkage espace-vector and
            inverter configurations, 232–234
        voltage space-vectors and control
            strategies, 234–237
    dynamic and steady-state models,
        see Dynamic and steady-state
        models, IM
    FOC, *see* Field-oriented control (FOC), IM
INFORM technique, *see* Indirect flux detection
        by on-line reactance measurement
        (INFORM) technique
Interior permanent magnet (IPM) motor,
        321–322

K

Kalman Filter (KF), 275
KF, *see* Kalman Filter (KF)
Kirchhoff's voltage law (KVL) equations,
        68–70

L

Least mean squares (LMS), 500
Least-squares and neural identification,
        electrical machines
    flux model adaptation, FOC and DTC IM
        drives
        block diagram, selection algorithm,
            442, 479
        description, 479
        FOC and DTC, 480, 481
        MATLAB-Simulink software,
            Mathworks®, 480
        real parameters, speed transient,
            480, 482
        rotor flux reference frame, 480
    flux model detuning, control performance
        amplitude= 0.6 Wb, 431, 432
        amplitude= 1.5 Wb, 431
        magnetization curve, 431
        sensitivity analysis, 431
        $t_e/\hat{t}_e$ *vs.* $\hat{R}_r/R_r$, 427, 432
    flux model, parameter variations
        sensitivity, current, 420–426
        sensitivity, voltage, 426–430
    IM parameters, standstill
        electrical parameters, 485
        electrical parameters and real ones,
            486, 487
        equations, 483
        estimated and true electrical
            parameters, 488
        estimated and true K-parameters, 488
        frequency domain, 483
        i_{sD} and i_{sQ} filtered stator currents,
            486, 487
        magnetomotive force (mmf), 486
        off-line, estimation techniques, 483
        phases, voltage references, 486
        remarks, 484
        schematics, supply, 485
        step voltage input, 485
        u_{sD} and u_{sQ}, filtered stator voltages, 486
    on-line estimation, IM parameters, *see*
            Ordinary least square (OLS) method
    on-line tracking, parameter variations

approaches, 432
classification, methodologies, 432, 433
drawbacks, 432–433
nonlinear formulation, 432
parameter estimation, IMs
description, 419
heating/cooling effects, 420
model, current and voltage, 419
on-line adaptation, 420
stator and rotor resistances, 419
TLS, 470–479
saturated and unsaturated conditions, IMs
constrained minimization, *see*
Constrained minimization, IMs
parameter estimation
equations, 450–451
gradient descent method and constrained
minimization, 451
paths, 451
Linear regression
DLS problem, 378–379
OLS problem, 378
TLS problem, 379
LMS, *see* Least mean squares (LMS)
Low-pass filters (LPFs)
algorithm 1
integration, 265, 266
locus, rotor flux linkage, 265, 266
algorithm 2
integration, 266, 267
locus, rotor flux linkage, 266, 267
algorithm 3
compensation scheme, 269, 270
flux linkage, 266
integration, 268
locus, rotor flux linkage, 268, 269
PCLPF, 266, 267
RNN, 268
stator flux linkage, locus, 269, 270
first-order, 264
integration methods, 265
integrator, z-domain, 265
z-domain transfer function, 265
LPFs, *see* Low-pass filters (LPFs)

M

Machine space-vector quantities
assumptions, 135–136
rotor
current space-vector, 138
flux-linkage space-vector,
139–140
stator
current space-vector, 137
flux-linkage space-vector, 138–139
phases, 136
reference frame, 140
three-phase machine, 135, 136
two-axis Park transformation, 137
zero-sequence stator voltages, 140
Magnetic flux-linkage reference frame,
153–154
Mathematical dynamic model
electromagnetic torque, 156, 158
iron path saturation effects, 156–157
magnetic linearity conditions, 156
magnetizing inductance, 154–155
modified rotor time constant, 155
static and incremental inductance, 155
vector equations, 155–156
MATLAB®–Simulink®, 162, 179, 503, 505
Maximum torque per ampere (MTPA)
normalized torque, 337
stator flux-linkage amplitude reference, 339
torque reference, 337–338
MCA EXIN neurons
convergence, first transient phase
null initial conditions, 383
ODE assumptions, limits, 381
stability analysis and phase
portrait, 382
theorem, 381
dynamic behaviour
analysis, 383
deviation, LUO, 384
divergences, 384
eigenvalue computation, 385
existence, sudden divergence, 384
FENG and FENG1 analysis, 385
weight vectors, 383
dynamic stability and learning rate
divergence phase, 386
weight vector, subspaces, 385, 386
gradient flow, RQ, 381
MCA EXIN + reduced-order observer
accuracy, low speed, 589, 590
computational complexity
comparison, 587–588
flops and IF-THEN, 588
dynamic performance
bandwidth, speed loop *vs.* reference
speed, 588, 589
speed reversal, 588, 590
speed step references, 588, 589
equations, 584–585

observer gain matrix
  defined, FPP, 586
  pole locus, amplitude *vs.* speed, 587
speed estimation, 585–586
zero-speed operation, 589, 591
Minor component analysis (MCA), *see* MCA
    EXIN + reduced-order observer
MLP, *see* Multilayer perceptron (MLP)
Model-based sensorless techniques
  adaptive systems
    classic MRAS, 279–282
    CL-MRAS, 282–283
  EKF, *see* Extended Kalman filter (EKF)
  estimators and observers, 275
  full-order Luenberger adaptive observer,
      284–289
  full-order sliding-mode observer
    estimation error, 290
    sliding hyperplane, 289
    speed estimation, 291
    switching signal, 289
  inverter nonlinearity
    direct and quadrature components,
        273, 274
    forward voltage compensation
        scheme, 272
    IGBT Semikron SMK 50 GB 123, 271
    stator voltage space-vector, 271, 272
    threshold voltage space-vector, 271
    voltage drop, 270
  machine parameter mismatch
    algorithm 1, 273
    algorithm 2, 273–274
    algorithm 3, 274–275
    stator and rotor resistances, 273
  open-loop integration
    DC biases, 263
    low speed, 264
    LPFs, *see* Low-pass filters (LPFs)
  open-loop speed estimators
    experimental time waveform, 277
    reference, measured and estimated
        speed, 278
    rotor flux linkage, 277, 278
    sensitivity analysis, 279
    slip speed, 276
    stator flux linkage, 276
  PMSMs
    description, 362
    emf-based estimators, open-loop, *see*
        Open-loop estimators, emf-based
    observer-based estimators, *see* Observer-
        based estimators, PMSMs

reduced-order adaptive observer
  equations, 291–292
  observer gain matrix, 292–295
  speed estimation, 295–296
Model reference adaptive system (MRAS)
  BPN-based, 533–538
  OLS-based, 538–563
Modified Euler neural adaptive model
  block diagram, MCA EXIN + MRAS speed
      observer, 557, 558
  defined, 557
  equation, 555
  matrix equation, 557
  power series expansion, 557
  simulation and prediction mode
    amplitude, poles, 558–560
    estimation and measurement, speeds,
        559–561
    integrator, 559
    relationships, 558
    rotor flux linkage, 558
    uses, 557
Modulation index, 21–22
MRAS, *see* Model reference adaptive system
    (MRAS)
MTPA, *see* Maximum torque per ampere (MTPA)
Multilayer perceptron (MLP), 532
Mutually coupled magnetic circuits
  description, 146
  equivalent circuit, 147
  parameter, 146–147

N

Neural adaptive filter based on linear neurons
    (ADALINEs)
  algorithm, LMS, 500
  band operation
    Bode diagram, NN filter, 503, 504
    description, 501
    grid fundamental frequencies, 503
  harmonic voltage fundamental frequency
      extraction, 501
  MATLAB®–Simulink® implementation,
      503, 505
  network transfer function, z-domain, 500–501
  NN band filter *vs.* PLL
    block diagram, 505, 506
    frequency response, 506, 507
    lower complexity and computational
        demand, 507
    transfer function, 506
  notch operation, 501, 502

outputs, signal and sinusoidal sequence, 499
schematic representation, NN, 499, 500
traditional digital filters, comparison
   Bode diagrams, 503, 505
   Butterworth and Chebyshev, 503
   performance, NN notch adaptive
     filter, 503
Neural adaptive integrator
  block diagram, integrator and filter,
    544, 545
  experimental results
   block diagram, 545, 547
   D, Q rotor flux components *vs.* time and
    rotor flux locus, 548–549
   steady-state locus, 548
   transient time waveform, 547
  feedback loop, 544
  frequency response, 545, 546
  law, 545
  primary and reference input, 544
Neural-based orthogonal regression
  acceleration techniques, 387
  ADALINE and least squares problems
   description, 377
   EIV, 378
   linear neuron network, 377, 378
   TLS and DLS approach, 378
  computational cost, 386
  data, PCA, 392
  GeMCA EXIN neuron
   central trajectories, 406–408
   error function analysis, GeTLS, 405–406
   qualitative analysis, error function, 405
  GeTLS EXIN neuron
   domain, convergence, 409–410
   MCAEXIN+, 412–414
   scheduling, 410–412
  linear
   DLS problem, 378–379
   OLS problem, 378
   TLS problem, 379
  linear least squares problems
   description, 402
   error function, 402
   generalized Rayleigh quotient, 403
   OLS and DLS spectra, 403
  MCA and MCA EXIN neuron
   applications, 380
   neural approach, 380–381
  MCA EXIN neuron
   convergence, first transient phase,
    381–383
   dynamic behaviour, 383–385

dynamic stability and learning rate,
   385–386
   gradient flow, RQ, 381
  quantization errors, 386–387
  signal processing, data analysis, 392
  simulations
   computation, eigenvalue, 387, 389, 390, 391
   divergence, MCA learning laws, 387, 390
   plot, squared weight moduli, 387, 391
   total cost, MCA learning laws, 387, 388
   zero mean Gaussian random vector, 387
  TLS EXIN neuron
   convergence domain, 397–399
   nongeneric TLS problem, 399–402
   stability analysis, 395–397
Neural-enhanced single-phase DG system, APF
   capability
  ADALINE, *see* Neural adaptive filter based
    on linear neurons (ADALINEs)
  compliance, international standards
   current THD, 525–526
   European standard, 525
   harmonic-by-harmonic analysis, 527
   IEEE standard, 523, 525
   nonlinear loads, 526
   steady-state grid currents, 520, 526
   steady-state inverter current, P_{ref} 522,
    526–527
  connection procedure
   electrical scheme active filtering
    capability, 527
   parallel connection, 527–528
  current reference building
   availability, grid harmonic voltage, 508
   block diagram, 499, 507
   components, 507–508
  defined, band configuration, 498
  description, grids, 497
  global stability analysis, 497
  insertion
   current waveform measurements, 516, 517
   electrical scheme, active filtering capacity,
    515, 516
   measurements, currents, 517–518
   parameters, electrical grid, 512, 515
   steady-state grid current, 517, 520
   steady-state inverter current, 517, 519
   transients, 516–517
  load variation
   convergence, multiresonant controller
    and notch adaptive filter, 520
   description, 519–520
   tests, numerical simulation, 520

multilayer perceptrons, BPN, 498
multiresonant current controller
    AC *vs.* DC control, 508
    block diagram, control loop, 509
    classical PI compensator, 508
    continuous Laplace domain, 508–509
    frequency response, 510
    Naslin polynomial, 509
NN filter *vs.* PLL
    grid connection capability, 520
    and steady-state grid voltage, 520,
        521, 524
NN filter *vs.* p-q theory
    currents, auxiliary nonlinear load,
        522, 525
    currents, nonlinear load, 522, 524
    description, 522
    nonlinear load, currents, 523, 525
PAF, 498
power reference insertion
    description, 518
    steady-state inverter current, 519, 522
    steady-state time waveforms, 519, 522
    step transient, 518, 521
principles
    block diagram, 498, 499
    grids, PCC, 498
    power units, 499
stability issues
    aging and fabrication tolerance, 515
    block diagram, continuous/discrete
        domain, 511
    description, 510
    parameters, electrical grid, 512
    poles and zeros, w(z), 512, 513
    position, poles, 512–514
    voltage equations, 510–511
test rig
    description, 515
    electrical scheme, 515, 516
Neural network (NN), *see* Neural adaptive filter
        based on linear neurons (ADALINEs)
Neural sensorless control, AC drives
    BPN-based MRAS speed observer, *see*
        BPN-based MRAS speed observer
    characteristics, schemes, 592
    description, test setup
        electric scheme, 597, 598
        factors, 596
        parameters, IM and DC machine,
            596, 597
        photograph, 597
        torque signal, 599

direct torque, IM
    block diagram, NN ST-DTC, 592, 594
    closed-loop control, 596
    inverter optimal switching table, 596
    stator flux reference, 592, 596
field-oriented-controlled, IM
    block diagram, FOC, 592, 593
    NN speed observer, 592
    quadrature axis, 592
MCA EXIN + reduced-order observer
    computational complexity, 587–588
    equations, 584–585
    experimental results, 588–591
    observer gain matrix, 586–587
    speed estimation, 585–586
NN-based
    ANNs, 532
    ASNFC, 532
    description, 531
    estimation, stator resistance, 533
    hard and soft computing, 531
    layers, feed-forward MLP, 532
    models, Hitachi and Yaskawa, 531
    sensitive *vs.* parameter variations, 531
    simple open-loop speed estimator, 532
OLS-based MRAS speed observer
    experimental results, 539–540
    MCA EXIN, 561–563
    TLS EXIN, 541–555
TLS EXIN full-order Luenberger adaptive
        observer
    adaptive speed, 564
    experimental comparative tests, 581–584
    experimental results, 570–581
    stability issues, 567–570
    state-space model, IM, 563–564
    TLS-based speed estimation, 564–567

O

Observer-based estimators, PMSMs
    extended Kalman filter, 363–364
    TLS EXIN neuron–based speed estimator,
        364–372
Observer gain matrix, 586–587
ODE, *see* Ordinary differential equation
        (ODE)
OLS-based MRAS speed observer
    block diagram, 539
    defined, "data matrix", 538
    experimental results
        DTC drive, 539, 540, 542
        FOC drive, 539–541

rotor flux linkage and i_{sy} current
component, 539, 541
stator flux linkage and electromagnetic
torque, 539, 540, 542
steady-state mean estimation error and
standard deviation, 539, 543
inputs and outputs, 538
MCA EXIN
DC drift, 561–562
description, 561
frequency response, 546, 561
rotor flux D-axis component, 561, 562
square-waveform reference, 562, 563
variation, learning factor, 561, 562
on-line training, 538
TLS EXIN
block diagram, 539, 543
description, 541
d-q-axis components, 542
with DTC drive, 552–555
with FOC drive, 549–552
modified Euler neural adaptive model,
555–561
neural adaptive integrator, 544–549
Open-loop estimators, emf-based
drift, 363
extended emf, 363
space-vector, 362–363
stator flux linkage, 363
Open-loop PWM
carrier-based
DPWM, 34–36
duty cycle, 37–38
modified space-vector modulation, 38
modified suboscillation method, 26–28
sampling techniques, 28–29
space-vector modulation, *see* Space-vector
modulation (SV-PWM)
suboscillation method, 25–26
synchronized carried modulation, 38
carrierless PWM, 38–39
categories, 24
control strategy, 25
experimental verification
duty cycles, 48, 51
loci, reference and average voltage space-
vector, 50
reference and measured phase
voltage, 49
simulation, logic command signals,
44, 45
spectrum, phase voltage, 47, 50
waveforms, reference voltage, 44, 45

optimized
high-power inverters, 43
optimal subcycle method, 44
overmodulation
angular position, 40
mode 1, 39
mode 2, 40–41
techniques divisions, 40
SV-PWM
CMV, 41–42
switching strategy, 42–43
Ordinary differential equation (ODE), 569–570
Ordinary least square (OLS) method
description, test setup
electric scheme diagram, 446
parameter estimation
algorithm, 446
photograph, 446, 447
transducers error, 447
VSI, 446
identification, 439–440
magnetization curve
flux-linkage amplitude *vs.* rotor current
amplitude, 438
nonlinear relationship, 439
static inductance, 439
$|\boldsymbol{\psi_r}|/|\hat{\boldsymbol{\psi}_r}|$ *vs.* ω_{sl}, 426, 439
parameters, IM, 433
RLS algorithm
block diagram, 442
constrained and unconstrained
minimization, 441
covariance, 440–441
data matrix, 440
electrical and K-parameters, 442
signal processing system
block diagram, 443
description, 442
FIR differentiator filters, 444
frequency response, differentiator FIR
filter, 444, 445
frequency response, low-pass analog
Bessel filter, 444
frequency response, low-pass FIR filter,
444, 445
simulation and experimental results
advantages, 449
capability, RLS algorithm, 447
drawbacks, 450
and IM, K-parameters, 423, 449
real and electrical parameters, motor, 447,
448, 450
rotor speed, i_{sD}, and i_{sQ} waveforms, 447–449

space-vector voltage equation
  constant/slow-varying signals, 438
  defined, K-parameters, 435
  description, 434
  general reference frame, 434
  linear and quadratic relationship, 435
  matrix coefficients, 437
  parameters, 435
  reference frame, stationary, 436
  scalar, 434
  sinusoidal steady-state, 438
  unconstrained optimization,
    437–438
  transient and sinusoidal steady-state, 434
Ordinary least squares (OLS), *see* OLS-based
    MRAS speed observer

P

Parallel active filters (PAFs), *see* Shunt active
    filters
Parallel passive filters (PPFs)
  description, 102
  electric schemes, 102, 103
  overall, frequency response, 103, 104
  parameters, components, 103, 104
  "quality factor of the inductor", 104
  resonance pulsation, 104–105
  single component, frequency
    response, 103
Parameter variations, flux model
  current sensitivity
    decomposition, 421, 422
    description, 420
    detuning, 421, 422
    heating, windings/flux reduction,
      421–422
    K-parameters, 422, 423
    $\hat{\rho}_r - \rho_r$ vs. $\hat{R}_r/R_r$ and ω_{sl}, 423, 425
    vs. steady-state curves, 422
    steady-state space-vector
      relationship, 421
    $|\psi_r|/|\hat{\psi}_r|$ vs. $\hat{R}_r/R_r$, 422
    $|\psi_r|/|\hat{\psi}_r|$ vs. $\hat{R}_r/R_r$ and ω_{sl}, 423, 424
    $|\psi_s|/|\hat{\psi}_s|$ vs. ω_r, 430
    $|\psi_r|/|\hat{\psi}_r|$ vs. ω_{sl}, 423–425
  variables, FOC and DTC, 420
  voltage sensitivity
    description, 426
    $\rho_r - \hat{\rho}_r$ vs. ω_r, 428–430
    scalar equations, 429
    $|\psi_r|/|\hat{\psi}_r|$ vs. ω_r, 428

Park transformation, 335
Passive filters, *see* Parallel passive
    filters (PPFs)
PCC, *see* Point of common coupling (PCC)
PCLPF, *see* Programmable cascaded low-pass
    filter (PCLPF)
Permanent magnet assisted reluctance (PMAR)
    motor, 321–322
Permanent magnet synchronous motors
    (PMSMs)
  AC brushless motors, 320–322
  applications, 319
  categories, 319–320
  DC brushless motors, 320
  direct torque control (DTC)
    control variables, 341–342
    description, 341, 347–348
    electromagnetic torque
      production, 343
    scheme, 347–348
    stator flux and the electromagnetic
      torque, 345–347
    stator flux-linkage reference, 348
    torque error, 347–348
    voltage space-vectors and control
      strategies, 343–344
  FOC
    current space-vector, 331
    electromagnetic torque, 332
    stator flux-linkage module, 333
    three-phase current, 331
    torque angle, 332–333
  PM
    coercive field, 322–323
    description, 322
    maximum energy product, 322
    properties, 323, 324
    Recoma 20, hysteresis curve, 323
    residual magnetization, 322–323
  sensorless control techniques
    anisotropy-based, 349–362
    description, 348
    model-based, 362–372
  space-vector model, *see* Space-vector model,
      PMSMs
  speed-controlled drives
    current waveforms, vector
      control, 340
    electromagnetic power, 340
    parameters, 341, 342
    polar coordinates, 339–340
    reference, 340
    vector control, 341

test setup, 372
torque-controlled drives, *see* Torque-controlled drives, PMSMs
Phase equations, IM, 141–142
Phase locked loop (PLL)
  grid fundamental voltage extraction, 498
  vs. NN band filter, 505–507
  vs. NN filter, 520–521
Phase-locked loop (PLL), 300
PLL, *see* Phase locked loop (PLL); Phase-locked loop (PLL)
PMSMs, *see* Permanent magnet synchronous motors (PMSMs)
Point of common coupling (PCC)
  DG systems, APF capability, 498
  effect, DG and nonlinear load, 525
  renewable source-based system, 523
  transfer function, 506
Power quality
  APFs, *see* Active power filters (APFs)
  harmonic propagation
    description, 97, 100
    groups, 100
    industrial network, linear and nonlinear loads., 101
    interharmonics, 100
    iteration method, 102
    linear and nonlinear load, 101
    network equation, 102
    PWM, 100
    real distribution system, 101
  nonlinear loads
    harmonic current sources, 97, 98
    harmonic voltage sources, 97, 99
  PPFs, *see* Parallel passive filters (PPFs)
Programmable cascaded low-pass filter (PCLPF), 266, 267
Pulsating carrier voltage carrier sensorless technique
  description, 357–359
  load test results
    high-speed, 359–361
    low-speed, 357–359
  speed estimation error, 359
Pulsewidth modulation (PWM)
  closed-loop control, *see* Closed-loop control, VSIs
  converters
    current source converter (CSC), electric scheme, 113, 114
    description, 113
    H bridge configuration, 113–114

open-loop
  carrier-based, 25–38
  carrierless, 38–39
  categories, 24
  classification, 25
  optimized, 43–44
  overmodulation, 39–41
  SV-PWM, 41–43
  verification, 44–51

R

Reduced-order adaptive observer (ROO)
  equations
    space-vectors, 292
    voltage error, 291
  gain matrix, choices
    constant amplitude, 292–293
    "current" and "voltage" model, 294, 295
    imaginary poles, observer, 293–294
    observer pole locus, 292, 293
    pole amplitude variation, 295
    rotor speed, 294
  speed estimation, 295, 296
Residual magnetization, 322–323
ROO, *see* Reduced-order adaptive observer (ROO)
Rotor flux-linkage reference frame, space-vector equations
  definition, 147–148
  dynamical model, 149, 150
  electromagnetic torque, 149
  magnetizing current space-vector, 148
Rotor flux-oriented control
  DFOC, 203
  electromagnetic torque, 202
  flux-linkage space-vector and the stator current, 203
  impressed currents
    AC/DC converter, 209
    constant frequency, 210
    control system, constant switching frequency, 210
    current-controlled cycloconverter, 209
    "flux model", 212–213
    hysteresis comparators, 209–210
    phase stator currents, 209
    rotor flux orientation, 211
    speed and tracking error, 212
  impressed voltages
    controllers, 216
    decoupling circuit, 214

IM drive, 214, 215
    maximum switching frequency, 213
    quadrature component, 216
    stator voltages and currents, field-
        oriented reference frame, 213
    voltage components, 214
  rotor space-vector equation, 202
Rotor slotting effects, IM model
  space-state model
    coefficients, matrix A and vector B,
        174–175
    "current" model, 174
    eigenvalue locus, matrix A, 175
  space-vector state model
    "current" flux model, 172–174
    resistance, stator winding, 171, 172
    stator reference frame, 171
    "voltage" flux model, 172

S

Scalar control, IMs
  impressed currents
    closed-loop scalar control, 199
    stator current amplitude *vs.* slip speed
        curves, 198
    stator pulsation reference, 200
    stator resistance and leakage
        reactance, 196
    steady-state torque, 198
  impressed voltages
    air-gap electromagnetic torque, 190–191
    closed-loop scalar control, 193, 194
    frequency variations, 192
    ideal steady-state speed *vs.* torque
        characteristics, 195
    open-loop scalar control, 192, 193,
        195–196
    stator flux amplitude *vs.* rotor speed, 194
    steady-state speed *vs.* torque
        characteristics, 192
    three-phase stator currents, 196, 197
Second constrained minimization method, IMs
    parameter estimation
  description, 456–457
  experimental results
    advantages, 467
    L_s and L_m *vs.* $|i_{mr}|$, 467–469
    no-load and locked-rotor tests, 464–467
    rotor speed and stator current, 464, 465
  matrix inverse, 457
  nonlinear numerical method, 459
  parametric error index, 460

simulation results
  on electrical parameters, 462, 463
  estimated and real electrical parameters,
    460, 461
  K-parameters, 462, 463
  percent estimation errors, RLS, 462
  rotor speed and stator current, 460, 461
  steady-state magnetization, 460
Sensorless control, IM drives
  anisotropy-based
    excitation signals, 263
    INFORM method, 263
    magnetic, 262–263
    techniques, *see* Signal injection techniques
  model-based
    ROOs, 262
    rotor speed, 261
    techniques, *see* Model-based sensorless
        techniques
  techniques, 261, 262
Series active filters (SAF)
  description, 117
  harmonic current loads
    description, 117–118
    impedance, 119
    load voltage, 118
    voltage equation, 118
  p-q theory
    Clarke transformation, 121
    control scheme, 121, 122
    control strategy, 121
  voltage harmonic loads
    description, 119
    grid current, 120
    reference voltage, 120
    source current, 121
Shunt active filters
  current source compensation
    expression, 108–109
    harmonic current compensation *vs.* active
        filter gain, 110, 111
    harmonic of order, 109–110
    load impedance, 110–111
  p-q theory
    active filter controllers, 114–115
    constant instantaneous power control
        strategy, 115
    description, 112, 113
    generalized Fryze current control
        strategy, 116–117
    harmonic currents, 112
    PWM converters, 113–114
    sinusoidal current control strategy, 116

principle
  current load, 108, 109
  voltage load, 108, 109
  voltage source compensation, 111–112
Signal injection techniques
  high-frequency excitation
    INFORM technique, 311–312
    instantaneous rotor position
      measurement, 313–314
    transient current, 311
    transient flux-linkage space-vector, 310
    transient voltage, 310
  IM saliency analysis
    bar cross-section area and slot depth, 307
    closed and open rotor slots, unskewed
      bars, 304
    L_{hexc} vs. $a/a_{prot.}$, 307
    L_{hexc} vs. carrier frequency, 305
    L_{hexc} vs. $c/c_{prot.}$, 307
    L_{hexc} vs. load torque, 306
    locus, stator inductance space-vector, 305
    magnetic flux density lines, 304
    rotor structure modifications, 306
    space-vector, stator inductance, 304
    synthesis, 306, 308
  pulsating carrier, 308–309
  revolving carrier
    closed-slot rotor bridges, 300
    current control, 298
    direct and quadrature leakage inductance
      components, 299
    FFT, stator current, 303
    finite-element analysis (FEA), 301
    leakage inductance tensor, 299
    locus, stator current, 302
    polyphase rotating carrier, 298
    rotor position estimator, 300, 302
    saturation current component, 301
    synchronization signal, 300
    ZST, 302
Slotting effects, IM model
  description, 165, 166
  principal slot harmonics (PSH), 167
  rotor, *see* Rotor slotting effects, IM model
  rotor slot harmonics (RSHs), 168–169
  stator and rotor, *see* Stator and rotor slotting
    effects, IM model
  stator slot harmonics (SSHs), 168
Space-vector analysis
  amplitude and spatial orientation, 2
  3→2 and 2→3 transformations
    description, 5
    non-power-invariant, form ver.1, 5–6

  non-power-invariant, form ver.2, 7
  power-invariant form, 6
coordinate transformation
  description, 7
  generalized reference frame, 7, 8
  nonlinear, 8
  real and imaginary components, 8
current density distribution, 3
defined, three-phase quantities, 5
description, Galileo Ferraris' field, 4
electric machine theory, 2
graphical description, quantity, 4
real and imaginary powers
  Clark transformation, 11
  components, current, 12
  defined, instantaneous complex, 9
  definition, "q", 10
  physical meaning, 13
  power-and non-power-invariant
    form, 10–11
  rectangular components, 9
  separation, 12
  sinusoidal steady-state, 9–10
  three-phase doubly fed electric
    machine, 13
resultant current density distribution,
  phases, 3–4
space-vector theory, 1
stator current, defined, 3
symmetrical and asymmetrical forms, 4
three-phase winding, 2
Space-vector equations
  generalized reference frame
    description, 144–146
    magnetic flux-linkage, 153–154
    mutually coupled magnetic circuits,
      146–147
    rotor flux-linkage, 147–149
    stator flux-linkage, 149–153
    stator voltage and flux-linkage, 144–145
    vector diagram, 144
  rotor reference frame
    electromagnetic torque, 144
    and stator, differential equations, 143
  stator reference frame
    differential equations, 142
    electromagnetic torque, 143
    and rotor flux linkages, 142
Space-vector model
  IM
    linear and nonlinear magnetization
      curve, 165, 167
    machine speed, 163

magnetization curve, 163, 164
MATLAB®-Simulink®, 162
parameters, 22 kW IM, 163, 164
stator and rotor slotting effects, *see* Stator and rotor slotting effects, IM model
steady-state, *see* Steady-state space-vector model
time simulated waveforms, 165, 166
torque *vs.* speed characteristics, 163–165, 167
waveforms *vs.* rotor magnetizing current, 163, 165
PMSMs
  direct and quadrature-axis inductances, 329–330
  Euler's formula, 327
  instantaneous complex power and real power, 330
  magnetizing inductance, A winding, 325
  matrix formulation, 326–327
  mutual inductance, 325–326
  reluctance torque, 330
  stator flux, 327–329
  stator voltage, 329
  vector scheme, 323, 324
Space-vector modulation (SV-PWM)
  decomposition, reference voltage, 30
  duty cycle, phase-n leg, 33
  nonnull space voltage vectors, 31
  operation time, null vector, 31
  time domain waveforms, driver signals, 32
  vector modulation, 33–34
  voltage space-vector, 29
  weighing time intervals, 30
Squirrel-cage IM, 151–152
Stator and rotor slotting effects, IM model
  space-state model
    coefficients, matrix A and vector B, 177
    "current model", 176
    eigenvalue locus, matrix A, 177–178
    parameters, 176–177
    "voltage model", 176
  space-vector model
    AC motor inductances, 169
    FFT, 181, 182
    filtering, stator current components, 180, 181
    flux-linkage equation, 170–171
    motors 2 and 3, measured and filtered stator current, 184, 186
    motors 2 and 3, steady-state current components, 184, 185

motor 1, stator current waveform and spectrum, 182, 183
motor 2, stator current waveform and spectrum, 182, 183
motor 3, stator current waveform and spectrum, 182, 184
mutual inductance matrices, 169–170
no load and 6.37% load slip, stator current, 184, 185
parameters, 2.2 kW IM, 180
stator current harmonics, 178–179
stator current waveform and spectrum, 181, 182
steady-state stator current spectrum, 180
structure, rotors, 179, 180
Stator flux and electromagnetic torque, PMSMs
  DC link voltage, 346
  electromagnetic torque, 347
  Laplace domain, 345–346
  similarities, 345
  stationary reference frame, 346
  stator linkage reference frame, estimation, 346, 347
Stator flux-linkage-oriented control, IM
  decoupling circuit, 220, 221
  electromagnetic torque, 218
  impressed currents, 221, 222
  impressed voltages, 223, 224
  rotor space-vector equation, 220
  slip pulsation, 221
  stator flux-linkage acquisition, 223–225
  stator magnetizing current space-vector, 219
  voltage generator, 221
Stator flux-linkage reference frame
  angular slip frequency, 153
  description, 149, 151
  doubly fed IM, 152
  electromagnetic torque, 152
  squirrel-cage IM, 151–152
Steady-state space-vector model
  description, 158
  electromagnetic torque, 159
  equivalent circuit, 158–159
  2.2 kW IM
    efficiency *vs.* speed, 161, 162
    Heyland diagram, 161, 162
    parameters, 160–161
    stator current *vs.* speed, 161
    torque *vs.* slip, 160–161
Suboscillation current control, 54–55
SV-PWM, *see* Space-vector modulation (SV-PWM)
"Symmetrical regular sampling", 29

T

THD, *see* Total harmonic distortion (THD)
TLS EXIN full-order Luenberger adaptive
observer
accuracy, low-speed range, 573–575
adaptive speed, 564
behaviour, speed observers, 581
comparative analysis, 583, 584
description, 563
dynamic performance
rotor flux-linkage amplitude and
electromagnetic torque, 571, 572
speed and speed estimation error, 571
speeds and electromagnetic torque,
571, 572
speed step reference, 571, 573
square-waveform reference, 573, 574
field-weakening operation, 579
open-loop estimator, 582–583
regenerative mode
bias error, 579
estimation error, constant speed
reference, 577, 578
robustness, load perturbations
electromagnetic and load torque, 576, 577
load torque sets, 577
speed estimation
block diagram, 566
defined, observation vector, 565
digital implementation, 564
equations, 565
possibilities, 566
rotor flux linkage and rotor speed,
566, 567
speeds and speed estimation error, 581, 582
stability issues
formula, 568–569
Lyapunov function, 569
ODE, 569–570
relations, 567–568
state-space model, IM, 563–564
very low speed, accuracy
constant reference, 573, 576
estimation error, 573, 575
mean estimation error percentage, 576
zero-speed operation, 579–581, 583
TLS EXIN neuron–based speed estimator
data and observation vector, 365
description, 364
matrix linear equation, 365
PMSM stator voltage space-vector equation,
364–365

sensorless scheme, 365
speed reversal, high speed with no load and
with load, 366–368
speed step reference, low speed with no
load, 367, 371
triangular speed reference, no load and with
load, 367, 369–370
TLS EXIN neurons; *see also* TLS EXIN
neuron–based speed estimator
convergence domain
barriers, 397
maximum locus tangent, 398
saddle cone projections, 398
saddle-maximum hypercrest
projection, 398
theorem, 398–399
convergence speed, 394
cost function $E_{TLS}(x)$, 392
cost landscape stability analysis, 393
fundamental problems, 393
linear neural networks, 273
minimum singular value, 392
nongeneric TLS problem
data errors, 399–400
description, 399
evolution map, 400, 401
formulas, 400
solution/center locus, 400, 401
reverse ridge regression, 394
saddle direction, 394
stability analysis
benchmark overdetermined system,
396, 397
convergence keys, 396
cost landscape and equilevel curves,
396, 397
description, 395
positive quadratic forms, 395
Torque-controlled drives, PMSMs
description, 333
feed-forward control, 339
FOC, surface-mounted
electromagnetic torque value, 333–334
Park transformation, 335
polar coordinates, torque control, 334
rectangular coordinates, torque
control, 335
torque angle, 334
voltage torque control, rectangular
coordinates, 336
interior-mounted
description, 336–337
MTPA, 337–339

stator flux-linkage amplitude
reference, 339
stator linkage reference frame, 337–338
vector diagram, 337
Total harmonic distortion (THD)
current waveform, nonlinear load and
spectrum, 516, 517
harmonic limits, 525–526
multiresonant controller, 519
Total least square (TLS) method
constrained minimization, 472
defined, reduced matrix, 472
description, 470
error function, 473
error function *vs.* K_1, K_2, 471, 472
EXIN neuron, 470–471
nonlinear minimization method, 472
OLS error function *vs.* K_1, K_2 and K
trajectory, 471
simulation and experimental results
contours and critical points, 476, 477
3-D error surfaces, 474–476
estimated K-parameters, 474, 475
flatness, 476
MATLAB and Simulink, 473
nonlinear constrained minimization
algorithm, 479
real and electrical parameters, 477, 478
relative errors, K-parameters, 477, 479
robustness *vs.* noise, 476
rotor speed, i_{sD}, and i_{sQ} waveforms, 474
virtual instruments, 473–474
Two-phase SV-PWM, *see* Discontinuous PWM
(DPWM)

V

VOC, *see* Voltage-oriented control (VOC)
"Voltage" flux model, 172, 204–206
Voltage-oriented control (VOC)
active rectifier
circuital analysis, 67–68
operating condition, 67

LCL filter, 66, 67
reference frame analysis
stationary, 68–69
synchronous, 69–71
single-phase representation, 66, 67
voltage-controlled bridge converter, 66, 67
Voltage source inverters (VSI)
characteristic hexagon, 19, 20
closed-loop control, VSIs, *see* Closed-loop
control, VSIs
CMV, *see* Common-mode voltage (CMV)
current harmonics
copper loss, 21
rms harmonic current, 20
space-vector trajectory, 20
energy-flow direction, 18
harmonic spectrum, 21
maximum modulation index, 21–22
ON and OFF state, 17–18
PWM, *see* Pulsewidth modulation (PWM)
switching frequency and switching
losses
energy loss, transition, 23
mean power, 22
mean power loss, 23
time domain waveform, 23
torque harmonics, 22
values, phase voltage, 18
voltage space-vector, 19
Voltage space-vectors
cases, stator flux amplitude and
electromagnetic torque, 234–235
effects, 234
PMSMs
decoupling, 343
optimal inverter, 344
plane, 343–344
sampling time interval, 344
stator flux space-vector, 236, 237
strategy choice, 235–236
switching frequency, 237
torque and stator flux control, 235
VSI, *see* Voltage source inverters (VSI)